Mathematical Methods in Chemistry and Physics

Mathematical Methods in Chemistry and Physics

Michael E. Starzak
State University of New York at Binghamton
Binghamton, New York

Plenum Press • New York and London

Library of Congress Cataloging in Publication Data

Starzak, Michael E.
Mathematical methods in chemistry and physics / Michael E. Starzak.
p. cm.
Includes bibliographical references and index.
ISBN 0-306-43066-5
1. Chemistry—Mathematics. 2. Physics—Mathematics. I. Title.
QD39.3.M3S73 1989 88-32133
510′.2454—dc19 CIP

A Division of Plenum Publishing Corporation
233 Spring Street, New York, N.Y. 10013

Printed in the United States of America

Preface

Mathematics is the language of the physical sciences and is essential for a clear understanding of fundamental scientific concepts. The fortunate fact that the same mathematical ideas appear in a number of distinct scientific areas prompted the format for this book. The mathematical framework for matrices and vectors with emphasis on eigenvalue–eigenvector concepts is introduced and applied to a number of distinct scientific areas. Each new application then reinforces the applications which preceded it.

Most of the physical systems studied involve the eigenvalues and eigenvectors of specific matrices. Whenever possible, I have selected systems which are described by 2×2 or 3×3 matrices. Such systems can be solved completely and are used to demonstrate the different methods of solution. In addition, these matrices will often yield the same eigenvectors for different physical systems, to provide a sense of the common mathematical basis of all the problems. For example, an eigenvector with components $(1, -1)$ might describe the motions of two atoms in a diatomic molecule or the orientations of two atomic orbitals in a molecular orbital. The matrices in both cases couple the system components in a parallel manner.

Because I feel that 2×2, 3×3, or soluble $N \times N$ matrices are the most effective teaching tools, I have not included numerical techniques or computer algorithms. A student who develops a clear understanding of the basic physical systems presented in this book can easily extend this knowledge to more complicated systems which may require numerical or computer techniques.

The book is divided into three sections. The first four chapters introduce the mathematics of vectors and matrices. In keeping with the book's format, simple examples illustrate the basic concepts. Chapter 1 introduces finite-dimensional vectors and concepts such as orthogonality and linear independence. Bra–ket notation is introduced and used almost exclusively in subsequent chapters. Chapter 2 introduces function space vectors. To illustrate the strong parallels between such spaces and N-dimensional vector spaces, the concepts of Chapter 1, e.g., orthogonality and linear independence, are developed for function space vectors. Chapter 3 introduces matrices, beginning with basic matrix operations and concluding with an introduction to eigenvalues and eigenvectors and their properties. Chapter 4 introduces practical techniques for the solution of matrix algebra and calculus problems. These include similarity transforms and projection operators. The chapter concludes with some finite difference techniques for determining eigenvalues and eigenvectors for $N \times N$ matrices.

Chapters 5–8 apply the mathematics to the major areas of normal mode analysis, kinetics, statistical mechanics, and quantum mechanics. The examples in the chapter demonstrate the parallels between the one-dimensional systems often introduced in introductory courses and multidimensional matrix systems. For example, the single vibrational frequency of a one-dimensional harmonic oscillator introduces a vibrating molecule where the vibrational frequencies are related to the eigenvalues of the matrix for the coupled system. In each chapter, the eigenvalues and eigenvectors for multicomponent coupled systems are related to familiar physical concepts.

The final three chapters introduce more advanced applications of matrices and vectors. These include perturbation theory, direct products, and fluctuations. The final chapter introduces group theory with an emphasis on the nature of matrices and vectors in this discipline.

The book grew from a course in matrix methods I developed for juniors, seniors, and graduate students. Although the book was originally intended for a one-semester course, it grew as I wrote it. The material can still be covered in a one-semester course, but I have arranged the topics so chapters can be eliminated without disturbing the flow of information. The material can then be covered at any pace desired. This material, with additional numerical and programming techniques for more complicated matrix systems, could provide the basis for a two-semester course. Since the book provides numerous examples in diverse areas of chemistry and physics, it can also be used as a supplemental text for courses in these areas.

Each chapter concludes with problems to reinforce both the concepts and the basic examples developed in the chapter. In all cases, the problems are directed to applications.

I wish to thank my wife Anndrea and my daughters Jocelyn and Alissa for their support throughout this project and Alissa for converting my pencil sketches into professional line drawings. I am grateful to the students whose comments and suggestions aided me in determining the most effective way to present the material. I also wish to thank my readers in advance for their suggestions for improvement.

Michael E. Starzak

Binghamton, New York

Contents

1

Vectors

1.1. Vectors

Vectors are used when both the magnitude and the direction of some physical quantity are required. A force applied to an object on a frictionless table (a two-dimensional system) can be any magnitude and it may pull the object along any direction on the table (Figure 1.1). This force is represented by a line with length proportional to the magnitude of the force. This line lies along the direction in which the force is applied. The line normally begins from the object and terminates with an arrow (Figure 1.1). If it acts on a rigid body, the force could be applied at any point on the body, i.e., the physical location of the vector is less important than its magnitude and direction. If the body is elastic, a force vector applied to different parts of the body may give a different response. In such cases, the vector cannot be separated from its location on the body.

For a rigid body, two forces of different magnitude which act in exactly the same direction will produce a net force equal to the sum of the two constituent forces:

$$\mathbf{F}_t = \mathbf{F}_1 + \mathbf{F}_2 \tag{1.1.1}$$

To translate into a vector format, either vector is moved so it starts from the terminus of the second vector. The resultant vector, $\mathbf{F}_t$, is a single vector which starts from the origin and ends at the terminus of the second vector; it has the same direction as the original two vectors. This resultant vector is found by arranging vectors in head-to-tail fashion and connecting the first tail to the final head.

This head-to-tail vector addition is valid even when the vectors have different directions. Two forces are oriented at a right angle in Figure 1.2. The total force is found by transposing either vector to the head of the other (Figure 1.3). The resultant vector then connects the initial tail and final head. The force from the two vectors is equivalent to a single force directed horizontally. Its magnitude can be found geometrically since the transposed vector is perpendicular to the initial vector creating a right triangle. The resultant (hypotenuse) is

$$(F_1^2 + F_2^2)^{1/2} \tag{1.1.2}$$

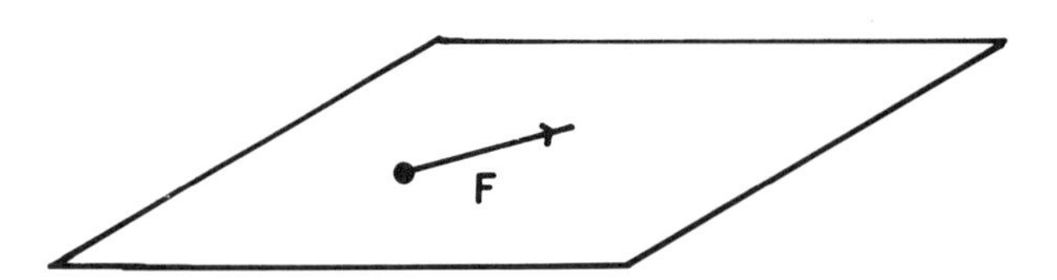

Figure 1.1. The force on an object on a table expressed as the magnitude of a directed line in the x–y plane.

If the two vectors make an angle θ rather than 90°, the law of cosines can be used:

$$F_t = (F_1^2 + F_2^2 - 2F_1F_2 \cos(\theta))^{1/2} \tag{1.1.3}$$

Vectors can also be used to locate positions in space. Under such circumstances, the vector represents the distance and direction from one point in space to another. Although most vectors in space involve three dimensions, two-dimensional systems can illustrate the concept. The vector **r** in Figure 1.4 represents the motion from the origin to a point $\sqrt{2}$ distant at a +45° angle. The vector **s** represents a motion from the origin of $\sqrt{2}$ at a −45° angle. The addition of these two vectors by connecting the head of **r** to the tail of **s** is like the addition of forces. In this case, the first vector changes the location in space from the tail to the head of the vector. The head of the first vector then serves as the origin for the second vector. The head of this vector is the final spatial location. The vectors are arranged in sequence to determine the final position. The order of the vectors is not important in this case.

The subtraction of two vectors requires only a change in the direction of the second vector. In Figure 1.5, the operation

$$\mathbf{r} - \mathbf{s} \tag{1.1.4}$$

involves the translation to the head of **r** as its first step. The position of the +**s** vector is shown as a dashed line. The subtraction is performed by reversing the direction of the **s** vector as shown. The resultant then connects the initial tail to the head of the negated vector.

Any number of vectors can be added in this fashion to produce a net resultant. For example, there is no reason that the vectors of Figure 1.4 be located at some origin. An origin may be defined at some other point in space. In such a case, a third vector might be used to bring an observer from this origin to

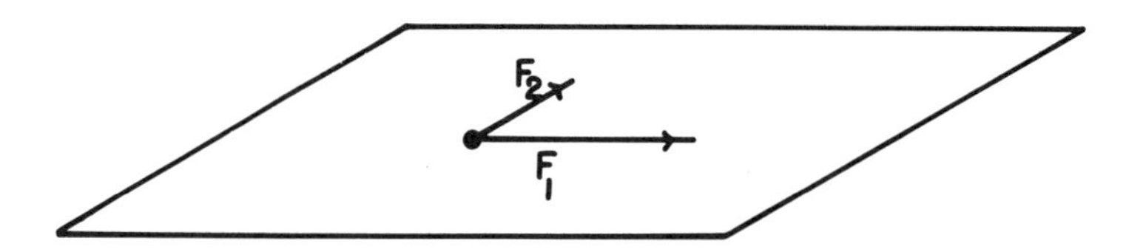

Figure 1.2. Two perpendicular forces F_1 and F_2 in a plane.

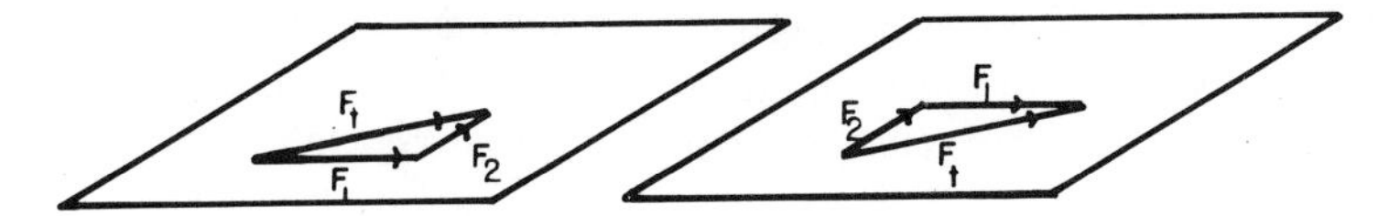

Figure 1.3. Graphical addition of the two vectors of Figure 1.2.

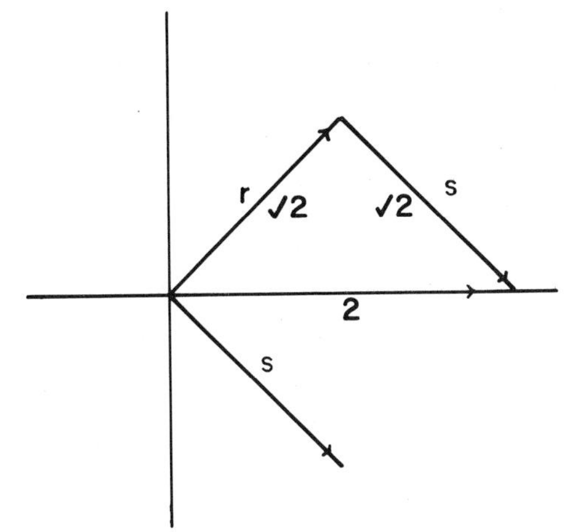

Figure 1.4. Two perpendicular vectors with magnitudes $\sqrt{2}$.

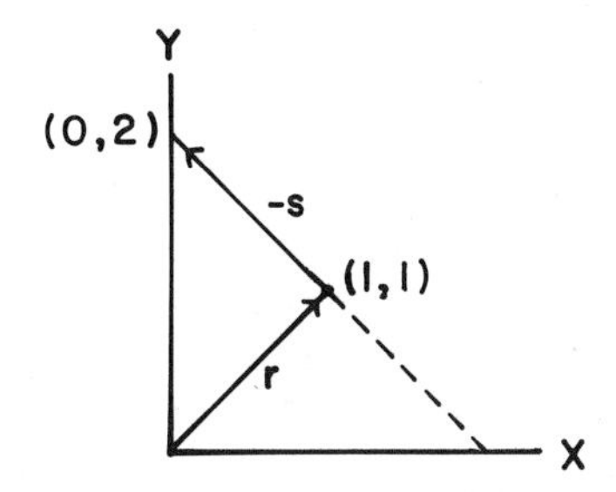

Figure 1.5. The graphical difference, $\mathbf{r}-\mathbf{s}$, of the vectors of Figure 1.4.

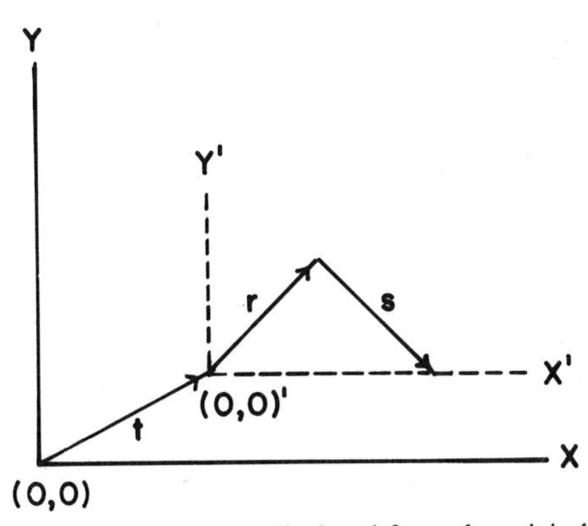

Figure 1.6. The vector sum $\mathbf{r}+\mathbf{s}$ displaced from the origin by a vector $\mathbf{t}$.

the tail of vector **r**, as shown in Figure 1.6. The reverse situation is often more important. Since the addition of interest is $\mathbf{r}+\mathbf{s}$, a system defined with the origin of Figure 1.6 could be converted to a simpler system by subtracting the vector **t**:

$$\mathbf{r}+\mathbf{s}-\mathbf{t} \tag{1.1.5}$$

Since the vector $-\mathbf{t}$ translates the point labeled $(0, 0)'$ to the tail of the **t** vector, this subtraction places the tail of the **r** vector at the origin of the coordinate system. If a system is located at the point $(0, 0)'$, this point can be defined as the new origin by subtracting the vector (**t**) which led to this point in the original coordinate system.

1.2. Vector Components

The most common vectors have a magnitude and direction in three-dimensional space. However, some situations may require a different number of dimensions. If forces are restricted to a plane, only two dimensions are required. If time is included as a variable, a fourth dimension may be required.

The vector sums of Section 1.1 required transposition of some vectors to generate the resultant vector. While this was a simple procedure for a one-dimensional system, it becomes increasingly difficult as the total dimension of the space increases. For example, the four-dimensional system would be impossible to draw for a determination of the resultant vector.

If two vectors are confined to a single dimension in space, the addition or subtraction of such vectors is simply the addition or subtraction of their scalar magnitudes. For this reason, it is useful to resolve multidimensional vectors into a set of scalar components. The vectors **r** and **s** of Figure 1.4 each have a single spatial direction. However, both these vectors could have a finite projection on a horizontal axis (an arbitrary selection). Both **r** and **s** have projections of 1 unit on this coordinate (Figure 1.7). The resultant vector for these two components also lies on this axis with a magnitude of 2. The two projections on this axis are now scalars and they can be added to give the resultant component. The difference of these scalar components r_1 and s_1 is

$$r_1 - s_1 = 1 - 1 = 0 \tag{1.2.1}$$

since both vectors have the same projection on this axis.

Since the vectors lie in a two-dimensional space, a second component is needed to completely describe **r** and **s**. The second axis is selected perpendicular to the first. This choice of perpendicular axes is extremely convenient. However, any axis which is not parallel to the first coordinate axis can be chosen. The projections on axis 2 are $+1$ and -1 for the **r** and **r** vectors, respectively (Figure 1.7). The component of the resultant $\mathbf{r}+\mathbf{s}$ on this axis is zero:

$$+1+(-1)=0 \tag{1.2.2}$$

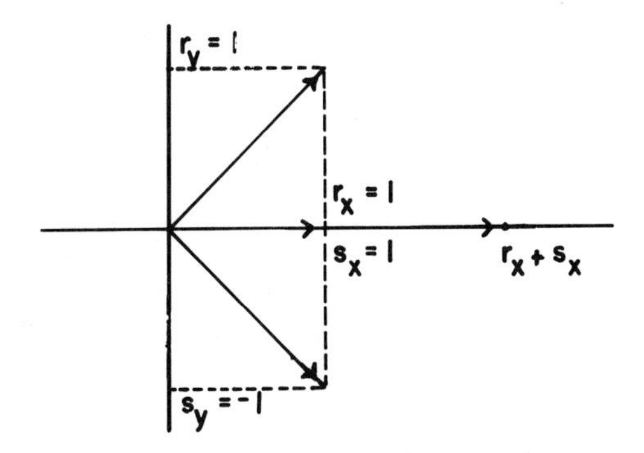

Figure 1.7. Projections of the vectors **r** and **s** on defined x and y coordinate axes.

The resultant vector can now be generated from its components, i.e., its projections on the first and second axes. For $\mathbf{r}+\mathbf{s}$, these are components are $+2$ and 0, and the resultant vector lies entirely on the first axis (Figure 1.7). By convention, this horizontal axis is called the x axis. The vertical axis is the y axis, and the projections of **r** on x and y are r_x and r_y, respectively.

Although each vector must be resolved into components on each of the defined coordinates, this initial work is compensated by the ease of manipulating the vectors. For example, the sum **u** of vectors **r** and **s** with x and y components

$$\begin{aligned} r_x &= 1; \qquad r_y = 2 \\ s_x &= 3; \qquad s_y = 1 \end{aligned} \tag{1.2.3}$$

has components

$$u_x = 3 + 1 = 4; \qquad u_y = 2 + 1 = 3 \tag{1.2.4}$$

These components of the resultant vector define a point in the x–y plane for the tip of the resultant vector, i.e., the head of the vector is located by a motion of 4 units on the x axis followed by a motion of 3 units in the vertical (y) direction (Figure 1.8).

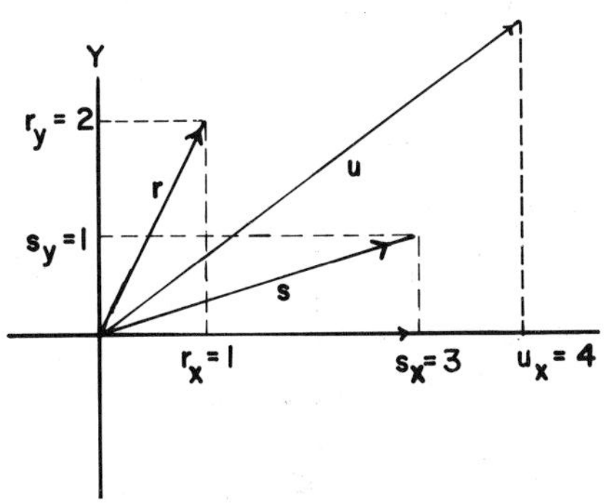

Figure 1.8. The resultant vector **u** reconstructed from the sums of components of **r** and **s** on the x and y coordinate axes.

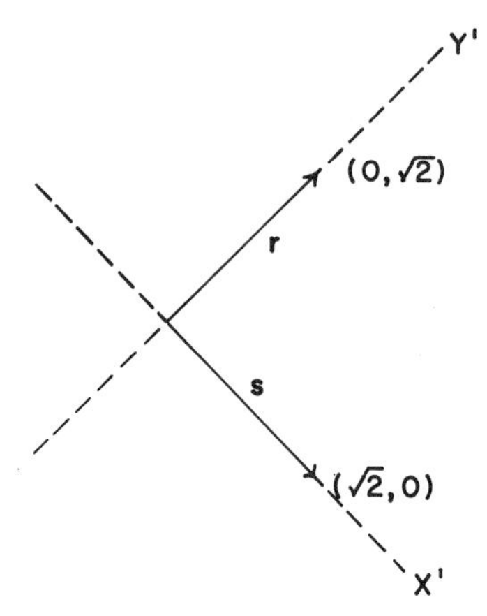

Figure 1.9. Coordinate axes x' and y' selected to coincide with the directions of the vectors **s** and **r**, respectively.

Although horizontal and vertical (x and y) axes are commonly used, this is a choice of convenience since it is relatively easy to project onto perpendicular axes. The axes do not have to be perpendicular and do not have to be oriented in the horizontal and vertical directions. For the vectors **r** and **s** of Figure 1.4, perpendicular axes might be selected to coincide with the **r**(y') and **s**(x') vector directions (Figure 1.9). The **r** vector will then have a component of $\sqrt{2}$ on the y' axis and a component of 0 on the x' axis. The **s** vector will have a component of $\sqrt{2}$ on the x' axis and a component of 0 on the y' axis. The resultant will again lie on the horizontal, but this horizontal is now constructed from the components on the x' and y' axes (Problem 1.17).

The resolution of a vector into coordinate projections or components is particularly effective for systems with more dimensions. Two three-dimensional vectors could be summed by first resolving each vector into its three components on each of the mutually perpendicular axes. The components for each direction are added to find a final component. This can be illustrated for a simple tetrahedral system shown in Figure 1.10. A carbon atom in the center of the cube

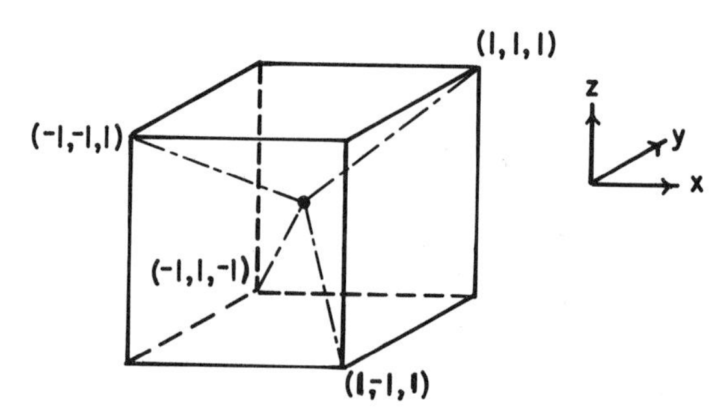

Figure 1.10. Location of H atoms in a tetrahedral system with a C atom at the origin.

defines the origin. For methane, H atoms would be located at relative distances having the x–y–z components

$$\underset{\text{i}}{(1, 1, 1)}, \quad \underset{\text{ii}}{(1, -1, -1)}, \quad \underset{\text{iii}}{(-1, -1, 1)}, \quad \underset{\text{iv}}{(-1, 1, -1)} \tag{1.2.5}$$

The vector distance between the two hydrogens above the C atom is determined by subtracting vector iii from vector i component by component. The resultant is

$$\{[1-(-1)], [1-(-1)], [1-(1)]\} = (2, 2, 0) \tag{1.2.6}$$

There is no z component since both atoms lie in the same x–y plane. The three components define the tip of the vector which would be formed by head-to-tail combination of $\mathbf{r}_1$ and $-\mathbf{r}_2$. This vector "point" is related to the separation between the atoms, as can be seen (i) and $-$(iii) by examining the x–y plane of the two atoms (Figure 1.11). The difference moves the tail of the vector to the origin so the tip defines the actual difference of $2\sqrt{2}$.

The separation between the two atoms can also be found using the Pythagorean theorem for this orthogonal (perpendicular axis) coordinate system. The theorem is applied twice for the three-dimensional system to give the magnitude of the resultant difference vector with components r_x, r_y, and r_z (Problem 1.1):

$$r^2 = r_x^2 + r_y^2 + r_z^2 \tag{1.2.7}$$

The two hydrogen atoms above C are then separated by a distance

$$r^2 = (2)^2 + (2)^2 + (0)^2 = 8$$
$$r = 2\sqrt{2} \tag{1.2.8}$$

when the distance from the C atom to any H atom is 1. This result can be verified with the law of cosines (Problem 1.2). With a Cartesian coordinate system, the absolute length of any vector can always be determined using the Pythagorean theorem. This absolute distance for the vector is called its norm.

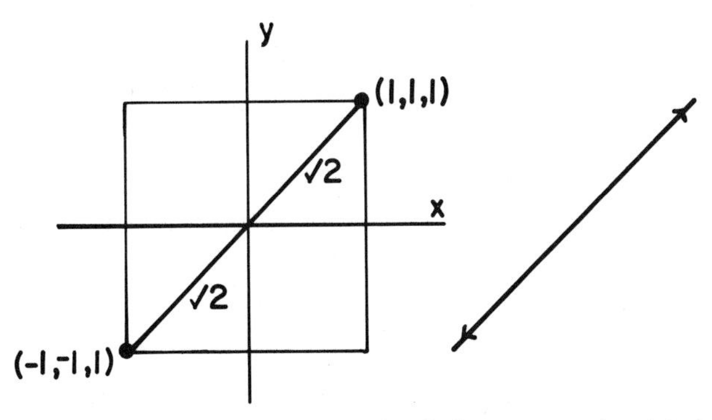

Figure 1.11. The x–y plane containing the upper two hydrogen atoms in a tetrahedral structure.

Several different types of notation can be used to describe a vector in terms of its components. In a Cartesian coordinate system, the x, y, and z coordinate axes are described by unit vectors **i**, **j**, and **k**, respectively. The unit vectors have a value of one with the appropriate units. For example, a force unit vector would be 1 N. This unit vector can be multiplied by a scalar to prrroduce the actual projection on this axis for a given vector. For example, the vector

$$\mathbf{r} = \mathbf{i}1 + \mathbf{j}2 + \mathbf{k}3 \tag{1.2.9}$$

has projections of 1 unit on the x axis, 2 units on the y axis, and 3 units on the z axis. To add two vectors, the scalars for **i** are added to form the **i** component of the resultant, the scalars for **j** are added to form the resultant **j** component, etc. The unit vectors serve as "markers" to distinguish the different projections for the vector.

Since the vector components are generally ordered as r_x, r_y, and r_z, the **i**, **j**, and **k** unit vector markers can be deleted; the position of the scalar components determines their coordinate axis. The vector $\mathbf{r} = \mathbf{i}r_x + \mathbf{j}r_y + \mathbf{k}r_z$ becomes the ordered set of scalars (r_x, r_y, r_z). The vector of Equation 1.2.9 is

$$(1, 2, 3) \tag{1.2.10}$$

To add this vector to a second vector **s** with components

$$(2, 2, 2) \tag{1.2.11}$$

the numbers in the appropriate positions in each vector are added to give the resultant

$$(3, 4, 5) \tag{1.2.12}$$

The actual resultant vector is generated from these components on their proper coordinate axes.

The use of consistent component order provides a convenient vector notation which can be generalized to vector systems of any dimension. For example, a particle in three-dimensional space must actually be described by six different entities; there are three position coordinates and three momentum or velocity components for the particle in a three-dimensional space. These six independent components constitute a coordinate system which has six perpendicular axis. Although such a phase space cannot be described graphically in a three-dimensional space, this is not an obstruction to writing the vector in its components. The vector would be

$$(r_x, r_y, r_z, p_x, p_y, p_z) \tag{1.2.13}$$

with the appropriate scalar values inserted in their proper positions. If the position and momentum of the particle changed with time, the position of the vector in this phase space could be monotored by observing the temporal evolution of each scalar component:

$$[r_x(t), r_y(t), r_z(t), p_x(t), p_y(t), p_z(t)] \tag{1.2.14}$$

When vectors were restricted in our discussion to a two-dimensional space, two coordinate directions were necessary to locate the head of the vector. In an N-dimensional space, N coordinate axes would be required. Moreover, none of the selected coordinate axes can be parallel since parallel axes will provide the same information on the total vector. For the two-and three-dimensional systems, mutually orthogonal (perpendicular) coordinate axes provided a full characterization of each vector. These coordinate axes spanned the two- or three-dimensional space. A choice of two coordinate directions along x and a third direction along z would not span a three-dimensional space since there would be no way to characterize the "y" direction. A set of vector coordinates which do span the space constitute a basis set of vectors for that space. These concepts can be expanded for vector spaces of dimension N. For such systems, special procedures are required to determine mutually orthogonal coordinate axes and to establish the completeness of this basis set, i.e., that it can describe any N-dimensional vector in the space.

1.3. The Scalar Product

Although the vector concept provides a convenient way to combine several directed quantities with the same units, there are situations which involve the product of vectors with different units. Such products are most effectively described using the component decomposition of these vectors.

A force in a one-dimensional system will do work if it acts over a certain distance. In other words, the product of a force and a distance in a one-dimensional system is equivalent to a work or an energy:

$$E = W = Fd \tag{1.3.1}$$

In the one-dimensional system, both the force and the distance of application are vectors. However, their product gives an energy, which is a scalar quantity. If the applied force and the direction of motion are parallel, the system will have a positive energy. If the force and direction of motion are opposed, this energy is negative. However, the energy is generated during the motion and has no intrinsic direction.

If a vector has units, each of its components must have the same units. In the force–distance example, force and distance, with their different units, are projected onto a common set of spatial directions, e.g., x, y, and z. These projections will give the magnitudes of the force and distance components in that spatial direction.

The force and direction produced a net energy only if both are directed along the same axis. In Figure 1.12, force and direction vectors are constrained to the x–y plane and oriented in different directions. This arrangement would not generate energy as effectively as the case where both vectors were parallel, i.e., a one-dimensional system.

To generate a scalar energy, both force and direction must be parallel. In Figure 1.12, some of the force can be resolved into a component parallel to the vector $\mathbf{r}$ by forming a projection of the vector $\mathbf{F}$ on the $\mathbf{r}$ vector. The projection reduces the system to the one-dimensional system needed to produce the scalar energy. If the angle between the two vectors is θ, the projection of $\mathbf{F}$ on $\mathbf{r}$ is

$$\mathbf{F}_{\mathbf{r}} = F \cos\theta \tag{1.3.2}$$

In this case, the "scalar" product is

$$\mathbf{r} \cdot \mathbf{F}_{\mathbf{r}} = Fr \cos\theta \tag{1.3.3}$$

The direction of the force could also be selected as the one-dimensional axis. In this case, the projection of $\mathbf{r}$ on the force vector is

$$\mathbf{r}_{\mathbf{F}} = r \cos\theta \tag{1.3.4}$$

and the scalar product is again Equation 1.3.3. In general, the scalar product will be independent of the coordinate axes selected for projections. Equation 1.3.3 is also valid in three-dimensional systems. The two vectors form a plane in the three-dimensional space. The angle is simply the angle between those two vectors in that plane.

Although the coordinates for the scalar product were selected to coincide with one of the vectors, this is not a requirement. Figure 1.13 shows the two vectors in a conventional (x–y) Cartesian coordinate system. The vectors are assigned the absolute values

$$F = 1\text{N}, \qquad r = 2m \tag{1.3.5}$$

and an angle of 45° so that the scalar product is

$$(1\text{N})(2\text{m}) \cos 45^\circ = \sqrt{2}\,\text{Nm} \tag{1.3.6}$$

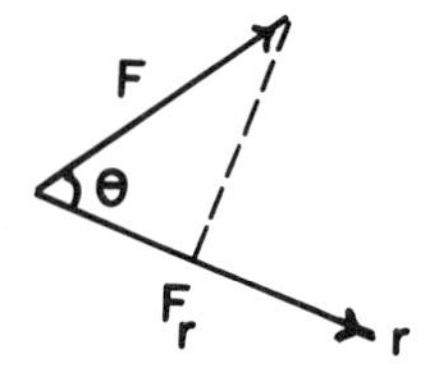

Figure 1.12. The projection of a force vector **F** on a distance vector **r**

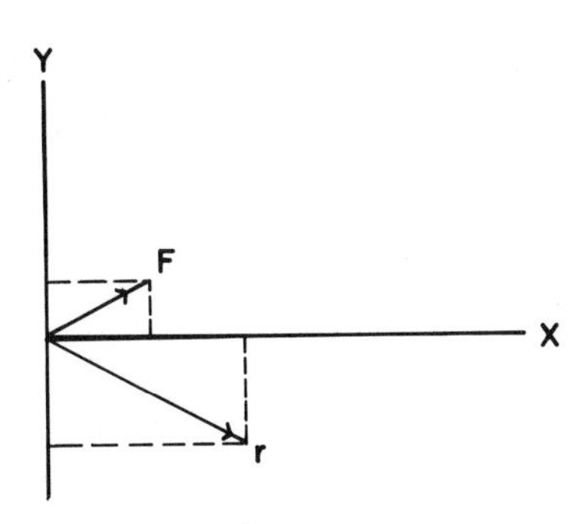

Figure 1.13. Scalar product using the x and y components of the force and distance vectors.

To determine a scalar product using the x and y coordinates, both the force and the direction must be projected onto the coordinate axes. In this case, the F_x and r_x components are

$$F_x = F \cos(22.5°) = (1)(0.9239)$$
$$r_x = r \cos(22.5°) = (2)(0.9239) = 1.8478 \qquad (1.3.7)$$

and the scalar product for the x coordinate is

$$F_x r_x = (0.9239)(1.8478) = 1.707 \qquad (1.3.8)$$

This represents only the energy generated with respect to the x coordinate, but it is still a scalar quantity. The remaining energy must be determined from the components on the y axis. These are

$$F_y = F \sin(22.5°) = (1)(0.3827)$$
$$r_y = r \sin(-22.5°) = -(2)(0.3827) = -0.7654 \qquad (1.3.9)$$

and the scalar product is

$$F_y r_y = -0.2929 \qquad (1.3.10)$$

The total energy generated is the sum of these two products,

$$F_x r_x + F_y r_y = 1.707 - 0.2929 = 1.414 = \sqrt{2} \qquad (1.3.11)$$

which is identical to the result generated by choosing **F** or **r** as the coordinate vector for projection.

Although only projections on two axes were considered, the component approach to scalar products is general. The two vectors are resolved into components on some preselected coordinate system. The scalar product is then formed by multiplying the respective components from each vector and adding these products. For example, the scalar product of vectors **a** and **b** in a three-dimensional space with x–y–z Cartesian coordinates is

$$\mathbf{a} \cdot \mathbf{b} = a_x b_x + a_y b_y + a_z b_z \tag{1.3.12}$$

The dot between the two vectors is often used to symbolize a scalar product. For this reason, the scalar product is often called the "dot" product.

The component form of the scalar product suggests the generalization to spaces with more than three dimensions. If there are N orthogonal coordinates and the projections of vectors **a** and **b** on the ith coordinate are a_i and b_i, respectively, then the scalar product is

$$\sum_i^N a_i b_i \tag{1.3.13}$$

Because the scalar product involves the product of components for a given product direction, the use of ordered coordinates is compatible with the generation of scalar products. For the ordered components

$$\begin{gathered} (a_x, a_y, a_z) \\ (b_x, b_y, b_z) \end{gathered} \tag{1.3.14}$$

the first components of each vector are multiplied, the second components are multiplied, etc., and the products are added. The scalar product of the vectors

$$(1, 2, 3) \quad \text{and} \quad (2, 3, 4) \tag{1.3.15}$$

is

$$(1 \times 2) + (2 \times 3) + (3 \times 4) = 20 \tag{1.3.16}$$

For the scalar products, the position of each set of vector components is not important. The components from either vector can be selected as the first term in each product; the product in each case will be the same. When matrix operations are introduced, the location of the vector relative to the matrix will play a role in the operation. For this reason, a scalar product

$$\mathbf{a} \cdot \mathbf{b} \tag{1.3.17}$$

must be arranged so that the vector on the left (**a**) can be distinguished from the vector on the right (**b**). This is done by arranging the components of the left-hand vector in a row with the proper coordinate order. The components of the right-hand vector are arranged in a column. The scalar product (Equation 1.3.17) is

$$(a_x \quad a_y \quad a_z) \begin{pmatrix} b_x \\ b_y \\ b_z \end{pmatrix} \tag{1.3.18}$$

The first component in the row is multiplied by the first component in the column, and so on.

The operation in reverse order,

$$\begin{pmatrix} b_x \\ b_y \\ b_z \end{pmatrix} (a_x \quad a_y \quad a_z) \tag{1.3.19}$$

is a matrix (Chapter 3). Scalar products are only possible with a row on the left and a column on the right.

Although the use of rows and columns is clear, it is convenient to have a notation which distinguishes row and column vectors without writing the full sets of components. The notation which will be used is the bra–ket notation. The row vector (**a** in this case) is placed in a bracket which points to the left, i.e.,

$$\langle a| = (a_x \quad a_y \quad a_z) \tag{1.3.20}$$

while the column vector (**b** in this case) is described by a bracket which faces right, i.e.,

$$|b\rangle = \begin{pmatrix} b_x \\ b_y \\ b_z \end{pmatrix} \tag{1.3.21}$$

The scalar product is now formed by combining these two brackets as

$$\langle a|b\rangle \tag{1.3.22}$$

The reason for the names bra and ket is now clear. They form the two portions of the word "bracket":

$$\begin{matrix} \langle a| & & |b\rangle \\ \text{bra} & \text{c} & \text{ket} \end{matrix} \tag{1.3.23}$$

The index within the bra and ket vectors may label the vector, as it does here. The scalar product must always form a closed combination bracket.

Imaginary numbers will appear often in quantum mechanics. However, there are many situations where the scalar product of two vectors which have some imaginary components must be a real number. Under such circumstances, the elements of the row vector will be complex conjugates of the components of the column vector. For example, consider a vector

$$|a\rangle = \begin{pmatrix} 1+i \\ i \\ 2-i \end{pmatrix} \tag{1.3.24}$$

If the scalar product

$$\langle a | a \rangle \tag{1.3.25}$$

is to be a real number, each component in $\langle a|$ must be the complex conjugate of the corresponding element in $|a\rangle$. The complex conjugate is formed by replacing each i with $-i$. The components of the row $\langle a|$ vector are

$$\langle a| = (1-i, \quad -i \quad 2+i) \tag{1.3.26}$$

The scalar product is

$$\langle a|a\rangle = (1-i \quad -i \quad 2+i)\begin{pmatrix} 1+i \\ i \\ 2-i \end{pmatrix}$$

$$(1-i)(1+i)+(-i)(i)+(2+i)(2-i) = 2+1+5 = 8 \tag{1.3.27}$$

The scalar product is real as it must be when row and column components are complex conjugates.

1.4. Scalar Product Applications

Since the scalar product can be determined either by a projection of one of the two vectors on the other or by a summation of projections on the coordinate axes of some preselected coordinate system, the two techniques can be melded in special ways. The norm of the vector **r**, for example, is just the absolute length of the vector. In Section 1.2, the norm of a vector was determined by finding its components in some Cartesian coordinate system and applying the Pythagorean theorem. The norm is

$$|\mathbf{r}|^2 = r_x^2 + r_y^2 + r_z^2 \tag{1.4.1}$$

$$|\mathbf{r}| = (r_x^2 + r_y^2 + r_z^2)^{1/2} \tag{1.4.2}$$

This result can be obtained directly by forming the scalar product of **r** with itself. In the Cartesian coordinate system of Equation 1.4.2, the scalar product is

$$\begin{aligned} \langle r|r\rangle &= |\mathbf{r}|^2 \\ &= (r_x \quad r_y \quad r_z)\begin{pmatrix} r_x \\ r_y \\ r_z \end{pmatrix} \\ &= r_x^2 + r_y^2 + r_z^2 \end{aligned} \tag{1.4.3}$$

The result still contains a sum of squared components which is identical to the scalar product. The procedure could be extended to an N-component vector for an N-dimensional space. The norm is still the sum of the squares of the real components of the vector if the basis set is orthogonal. Any orthogonal basis set in the N-dimensional space will give the same norm since the length of the vector remains the same.

Vector norms can be used to determine the angle θ between two vectors in the space. The scalar product of the vectors $\langle r|$ and $|s\rangle$ can be written in two ways. The scalar product can be determined in component form for any coordinate system, i.e.,

$$\langle r|s\rangle = (r_1 \quad r_2 \quad r_3)\begin{pmatrix} s_1 \\ s_2 \\ s_3 \end{pmatrix}$$

$$= r_1 s_1 + r_2 s_2 + r_3 s_3 \tag{1.4.4}$$

where the indices 1, 2, and 3 are used to show that the three coordinate axes need not be the standard x, y, and z axes.

The second definition of scalar product requires the absolute lengths, i.e., the norms of the vectors $\mathbf{r}$ and $\mathbf{s}$,

$$\langle r|s\rangle = rs\cos(\theta) = |\mathbf{r}|\;|\mathbf{s}|\cos\theta \tag{1.4.5}$$

which are determined by forming their scalar products using a coordinate system:

$$|\mathbf{r}|^2 = r_1^2 + r_2^2 + r_3^2 \tag{1.4.6}$$

$$|\mathbf{s}|^2 = s_1^2 + s_2^2 + s_3^2 \tag{1.4.7}$$

The scalar product for the vectors

$$|r\rangle = \begin{pmatrix} 1 \\ 0 \\ 0 \end{pmatrix} \qquad |s\rangle = \begin{pmatrix} 1 \\ 1 \\ 1 \end{pmatrix} \tag{1.4.8}$$

is

$$\langle r|s\rangle = (1 \quad 0 \quad 0)\begin{pmatrix} 1 \\ 1 \\ 1 \end{pmatrix} = 1 \tag{1.4.9}$$

The norms for $|r\rangle$ and $|s\rangle$ are

$$\langle r|r\rangle = (1 \quad 0 \quad 0)\begin{pmatrix} 1 \\ 0 \\ 0 \end{pmatrix} = 1 \tag{1.4.10}$$

$$\langle s|s\rangle = (1 \quad 1 \quad 1)\begin{pmatrix}1\\1\\1\end{pmatrix} = 3 \tag{1.4.11}$$

$$|\mathbf{r}| = 1, \qquad |\mathbf{s}| = \sqrt{3} \tag{1.4.12}$$

$\cos(\theta)$ is determined as

$$\langle r|s\rangle = |\mathbf{r}|\ |\mathbf{s}| \cos\theta \tag{1.4.13}$$

$$\begin{aligned}\cos\theta &= \langle r|s\rangle/|\mathbf{r}|\ |\mathbf{s}|\\ &= 1/(1)(\sqrt{3})\end{aligned} \tag{1.4.14}$$

and

$$\theta = 54.74^\circ \tag{1.4.15}$$

The technique can be used to determine the tetrahedral bond angle for the HCH bond in methane. If carbon is located at the origin, then the two H atoms above the carbon have the coordinates

$$|1\rangle = \begin{pmatrix}1\\1\\1\end{pmatrix} \qquad |2\rangle = \begin{pmatrix}-1\\-1\\1\end{pmatrix} \tag{1.4.16}$$

Since the cosine will be defined as a ratio of lengths, the length of each C–H bond is equated to 1. Then

$$\begin{aligned}\cos\theta &= \langle 1|2\rangle/(\langle 1|1\rangle\langle 2|2\rangle)^{1/2}\\ &= (-1)/3\end{aligned} \tag{1.4.17}$$

and

$$\theta = 109.5^\circ \tag{1.4.18}$$

which is the well-known tetrahedral angle.

The scalar product also defines relationships between unit vectors along the coordinate axes. Unit vectors on Cartesian x, y, and z coordinate axes are

$$|x\rangle = \begin{pmatrix}1\\0\\0\end{pmatrix} \qquad |y\rangle = \begin{pmatrix}0\\1\\0\end{pmatrix} \qquad |z\rangle = \begin{pmatrix}0\\0\\1\end{pmatrix} \tag{1.4.19}$$

Since the vectors are perpendicular to each other, they have no projections onto each other. The scalar product between any pair of different unit vectors must be zero. For example,

$$\langle x|y\rangle = (1 \quad 0 \quad 0)\begin{pmatrix}0\\1\\0\end{pmatrix} = 0 \tag{1.4.20}$$

Two vectors are orthogonal when their scalar product is zero. In this case, the three coordinate vectors which span the space are all orthogonal.

The norms of the three chosen vectors are all unity. For example,

$$\langle x|x\rangle = (1 \quad 0 \quad 0)\begin{pmatrix}1\\0\\0\end{pmatrix} = 1 \tag{1.4.21}$$

When the norms of each vector in a set of orthogonal vectors all are unity, the vectors are called an orthonormal set.

Any set of mutually orthogonal vectors can be converted into an orthonormal set. For example, the set of vectors

$$|1\rangle = \begin{pmatrix}1\\1\\0\end{pmatrix} \qquad |2\rangle = \begin{pmatrix}1\\-1\\0\end{pmatrix} \qquad |3\rangle = \begin{pmatrix}0\\0\\1\end{pmatrix} \tag{1.4.22}$$

are mutually orthogonal (Problem 1.5) but the norms of vectors $|1\rangle$ and $|2\rangle$ are each $\sqrt{2}$. The vectors become orthonormal if the components of each vector are divided by the norm of the vector. The orthonormal vectors are

$$|1\rangle = \begin{pmatrix}\dfrac{1}{\sqrt{2}}\\ \dfrac{1}{\sqrt{2}}\\ 0\end{pmatrix} \qquad |2\rangle = \begin{pmatrix}\dfrac{1}{\sqrt{2}}\\ -\dfrac{1}{\sqrt{2}}\\ 0\end{pmatrix} \qquad |3\rangle = \begin{pmatrix}0\\0\\1\end{pmatrix} \tag{1.4.23}$$

The scalar products of this set of vectors will now be 1 (for the same vector) or 0 (for different vectors).

For convenience, the common normalization factor is often presented as a factor with the vector. For example, $|2\rangle$ can be written as

$$|2\rangle = \left(\frac{1}{\sqrt{2}}\right)\begin{pmatrix}1\\-1\\0\end{pmatrix} \tag{1.4.24}$$

In situations where the two vectors will ultimately be used to generate a scalar, the normalization factors from the bra and ket vectors can be combined and associated with the bra vector. $\langle 2|$ and $|2\rangle$ are then

$$\langle 2| = \left(\frac{1}{2}\right)(1 \quad -1 \quad 0) \qquad |2\rangle = \begin{pmatrix}1\\-1\\0\end{pmatrix} \tag{1.4.25}$$

The orthogonality of vectors can be used to develop some geometric properties and equations in space. The linear equation

$$y = 2x - 3 \tag{1.4.26}$$

can be written as a sum of x and y factors:

$$2x - 1y = 3 \tag{1.4.27}$$

The actual direction of the line involves a motion of 2 units on x for every one unit on y. A vector for this line can then be written using 2 as the x component and -1 as the y component:

$$|r\rangle = \begin{pmatrix} 2 \\ -1 \end{pmatrix} \tag{1.4.28}$$

A vector $\langle s|$ orthogonal to $|r\rangle$ must have components such that

$$\langle s|r\rangle = 0 \tag{1.4.29}$$

In component form with unknown $\langle s|$ components s_x and s_y, this condition is

$$(s_x \ \ s_y)\begin{pmatrix} 2 \\ -1 \end{pmatrix} = 0 \tag{1.4.30}$$

To satisfy this condition, the $\langle s|$ vector must have components proportional to

$$\langle s| = (1 \quad 2) \tag{1.4.31}$$

The linear equation perpendicular to $|r\rangle$ then has the form

$$1x + 2y = c \tag{1.4.32}$$

where c will be determined by the point (x, y) where the two lines intersect. For example, if the point $(1, -1)$ on the first line is chosen as the point of intersection, the orthogonal linear equation is

$$x + 2y = 1(1) + 2(-1) = -1 = c \tag{1.4.33}$$

The linear equation is

$$\begin{aligned} x + 2y &= -1 \\ 2y &= -x - 1 \end{aligned} \tag{1.4.34}$$

The two linear plots are shown in Figure 1.14.

While equations of the form

$$ax + by = c \tag{1.4.35}$$

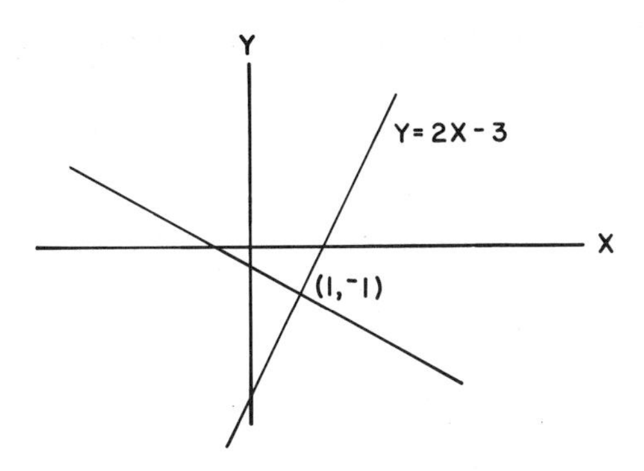

Figure 1.14. The linear equation $2y = -x - 1$ perpendicular to the equation $2x - y = 3$ through the point $(1, -1)$.

define a line in the three-dimensional space, an equation of the form

$$ax + by + cz = d \tag{1.4.36}$$

defines a plane in the three-dimensional space. x and y must be specified to find z. A vector normal to the plane is determined and used to generate the equation for the plane. The vector

$$|n\rangle = \begin{pmatrix} a \\ b \\ c \end{pmatrix} \tag{1.4.37}$$

must be perpendicular to any line in the plane. If the normal is to be defined at the point (x_0, y_0, z_0) in the plane, then a line in the plane to a point (x, y, z) has the components

$$(x - x_0, y - y_0, z - z_0) \tag{1.4.38}$$

and the scalar product of this vector in the plane and the normal vector must be zero:

$$(x - x_0 \quad y - y_0 \quad z - z_0) \begin{pmatrix} a \\ b \\ c \end{pmatrix} = 0$$

$$a(x - x_0) + b(y - y_0) + c(z - z_0) = 0 \tag{1.4.39}$$

This equation can be converted to

$$ax + by + cz = d = ax_0 + by_0 + cz_0 \tag{1.4.40}$$

If the equation of a plane is written in this form, the normal to this plane will always be parallel to a vector with components (a, b, c).

1.5. Other Vector Combinations

The scalar product multiplies the components of two vectors to form a scalar. Two vectors can be combined by addition to form a new resultant vector. These two ideas can be combined into an operation which multiplies two vectors to produce a new vector. This is the vector or cross product.

If a particle of mass m and velocity $\mathbf{v}$ is constrained to a circular path of radius r, its linear momentum is

$$\mathbf{p} = m\mathbf{v} \tag{1.5.1}$$

and its angular momentum is the product of this linear momentum and the radius:

$$\mathbf{L} = r\mathbf{p} \tag{1.5.2}$$

This angular momentum is expressed in vector components by examining the system at some instant of time and introducing a coordinate system. The appropriate Cartesian coordinate system is shown in Figure 1.15. Because the linear momentum is always tangent to the circle of motion, it is perpendicular to the radius from the center. Using this system, the angular momentum at this instant is the product

$$r_x p_y \tag{1.5.3}$$

The vector components have different axis subscripts.

If the angular momentum is generated by $\mathbf{r}$ and $\mathbf{p}$ components along x and y, respectively, what is the direction of the $\mathbf{L}$ vector? In Figure 1.15, the angular momentum vector has an origin at the center of the circle and has a magnitude proportional to the product of $\mathbf{r}$ and $\mathbf{p}$. However, its direction must be perpendicular to both $\mathbf{r}$ and $\mathbf{p}$, i.e., normal to plane of rotation. This is a right-handed coordinate system where the counterclockwise motion of the particle describes an angular momentum vector in a positive z direction, i.e., above the plane of the ring. The L_z vector is located at the center of the circle and has a magnitude equal to $r_x p_y$.

The product $r_x p_y$ defines only part of L_z. For a full vector product, the three components of $\mathbf{r}(r_x, r_y, r_z)$ and $\mathbf{p}(p_x, p_y, p_z)$ must be combined to produce the three components of angular momentum. The vector product multiplied orthogonal components of $\mathbf{r}$ and $\mathbf{p}$. If the actual $\mathbf{r}$ and $\mathbf{p}$ vectors in space are not orthogonal, their vector product is formed by selecting a component on one

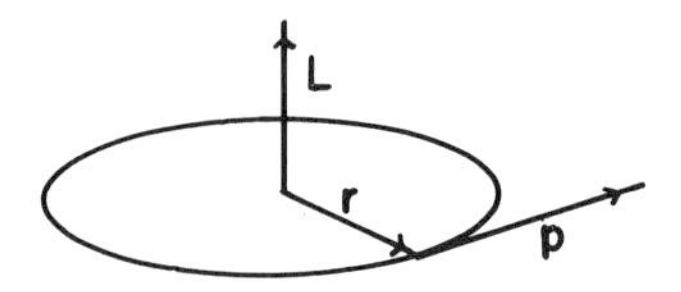

Figure 1.15. The components r_x, p_y, and L_z for a rotating particle.

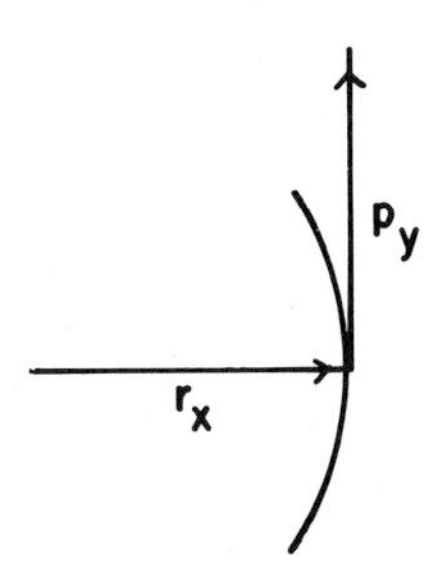

Figure 1.16. The vector product defined by selecting a vector component of **p** perpendicular to vector **r**.

vector which is perpendicular to the other vector (Figure 1.16). The component of **p** perpendicular to **r** is

$$p \sin \theta \tag{1.5.4}$$

and the vector product is

$$\mathbf{r} \times \mathbf{p} = rp \sin \theta \tag{1.5.5}$$

with a direction perpendicular to the plane formed by the two vectors.

The use of perpendicular components suggests that the vector prroduct can be defined using the components of the vectors in an arbitrary orthogonal coordinate system. The vectors **r** and **p** have already been resolved into components to form the product

$$r_x p_y \tag{1.5.6}$$

which gave the L_z component. However, the components p_x and r_y are also in this plane and perpendicular to each other. Their contribution to L_z is illustrated in Figure 1.17. To produce the counterclockwise motion of the particle which defines a positive L_z, the component of the **r** vector on the y axis must be negative. If a positive r_y was selected, then p_x would be negative as shown in the figure. The two contributions to L_z have opposite signs, and the total z component of angular momentum is

$$L_z = r_x p_y - p_x r_y \tag{1.5.7}$$

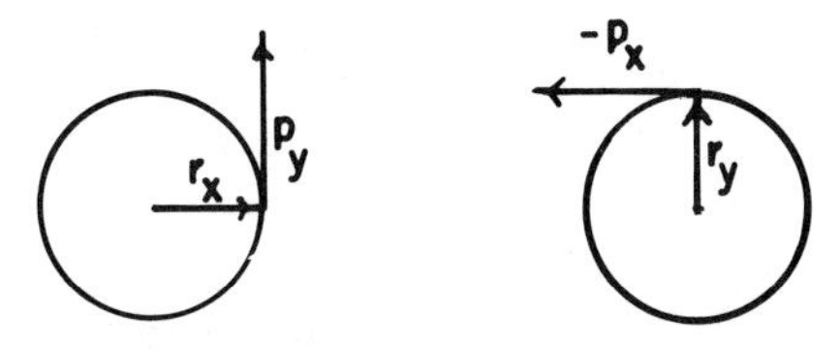

Figure 1.17. The two component contributions to the angular momentum vector component in the z direction.

It is easy to show (Problem 1.6) that the remaining two components of angular momentum are

$$L_x = r_y p_z - p_y r_z \tag{1.5.8}$$

$$L_y = r_z p_x - p_z r_x \tag{1.5.9}$$

The order of coordinates is a characteristic of the right-handed coordinate system which was chosen.

Since each component does require the difference of the products of the vector components, it is convenient to introduce a framework to generate these components. Determinants are discussed in Chapter 3. However, a 3×3 determinant is introduced here to provide a framework for component generation. If **i**, **j**, and **k** are defined as the unit vectors for the x, y, and z directions, respectively, these three vectors are listed as the first row of a 3×3 array. The second row contains the r_x, r_y, and r_z components of **r** and the third row contains the p_x, p_y, and p_z components of **p**. In this form, the first column contains only x components, the second column contains only y components, and the third column contains only z components. The determinant is

$$\begin{vmatrix} \mathbf{i} & \mathbf{j} & \mathbf{k} \\ r_x & r_y & r_z \\ p_x & p_y & p_z \end{vmatrix} \tag{1.5.10}$$

The L_x component is found by crossing out the row and column containing **i**. The 2×2 array which remains is evaluated as the products of pairs of diagonal elements. The product $r_y p_z$ is assigned a positive sign while the product $p_y r_z$ is assigned a negative sign to give the result

$$\mathbf{i}(r_y p_z - p_y r_z) \tag{1.5.11}$$

The sign for the vector **i** is determined by its position in the array. Its position in the first row and first column gives a positive sign:

$$(-1)^{1+1} = +1 \tag{1.5.12}$$

The unit vector **j** must be preceded by a negative sign since it lies in the first row and second column (12 position):

$$(-1)^{1+2} = -1 \tag{1.5.13}$$

The remaining components are determined in the same manner (Problem 1.18).

The vector product in three-dimensional space will always generate a third vector normal to the plane of the first two vectors. The plane of the first two vectors and the third, normal vector define a coordinate system in the space. Even if the two vectors in the plane are not orthogonal, the three vectors can constitute a three-dimensional basis set. The technique fails if the two vectors in the plane are parallel.

Two operations determine if three vectors **a**, **b**, and **c** in the three-dimensional space span this space. Two of the vectors, e.g., **b** and **c**, are selected as the elements of a vector product; the vector product gives a new vector **d** which is orthogonal to both **b** and **c**. If the final vector (**a**) of the set has a component on the **d** vector, a nonzero scalar product is obtained, i.e.,

$$\mathbf{a} \cdot \mathbf{d} \neq 0 \tag{1.5.14}$$

The vectors **a**, **b**, and **c** will have components in all three spatial directions, i.e., they will span the space, if

$$\mathbf{a} \cdot (\mathbf{b} \times \mathbf{c}) \neq 0 \tag{1.5.15}$$

The combination of scalar and vector product can be generated using any arrangement of the three vectors (Problem 1.8).

The combination $\mathbf{a} \cdot (\mathbf{b} \times \mathbf{c})$ is called a triple scalar product, and it assumes a convenient form when written in terms of components. The x component of the vector product $\mathbf{b} \times \mathbf{c}$ is

$$(\mathbf{b} \times \mathbf{c})_x = \mathbf{i}(b_y c_z - b_z c_y) \tag{1.5.16}$$

The scalar product of this vector with **a** will then give a scalar term associated with the i components:

$$\begin{aligned}(\mathbf{i}a_x) \cdot (b_y c_z - b_z c_y) &= (\mathbf{i} \cdot \mathbf{i})(a_x b_y c_z - a_x b_z c_y) \\ &= a_x b_y c_z - a_x b_z c_y \end{aligned} \tag{1.5.17}$$

The scalar terms for the remaining two vector components are

$$-[a_y b_x c_z - a_y c_x b_z] \tag{1.5.18}$$

$$a_z b_x c_y - a_z c_x b_y \tag{1.5.19}$$

The full triple scalar product is the sum of these nine component products. However, this scalar result can be generated using the 3×3 determinant (Problem 1.7)

$$\begin{vmatrix} a_x & a_y & a_z \\ b_x & b_y & b_z \\ c_x & c_y & c_z \end{vmatrix} \tag{1.5.20}$$

If this determinant gives a nonzero result, the vectors **a**, **b**, and **c** span the three-dimensional space.

The vector product can generate geometric equations in the three-dimensional space. The equation for the plane which passes through three H atoms in a tetrahedral methane atom constitutes a good example. The three atoms are found at the Cartesian coordinates

$$
\begin{gathered}
(1, 1, 1) \\
(-1, -1, 1) \\
(-1, 1, -1)
\end{gathered}
\tag{1.5.21}
$$

These three points can be used to create two vectors by subtracting components point by point:

$$
\begin{aligned}
(1, 1, 1) - (-1, -1, 1) &= (2, 2, 0) \\
(1, 1, 1) - (-1, 1, -1) &= (2, 0, 2)
\end{aligned}
\tag{1.5.22}
$$

The vector product of these two vectors will give a vector which is normal to the plane of the two vectors:

$$
\begin{vmatrix} \mathbf{i} & \mathbf{j} & \mathbf{k} \\ 2 & 2 & 0 \\ 2 & 0 & 2 \end{vmatrix} = \mathbf{i}4 - \mathbf{j}4 - \mathbf{k}4 \tag{1.5.23}
$$

Since this vector is normal to the plane, the equation for the plane is the dot product of this vector and the vector

$$
\mathbf{i}(x-1) + \mathbf{j}(y-1) + \mathbf{k}(z-1) \tag{1.5.24}
$$

through the point (1, 1, 1) in the plane. The equation of the plane is

$$
\begin{gathered}
4(x-1) - 4(y-1) - 4(z-1) = 0 \\
x - y - z = -1
\end{gathered}
\tag{1.5.25}
$$

The determinant format gives the components of the final vector but does not give the norm of the vector explicitly. This norm can be determined from the scalar products of the initial vectors. If the vectors are **r** and **p**, the vector product is

$$
|\mathbf{L}| = |\mathbf{r} \times \mathbf{p}| = |\mathbf{r}|\ |\mathbf{p}| \sin\theta \tag{1.5.26}
$$

Squaring each side and applying the identity

$$
\sin^2\theta = 1 - \cos^2\theta \tag{1.5.27}
$$

gives

$$
\begin{aligned}
|\mathbf{L}|^2 &= |\mathbf{r}|^2\ |\mathbf{p}|^2 \sin^2\theta \\
&= |\mathbf{r}|^2\ |\mathbf{p}|^2 - |\mathbf{r}|^2\ |\mathbf{p}|^2 \cos^2\theta
\end{aligned}
\tag{1.5.28}
$$

Since the second term is the scalar product of **r** and **p**, the equation is

$$\langle L|L\rangle = \langle r|r\rangle\langle p|p\rangle - \langle r|p\rangle^2 \tag{1.5.29}$$

Although the vector and scalar products are the more common vector products, a third product does exist. This is the direct product of two vectors. When two particles exist in a space and do not interact with each other, then it is necessary to specify the position coordinates of each particle to locate them both. This would create a six-dimensional space, and it would be necessary to examine each component individually. This format creates difficulties if operations must be performed on one of these particles (Chapter 10). If the two vectors are

$$|r_1\rangle \quad \text{and} \quad |r_2\rangle \tag{1.5.30}$$

for particles 1 and 2, then an operation on $|r_1\rangle$ should not disturb $|r_2\rangle$. This leads to the notation for forming a direct product of the two vectors. The direct product can be written

$$|r_1\rangle|r_2\rangle \tag{1.5.31}$$

or

$$|r_1\rangle \otimes |r_2\rangle \tag{1.5.32}$$

An operation $\mathbf{O}(\mathbf{r}_2)$ which acts only on the coordinates of $|r_2\rangle$ then ignores $|r_1\rangle$ in the product:

$$O(\mathbf{r}_2)|r_1\rangle|r_2\rangle = |r_1\rangle 0(\mathbf{r}_2)|r_2\rangle \tag{1.5.33}$$

In component notation, each component of $|r_2\rangle$ is multiplied by the vector $|r_1\rangle$:

$$|r_1\rangle|r_2\rangle = \begin{pmatrix} |r_1\rangle r_{2x} \\ |r_1\rangle r_{2y} \\ |r_1\rangle r_{2z} \end{pmatrix} \tag{1.5.34}$$

If $|r_1\rangle$ is resolved into its components, a nine-component vector results:

$$\begin{pmatrix} r_{1x}r_{2x} \\ r_{1y}r_{2x} \\ r_{1z}r_{2x} \\ r_{1x}r_{2y} \\ r_{1y}r_{2y} \\ r_{1z}r_{2y} \\ r_{1x}r_{2z} \\ r_{1y}r_{2z} \\ r_{1z}r_{2z} \end{pmatrix} \tag{1.5.35}$$

Each r_2 coordinate appears with each r_1 coordinate in the nine products.

The direct product appears frequently in quantum mechanics for systems which have both spatial and spin coordinates. These parameters have distinct coordinate systems, and it is common to write the total wavefunction as the direct product of a spatial wavefunction $|L\rangle$ and a spin wavefunction $|S\rangle$:

$$|L\rangle |S\rangle \tag{1.5.36}$$

An operator involving spatial coordinates will then have no effect on $|S\rangle$ but will operate on $|L\rangle$. The spin wavefunction will be present throughout the operation.

1.6. Orthogonality and Biorthogonality

It has been convenient to resolve vectors into their components to perform operations like the scalar and vector products. In the cases considered, these components were determined for a "standard" Cartesian coordinate system. The x, y, and z axes were chosen as the coordinate system, and all vectors in the space were projected onto these axes. However, any set of mutually orthogonal axes could be used as the coordinate system for the three-dimensional space. Care is necessary since these new coordinate axes are defined in terms of the original x, y, and z axes. A vector $\mathbf{r}$ will project onto coordinate axes which, in turn, project onto the original axes such as the x, y, and z axes in three-dimensional space. These new coordinate systems need not be orthogonal.

The original Cartesian coordinate system to describe vectors in three-dimensional space is oriented on x, y, and z axes (Figure 1.18). Unit vectors for this system are defined as a single unit on the appropriate axis, i.e., the x-coordinate unit vector will have a value of 1 on the x axis and values of zero on the y and z

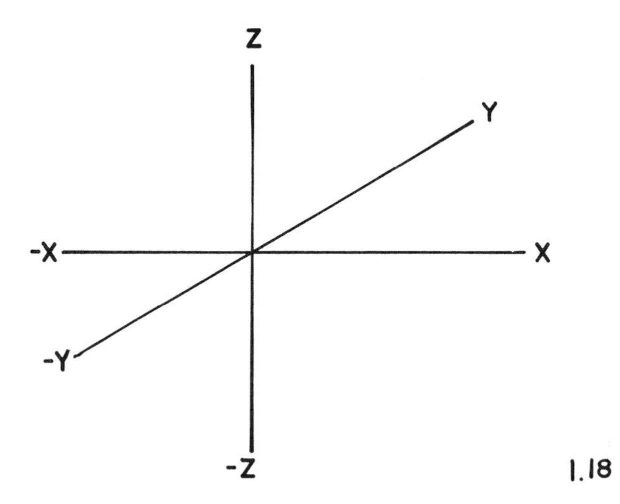

Figure 1.18. A right-handed x–y–z coordinate system.

axes. The unit vector on the y axis will have a value of unity for its y component and zeros for the x and z components. The three unit vectors are written as three-component column vectors:

$$|x\rangle = \begin{pmatrix} 1 \\ 0 \\ 0 \end{pmatrix} \qquad |y\rangle = \begin{pmatrix} 0 \\ 1 \\ 0 \end{pmatrix} \qquad |z\rangle = \begin{pmatrix} 0 \\ 0 \\ 1 \end{pmatrix} \tag{1.6.1}$$

These coordinate vectors are orthonormal. For example,

$$\langle y|x\rangle = (0 \quad 1 \quad 0) \begin{pmatrix} 1 \\ 0 \\ 0 \end{pmatrix} = 0 \tag{1.6.2}$$

The orthogonality of the three vectors confirms that they span the three-dimensional space. This is verified with the determinant of their components, which is not zero:

$$\begin{vmatrix} 1 & 0 & 0 \\ 0 & 1 & 0 \\ 0 & 0 & 1 \end{vmatrix} = 1 \tag{1.6.3}$$

In many situations, a coordinate system which is orthogonal but different from the standard coordinate system will be preferred. The coordinate system of Figure 1.19 retains the same z axis as the standard system but the axes in the x–y plane are oriented at 45° to the original axes. These original axes could now be erased and the new axes defined as x and y, respectively. These new vectors are orthogonal to each other and z, and any vector could be projected onto them in exactly the same way as they are projected onto the standard axes.

The standard x and y axes are familiar, and it is convenient not to eliminate them completely. Instead, the new coordinate axes are expressed in terms of their components on the standard axes. Figure 1.19 shows that the new x axis (x') will have a unit vector with projections of $\frac{1}{\sqrt{2}}$ on the original x and y axes and no projection on the z axis. The new y axis (y') will have projections of $\frac{1}{\sqrt{2}}$ and $\frac{-1}{\sqrt{2}}$ on the x and y axes and no projection on the z axis. The z axis is unchanged. The set of unit vectors for the new coordinate system is

$$|x'\rangle = \begin{pmatrix} \frac{1}{\sqrt{2}} \\ \frac{1}{\sqrt{2}} \\ 0 \end{pmatrix} \qquad |y'\rangle = \begin{pmatrix} \frac{1}{\sqrt{2}} \\ \frac{-1}{\sqrt{2}} \\ 0 \end{pmatrix} \qquad |z\rangle = \begin{pmatrix} 0 \\ 0 \\ 1 \end{pmatrix} \tag{1.6.4}$$

Scalar products of these new coordinate axes establish their orthogonality. For example,

$$\langle y'|x'\rangle = \begin{pmatrix} \frac{1}{\sqrt{2}} & \frac{-1}{\sqrt{2}} & 0 \end{pmatrix} \begin{pmatrix} \frac{1}{\sqrt{2}} \\ \frac{1}{\sqrt{2}} \\ 0 \end{pmatrix}$$

$$= \frac{1}{2} - \frac{1}{2} + 0 = 0 \tag{1.6.5}$$

The determinant of vector components is nonzero,

$$\begin{vmatrix} \frac{1}{\sqrt{2}} & \frac{1}{\sqrt{2}} & 0 \\ \frac{1}{\sqrt{2}} & \frac{-1}{\sqrt{2}} & 0 \\ 0 & 0 & 1 \end{vmatrix} = -1 \tag{1.6.6}$$

which confirms that these vectors do span the space.

Coordinate systems are not restricted to a three-dimensional space. In the cyclobutadiene molecule, each of the four carbon atoms contributes an electron to a molecular orbital for the molecule. These four atomic orbitals each represent

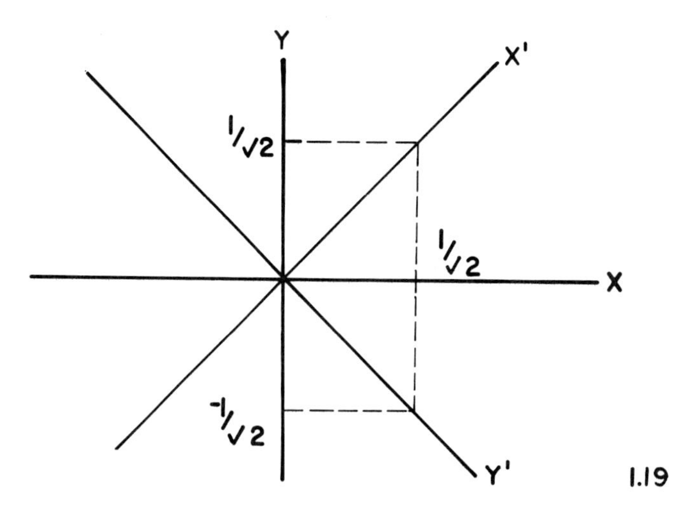

Figure 1.19. A three-dimensional coordinate system with x' and y' axes oriented 45° to the standard x and y axes. The z axis is perpendicular to the paper.

an independent axis in a four-dimensional space. The atomic orbital coordinates are then defined as

$$|1\rangle = \begin{pmatrix} 1 \\ 0 \\ 0 \\ 0 \end{pmatrix} \quad |2\rangle = \begin{pmatrix} 0 \\ 1 \\ 0 \\ 0 \end{pmatrix} \quad |3\rangle = \begin{pmatrix} 0 \\ 0 \\ 1 \\ 0 \end{pmatrix} \quad |4\rangle = \begin{pmatrix} 0 \\ 0 \\ 0 \\ 1 \end{pmatrix} \tag{1.6.7}$$

i.e., the atomic orbitals φ_1, φ_2, φ_3, and φ_4 are coordinate axes labeled as the first, second, third, and fourth components of the four-dimensional vector.

The simplest molecular orbital for cyclobutadiene is a sum of all four atomic orbitals:

$$\psi = \varphi_1 + \varphi_2 + \varphi_3 + \varphi_4 \tag{1.6.8}$$

This molecular orbital defines a new coordinate direction in the four-dimensional space. The new coordinate has relative components

$$|1'\rangle = \begin{pmatrix} 1 \\ 1 \\ 1 \\ 1 \end{pmatrix} \tag{1.6.9}$$

For an orthogonal set of molecular orbitals in this four-dimensional space, three other coordinate vectors which are orthogonal to both $|1'\rangle$ and each other are required (Problem 1.9), e.g.,

$$|2'\rangle = \left(\frac{1}{2}\right) \begin{pmatrix} 1 \\ -1 \\ -1 \\ 1 \end{pmatrix} \quad |3'\rangle = \left(\frac{1}{2}\right) \begin{pmatrix} 1 \\ 1 \\ -1 \\ -1 \end{pmatrix} \quad |4'\rangle = \left(\frac{1}{2}\right) \begin{pmatrix} 1 \\ -1 \\ 1 \\ -1 \end{pmatrix} \tag{1.6.10}$$

These vector components give the contribution of each atomic orbital to the final molecular orbital. The atomic orbitals each have the same shape but their coefficients may differ to produce maxima and minima (nodes) for the total wavefunction.

The coordinate axes do not have to be orthogonal. The coordinates of Figure 1.20 have a positive z axis above the page to permit examination of the x–y plane. The two vectors in the original x–y–z (standard) component system are

$$|1\rangle = \begin{pmatrix} 1 \\ 1 \\ 0 \end{pmatrix} \quad |2\rangle = \begin{pmatrix} 1 \\ -2 \\ 0 \end{pmatrix} \tag{1.6.11}$$

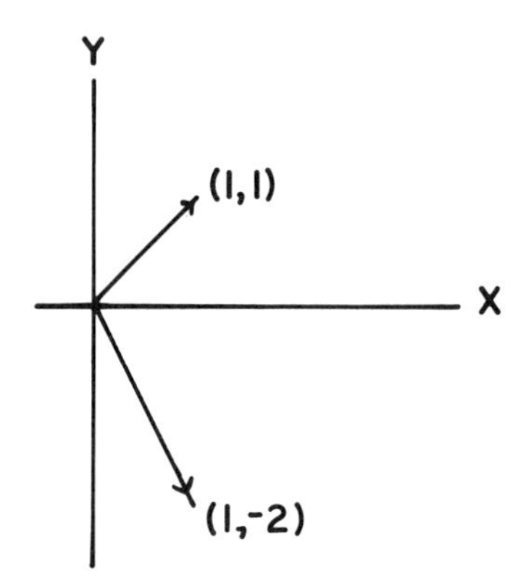

Figure 1.20. Two nonorthogonal, linearly independent coordinate vectors in a two-dimensional space.

With the unaltered z coordinate, the three vectors span the three-dimensional space:

$$\begin{vmatrix} 1 & 1 & 0 \\ 1 & -2 & 0 \\ 0 & 0 & 1 \end{vmatrix} = -3 \tag{1.6.12}$$

Since the z axis remains unchanged, only the two-dimensional vectors for components in the x–y plane are analyzed.

Although the components of the two-dimensional ket vectors for the coordinates are known, the bra vectors required for mutual orthogonality may not have the same components. If they were assigned the same components,

$$\langle 1| = (1 \quad 1) \qquad \langle 2| = (1 \quad -2) \tag{1.6.13}$$

their scalar products are not zero:

$$\begin{aligned} \langle 1|2\rangle &= (1 \quad 1)\begin{pmatrix} 1 \\ -2 \end{pmatrix} = -1 \\ \langle 2|1\rangle &= (1 \quad -2)\begin{pmatrix} 1 \\ 1 \end{pmatrix} = -1 \end{aligned} \tag{1.6.14}$$

Although bra and ket vectors with the same components are not orthogonal for this case, it is possible to find a set of bra vectors which are mutually orthogonal to the ket vectors. For example,

$$\begin{aligned} \langle 1| &= (2 \quad 1) \\ \langle 2| &= (1 \quad -1) \end{aligned} \tag{1.6.15}$$

give zero scalar products:

$$\langle 1|2\rangle = (2 \quad 1)\begin{pmatrix} 1 \\ -2 \end{pmatrix} = 0$$
$$\langle 2|1\rangle = (1 \quad -1)\begin{pmatrix} 1 \\ 1 \end{pmatrix} = 0 \tag{1.6.16}$$

The scalar products $\langle 1|1\rangle$ and $\langle 2|2\rangle$ are finite and the vectors can be normalized:

$$\langle 1|1\rangle = (2 \quad 1)\begin{pmatrix} 1 \\ 1 \end{pmatrix} = 3$$
$$\langle 2|2\rangle = (1 \quad -1)\begin{pmatrix} 1 \\ -2 \end{pmatrix} = 3 \tag{1.6.17}$$

It is convenient to place the normalization factor with the bra vectors:

$$\langle 1| = \left(\frac{1}{3}\right)(2 \quad 1)$$
$$\langle 2| = \left(\frac{1}{3}\right)(1 \quad -1) \tag{1.6.18}$$

The set of ket vectors and the set of bra vectors constitute a biorthogonal set of vectors. The vectors belong to different two-dimensional spaces. These two spaces can be paired to produce the scalar product. The two vector spaces and basis vectors are shown in Figure 1.21. The $(1 \quad -1)$ vector in the ket space is obviously orthogonal to the $(1 \quad 1)$ vector in the bra space. The same is true for the $(2 \quad 1)$ and $(1 \quad -2)$ vectors.

Biorthogonality appears for vector systems with imaginary components. For example, the ket vectors

$$|1\rangle = \begin{pmatrix} i \\ 1 \end{pmatrix} \qquad |2\rangle = \begin{pmatrix} -i \\ 1 \end{pmatrix} \tag{1.6.19}$$

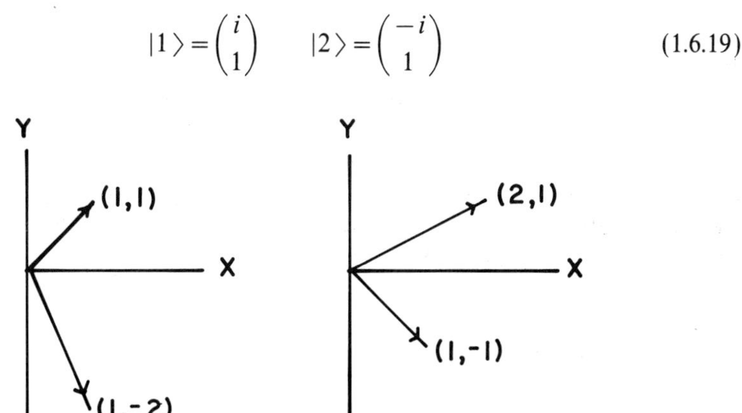

Figure 1.21. Comparison of vector coordinates in the bra and ket space to illustrate biorthogonality.

are not orthogonal without some change in the components of the bra vectors. In this case, the bra vectors must have components which are the complex conjugates of the ket vector components. These vectors are

$$\begin{aligned} \langle 1| &= (-i, \quad 1) \\ \langle 2| &= (i \quad 1) \end{aligned} \tag{1.6.20}$$

The vectors are orthogonal:

$$\begin{aligned} \langle 1|2\rangle &= (-i \quad 1)\begin{pmatrix} -i \\ 1 \end{pmatrix} = i^2 + 1 = 0 \\ \langle 2|1\rangle &= (i \quad 1)\begin{pmatrix} i \\ 1 \end{pmatrix} = i^2 + 1 = 0 \end{aligned} \tag{1.6.21}$$

and give real norms:

$$\begin{aligned} \langle 1|1\rangle &= (-i \quad 1)\begin{pmatrix} i \\ 1 \end{pmatrix} = -i^2 + 1 = 2 \\ \langle 2|2\rangle &= (i \quad 1)\begin{pmatrix} -i \\ 1 \end{pmatrix} = -i^2 + 1 = 2 \end{aligned} \tag{1.6.22}$$

Bra and ket vectors whose components are complex conjugates are closely related to quantum mechanical integrals such as

$$\int \psi^* \psi \, dV \tag{1.6.23}$$

Such integrals give real values (Problem 1.10) and the product

$$\psi^* \psi \tag{1.6.24}$$

is always a real probability density.

Sets of vectors in the ket space which are orthogonal to each other produce a set of bra vectors with the same components. In this case, the vectors within the space were mutually orthogonal already. Their components were simply reassigned to the bra vectors to form that vector space. When the ket vectors were not mutually orthogonal, it was necessary to find a set of vectors for the bra space which would provide consistent orthogonality relations.

1.7. Projection Operators

The resolution of a vector into its projections on some coordinate system axes facilitates the determination of vector-related quantities such as the scalar and vector products. It also permits representation of the vector as an ordered set of scalars. The direction associated with each of the scalars is contained explicitly in the order. However, there are some cases where both the scalar projection of a vector and the direction of the component in the space must be known.

The standard x–y–z Cartesian coordinate system is characterized by the ordered vectors

$$|x\rangle = \begin{pmatrix} 1 \\ 0 \\ 0 \end{pmatrix} \qquad |y\rangle = \begin{pmatrix} 0 \\ 1 \\ 0 \end{pmatrix} \qquad |z\rangle = \begin{pmatrix} 0 \\ 0 \\ 1 \end{pmatrix} \tag{1.7.1}$$

$$\langle x| = (1 \quad 0 \quad 0) \qquad \langle y| = (0 \quad 1 \quad 0) \qquad \langle z| = (0 \quad 0 \quad 1)$$

If a vector $\langle r|$ is resolved into components r_x, r_y, and r_z on the coordinate axes, the scalar product of $|r\rangle$ with the unit vectors $\langle i|$, $i = x, y,$ and z, gives these components:

$$\langle x|r\rangle = (1 \quad 0 \quad 0) \begin{pmatrix} r_x \\ r_y \\ r_z \end{pmatrix} = r_x \tag{1.7.2}$$

The operations $\langle y|r\rangle$ and $\langle z|r\rangle$ will give the components r_y and r_z, respectively.

The scalar product gives the magnitudes of $\mathbf{r}$ in each of the coordinate directions. These magnitudes lie along their respective coordinate directions so the unit vectors $\mathbf{i}$, $\mathbf{j}$, and $\mathbf{k}$ for the x, y, and z coordinate directions must be included:

$$\mathbf{i}r_x + \mathbf{j}r_y + \mathbf{k}r_z \tag{1.7.3}$$

Using $|x\rangle$, $|y\rangle$, and $|z\rangle$ as the unit vectors, Equation 1.7.3 becomes

$$|x\rangle\, r_x + |y\rangle\, r_y + |z\rangle\, r_z \tag{1.7.4}$$

Since the components of $|r\rangle$ are scalar products on $\langle x|$, $\langle y|$, and $\langle z|$, these scalar products are substituted for the components, r_i, in Equation 1.7.4 to give

$$|r\rangle = |x\rangle\langle x|r\rangle + |y\rangle\langle y|r\rangle + |z\rangle\langle z|r\rangle \tag{1.7.5}$$

The bra and ket unit vectors appear as

$$|i\rangle\langle i| \tag{1.7.6}$$

and cannot be a scalar product. The vector $|r\rangle$ is common to all three terms and can be separated to give

$$|r\rangle = (|x\rangle\langle x| + |y\rangle\langle y| + |z\rangle\langle z|)\,|r\rangle \tag{1.7.7}$$

Since $|r\rangle$ appears on each side of the equation, the factor in brackets must be unity:

$$|x\rangle\langle x| + |y\rangle\langle y| + |z\rangle\langle z| = 1 \tag{1.7.8}$$

The "reversed" ket–bra pairs are called projection operators. They operate on a vector in the space to determine both the scalar projection on a preselected coordinate and the spatial direction of this coordinate. The $|r\rangle$ vector

$$\begin{pmatrix} 1 \\ 2 \\ 3 \end{pmatrix} \tag{1.7.9}$$

will have the projection

$$|x\rangle\langle x|r\rangle \tag{1.7.10}$$

on the x coordinate axis, i.e.,

$$\begin{pmatrix} 1 \\ 0 \\ 0 \end{pmatrix} (1 \quad 0 \quad 0) \begin{pmatrix} 1 \\ 2 \\ 3 \end{pmatrix} = \begin{pmatrix} 1 \\ 0 \\ 0 \end{pmatrix} (1) \tag{1.7.11}$$

The projection on the z coordinate is

$$\begin{pmatrix} 0 \\ 0 \\ 1 \end{pmatrix} (0 \quad 0 \quad 1) \begin{pmatrix} 1 \\ 2 \\ 3 \end{pmatrix} = \begin{pmatrix} 0 \\ 0 \\ 1 \end{pmatrix} (3) \tag{1.7.12}$$

The unit ket vector is not involved in the scalar product and remains to give the coordinate direction. If **r** a bra vector ($\langle r|$), the vector gives a scalar product with the column vector while the row vector indicates the coordinate direction.

The utility of the projection operators becomes clear when the coordinate system is not the standard x–y–z coordinate system. For example, in the previous section, the three vectors

$$|x'\rangle = \left(\frac{1}{\sqrt{2}}\right) \begin{pmatrix} 1 \\ 1 \\ 0 \end{pmatrix} \qquad |y'\rangle = \left(\frac{1}{\sqrt{2}}\right) \begin{pmatrix} 1 \\ -1 \\ 0 \end{pmatrix} \qquad |z\rangle = \begin{pmatrix} 0 \\ 0 \\ 1 \end{pmatrix} \tag{1.7.13}$$

were developed as an orthonormal coordinate system. If these vectors are used as elements of the projection operator, the projectors

$$|x'\rangle\langle x'| \qquad |y'\rangle\langle y'| \qquad |z'\rangle\langle z'| \tag{1.7.14}$$

will select the scalar component of the vector $|r\rangle$ on a unit primed vector and indicate the spatial direction of this vector. The components of both **r** and the unit vectors are expressed relative to the standard x–y–z coordinate system.

The technique is illustrated with the $|r\rangle$ vector of Equation 1.7.9. The projection on the coordinate x' is

$$|x'\rangle\langle x|r\rangle = \left(\frac{1}{\sqrt{2}}\right)\begin{pmatrix}1\\1\\0\end{pmatrix}\left(\frac{1}{\sqrt{2}}\right)(1 \quad 1 \quad 0)\begin{pmatrix}1\\2\\3\end{pmatrix}$$

$$= \left(\frac{1}{\sqrt{2}}\right)\begin{pmatrix}1\\1\\0\end{pmatrix}\left(\frac{3}{\sqrt{2}}\right)$$

$$= |x'\rangle\left(\frac{3}{\sqrt{2}}\right) \tag{1.7.15}$$

This scalar component could be difficult to determine via a graphical projection of **r** onto **x**$'$. However, expressing both these vectors in terms of their standard x–y–z components simplifies the problem. The vector $|r\rangle$ resolves into the primed component vector sum as (Problem 1.11)

$$|r\rangle = |x'\rangle\left(\frac{3}{\sqrt{2}}\right) + |y'\rangle\left(\frac{-1}{\sqrt{2}}\right) + |z'\rangle(3) \tag{1.7.16}$$

Projection operator techniques will generate the standard x–y–z coordinates from the primed coordinates. The two-dimensional coordinate axes x' and y' in Figure 1.19 are expressed in x–y components as

$$|1\rangle = \left(\frac{1}{\sqrt{2}}\right)\begin{pmatrix}1\\1\end{pmatrix} \qquad |2\rangle = \left(\frac{1}{\sqrt{2}}\right)\begin{pmatrix}1\\-1\end{pmatrix} \tag{1.7.17}$$

and the x coordinate in this coordinate system is

$$|x\rangle = |1\rangle\langle 1|x\rangle + |2\rangle\langle 2|x\rangle$$

$$= \left(\frac{1}{\sqrt{2}}\right)\begin{pmatrix}1\\1\end{pmatrix}\left(\frac{1}{\sqrt{2}}\right)(1 \quad 1)\begin{pmatrix}1\\0\end{pmatrix} + \left(\frac{1}{\sqrt{2}}\right)\begin{pmatrix}1\\-1\end{pmatrix}\left(\frac{1}{\sqrt{2}}\right)(1 \quad -1)\begin{pmatrix}1\\0\end{pmatrix}$$

$$= |1\rangle\left(\frac{1}{\sqrt{2}}\right) + |2\rangle\left(\frac{1}{\sqrt{2}}\right) \tag{1.7.18}$$

The factors of $\dfrac{1}{\sqrt{2}}$ are the projections of the unit **x** vector on the x' and y' axes (Figure 1.19). The unit **y** vector is

$$|y\rangle = |1\rangle\langle 1|y\rangle + |2\rangle\langle 2|y\rangle$$
$$= |1\rangle\left(\frac{1}{\sqrt{2}}\right) + |2\rangle\left(\frac{-1}{\sqrt{2}}\right) \tag{1.7.19}$$

which is easily verified with Figure 1.19.

In the cases selected, the coordinate systems were orthogonal and the components of the vectors could be used in either row or column vectors. The projection operators must be modified for biorthogonal systems in which the sets of bra and ket vectors can be different. For the biorthogonal vector pairs,

$$\langle 1| = \left(\frac{1}{3}\right)(2 \quad 1) \qquad \langle 2| = \left(\frac{1}{3}\right)(1 \quad -1) \tag{1.7.20}$$

$$|1\rangle = \begin{pmatrix} 1 \\ 1 \end{pmatrix} \qquad |2\rangle = \begin{pmatrix} 1 \\ -2 \end{pmatrix} \tag{1.7.21}$$

the projections operators are prepared like the orthogonal projection operators,

$$|1\rangle\langle 1| \quad \text{and} \quad |2\rangle\langle 2| \tag{1.7.22}$$

but the components in the row and column vectors are different.

The resolution of the vector

$$|s\rangle = \begin{pmatrix} 2 \\ 3 \end{pmatrix} \tag{1.7.23}$$

for this coordinate system is

$$|s\rangle = |1\rangle\langle 1|s\rangle + |2\rangle\langle 2|s\rangle$$
$$= |1\rangle\left(\frac{1}{3}\right)(2 \quad 1)\begin{pmatrix} 2 \\ 3 \end{pmatrix} + |2\rangle\left(\frac{1}{3}\right)(1 \quad -1)\begin{pmatrix} 2 \\ 3 \end{pmatrix}$$
$$= |1\rangle\left(\frac{7}{3}\right) + |2\rangle\left(\frac{-1}{3}\right) \tag{1.7.24}$$

The ket vectors $|1\rangle$ and $|2\rangle$ have no normalization facttor. Since the two spaces are required for the normalization, the normalization factor was included in the bra space for convenience.

Biorthogonal projection operators operate differently on row and column vectors. The bra vector with the same components,

$$\langle s| = (2 \quad 3) \tag{1.7.25}$$

resolves as

$$\langle s| = \langle s|1\rangle\langle 1| + \langle s|1\rangle\langle 2|$$
$$= \langle 1|(5) + \langle 2|(-4) \tag{1.7.26}$$

These are the scalar components required in the bra coordinate system.

1.8. Linear Independence and Dependence

The development of sets of unit coordinate vectors and their projection operators had one major requirement. The unit vectors had to span the space so that any vector in the space could be resolved into components on them. If just one unit vector were absent, it would be impossible to know if the description of the vector was accurate since its component on the missing coordinate direction would be unknown. In the three–dimensional space, three arbitrary vectors could span the space. However, if any two of these vectors are parallel, the three vectors cannot span the space. Vectors which do span the space are linearly independent while those which do not are linearly dependent.

Since any two nonparallel vectors will define a plane in a three-dimensional space, two dimensions are spanned if the two vectors give a vector product which is nonzero:

$$\mathbf{a} \times \mathbf{b} \neq 0 \tag{1.8.1}$$

The vector generated by the vector product will be perpendicular to the plane. If the third vector in the trial set has some component on this resultant vector, then the vectors **a**, **b**, and **c** will span the space. This was expressed using the determinant:

$$\begin{vmatrix} a_x & a_y & a_z \\ b_x & b_y & b_z \\ x_x & c_y & c_z \end{vmatrix} \tag{1.8.2}$$

This determinant is evaluated as (Chapter 3)

$$\begin{aligned} &a_x \begin{vmatrix} b_y & b_z \\ c_y & c_z \end{vmatrix} - a_y \begin{vmatrix} b_x & b_z \\ c_x & c_z \end{vmatrix} + a_z \begin{vmatrix} b_x & b_y \\ c_x & c_y \end{vmatrix} \\ &= a_x(b_y c_z - c_y b_z) - a_y(b_x c_z - c_x b_z) + a_z(b_x c_y - c_x b_y) \end{aligned} \tag{1.8.3}$$

If the determinant is zero, then the three vectors do not span the space.

The three vectors

$$|1\rangle = \begin{pmatrix} 1 \\ 0 \\ 0 \end{pmatrix} \qquad |2\rangle = \begin{pmatrix} 0 \\ 1 \\ 0 \end{pmatrix} \qquad |3\rangle = \begin{pmatrix} 1 \\ 1 \\ 0 \end{pmatrix} \tag{1.8.4}$$

do not span the space since

$$\begin{vmatrix} 1 & 0 & 0 \\ 0 & 1 & 0 \\ 1 & 1 & 0 \end{vmatrix} = 0 \tag{1.8.5}$$

When the coordinate vectors are plotted (Figure 1.22), all three vectors lie in the x–y plane. They cannot be used to describe a three-dimensional vector which includes a z component. The abence of a z component appears in the determinant; its final column contains only zeros.

The use of a determinant of components to determine if a set of vectors span the space is not confined to three-dimensional systems. For example, in a two-dimensional space, the two vectors

$$|1\rangle = \begin{pmatrix} 1 \\ 1 \end{pmatrix} \qquad |2\rangle = \begin{pmatrix} 1 \\ -1 \end{pmatrix} \tag{1.8.6}$$

were definitely orthogonal (Figure 1.19). The 2×2 determinant formed with these vector components is

$$\begin{vmatrix} 1 & 1 \\ 1 & -1 \end{vmatrix} = -2 \tag{1.8.7}$$

The nonzero result confirms the linear independence of the two vectors.

For four dimensions, the four orthogonal vectors for cyclobutadiene (Section 1.6) generate the components of a 4×4 determinant:

$$\begin{vmatrix} 1 & 1 & 1 & 1 \\ 1 & -1 & -1 & 1 \\ 1 & 1 & -1 & -1 \\ 1 & -1 & 1 & -1 \end{vmatrix} \tag{1.8.8}$$

Although it cannot be evaluated with the techniques introduced thus far, this determinant has a nonzero value and the four vectors span the four-dimensional space.

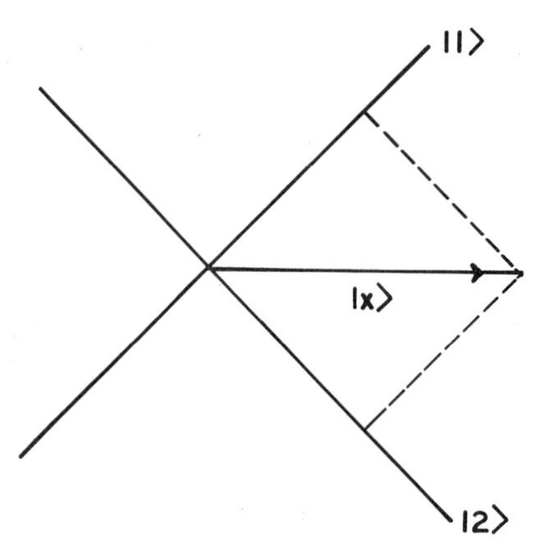

Figure 1.22. A set of three three-component vectors which all lie in the x–y plane. One vector will be linearly dependent on the remaining two in the plane.

N vectors in an N-dimensional space which span this space are linearly independent. Each projects onto some region of the total space. This notion is elucidated using a linearly dependent set in the two-dimensional space. The two vectors

$$|r\rangle=\begin{pmatrix}1\\0\end{pmatrix} \quad \text{and} \quad |s\rangle=\begin{pmatrix}2\\0\end{pmatrix} \tag{1.8.9}$$

have the same direction but different magnitudes and they are linearly dependent:

$$\begin{vmatrix}1 & 0\\2 & 0\end{vmatrix}=0 \tag{1.8.10}$$

The two vectors do not span the two-dimensional space since they are related to each other. The vector $|s\rangle$ can be generated by multiplying $|r\rangle$ by a factor of 2:

$$|s\rangle=2|r\rangle \tag{1.8.11}$$

$$\begin{pmatrix}2\\0\end{pmatrix}=2\begin{pmatrix}1\\0\end{pmatrix} \tag{1.8.12}$$

The equation can be rewritten as

$$2|r\rangle-1|s\rangle=0 \tag{1.8.13}$$

It was possible to find two vector coefficients (2 and -1) which produce a vector difference of zero. In general, a pair of linearly dependent vectors in two-dimensional space will satisfy an equation of the form:

$$c_1|r\rangle+c_2|s\rangle=0 \tag{1.8.14}$$

The coefficients c_i cannot equal zero. If one vector is known, the second is found using Equation 1.8.14, i.e., it is linearly dependent on the first vector. Only one vector is required to span this portion of the space. A new second vector is required for the second dimension in this space.

The orthogonal vectors of Equation 1.8.6 are linearly independent. If they are related using

$$c_1|1\rangle+c_2|2\rangle=0 \tag{1.8.15}$$

the coefficients c_i must be zero. The two-dimensional vectors give two linear equations in the coefficients:

$$c_1\begin{pmatrix}1\\1\end{pmatrix}+c_2\begin{pmatrix}1\\-1\end{pmatrix}=0$$

$$\begin{aligned}(1)\,c_1+(1)\,c_2&=0\\(1)\,c_1+(-1)\,c_2&=0\end{aligned} \tag{1.8.16}$$

Adding the two equations gives

$$2c_1 = 0; \qquad c_1 = 0 \tag{1.8.17}$$

while their subtraction gives

$$2c_2 = 0; \qquad c_2 = 0 \tag{1.8.18}$$

When the two vectors are linearly independent, the only coefficients which satisfy Equation 1.8.14 are zero. There is no way to establish a nonzero linear relationship between two vectors in the two-dimensional space which are linearly independent. The two vectors are directed to different subspaces of the space, and one cannot be used to describe the other.

The arguments can be extended to a space of N dimensions. If a set of N N-component vectors are linearly dependent, then at least two of the coefficients c_i in the equation

$$c_1|1\rangle + c_2|2\rangle + \cdots + c_N|N\rangle = 0 \tag{1.8.19}$$

are nonzero. A set of linearly independent vectors must havc all coefficients equal to zero to satisfy the equation.

A set of vectors may be linearly independent although they are not mutually orthogonal. The set of linearly independent vectors will provide some component on every possible direction in the space. A set of nonorthogonal coordinate vectors which are linearly independent can be converted into an orthogonal coordinate system by selecting vector projections which are orthogonal.

1.9. Orthogonalization of Coordinates

A set of N linearly independent vectors in an N-dimensional space will span that space even if they are not mutually orthogonal. However, it is generally more convenient to use an orthogonal coordinate system. In such a case, the vector components for both the bra and ket vectors are the same, and both the vectors and projection operators are easy to generate. The linearly independent vectors can always be used to generate a new set of vectors which are mutually orthogonal.

The Gram–Schmidt orthogonalization is applicable in any N-dimensional space. For a two-dimensional space, the vectors

$$|1\rangle = \begin{pmatrix} 1 \\ 0 \end{pmatrix} \qquad |2\rangle = \begin{pmatrix} 1 \\ 1 \end{pmatrix} \tag{1.9.1}$$

are linearly independent since the determinant

$$\begin{vmatrix} 1 & 0 \\ 1 & 1 \end{vmatrix} \tag{1.9.2}$$

is nonzero. Although the vectors are not orthogonal (Figure 1.23), it is clear that $|2\rangle$ has a projection on the orthogonal y axis which could be selected as the new coordinate axis.

One axis is always selected as the starting axis. In this case, $|1\rangle$ is selected as this axis and the vector is normalized to create a unit vector. $|2\rangle$ is not orthogonal to $|1\rangle$ so it has a finite projection on $|1\rangle$. If a new $|2'\rangle$ is orthogonal to $|1'\rangle = |1\rangle$, any projections of $|2\rangle$ on $|1'\rangle$ must be eliminated from the vector by subtracting their magnitude $\langle 1'|2\rangle$ and direction $|1'\rangle$ from the original vector:

$$|2'\rangle = |2\rangle - |1\rangle\langle 1|2\rangle \tag{1.9.3}$$

$$= \begin{pmatrix} 1 \\ 1 \end{pmatrix} - \begin{pmatrix} 1 \\ 0 \end{pmatrix} (1 \quad 0) \begin{pmatrix} 1 \\ 1 \end{pmatrix}$$

$$= \begin{pmatrix} 1 \\ 1 \end{pmatrix} - \begin{pmatrix} 1 \\ 0 \end{pmatrix} = \begin{pmatrix} 0 \\ 1 \end{pmatrix} \tag{1.9.4}$$

The new vector $|2'\rangle$ is the unit vector on the y axis since the x axis projection has been subtracted from the original vector $|2\rangle$.

The technique also works if the vector $|2\rangle$ is selected as the initial vector. After normalization, its projection operator is

$$|2\rangle\langle 2| = \left(\frac{1}{\sqrt{2}}\right) \begin{pmatrix} 1 \\ 1 \end{pmatrix} \left(\frac{1}{\sqrt{2}}\right) (1 \quad 1) \tag{1.9.5}$$

The new vector $|1'\rangle$ orthogonal to this vector is

$$|1'\rangle = |1\rangle - |2\rangle\langle 2|1\rangle$$

$$= \begin{pmatrix} 1 \\ 0 \end{pmatrix} - \begin{pmatrix} 1 \\ 1 \end{pmatrix} \left(\frac{1}{2}\right) (1 \quad 1) \begin{pmatrix} 1 \\ 0 \end{pmatrix} \tag{1.9.6}$$

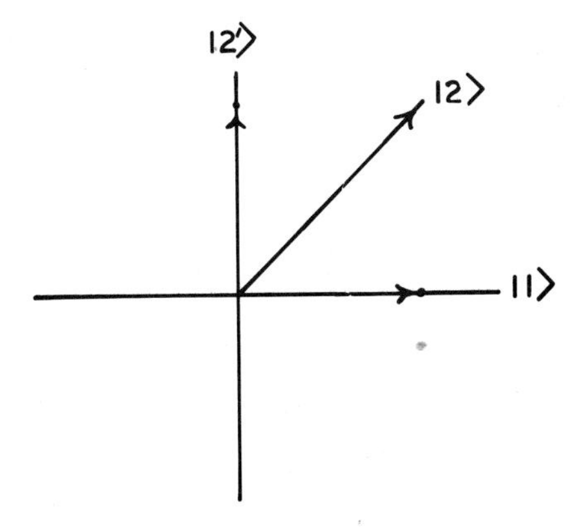

Figure 1.23. The component of a linearly independent vector which is subtracted from this vector in a Gram–Schmidt orthogonalization.

$$\begin{pmatrix}1\\0\end{pmatrix}-\begin{pmatrix}1\\1\end{pmatrix}\begin{pmatrix}1\\\overline{2}\end{pmatrix}=\begin{pmatrix}\dfrac{1}{2}\\-\dfrac{1}{2}\end{pmatrix} \tag{1.9.7}$$

The $|1'\rangle$ vector is normalized as

$$|1'\rangle=\begin{pmatrix}\dfrac{1}{\sqrt{2}}\\\dfrac{-1}{\sqrt{2}}\end{pmatrix} \tag{1.9.8}$$

Since the initial vector was $|2\rangle$,

$$\begin{aligned}|2'\rangle&=|2\rangle\\&=\begin{pmatrix}\dfrac{1}{\sqrt{2}}\\\dfrac{1}{\sqrt{2}}\end{pmatrix}\end{aligned} \tag{1.9.9}$$

Since vectors $|1'\rangle$ and $|2'\rangle$ are oriented $+45°$ and $-45°$ with respect to the x axis, they are separated by $90°$ and are orthogonal.

For an N-dimensional space, $N-1$ vector projections might be subtracted from the selected initial vector. For a four-dimensional system,

$$|1'\rangle=|1\rangle \tag{1.9.10}$$

is selected as the initial vector and subtracted from a second vector $|2\rangle$ to give a $|2'\rangle$ orthogonal to $|1'\rangle$. The normalized $|1'\rangle$ and $|2'\rangle$ vector projections are then subtracted from a third vector $|3\rangle$ to give a $|3'\rangle$ orthogonal to both $|1'\rangle$ and $|2'\rangle$. Finally, the projections on $|1'\rangle$, $|2'\rangle$, and $|3'\rangle$ are subtracted from the remaining vector. This vector is normalized to produce the final orthogonal vector required to span the space.

The technique is illustrated by selecting the four vectors

$$\begin{aligned}|1\rangle&=\begin{pmatrix}1\\1\\1\\1\end{pmatrix} \quad |2\rangle=\begin{pmatrix}2\\2\\1\\1\end{pmatrix}\\|2\rangle&=\begin{pmatrix}3\\2\\2\\1\end{pmatrix} \quad |4\rangle=\begin{pmatrix}1\\1\\1\\3\end{pmatrix}\end{aligned} \tag{1.911}$$

The determinant of the four vectors is

$$\begin{vmatrix} 1 & 1 & 1 & 1 \\ 2 & 2 & 1 & 1 \\ 3 & 2 & 2 & 1 \\ 1 & 1 & 1 & 3 \end{vmatrix} = -2 \tag{1.9.12}$$

so they are linearly independent. $|1\rangle$ is selected as the starting vector and normalized to give

$$|1'\rangle = \left(\frac{1}{2}\right)\begin{pmatrix} 1 \\ 1 \\ 1 \\ 1 \end{pmatrix} \tag{1.9.13}$$

The unnormalized $|2'\rangle$ is determined as

$$|2'\rangle = |2\rangle - |1'\rangle\langle 1'|2\rangle$$

$$= \begin{pmatrix} 2 \\ 2 \\ 1 \\ 1 \end{pmatrix} - \begin{pmatrix} 1 \\ 1 \\ 1 \\ 1 \end{pmatrix}\left(\frac{1}{4}\right)(1 \quad 1 \quad 1 \quad 1)\begin{pmatrix} 2 \\ 2 \\ 1 \\ 1 \end{pmatrix} \tag{1.9.14}$$

$$= \begin{pmatrix} 2 \\ 2 \\ 1 \\ 1 \end{pmatrix} - \begin{pmatrix} \frac{3}{2} \\ \frac{3}{2} \\ \frac{3}{2} \\ \frac{3}{2} \end{pmatrix} = \begin{pmatrix} \frac{1}{2} \\ \frac{1}{2} \\ -\frac{1}{2} \\ -\frac{1}{2} \end{pmatrix} \tag{1.9.15}$$

The vector $|2'\rangle$ is already normalized since

$$\langle 2'|2'\rangle = 1 \tag{1.9.16}$$

Two projections must be subtracted from $|2\rangle$ to produce $|3'\rangle$:

$$|3'\rangle = |3\rangle - |1'\rangle\langle 1'|3\rangle - |2'\rangle\langle 2'|3\rangle \tag{1.9.17}$$

$$= \begin{pmatrix} 3 \\ 2 \\ 2 \\ 1 \end{pmatrix} - \begin{pmatrix} 1 \\ 1 \\ 1 \\ 1 \end{pmatrix} \left(\frac{1}{4}\right) (1 \quad 1 \quad 1 \quad 1) \begin{pmatrix} 3 \\ 2 \\ 2 \\ 1 \end{pmatrix}$$

$$- \begin{pmatrix} 1 \\ 1 \\ -1 \\ -1 \end{pmatrix} \left(\frac{1}{4}\right) (1 \quad 1 \quad -1 \quad -1) \begin{pmatrix} 3 \\ 2 \\ 2 \\ 1 \end{pmatrix} \tag{1.9.18}$$

$$= \begin{pmatrix} 3 \\ 2 \\ 2 \\ 1 \end{pmatrix} - \begin{pmatrix} 2 \\ 2 \\ 2 \\ 2 \end{pmatrix} - \begin{bmatrix} \frac{1}{2} \\ \frac{1}{2} \\ -\frac{1}{2} \\ -\frac{1}{2} \end{bmatrix} = \begin{bmatrix} \frac{1}{2} \\ -\frac{1}{2} \\ \frac{1}{2} \\ -\frac{1}{2} \end{bmatrix} = \left(\frac{1}{2}\right) \begin{pmatrix} 1 \\ -1 \\ 1 \\ -1 \end{pmatrix} \tag{1.9.19}$$

This vector is also normalized. Its scalar product with either $|1'\rangle$ or $|2'\rangle$ is zero.

The final orthogonal vector is determined from the equation

$$|4'\rangle = |4\rangle - |1'\rangle\langle 1'|4\rangle - |2'\rangle\langle 2'|4\rangle - |3'\rangle\langle 3'|4\rangle \tag{1.9.20}$$

and is (Problem 1.13)

$$|4'\rangle = \left(\frac{1}{2}\right) \begin{pmatrix} 1 \\ -1 \\ -1 \\ 1 \end{pmatrix} \tag{1.9.21}$$

For Gram–Schmidt orthogonalization, the initial orthogonalization steps are easiest since fewer components must be eliminated. Although the process becomes more difficult as the number of components increases, the method is still tractable for arbitrary N-dimensional spaces.

Other methods can be used for vectors in two- and three-dimensional spaces. For example, if two orthogonal vectors are known in a three-dimensional space, the third vector is known using the vector product. The two vectors

$$|1\rangle = \begin{pmatrix} 1 \\ 1 \\ 1 \end{pmatrix} \qquad |2\rangle = \begin{pmatrix} 1 \\ -2 \\ 1 \end{pmatrix} \tag{1.9.22}$$

are orthogonal since

$$\langle|1\rangle = (1 \quad -2 \quad 1)\begin{pmatrix}1\\1\\1\end{pmatrix} = 0 \tag{1.9.23}$$

Their vector product is

$$\begin{vmatrix}\mathbf{i} & \mathbf{j} & \mathbf{k}\\ 1 & 1 & 1\\ 1 & -2 & 1\end{vmatrix} = 3\mathbf{i} + 0\mathbf{j} - 3\mathbf{k} \tag{1.9.24}$$

The normalized ket vector is

$$|3\rangle = \left(\frac{1}{\sqrt{2}}\right)\begin{pmatrix}1\\0\\-1\end{pmatrix} \tag{1.9.25}$$

The problem can also be solved by using the two orthogonal vectors to generate two equations in two unknowns. The vector $|3\rangle$ is assigned the coordinates

$$\begin{pmatrix}1\\a\\b\end{pmatrix} \tag{1.9.26}$$

The first component is assigned a value of 1 since the vector represents the relative magnitudes of the components. A normalization provides a third relationship,

$$\langle 3|3\rangle = 1 \tag{1.9.27}$$

The scalar product $\langle 1|3\rangle$ is

$$(1 \quad 1 \quad 1)\begin{pmatrix}1\\a\\b\end{pmatrix} = 1 + a + b = 0 \tag{1.9.28}$$

while $\langle 2|3\rangle$ is

$$(1 \quad -2 \quad 1)\begin{pmatrix}1\\a\\b\end{pmatrix} = 1 - 2a + b = 0 \tag{1.9.29}$$

The two equations in two unknowns are

$$\begin{aligned} a + b &= -1\\ -2a + b &= -1 \end{aligned} \tag{1.9.30}$$

with solution

$$a=0 \qquad b=-1 \tag{1.9.31}$$

The orthogonal vector is

$$|3\rangle=\begin{pmatrix}1\\0\\-1\end{pmatrix} \tag{1.9.32}$$

As the number of dimensions increases, the number of unknowns increases and the Gram–Schmidt orthogonization is preferable.

1.10. Vector Calculus

When a vector is resolved into its components along coordinate axes, these coordinate directions function as distinct one-dimensional vector systems. Each component can be observed individually.

The resolution into components is very useful when the vector changes with time or position. The change of $|r\rangle$ with time is written formally as

$$d|r\rangle/dt \tag{1.10.1}$$

If $|r\rangle$ changes with time, each of its components on a set of coordinate axes can also change with time. For standard x–y–z coordinates, the vector derivative is

$$d|r\rangle/dt=d/dt\begin{pmatrix}r_x\\r_y\\r_z\end{pmatrix} \tag{1.10.2}$$

which is a vector of three separate derivatives:

$$d|r\rangle/dt=\begin{pmatrix}dr_x/dt\\dr_y/dt\\dr_z/dt\end{pmatrix} \tag{1.10.3}$$

The time evolution of the vector can be reconstructed from the components at any time.

Derivatives with respect to position can be performed in the same way. The derivative with respect to x, for example, can operate on all three components, i.e.,

$$d|r\rangle/dx=\begin{pmatrix}\partial r_x/\partial x\\\partial r_y/\partial x\\\partial r_z/\partial x\end{pmatrix} \tag{1.10.4}$$

The derivative $\partial r_y/\partial x$ indicates that the component r_y may be a function of its location in space, i.e., its x, y, and z coordinates.

If a vector changes with respect to x, y, and z, there are several ways to define differentiation for this vector. The function $f(x, y)$ in Figure 1.24 has a very small slope in the x direction and a large slope in the y direction. A small change if f, df, must be related to changes in the x and y coordinates. Since f can change as either x or y is changed, both types of change must be included in a description of df.

If the change in f is infinitesimal, i.e., df, the system can be resolved into components along the two orthogonal axes. The change df for the change along the x axis is given by

$$df = (\partial d/\partial x)\, dx \tag{1.10.5}$$

for a distance dx. Since the change in f is infinitesimal, the same function f, i.e., at position x and y, can be used to determine the change in f along the y axis:

$$f(x + dx, y) \sim f(x, y) \tag{1.10.6}$$

The change in f for a change of dy,

$$(\partial f/\partial y)\, dy \tag{1.10.7}$$

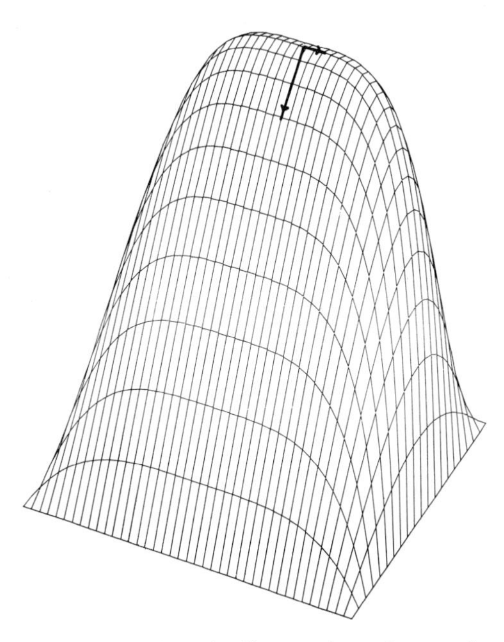

Figure 1.24. A function $f(x, y)$ defined as the distance above the x–y plane. The function has a steep slope along the y axis and a gentle slope along the x axis.

is added to the change for dx to give the full change in df:

$$df = (\partial f/\partial x)_y \, dx + (\partial f/\partial y)_x \, dy \tag{1.10.8}$$

The changes (dx, dy) in the x–y plane are represented by a vector,

$$|dr\rangle = \begin{pmatrix} dx \\ dy \end{pmatrix} \tag{1.10.9}$$

Although $|dr\rangle$ is a vector quantity, the final result, df, is not a vector; it is a scalar quantity describing some change in a function f. A scalar is generated only via a scalar product of $|dr\rangle$ with a second vector. Equation 1.10.8 can be generated only by creating a vector of derivatives in x and y, i.e.,

$$(\partial f/\partial x \quad \partial f/\partial y) \tag{1.10.10}$$

to give a scalar product,

$$\begin{aligned} \langle \mathrm{grad}\, f \,|dr\rangle &= (\partial f/\partial x \quad \partial f/\partial y) \begin{pmatrix} dx \\ dy \end{pmatrix} \\ &= (\partial f/\partial x)\, dx + (\partial f/\partial y)\, dy \end{aligned} \tag{1.10.11}$$

The row vector contains components which are derivatives with respect to the independent variables. In this case, the derivative in x acts only along the x coordinate direction while the derivative in y acts only along the y coordinate direction. The vector in Cartesian coordinates is

$$\nabla = \mathbf{i}\, \partial x + \mathbf{j}\, \partial/\partial y \tag{1.10.12}$$

for two dimensions and

$$\nabla = \mathbf{i}\, \partial/\partial x + \mathbf{j}\, \partial/\partial y + \mathbf{k}\, \partial/\partial z \tag{1.10.13}$$

for three dimensions. The sum of differential operations is the gradient. The symbols grad f or ∇f describe this vector operation.

The scalar product $\langle \mathrm{grad}\, f\, |dr\rangle$ defined a change df at some point (x, y). The gradient portion of this product defines the slopes of the function for two (or three) orthogonal directions in the space. The function of Figure 1.24 has distinctly different slopes at some point (x, y) depending on whether one is examining the x or y direction. For orientations intermediate to these two directions, the net change in the function is a linear combination of the two orthogonal changes in x and y.

The gradient is extremely useful for conservative systems in which the same result is obtained for any path between two points. In thermodynamics, the free energy, dG, is

$$dG = (\partial G/\partial T)_P \, dT + (\partial G/\partial P)_T \, dP \tag{1.10.14}$$

The interpretation is identical to that for a change of function $f(x, y)$ in space. In this case, the gradient is

$$(\partial G/\partial T \quad \partial G/\partial P) \tag{1.10.15}$$

Since the total free energy between (T_1, P_1) and (T_2, P_2) is independent of path, the gradient is used to select a convenient path. It is most convenient to select a path where P_1 is held constant and the term

$$(\partial G/\partial T)_{P_1}\, dT \tag{1.10.16}$$

is integrated from T_1 to T_2. The temperature is then held constant at T_2 and the function is integrated from P_1 to P_2. The two steps are added to give the free energy for the change,

$$dG = (\partial G/\partial T)_{P_1}\, dT + (\partial G/\partial P)_{T_2}\, dP \tag{1.10.17}$$

The change in a function, *df*, should be independent of the coordinate system chosen since any coordinate system should give the same scalar product,

$$\langle \mathrm{grad}\, f \,|\, dr \rangle \tag{1.10.18}$$

For this reason, the gradient operator is written in the coordinate system which is most convenient for the problem under consideration. A system which has spherical symmetry, i.e., no explicit dependence on the variables θ and φ, should have a gradient in spherical coordinates. The total gradient in spherical coordinates is

$$\mathrm{grad} = (\partial/\partial r,\ (1/r)\, \partial/\partial\theta,\ [1/r \sin(\theta)]\, \partial/\partial\varphi) \tag{1.10.19}$$

For polar coordinates in two dimensions, the gradient is

$$\mathrm{grad} = (\partial/\partial r \quad (1/r)\, \partial/\partial\theta) \tag{1.10.20}$$

which is easily extended to three dimensions by adding a third variable z to create cylindrical coordinates:

$$\mathrm{grad} = (\partial/\partial r \quad (1/r)\, \partial/\partial\theta \quad \partial/\partial z) \tag{1.10.21}$$

For the system with spherical symmetry, the function $f(r, \theta, \varphi)$ will not change when θ or φ is changed so that

$$\partial f/\partial\theta = \partial f/\partial\varphi = 0 \tag{1.10.22}$$

and the system need only be solved for one dimension, i.e., with respect to dr:

$$\mathrm{grad}\, f = (\partial f/\partial r)\, dr \tag{1.10.23}$$

It was seen above that *df* became a scalar because it was a scalar product between the vector grad(f) and the vector $|dr\rangle$. In other cases, the directed derivatives, e.g., $\mathbf{i}\ \partial/\partial x$, can define a scalar product. For example, a region of cross-sectional area A and thickness dx contains a concentration c of molecules. If molecules enter this region along x from the left, some may be adsorbed within the system so the number which exit the system to the right is smaller (Figure 1.25). This difference between entering and leaving molecules requires a change of concentration within the system, and this can be determined as the difference between the input and output fluxes along x. For an input flux J in units of moles per area per second, the output flux must change over the interval dx and can be expressed as a Taylor expansion of J,

$$J + (\partial J/\partial x)\, dx \tag{1.10.24}$$

where (dJ/dx) is the change in flux over the interval dx. The net change in flux is

$$\mathbf{J} - [\mathbf{J} + (\partial J/\partial x)\, dx] = -(\partial J/\partial x)\, dx \tag{1.10.25}$$

and this must describe the total change in concentration with time in the interval dx:

$$\begin{aligned}(\partial c/\partial t)\, dx &= -(\partial J/\partial x)\, dx \\ \partial c/\partial t &= -\partial J/\partial x\end{aligned} \tag{1.10.26}$$

This is the equation of continuity in one dimension. If the flux is not constant with respect to a change in x, the system is gaining or losing molecules with time in this region. If molecules disappear, the region acts as a sink; if they appear, it acts as a source.

Equation 1.10.26 describes divergence. The change in concentration with time is a scalar but it is equal to some operation $(\partial/\partial x)$ on a vector quantity, $\mathbf{J}$, which describes the vector flow of molecules on x. For consistency, the derivative $\partial/\partial x$ must also be a vector quantity; the flux is changing as the position on a coordinate axis x changes. Thus, $\partial/\partial x$ and $\mathbf{J}$ are vector components which combine to create a scalar product. For a two-dimensional system, the net flux is

$$\begin{aligned}J = \operatorname{div} \mathbf{J} &= (\partial/\partial x \quad \partial/\partial y)\begin{pmatrix} J_x \\ J_y \end{pmatrix} \\ &= \partial J_x/\partial x + \partial J_y/\partial y\end{aligned} \tag{1.10.27}$$

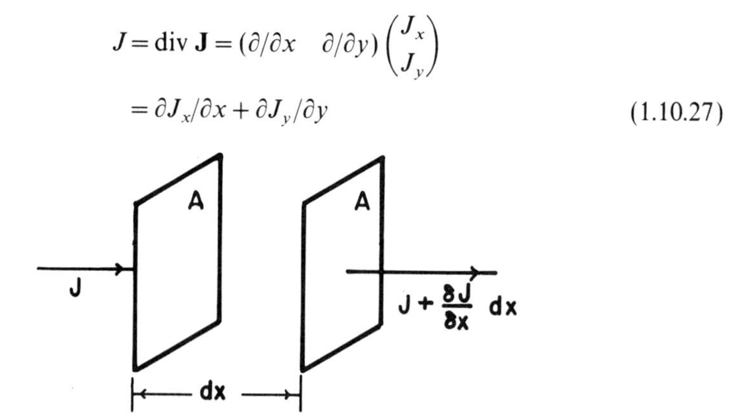

Figure 1.25. Flux J into a region of thickness $x + dx$ with an exit flux of $J + (\partial J/\partial x)\, dx$.

The derivative $\partial/\partial x$ operates only on the component J_x. The divergence monitors the change in flux in distinct coordinate directions to determine the net flux for the system. In three dimensions, the divergence is

$$\begin{aligned} J &= \text{div } \mathbf{J} \\ &= (\partial/\partial x \quad \partial/\partial y \quad \partial/\partial z) \begin{pmatrix} J_x \\ J_y \\ J_z \end{pmatrix} \\ &= \partial J_x/\partial x + \partial J_y/\partial y + \partial J_z/\partial z \end{aligned} \tag{1.10.28}$$

Since the divergence will monitor the net loss or gain of molecules in some three-dimensional region via a scalar product, the choice of coordinate system will have no effect on the final result. Again, a spherical volume is best described with spherical coordinates. The divergence in this case has components

$$((1/r)\, \partial/\partial r(r^2),\ [1/(r \sin(\theta)]\, \partial/\partial\theta(\sin(\theta),\ [1/r \sin(\theta)]\, \partial/\partial\varphi) \tag{1.10.29}$$

These operations are best illustrated with the divergence of $\mathbf{J}$ with respect to its three flux components, J_r, J_θ, and J_φ,

$$(1/r)\, \partial(r^2 J_r)/\partial r + [1/r \sin(\theta)]\, \partial(\sin\theta\, J_\theta)/\partial\theta + [1/r \sin(\theta)]\, \partial J_\varphi/\partial\varphi \tag{1.10.30}$$

The divergence for a two-dimensional system in polar coordinates is

$$(1/r)\, \partial(rJ_r)/\partial r + (1/r)\, \partial J_\theta/\partial\theta \tag{1.10.31}$$

and the divergence for cylindrical coordinates is

$$(1/r)\, \partial(rJ_r)/\partial r + (1/r)\, \partial J_\theta/\partial\theta + \partial J_z/\partial z \tag{1.10.32}$$

The equation of continuity in three dimensions is

$$\text{div}\, \mathbf{J} = -dc/dt \tag{1.10.33}$$

The total concentration change in a three-dimensional region is dictated by the net flow of molecules across the surface which surrounds it. The total change of concentration in some volume with time is determined by integrating dc/dt over the entire volume,

$$\int (dc/dt)\, dV \tag{1.10.34}$$

which is equivalent to

$$-\int \text{div}\, \mathbf{J}\, dV \tag{1.10.35}$$

This change in concentration is also equal to the net flow across all surfaces. This is equal to the scalar product of $|J\rangle$ and a normal to the surface $\langle n|$,

$$\langle n|J\rangle \tag{1.10.36}$$

which is integrated over the entire surface,

$$-\int \langle n|J\rangle\, dS \tag{1.10.37}$$

where dS is an infinitesimal surface area. Since this loss of molecules also describes the net change in the concentration within the volume, the two equations (1.10.35 and 1.10.37) are equated to give

$$\int \langle n|J\rangle\, dS = \int \operatorname{div} \mathbf{J}\, dV \tag{1.10.38}$$

Although this equation is developed for the flux **J**, it is valid for any vector quantity and is called the divergence theorem. The theorem permits a change from a volume integral to a surface integral or vice versa. Problems which might be difficult in one format can often be solved more easily in the alternate format.

Problems

1.1. Determine the equation of the plane formed by the two upper H atoms and the C atom in the methane molecule when C is centered at the origin (Figure 1.10).

1.2. Verify the distance between two H atoms in a methane molecule using the law of cosines,

$$r^2 = a^2 + b^2 - 2ab\cos\theta$$

where $\theta = 109°$.

1.3. A particle confined to the x axis with a momentum p_x vibrates about a fixed point. Show that the trajectory of the particle in the two-dimensional phase space is an ellipse.

1.4. The force and distance vectors of Figure 1.13 make an angle of 60°. Show that the scalar product is unchanged for an orthogonal coordinate system in which the y' axis lies $+45°$ with respect to the horizontal x axis and the x' axis lies $-45°$ with respect to this axis.

1.5. Show that the vectors of Equation 1.4.22 are mutually orthogonal.

1.6. Prove that the angular momentum components L_x and L_y are

$$L_x = r_y p_z - r_z p_y$$

$$L_y = r_z p_x + r_x p_z$$

1.7. Show that the operation

$$\mathbf{a} \cdot (\mathbf{b} \times \mathbf{c})$$

leads to a determinant

$$\begin{vmatrix} a_x & a_y & a_z \\ b_x & b_y & b_z \\ c_x & c_y & c_z \end{vmatrix}$$

1.8. Show that the triple scalar products

$$\mathbf{a} \cdot (\mathbf{b} \times \mathbf{c}) \qquad \mathbf{c} \cdot (\mathbf{a} \times \mathbf{b}) \qquad \mathbf{b} \cdot (\mathbf{c} \times \mathbf{a})$$

give the same result while

$$\mathbf{b} \cdot (\mathbf{a} \times \mathbf{c})$$

gives the negative of this result.

1.9. Show that the four cyclobutadiene molecular orbital vectors (Equations 1.6.9 and 1.6.10) are orthogonal.

1.10. Verify that the scalar product

$$\langle 1| = (a + ib \quad c - id) \begin{pmatrix} a - ib \\ c + id \end{pmatrix}$$

is real.

1.11. Determine the projections of the vector of Equation 1.7.9 on $|y'\rangle$ and $|z'\rangle$ (Equation 1.7.13).

1.12. Show that the vector $|3'\rangle$ (Equation 1.9.19) is orthogonal to vectors $|1'\rangle$ (Equation 1.9.13) and $|2'\rangle$ (Equation 1.9.15).

1.13. Perform the Gram–Schmidt orthogonalization to verify $|4'\rangle$ of Equation 1.9.21.

1.14. Show that the determinant (Equation 1.9.12) equals -2.

1.15. Verify that the determinant constructed from the components of the four orthogonalized vectors $|1'\rangle$, $|2'\rangle$, $|3'\rangle$, and $|4'\rangle$ (Equations 1.9.13, 1.9.15, 1.9.19, and 1.9.21) has a nonzero value.

1.16. a. Show that the vectors $(1 \quad 0 \quad -1)$ and $(1 \quad \sqrt{2} \quad 1)$ are orthogonal.
b. Find a third vector with components (a, b, c) which is orthogonal to both vectors in (a).
c. Normalize the vector determined in (b).

1.17. Use the coordinates defined by Figure 1.9 to determine the vector sum $\mathbf{r} + \mathbf{s}$ and show that the resultant vector has the proper orientation.

1.18. Determine expressions for angular momentum components L_y and L_z using the matrix of Equation 1.5.10.

2

Function Spaces

2.1. The Function as a Vector

A vector in an N-dimensional space can always be resolved into components on a set of basis vectors. These basis vectors are linearly independent and span the N-dimensional space. Many of the properties of such finite-dimensional spaces can be extended to a study of vector spaces with an infinite number of dimensions. These spaces utilize continuous functions as basis vectors.

In an N-dimensional space, as many as N basis vectors might be required to completely specify an arbitrary vector. The complicated function of Figure 2.1 on the interval $-1 \leqslant x \leqslant 1$ is not a standard function. A table of $f(x)$ for each value of x on the interval is required to reproduce it. Of course, there are infinite values of x on the interval. If each x defines one component of an infinite-dimensional vector, then the value of this component will be $f(x)$ since this is the value which reproduces the shape of the function. The x defines the position in the vector just as the basis coordinates defined the position in a finite-dimensional column vector. The number in the vector at that position described that component's contribution to the total vector. Taking each value of $f(x)$ from the "vector" simply reproduces the function on the interval.

Although it is impossible to develop an infinite table of function values on the interval, it is not necessary if the function is known and well behaved. For a continuous function such as

$$f(x) = x^2 \tag{2.1.1}$$

$f(x)$ can be determined for any value of x by substituting that value of x into Equation 2.1.1. In such a case, the function defines the entire vector, and it is not necessary to tabulate the infinity of possible $f(x)$.

If the function $f(x)$ is interpreted as an infinite-dimensional vector, it must satisfy some of the properties of the finite-dimensional vector spaces. In fact, there are many parallels. For example, the sum of two finite vectors is simply the addition of their corresponding components. For **r** and **s**, r_x and s_x add to give the resultant on x, etc. The sum of two functions, $f(x)$ and $g(x)$,

$$f(x) + g(x) \tag{2.1.2}$$

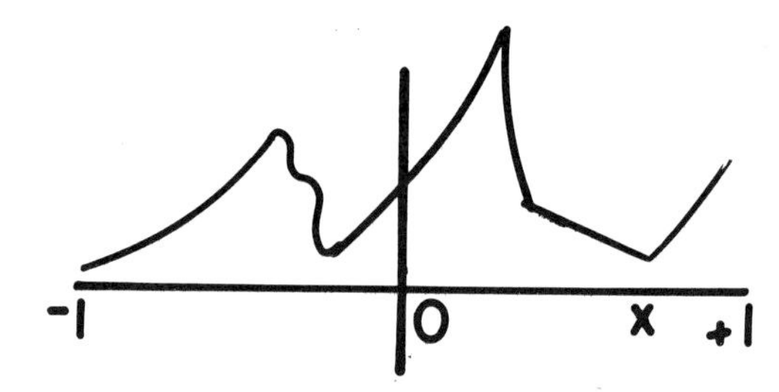

Figure 2.1. An arbitrary function on the interval $-1 \leqslant x \leqslant 1$.

is really a sum of f and g for each separate value of x. The function notation permits the "shorthand" notation of Equation 2.1.2. A scalar a multiplies every component of a vector $|r\rangle$:

$$a\,|r\rangle = \begin{pmatrix} ar_x \\ ar_y \\ ar_z \end{pmatrix} \tag{2.1.3}$$

while the product of a and the function $f(x)$,

$$af(x) \tag{2.1.4}$$

really multiplies every $f(x)$ in the interval by a. The function notation reduces these multiplications to a single statement.

Since there are an infinite set of x on the interval, the dimension of the space is infinite. A basis vector for a specific x would be a column with a single 1 (at the location of this specific x) and the remainder zero. As such, it corresponds to the x, y, and z basis vectors in three-dimensional space. For example, $|x\rangle$ is

$$\begin{pmatrix} 1 \\ 0 \\ 0 \end{pmatrix} \tag{2.1.5}$$

However, in three-dimensional space, it was possible to select other basis vectors, e.g.,

$$\begin{pmatrix} 1 \\ 1 \\ 0 \end{pmatrix} \tag{2.1.6}$$

which had components on several axes. The analogy is extended to the infinite-dimension vectors by noting that any function constitutes a vector in this space. Every nonzero value of $f(x)$ will appear as a nonzero component in the vector. This function might then serve as a basis vector in the space. An arbitrary function might then be "projected" onto this basis function to determine the contribution of this functional basis vector to the total function. This technique of resolving a function into a summation of simpler basis functions is the basis for function approximation techniques such as Fourier series. Infinite-dimensional function spaces are called Hilbert spaces.

The parallels between finite- and infinite-component vectors extend to most of the operations between vectors introduced in Chapter 1. A function scalar product will extend the notions of orthogonality and normalization. Linear independence will permit development of an orthogonal basis set of functions using a version of Gram–Schmidt orthogonalization. Finally, projection techniques will permit the description of any function in the space as a sum of orthogonal basis functions. In each case, the techniques of Chapter 1 serve as a heuristic guide.

The description of a function as a summation of simpler functions can be illustrated with a Taylor series expansion. The expansion about $x=0$ is

$$f(x)=a_0+a_1x+a_2x^2+\cdots \qquad (2.1.7)$$

where the a_i are constants. The Taylor series has described $f(x)$ in terms of the basis set of functions, x^i:

$$1, x, x^2,\ldots x^n,\ldots \qquad (2.1.8)$$

This series of single powers of x does constitute a basis set with linear independence. However, if they were used for the full range of x, the higher powers would dominate. Thus, it will often be necessary to restrict the interval over which a set of functions is applicable. The functions of Equation 2.1.8 will be restricted to the interval

$$-1\leqslant x\leqslant 1 \qquad (2.1.9)$$

for many of the examples which follow.

2.2. Function Scalar Products and Orthogonality

The scalar product for a finite-dimensional space provided a technique to determine the product of two vectors along the same spatial direction. It did so by establishing the projection of one vector parallel to the second vector. The same result was obtained by resolving each vector into components on a common set of axes and multiplying the vector components for each coordinate axis. These products were then summed to give the scalar product. The scalar product is extended to functions using the simple power functions

$$1 \quad x \quad x^2 \quad \cdots \qquad (2.2.1)$$

on the interval

$$-1 \leqslant x \leqslant 1 \tag{2.2.2}$$

to demonstrate the technique.

Although there are an infinite number of values of x on the interval, selected values of x can be selected to form a function "component" table. For the functions

$$f_0 = 1 \qquad f_1 = x \tag{2.2.3}$$

the table is

x	$f_0(x) = 1$	$f_1(x)$
-1	1	-1
-0.5	1	-0.5
0	1	0
0.5	1	0.5
1	1	1

If these were the only components for the function, the scalar product of the two functions could be formed by taking the product of f_0 and f_1 at each value of x and then adding them, i.e.,

$$\langle f_0 | f_1 \rangle = \sum f_0(x) f_1(x) \tag{2.2.4}$$

Actually, the infinite set of products

$$f_0(x) f_1(x) \tag{2.2.5}$$

on the interval must be summed in infinitesimal steps. The summation of Equation 2.2.4 is replaced by an integration in the variable dx. The scalar product for the function vectors is

$$\langle f_0 | f_1 \rangle = \int_{-1}^{1} f_0(x) f_1(x)\, dx \tag{2.2.6}$$

The bra–ket notation is used for this scalar product. However, for these spaces, the vectors are functions and the summation becomes an integration. Because each component is defined by the functions, integration of the function product completely eliminates any need for an infinite tabulation. However, if the functions $f_i(x)$ in the scalar product were quite complicated and impossible to describe in functional form, the tabulation might be necessary.

Any two functions on the selected interval can form a scalar product. The two arbitrary functions $g(x)$ and $h(x)$ will have a scalar product

$$\langle g \mid h \rangle = \int_{-1}^{1} g(x)\, h(x)\, dx \tag{2.2.7}$$

In general, any interval can be selected as long as the integral of the functions is defined. For example, many integrals with limits of $\pm\infty$ give infinite integrals. The range of x is normally selected so that the functions give finite integrals for their scalar products.

In a finite space, two vectors **r** and **s** which gave a scalar product of zero,

$$\langle r \mid s \rangle = 0 \tag{2.2.8}$$

were orthogonal to each other. In a two- or three-dimensional space, the two vectors were actually perpendicular. In higher-dimensional spaces, a scalar product of zero was again used to define orthogonality even though it was impossible to describe the orthogonal behavior in any graphical way. Since the graphical format is not necessary to define orthogonality, the concept can be extended to the infinite-dimensional vectors in function space. Any two functions, e.g., $g(x)$ and $h(x)$, which give a scalar product of zero on the interval are orthogonal in the Hilbert space:

$$\langle g \mid h \rangle = \int_{-1}^{1} g(x)\, h(x)\, dx = 0 \tag{2.2.9}$$

An infinite set of mutually orthogonal functions would then constitute a coordinate system in the Hilbert space.

The orthogonality property is illustrated with the vectors of Equation 2.2.1 on the interval $-1 \leqslant x \leqslant 1$. The scalar product of f_0 and f_1 is

$$\begin{aligned} \langle f_0 \mid f_1 \rangle &= \int_{-1}^{1} (1)(x)\, dx \\ &= (x^2/2)\Big|_{-1}^{1} = 0 \end{aligned} \tag{2.2.10}$$

These two functions are orthogonal.

The functions are not orthogonal in the sense of being perpendicular to each other. All components in f_0 are equal for this constant function. For each $f_1(+x)$ on the interval $0 \leqslant x \leqslant 1$, there is an equal and opposite $f_1(-x)$ from the interval $-1 \leqslant x \leqslant 0$ which produces a net "sum" of zero.

Since the first two members, 1 and x, of the function set are orthogonal, additional functions in the set might also be mutually orthogonal. However, the next function,

$$f_2(x) = x^2 \tag{2.2.11}$$

is orthogonal to f_1 but not to f_0:

$$\langle f_2 | f_0 \rangle = \int_{-1}^{1} (x^2)(1)\, dx = +\frac{2}{3} \tag{2.2.12}$$

$$\langle f_2 | f_1 \rangle = \int_{-1}^{1} (x^2)(x)\, dx = 0 \tag{2.2.13}$$

In general, the integral will be nonzero if the power for x in the product of the two functions is even; it will be zero when the power is odd.

The set of x^i are linearly independent, and it is possible to find functions orthogonal to 1 and x. The next orthogonal function is (Problem 2.6)

$$f'_2 = 3x^2 - 1 \tag{2.2.14}$$

Each function $f(x)$ is normalized using the scalar product. The norm is formed from the scalar product $\langle f(x) | f(x) \rangle$. For example, $\langle f_0 | f_0 \rangle$ has a norm

$$\langle f_0 | f_0 \rangle = \int_{-1}^{1} (1)(1)\, dx = 2 \tag{2.2.15}$$

Each function is divided by the square root of this norm to give the normalized functions:

$$\langle f_0 | = \frac{1}{\sqrt{2}} \tag{2.2.16}$$

$$| f_0 \rangle = \frac{1}{\sqrt{2}} \tag{2.2.17}$$

The components for every value of x are equal to $\frac{1}{\sqrt{2}}$. There is no x dependence in the vectors.

The normalization for f_1 uses the scalar product

$$\langle f_1 | f_1 \rangle = \int_{-1}^{1} (x)(x)\, dx$$

$$= (x^3/3)\Big|_{-1}^{1} = \frac{2}{3} \tag{2.2.18}$$

and the normalized function vectors are

$$\langle f_1 | = \sqrt{\frac{3}{2}}\, x$$
$$| f_1 \rangle = \sqrt{\frac{3}{2}}\, x \tag{2.2.19}$$

The scalar product for $f_2 = 3x^2 - 1$ is $\frac{8}{5}$, which gives the orthonormal function (Problem 2.20)

$$|f_2\rangle = \sqrt{\frac{5}{8}}\,(3x^2 - 1) \tag{2.2.20}$$

These scalar products are formed from real functions. If the functions are complex, the scalar product requires functions from a dual space. The function is paired with its complex conjugate to form the scalar product. For a ket vector,

$$|g(x)\rangle = \exp(im\varphi) \tag{2.2.21}$$

on the interval

$$0 \leqslant \varphi \leqslant 2\pi \tag{2.2.22}$$

the bra vector in the dual space is

$$\langle g(\varphi)| = [\exp(im\varphi)]^* = \exp(-im\varphi) \tag{2.2.23}$$

The scalar product is

$$\begin{aligned}\langle g | g\rangle &= \int_0^{2\pi} g^*(\varphi)\, g(\varphi)\, d\varphi \\ &= \int_0^{2\pi} \exp(-im\varphi)\exp(im\varphi)\, d\varphi = \int_0^{2\pi} d\varphi = 2\pi\end{aligned} \tag{2.2.24}$$

and the normalized functions are

$$\begin{aligned}\langle g(\varphi)| &= \sqrt{1/(2\pi)}\,\exp(-im\varphi) \\ |g(\varphi)\rangle &= \sqrt{1/(2\pi)}\,\exp(im\varphi)\end{aligned} \tag{2.2.25}$$

Two exponential functions with different m will be orthogonal. For

$$\begin{aligned}\langle h| &= \exp(-im'\varphi) \\ |g\rangle &= \exp(im\varphi)\end{aligned} \tag{2.2.26}$$

with $m \neq m'$, the scalar product is

$$\begin{aligned}\langle h | g\rangle &= \int_0^{2\pi} \exp(-im'\varphi)\exp(im\varphi)\, d\varphi \\ &= \int_0^{2\pi} \exp[i(m-m')\varphi]\, d\varphi = [1/i(m-m')]\exp[i(m-m')\varphi]\Big|_0^{2\pi} = 0\end{aligned} \tag{2.2.27}$$

These orthogonal functions are the wavefunctions for a particle confined to a ring in quantum mechanics.

Although functions in a single variable, i.e., one dimension, have been used as examples, the extension to a three-dimensional space is straightforward. Each point in the three-dimensional space will define one function vector component. To form a scalar product, one must integrate over the three-dimensional space. The scalar product in three dimensions can be written formally in terms of a vector **r**,

$$\langle g(\mathbf{r}) | f(\mathbf{r}) \rangle = \int_a^b g(\mathbf{r}) f(\mathbf{r}) \, d\mathbf{r} \tag{2.2.28}$$

where a and b specify limits for the volume in the three-dimensional space. In Cartesian coordinates, Equation 2.2.28 evolves into a triple integral over the three coordinate directions of the three-dimensional space, i.e.,

$$\langle g(\mathbf{r}) | f(\mathbf{r}) \rangle = \int_{a_x}^{b_x} \int_{a_y}^{b_y} \int_{a_z}^{b_z} g(x, y, z) f(x, y, z) \, dx \, dy \, dz \tag{2.2.29}$$

where the limits a_x, b_x, etc. set the limits for the volume selected. If the triple integral is equal to zero, then the functions g and f are orthogonal.

The three-dimensional scalar products can be evaluated using any three-dimensional coordinate, e.g., Cartesian, spherical. In each case, the integrals span the entire defined space and the result will be the same. It is often convenient, therefore, to choose a coordinate system which requires the minimum integration. For example, if two functions are spherically symmetric, the functions have the form

$$\begin{aligned} f(r, \theta, \varphi) &= f(r) \\ g(r, \theta, \varphi) &= g(r) \end{aligned} \tag{2.2.30}$$

The functions have no explicit dependence on θ or φ. The scalar product using a spherical coordinate system would be

$$\langle f | g \rangle = \int_0^\infty \int_0^\pi \int_0^{2\pi} f(r) \, g(r) \, r^2 \sin\theta \, dr \, d\theta \, d\varphi \tag{2.2.31}$$

Since the functions depend only on r, the integrals separate into the product of three integrals

$$\int_0^\infty f(r) \, g(r) \, r^2 \, dr \int_0^\pi \sin\theta \, d\theta \int_0^{2\pi} d\varphi \tag{2.2.32}$$

The last two integrals are finite with values of 2 and 2π, respectively. The radial integral dictates whether the two functions are orthogonal for this spherically symmetric example.

In some cases, the orthogonality property is mediated by an additional weighting function, $w(x)$, in the scalar product integral, i.e.,

$$\langle f \mid g \rangle = \int_{-\infty}^{\infty} f(x)\, g(x)\, w(x)\, dx \tag{2.2.33}$$

The Hermite polynomials, $H_n(x)$, which describe a harmonic oscillator in quantum mechanics, require such a weighting factor. The first two Hermite polynomials

$$H_0 = 1 \qquad H_1 = 2x \tag{2.2.34}$$

require a weighting factor

$$\exp(-x^2) \tag{2.2.35}$$

to give a weighted scalar product (Problem 2.9):

$$\begin{aligned} \langle H_1 \mid H_0 \rangle &= \int_{-\infty}^{\infty} (1)(2x) \exp(-x^2)\, dx \\ &= -\exp(-x^2) \Big|_{-\infty}^{\infty} = 0 \end{aligned} \tag{2.2.36}$$

Although the first two Hermite polynomials are 1 and x which are orthogonal on the interval $-1 \leqslant x \leqslant 1$, the weighting factor permits a scalar product on the full x axis.

2.3. Linear Independence

Two functions are orthogonal if their scalar product is zero. Even if two functions are not orthogonal, they may be linearly independent. The linear independence of functions on a given interval must be established by procedures which are similar, but not identical, to the techniques developed for finite-component vectors.

The linear independence or dependence of a set of vectors in an N-dimensional space was established by ranging the components of each vector in successive rows of a determinant. If the determinant was nonzero, the vectors were linearly independent. The identical procedure for function vectors would require a infinite determinant since the function has infinite components. However, each N-dimensional vector represented projections onto each coordinate axis to generate the components. For an arbitrary set of functions, the functions themselves are selected as the coordinate axes. If two functions project onto the remaining functions in exactly the same way, they will be linearly dependent.

The scalar product of two functions gives the projection of one function on the other. Thus, the components for function f_1 are its scalar products with the remaining functions,

$$\langle f_1 | f_2 \rangle, \langle f_1 | f_3 \rangle, \ldots \tag{2.3.1}$$

and these will become the off-diagonal elements for the first row of the determinant. The diagonal element will be $\langle f_1 | f_1 \rangle$. The second row will have elements

$$\langle f_2 | f_i \rangle \qquad i = 1, 2, \ldots \tag{2.3.2}$$

The Gram determinant which is formed cannot include the infinity of functions which are required to span the Hilbert space. It simply establishes whether some finite set of functions can span mutually exclusive regions of this space; i.e., it determines whether two functions are "parallel" to each other via the scalar product.

The Gram determinant is illustrated for three functions which are mutually orthogonal:

$$\begin{aligned} \langle f_i | f_j \rangle &= 1 \qquad \text{for} \quad i = j \\ \langle f_i | f_j \rangle &= 0 \qquad \text{for} \quad i \neq j;\ i, j = 1, 2, 3 \end{aligned} \tag{2.3.3}$$

The Gram determinant is formed by determining all the scalar products of a function f_1 with the remaining functions and ranging these products in the first row; f_2 forms scalar products with all the remaining functions to produce the components of the second row; etc. The Gram determinant is

$$\begin{vmatrix} \langle f_1 | f_1 \rangle & \langle f_1 | f_2 \rangle & \langle f_1 | f_3 \rangle \\ \langle f_2 | f_1 \rangle & \langle f_2 | f_2 \rangle & \langle f_2 | f_3 \rangle \\ \langle f_3 | f_1 \rangle & \langle f_3 | f_2 \rangle & \langle f_3 | f_3 \rangle \end{vmatrix} \tag{2.3.4}$$

The orthogonal functions give the diagonal determinant

$$\begin{vmatrix} 1 & 0 & 0 \\ 0 & 1 & 0 \\ 0 & 0 & 1 \end{vmatrix} \tag{2.3.5}$$

with a nonzero value of 1. The functions are linearly independent. Since each nonzero element is located in a separate column, there is no "coupling" between the different functions.

Linearly independent functions will give a nonzero Gram determinant even though the determinant has nondiagonal elements. The functions

$$f_1 = 1 \qquad f_2 = x \qquad f_3 = x^2 \tag{2.3.6}$$

on the interval $-1 \leqslant x \leqslant 1$ have a Gram determinant

$$\begin{vmatrix} 2 & 0 & 0 \\ 0 & \frac{2}{3} & 0 \\ \frac{2}{3} & 0 & \frac{2}{5} \end{vmatrix} = \frac{8}{15} \tag{2.3.7}$$

The functions are linearly independent; they cannot be combined to express one function as a sum of the remaining two.

The three functions

$$\begin{aligned} f_1 &= \sin^2 \theta \\ f_2 &= \cos^2 \theta \\ f_3 &= 1 \end{aligned} \tag{2.3.8}$$

on the interval $0 \leqslant \theta \leqslant 2\pi$ are linearly dependent since a particular set of coefficients permits a sum of zero. The linear combination is

$$(1) \sin^2 \theta + (1) \cos^2 \theta + (-1)1 = 1 - 1 = 0 \tag{2.3.9}$$

The Gram determinant is (Problem 2.19):

$$\begin{vmatrix} \langle f_1 | f_1 \rangle & \langle f_1 | f_2 \rangle & \langle f_1 | f_3 \rangle \\ \langle f_2 | f_1 \rangle & \langle f_2 | f_2 \rangle & \langle f_2 | f_3 \rangle \\ \langle f_3 | f_1 \rangle & \langle f_3 | f_2 \rangle & \langle f_3 | f_3 \rangle \end{vmatrix} = \begin{vmatrix} 2 & 1 & 1 \\ 1 & \frac{3}{4} & \frac{1}{4} \\ 1 & \frac{1}{4} & \frac{3}{4} \end{vmatrix} = 0 \tag{2.3.10}$$

Linear independence is also related to the Taylor series expansions for the functions. The functions $f(x)$ and $g(x)$ can be expanded in Taylor series to give

$$\begin{aligned} f(x) &= f(x_0) + (df/dx)_{x0}(x - x_0) + (1/2!)(d^2f/dx^2)_{x0}(x - x_0)^2 + \cdots \\ g(x) &= g(x_0) + (dg/dx)_{x0}(x - x_0) + (1/2!)(d^2g/dx^2)_{x0}(x - x_0)^2 + \cdots \end{aligned} \tag{2.3.11}$$

If each term in x^i in the expansions is considered a component of the complete function, then a row or column of the determinant can be formed from the function, its first derivative, its second derivative, etc. The total number of such functions will equal the total number of functions considered to produce a square determinant, the Wronskian. For two functions $f(x)$ and $g(x)$, the elements of the Wronskian are $f(x)$, $df(x)/dx$, $g(x)$, and $dg(x)/dx$. Since the determinant gives the same result if the components are ranged as rows or columns, the determinant is written

$$\begin{vmatrix} f(x) & g(x) \\ df/dx & dg/dx \end{vmatrix} \tag{2.3.12}$$

The functions are linearly dependent only if the determinant is zero.

The Wronskian is also generated by noting that two linearly dependent functions can sum to zero with the proper scalar constants a and b:

$$af(x) + bg(x) = 0 \tag{2.3.13}$$

This linear combination can be differentiated with respect to x and the sum is still zero:

$$a(df/dx) + b(dg/dx) = 0 \tag{2.3.14}$$

For linear dependence, the coefficients a and b must be nonzero. This occurs only if the determinant

$$\begin{vmatrix} f & g \\ df/dx & dg/dx \end{vmatrix} \tag{2.3.15}$$

is zero. Higher-order derivatives are used as the number of functions increases. For three functions f, g, and h, the Wronskian is

$$\begin{vmatrix} f & g & h \\ df/dx & dg/dx & dh/dx \\ d^2f/dx^2 & d^2g/dx^2 & d^2h/dx^2 \end{vmatrix} \tag{2.3.16}$$

The linear independences of 1, x, and x^2 can be confirmed with the Wronskian. The determinant is

$$\begin{vmatrix} 1 & x & x^2 \\ 0 & 1 & 2x \\ 0 & 0 & 2 \end{vmatrix} \tag{2.3.17}$$

Since the diagonal elements are all nonzero and the elements below the diagonal are all zero, the determinant is just the product of the three diagonal elements, i.e., 2. The Wronskian determinant is nonzero and the three functions are linearly independent.

The general sequence $1, x, \ldots, x^i, \ldots$ must also be linearly independent since each subsequent row in the determinant will add zero elements below the diagonal. The determinant will always be the product of the diagonal elements and is always nonzero.

The Wronskian determinant can be used to check the linear independence of sine and cosine functions which will be used as a function basis set. For example, $\cos\theta$ and $\sin\theta$ give a Wronskian

$$\begin{vmatrix} \sin\theta & \cos\theta \\ \cos\theta & -\sin\theta \end{vmatrix} = -\sin^2\theta - \cos^2\theta = -1 \tag{2.3.18}$$

The functions are linearly independent. The set of sine functions

$$\sin\theta, \sin 2\theta, \ldots, \sin m\theta, \ldots \tag{2.3.19}$$

constitute a basis set. For the first two functions, the Wronskian is

$$\begin{vmatrix} \sin\theta & \sin 2\theta \\ \cos\theta & 2\cos 2\theta \end{vmatrix} = 2\sin^3\theta \tag{2.3.20}$$

The functions are linearly independent.

The three functions

$$1 \qquad \sin^2\theta \qquad \cos^2\theta \tag{2.3.21}$$

are linearly dependent since

$$\begin{vmatrix} 1 & \sin^2\theta & \cos^2\theta \\ 0 & 2\sin\theta\cos\theta & -2\cos\theta\sin\theta \\ 0 & 2(\cos^2\theta-\sin^2\theta) & -2(\cos^2\theta-\sin^2\theta) \end{vmatrix}$$
$$= 1(4\sin\theta\cos\theta)[(\sin^2\theta-\cos^2\theta)-(\sin^2\theta-\cos^2\theta)]=0 \tag{2.3.22}$$

2.4. Orthogonalization of Basis Functions

A set of linearly independent functions in a Hilbert space do not have to be orthogonal. However, any linearly independent set of functions can be converted to an orthogonal set of function vectors. The techniques for producing this vector set parallel those developed for vectors with finite components. A function f_1 is selected as the initial function. The projection of f_2 on f_1 is then subtracted from f_2 to create a new function f'_2 which is orthogonal to f_1. Projections on f_1 and f_2 are then subtracted from a third function f_3 to produce the orthogonal f'_3, etc.

If a vector $|f_0\rangle$ is selected as the initial coordinate vector in the Hilbert space, the scalar product determines the projection of a second function $|f_1\rangle$ on it:

$$\langle f_0 | f_1 \rangle \tag{2.4.1}$$

Since the "orientation" of $|f_0\rangle$ must be included, the projection on $|f_0\rangle$ which must be subtracted from $|f_2\rangle$ is

$$|f_0\rangle\langle f_0|f_1\rangle \tag{2.4.2}$$

The interpretation of this projection is different in function space. The function f_0 is the vector. It simply represents the infinity of components along the interval. However, on this interval, the projection technique is locating the portion of the function f_1 which looks like the function f_0. The projection of Equation 2.4.2 eliminates this portion of the function from f_1.

The Gram–Schmidt orthogonalization procedure for functions subtracts the projection of f_1 on f_0 from f_1 to give an f_1' orthogonal to f_0:

$$|f_1'\rangle = |f_1\rangle - |f_0\rangle\langle f_0|f_1\rangle \tag{2.4.3}$$

A third function $|f_2\rangle$ becomes mutually orthogonal to the first two function vectors when the projections on both these functions are eliminated. The second step in the Gram–Schmidt orthogonalization is

$$|f_2'\rangle = |f_2\rangle - |f_0\rangle\langle f_0|f_2\rangle - |f_1\rangle\langle f_1|f_2\rangle \tag{2.4.4}$$

For the nth function in the infinite set of functions required, the projections for the $n-1$ mutually orthogonal vectors must be subtracted from $|f_n\rangle$:

$$|f_n'\rangle = |f_n\rangle - \sum_{i}^{n-1} |f_i\rangle\langle f_i|f_n\rangle \tag{2.4.5}$$

It is obviously impossible to produce the full set of vectors in this manner since the total is infinite. However, a finite set of such vectors is often quite useful.

The orthogonalization procedure can be illustrated with the series of functions

$$1 \qquad x \qquad x^2 \qquad x^3 \tag{2.4.6}$$

on the interval $-1 \leqslant x \leqslant 1$. The normalized vectors, $f_0 = \frac{1}{\sqrt{2}}$ and $f_1 = \sqrt{\left(\frac{3}{2}\right)}\, x$, are already orthogonal (Section 2.2). The third function, $f_2 = x^2$, must be orthogonalized by subtracting these two normalized function vectors,

$$|f_2'\rangle = |f_2\rangle - |f_0\rangle\langle f_0|f_2\rangle - |f_1\rangle\langle f_1|f_2\rangle \tag{2.4.7}$$

to give

$$\begin{aligned} |f_2'\rangle &= x^2 - \frac{1}{2}\int_{-1}^{1} (1)(x^2)\,dx - \frac{3}{2}x\int_{-1}^{1} (x)(x^2)\,dx \\ &= x^2 - \left(\frac{1}{2}\right)\left(\frac{2}{3}\right) - \left(\frac{3}{2}\right)(0) = x^2 - \frac{1}{3} \end{aligned} \tag{2.4.8}$$

In the third term, one x remains outside the integral; this is the directional portion ($|f_1\rangle$) of this projection. The orthogonal function is x^2 is now proportional to

$$3x^2 - 1 \tag{2.4.9}$$

The orthogonal function $|f_2\rangle$ must be normalized (Problem 2.20),

$$|f_2'\rangle = \sqrt{\frac{5}{8}}\,(3x^2 - 1) \tag{2.4.10}$$

to permit determination of the orthogonal vector containing x^3:

$$|f_3'\rangle = |f_3\rangle - |f_0\rangle\langle f_0|f_3\rangle - |f_1\rangle\langle f_1|f_3\rangle - |f_2\rangle\langle f_2|f_3\rangle \tag{2.4.11}$$

The values of the integral in these terms will depend on the power of x which appears in the integral. If the net power is odd, the integral will have an even power and will be zero. Since the powers correspond to the function subscripts, only the projection $\langle f_1|f_3\rangle$ will be nonzero, and this projection is

$$\begin{aligned} |f_1\rangle\langle f_1|f_3\rangle &= \frac{3}{2}x\int_{-1}^{1}(x)(x^3)\,dx \\ &= \left(\frac{3}{2}\right)(x)\left(\frac{2}{5}\right) = \frac{3}{5}x \end{aligned} \tag{2.4.12}$$

The orthogonal function is

$$x^3 - \frac{3}{5}x \tag{2.4.13}$$

or

$$|f_3'\rangle = \sqrt{\frac{7}{8}}\,(5x^3 - 3x) \tag{2.4.14}$$

The orthogonal function set developed in this manner is the set of Legendre polynomials. These polynomials appear as solutions for the $\Theta(\theta)$ portion of the solution for the hydrogen atom in quantum mechanics. In that case, x is replaced by $\cos\theta$, and the first Legendre polynomials on the interval $0 \leqslant \theta \leqslant \pi$ are

$$\begin{aligned} P_0(\cos\theta) &= \sqrt{\frac{1}{2}} \\ P_1(\cos\theta) &= \sqrt{\frac{3}{2}}\cos\theta \\ P_2(\cos\theta) &= \sqrt{\frac{5}{8}}\,(3\cos^2\theta - 1) \\ P_3(\cos\theta) &= \sqrt{\frac{7}{8}}\,(5\cos^3\theta - 3\cos\theta) \end{aligned} \tag{2.4.15}$$

These functions describe the θ portion of the s, p, d, and f orbitals, respectively. In other words, the functional solutions to the Schrödinger wave equation in this case are an orthogonal set of function vectors.

Scalar products in the $x = \cos\theta$ format require the integration variable

$$dx = -\sin\theta \, d\theta \tag{2.4.16}$$

$\sin\theta$ appears as a weighting factor in the integration. The minus sign is eliminated by reversing the order of integration since

$$\begin{aligned} x &= \cos\pi = -1 \qquad \theta = \pi \\ x &= \cos 0 = +1 \qquad \theta = 0 \end{aligned} \tag{2.4.17}$$

The scalar product in the variable θ is

$$-\int_{\pi}^{0} P_i P_j \sin\theta \, d\theta = +\int_{0}^{\pi} P_i P_j \sin\theta \, d\theta \tag{2.4.18}$$

Orthogonal functions were developed from polynomial functions, x^i, on the interval $-1 \leqslant x \leqslant 1$. However, there are an extremely large number of ways in which such functions may be generated. The limits may be changed; an additional function may be required in the scalar product as it was for the Hermite polynomials; etc. The development of orthogonal function sets can be quite tedious using the Gram–Schmidt orthogonalization. For this reason, alternate methods of creating such sets are desirable. Different orthogonal sets of functions can be solutions of specific differential equations. Each specific type of differential equation produces a unique set of orthogonal function vectors.

2.5. Differential Operators

A vector function in Hilbert space represents an orthogonal "direction" in that space. For the finite-dimension vectors, multiplication of each component of that vector by the same scalar did not change its direction. In Hilbert space, the function

$$2f_0 \tag{2.5.1}$$

has the same direction as f_0 since each component in the vector is changed by the same factor. The vector is longer but points in the same direction.

An N-dimensional vector could be differentiated with respect to some argument x by differentiating each component of the vector with respect to x. For a three-dimensional vector, this means

$$d\,|a\rangle/dx = \begin{pmatrix} da_x/dx \\ da_y/dx \\ da_z/dx \end{pmatrix} \tag{2.5.2}$$

The process of differentiation is even easier for a function vector, $f(x)$. The vector components of $f(x)$ are its values for each x on the interval. Differentiation of each of the components is equivalent to finding the derivative of $f(x)$. This quantity could then be evaluated for each x to generate all the components of the derivative. The vector differentiation is equivalent to "normal" differentiation of a function,

$$d\,|f(x)\rangle/dx = df(x)/dx \tag{2.5.3}$$

In most cases, differentiation of the function will generate a new function or combination of functions. For example, the derivative of the Legendre polynomial

$$dP_1(x)/dx = x \tag{2.5.4}$$

is

$$dP_1(x)/dx = 1 = P_0(x) \tag{2.5.5}$$

Differentiation has changed the "orientation" in the space.

Differentiation can also produce a linear combination of new function vectors. For example, the x derivative of $P_3(x)$ is

$$\begin{aligned} dP_3/dx &= d/dx(5x^3 - 3x) \\ &= 15x^2 - 3 \end{aligned} \tag{2.5.6}$$

which is a linear combination of P_2 and P_0 (Problem 2.7).

In some cases, a derivative will reproduce the original function. For example, if

$$|f(x)\rangle = \exp(-2x) \tag{2.5.7}$$

the derivative is

$$d\,|f(x)\rangle/dx = d/dx[\exp(-2x)] = -2\exp(-2x) \tag{2.5.8}$$

The derivative regenerates the same function vector multiplied by a scalar, i.e., its magnitude but not its direction is changed by the derivative. Other functions may be reproduced by higher-order derivatives. The function

$$|g(x)\rangle = \sin 2x \tag{2.5.9}$$

is reproduced with a scalar factor by a second derivative in x,

$$d^2 \sin 2x/dx^2 = -4 \sin 2x \tag{2.5.10}$$

The derivatives applied to the function vectors are examples of operations on these vectors. The differential operator operates on the function by differentiation or multiplication with other functions or scalars. The result can be a linear combination of function vectors. In other cases of more interest here, the operators acts on the function vector and reproduces this vector multiplied by some scalar. Such operations maintain the system in a single "vectorial" direction.

The differential operations which reproduce the original function are actually differential equations for this function. For example, the differential operator

$$\mathbf{O} = (1 - x^2)\, d^2/dx^2 - 2x\, d/dx \tag{2.5.11}$$

acts on the function $f(x) = x$ to give x and a scalar factor of -2:

$$\begin{aligned}(1 - x^2)\, d^2 f(x)/dx^2 - 2x\, df/dx &= (1 - x^2)\, d^2(x)/dx^2 - 2x\, dx/dx \\ &= 0 - 2x = -2x = -2f(x)\end{aligned} \tag{2.5.12}$$

$f(x) = x$ is a solution of the differential equation. Since it appears on the right, the complicated differential operator has not changed its direction in function space.

Some mutually orthogonal functions will be regenerated and multiplied by a scalar under the differential operation. This is written formally as

$$\mathbf{O}\,|\,f\rangle = \lambda\,|\,f\rangle \tag{2.5.13}$$

where λ is a scalar. For the differential operator of Equation 2.5.11, the appropriate functions are the Legendre polynomials. The ith Legendre polynomial, $P_i(x)$, will give a scalar factor

$$-i(i+1) \tag{2.5.14}$$

i.e.,

$$\mathbf{O}P_i(x) = -i(i+1)\, P_i(x) \tag{2.5.15}$$

The operation has been verified for $P_1(x) = x$, i.e.,

$$[(1 - x^2)\, d^2/dx^2 - 2x\, d/dx]\, x = (1 - x^2)(0) - 2x(1) = -2x \tag{2.5.16}$$

$P_1 = x$ is reproduced by the operation times a factor of -2 which is equal to

$$-i(i+1) = -1(1+1) = -2 \tag{2.5.17}$$

The orientation of each Legendre polynomial is not disturbed by the differential operation. The entire operation is confined to a single component in the Hilbert space.

If the Legendre differential operaor is rewritten in terms of $\cos\theta$ and multiplied by $-\hbar^2$, it becomes the angular momentum operator of quantum mechanics. It operates on an arbitrary Legendre function, P_i, to regenerate the function and a scalar equal to the angular momentum associated with that "wave" function,

$$-\hbar^2\{d/dx[(1-x^2)\, d/dx]\}\, P_i = \hbar^2 i(i+1)\, P_i \tag{2.5.18}$$

or

$$(-\hbar^2/\sin\theta)\{d/d\theta[(\sin\theta)\, d/d\theta]\}\, P_i(\theta) = \hbar^2 i(i+1)\, P_i(\theta) \tag{2.5.19}$$

The wavefunctions of quantum mechanics for simple systems will be orthogonal functions which satisfy an operator equation,

$$\mathbf{H}\, |\psi\rangle = E\, |\psi\rangle \tag{2.5.20}$$

where **H** is the Hamiltonian differential operator and E is a scalar equal to the system energy.

The Legendre polynomials constitute a single example of orthogonal functions which are solutions of differential equations of the form

$$\mathbf{O}\, |f\rangle = \lambda\, |f\rangle \tag{2.5.21}$$

Other orthogonal sets serve as solutions for other differential equations. For example, sine functions

$$\psi(x) = \sin(n\pi x/L) \tag{2.5.22}$$

serve as solutions of the operator equation

$$(d^2/dx^2)\, |\psi(x)\rangle = \lambda\, |\psi(x)\rangle \tag{2.5.23}$$

If the operator is multiplied by

$$-\hbar^2/2m \tag{2.5.24}$$

it becomes the Hamiltonian for a particle of mass m in a one-dimensional box of length L. The scalar factor generated by the differential operation is the energy of the particle, i.e.,

$$\mathbf{H}\, |\psi(x)\rangle = E\, |\psi(x)\rangle \tag{2.5.25}$$

$$(-\hbar^2/2m)\, d^2[\sin(n\pi x/L)]/dx^2 = (\hbar n\pi/L)^2\, (1/2m)\sin(n\pi x/L) = E\sin(n\pi x/L) \tag{2.5.26}$$

The second-order differential equation in x is generalized to three dimensions by forming a differential operator from the divergence of the gradient, which is the Laplacian or "del-squared" operator:

$$\nabla^2 = \text{div}(\text{grad}) \tag{2.5.27}$$

In Cartesian coordinates, the operation is

$$(\partial/\partial x \quad \partial/\partial y \quad \partial/\partial z) \begin{pmatrix} \partial/\partial x \\ \partial/\partial y \\ \partial/\partial z \end{pmatrix} = \partial^2/\partial x^2 + \partial^2/\partial y^2 + \partial^2/\partial z^2 \tag{2.5.28}$$

The operation is a sum of second derivatives for Cartesian coordinates. However, the actual differential operator will be dependent on the particular coordinate system chosen. The Schrödinger equation

$$(-\hbar^2/2m)\nabla^2 |\psi\rangle = E|\psi\rangle \tag{2.5.29}$$

is valid for any coordinate system. However, the choice of coordinate system will dictate the orthogonal functions, $|\psi\rangle$, which satisfy Equation 2.5.29.

Several sets of orthogonal functions and their appropriate differential equations appear frequently. The associated Legendre polynomials appear in two-dimensional systems described by the polar and azimuthal angles, θ and φ. The differential equation resolves into two parts. The equation for φ is

$$d^2\Phi(\varphi)/d\varphi^2 = m\Phi(\varphi) \qquad m = 0, 1, 2, \ldots \tag{2.5.30}$$

with solutions

$$\Phi(\varphi) = \exp(im\varphi) \quad \text{and} \quad \Phi(\varphi) = \exp(-im\varphi) \tag{2.5.31}$$

The integer m must appear in the differential equation for Θ. For convenience, this equation can be written as an equation in x where

$$x = \cos\theta \tag{2.5.32}$$

to give

$$\{(1-x^2)\, d^2/dx^2 - 2x\, d/dx - m^2/(1-x^2)\}\, P_{lm}(x) = l(l+1)\, P_{lm} \tag{2.5.33}$$

The differential equation has an additional term in m. However, the associated Legendre polynomials can be determined from the "one-dimensional" Legendre polynomials by performing m derivatives using the equation

$$P_{lm}(x) = (1-x^2)^{m/2}\, d^m P_l/dx^m \tag{2.5.34}$$

The derivative ensures that solutions for the associated Legendre polynomials are valid only when (Problem 2.12)

$$m \leqslant l \tag{2.5.35}$$

The product of $P_{lm}(\cos\theta)$ and $\Phi(\varphi)$ is the spherical harmonic function Y_{lm} which represents the two-dimensional solution of the differential equation. The function can be illustrated by determining Y_{21}. Since $m=1$, the Legendre polynomial for $l=2$,

$$P_2(x)=3x^2-1 \tag{2.5.36}$$

must be differentiated once to obtain the associated Legendre polynomial,

$$\begin{aligned} P_{21}(x) &= (1-x^2)^{1/2}\, d(3x^2-1)/dx \\ &= (1-x^2)^{1/2}\,(6x) \end{aligned} \tag{2.5.37}$$

Using $x=\cos\theta$ and

$$1-\cos^2\theta=\sin^2\theta \tag{2.5.38}$$

this equation becomes

$$6\cos\theta\sin\theta \tag{2.5.39}$$

and Y_{21} is

$$Y_{21}(\theta,\varphi)=\cos\theta\sin\theta\exp[i(1)\varphi] \tag{2.5.40}$$

Bessel functions are the solution of Bessel's differential equation,

$$(x^2d^2/dx^2+x\,d/dx+x^2)\,J_n=n^2J_n \tag{2.5.41}$$

where n is not restricted to integer values. The Bessel functions J_0 and J_1 are written as infinite series in powers of x:

$$\begin{aligned} J_0(x) &= 1-x^2/2^2+x^4/(2\cdot 4)^2-x^6/(2\cdot 4\cdot 6)^2+\cdots \\ J_1(x) &= (x/2)\{1-x^2/(2\cdot 4)+x^4/(2\cdot 4^2\cdot 6)-x^6/[2\cdot(4\cdot 6)^2\cdot 8]+\cdots\} \end{aligned} \tag{2.5.42}$$

The two functions are shown in Figure 2.2. For small x, J_0 falls as x^2 while J_1 is

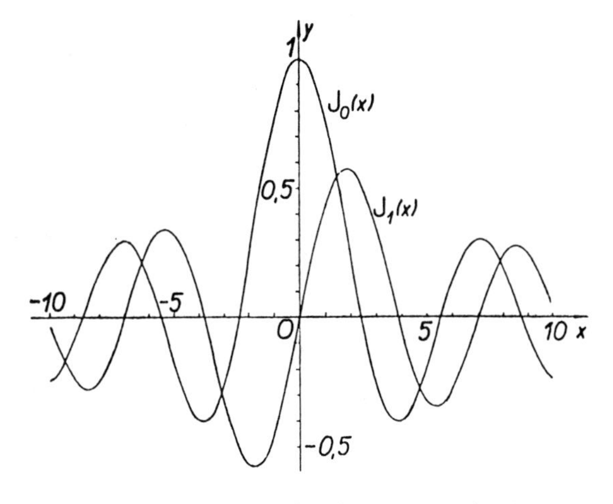

Figure 2.2. The Bessel functions $J_0(x)$ and $J_1(x)$.

dominated by x^1. The Bessel functions oscillate with decreasing amplitude and can describe standing waves on a circular surface such as a drumhead.

When $n = \frac{1}{2}$ or $-\frac{1}{2}$, the oscillation is proportional to sin x:

$$\begin{aligned} J_{1/2} &= (2/\pi x) \sin x \\ J_{-1/2} &= (2/\pi x) \cos x \end{aligned} \tag{2.5.43}$$

The amplitudes of the oscillations decrease with increasing x because of the x^{-1} dependence of the function.

The Hermite polynomials are solutions of the equation

$$(d^2/dx^2 - 2x\, d/dx)\, H_n = -2nH_n \tag{2.5.44}$$

The first four Hermite polynomials are

$$\begin{aligned} H_0(x) &= 1 \\ H_1(x) &= 2x \\ H_2(x) &= 4x^2 - 2 \\ H_3(x) &= 8x^3 - 12x \end{aligned} \tag{2.5.45}$$

The first three of these are plotted in Figure 2.3.

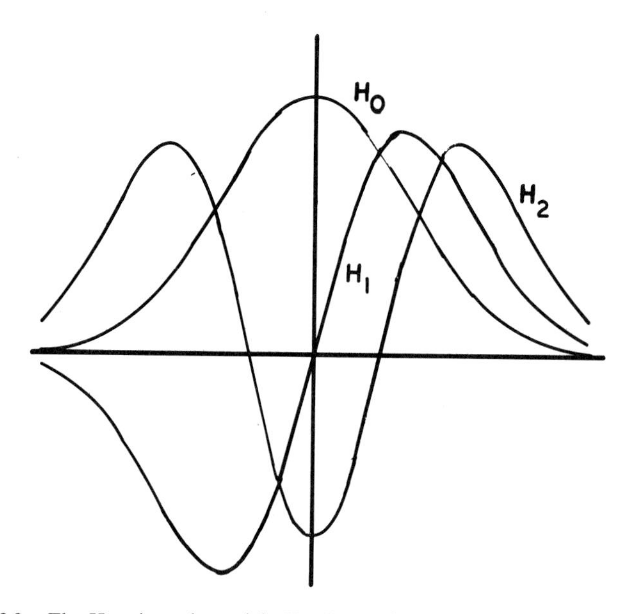

Figure 2.3. The Hermite polynomials H_0, H_1, and H_2 on the interval $-\infty < x < \infty$.

The general formula for H_n is

$$H_n(x) = (2x)^n - [n(n-1)/1!](2x)^{n-2} + [n(n-1)(n-2)(n-3)/2!](2x)^{n-4} - \cdots \quad (2.5.46)$$

where the final term is proportional to x^0 for even n and x^1 for odd n. In quantum mechanics, this basis set describes wavefunctions for a harmonic oscillator.

The Laguerre polynomials satisfy the differential equation

$$[x\, d^2/dx^2 + (1-x)\, d/dx]\, L_n(x) = -nL_n(x) \quad (2.5.47)$$

The first four Laguerre polynomials are

$$\begin{aligned} L_0 &= 1 \\ L_1(x) &= 1 - x \\ L_2(x) &= 1 - 2x + x^2/2 \\ L_3(x) &= 1 - 3x + 3x^2/2 - x^3/6 \end{aligned} \quad (2.5.48)$$

The nth Laguerre polynomial is

$$L_n(x) = (-1)^n\{x^n - [n^2/1!]\, x^{n-1} + [n^2(n-1)^2/2!]\, x^{n-2} - \cdots + (-1)^n n!\} \quad (2.5.49)$$

where the sequence continues until the power of x is zero. These basis functions give the radial portion of the wavefunction for the hydrogen atom in quantum mechanics.

These function basis sets are selected since their associated differential equations appear frequently in descriptions of physical systems. However, many other basis functions and differential equations have been developed and can be found in any textbook of mathematical physics.

All basis sets share some basic properties. They are usually generated as the solution to some differential equation. They can be normalized although some require a weighting factor. They can be used to approximate an arbitrary function in the same domain.

2.6. Generation of Special Functions

The special functions introduced in Section 2.5 can be generated as solutions of differential equations. Although many systems will require different differential equations, it is often possible to recast them in the form of the "tabulated" differential equations so they can be solved rapidly. For this reason, the solutions for these standard differential equations appear frequently, and several techniques have been developed to generate their solutions directly. The most important generating techniques are outlined in this section.

Although the Legendre polynomials are generated via a Gram–Schmidt orthogonalization or via solution of the Legendre differential equation, they can be generated directly using the summation

$$P_l(x) = (2l)!/(2^l l!^2)\left[x^l - \frac{l(l-1)\,x^{l-2}}{2(2l-1)} + \frac{l(l-1)(l-2)(l-3)\,x^{l-4}}{2(4)(2l-1)(2l-3)} + \cdots\right] \qquad (2.6.1)$$

The integer l for P_l appears as the highest power of x. The series terminates when the power of x is less than 0. For example, the polynomial P_4 is

$$\begin{aligned} P_4 &= [8!/16(4!)^2]\{x^4 - [(4)(3)/(2)(7)]\,x^2 + [4!/(8)(7)(5)]\} \\ &= \frac{35}{8}\left(x^4 - \frac{6}{7}x^2 + \frac{3}{35}\right) \\ &= \frac{35}{8}x^4 - \frac{15}{4}x^2 + \frac{3}{8} \end{aligned} \qquad (2.6.2)$$

The Legendre polynomials can also be generated with Rodrigues' formula:

$$P_l(x) = (1/2^l l!)\, d^l(x^2-1)^l/dx^l \qquad (2.6.3)$$

$P_2(x)$ is

$$\begin{aligned} P_2 &= \left(\frac{1}{8}\right) d^2(x^2-1)^2/dx^2 \\ &= \left(\frac{1}{8}\right) d^2(x^4 - 2x^2 + 1)/dx^2 = \frac{1}{8}(12x^2 - 4) \\ &= \frac{1}{2}(3x^2 - 1) \end{aligned} \qquad (2.6.4)$$

P_i can be related to other P_j using recursion relations. The most useful recursion relation for the Legendre polynomials is

$$(2i+1)\,xP_i = (i+1)\,P_{i+1} + iP_{i-1} \qquad (2.6.5)$$

P_i is related to P_{i+1} and P_{i-1}. This recursion relation coupled with the orthogonality of the P_i can be used to evaluate some integrals efficiently. For example, an integral

$$\int_{-1}^{1} P_j(x)\, xP_i(x)\, dx \qquad (2.6.6)$$

can involve some complicated function products. However, instead of listing the functions in detail, xP_j is replaced using Equation 2.6.5:

$$[1/(2j+1)]\left[\int_{-1}^{1} (j+1)\, P_{j+1} P_i\, dx + \int_{-1}^{1} jP_{j-1} P_i\, dx\right] \tag{2.6.7}$$

The integrals are nonzero only if the subscripts on both functions are identical. The first integral is nonzero if $j=i-1$. The second integral is nonzero if $j=i+1$. For the first case, the integral is

$$\{1/[2(i-1)+1]\}(i-1+1)\int_{-1}^{1} P_i P_i\, dx = i/(2i-1) \tag{2.6.8}$$

The recursion relations provide a simple explanation for selection rules in quantum mechanics. A transition between electronic states described by P_i and P_j depends on the dipole moment integral where the dipole is proportional to x:

$$\int_{-1}^{1} P_j x P_i\, dx \tag{2.6.9}$$

The recursion relation shows that this integral is nonzero only when $j=i+1$ or $j=i-1$. A dipole-induced transition can occur only between wavefunctions with quantum numbers separated by one unit, i.e., nearest-neighbor transitions.

The Legendre polynomials are connected by other recursion relations. For example,

$$(2i+1)\, P_i = d/dx(P_{i+1} - P_{i-1}) \tag{2.6.10}$$

$$(i+1)\, P_i = dP_{i+1}/dx - x\, dP_i/dx \tag{2.6.11}$$

The associated Legendre polynomials are generated from the Legendre polynomials via m differentiations. The associated Legendre polynomials satisfy the recursion relations

$$(2i+1)(1-x^2)^{1/2}\, P_{im} = P_{i+1,m+1} - P_{i-1,m+1} \tag{2.6.12}$$

and

$$(2i+1)(1-x^2)^{1/2}\, P_{im} = (i+m)(i+m-1)\, P_{i-1,\geqslant m-1} - (i-m+1)(i-m+2)\, P_{i+1,m-1} \tag{2.6.13}$$

The Hermite polynomials are given by the general expression

$$H_n = (2x)^n - [n(n-1)/1!](2x)^{n-1} + [n(n-1)(n-2)(n-3)/2!](2x)^{n-4} - \cdots \tag{2.6.14}$$

expressed compactly as

$$H_n(x) = \sum_k^h (-1)^k [n!/k!(n-2k)!](2x)^{n-2k} \qquad (2.6.15)$$

where h, the maximum value of k in the sum, is $n/2$ when n is even and $(n-1)/2$ when n is odd.

The Hermite polynomials are generated by

$$H_n(x) = (-1)^n \exp(x^2)\, d^n[\exp(-x^2)]/dx^n \qquad (2.6.16)$$

$H_2(x)$ is

$$\begin{aligned} H_2 &= (-1)^2 \exp(x^2)\, d^2[\exp(-x^2)]/dx^2 \\ &= \exp(x^2)[4x^2 \exp(-x^2) - 2\exp(-x^2)] \\ &= 4x^2 - 2 \end{aligned} \qquad (2.6.17)$$

The recursion relation for the Hermite polynomials,

$$xH_n(x) = nH_{n-1} + \frac{1}{2} H_{n+1} \qquad (2.6.18)$$

can also give quantum mechanical selection rules. The dipole moment integral for a transition between n and $n+1$ is given as

$$\int_{-\infty}^{\infty} H_{n+1}(x)\, xH_n(x) \exp(-x^2)\, dx \qquad (2.6.19)$$

since the scalar product for the Hermite polynomials must include the weighting factor

$$\exp(-x^2) \qquad (2.6.20)$$

Replacing $xH_n(x)$ by the right-hand side of Equation 2.6.18 gives

$$\int_{-\infty}^{\infty} nH_{n+1} nH_{n-1} \exp(-x^2)\, dx + \frac{1}{2}\int_{-\infty}^{\infty} H_{n+1} H_{n+1} \exp(-x^2)\, dx \qquad (2.6.21)$$

The second integral is the norm of H_{n+1} so the total integral in this case is $\frac{1}{2}$. The Hermite polynomials, which describe a harmonic oscillator in quantum mechanics, also obey nearest-neighbor selection rules. The integrations are deduced directly from the recursion relations and the orthogonality properties of the polynomials.

The Laguerre polynomials are defined by the general summation

$$L_n(x) = \sum_k^h [(-1)^k/k!][n!/k!(n-k)!]\, x^k \qquad (2.6.22)$$

L_2 in this case is

$$L_2 = [2!/0!2!]\, x^0 - [1/1!][2!/1!1!]\, x^1 + [1/2!][2!/2!(2-2)!]\, x^2$$
$$= 1 - 2x + \frac{1}{2} x^2 \tag{2.6.23}$$

L_n is generated using the differential expression

$$L_n(x) = [\exp(x)/n!]\, d^n[x^n \exp(-x)]/dx^n \tag{2.6.24}$$

for $n > 0$; $L_0 = 1$. L_1 determined with this formula is

$$L_1 = [\exp(x)/1!]\, d[x \exp(-x)]/dx$$
$$= [\exp(+x)][\exp(-x) - x \exp(-x)]$$
$$= 1 - x \tag{2.6.25}$$

The scalar product for the Laguerre polynomials must include the weighting factor

$$\exp(-x) \tag{2.6.26}$$

i.e.,

$$\langle L_i | L_j \rangle = \int_0^\infty \exp(-x)\, L_i(x)\, L_j(x)\, dx \tag{2.6.27}$$

The Laguerre polynomials are solutions for the differential equation

$$xd^2L/dx^2 + (1-x)\, dL/dx = -nL \tag{2.6.28}$$

The associated Laguerre polynomials, L_{nm}, satisfy the differential equation

$$x\, d^2L_{nm}/dx^2 + (m+1-x)\, dL_{nm}/dx + (n-m)\, L_{nm} = 0 \tag{2.6.29}$$

The function L_{nm} is derived directly from the Laguerre polynomial by differentiating it with respect to x m times. The associated Laguerre polynomial, L_{21}, is then

$$dL_{21}/dx = d/dx\left(1 - 2x + \frac{1}{2} x^2\right)$$
$$= -2 + x \tag{2.6.30}$$

This result is verified by substitution in Equation 2.6.29 with $n = 2$ and $m = 1$ (Problem 2.14).

The scalar product for the associated Laguerre polynomials must include an additional weighting factor. The scalar product for L_{im} and L_{jm} is

$$\int_0^\infty \exp(-x)\, x^m L_{im}(x)\, L_{jm}(x)\, dx \tag{2.6.31}$$

and the associated Laguerre polynomials form an orthogonal function set.

The orthogonal function sets used in this section were solutions of second-order differential equations. These differential equations differed in the coefficients for each term but all of these equations could be written in the general form

$$d/dx[a(x)\, dy/dx] + [b(x) + \lambda c(x)]\, y = 0 \tag{2.6.32}$$

This general equation is called a Sturm–Liouville equation and each of the equations discussed here can be rewritten in this form. For example, a Sturm–Liouville equation with

$$a(x) = 1 - x^2 \qquad b(x) = 0 \qquad c(x) = 1 \tag{2.6.33}$$

gives

$$\begin{aligned} d/dx[(1 - x^2)\, dy/dx] + \lambda y &= 0 \\ (1 - x^2)\, d^2y/dx^2 - 2x\, dy/dx + \lambda y &= 0 \end{aligned} \tag{2.6.34}$$

This is the Legendre differential equation.

Although the Legendre equation converts easily to a Sturm–Liouville format, the Hermite differential equation,

$$d^2H/dx^2 - 2x\, dH/dx + 2nH = 0 \tag{2.6.35}$$

must be manipulated to reach the proper form. The equation is multiplied by

$$\exp(-x^2) \tag{2.6.36}$$

to give

$$\exp(-x^2)\, d^2H/dx^2 - 2x \exp(-x^2)\, dH/dx + 2n \exp(-x^2) H = 0 \tag{2.6.37}$$

The equation is equivalent since the exponential factor could be factored from each term. However, in this form, the differential equation is equivalent to a Sturm–Liouville form (Problem 2.22),

$$d/dx[\exp(-x^2)\, dH/dx] + 2n \exp(-x^2) H = 0 \tag{2.6.38}$$

For this equation,

$$a(x) = c(x) = \exp(-x^2) \qquad b(x) = 0 \tag{2.6.39}$$

The exponential factor, $\exp(-x^2)$, appeared as the weighting factor in the scalar product for the Hermite polynomials. For Sturm–Liouville equations, the factor $c(x)$ will always appear as the weighting factor. For H, the scalar product is

$$\langle H_i | H_j \rangle = \int_{-\infty}^{\infty} H_i H_j \exp(-x^2)\, dx \tag{2.6.40}$$

For the Legendre differential equation in the Sturm–Liouville format, $c(x) = 1$. There is no weighting factor in the scalar product for the Legendre polynomials.

The function $a(x)$ determines the limits where the functions are orthogonal. For the Legendre polynomials,

$$a(x) = 1 - x^2 \tag{2.6.41}$$

At the limits $x = +1$ and $x = -1$, $a(x)$ is zero. This is one of a number of possible conditions for the existence of a valid set of orthogonal functions on the interval.

The Sturm–Liouville equation is an example of an eigenvalue–eigenvector equation. The equation is written as

$$d/dx[a(x)\, dy/dx] + [b(x)]\, y = -\lambda c(x)\, y \tag{2.6.42}$$

The differential operations on y reproduce the eigenfunction (or eigenvector) y multiplied by a scalar, λ (the eigenvalue), and $c(x)$. Eigenvalue–eigenvector equations have special properties which will be discussed in subsequent chapters. In this case, the eigenfunction exists in the Hilbert space. Eigenvectors and eigenvalues will also be observed via operations in finite-dimensional spaces.

2.7. Function Resolution in a Set of Basis Functions

An arbitrary vector $|r\rangle$ in a finite-dimensional vector space could always be resolved into a sum of projections on a complete set of coordinate directions in the space. The projection of $|r\rangle$ on the unit coordinate $|1\rangle$ was

$$|1\rangle\langle 1 | r\rangle \tag{2.7.1}$$

where $\langle 1 | r\rangle$ gave the scalar magnitude of the projection and $|1\rangle$ indicated the direction of this component. The total vector $|r\rangle$ was the sum of projections on all possible coordinate directions and could be written

$$|r\rangle = \sum_i |i\rangle\langle i | r\rangle \tag{2.7.2}$$

where the summation extends over the full set of vectors in the finite-dimensional space.

A basis set in function space requires an infinite set of orthogonal functions. These functions are characterized by their "shape" on the interval of interest. Since each function will have a distinct "shape," i.e., a distinct coordinate direction in the space, any function in the interval should be described as a combination of the basis function vectors. A function $f(x)$ on a one-dimensional interval, $a \leqslant x \leqslant b$, is described by a set of basis functions $|i\rangle$ by analogy with the finite-dimensional example of Equation 2.7.2:

$$|f(x)\rangle = \sum_{i}^{\infty} |i\rangle\langle i|f(x)\rangle \tag{2.7.3}$$

The summation extends to infinity in this case. The scalar component of the projection of $f(x)$ on the functional direction $|i\rangle$ is just

$$\langle i|f(x)\rangle = \int b_i(x)\, f(x)\, w(x)\, dx \tag{2.7.4}$$

where $b_i(x)$ is the ith function vector for a basis set such as the Legendre or Hermite polynomials, and $w(x)$ is the weighting function for $b_i(x)$ if it is necessary for the scalar product.

Decomposition of the function into a sum of basis functions can be illustrated by considering the function

$$f(x) = x^2 \tag{2.7.5}$$

on the interval $-1 \leqslant x \leqslant 1$. The function must be described as a summation of mutually orthogonal Legendre polynomials. P_0 and P_2 provide the best approximation for this simple function,

$$P_0 = \sqrt{\frac{1}{2}} \qquad P_2 = \sqrt{\frac{5}{8}}\,(3x^2 - 1) \tag{2.7.6}$$

since $P_1 = \sqrt{\frac{3}{2}}\,x$ will produce a scalar product with an odd power of x and a zero projection. The projections are

$$\begin{aligned} \langle P_0|x^2\rangle &= \sqrt{\frac{1}{2}} \int_{-1}^{1} (1)(x^2)\, dx = \sqrt{\frac{1}{2}}\left(\frac{2}{3}\right) \\ \langle P_2|x^2\rangle &= \sqrt{\frac{5}{8}} \int_{-1}^{1} (3x^2 - 1)(x^2)\, dx \\ &= \sqrt{\frac{5}{8}}\,(3x^5/5 - x^3/3)\Big|_{-1}^{1} = \sqrt{\frac{5}{8}}\left(\frac{8}{15}\right) \end{aligned} \tag{2.7.7}$$

The function x^2 is now the linear combination of P_0 and P_1 weighted by these scalars:

$$x^2 = |P_0\rangle \sqrt{\frac{1}{2}}\left(\frac{2}{3}\right) + |P_2\rangle \sqrt{\frac{5}{8}}\left(\frac{8}{15}\right) \tag{2.7.8}$$

Although this function appears more complicated than x^2, it is only a resolution of x^2 on the two orthogonal basis functions. The polynomial functions P_0 and P_2 are inserted in functional form to give

$$x^2 = \sqrt{\frac{1}{2}}\sqrt{\frac{1}{2}}\left(\frac{2}{3}\right) + \sqrt{\frac{5}{8}}(3x^2-1)\sqrt{\frac{5}{8}}\left(\frac{8}{15}\right)$$

$$= \frac{1}{3} + \frac{1}{3}(3x^2-1) = x^2 \tag{2.7.9}$$

The function x^2 is not orthogonal to P_0 and P_2; a combination of these basis functions is required to regenerate it.

The two basis functions and their coefficients for x^2 are illustrated in Figure 2.4. The P_0 component is a horizontal line through $y = +\frac{1}{3}$. P_2 is equal to

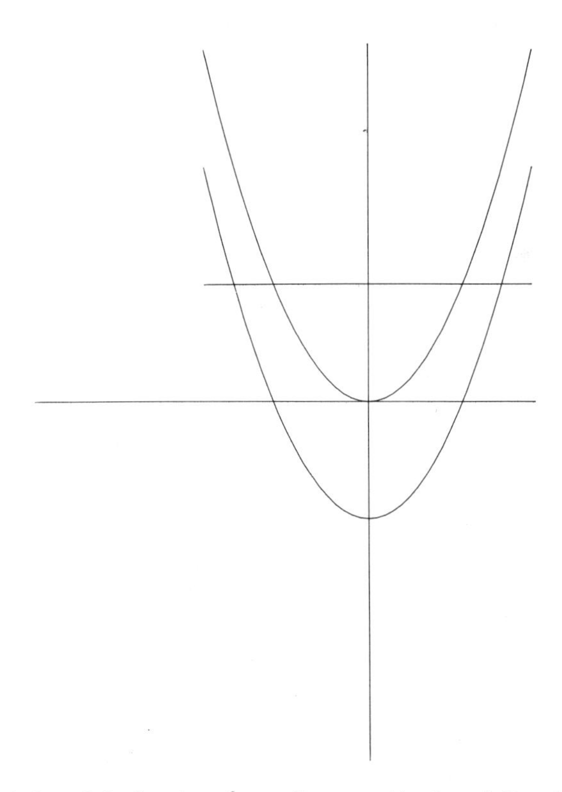

Figure 2.4. Description of the function x^2 as a linear combination of P_0 and P_1 on the interval $-1 \leqslant x \leqslant 1$.

$-\frac{1}{3}$ at $x=0$. P_0 translates P_2 up by $+\frac{1}{3}$ so that $y=0$ at $x=0$. This reproduces the function x^2.

The real power of expansions in orthogonal functions is realized for more complicated functions. For example, the function of Figure 2.5 consists of two line segments. The function is 0 from -1 to 0 and rises with slope $+1$ on the interval 0 to 1, i.e., $y=x$ in the positive x interval. Because there are two different line segments, the function does not correspond to any of the regular functions in the basis set. However, the function can still be constructed from these regular basis functions.

The decomposition of the function is illustrated for the first few basis functions. The normalization factors will be incorporated entirely in the ket basis vectors for convenience. The projections are determined by integrating the function over each of its segments. The scalar product $\langle P_0 | f \rangle$ is

$$\int_{-1}^{0} (1)(0)\, dx + \int_{0}^{1} (1)(x)\, dx = (x^2/2)\Big|_0^1 = \frac{1}{2} \tag{2.7.10}$$

The integral for the negative portion of the interval is zero since the function is zero in this range. The integral for $\langle P_1 | f \rangle$ is

$$\int_{-1}^{0} (x)(0)\, dx + \int_{0}^{1} (x)(x)\, dx = (x^3/3)\Big|_0^1 = \frac{1}{3} \tag{2.7.11}$$

The next projection is

$$\begin{aligned} \langle P_2 | f \rangle &= \int_0^1 (3x^2 - 1)(x)\, dx \\ &= (3x^4/4 - x^2/2)\Big|_0^1 = \frac{1}{4} \end{aligned} \tag{2.7.12}$$

The process must be continued for additional basis functions. At this point, the function is approximated by the sum

$$f(x) = |P_0\rangle(1) + |P_1\rangle\left(\frac{1}{2}\right) + |P_2\rangle\left(\frac{1}{3}\right) \tag{2.7.13}$$

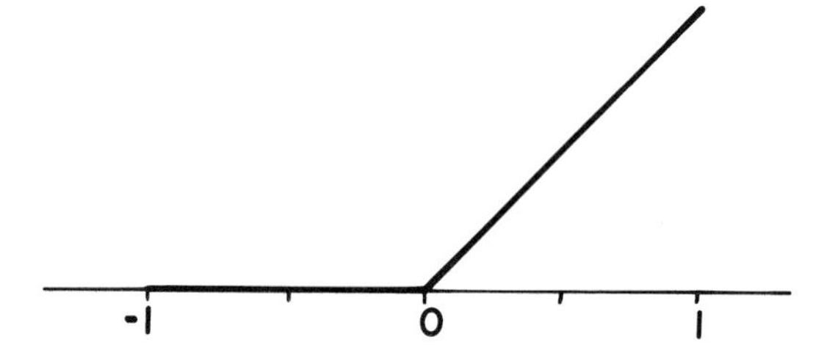

Figure 2.5. A function consisting of two line segments of different slope.

Inserting the values of the ket functions gives

$$f(x)=\frac{1}{2}+\left(\frac{3}{2}\right)\left(\frac{1}{2}\right)x+\left(\frac{5}{8}\right)\left(\frac{1}{3}\right)(3x^2-1) \tag{2.7.14}$$

The increased accuracy of the fit is observable in Figure 2.6 for two, four, and five projections. The functions add and subtract on the negative interval while they always add on the positive interval. As the number of functions in the summation is increased, the approximation to the function gets better and better. The worst fit is observed near $x=0$, where the function changes slope.

For finite-dimensional systems, the addition of contributions for all basis vectors in the space had to reproduce a given vector. Every direction in the space was counted. For a function in the Hilbert space, the summation of function vectors could continue indefinitely. However, the representation of the function will get better as more components are added. The Bessel inequality is used to provide a measure of the effectiveness of fit for a finite sum of function vectors. The norm of $f(x)$, i.e., the scalar product of $f(x)$ with itself,

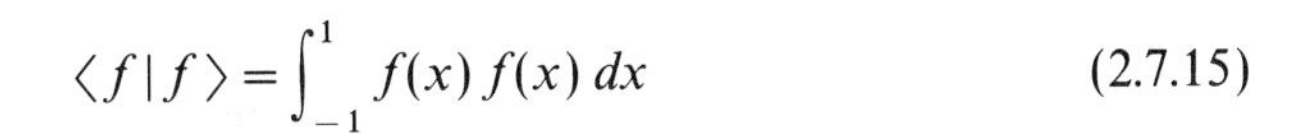

$$\langle f \mid f\rangle=\int_{-1}^{1} f(x)\, f(x)\, dx \tag{2.7.15}$$

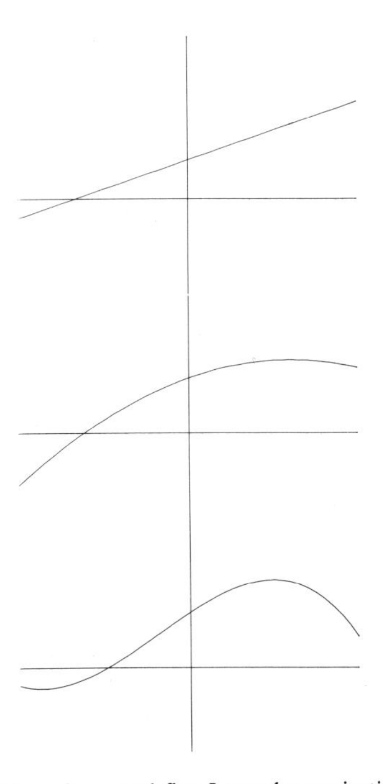

Figure 2.6. The summation of two, four, and five Legendre projections to describe the function of Figure 2.5.

plays a role parallel to that of the norm in finite-dimensional spaces; it gives the square of the absolute "length" of the function vector with no directional information. Since the function is resolved into a sum of basis function vectors with scalar coefficients,

$$|f(x)\rangle = \sum_i c_i \, |i\rangle \tag{2.7.16}$$

the scalar product $\langle f \mid f \rangle$ is also equal to

$$\langle f \mid f \rangle = \sum_i c_i^2 \langle i \mid i \rangle = \sum c_i^2 \tag{2.7.17}$$

since all product terms $\langle i \mid j \rangle$, $i \neq j$, will be zero for this orthogonal basis set. The scalar norm is just equal to the sum of the squares of all the scalar factors, c_i. However, all possible function vectors in the Hilbert space cannot be determined. If the set is incomplete, the sum of c_i^2 must be less than $\langle f \mid f \rangle$. This is the Bessel inequality,

$$\langle f \mid f \rangle \geqslant \sum c_i^2 \tag{2.7.18}$$

where the equality is approached as the number of basis functions is increased.

The inequality can be illustrated using three coefficients for the expansion of the function of Figure 2.5. The normalized basis functions (kets) must be factored from the expressions so that

$$c_0 = \left(\frac{1}{2}\right)\left(\frac{1}{\sqrt{2}}\right) \qquad c_0^2 = \frac{1}{8} \tag{2.7.19}$$

$$c_1 = \left(\frac{1}{3}\right)\sqrt{\frac{3}{2}} \qquad c_1^2 = \frac{1}{6} \tag{2.7.20}$$

$$c_2 = \left(\frac{1}{4}\right)\sqrt{\frac{5}{8}} \qquad c_2^2 = \frac{5}{128} \tag{2.7.21}$$

The sum of the three squared coefficients is 0.331. This is contrasted with the function norm

$$\langle f \mid f \rangle = \int_0^1 (x)(x)\, dx = (x^3/3)\Big|_0^1 = \frac{1}{3} = 0.333 \tag{2.7.22}$$

The first three terms have generated 99.9% of the norm. This phenomenon will be observed frequently. If the function is reasonably well behaved, the first terms in the series will do an effective approximation. Each term contributes a smaller c_i^2 and more terms must be used for subsequent enhancement of the function fit.

The Legendre polynomials on the interval $-1 \leqslant x \leqslant 1$ provide a good example of the simplifications possible if the function has symmetry about an

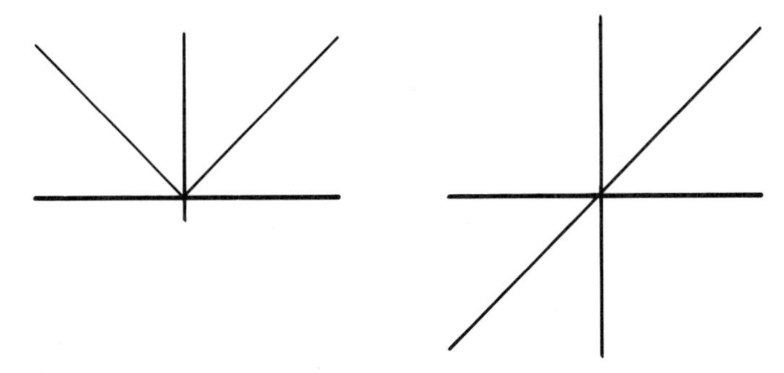

Figure 2.7. (a) The even function $f(x)=|x|$ on the interval $-1 \leqslant x \leqslant 1$. (b) The odd function $f(x)=x$ on the interval $-1 \leqslant x \leqslant 1$.

axis perpendicular to $x=0$. An even function is illustrated in Figure 2.7a. If the function along the positive x axis is rotated about the perpendicular $x=0$ axis, it will coincide with the function on the negative portion of the interval. The odd version of this function is illustrated in Figure 2.7b. The function on the positive x axis must be rotated and then inverted to coincide with the function on the negative x axis.

The basic symmetry of the functions determines the symmetry of basis functions which must be used to describe this function. The Legendre functions P_1, P_2, P_3, and P_4 are graphed in Figure 2.8. The even function of Figure 2.7a

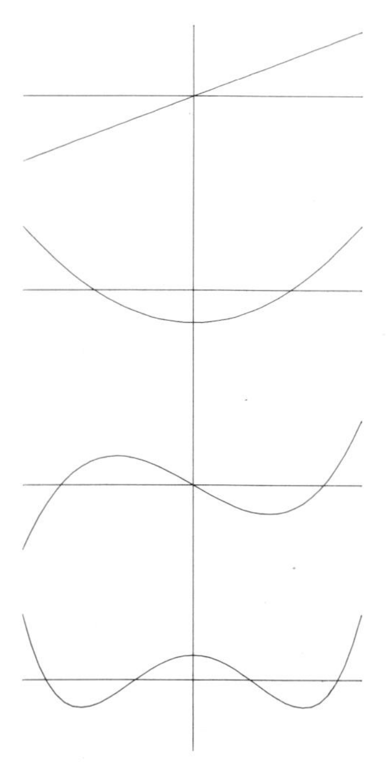

Figure 2.8. The Legendre functions P_1, P_2, P_3, and P_4 on the interval $-1 \leqslant x \leqslant 1$.

requires even basis functions, which is verified by determining the projections on the first four basis functions. These are

$$\langle P_0|f\rangle = \int_{-1}^{0} (1)(-x)\,dx + \frac{1}{\sqrt{2}}\int_0^1 (1)(x)\,dx$$

$$= (-x^2/2)\Big|_{-1}^{0} + (x^2/2)\Big|_0^1 = 1 \tag{2.7.23}$$

$$\langle P_1|f\rangle = \int_{-1}^{0} (x)(-x)\,dx + \int_0^1 (x)(x)\,dx$$

$$= (-x^3/3)\Big|_{-1}^{0} + (x^3/3)\Big|_0^1 = 0 \tag{2.7.24}$$

$$\langle P_2|f\rangle = \int_{-1}^{0} (3x^2-1)(-x)\,dx + \int_0^1 (3x^2-1)(x)\,dx$$

$$= (-3x^4/4 + x^2/2)\Big|_{-1}^{0} + (3x^4/4 - x^2/2)\Big|_0^1 = \frac{1}{2} \tag{2.7.25}$$

$$\langle P_3|f\rangle = \int_{-1}^{0} (5x^3-3x)(-x)\,dx + \int_0^1 (5x^3-3x)(x)\,dx$$

$$= -(x^5 - x^3)\Big|_{-1}^{0} + (x^5 - x^3)\Big|_0^1 = 0 \tag{2.7.26}$$

The even Legendre functions give nonzero components for the expansion of the even $f(x)$ since only these functions have the proper "shape" to describe the even function.

The symmetry of the function which must be expanded in the basis set dictates which basis functions will contribute to the sum. An even function will produce a sum of even function vectors while an odd function will produce a sum of odd function vectors. If the symmetry is included, half the integrals need not be evaluated since they will give a zero contribution. This can result in a considerable saving of time when the basis set of function vectors is more complicated than the Legendre polynomials used here.

2.8. Fourier Series

The Legendre functions were polynomials with orthogonality defined on a preselected interval. There was no interest in dealing with the function outside this range. However, it is often necessary to reproduce periodic functions using an expansion in some basis set. The even function of Figure 2.9 is periodic with cycles of 2π. Because the function changes slope abruptly at the maxima and

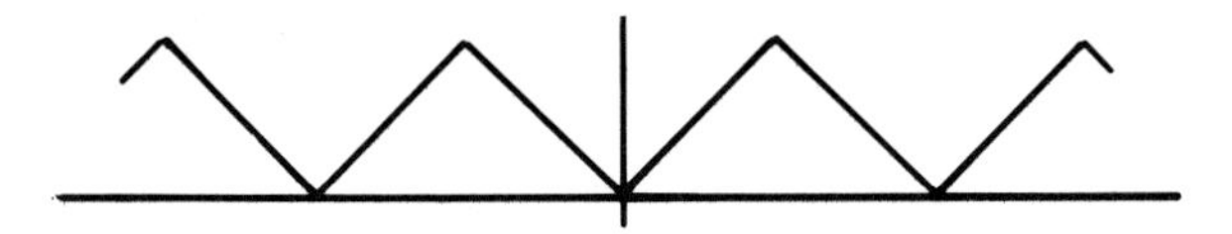

Figure 2.9. The periodic continuation of the even function $f(x) = |x|$.

minima, it will involve some combination of well-behaved functions. The sine and cosine functions

$$\sin n\theta \quad \text{and} \quad \cos n\theta \qquad n = 1, 2, 3, \ldots \tag{2.8.1}$$

are orthogonal and periodic and represent the ideal choice to describe this type of function.

Since the function of Figure 2.9 is minimal (zero) at the origin, it is even. The sine and cosine functions of Figure 2.10 are odd and even, respectively. The basis functions for the even function must be even, i.e.,

$$\cos n\theta \qquad n = 0, 1, 2, \ldots \tag{2.8.2}$$

for the range $0 \leqslant \theta \leqslant 2\pi$.

Although the x axis has units of degrees in this case, it is easily converted to units of length. If the half-wavelength from minimum to minimum of the wave is L, the function

$$\cos (n\pi x/L) \tag{2.8.3}$$

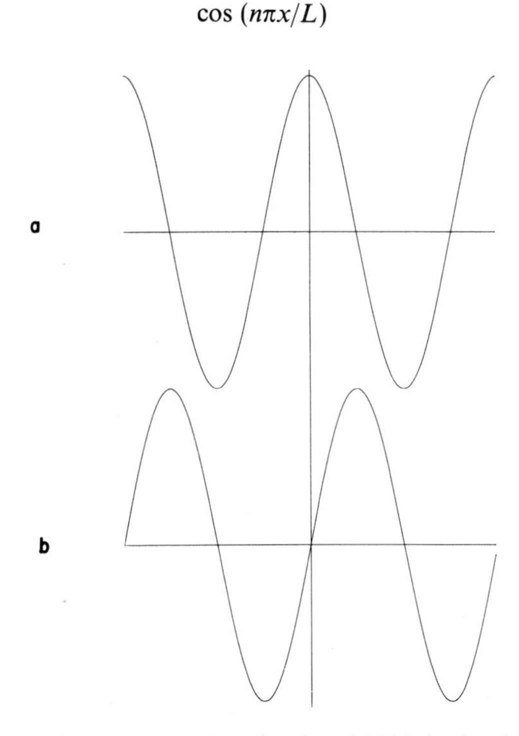

Figure 2.10. (a) The cosine function as an even function. (b) The sine function as an odd function.

will have a maximum at $x=0$ and a minimum at $x=L$. The shape of the function has not changed; it has been reexpressed in terms of a distance x on the axis rather than an angle θ.

When $n=0$,

$$\cos 0\theta = 1 \tag{2.8.4}$$

The first term in the cosine series is a constant which is normalized on the interval $0 \leqslant \theta \leqslant 2\pi$ as

$$\langle 0 | 0 \rangle = \int_0^{2\pi} (1)(1)\, d\theta = 2\pi \tag{2.8.5}$$

The expansion term will contain both the scalar product with the function and the directional function vector. Since they will be combined in the final summation, the full normalization is introduced as a factor in the expression. The first term in the expansion for the function of Figure 2.9 is

$$\begin{aligned} |0\rangle\langle 0 | f \rangle &= (1/2\pi)\left[\int_0^{\pi} (1)(\theta)\, d\theta + \int_{\pi}^{2\pi} (1)(2\pi - \theta)\, d\theta\right] \\ &= (1/2\pi)\left[\theta^2/2\Big|_0^{\pi} + (2\pi\theta - \theta^2/2)\Big|_{\pi}^{2\pi}\right] \\ &= (1/2\pi)[\pi^2/2 + (2\pi)(\pi) - (2\pi^2 - \pi^2/2)] = \pi^2/2\pi = \pi/2 \end{aligned} \tag{2.8.6}$$

which is independent of θ. All other cosine functions in the series will have equal amplitudes above and below the horizontal line defined by this term. Since the maximum amplitude of the function is π, the constant term moves the horizontal axis to the average value of the function (Figure 2.9).

The maximum amplitude of this function is π. Any maximal amplitude is produced by multiplying this function by an appropriate constant A,

$$f'(x) = Af(x) \tag{2.8.7}$$

In this case, every term in the summation will be multiplied by the factor A. The scalar factor will have no effect on the period of the function.

Although $|0\rangle$ is normalized with a factor of $1/2\pi$, the remaining cosine (and sine) functions will have a normalization factor of $1/\pi$. The normalization for $\cos n\theta$ with $n=1$ is

$$\langle 1 | 1 \rangle = \int_0^{2\pi} \cos\theta \cos\theta\, d\theta \tag{2.8.8}$$

This integral can be evaluated using the identities

$$\cos^2 n\theta = \frac{1}{2}(1 + \cos 2n\theta) \tag{2.8.9}$$

(Problem 2.16) or

$$\cos n\theta = \frac{1}{2}[\exp(in\theta) + \exp(-in\theta)] \tag{2.8.10}$$

for any n. Equation 2.8.8 becomes

$$\begin{aligned}\langle n|n\rangle &= \frac{1}{4}\int_0^{2\pi} [\exp(in\theta) + \exp(-in\theta)]^2\, d\theta \\ &= \frac{1}{4}\int_0^{2\pi} [\exp(2in\theta) + 2 + \exp(-2in\theta)]\, d\theta \\ &= \frac{1}{4}(1/2in)[\exp(2in\theta) + \exp(-2in\theta)]\Big|_0^{2\pi} + \left(\frac{1}{4}\right) 2\theta\Big|_0^{2\pi} = 0 + \pi \end{aligned} \tag{2.8.11}$$

The normalization factor for the cos $n\theta$ summation term is $1/\pi$. The normalized functions are then

$$(1/2\pi)(\cos\theta)/\sqrt{\pi}, \qquad (\cos 2\theta)/\sqrt{\pi}, \qquad (\cos 3\theta)/\sqrt{\pi}, \ldots \tag{2.8.12}$$

The sine functions have no constant term and each has the form (Problem 2.24)

$$(\sin n\theta)/\sqrt{\pi} \tag{2.8.13}$$

Since each function in this even Fourier series is a cosine, a single projection of the form

$$\langle n|f\rangle = \int_0^{2\pi} \cos n\theta\, f(\theta)\, d\theta \tag{2.8.14}$$

defines each projection in terms of the index n. For convenience, the normalization and ket vector are not included. When the projection $\langle n|f\rangle$ is found, it must be multiplied by

$$|f\rangle = (\cos n\theta)/\pi \tag{2.8.15}$$

to complete the term in the series.

Equation 2.8.14 resolves into two integrals for $y = +\theta$ and $y = 2\pi - \theta$:

$$\langle n|f\rangle = \int_0^{\pi} (\cos n\theta)(+\theta)\, d\theta + \int_{\pi}^{2\pi} (\cos n\theta)(2\pi - \theta)\, d\theta \tag{2.8.16}$$

The first integral is solved by integration by parts:

$$\begin{aligned}\int_0^{\pi} (\cos n\theta)(+\theta)\, d\theta &= (1/n)[(\sin n\theta)(\theta)]\Big|_0^{\pi} + (1/n)\int_0^{\pi} -(\sin n\theta)\, d\theta \\ &= 0 + (1/n^2)\cos n\theta\Big|_0^{\pi} \\ &= (1/n^2)[(-1)^n - 1]\end{aligned} \tag{2.8.17}$$

The second integral is

$$\int_{\pi}^{2\pi} 2\pi \cos n\theta \, d\theta - \int_{\pi}^{2\pi} \theta \cos n\theta \, d\theta$$

$$= (2\pi/n) \sin n\theta \Big|_{\pi}^{2\pi} - [(\cos n\theta)/n^2 + \theta(\sin n\theta)/n] \Big|_{\pi}^{2\pi}$$

$$= (1/n^2)[(-1)^n - 1] \tag{2.8.18}$$

This could be anticipated since the areas under each half of the function are identical. The two parts are summed to give the coefficients for the Fourier series,

$$\langle n | f \rangle = (2/n^2)[(-1)^n - 1] \tag{2.8.19}$$

The coefficients will be zero for even n. The expansion will contain only the terms with odd n. With its normalized ket vector, the nth term in the series is

$$|n\rangle\langle n | f \rangle = -(4/n^2)(1/\pi) \cos n\theta \tag{2.8.20}$$

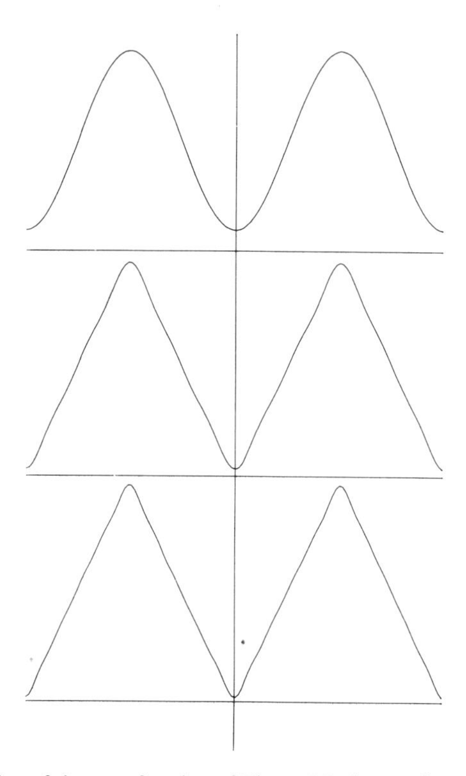

Figure 2.11. Description of the even function of Figure 2.9 via two, four, and six Fourier terms.

The Fourier series is the sum of these terms,

$$|f\rangle = \sum |n\rangle\langle n|f\rangle$$
$$= (\pi/2) - \sum_n (4/\pi)[(\cos n\theta)/n^2] \quad (2.8.21)$$

The first several terms are

$$(\pi/2) - (4/\pi)[(\cos\theta)/1^2 + (\cos 3\theta)/9 + (\cos 5\theta)/25 + \cdots] \quad (2.8.22)$$

The factor of n^2 in the denominator reduces the magnitude of terms with larger n rapidly. The series converges rapidly toward the function. The summations of two, four, and six terms, respectively, are illustrated in Figure 2.11. The worst fit is observed near the maxima and minima where the function changes abruptly.

Infinite series can be evaluated using such Fourier expansions. If $\theta = 0$ in Equation 2.8.21, $f(0) = 0$ since the function is zero at the origin. The right-hand side of the equation becomes

$$0 = (\pi/2) - \sum (4/\pi) 1/n^2 \quad (2.8.23)$$

The equation is now solved to give a value for the summation:

$$(\pi/2)(\pi/4) = (\pi^2/8) = \sum_{n=1}^{\infty} (1/n^2) \quad (2.8.24)$$

The function of Figure 2.9 becomes odd if the origin is shifted as shown in Figure 2.12. The function is $f(\theta) = \theta$ on the interval $-\pi \leqslant \theta \leqslant \pi$. At $\theta = \pi$ and $\theta = -\pi$, the function is discontinuous. At $\theta = \pi$, the function jumps to $-\pi$; at $\theta = -\pi$, it jumps to $+\pi$.

Since the function is odd, it is described by an odd set of basis functions, i.e., the sine functions

$$(\sin n\theta)/\pi \quad (2.8.25)$$

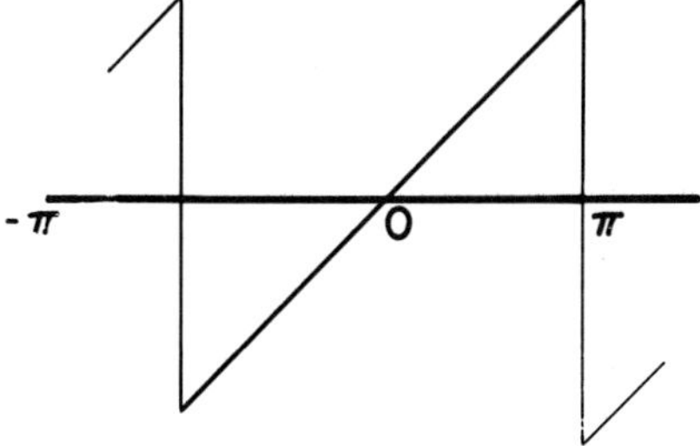

Figure 2.12. The odd function $f(x) = x$ on the interval $-\pi < x < \pi$.

The scalar projection requires a single integral on the range from $-\pi$ to $+\pi$:

$$\begin{aligned}\langle n|f\rangle &= \int_{-\pi}^{\pi} (\sin n\theta)(+\theta)\, d\theta \\ &= [(\sin n\theta)/n^2 - (\theta)(\cos n\theta)/n]\Big|_{-\pi}^{\pi} \\ &= [(\pi)(\cos n\pi)/n - (-\pi)(\cos -n\pi)/n] \\ &= (\pi/n)[(-1)^n + (-1)^n] = (2\pi/n)(-1)^n \end{aligned} \tag{2.8.26}$$

For this odd function, all indices n appear in the series and their signs alternate. The ket and normalization factor are introduced to produce the terms for the series,

$$(-1)^{n+1}\,(2/n)\sin n\theta \tag{2.8.27}$$

The first several terms of the Fourier series are

$$f(\theta) = 2\left(\sin\theta - \frac{1}{2}\sin 2\theta + \frac{1}{3}\sin 3\theta + \cdots\right) \tag{2.8.28}$$

The contributions of successive terms toward a description of the function are illustrated in Figure 2.13.

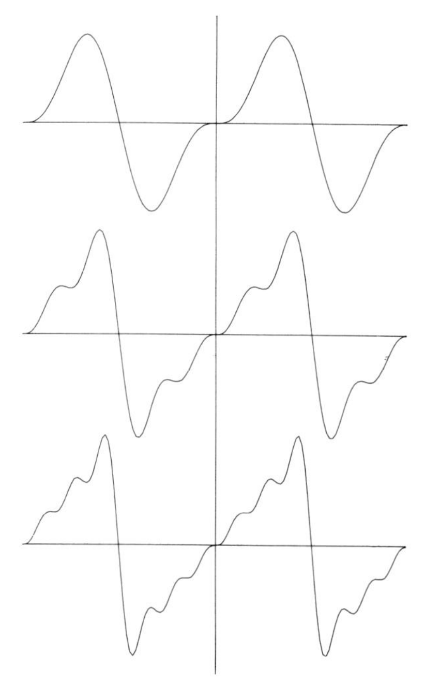

Figure 2.13. Reproduction of the function of Figure 2.12 with sums of the first two, four, and six Fourier terms.

The basis functions also describe functions with discontinuities within the periodic range. The step function of Figure 2.14 is discontinuous at $x=0$. This odd function requires the sine basis functions

$$(\sin n\theta)/\pi \tag{2.8.29}$$

Since the function has no negative region, a constant term is required to center the sinusoids. This is

$$\begin{aligned}\langle 0|f\rangle &= (1/2\pi)\int_{-\pi}^{0}(1)(0)\,d\theta + \int_{0}^{\pi}(1)(1)\,d\theta\\ &= (\pi/2\pi) = \frac{1}{2}\end{aligned} \tag{2.8.30}$$

Since the limits must cover a distance 2π along the axis, a choice of $-\pi \leqslant \theta \leqslant \pi$ or $0 \leqslant \theta < 2\pi$ will give the same results. The constant value of $+\frac{1}{2}$ moves the x axis to the center of the function.

For limits of $-\pi \leqslant \theta \leqslant \pi$, the expansion coefficients are

$$\langle n|f\rangle = \int_{-\pi}^{0}(0)\sin n\theta\,d\theta + \int_{0}^{\pi}(1)\sin n\theta\,d\theta \tag{2.8.31}$$

The second integral is nonzero and is equal to

$$(-1/n)(\cos n\theta)\Big|_{0}^{\pi} = (-1/n)[(-1)^{n}-1] \tag{2.8.32}$$

Again, even coefficients are zero. The series is formed from the product of the odd coefficients and $\sin n\theta$ as

$$f(x) = \frac{1}{2} + (4/\pi)[\sin\theta + (\sin 3\theta)/3 + (\sin 5\theta)/5 + \cdots] \tag{2.8.33}$$

The indices appear in the denominator to the first power in this case. The contributions to the function are illustrated in Figure 2.15.

This function was chosen because of its discontinuity at $x=0$. How does the Fourier series resolve this problem? The function at 0, $f(0)$, can be evaluated by

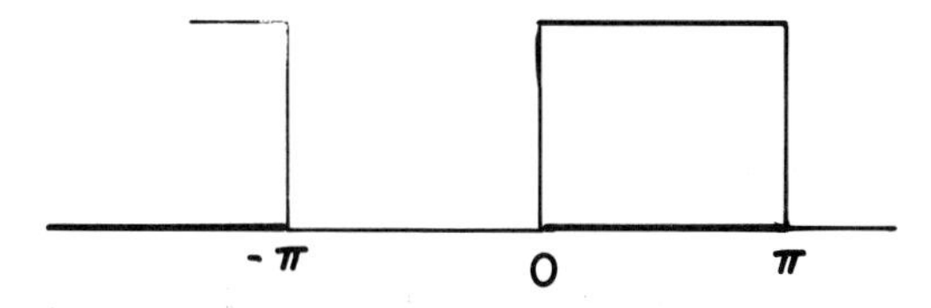

Figure 2.14. A unit step function on the interval from $-\pi < x < \pi$.

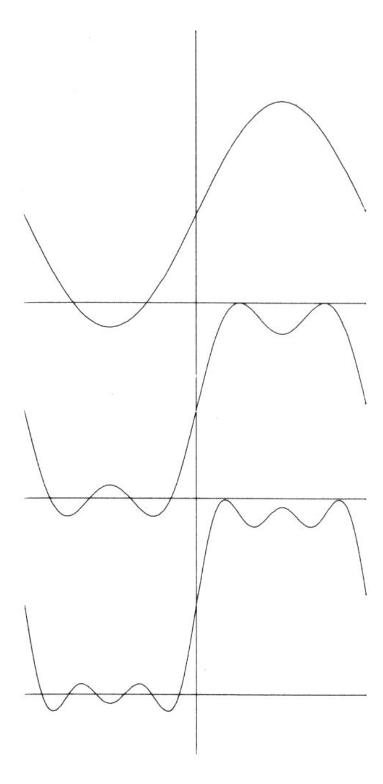

Figure 2.15. The unit step function described as a sum of two, four, and six Fourier terms.

summing all terms in the series. However, each term is sin $n\theta$ and must be zero. Only the constant term remains at $x=0$. Thus,

$$f(0)=\frac{1}{2}=\left(\frac{1}{2}\right)[f_+(0)+f_-(0)] \qquad (2.8.34)$$

where $f_+(0)$ is the value of the function at zero on approach from the positive axis and $f_-(0)$ is its value when approached from the negative axis. This example illustrates a general property of Fourier series expansions. When the function has a discontinuity and is therefore double-valued at that point, the continuous basis functions seek a value which is the average of the two values. This can be checked using the Fourier series for the odd function $f(x)=x$ (Equation 2.8.28) at π.

Problems

2.1. a. Using the Legendre polynomials, P_i, as a basis set,

$$\left[\frac{1}{\sqrt{2}}, \frac{\sqrt{3}}{2}\cos\theta, \frac{\sqrt{8}}{5}(3\cos^2\theta-1), 5\cos^3\theta-3\cos\theta\right]$$

describe a p_z orbital which has electron density only in the upper hemisphere, i.e., $P(\theta) = \cos\theta$ for $0 \leqslant \theta \leqslant 90°$; $P(\theta) = 0$ for $90° \leqslant \theta < 180°$.

b. Since

$$\mathbf{H}P_l = [l(l+1)\,\hbar^2/2I]\,P_l = EP_l$$

determine the energy of your new orbital. Is it higher or lower than that of a normal p orbital?

2.2. Calculate the Legendre polynomial for $l = 4$ (a g orbital) using the generating function. How many angular nodes are present?

2.3. Generate the third Hermite polynomial $H_2(x)$ using the generating function

$$H_n(x) = (-1)^n \exp(x^2)\, d^n/dx^n[\exp(-x^2)]$$

2.4. $\psi_n(x) = H_n \exp(-x^2/2)$ is the wavefunction for a harmonic oscillator. Show that $\psi_3(x)$ is a solution of the Schrödinger equation for a harmonic oscillator.

2.5. a. Find the Hermite polynomial, $H_1(x)$, using a generating function.
b. Use this polynomial to find $\psi(x)$ for a harmonic oscillator and find the energy eigenvalue associated with this $\psi(x)$.
c. Evaluate the delta function $\delta(x-0)$ in terms of the first three Legendre polynomials.

2.6. a. Determine the first three orthogonal vectors which can be constructed from the functions 1, x, x^2, and x^3 on the interval $0 \leqslant x \leqslant 1$.
b. Describe the function $\exp(x)$ on the interval $0 \leqslant x \leqslant 1$ by determining the projections on each of the orthogonal functions.

2.7. Determine the coefficients for P_0 and P_2 required to describe the function $f(x) = x^2$ algebraically by equating terms of the same power on each side of the equation

$$x^2 = c_0 P_0 + c_2 P_2$$

2.8. Show that the functions $\sin x$, $\sin 2x$, and $\sin 3x$ are linearly independent.

2.9. Prove

a. $$\int_{-\infty}^{\infty} H_1(x)\, H_0(x) \exp(-x^2)\, dx = 0$$

b. $$\int_{-\infty}^{\infty} H_0(x)\, H_0(x) \exp(-x^2)\, dx = 1$$

2.10. Normalize the function

$$3\cos^2\theta - 1$$

using spherical coordinates.

2.11. Show that the first derivative of the function

$$P_3 = 5x^3 - 3x$$

is a linear combination of Legendre functions.

2.12. Prove that $m < n$ for a nonzero associated Legendre polynomial P_{nm}.

2.13. a. Determine the third Laguerre function, L_3, using the generating function for these polynomials.
b. Determine the associated Laguerre polynomial, L_{32}.

2.14. Show that the Laguerre function

$$L_2 = x - 2$$

is a solution to the Laguerre differential equation.

2.15. Prove that the Hermite polynomials, H_n, must be normalized with a weighting factor $w(x) = \exp(-x^2)$.

2.16. Normalize the functions $\cos n\theta$ with an index n ($n = 0, 1, 2, ...$) using the conversion

$$\cos^2 x = \frac{1}{2}(1 + \cos 2x)$$

2.17. Prove the identity

$$\cos^2 x = \frac{1}{2}(1 + \cos 2x)$$

2.18. Determine the energy for an electron described by the Fourier expansion in a one-dimensional box of length L

$$f(x) = \frac{1}{2} + (4/\pi)[\sin(\pi x/L) + \sin(3\pi x/L)/3 + \cdots]$$

using the differential operator

$$(-\hbar^2/2m)\, d^2f(x)/dx^2 = Ef(x)$$

and compare this energy with that obtained from an operation on the $n = 3$ basis function, $\sin(3\pi x/L)$.

2.19. Show that the Gram determinant, G, for the three functions 1, $\sin^2\theta$, and $\cos^2\theta$ is zero, i.e., the functions are linearly dependent.

2.20. Normalize the second Legendre polynomial

$$P_2 = 3x^2 - 1$$

2.21. Show the associated Legendre polynomial, P_{21}, is a solution of the Legendre differential equation (Equation 2.5.18).

2.22. Show that the modified Hermite differential equation (Equation 2.6.37) can be cast in a Sturm–Liouville format and determine $a(x)$, $b(x)$, and $c(x)$.

2.23. Show that the normalization for $\sin nx$ on the interval $-\pi \leqslant x < \pi$ is $1/\sqrt{\pi}$.

3

Matrices

3.1. Vector Rotations

A fixed vector in space can always be resolved into components on some coordinate system in this space. The choice of coordinate system is completely arbitrary although there are some distinct advantages in the choice of a set of mutually orthogonal coordinates. In some situations, the orientation of the vector in space must be changed. The simplest change involves a rotation of the vector to a new orientation so that its direction, but not its length, is changed. The process is illustrated in Figure 3.1 for a rotation of $+135°$. The original vector with components of $(+1, +1)$ on the x and y axes, respectively, now lies on the negative x axis and has a length of $\sqrt{2}$, which is equivalent to the length of the original vector:

$$\langle r | r \rangle = (1 \quad 1) \begin{pmatrix} 1 \\ 1 \end{pmatrix} = |2|^2$$
$$|r| = \sqrt{2} \tag{3.1.1}$$

The rotation changes only the orientation, not the absolute magnitude, of the vector.

The rotation of the vector in space produces a new set of coordinate components. Consider the vector $|r\rangle$ in two-dimensional space with components $(2, 1)$ on the x and y axes, respectively, as shown in Figure 3.2. The vector is now rotated $90°$ counterclockwise $(+90°)$ to give a vector $|s\rangle$. The components of $|s\rangle$ on the x and y axes are now -1 and $+2$, respectively. The components undergo both sign and magnitude changes as a result of the rotation. The rotation of the vector can now be described by an operator $\mathbf{R}$ which acts on the vector $|r\rangle$ to change it to the vector $|s\rangle$:

$$|s\rangle = \mathbf{R} |r\rangle \tag{3.1.2}$$

Since this operation changes the components of $|r\rangle$ to the components of $|s\rangle$ in

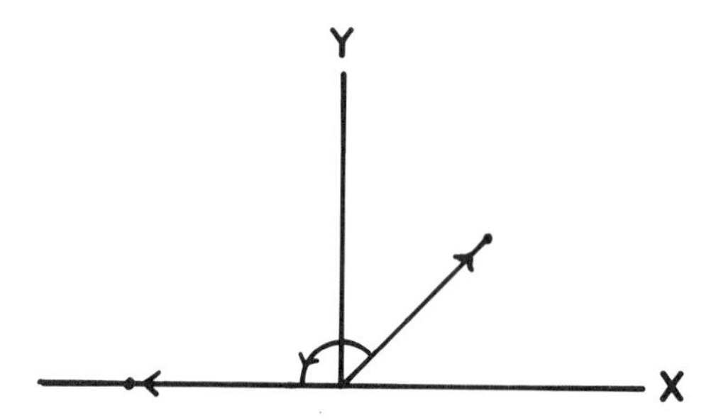

Figure 3.1. The rotation of a vector with components $(\sqrt{2}, \sqrt{2})$ counterclockwise through 135°.

the two-dimensional space, the operator must act on the individual components of the vector to produce the change:

$$\begin{pmatrix} 2 \\ 1 \end{pmatrix} \rightarrow \begin{pmatrix} -1 \\ 2 \end{pmatrix} \tag{3.1.3}$$

This component operator will be a matrix.

The operator **R** must transform each component of the original $|r\rangle$ to produce $|s\rangle$. The component of $|r\rangle$ on the x axis, r_x, must become s_x, the component of $|s\rangle$ on the x axis. Since the vector $|r\rangle$ can be resolved into vectors on the x and y axes, a rotation of the vector is equivalent to a rotation of each of these component vectors as illustrated in Figure 3.3. The rotation of 90° places r_x on the $+y$ axis where it has no projection on the x axis and cannot contribute to s_x. The r_y component ends on the negative x axis and thus contributes -1 to the s_x component. The total contribution to s_x is

$$s_x = (0)\, r_x + (-1) r_y \tag{3.1.4}$$

where r_x and r_y are numbers, the original vector components, and the factors 0 and -1 give their projections on the x axis after the rotation.

The process can be repeated for the s_y projection. After rotation, r_x is

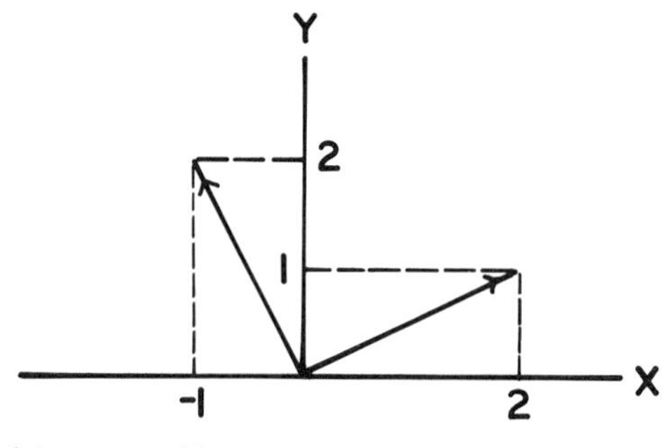

Figure 3.2. Rotation of the vector with components (2, 1) through 90° to produce a vector $|s\rangle$ with components $(-1, +2)$.

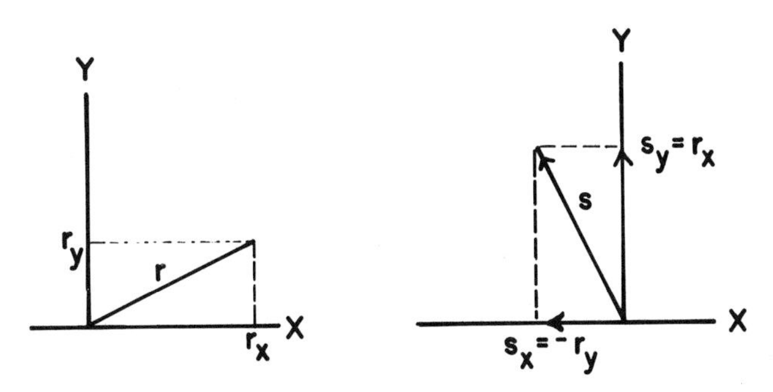

Figure 3.3. Rotation of the vector components of $|r\rangle$: (a) r_x rotation; (b) r_y rotation.

located on the $+y$ axis with its full length r_x. The r_y component now lies on the $-x$ axis and makes no contribution to s_y so this component is

$$s_y = (+1)r_x + (0)r_y \tag{3.1.5}$$

Contributions to s_x from both r_x and r_y are required since the rotation moves both these projections off their original axes. The expression for s_x has the form of a scalar product: r_x is multiplied by a factor (0) which gives its contribution to s_x while r_y is multiplied by the factor -1. The two are then summed to give a single scalar quantity, s_x. This scalar product is then

$$\begin{pmatrix} 0 & -1 \end{pmatrix}\begin{pmatrix} r_x \\ r_y \end{pmatrix} \tag{3.1.6}$$

The second projection s_y also contains contributions from r_x and r_y and must be described by a second scalar product:

$$\begin{pmatrix} +1 & 0 \end{pmatrix}\begin{pmatrix} r_x \\ r_y \end{pmatrix} \tag{3.1.7}$$

Since both the s_x and s_y components will be needed in any rotation, it is convenient to arrange the two row vectors as successive rows of an array. The position of the row determines the position of the component in the $|s\rangle$ vector. This array is a matrix. For the rotation of 90°, the transformation is

$$\begin{pmatrix} s_x \\ s_y \end{pmatrix} = \begin{pmatrix} 0 & -1 \\ 1 & 0 \end{pmatrix}\begin{pmatrix} r_x \\ r_y \end{pmatrix} \tag{3.1.8}$$

The rotation is illustrated by replacing the $|r\rangle$ components with their numerical values. The operation is

$$\begin{pmatrix} 0 & -1 \\ 1 & 0 \end{pmatrix}\begin{pmatrix} 2 \\ 1 \end{pmatrix} = \begin{pmatrix} (0)(2) + (-1)(1) \\ (1)(2) + (0)(1) \end{pmatrix} = \begin{pmatrix} -1 \\ 2 \end{pmatrix} \tag{3.1.9}$$

This procedure is applicable to matrices and vectors of any dimension. The number of elements in each row of the matrix must equal the number of the components in the vector to permit a scalar product. In most cases considered in this text, the matrix will be square, i.e., a matrix of N rows and N columns will operate to the right on a column vector of N components to give a new column vector of N components. The N components of the final vector are the N scalar products generated in order by the rows of the matrix.

The procedure outlined for a rotation of 90° is applicable to rotations through any angle θ. Such a rotation is illustrated in Figure 3.4 for the $|r\rangle$ vector components r_x and r_y. The value of s_x is

$$\begin{aligned} s_x &= \cos(\theta) r_x + [\cos(90° + \theta)]\, r_y \\ &= \cos(\theta) r_x - \sin(\theta)\, r_y \end{aligned} \tag{3.1.10}$$

while s_y is

$$\begin{aligned} s_y &= [\cos(90° - \theta)]\, r_x + \cos(\theta) r_y \\ &= \sin(\theta) r_x + \cos(\theta) r_y \end{aligned} \tag{3.1.11}$$

The matrix operation for the rotation of θ is

$$|s\rangle = \mathbf{R}|r\rangle$$

$$\begin{pmatrix} s_x \\ s_y \end{pmatrix} = \begin{pmatrix} \cos\theta & -\sin\theta \\ \sin\theta & \cos\theta \end{pmatrix} \begin{pmatrix} r_x \\ r_y \end{pmatrix} \tag{3.1.12}$$

For a rotation of the vector

$$|r\rangle = \begin{pmatrix} \dfrac{\sqrt{3}}{2} \\ \dfrac{1}{2} \end{pmatrix} \tag{3.1.13}$$

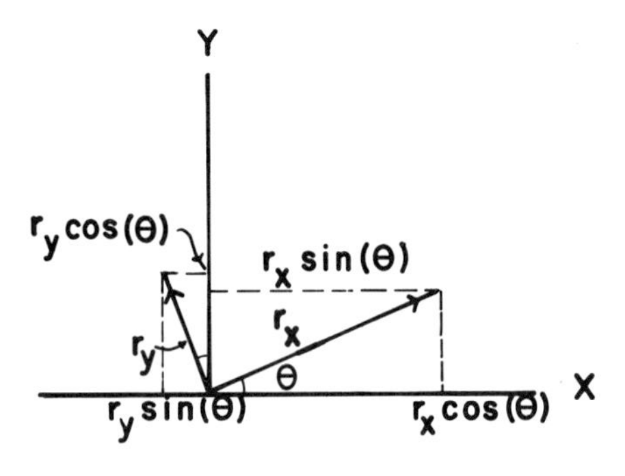

Figure 3.4. Rotation of vector components of $|r\rangle$ by an angle θ.

by 30°, the matrix operation is

$$\begin{pmatrix} \cos 30^\circ & -\sin 30^\circ \\ \sin 30^\circ & \cos 30^\circ \end{pmatrix} \begin{pmatrix} \dfrac{\sqrt{3}}{2} \\ \dfrac{1}{2} \end{pmatrix} = \begin{pmatrix} \dfrac{\sqrt{3}}{2} & \dfrac{-1}{2} \\ \dfrac{1}{2} & \dfrac{\sqrt{3}}{2} \end{pmatrix} \begin{pmatrix} \dfrac{\sqrt{3}}{2} \\ \dfrac{1}{2} \end{pmatrix} = \begin{pmatrix} \dfrac{1}{2} \\ \dfrac{\sqrt{3}}{2} \end{pmatrix} \tag{3.1.14}$$

The rotation with appropriate components is shown in Figure 3.5.

The choice of column vectors was entirely arbitrary. The matrix operation should work for row vectors as well. However, since the row vector must form scalar products with the matrix for rotations, the row vector must be placed to the left of the matrix. The scalar products are then those of the row vector with each column of the matrix. If the row vector

$$\langle r| = \left(\frac{\sqrt{3}}{2}, \ \frac{1}{2}\right) \tag{3.1.15}$$

is to be rotated $+30°$, it must again give the final components

$$\langle s| = \left(\frac{1}{2} \ \ \frac{\sqrt{3}}{2}\right) \tag{3.1.16}$$

However, the matrix developed for rotation of the column vectors does not give this result:

$$\left(\frac{\sqrt{3}}{2} \ \ \frac{1}{2}\right) \begin{pmatrix} \dfrac{\sqrt{3}}{2} & \dfrac{-1}{2} \\ \dfrac{1}{2} & \dfrac{\sqrt{3}}{2} \end{pmatrix} = (1 \ \ 0) \tag{3.1.17}$$

The matrix rotates the row vector clockwise through 30°, i.e., exactly opposite to the direction it rotated the equivalent column vector. The proper matrix for a rotation of a row vector by $+30°$ must place the elements of the first row of the

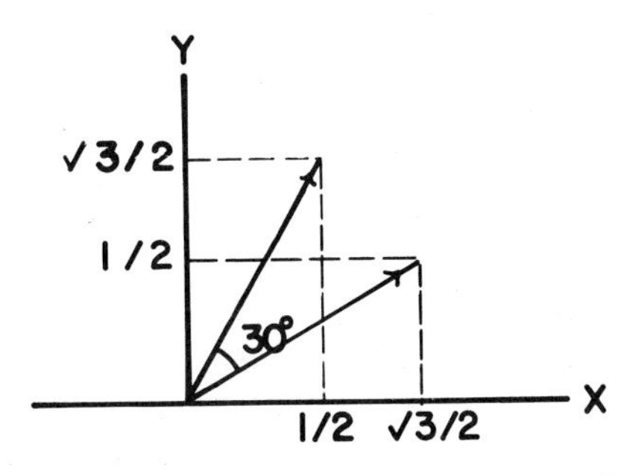

Figure 3.5. Rotation of the components of the vector $(\sqrt{3}/2, 1/2)$ by 30°.

R matrix in the first column. The second row of the **R** matrix becomes the second column of the **R**′ matrix for a $+30°$ rotation. The matrix

$$\mathbf{R}' = \begin{pmatrix} \frac{\sqrt{3}}{2} & \frac{1}{2} \\ \frac{-1}{2} & \frac{\sqrt{3}}{2} \end{pmatrix} \tag{3.1.18}$$

gives a rotation

$$\begin{pmatrix} \frac{\sqrt{3}}{2} & \frac{1}{2} \end{pmatrix} \begin{pmatrix} \frac{\sqrt{3}}{2} & \frac{1}{2} \\ \frac{-1}{2} & \frac{\sqrt{3}}{2} \end{pmatrix} = \begin{pmatrix} \frac{1}{2} & \frac{\sqrt{3}}{2} \end{pmatrix} \tag{3.1.19}$$

The matrix **R**′ is the transpose of matrix **R** since the rows of **R** are transposed into the columns of **R**′.

The **R** and **R**′ matrices rotate the vector but do not change its magnitude. The scalar product of the vector must remain constant. Since $\langle s|$ and $|s\rangle$ are determined from $\langle r|$ and $|r\rangle$ as

$$\begin{aligned} \langle s| &= \langle r|\mathbf{R}' \\ |s\rangle &= \mathbf{R}\,|r\rangle \end{aligned} \tag{3.1.20}$$

this equality is

$$\langle r\,|\,r\rangle = \langle s\,|\,s\rangle = \langle r|\,\mathbf{R}'\mathbf{R}\,|r\rangle \tag{3.1.21}$$

This suggests that the product of **R**′ and **R** must act as an identity operation. The product **R′R** for a rotation of 30° is

$$\mathbf{R}'\mathbf{R} = \begin{pmatrix} \frac{\sqrt{3}}{2} & \frac{1}{2} \\ \frac{-1}{2} & \frac{\sqrt{3}}{2} \end{pmatrix} \begin{pmatrix} \frac{\sqrt{3}}{2} & \frac{-1}{2} \\ \frac{1}{2} & \frac{\sqrt{3}}{2} \end{pmatrix} \tag{3.1.22}$$

Four distinct scalar products are possible. The top row of the left-hand matrix forms scalar products with the first and second columns of the right-hand matrix. These two scalar products become the first and second elements of the first row of the final product matrix. Similarly, the scalar products of the lower row of **R**′ with the two columns of **R** give the first and second elements of the second row of the product matrix. A matrix multiplication of two 2×2 matrices creates a new 2×2 product matrix. The matrix product of Equation 3.1.22 is

$$\begin{pmatrix} \left(\frac{\sqrt{3}}{2}\right)\left(\frac{\sqrt{3}}{2}\right)+\left(\frac{1}{2}\right)\left(\frac{1}{2}\right) & \left(\frac{\sqrt{3}}{2}\right)\left(\frac{-1}{2}\right)+\left(\frac{1}{2}\right)\left(\frac{\sqrt{3}}{2}\right) \\ \left(\frac{-1}{2}\right)\left(\frac{-\sqrt{3}}{2}\right)+\left(\frac{\sqrt{3}}{2}\right)\left(\frac{1}{2}\right) & \left(\frac{-1}{2}\right)\left(\frac{-1}{2}\right)+\left(\frac{\sqrt{3}}{2}\right)\left(\frac{\sqrt{3}}{2}\right) \end{pmatrix} = \begin{pmatrix} 1 & 0 \\ 0 & 1 \end{pmatrix} \quad (3.1.23)$$

A matrix with 1's on its "principal" diagonal like this is an identity matrix, **I**. It will not change a vector or matrix on which it acts and is equivalent to multiplication by a scalar 1. There are situations, however, when it is maintained in the matrix form.

The product $\mathbf{R}'\mathbf{R}$ gives the identity matrix for any rotation angle θ:

$$\begin{aligned} &\begin{pmatrix} \cos\theta & \sin\theta \\ -\sin\theta & \cos\theta \end{pmatrix}\begin{pmatrix} \cos\theta & -\sin\theta \\ \sin\theta & \cos\theta \end{pmatrix} \\ &= \begin{pmatrix} \cos^2\theta+\sin^2\theta & -\cos\theta\sin\theta+\cos\theta\sin\theta \\ -\sin\theta\cos\theta+\cos\theta\sin\theta & \sin^2\theta+\cos^2\theta \end{pmatrix} \\ &= \begin{pmatrix} 1 & 0 \\ 0 & 1 \end{pmatrix} \end{aligned} \quad (3.1.24)$$

The scalar product $\langle s|s\rangle$ is then equal to $\langle r|r\rangle$, as predicted:

$$\langle s|s\rangle = \langle r|\mathbf{R}'\mathbf{R}|r\rangle = \langle r|\mathbf{I}|r\rangle = \langle r|r\rangle \quad (3.1.25)$$

Rotation matrices which preserve the vector length are examples of unitary or orthogonal matrices (orthogonal matrices have only real elements). Their determinants are unitary. For example, the determinant of the elements of the **R** matrix is

$$\begin{vmatrix} \frac{\sqrt{3}}{2} & \frac{-1}{2} \\ \frac{1}{2} & \frac{\sqrt{3}}{2} \end{vmatrix} = \frac{3}{4} + \frac{1}{4} = 1 \quad (3.1.26)$$

However, a matrix with a determinant of unity is not necessarily a unitary matrix.

Any array of numbers constitutes a matrix. In general, such matrices can change both the direction and length of a vector. For example, the matrix operation

$$\begin{pmatrix} 2 & 1 \\ 1 & 1 \end{pmatrix}\begin{pmatrix} 1 \\ 1 \end{pmatrix} = \begin{pmatrix} 3 \\ 2 \end{pmatrix} \quad (3.1.27)$$

rotates the vector as illustrated in Figure 3.6. At the same time, the original norm of $\sqrt{2}$ increases to $\sqrt{13}$ as a result of this matrix operation. This non-unitary matrix has a determinant

$$\begin{vmatrix} 2 & 1 \\ 1 & 1 \end{vmatrix} = 2 - 1 = 1 \tag{3.1.28}$$

even though it changes the norm of the vector.

Matrix operations appear in a variety of physical systems. For example, a rotating body has angular velocities ω_1, ω_2, and ω_3 about axes 1, 2, and 3, respectively. These angular velocities produce angular momenta components L_1, L_2, and L_3. For a simple rotation about one axis, the angular momentum and angular velocity are related through the moment of inertia for this axis,

$$L = \mathscr{I}\omega \tag{3.1.29}$$

In a three-dimensional system, each of the three angular velocities may contribute to each component of the angular momentum. This relationship is expressed as a matrix vector operation:

$$|L\rangle = \mathscr{I}\,|\omega\rangle$$

$$\begin{pmatrix} L_1 \\ L_2 \\ L_3 \end{pmatrix} = \begin{pmatrix} \mathscr{I}_{11} & \mathscr{I}_{12} & \mathscr{I}_{13} \\ \mathscr{I}_{21} & \mathscr{I}_{22} & \mathscr{I}_{23} \\ \mathscr{I}_{31} & \mathscr{I}_{32} & \mathscr{I}_{33} \end{pmatrix} \begin{pmatrix} \omega_1 \\ \omega_2 \\ \omega_3 \end{pmatrix} \tag{3.1.30}$$

The diagonal elements $\mathscr{I}_{11}$, $\mathscr{I}_{22}$, and $\mathscr{I}_{33}$ are the moments of inertia. They determine the total angular momentum for the ith component produced by the ith component of angular velocity. The off-diagonal elements are the products of inertia and describe the effects of angular velocity in one direction on the angular momentum in a second direction. For example, $\mathscr{I}_{12}$ describes the contribution of angular velocity from the 2 coordinate direction on angular momentum in the 1 coordinate direction. With a careful choice of coordinates, the products of inertia can be eliminated. The choice of such proper coordinate systems will be a major concern throughout this book.

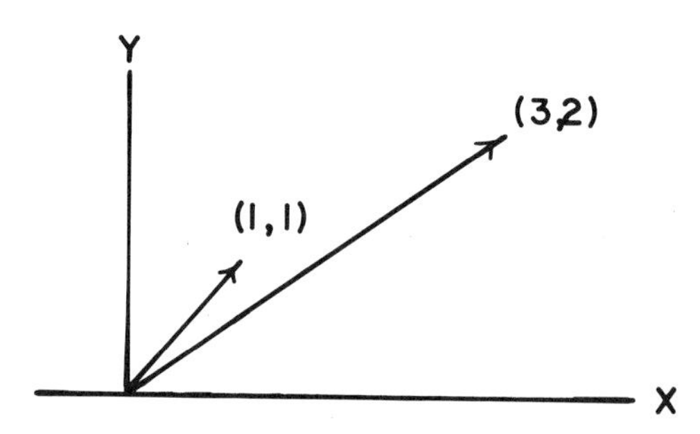

Figure 3.6. The rotation and elongation of a vector with a nonunitary matrix.

3.2. Special Matrices

Although an $N \times N$ matrix may contain N^2 unrelated elements, many matrices contain certain elements or arrangements of elements which impart special properties to the matrix. For example, in Section 3.1, the transpose matrix $\mathbf{R}'$ was generated from an $\mathbf{R}$ matrix via a simple transposition of rows and columns. Both the $\mathbf{R}$ and $\mathbf{R}'$ matrices were unitary and did not change the absolute length of the vector in the space.

An $N \times N$ identity matrix will contain only 1's on the diagonal. Since the full $N \times N$ array can be large although it contains little information, it is often convenient to introduce the notation

$$[0, 1, 0]_N \tag{3.2.1}$$

for $\mathbf{I}$. The center number in the bracket gives the 1 which appears for each diagonal element. The 0's indicate that zeros appear to the left and right of the diagonal elements and that all elements on either side of these elements are also zero.

A diagonal matrix has elements a_{ii} only along the diagonal while all other elements are zero, i.e.,

$$[0, a_{ii}, 0]_N \tag{3.2.2}$$

For example, the 3×3 diagonal matrix

$$\begin{pmatrix} a & 0 & 0 \\ 0 & b & 0 \\ 0 & 0 & c \end{pmatrix} \tag{3.2.3}$$

has diagonal elements a, b, and c. If $a = b = c$, a could be factored from each element of the matrix, leaving the identity matrix:

$$\begin{pmatrix} a & 0 & 0 \\ 0 & a & 0 \\ 0 & 0 & a \end{pmatrix} = a \begin{pmatrix} 1 & 0 & 0 \\ 0 & 1 & 0 \\ 0 & 0 & 1 \end{pmatrix} \tag{3.2.4}$$

In general, multiplication of a matrix by a scalar is equivalent to multiplying every element in the matrix by this scalar. The product $2\mathbf{A}$ for

$$\mathbf{A} = \begin{pmatrix} 2 & 1 \\ 1 & 3 \end{pmatrix} \tag{3.2.5}$$

is

$$2\mathbf{A} = \begin{pmatrix} 4 & 2 \\ 2 & 6 \end{pmatrix} \tag{3.2.6}$$

The product of two diagonal matrices is a matrix which is also diagonal. For example, the matrix product

$$\begin{pmatrix} 1 & 0 & 0 \\ 0 & 2 & 0 \\ 0 & 0 & 3 \end{pmatrix} \begin{pmatrix} 3 & 0 & 0 \\ 0 & 2 & 0 \\ 0 & 0 & 1 \end{pmatrix} = \begin{pmatrix} 3 & 0 & 0 \\ 0 & 4 & 0 \\ 0 & 0 & 3 \end{pmatrix} \tag{3.2.7}$$

has diagonal elements which are the products of the elements at the same location in the two multiplied matrices. Multiplication of two $N \times N$ diagonal matrices is equivalent to N independent products.

The diagonal matrix product provides an example of a commuting matrix pair. The matrix product is the same when the matrices are multiplied in either order. Reversing the order of the two matrices of Equation 3.2.7 gives

$$\begin{pmatrix} 3 & 0 & 0 \\ 0 & 2 & 0 \\ 0 & 0 & 1 \end{pmatrix} \begin{pmatrix} 1 & 0 & 0 \\ 0 & 2 & 0 \\ 0 & 0 & 3 \end{pmatrix} = \begin{pmatrix} 3 & 0 & 0 \\ 0 & 4 & 0 \\ 0 & 0 & 3 \end{pmatrix} \tag{3.2.8}$$

This commutative property is not satisfied for matrices in general. For example, the product of a 2×2 diagonal matrix, **D**, and a second matrix, **A**, in the order **DA** is

$$\begin{pmatrix} 2 & 0 \\ 0 & 1 \end{pmatrix} \begin{pmatrix} 1 & 1 \\ 1 & 1 \end{pmatrix} = \begin{pmatrix} 2 & 2 \\ 1 & 1 \end{pmatrix} \tag{3.2.9}$$

while the product **AD** is

$$\begin{pmatrix} 1 & 1 \\ 1 & 1 \end{pmatrix} \begin{pmatrix} 2 & 0 \\ 0 & 1 \end{pmatrix} = \begin{pmatrix} 2 & 1 \\ 2 & 1 \end{pmatrix} \tag{3.2.10}$$

The nonzero difference between the two products,

$$\mathbf{DA} - \mathbf{AD} \tag{3.2.11}$$

is

$$\begin{pmatrix} 2 & 2 \\ 1 & 1 \end{pmatrix} - \begin{pmatrix} 2 & 1 \\ 2 & 1 \end{pmatrix} = \begin{pmatrix} 0 & 1 \\ -1 & 0 \end{pmatrix} \tag{3.2.12}$$

where the difference of two matrices involves the differences between corresponding elements in each matrix. The difference operation is called a commutator and is often written symbolically with a bracket notation,

$$[\mathbf{D}, \mathbf{A}] = \mathbf{DA} - \mathbf{AD} \tag{3.2.13}$$

where the first matrix in the bracket always appears on the left in the first term. Commuting matrices,

$$[\mathbf{B}, \mathbf{A}] = \mathbf{BA} - \mathbf{AB} = 0 \tag{3.2.14}$$

can operate in either order to give the same result.

Many systems which have coupled adjacent units are described by tridiagonal matrices with nonzero elements only on the principal diagonal and the two adjacent "diagonals." Each row has the elements

$$[a_{i,i-1}, a_{ii}, a_{i,i+1}] \tag{3.2.15}$$

The 3×3 tridiagonal matrix is

$$\begin{pmatrix} a_{11} & a_{12} & 0 \\ a_{21} & a_{22} & a_{23} \\ 0 & a_{32} & a_{33} \end{pmatrix} \tag{3.2.16}$$

A matrix having only zero elements to the right or left of the principal diagonal is a triangular matrix, e.g.,

$$\begin{pmatrix} 2 & 0 & 0 \\ 1 & 3 & 0 \\ 1 & 1 & 2 \end{pmatrix} \tag{3.2.17}$$

The triangular matrices are useful when they appear in algebraic equations such as

$$\begin{pmatrix} 2 & 0 & 0 \\ 1 & 3 & 0 \\ 1 & 1 & 2 \end{pmatrix} \begin{pmatrix} u \\ v \\ w \end{pmatrix} = \begin{pmatrix} 2 \\ -2 \\ 0 \end{pmatrix} \tag{3.2.18}$$

where the vector components u, v, and w are unknown. Matrix multiplication of the first row gives

$$2u = 2 \tag{3.2.19}$$

with solution

$$u = 1 \tag{3.2.20}$$

The vector component generated by the matrix multiplication is equated to the corresponding vector component on the right. The second-row matrix operation gives the equation

$$1u + 3v = -2 = (1)(1) + 3v = -2$$

$$v = -1 \tag{3.2.21}$$

The third row of the matrix gives the final equation

$$u + v + 2w = 0 = 1 - 1 + 2w = 0$$

$$w = 0 \tag{3.2.22}$$

The vector soluttion of the matrix equation is

$$\begin{pmatrix} u \\ v \\ w \end{pmatrix} = \begin{pmatrix} 1 \\ -1 \\ 0 \end{pmatrix} \tag{3.2.23}$$

A number of special matrices are defined by relationships between the elements of the matrix. The transpose matrix is one example. Since the rows and columns in the original matrix are transposed in the transpose matrix, the elements in the transpose matrix are related to the elements in the original matrix. If the original elements are a_{ij} for the element located in the ith row and jth column, then this will be equal to the element located in the jth row and ith column of the transpose matrix, i.e.,

$$a'_{ji} = a_{ij} \tag{3.2.24}$$

For the 2×2 matrix $\mathbf{A}$ and its transpose $\mathbf{A}'$,

$$\mathbf{A} = \begin{pmatrix} 1 & 1 \\ 3 & 2 \end{pmatrix} \qquad \mathbf{A}' = \begin{pmatrix} 1 & 3 \\ 1 & 2 \end{pmatrix} \tag{3.2.25}$$

$$a'_{12} = 3 = a_{21} \tag{3.2.26}$$

Only the off-diagonal elements of $\mathbf{A}'$ are transposed. The rows in the $\mathbf{A}$ matrix become columns of the $\mathbf{A}'$ matrix.

The elements of a symmetric matrix satisfy the condition

$$a_{ij} = a_{ji} \tag{3.2.27}$$

The elements of the ith row and ith column are identical. A symmetric matrix will operate on row or column vectors with the same components to give final vectors with the same components. For example,

$$\begin{pmatrix} 2 & 1 & 0 \\ 1 & 2 & 1 \\ 0 & 1 & 2 \end{pmatrix} \begin{pmatrix} 1 \\ 2 \\ 3 \end{pmatrix} = \begin{pmatrix} 4 \\ 8 \\ 8 \end{pmatrix} \tag{3.2.28}$$

and

$$(1 \quad 2 \quad 3) \begin{pmatrix} 2 & 1 & 0 \\ 1 & 2 & 1 \\ 0 & 1 & 2 \end{pmatrix} = (4 \quad 8 \quad 8) \tag{3.2.29}$$

Once a set of column vector components is known, the row vector components are identical.

An antisymmetric matrix has elements which obey the relation

$$a_{ij} = -a_{ji} \tag{3.2.30}$$

e.g.,

$$\begin{pmatrix} 0 & -1 & 1 \\ 1 & 0 & -1 \\ -1 & 1 & 0 \end{pmatrix} \tag{3.2.31}$$

For a skew-symmetric matrix, $a_{ii} = -a_{ii}$. The diagonal elements of skew-symmetric matrices must be zero to satisfy this relation. The elements of a Hermitian matrix satisfy

$$a_{ij} = a_{ji}^* \tag{3.2.32}$$

where the asterisk indicates a complex conjugate. The matrix

$$\begin{pmatrix} 1 & -i \\ i & 1 \end{pmatrix} \tag{3.2.33}$$

is Hermitian since the 12 element is the complex conjugate of the 21 element, i.e., the i in a_{21} is replaced by $-i$ in a_{12}. While symmetric matrices operated identically on row or column vectors with the same components, a Hermitian matrix must operate on row and column vectors whose components are complex conjugates. The Hermitian matrix of Equation 3.2.33 operates on the column vector

$$\begin{pmatrix} 1 \\ i \end{pmatrix} \tag{3.2.34}$$

to give

$$\begin{pmatrix} 2 \\ 2i \end{pmatrix} \tag{3.2.35}$$

The row vector with complex conjugate components gives

$$(1 \quad -i)\begin{pmatrix} 1 & -i \\ i & 1 \end{pmatrix} = (2 \quad -2i) \tag{3.2.36}$$

The final vectors have corresponding components which are complex conjugates. The row and column vectors give a finite real norm:

$$(1 \quad -i)\begin{pmatrix} 1 \\ i \end{pmatrix} = 2 \tag{3.2.37}$$

A symmetric matrix is a Hermitian matrix with real elements.

Vectors whose components are complex conjugates of each other occur frequently for Hilbert space vectors in quantum mechanics. For example, the wavefunction

$$\psi = \exp(-im\varphi) \tag{3.2.38}$$

must be multiplied by its complex conjugate

$$\psi^* = \exp(im\varphi) \tag{3.2.39}$$

and integrated (summed) on the interval $0 \leqslant \varphi \leqslant 2\pi$ to give a finite scalar product. The product $\psi^*\psi$ gives

$$\int_0^{2\pi} \psi^*\psi \, d\varphi = \int_0^{2\pi} \exp(-0\varphi)\, d\varphi = 2\pi \tag{3.2.40}$$

Idempotent matrices are matrices whose product returns the original matrix. The matrix product **AA** equals **A**, e.g.,

$$\left(\frac{1}{2}\right)\begin{pmatrix} 1 & -1 \\ -1 & 1 \end{pmatrix}\left(\frac{1}{2}\right)\begin{pmatrix} 1 & -1 \\ -1 & 1 \end{pmatrix} = \left(\frac{1}{4}\right)\begin{pmatrix} 2 & -2 \\ -2 & 2 \end{pmatrix} = \left(\frac{1}{2}\right)\begin{pmatrix} 1 & -1 \\ -1 & 1 \end{pmatrix} \tag{3.2.41}$$

Projection matrices which select a single coordinate direction in an N-dimensional space will be idempotent.

3.3. Matrix Equations and Inverses

Matrices and vectors provide a convenient method of expressing sets of linear equations involving many variables. Since the sets of equations can be reduced to a common matrix notation, general solutions for such sets of equations can be expressed in a common matrix format. The pair of linear equations in the two unknowns x and y

$$\begin{aligned} 2x + y &= 4 \\ x + y &= 3 \end{aligned} \tag{3.3.1}$$

is easily solved by subtracting the two equations to eliminate y, leaving a solution $x = 1$. Either equation then gives $y = 2$. A matrix expression for this problem permits an alternative method of solution which is easily generalized to a system of N equations in N unknowns.

If x and y are selected as the components of a vector, the four coefficients on the left side of the equations become the elements of a 2×2 matrix:

$$\begin{pmatrix} 2 & 1 \\ 1 & 1 \end{pmatrix}\begin{pmatrix} x \\ y \end{pmatrix} = \begin{pmatrix} 4 \\ 3 \end{pmatrix} \tag{3.3.2}$$

For this matrix **A**, the x–y vector $|r\rangle$, and the 4–3 vector $|c\rangle$, this equation is

$$\mathbf{A}|r\rangle = |c\rangle \tag{3.3.3}$$

This compact notation yields no real advantages since the separate equation must still be generated from the matrix equation and solved for x and y. However, the matrix equation can be solved formally for $|r\rangle$ using an inverse ($\mathbf{A}^{-1}$) matrix which satisfies the condition

$$\mathbf{A}^{-1}\mathbf{A} = \mathbf{I} \tag{3.3.4}$$

Not all **A** matrices have an inverse. If such an inverse does exist, the formal solution to Equation 3.3.3 is found by operating on both sides of the equation with $\mathbf{A}^{-1}$ to give

$$\mathbf{A}^{-1}\mathbf{A}|r\rangle = \mathbf{A}^{-1}|c\rangle \tag{3.3.5}$$

$$\mathbf{I}|r\rangle = |r\rangle = \mathbf{A}^{-1}|c\rangle \tag{3.3.6}$$

For systems with two or three unknowns, the inverse can be determined using the identity:

$$\mathbf{A}^{-1}\mathbf{A} = \mathbf{I} \tag{3.3.7}$$

If the inverse matrix has elements a, b, c, and d, the equation

$$\begin{pmatrix} a & b \\ c & d \end{pmatrix}\begin{pmatrix} 2 & 1 \\ 1 & 1 \end{pmatrix} = \begin{pmatrix} 1 & 0 \\ 0 & 1 \end{pmatrix} \tag{3.3.8}$$

gives four equations in four unknowns:

$$\begin{aligned} &(11) \qquad 2a + 1b = 1 \\ &(12) \qquad 1a + 1b = 0 \\ &(21) \qquad 2c + 1d = 0 \\ &(22) \qquad 1c + 1d = 1 \end{aligned} \tag{3.3.9}$$

The first two equations determine a and b while the second two equations determine c and d. The inverse matrix

$$\begin{pmatrix} 1 & -1 \\ -1 & 2 \end{pmatrix} \tag{3.3.10}$$

gives $|r\rangle$ as

$$|r\rangle = \mathbf{A}^{-1}|c\rangle = \begin{pmatrix} 1 & -1 \\ -1 & 2 \end{pmatrix}\begin{pmatrix} 4 \\ 3 \end{pmatrix} = \begin{pmatrix} 1 \\ 2 \end{pmatrix} \tag{3.3.11}$$

The solution of a set of two linear equations in two unknowns,

$$\begin{aligned} a_{11}x + a_{12}y &= c_1 \\ a_{21}x + a_{22}y &= c_2 \end{aligned} \tag{3.3.12}$$

provides a pattern for the determination of matrix inverses. The equation can be solved for x by eliminating the y terms. The first equation is multiplied by a_{22} while the second is multiplied by a_{12}. The two equations are subtracted to give

$$(a_{22}a_{11} - a_{12}a_{21})\, x = c_1 a_{22} - c_2 a_{12} \tag{3.3.13}$$

which can be solved for x:

$$x = (c_1 a_{22} - c_2 a_{12})/(a_{22}a_{11} - a_{12}a_{21}) \tag{3.3.14}$$

Eliminating terms in x in Equations 3.3.12 gives a solution for y:

$$y = (a_{11}c_2 - a_{21}c_2)/(a_{22}a_{11} - a_{12}a_{21}) \tag{3.3.15}$$

The denominator is the same for each case and is generated from the determinant of all the elements in the matrix **A**, i.e.,

$$\begin{vmatrix} a_{11} & a_{12} \\ a_{21} & a_{22} \end{vmatrix} = a_{11}a_{22} - a_{12}a_{21} = D \tag{3.3.16}$$

The numerator of the solution for x can be generated from a determinant in which the components c_i are substituted in the first column of the matrix (the x column):

$$\begin{vmatrix} c_1 & a_{12} \\ c_2 & a_{22} \end{vmatrix} = c_1 a_{22} - c_2 a_{12} = D_x \tag{3.3.17}$$

The numerator for the y solution is formed by substituting the components c_i in the second column of the matrix:

$$\begin{vmatrix} a_{11} & c_1 \\ a_{21} & c_2 \end{vmatrix} = a_{11}c_2 - a_{21}c_1 = D_y \tag{3.3.18}$$

The solutions for x and y are then ratios of determinants:

$$\begin{aligned} x &= D_x/D \\ y &= D_y/D \end{aligned} \tag{3.3.19}$$

The above technique, known as Cramer's method, can be extended to systems with several unknowns. The procedure is the same. A determinant D is formed using all the matrix coefficients and appears as the denominator in each case. For the ith unknown, x_i, the ith column of the matrix is replaced with the $|c_i\rangle$

vector components in order. The determinant for this matrix is divided by D to give x_i (Problem 3.13). Some techniques for evaluating determinants are discussed in Section 3.4.

Solutions for the 2×2 matrix system suggest a technique for the evaluation of matrix inverses. The solution for the vector $|r\rangle$ for a matrix with elements a_{ij} is still

$$|r\rangle = \mathbf{A}^{-1}|c\rangle \tag{3.3.20}$$

However, the solutions for the components of $|r\rangle$ are now known from Cramer's technique. The two solutions are equated:

$$\begin{aligned} \mathbf{A}^{-1}|c\rangle &= \begin{pmatrix} D_x/D \\ D_y/D \end{pmatrix} \\ &= (1/D)\begin{pmatrix} a_{22}c_1 - a_{12}c_2 \\ -a_{21}c_1 + a_{11}c_2 \end{pmatrix} \end{aligned} \tag{3.3.21}$$

The right-hand side is expressed as a matrix–vector product,

$$(1/D)\begin{pmatrix} a_{22} & -a_{12} \\ -a_{21} & a_{11} \end{pmatrix}\begin{pmatrix} c_1 \\ c_2 \end{pmatrix} \tag{3.3.22}$$

so that the inverse matrix is

$$\mathbf{A}^{-1} = \begin{pmatrix} a_{22} & -a_{12} \\ -a_{21} & a_{11} \end{pmatrix} D^{-1} \tag{3.3.23}$$

The elements of the inverse are determined as the cofactors of the **A** matrix. To determine the 11 element, the first row and first column are eliminated, leaving only a_{22}. This element, divided by D, becomes the 11 element of the inverse matrix. If the first row and second column are eliminated, only $-a_{21}$ remains. After division by D, this becomes the 21 element of the inverse matrix.

The cofactor technique for the inverse can be extended to higher-order matrices. All elements are divided by the determinant of **A**. The units for the *ji* element (not the *ij* element) of the matrix are formed by eliminating the *i*th row and *j*th column from the matrix and evaluating the determinant which remains. The sign of this element is determined by the sum of the indices i and j via the expression

$$(-1)^{i+j} \tag{3.3.24}$$

The technique can be demonstrated for the 3×3 matrix

$$\begin{pmatrix} 2 & 1 & 0 \\ 1 & 2 & 1 \\ 0 & 1 & 2 \end{pmatrix} \tag{3.3.25}$$

The determinant for a 3×3 matrix can be determined using the technique outlined in Figure 3.7. Each product of three matrix elements contains only one element from each row and each column. The determinant for this 3×3 matrix is

$$\begin{aligned} D &= (2)(2)(2) + (1)(1)(0) + (1)(1)(0) - (0)(2)(0) - (1)(1)(2) - (2)(1)(1) \\ &= 8 - 2 - 2 = 4 \end{aligned} \tag{3.3.26}$$

a_{11} for the inverse is the determinant which remains after the elements of the first row and first column are eliminated, i.e.,

$$\begin{vmatrix} 2 & 1 \\ 1 & 2 \end{vmatrix} = 4 - 1 = 3 \tag{3.3.27}$$

with a positive sign, as given by $(-1)^{1+1}$. a_{12} is the negative of the determinant which remains when the second row and first column are eliminated, i.e.,

$$\begin{vmatrix} 1 & 0 \\ 1 & 2 \end{vmatrix} = 2 - 0 = 2 \tag{3.3.28}$$

The remaining cofactors are determined in the same manner to give the inverse matrix

$$\left(\frac{1}{4}\right) \begin{pmatrix} 3 & -2 & 1 \\ -2 & 4 & -2 \\ 1 & -2 & 3 \end{pmatrix} \tag{3.3.29}$$

(Problem 3.1) which satisfies

$$\mathbf{A}^{-1}\mathbf{A} = \mathbf{I} \tag{3.3.30}$$

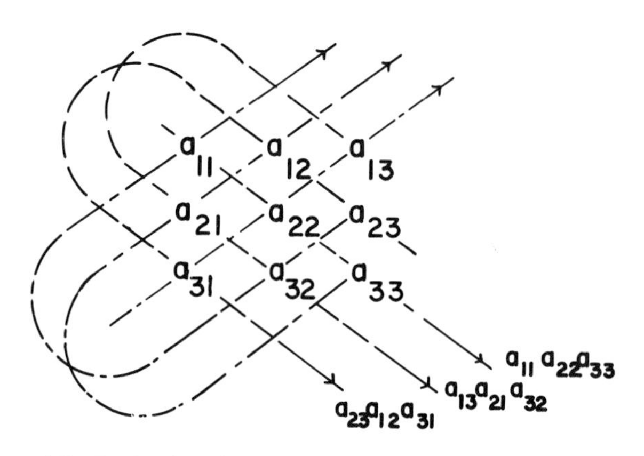

Figure 3.7. Evaluation of a 3×3 determinant with appropriate products.

Not every matrix has an inverse. The equations

$$\begin{aligned} 2x + 4y &= 6 \\ x + 2y &= 3 \end{aligned} \tag{3.3.31}$$

are linearly dependent since the second equation is exactly one-half the first equation. The inverse solution of these equations requires the determinant D,

$$D = \begin{vmatrix} 2 & 4 \\ 1 & 2 \end{vmatrix} = 0 \tag{3.3.32}$$

Since D appears in the denominator, it is impossible to define unique values of x and y for this pair of equations. However, a single equation is sufficient to define y in terms of x:

$$y = (3 - x)/2 \tag{3.3.33}$$

In this case, y is a dependent variable and x is the only independent variable. Any value of x can be substituted in Equation 3.3.33 to give a value of y. This occurs when x and y are linearly dependent. A matrix which has a determinant equal to zero is called a singular matrix.

The formal matrix solution provides an immediate solution to homogeneous equations where $|c\rangle = 0$. For example,

$$\begin{aligned} a_{11}x + a_{12}y &= 0 \\ a_{21}x + a_{22}y &= 0 \end{aligned} \tag{3.3.34}$$

must have a unique solution $x = y = 0$ since

$$|r\rangle = \mathbf{A}^{-1}|c\rangle = \mathbf{A}^{-1}|0\rangle = |0\rangle \tag{3.3.35}$$

when $\mathbf{A}$ is nonsingular.

3.4. Determinants

Determinants are used for matrix inverses and the solutions of sets of equations. They will also establish linear dependence or independence as required for the study of eigenvalue–eigenvector problems. For this reason, they are reviewed in this section.

The determinant of the general 3×3 matrix

$$\begin{pmatrix} a_{11} & a_{12} & a_{13} \\ a_{21} & a_{22} & a_{23} \\ a_{31} & a_{32} & a_{33} \end{pmatrix} \tag{3.4.1}$$

is found using the diagonal multiplication method (Figure 3.7) to be

$$a_{11}a_{22}a_{33}+a_{21}a_{32}a_{13}+a_{31}a_{12}a_{23}-a_{31}a_{22}a_{13}-a_{11}a_{32}a_{23}-a_{21}a_{12}a_{33} \tag{3.4.2}$$

Each product has been arranged so the column indices increase in the order 1, 2, 3. The row indices in each term also follow a pattern. The three positive products have row indices which increase in the order

$$1\ 2\ 3\ 1 \tag{3.4.3}$$

For example, the third product has the ordered row indices

$$3\ 1\ 2 \tag{3.4.4}$$

The fourth product with a negative coefficient has the row index order

$$3\ 2\ 1 \tag{3.4.5}$$

The three elements in each product are always elements from distinct rows and columns of the matrix. Each row and column index will appear only once in the product. The determinant has selected all possible ways of combining elements from distinct rows and columns in groups of three.

The nature of this ordering is shown for a permutation involving four objects, a green hat, a red hat, a green sock, and a red sock. To choose one hat and one sock with different colors, objects of the same color are placed in the same row while each object occupies a specific column. The determinant is

$$\begin{vmatrix} \mathrm{gh} & \mathrm{gs} \\ \mathrm{rh} & \mathrm{rs} \end{vmatrix} = (\mathrm{gh})(\mathrm{rs})-(\mathrm{rh})(\mathrm{gs}) \tag{3.4.6}$$

Only mismatched combinations are allowed. The determinant does not generate the case of a red hat and red sock.

A determinant generates combinations of product wavefunctions so that a single electron is associated with a single wavefunction. For example, two electrons (1 and 2) might be in a $+\frac{1}{2}$ or $\frac{-1}{2}$ electron spin state characterized by the wavefunctions α and β, respectively. For the case where one electron must have a β wavefunction and the other must have an α wavefunction, the determinant

$$\begin{vmatrix} \alpha(1) & \beta(1) \\ \alpha(2) & \beta(2) \end{vmatrix} = \alpha(1)\,\beta(2)-\alpha(2)\,\beta(1) \tag{3.4.7}$$

gives the full product wavefunction, which includes the case where electron 1 has the α wavefunction and electron 2 has β (the first term) and the case where electron 2 has the α wavefunction while electron 1 has β (the second term). Reversing the order reverses the sign.

When N electrons must be distributed as N distinct wavefunctions, Slater determinants are formed by placing the different wavefunctions in successive rows. Electron 1 for this wavefunction appears in the 11 position, electron 2 in the 12 position, etc.; i.e.,

$$\begin{vmatrix} \psi_1(1) & \psi_1(2) & \cdots & \psi_1(N) \\ \psi_2(1) & \psi_2(2) & \cdots & \psi_2(N) \\ & & \vdots & \\ \psi_N(1) & \psi_N(2) & \cdots & \psi_N(N) \end{vmatrix} \qquad (3.4.8)$$

The Slater determinant will generate every wavefunction product in which each wavefunction and each indexed electron appears only once.

The solution of 2×2 and 3×3 determinants can be achieved using simple multiplication schemes. Larger determinants must be solved by reducing them to products containing such smaller determinants. The basic reduction procedure can be inferred from the determinant for Equation 3.4.1, i.e., Equation 3.4.2. Each element in the first row of the matrix appears twice. Collecting terms with these elements gives

$$a_{11}(a_{22}a_{33}-a_{32}a_{23})+a_{12}(a_{31}a_{23}-a_{33}a_{21})+a_{13}(a_{21}a_{32}-a_{31}a_{22}) \qquad (3.4.9)$$

The factor for a_{11} is the determinant which results when the first row and first column of the matrix are deleted, i.e.,

$$\begin{vmatrix} a_{22} & a_{23} \\ a_{32} & a_{33} \end{vmatrix} = a_{22}a_{33}-a_{32}a_{23}=D_{11} \qquad (3.4.10)$$

The factor for $a_{12}(1+2=3)$ is the negative of the determinant formed by eliminating the first row and second column from the matrix:

$$\begin{vmatrix} a_{21} & a_{23} \\ a_{31} & a_{33} \end{vmatrix} = a_{21}a_{33}-a_{31}a_{23}=D_{12} \qquad (3.4.11)$$

Finally, the factor D_{13} for a_{13} $(1+3=4)$ is equal to the determinant formed by eliminating row 1 and column 3 from the matrix.

The full determinant for the 3×3 matrix can be resolved into a sum of three terms. The terms are the product of an element from a given row or column multiplied by the 2×2 determinant formed when the row and column associated with the element are eliminated. For expansion in the first row, the determinant is

$$D=(-1)^{1+1}a_{11}D_{11}+(-1)^{1+2}a_{12}D_{12}+(-1)^{1+3}a_{13}D_{13} \qquad (3.4.12)$$

The procedure can be used to expand a determinant of any size. For example, a 4×4 determinant can be expanded in any row or column to give a sum of each element in that row or column multiplied by a 3×3 determinant.

The 3×3 determinants could be further expanded in one of their rows or columns to give each of the relevant three elements times the appropriate 2×2 determinant.

The 4×4 matrix

$$\begin{pmatrix} 2 & 1 & 0 & 0 \\ 1 & 2 & 1 & 0 \\ 0 & 1 & 2 & 1 \\ 0 & 0 & 1 & 2 \end{pmatrix} \tag{3.4.13}$$

is expanded via the last column to give

$$D = (-1)^8 a_{44} D_{44} + (-1)^7 a_{34} D_{34} + 0 + 0 \tag{3.4.14}$$

The last two terms are zero since a_{24} and a_{14} are zero. D is

$$(+2)\begin{vmatrix} 2 & 1 & 0 \\ 1 & 2 & 1 \\ 0 & 1 & 2 \end{vmatrix} + (-1)\begin{vmatrix} 2 & 1 & 0 \\ 1 & 2 & 1 \\ 0 & 0 & 1 \end{vmatrix} \tag{3.4.15}$$

These 3×3 determinants can also be expanded. The first determinant is expanded in the first column while the second determinant is evaluated in the third row since this row contains only one nonzero element:

$$\begin{aligned} D &= (+2)\left[(+2)\begin{vmatrix} 2 & 1 \\ 1 & 2 \end{vmatrix} + (-1)\begin{vmatrix} 1 & 0 \\ 1 & 2 \end{vmatrix}\right] + (-1)(+1)\begin{vmatrix} 2 & 1 \\ 1 & 2 \end{vmatrix} \\ &= 2[2(3) - 1(2)] - 1(3) = 8 - 3 = 5 \end{aligned} \tag{3.4.16}$$

The determinant can be altered in some consistent ways which do not alter its final value. The 3×3 determinant

$$\begin{vmatrix} 2 & 1 & 0 \\ 1 & 2 & 1 \\ 0 & 1 & 2 \end{vmatrix} \tag{3.4.17}$$

equals 4. Rows 1 and 2 are now added and this result is substituted into the first row to give the determinant

$$\begin{vmatrix} 3 & 3 & 1 \\ 1 & 2 & 1 \\ 0 & 1 & 2 \end{vmatrix} \tag{3.4.18}$$

which also equals 4. The result will be the same if the two rows are added and the result is substituted for row 2. In general, the elements of any row can be multiplied by an arbitrary constant and added to a second row. The value of the determinant is unchanged. Likewise, the elements of any column can be multiplied by a scalar and added to the corresponding elements of a second column without changing the result.

A determinant can be resolved into a sum of determinants if all the elements of a row or column are separated into two parts. For example, Equation 3.4.17 becomes

$$\begin{vmatrix} 4 & -2 & 1 & 0 \\ -3 & +4 & 2 & 1 \\ 1 & -1 & 1 & 2 \end{vmatrix} \tag{3.4.19}$$

which separates into the sum

$$\begin{vmatrix} 4 & 1 & 0 \\ -3 & 2 & 1 \\ 1 & 1 & 2 \end{vmatrix} + \begin{vmatrix} -2 & 1 & 0 \\ 4 & 2 & 1 \\ -1 & 1 & 2 \end{vmatrix} \tag{3.4.20}$$

with the same value,

$$(16+0+1-0-4+6)+(-8+0-1+0+2-8)=19-15=4 \tag{3.4.21}$$

Both these techniques are normally used to evaluate determinants with the maximal number of zero elements. The determinant

$$\begin{vmatrix} 1 & 1 & 1 \\ 1 & 3 & 3 \\ 0 & 1 & 1 \end{vmatrix} \tag{3.4.22}$$

can be simplified if the elements of row 3 are subtracted from those of row 1 to give

$$\begin{vmatrix} 1 & 0 & 0 \\ 1 & 3 & 3 \\ 0 & 1 & 1 \end{vmatrix} \tag{3.4.23}$$

The center row is now distributed between two determinants to maximize zeros:

$$\begin{vmatrix} 1 & 0 & 0 \\ 1 & 3 & 0 \\ 0 & 1 & 1 \end{vmatrix} + \begin{vmatrix} 1 & 0 & 0 \\ 0 & 0 & 3 \\ 0 & 1 & 1 \end{vmatrix} \tag{3.4.24}$$

This gives a determinant of $3-3=0$. The same result is obtained by multiplying the bottom row by -3 and adding it to row 2.

The determinant in Equation 3.4.22 demonstrates another important property of determinants. Row 1 can be subtracted from row 2 to give

$$\begin{vmatrix} 1 & 1 & 1 \\ 0 & 2 & 2 \\ 0 & 1 & 1 \end{vmatrix} \tag{3.4.25}$$

Row 3 can now be multiplied by -2 and added to row 2 to give

$$\begin{vmatrix} 1 & 1 & 1 \\ 0 & 0 & 0 \\ 0 & 1 & 1 \end{vmatrix} \tag{3.4.26}$$

The determinant is zero as expected. This can be determined immediately if any row or column contains all zero elements. In chemical kinetics, matrices of rate constants for changes in a conservative system will always have columns which sum to zero:

$$\begin{pmatrix} -k_1 & k_2 \\ k_1 & -k_2 \end{pmatrix} \tag{3.4.27}$$

The determinant of such matrices is always zero.

If all the elements in any row or column of the matrix are multiplied by some scalar c, the value of the determinant is also multiplied by c. For example,

$$\begin{vmatrix} 2\times 2 & 1 & 0 \\ 2\times 1 & 2 & 1 \\ 2\times 0 & 1 & 2 \end{vmatrix} = 16+0+0-0-4-4=8=2(4) \tag{3.4.28}$$

where 4 is the value of the determinant without the factor of 2.

If any two rows or columns of a determinant are interchanged, the value of the determinant is negated. For example, if

$$\begin{vmatrix} 2 & 1 \\ 1 & 1 \end{vmatrix} \tag{3.4.29}$$

is changed to

$$\begin{vmatrix} 1 & 2 \\ 1 & 1 \end{vmatrix} \tag{3.4.30}$$

the determinant changes from $+1$ to -1. For the Slater determinant,

$$\begin{vmatrix} \alpha(1) & \alpha(2) \\ \beta(1) & \beta(2) \end{vmatrix} \tag{3.4.31}$$

the spin wavefunction

$$\psi = \alpha(1)\beta(2) - \alpha(2)\beta(1) \tag{3.4.32}$$

becomes

$$\psi' = \alpha(2)\,\beta(1) - \alpha(1)\,\beta(2) = -\psi \tag{3.4.33}$$

when electrons 1 and 2 are interchanged, i.e.,

$$\begin{vmatrix} \alpha(2) & \alpha(1) \\ \beta(2) & \beta(1) \end{vmatrix} \tag{3.4.34}$$

Wavefunctions generated with the determinants will always be antisymmetric, i.e., will change sign, for an interchange of a pair of electrons in the system.

3.5. Rotation of Coordinate Systems

A unitary matrix will rotate a vector in space without changing its norm. It replaces the original components on the coordinate axes with a new set of components on these same axes.

A two-dimensional rotation operation is illustrated in Figure 3.8 for a two-dimensional space. A rotation of the vector with components (1, 1) through $+90°$ produces a new vector with components $(-1, +1)$. The components are transformed using a unitary matrix **R**:

$$\mathbf{R} = \begin{pmatrix} \cos\theta & -\sin\theta \\ \sin\theta & \cos\theta \end{pmatrix} = \begin{pmatrix} 0 & -1 \\ 1 & 0 \end{pmatrix} \tag{3.5.1}$$

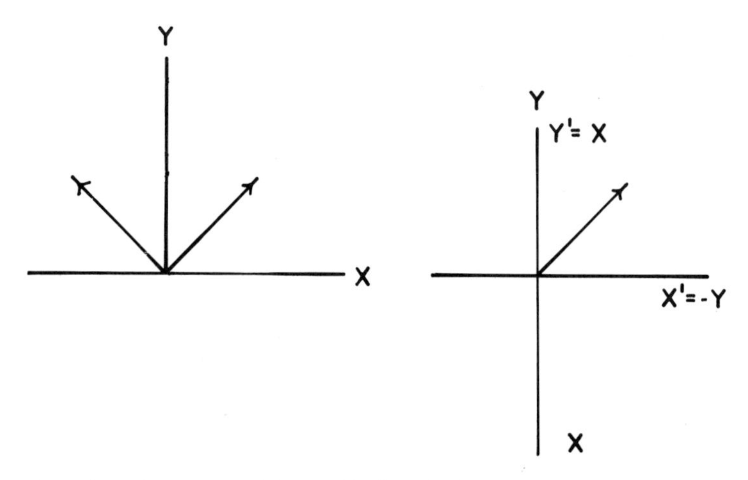

Figure 3.8. Rotation of a vector $\langle r| = (1 \quad 1)$ through 90° to the new vector $\langle s| = (-1 \quad 1)$: (a) rotation of the vector; (b) rotation of the coordinate system in the opposite direction.

The rotation of the vector $|r\rangle$ is a rotation relative to fixed x and y coordinate axes. However, the final vector $|s\rangle$ can assume its correct final position relative to the x and y axes if it remains fixed in space while the coordinate system is rotated in space through $-90°$ (Figure 3.8). The rotated y axis now lies on the position of the original x axis while the rotated x axis lies along the original negative y axis. Although the vector $|r\rangle$ has not moved, its projections on the new x and y axes are identical to those from a rotation of the vector against fixed axes.

Since the coordinate axes must rotate through $-\theta$ to achieve the effect of a vector rotation of $+\theta$, the appropriate axis rotation matrix is

$$\begin{pmatrix} \cos(-\theta) & -\sin(-\theta) \\ \sin(-\theta) & \cos(-\theta) \end{pmatrix} = \begin{pmatrix} \cos\theta & \sin\theta \\ -\sin\theta & \cos\theta \end{pmatrix} \tag{3.5.2}$$

i.e., the transpose of **R**. It is also the inverse of this matrix since a rotation of $+\theta$ is canceled by the second rotation of $-\theta$:

$$\mathbf{R}(-\theta) = \mathbf{R}' = \mathbf{R}^{-1} \tag{3.5.3}$$

A unitary matrix with real elements is also called an orthogonal matrix. If elements of the unitary matrix are imaginary, the transpose matrix elements are replaced by their complex conjugates to produce this inverse:

$$\mathbf{A}^{-1} = [\mathbf{A}']^* \tag{3.5.4}$$

The inverse matrix operates on unit vectors $|x\rangle$ and $|y\rangle$. $\mathbf{R}^{-1}$ rotates each of these vectors to generate the new axes:

$$\mathbf{R}^{-1}|x\rangle = \begin{pmatrix} 0 & 1 \\ -1 & 0 \end{pmatrix} \begin{pmatrix} 1 \\ 0 \end{pmatrix} = \begin{pmatrix} 0 \\ -1 \end{pmatrix} \tag{3.5.5}$$

$$\mathbf{R}^{-1}|y\rangle = \begin{pmatrix} 0 & 1 \\ -1 & 0 \end{pmatrix} \begin{pmatrix} 0 \\ 1 \end{pmatrix} = \begin{pmatrix} 1 \\ 0 \end{pmatrix} \tag{3.5.6}$$

The new x' and y' axes are oriented on the original $-y$ and $+x$ axes, respectively. The vectors in Equations 3.5.5 and 3.5.6 are unit vectors expressed in terms of the original x and y (first and second component) coordinate directions.

When $|r\rangle$ was rotated through $+90°$, the new vector $|s\rangle$ had components

$$|s\rangle = \begin{pmatrix} -1 \\ 1 \end{pmatrix} \tag{3.5.7}$$

The rotated coordinate system must give the same final projections on the original x and y axes.

The $|r\rangle$ vector has not moved, so its components with respect to the original x and y are

$$|r\rangle = \begin{pmatrix} 1 \\ 1 \end{pmatrix} \tag{3.5.8}$$

This vector must now be projected onto the new coordinate axes. The projection on the new x axis (x') is

$$|x'\rangle\langle x'|r\rangle = \begin{pmatrix} 0 \\ -1 \end{pmatrix} (0 \quad -1) \begin{pmatrix} 1 \\ 1 \end{pmatrix} = (-1) \begin{pmatrix} 0 \\ -1 \end{pmatrix} \tag{3.5.9}$$

The projection on the new y axis (y') is

$$|y'\rangle\langle y'|r\rangle = \begin{pmatrix} 1 \\ 0 \end{pmatrix} (1 \quad 0) \begin{pmatrix} 1 \\ 1 \end{pmatrix} = (1) \begin{pmatrix} 1 \\ 0 \end{pmatrix} \tag{3.5.10}$$

The projections onto these two axes are -1 and 1, respectively, and are exactly the projections for an $|r\rangle$ vector rotation in the opposite direction.

Either procedure describes rotations in two or more dimensions. Rotations of θ about a z axis perpendicular to the x–y plane require the matrix

$$\begin{pmatrix} \cos\theta & -\sin\theta & 0 \\ \sin\theta & \cos\theta & 0 \\ 0 & 0 & 1 \end{pmatrix} \tag{3.5.11}$$

The $\mathbf{R}^{-1}$ for rotation of the axes is the transpose of this matrix (Problem 3.16). For a rotation of 180° about the z axis and a reflection through the x–y plane, the appropriate matrix is

$$\begin{pmatrix} \cos 180^\circ & -\sin 180^\circ & 0 \\ \sin 180^\circ & \cos 180^\circ & 0 \\ 0 & 0 & -1 \end{pmatrix} = \begin{pmatrix} -1 & 0 & 0 \\ 0 & -1 & 0 \\ 0 & 0 & -1 \end{pmatrix} \tag{3.5.12}$$

This inversion matrix negates all three components of the vector (Figure 3.9). The vector direction in space is reversed.

For a three-dimensional rotation, it is convenient to adopt some common method for the rotation of a vector to a new orientation. The Eulerian angles of rotation are illustrated in Figure 3.10. The first operation is a rotation by an angle φ about the z axis described by the matrix

$$\begin{pmatrix} \cos\varphi & \sin\varphi & 0 \\ -\sin\varphi & \cos\varphi & 0 \\ 0 & 0 & 1 \end{pmatrix} \tag{3.5.13}$$

This operation creates a new set of unit vectors x', y', and z'. The next rotation

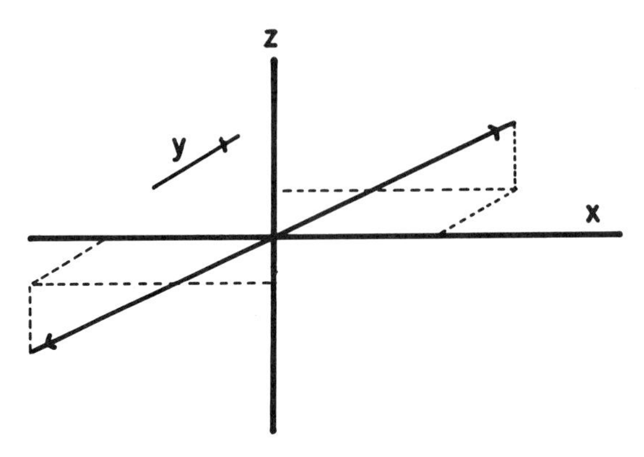

Figure 3.9. Vector inversion generated by the inversion matrix of Equation 3.5.12.

takes place about the y' (not the original y) axis through an angle θ so the second matrix operation is

$$\begin{pmatrix} \cos\theta & 0 & \sin\theta \\ 0 & 1 & 0 \\ -\sin\theta & 0 & \cos\theta \end{pmatrix} \tag{3.5.14}$$

Since rotation takes place about y', other elements in the second row and column are zero. The final Eulerian rotation involves the z'' axis created by the second rotation. The rotation of ψ about this axis requires the matrix

$$\begin{pmatrix} \cos\psi & \sin\psi & 0 \\ -\sin\psi & \cos\psi & 0 \\ 0 & 0 & 1 \end{pmatrix} \tag{3.5.15}$$

The rotation of a vector $|r\rangle$ in the three-dimensional space must proceed only in the order outlined. The full rotation is

$$\mathbf{R}_{z''}(\psi)\mathbf{R}_{y'}(\theta)\mathbf{R}_{z}(\varphi)|r\rangle \tag{3.5.16}$$

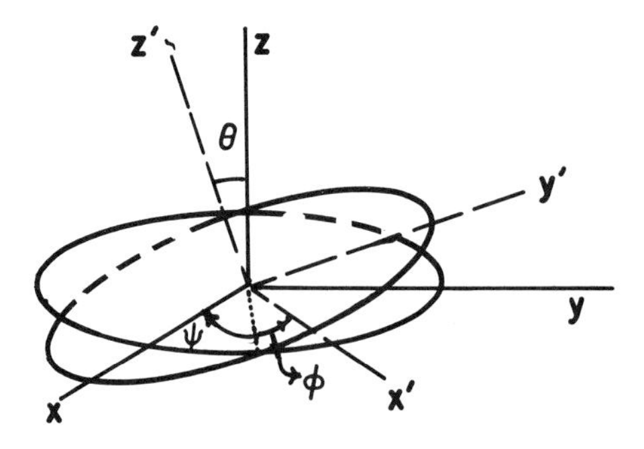

Figure 3.10. The Eulerian angles in three-dimensional space.

The matrix formulation will generate the three components of the final vector and is quite convenient for this reason. The net rotation for three Eulerian rotations of 90° is shown in Figure 3.11.

Although rotations of the vector or the coordinate system give the same final result, rotation of the coordinate system is preferable when functions must be rotated. The function

$$f(x, y) = 1x + 0y \tag{3.5.17}$$

describes a plane which is oriented 45° above the x–y plane and passes through the y axis. If the function is rotated $+90°$ in the x–y plane, the function intersects the x axis and the function is positive for positive y, i.e.,

$$g(x, y) = 0x + 1y \tag{3.5.18}$$

Since the entire function was rotated, a rotation operation $\mathbf{P}_R(+90)$ is defined as

$$\mathbf{P}_R f(x, y) \tag{3.5.19}$$

$\mathbf{P}_R$ is an operation since it must rotate the function for each x and y.

For any $f(x, y)$, this rotation is performed more easily using a rotation of the coordinate system in the opposite direction. The coordinate x and y axes are rotated to define new axes x' and y' which then replace x and y in the original function $f(x, y)$. Since both techniques give the same final result,

$$\mathbf{P}_R f(x, y) = f(\mathbf{R}^{-1}|r\rangle) \tag{3.5.20}$$

For functions, the rotation $\mathbf{R}^{-1}$ rotates every value of x on the x axis. The variable x is the first vector component; y is the second. A 90° rotation of these variables is

$$\begin{pmatrix} 0 & 1 \\ -1 & 0 \end{pmatrix} \begin{pmatrix} x \\ y \end{pmatrix} = \begin{pmatrix} y \\ -x \end{pmatrix} = \begin{pmatrix} x' \\ y' \end{pmatrix} \tag{3.5.21}$$

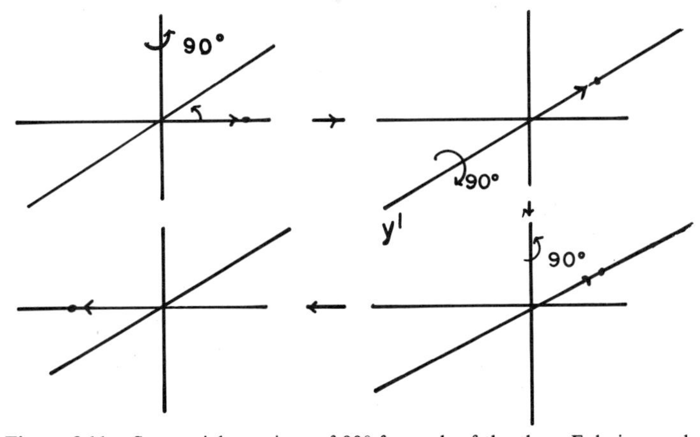

Figure 3.11. Sequential rotations of 90° for each of the three Eulerian angles.

The function $f(x, y)$ becomes $g(x, y)=f(x', y')$, i.e., x' and y' from Equation 3.5.21 replace x and y. The plane of Equation 3.5.17 becomes

$$\begin{aligned}\mathbf{R}f(x, y) &= f(\mathbf{R}^{-1}|r\rangle) \\ &= 1(+y)+0(-x)=1y-0x\end{aligned} \tag{3.5.22}$$

which is the function generated by the direct rotation of the original $f(x, y)$. For a rotation of θ in two dimensions, the matrix $\mathbf{R}^{-1}$,

$$\begin{pmatrix} \cos\theta & \sin\theta \\ -\sin\theta & \cos\theta \end{pmatrix} \tag{3.5.23}$$

gives

$$\begin{aligned} x' &= x\cos\theta + y\sin\theta \\ y' &= -x\sin\theta + y\cos\theta \end{aligned} \tag{3.5.24}$$

The equation for the plane after a rotation of θ is

$$g(x, y)=1x'+0y'=x\cos\theta+y\sin\theta \tag{3.5.25}$$

The technique is applicable to three dimensions as well. The function

$$f(x, y, z)=x^2y+yz+xz^3 \tag{3.5.26}$$

can be rotated 30° in the x–y plane using the three-dimensional rotation matrix for $-30°$:

$$\begin{pmatrix} \frac{\sqrt{3}}{2} & \frac{1}{2} & 0 \\ -\frac{1}{2} & \frac{\sqrt{3}}{2} & 0 \\ 0 & 0 & 1 \end{pmatrix}\begin{pmatrix} x \\ y \\ z \end{pmatrix} = \begin{pmatrix} \frac{\sqrt{3}}{2}x+\frac{1}{2}y+0z \\ -\frac{1}{2}x+\frac{\sqrt{3}}{2}y+0z \\ 0x+0y+1z \end{pmatrix} \tag{3.5.27}$$

Substitution of x', y', and z' in $f(x, y, z)$ gives the rotated function,

$$\begin{aligned}\mathbf{P}_R f(x, y, z) &= (\sqrt{3}x/2-y-2)^2(-x/2+\sqrt{3}y/2) \\ &\quad +(-x/2+\sqrt{3}y/2)(z)+(\sqrt{3}x/2+y/2)(z^3)\end{aligned} \tag{3.5.28}$$

The coordinate axes rotation technique is particularly valuable in quantum mechanics where rotation operations generate new orbitals from a single orbital. A p_x orbital, for example, generates both the p_y and p_z orbitals. In Cartesian coordinates, the p_x orbital is

$$p_x = x/r = x/(x^2+y^2+z^2)^{1/2} \tag{3.5.29}$$

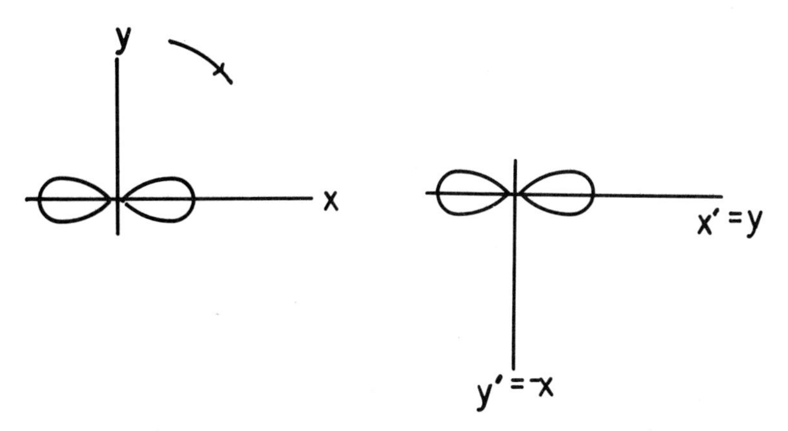

Figure 3.12. Rotation of a p_x orbital to the y axis in three-dimensional space.

The wave has positive amplitude on the positive x axis but extends throughout the three-dimensional space. To rotate this orbital $+90^\circ$ to the y axis, the coordinate system is rotated through -90° to give (Figure 3.12)

$$\begin{pmatrix} 0 & 1 & 0 \\ -1 & 0 & 0 \\ 0 & 0 & 1 \end{pmatrix} \begin{pmatrix} x \\ y \\ z \end{pmatrix} = \begin{pmatrix} y \\ -x \\ z \end{pmatrix} = \begin{pmatrix} x' \\ y' \\ z' \end{pmatrix} \tag{3.5.30}$$

The p_y orbital generated in this manner is

$$\begin{aligned} p_y = \mathbf{R}(+90)p_x &= y/[(y)^2 + (-x)^2 + (z)^2]^{1/2} \\ &= y/(x^2 + y^2 + z^2)^{1/2} = y/r \end{aligned} \tag{3.5.31}$$

Rotation of a p_x orbital to generate a p_z orbital is illustrated in Figure 3.13. The p_x orbital remains in place while the coordinate system is rotated about y to

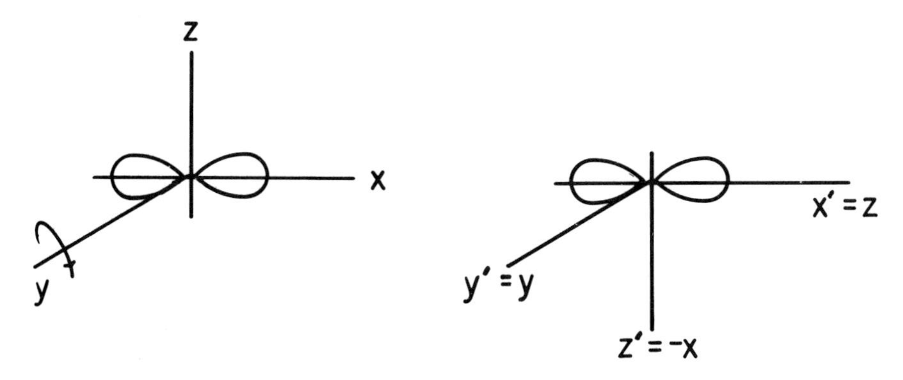

Figure 3.13. Rotation of a p_x orbital to the z axis in three-dimensional space.

bring the positive z axis to the positive x axis as defined in this right-handed coordinate system. The coordinate transform gives

$$\begin{pmatrix} 0 & 0 & 1 \\ 0 & 1 & 0 \\ -1 & 0 & 0 \end{pmatrix} \begin{pmatrix} x \\ y \\ z \end{pmatrix} = \begin{pmatrix} z \\ y \\ -x \end{pmatrix} = \begin{pmatrix} x' \\ y' \\ z' \end{pmatrix} \tag{3.5.32}$$

This result is substituted into p_x to give the p_z orbital:

$$\mathbf{P}_R p_x = (z)/[(-x)^2 + (y)^2 + (z)^2] \tag{3.5.33}$$

The coordinate rotation technique is ideal for more complex wavefunctions such as d orbitals. For example, the d_{xz} orbital is

$$d_{xz} = Axz/r^2 \tag{3.5.34}$$

with

$$r^2 = x^2 + y^2 + z^2 \tag{3.5.35}$$

and constant A. This wavefunction can be rotated by 120° about the z axis to create a new wavefunction which can be described as some linear combination from the set of five d orbitals.

The $\mathbf{R}^{-1}$ matrix is

$$\begin{pmatrix} \cos 120° & \sin 120° & 0 \\ -\sin 120° & \cos 120° & 0 \\ 0 & 0 & 1 \end{pmatrix} = \begin{pmatrix} \dfrac{-1}{2} & \dfrac{-\sqrt{3}}{2} & 0 \\ \dfrac{\sqrt{3}}{2} & \dfrac{-1}{2} & 0 \\ 0 & 0 & 1 \end{pmatrix} \tag{3.5.36}$$

The new coordinate vector is

$$\begin{pmatrix} \dfrac{-1}{2} & \dfrac{-\sqrt{3}}{2} & 0 \\ \dfrac{\sqrt{3}}{2} & \dfrac{-1}{2} & 0 \\ 0 & 0 & 1 \end{pmatrix} \begin{pmatrix} x \\ y \\ z \end{pmatrix} = \begin{pmatrix} \dfrac{-x}{2} - \dfrac{\sqrt{3}y}{2} \\ \dfrac{\sqrt{3}x}{2} - \dfrac{y}{2} \\ z \end{pmatrix} \tag{3.5.37}$$

and these new components are substituted into the original d_{xz} orbital to form the orbital which results from the rotation:

$$\begin{aligned} \mathbf{P}_R d_{xz} &= (A/r^2)(-x/2 - \sqrt{3}y/2)(z) \\ &= (A/r^2)\left(-xz/2 - \left(\frac{\sqrt{3}}{2}\right) yz\right) \end{aligned} \tag{3.5.38}$$

r^2 is not changed by the transform:

$$\begin{aligned} r^2 &= (-x/2 - \sqrt{3}y/2)^2 + (\sqrt{3}x/2 - y/2)^2 + z^2 \\ &= (x^2|4 + \sqrt{3}xy/2 + 3y^2/4) + (3x^2/4 - \sqrt{3}xy/4 + y^2/4) + z^2 \\ &= x^2 + y^2 + z^2 \end{aligned} \tag{3.5.39}$$

The function generated by the rotation is a weighted sum of the d_{xz} and d_{yz} orbitals,

$$\frac{-1}{2}d_{xz} + \frac{\sqrt{3}}{2}d_{yz} \tag{3.5.40}$$

The rotation of the function through 120° (a C_{3v} rotation) generates a new wavefunction using d orbitals from the set of five d orbitals. The effects of symmetry operations in addition to rotations can be determined in the same way.

Since the z coordinate is not changed by the rotation of the function, the d_{z^2} orbital is unchanged:

$$(A/r^2)(3z^2 - 1) = (A/r^2)(3(z)^2 - 1) \tag{3.5.41}$$

The rotation of d_{xz} generated a combination of two d orbitals; the rotation of d_{z^2} regenerated only itself. Orbitals or combinations of orbitals will ultimately be separated from each other based on such symmetry operations.

3.6. Principal Axes

Vectors in space can always be defined by choosing a coordinate system in the space and finding the projections of the vector on these coordinates. Although any coordinate system can be chosen, each matrix will have one coordinate system which separates the physical problem into distinct scalar systems. These coordinates are the principal axes or eigenvectors for the matrix.

The equation of a circle of radius 1 in a two-dimensional (x–y) space is

$$x^2 + y^2 = 1 \tag{3.6.1}$$

This quadratic equation is the scalar product of row and column vectors in the variables x and y:

$$(x \quad y)\begin{pmatrix} x \\ y \end{pmatrix} = x^2 + y^2 \tag{3.6.2}$$

$$(x \quad y)\begin{pmatrix} 1 & 0 \\ 0 & 1 \end{pmatrix}\begin{pmatrix} x \\ y \end{pmatrix} \tag{3.6.3}$$

If the identity matrix is replaced by a diagonal matrix

$$\begin{pmatrix} 1 & 0 \\ 0 & 4 \end{pmatrix} \tag{3.6.4}$$

the equation becomes

$$(x \quad y)\begin{pmatrix} 1 & 0 \\ 0 & 4 \end{pmatrix}\begin{pmatrix} x \\ y \end{pmatrix} \tag{3.6.5}$$

$$x^2 + 4y^2 = 1 \tag{3.6.6}$$

This is the equation of an ellipse oriented so that its major axis lies along the x axis with length 1. The minor axis with length $\frac{1}{2}$ is located on the y axis as shown in Figure 3.14. The ellipse can be elongated along x by increasing the a_{22} diagonal element. An increase in the a_{11} element relative to a_{22} produces an ellipse with its major axis along the y coordinate.

The equation for an ellipse oriented on the x and y coordinate axes contained only x^2 and y^2. When the ellipse is oriented off the x and y axes, xy product terms appear. The ellipse of Figure 3.15 is oriented along axes of $+45°$ and $+135°$ with respect to the x axis and is described by the equation

$$\frac{5}{2}x^2 - 3xy + \frac{5}{2}y^2 = 1 \tag{3.6.7}$$

The consistent matrix format is

$$(x \quad y)\begin{pmatrix} \frac{5}{2} & \frac{-3}{2} \\ \frac{-3}{2} & \frac{5}{2} \end{pmatrix}\begin{pmatrix} x \\ y \end{pmatrix} = 1 \tag{3.6.8}$$

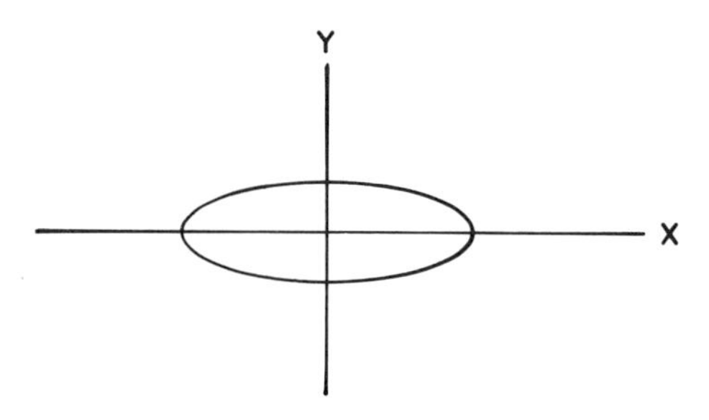

Figure 3.14. An ellipse with its major axis (1) on the x axis and its minor axis $\left(\frac{1}{2}\right)$ on the y axis.

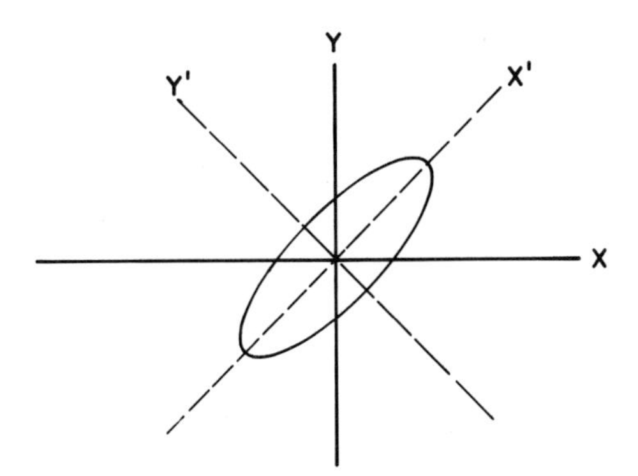

Figure 3.15. The ellipse with major and minor axes oriented 45° to the x and y coordinate axes.

The ellipse of Figure 3.14 has exactly the same shape as the ellipse of Figure 3.15. Both have major axes of 1 and minor axes of $\frac{1}{2}$. The matrix of Equation 3.6.8 with its off-diagonal elements appeared when the major and minor axes of the ellipse did not align with the coordinate axes. These axes could be aligned if the coordinate system was rotated counterclockwise through 45°. The rotation matrix to transform the coordinate axes $|r\rangle$ to $|r'\rangle$ is

$$\begin{pmatrix} \cos 45^\circ & \sin 45^\circ \\ -\sin 45^\circ & \cos 45^\circ \end{pmatrix} = \begin{pmatrix} \frac{1}{\sqrt{2}} & \frac{1}{\sqrt{2}} \\ \frac{-1}{\sqrt{2}} & \frac{1}{\sqrt{2}} \end{pmatrix} \tag{3.6.9}$$

The matrix will produce the new coordinate axes

$$\begin{pmatrix} x' \\ y' \end{pmatrix} = \begin{pmatrix} \frac{1}{\sqrt{2}} & \frac{1}{\sqrt{2}} \\ \frac{-1}{\sqrt{2}} & \frac{1}{\sqrt{2}} \end{pmatrix} \begin{pmatrix} x \\ y \end{pmatrix} = \begin{pmatrix} \frac{x}{\sqrt{2}} + \frac{y}{\sqrt{2}} \\ \frac{-x}{\sqrt{2}} + \frac{y}{\sqrt{2}} \end{pmatrix} \tag{3.6.10}$$

i.e., 45° with respect to the original x and y.

The combination of a row and a column vector bracketing a matrix is a quadratic form. The quadratic form of Equation 3.6.8 will assume a new form if the x and y components are replaced by x' and y' along the major and minor axes of the ellipse. The actual equation of the ellipse must not change although its coordinate system will. To maintain the quadratic form,

$$\langle r | \mathbf{A} | r \rangle \tag{3.6.11}$$

while rotating the coordinate system,

$$\mathbf{I} = \mathbf{R}\mathbf{R}^{-1} \tag{3.6.12}$$

must be inserted between **A** and the $|r\rangle$ vector:

$$\langle r|\mathbf{A}|\mathbf{R}\mathbf{R}^{-1}|r\rangle \qquad (3.6.13)$$

Since rotation of $\langle r|$ requires **R**, the identity

$$\mathbf{I}=\mathbf{R}\mathbf{R}^{-1} \qquad (3.6.14)$$

is inserted between $\langle r|$ and **A** to produce an equivalent quadratic form,

$$\langle r|\mathbf{A}|r\rangle=\langle r|\mathbf{R}\mathbf{R}^{-1}|\mathbf{A}|\mathbf{R}\mathbf{R}^{-1}|r\rangle \qquad (3.6.15)$$

Introducing the transformed vectors $\langle r'|$ and $|r'\rangle$ gives

$$\langle r'|\mathbf{R}^{-1}\mathbf{A}\mathbf{R}|r'\rangle \qquad (3.6.16)$$

The transformed vectors require a transformed matrix, $\mathbf{R}^{-1}\,\mathbf{A}\,\mathbf{R}$,

$$\begin{pmatrix} \frac{1}{\sqrt{2}} & \frac{1}{\sqrt{2}} \\ \frac{-1}{\sqrt{2}} & \frac{1}{\sqrt{2}} \end{pmatrix}\begin{pmatrix} \frac{5}{2} & \frac{-3}{2} \\ \frac{-3}{2} & \frac{5}{2} \end{pmatrix}\begin{pmatrix} \frac{1}{\sqrt{2}} & \frac{-1}{\sqrt{2}} \\ \frac{1}{\sqrt{2}} & \frac{1}{\sqrt{2}} \end{pmatrix}=\begin{pmatrix} 1 & 0 \\ 0 & 4 \end{pmatrix} \qquad (3.6.17)$$

The equation in the x'–y' coordinate system is

$$(x' \quad y')\begin{pmatrix} 1 & 0 \\ 0 & 4 \end{pmatrix}\begin{pmatrix} x' \\ y' \end{pmatrix}=x'^2+4y'^2=1 \qquad (3.6.18)$$

The equation in the primed coordinates is identical to that obtained when this ellipse was oriented on the x and y axes. Here x' and y' are the principal axes for the ellipse. They coincide with the major and minor axes of the ellipse.

If coordinate axes are selected so that mixed terms xy exist, the matrix for the quadratic form will be nondiagonal. However, it is always possible to rotate the coordinate system so that it coincides with the major and minor axes. The matrix must also be changed and will become a diagonal matrix when the rotation matrices describe an orientation on the ellipse axes. The principal axes are easily located for the ellipse in two dimensions. However, systems of any dimension can have principal axes.

The diatomic molecule of Figure 3.16 is aligned with the x axis and centered at the origin. For rotation in the x–y plane, the moment of inertia, $\mathcal{I}_{zz}$, is

$$\mathcal{I}_{zz}=\sum m_i(x_i^2+y_i^2) \qquad (3.6.19)$$

where the summation extends over the two atoms. Since there is no y component for this orientation, the moment of inertia is

$$\mathcal{I}_{zz}=m(d/2)^2+m(d/2)^2=md^2/2 \qquad (3.6.20)$$

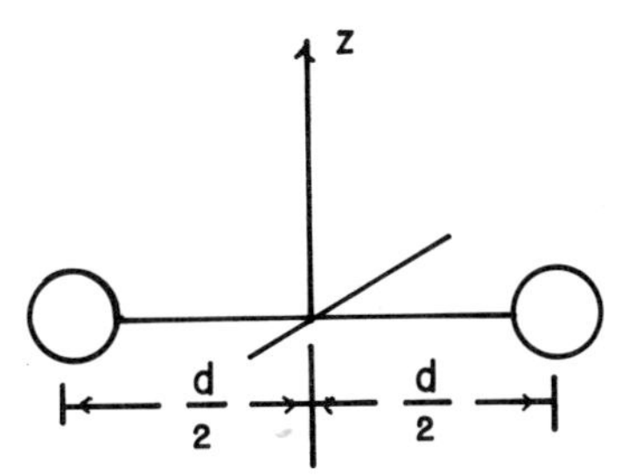

Figure 3.16. A diatomic molecule with the interatomic bond oriented on the x axis in space.

The second moment of inertia, $\mathscr{I}_{yy}$, is identical:

$$\mathscr{I}_{yy} = md^2/2 \tag{3.6.21}$$

These moments of inertia are identical to the moment of inertia for a single particle of reduced mass

$$1/\mu = 1/m_1 + 1/m_2 \tag{3.6.22}$$

located a distance d from a fixed rotation center.

The moment of inertia $\mathscr{I}_{zz}$ and the angular velocity ω_z about z determine the angular momentum L_z about the z axis:

$$L_z = \mathscr{I}_{zz}\omega_z \tag{3.6.23}$$

This equation is linear as is that for L_y:

$$L_y = \mathscr{I}_{yy}\omega_y \tag{3.6.24}$$

The two equations are independent. A molecule rotating about the z axis will have no influence on events involving the y rotation axis.

The products of inertia contain two coordinate directions. For example, the product of inertia $\mathscr{I}_{xy}$ is

$$\mathscr{I}_{xy} = -\sum m_i x_i y_i \tag{3.6.25}$$

Since all y_i and z_i are zero for the orientation of Figure 3.16, all products of inertia are zero.

When the diatomic molecule is oriented 45° to the x and y axes (Figure 3.17) in the x–y plane, both moments of inertia and products of inertia must be calculated. The moments of inertia are now

$$\begin{aligned}\mathscr{I}_{zz} &= m[(d/2\sqrt{2})^2 + (d/2\sqrt{2})^2] + m[(d/2\sqrt{2})^2 + (d/2\sqrt{2})^2] \\ &= md^2/4 + md^2/4 = md^2/2 \end{aligned} \tag{3.6.26}$$

$$\mathscr{I}_{yy} = m(d/2\sqrt{2})^2 + m(d/2\sqrt{2})^2$$
$$= md^2/8 + md^2/8 = md^2/4 \tag{3.6.27}$$
$$\mathscr{I}_{xx} = md^2/8 + md^2/8 = md^2/4 \tag{3.6.28}$$

The products of inertia are

$$\mathscr{I}_{xy} = \mathscr{I}_{yx} = -m(d/2\sqrt{2})(d/2\sqrt{2}) - m(d/2\sqrt{2})(d/2\sqrt{2})$$
$$= -md^2/4 \tag{3.6.29}$$

and

$$\mathscr{I}_{xz} = \mathscr{I}_{zx} = \mathscr{I}_{yz} = \mathscr{I}_{zy} = 0 \tag{3.6.30}$$

since all z components are zero.

The three moments of inertia and six products of inertia are the elements of a matrix,

$$\begin{pmatrix} \mathscr{I}_{xx} & \mathscr{I}_{xy} & \mathscr{I}_{xz} \\ \mathscr{I}_{yx} & \mathscr{I}_{yy} & \mathscr{I}_{yz} \\ \mathscr{I}_{zx} & \mathscr{I}_{zy} & \mathscr{I}_{zz} \end{pmatrix} \tag{3.6.31}$$

$$= (md^2/4) \begin{pmatrix} 1 & -1 & 0 \\ -1 & 1 & 0 \\ 0 & 0 & 2 \end{pmatrix} \tag{3.6.32}$$

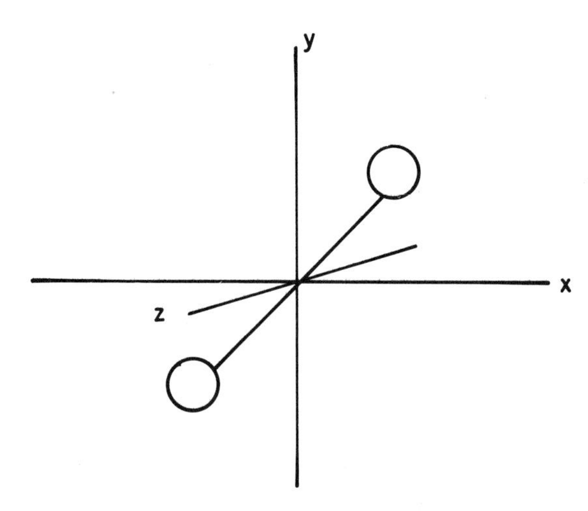

Figure 3.17. A diatomic molecule with its interatomic bond oriented 45° with respect to the x coordinate axis.

The effect of the 45° coordinate axes is apparent when the matrix operates on angular velocity components about the x, y, and z axes. The angular momentum components about these axes are

$$\begin{pmatrix} L_x \\ L_y \\ L_z \end{pmatrix} = (md^2/8) \begin{pmatrix} 1 & -1 & 0 \\ -1 & 1 & 0 \\ 0 & 0 & 2 \end{pmatrix} \begin{pmatrix} \omega_x \\ \omega_y \\ \omega_z \end{pmatrix}$$

$$= (md^2/8) \begin{pmatrix} \omega_x - \omega_y \\ -\omega_x + \omega_y \\ 2\omega_z \end{pmatrix} \tag{3.6.33}$$

The angular momentum about the z axis is unaffected by the new orientation since it describes a rotation in the x–y plane. The orientation of 45° changes the phase of this rotation. The angular momentum about the x axis receives contributions from both ω_x and ω_y. $\mathscr{I}_{xy}$ gives the contribution of ω_y to L_x.

When the coordinate axes do not coincide with the principal axes of the molecule, the $\mathscr{I}$ matrix is nondiagonal. Both $|L\rangle$ and $|\omega\rangle$ must be rotated 45° using the $\mathbf{R}^{-1}$ rotation matrix. Introducing

$$\mathbf{I} = \mathbf{R}\mathbf{R}^{-1} \tag{3.6.34}$$

between $\mathscr{I}$ and $|\omega\rangle$ gives

$$|L'\rangle = \mathbf{R}^{-1}|L\rangle = \mathbf{R}^{-1}\mathscr{L}\mathbf{R}\mathbf{R}^{-1}|\omega\rangle = \mathbf{R}^{-1}\mathscr{I}\mathbf{R}|\omega'\rangle \tag{3.6.35}$$

Then,

$$\begin{pmatrix} L'_x \\ L'_y \\ L'_z \end{pmatrix} = \begin{pmatrix} \frac{1}{\sqrt{2}} & \frac{1}{\sqrt{2}} & 0 \\ \frac{-1}{\sqrt{2}} & \frac{1}{\sqrt{2}} & 0 \\ 0 & 0 & 1 \end{pmatrix} \begin{pmatrix} L_x \\ L_y \\ L_z \end{pmatrix} = \begin{pmatrix} \frac{1}{\sqrt{2}}(L_x + L_y) \\ \frac{1}{\sqrt{2}}(-L_x + L_y) \\ L_z \end{pmatrix} \tag{3.6.36}$$

and

$$\begin{pmatrix} \omega'_x \\ \omega'_y \\ \omega'_z \end{pmatrix} = \begin{pmatrix} \frac{1}{\sqrt{2}}(\omega_x + \omega_y) \\ \frac{1}{\sqrt{2}}(-\omega_x + \omega_y) \\ \omega_z \end{pmatrix} \tag{3.6.37}$$

The matrix for the rotated coordinate system is

$$\mathbf{R}^{-1}\mathfrak{J}\mathbf{R} = \left(\frac{1}{2}\right)(md^2/4)\begin{pmatrix} 1 & 1 & 0 \\ -1 & 1 & 0 \\ 0 & 0 & 1 \end{pmatrix}\begin{pmatrix} 1 & -1 & 0 \\ 1 & 1 & 0 \\ 0 & 0 & 2 \end{pmatrix}\begin{pmatrix} 1 & -1 & 0 \\ 1 & 1 & 0 \\ 0 & 0 & 1 \end{pmatrix}$$

$$= (md^2/2)\begin{pmatrix} 0 & 0 & 0 \\ 0 & 1 & 0 \\ 0 & 0 & 1 \end{pmatrix} \tag{3.6.38}$$

The coordinate transform regenerates the two moments of inertia observed when the coordinate system was aligned with the principal axes of the diatomic molecule. If these axes had been selected initially, L_x, L_y, and L_z would be three independent (uncoupled) equations.

Matrices provide coupled linear equations for a variety of systems. For example, the three electric field components E_x, E_y, and E_z may all contribute to the induced dipole moment in the x direction. The simple scalar equation for one dimension,

$$P = \alpha E \tag{3.6.39}$$

where P is the polarizability becomes the matrix equation

$$\begin{pmatrix} P_x \\ P_y \\ P_z \end{pmatrix} = \begin{pmatrix} \alpha_{xx} & \alpha_{xy} & \alpha_{xz} \\ \alpha_{yx} & \alpha_{yy} & \alpha_{yz} \\ \alpha_{zx} & \alpha_{zy} & \alpha_{zz} \end{pmatrix}\begin{pmatrix} E_x \\ E_y \\ E_z \end{pmatrix} \tag{3.6.40}$$

The magnetization produced by a magnetic field H with components H_x, H_y, and H_z is

$$\begin{pmatrix} M_x \\ M_y \\ M_z \end{pmatrix} = \begin{pmatrix} \chi_{xx} & \chi_{xy} & \chi_{xz} \\ \chi_{yx} & \chi_{yy} & \chi_{yz} \\ \chi_{zx} & \chi_{zy} & \chi_{zz} \end{pmatrix}\begin{pmatrix} H_x \\ H_y \\ H_z \end{pmatrix} \tag{3.6.41}$$

In each of these cases, a coordinate system with the principal axes of the system can produce a diagonal matrix, i.e., scalar equations for each principal axis.

3.7. Eigenvalues and the Characteristic Polynomial

The principal axes of Section 3.6 were generally coincident with major axes for the system. For systems where such principal axes are not apparent, they must be determined via mathematical analysis.

The inertia matrix for the diatomic molecule aligned 45° with respect to the x axis (Figure 3.17) was

$$(md^2/4)\begin{pmatrix} 1 & -1 & 0 \\ -1 & 1 & 0 \\ 0 & 0 & 2 \end{pmatrix} \tag{3.7.1}$$

One principal axis for this diatomic molecule is oriented −45° with respect to the x axis. In terms of the x, y, and z coordinates, it is

$$\left(\frac{1}{\sqrt{2}}\right)\begin{pmatrix} 1 \\ -1 \\ 0 \end{pmatrix} \tag{3.7.2}$$

When the matrix $\mathscr{I}$ operates on this vector, the vector retains the same relative components

$$(md^2/4)\begin{pmatrix} 1 & -1 & 0 \\ -1 & 1 & 0 \\ 0 & 0 & 2 \end{pmatrix}\begin{pmatrix} 1 \\ -1 \\ 0 \end{pmatrix}\left(\frac{1}{\sqrt{2}}\right) = \left(\frac{1}{\sqrt{2}}\right)\begin{pmatrix} 1 \\ -1 \\ 0 \end{pmatrix}(2)(md^2/4) \tag{3.7.3}$$

multiplied by two to give a scalar moment of inertia

$$md^2/2 \tag{3.7.4}$$

identical to the moment of inertia about this principal axis.

The diatomic molecule lies along a principal axis with vector components

$$\left(\frac{1}{\sqrt{2}}\right)\begin{pmatrix} 1 \\ 1 \\ 0 \end{pmatrix} \tag{3.7.5}$$

Operation with the matrix returns this vector and a moment of inertia of zero:

$$(md^2/4)\begin{pmatrix} 1 & -1 & 0 \\ -1 & 1 & 0 \\ 0 & 0 & 2 \end{pmatrix}\begin{pmatrix} 1 \\ 1 \\ 0 \end{pmatrix}\left(\frac{1}{\sqrt{2}}\right) = \begin{pmatrix} 1 \\ 1 \\ 0 \end{pmatrix}(0)(md^2/4) \tag{3.7.6}$$

The nondiagonal matrix operates on a principal axis vector to give the moment of inertia for that principal axis. This is confirmed for the third and final principal axis,

$$(md^2/4)\begin{pmatrix} 1 & -1 & 0 \\ -1 & 1 & 0 \\ 0 & 0 & 2 \end{pmatrix}\begin{pmatrix} 0 \\ 0 \\ 1 \end{pmatrix} = \begin{pmatrix} 0 \\ 0 \\ 1 \end{pmatrix}(2)(md^2/4) \tag{3.7.7}$$

If the vectors for the principal axes are $|i\rangle$, each vector satisfies

$$\mathbf{A}\,|i\rangle = \lambda_i\,|i\rangle \tag{3.7.8}$$

The scalar λ_i is the moment of inertia about the principal axis $|i\rangle$. The λ_i are the eigenvalues for the moment of inertia matrix while the $|i\rangle$ are its eigenvectors.

The eigenvectors $|i\rangle$ for

$$\begin{pmatrix} 2 & 1 & 0 \\ 1 & 2 & 1 \\ 0 & 1 & 2 \end{pmatrix} \tag{3.7.9}$$

are

$$\begin{pmatrix} 1 \\ \sqrt{2} \\ 1 \end{pmatrix} \qquad \begin{pmatrix} 1 \\ 0 \\ -1 \end{pmatrix} \qquad \begin{pmatrix} 1 \\ -\sqrt{2} \\ 1 \end{pmatrix} \tag{3.7.10}$$

The matrix operations give

$$\begin{pmatrix} 2 & 1 & 0 \\ 1 & 2 & 1 \\ 0 & 1 & 2 \end{pmatrix} \begin{pmatrix} 1 \\ \sqrt{2} \\ 1 \end{pmatrix} = (2+\sqrt{2}) \begin{pmatrix} 1 \\ \sqrt{2} \\ 1 \end{pmatrix} \tag{3.7.11}$$

$$\begin{pmatrix} 2 & 1 & 0 \\ 1 & 2 & 1 \\ 0 & 1 & 2 \end{pmatrix} \begin{pmatrix} 1 \\ 0 \\ -1 \end{pmatrix} = (2) \begin{pmatrix} 1 \\ 0 \\ -1 \end{pmatrix} \tag{3.7.12}$$

$$\begin{pmatrix} 2 & 1 & 0 \\ 1 & 2 & 1 \\ 0 & 1 & 2 \end{pmatrix} \begin{pmatrix} 1 \\ -\sqrt{2} \\ 1 \end{pmatrix} = (2-\sqrt{2}) \begin{pmatrix} 1 \\ -\sqrt{2} \\ 1 \end{pmatrix} \tag{3.7.13}$$

The three vectors for this symmetric matrix are orthogonal:

$$\langle \lambda_i | \lambda_j \rangle = 0 \qquad i \neq j \tag{3.7.14}$$

The same components are used for row and column eigenvectors.

In practice, the eigenvectors are determined once the eigenvalues are known. The 2×2 matrix

$$\mathbf{A} = \begin{pmatrix} 1 & 1 \\ 1 & 1 \end{pmatrix} \tag{3.7.15}$$

has two eigenvectors with unknown components a and b. When the matrix operates on this vector, it reproduces this vector and an eigenvalue:

$$\mathbf{A}|\lambda\rangle = \lambda|\lambda\rangle \tag{3.7.16}$$

$$\begin{pmatrix} 1 & 1 \\ 1 & 1 \end{pmatrix} \begin{pmatrix} a \\ b \end{pmatrix} = \lambda \begin{pmatrix} a \\ b \end{pmatrix} \tag{3.7.17}$$

The identity operator **I** is introduced on the right so both sides are in matrix format. $\mathbf{I}|\lambda\rangle$ can be subtracted from both sides to give

$$\begin{pmatrix} 1 & 1 \\ 1 & 1 \end{pmatrix}\begin{pmatrix} a \\ b \end{pmatrix} - \begin{pmatrix} \lambda & 0 \\ 0 & \lambda \end{pmatrix}\begin{pmatrix} a \\ b \end{pmatrix} = 0$$
$$\begin{pmatrix} 1-\lambda & 1 \\ 1 & 1-\lambda \end{pmatrix}\begin{pmatrix} a \\ b \end{pmatrix} = 0 \tag{3.7.18}$$

Since λ is a scalar, the equation is now a homogeneous equation where the vector $|c\rangle = 0$. The only possible solution for such an equation is

$$a = b = 0 \tag{3.7.19}$$

(Section 3.3). A nontrivial solution is possible only if the equations are linearly dependent. Then a and b are relative values.

The matrix equation is a set of linear equations. The linear dependence of such a system can be verified if the determinant formed from the coefficients in the equation is equal to zero:

$$\begin{vmatrix} 1-\lambda & 1 \\ 1 & 1-\lambda \end{vmatrix} = 0 \tag{3.7.20}$$

The eigenvalues are now the scalar quantities which give a zero determinant when inserted into Equation 3.7.20. These values are determined by expanding and solving the determinant to give

$$(1-\lambda)(1-\lambda) - 1 = 0 = -2\lambda + \lambda^2 \tag{3.7.21}$$

The determinant (Equation 3.7.20) is called the characteristic determinant; the polynomial in λ generated when this determinant is expanded is called the characteristic polynomial. The roots of this characteristic polynomial are the two eigenvalues for the matrix,

$$\lambda = 0; \qquad \lambda = +2 \tag{3.7.22}$$

The procedure is applicable to any $N \times N$ matrix. The formal equation

$$\mathbf{A}|\lambda\rangle = \lambda\mathbf{I}|\lambda\rangle \tag{3.7.23}$$

becomes

$$(\mathbf{A} - \lambda\mathbf{I})\,|\lambda\rangle = |0\rangle \tag{3.7.24}$$

and the characteristic determinant is equated to zero,

$$|\mathbf{A} - \lambda\mathbf{I}| = 0 \tag{3.7.25}$$

The Nth order polynomial generated from the determinant is evaluated to give the N eigenvalues.

An $N \times N$ matrix has a rank of N if its determinant is not equal to zero. The introduction of eigenvalues which give a determinant of zero reduces the rank of the matrix to at least $N-1$. This states that the N components of the vector are not independent. They are all determined as ratios of one of the vector components.

The characteristic polynomial permits determination of the eigenvalues for the matrix

$$\begin{pmatrix} 2 & 1 & 0 \\ 1 & 2 & 1 \\ 0 & 1 & 2 \end{pmatrix} \tag{3.7.26}$$

The characteristic determinant is generated by subtracting λ from each diagonal element of the matrix,

$$\begin{vmatrix} 2-\lambda & 1 & 0 \\ 1 & 2-\lambda & 1 \\ 0 & 1 & 2-\lambda \end{vmatrix} = 0 \tag{3.7.27}$$

and expanded in the first row to give

$$(2-\lambda)\begin{vmatrix} 2-\lambda & 1 \\ 1 & 2-\lambda \end{vmatrix} - (1)\begin{vmatrix} 1 & 1 \\ 0 & 2-\lambda \end{vmatrix} + (0)\begin{vmatrix} 1 & 2-\lambda \\ 0 & 1 \end{vmatrix} = 0$$

$$(2-\lambda)[(2-\lambda)^2+1]-(2-\lambda)=0 \tag{3.7.28}$$

$(2-\lambda)$ is common to both terms and can be factored to give

$$(2-\lambda)(4-4\lambda+\lambda^2-1-1)=(2-\lambda)(2-4\lambda+\lambda^2) \tag{3.7.29}$$

One root (eigenvalue) is $\lambda = 2$ while the remaining two roots are determined from the quadratic formula

$$\lambda_{\pm} = +2 \pm \sqrt{16-8}/2 = +2 \pm \sqrt{2} \tag{3.7.30}$$

These are the three eigenvalues which resulted from application of the matrix to the principal axis vectors. However, the characteristic polynomial gives the eigenvalues without prior knowledge of the eigenvectors.

The notion of eigenvalues and eigenvectors extends to function vectors and operators as well. The Schrödinger equation

$$\mathbf{H}\psi = E\psi \tag{3.7.31}$$

states that the Hamiltonian operator $\mathbf{H}$ operates on the wavefunction ψ to regenerate the wavefunction and an eigenvalue E for that wavefunction. For example, a particle in a one-dimensional box of length L has the wavefunction

$$\psi(x) = \sin\,[n\pi x/L] \qquad n = 1, 2, \ldots \tag{3.7.32}$$

The Hamiltonian contains only the operator for the kinetic energy of the particle,

$$\mathbf{H} = -\hbar^2/2m\, d^2/dx^2 \tag{3.7.33}$$

where

$$\hbar = h/2\pi \tag{3.7.34}$$

The operation $\mathbf{H}\psi(x)$ gives

$$(-\hbar^2/2m)\, d^2[\sin\,(n\pi x/L)]/dx^2 = (-\hbar^2/2m)(-1)(n\pi/L)^2 \sin\,(n\pi x/L) \tag{3.7.35}$$

The wavefunction is regenerated with a factor which must equal the energy,

$$E = +(\hbar n\pi)^2/2mL^2 \tag{3.7.36}$$

or

$$E = h^2 n^2/8mL^2 \tag{3.7.37}$$

Eigenvalues and eigenfunctions are often determined via a full solution of the differential equation generated by the Hamiltonian operator and the Schrödinger equation. The solution of the differential equation gives the wavefunction and the energy eigenvalue. A Hermitian or symmetric matrix will always have real eigenvalues.

3.8. Eigenvectors

Eigenvalues for a matrix are the roots of the characteristic polynomial of this matrix. For an $N \times N$ matrix, N eigenvalues are associated with N N-component eigenvectors. The matrix

$$\begin{pmatrix} 1 & 1 \\ 1 & 1 \end{pmatrix} \tag{3.8.1}$$

has eigenvalues $\lambda = 0$ and $\lambda = 2$. Each of these eigenvalues generates a single eigenvector. The original matrix equation

$$\mathbf{A}|\lambda\rangle = \lambda|\lambda\rangle \tag{3.8.2}$$

generates a linearly dependent solution for the components a and b of $|\lambda\rangle$. For an eigenvalue λ, the characteristic matrix equation is

$$(\mathbf{A}-\lambda\mathbf{I})\,|\lambda\rangle=0 \tag{3.8.3}$$

$$\begin{gathered}\left[\begin{pmatrix}1 & 1\\ 1 & 1\end{pmatrix}-\begin{pmatrix}\lambda & 0\\ 0 & \lambda\end{pmatrix}\right]\begin{pmatrix}a\\ b\end{pmatrix}=0\\ \begin{pmatrix}1-\lambda & 1\\ 1 & 1-\lambda\end{pmatrix}\begin{pmatrix}a\\ b\end{pmatrix}=0\end{gathered} \tag{3.8.4}$$

To determine $|0\rangle$ for the eigenvalue $\lambda=0$, $\lambda=0$ is substituted into Equation 3.8.4 and the resultant equations are solved for a and b:

$$\begin{gathered}\begin{pmatrix}1 & 1\\ 1 & 1\end{pmatrix}\begin{pmatrix}a\\ b\end{pmatrix}=0\\ (1)a+(1)b=0\\ (1)a+(1)b=0\end{gathered} \tag{3.8.5}$$

The two equations are identical since a and b are linearly dependent. The coefficient b is determined with respect to a:

$$b=-a \tag{3.8.6}$$

Since a can have any value, it can be set equal to 1. Then, $b=-1$ and the eigenvector for $\lambda=0$ is

$$|0\rangle=\begin{pmatrix}1\\ -1\end{pmatrix} \tag{3.8.7}$$

This vector can be normalized to produce a unit vector in this coordinate direction:

$$|0\rangle=\left(\frac{1}{\sqrt{2}}\right)\begin{pmatrix}1\\ -1\end{pmatrix} \tag{3.8.8}$$

The second eigenvector for $\lambda=2$ is determined by substituting $\lambda=2$ into the matrix equation and solving for a and b:

$$\begin{pmatrix}1-2 & 1\\ 1 & 1-2\end{pmatrix}\begin{pmatrix}a\\ b\end{pmatrix}=0 \tag{3.8.9}$$

The two equations in a and b are again linearly dependent:

$$\begin{gathered}(-1)a+(1)b=0\\ (1)a+(-1)b=0\end{gathered} \tag{3.8.10}$$

The solution for b in terms of a is

$$b = a \tag{3.8.11}$$

and the eigenvector is

$$|2\rangle = \left(\frac{1}{\sqrt{2}}\right)\begin{pmatrix}1\\1\end{pmatrix} \tag{3.8.12}$$

The two eigenvectors for this symmetric matrix are orthogonal:

$$\langle 2|0\rangle = (1 \quad 1)\begin{pmatrix}1\\-1\end{pmatrix} = 0 \tag{3.8.13}$$

The 3×3 matrix

$$\begin{pmatrix}2 & 1 & 0\\1 & 2 & 1\\0 & 1 & 2\end{pmatrix} \tag{3.8.14}$$

will produce a three-component vector where two of the components can be written relative to the third component. The eigenvalues for this matrix were determined in Section 3.7. The eigenvalue $\lambda = 2$ gives an equation

$$\begin{gathered}\begin{pmatrix}2-2 & 1 & 0\\1 & 2-2 & 1\\0 & 1 & 2-2\end{pmatrix}\begin{pmatrix}a\\b\\c\end{pmatrix} = 0\\ \begin{pmatrix}0 & 1 & 0\\1 & 0 & 1\\0 & 1 & 0\end{pmatrix}\begin{pmatrix}a\\b\\c\end{pmatrix} = 0\end{gathered} \tag{3.8.15}$$

The first and last equations are

$$(1)b = 0 \tag{3.8.16}$$

so b must equal zero. The remaining two components appear only in the second equation,

$$(1)a + (1)c = 0 \tag{3.8.17}$$

so c is expressed relative to a:

$$c = -a \tag{3.8.18}$$

When $a = 1$, the unnormalized eigenvector is

$$|2\rangle = \begin{pmatrix}1\\0\\-1\end{pmatrix} \tag{3.8.19}$$

The eigenvector for $\lambda = 2 - \sqrt{2}$ is found when this eigenvalue is inserted into the matrix to give

$$\begin{pmatrix} \sqrt{2} & 1 & 0 \\ 1 & \sqrt{2} & 1 \\ 0 & 1 & \sqrt{2} \end{pmatrix} \begin{pmatrix} a \\ b \\ c \end{pmatrix} = 0 \tag{3.8.20}$$

The matrix equation generates the three equations

$$\begin{aligned} \sqrt{2}a + 1b &= 0 \\ 1a + \sqrt{2}b + 1c &= 0 \\ 1b + \sqrt{2}c &= 0 \end{aligned} \tag{3.8.21}$$

Using the first and third equations, a and c are determined in terms of b:

$$\begin{aligned} b &= -\sqrt{2}a \\ b &= -\sqrt{2}c \end{aligned} \tag{3.8.22}$$

Since both equations equal b,

$$a = c \tag{3.8.23}$$

Setting

$$a = c = 1 \tag{3.8.24}$$

and substituting them into the second of Equations 3.8.21 gives

$$1 + \sqrt{2}b + 1 = 0 \tag{3.8.25}$$

which yields

$$b = -2/\sqrt{2} = -\sqrt{2} \tag{3.8.26}$$

and the eigenvector

$$|2 - \sqrt{2}\rangle = \begin{pmatrix} 1 \\ -\sqrt{2} \\ 1 \end{pmatrix} \tag{3.8.27}$$

The process is repeated for $\lambda = 2 + \sqrt{2}$ to give

$$|2 + \sqrt{2}\rangle = \begin{pmatrix} 1 \\ \sqrt{2} \\ 1 \end{pmatrix} \tag{3.8.28}$$

This final eigenvector has components of the same sign. $|2\rangle$ has one component of zero and a single change of sign while $|2-\sqrt{2}\rangle$ changes sign between the first and second components and the second and third components.

The components of the eigenvectors can be determined as cofactors of any column of the matrix. For $\lambda=2$, the first column of the characteristic matrix of Equation 3.8.15 gives the components

$$\begin{vmatrix} 0 & 1 \\ 1 & 0 \end{vmatrix} = -1 \qquad \begin{vmatrix} 1 & 0 \\ 1 & 0 \end{vmatrix} = 0 \qquad \begin{vmatrix} 1 & 0 \\ 0 & 1 \end{vmatrix} = 1 \tag{3.8.29}$$

A nonsymmetric matrix has different row and column eigenvector components. The matrix

$$\begin{pmatrix} 1 & 2 \\ 1 & 2 \end{pmatrix} \tag{3.8.30}$$

has eigenvalues $\lambda=0$ and $\lambda=3$. The column eigenvector for $\lambda=0$ is determined from the matrix equation

$$\begin{pmatrix} 1-0 & 2 \\ 1 & 2-0 \end{pmatrix} \begin{pmatrix} a \\ b \end{pmatrix} = 0 \tag{3.8.31}$$

with solution

$$a = -2b \tag{3.8.32}$$

In this case, it is more convenient to select $b=1$ so that $a=-2$ and the eigenvector is

$$|0\rangle = \begin{pmatrix} -2 \\ 1 \end{pmatrix} \tag{3.8.33}$$

The row eigenvector for $\lambda=0$ is determined from

$$(a \quad b) \begin{pmatrix} 1-0 & 2 \\ 1 & 2-0 \end{pmatrix} = 0 \tag{3.8.34}$$

as

$$\langle 0| = (1 \quad -1)$$

The $\lambda=3$ eigenvectors are

$$\langle 3| = (1 \quad 2) \qquad |3\rangle = \begin{pmatrix} 1 \\ 1 \end{pmatrix} \tag{3.8.35}$$

Although the components for the row and column eigenvectors are different for this nonsymmetric matrix, these eigenvectors are biorthogonal:

$$\langle 0|3\rangle = (1 \quad -1)\begin{pmatrix}1\\1\end{pmatrix} = 0 \tag{3.8.36}$$

$$\langle 3|0\rangle = (1 \quad 2)\begin{pmatrix}-2\\1\end{pmatrix} = 0 \tag{3.8.37}$$

The Hermitian matrix

$$\begin{pmatrix}1 & i\\-i & 1\end{pmatrix} \tag{3.8.38}$$

has the characteristic polynomial

$$(1-\lambda)^2 - (i)(-i) = \lambda^2 - 2\lambda = 0 \tag{3.8.39}$$

with eigenvalues $\lambda = 0$ and $\lambda = 2$. Although some elements of the Hermitian matrix are imaginary, the eigenvalues are real. The eigenvalues for all Hermitian and symmetric matrices will always be real.

For $\lambda = 0$, the matrix equation is

$$\begin{pmatrix}1 & i\\-i & 1\end{pmatrix}\begin{pmatrix}a\\b\end{pmatrix} = 0 \tag{3.8.40}$$

The two solutions,

$$\begin{aligned} a + ib &= 0 \\ -ia + 1b &= 0 \end{aligned} \tag{3.8.41}$$

are identical. This is verified by multiplying the second equation by $-i$. The eigenvector is

$$|0\rangle = \begin{pmatrix}-i\\1\end{pmatrix} \tag{3.8.42}$$

The row eigenvector for $\lambda = 0$ is

$$\langle 0| = (i \quad 1) \tag{3.8.43}$$

The components of the row eigenvector are the complex conjugates of the corresponding components of the column eigenvector. The vector norm is real.

The row and column eigenvectors for $\lambda = 2$ are

$$\langle 2| = (-i \quad 1) \qquad |2\rangle = \begin{pmatrix}i\\1\end{pmatrix} \tag{3.8.44}$$

These eigenvectors are orthogonal in complex space:

$$\langle 0|2\rangle = (i \quad 1)\begin{pmatrix} i \\ 1 \end{pmatrix} = 0 \qquad \langle 2|0\rangle = (-i \quad 1)\begin{pmatrix} -i \\ 1 \end{pmatrix} \tag{3.8.45}$$

An Argand diagram for the two column eigenvectors appears in Figure 3.18.

An eigenvector selects a specific direction in the space. If the matrix operates on this eigenvector, it always generates a scalar which represents a scalar distance along this one direction in space. The eigenvectors are particularly important since the matrix can be applied an unlimited number of times. Each matrix operation returns the scalar and the eigenvector. If the matrix is applied M times,

$$\mathbf{A}^M|\lambda_i\rangle = \lambda_i^M|\lambda_i\rangle \tag{3.8.46}$$

For operations on function vectors, the eigenvalue and eigenvector are normally found simultaneously via a solution of a differential equation. For example, the Hamiltonian operator for a system which depends only on the polar angle, θ, is

$$\mathbf{H} = (-\hbar^2/2m)(1/\sin(\theta)\ d/d\theta(\sin\theta\ d/d\theta) \tag{3.8.47}$$

The Schrödinger equation for a wavefunction $\psi(\theta)$ is

$$\mathbf{H}\psi(\theta) = E\psi(\theta) \tag{3.8.48}$$

$$(-\hbar^2/2m)(1/\sin\theta)\ d/d\theta(\sin\theta\ d/d\theta)\ \psi(\theta) = E\psi(\theta) \tag{3.8.49}$$

The equation is similar to the standard differential equation for Legendre polynomials,

$$(1/\sin\theta)\ d/d\theta(\sin\theta\ d/d\theta)\ P_l(\theta) + l(l+1)\ P_l(\theta) = 0 \tag{3.8.50}$$

where l must be an integer to generate the Legendre functions. The Legendre equation has the form of an eigenvalue–eigenvector equation,

$$(1/\sin\theta)\ d/d\theta(\sin\theta\ d/d\theta)\ P_l(\theta) = -l(l+1)\ P_l(\theta) \tag{3.8.51}$$

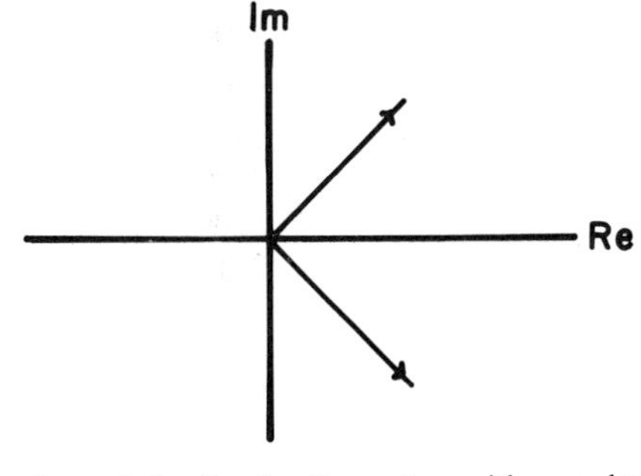

Figure 3.18. Argand diagram for the vectors with complex components.

and the energy for the Schrödinger equation is found by multiplying both sides of Equation 3.8.51 by $(-\hbar^2/2m)$:

$$(-\hbar^2/2m)(1/\sin\theta)\, d/d\theta(\sin\theta\, d/d\theta)\, P_l(\theta) = (\hbar^2/2m)[l(l+1)]\, P_l(\theta) = EP_l(\theta) \tag{3.8.52}$$

The scalar energy for the eigenfunction $P_l(\theta)$ must then be

$$E = \hbar^2[l(l+1)]/2m \tag{3.8.53}$$

3.9. Properties of the Characteristic Polynomial

The characteristic polynomial, $P(\lambda)$, for an $N \times N$ matrix is

$$\lambda^N + a_{N-1}\lambda^{N-1} + \cdots + a_1\lambda^1 + a_0 = 0 \tag{3.9.1}$$

Since each eigenvalue λ_i is a root of this equation, the entire scalar polynomial $P(\lambda_i)$ can operate on the eigenvector $|i\rangle$ and the result must still be zero since $P(\lambda_i) = 0$:

$$P(\lambda_i)\,|i\rangle = 0 \tag{3.9.2}$$

The jth term in the polynomial is

$$a_j\lambda_i^j\,|i\rangle \tag{3.9.3}$$

Since

$$\mathbf{A}\,|i\rangle = \lambda_i\,|i\rangle \tag{3.9.4}$$

and

$$\mathbf{A}^j\,|i\rangle = \lambda_i^j\,|i\rangle \tag{3.9.5}$$

each λ_i in Equation 3.9.2 is replaced by the matrix $\mathbf{A}$ to give a polynomial in powers of $\mathbf{A}$:

$$P(\mathbf{A})|i\rangle = (\mathbf{A}^N + a_{N-1}\mathbf{A}^{N-1} + \cdots + a_1\mathbf{A} + a_0\mathbf{I})|i\rangle = 0 \tag{3.9.6}$$

Since $|i\rangle \neq 0$, the polynomial must be zero:

$$P(\mathbf{A}) = 0 \tag{3.9.7}$$

This is the Hamilton–Cayley theorem. A matrix satisfies its own characteristic equation.

The Hamilton–Cayley theorem can be verified with the 2×2 matrix

$$\begin{pmatrix} 2 & 1 \\ 1 & 2 \end{pmatrix} \tag{3.9.8}$$

which has characteristic polynomial

$$\lambda^2 - 4\lambda + 3 = 0 \tag{3.9.9}$$

The matrix now replaces λ in this polynomial,

$$\mathbf{A}^2 - 4\mathbf{A} + 3\mathbf{I} = 0 \tag{3.9.10}$$

and

$$\begin{aligned} &\begin{pmatrix} 2 & 1 \\ 1 & 2 \end{pmatrix}\begin{pmatrix} 2 & 1 \\ 1 & 2 \end{pmatrix} - 4\begin{pmatrix} 2 & 1 \\ 1 & 2 \end{pmatrix} + 3\begin{pmatrix} 1 & 0 \\ 0 & 1 \end{pmatrix} \\ &\quad = \begin{pmatrix} 5 & 4 \\ 4 & 5 \end{pmatrix} - \begin{pmatrix} 8 & 4 \\ 4 & 8 \end{pmatrix} + \begin{pmatrix} 3 & 0 \\ 0 & 3 \end{pmatrix} = 0 = \begin{pmatrix} 0 & 0 \\ 0 & 0 \end{pmatrix} \end{aligned} \tag{3.9.11}$$

The Hamilton–Cayley theorem is valid for both symmetric and nonsymmetric matrices.

The Hamilton–Cayley theorem can generate matrix powers and polynomials. For example, the theorem generates inverses. The nonsingular 3×3 matrix

$$\begin{pmatrix} 2 & 1 & 0 \\ 1 & 2 & 1 \\ 0 & 1 & 2 \end{pmatrix} \tag{3.9.12}$$

has a characteristic polynomial

$$4 - 10\lambda + 6\lambda^2 - \lambda^3 = 0 \tag{3.9.13}$$

By the Hamilton–Cayley theorem,

$$4\mathbf{I} - 10\mathbf{A} + 6\mathbf{A}^2 - \mathbf{A}^3 = 0 \tag{3.9.14}$$

Applying $\mathbf{A}^{-1}$ to each term gives

$$\mathbf{A}^{-1}[4\mathbf{I} - 10\mathbf{A} + 6\mathbf{A}^2 - \mathbf{A}^3] = 4\mathbf{A}^{-1} - 10\mathbf{I} + 6\mathbf{A} - \mathbf{A}^2 = 0 \tag{3.9.15}$$

which is solved for $\mathbf{A}^{-1}$:

$$\mathbf{A}^{-1} = \frac{1}{4}(10\mathbf{I} - 6\mathbf{A} + \mathbf{A}^2) \tag{3.9.16}$$

The matrix inverse is now found as a sum of matrix products and scalars. Substituting the matrices $\mathbf{A}$ (Equation 3.9.12) and $\mathbf{I}$ gives

$$\left(\frac{1}{4}\right)\left(\begin{pmatrix}10&0&0\\0&10&0\\0&0&10\end{pmatrix}-6\begin{pmatrix}2&1&0\\1&2&1\\0&1&2\end{pmatrix}+\begin{pmatrix}2&1&0\\1&2&1\\0&1&2\end{pmatrix}\begin{pmatrix}2&1&0\\1&2&1\\0&1&2\end{pmatrix}\right)$$

$$=\left(\frac{1}{4}\right)\left(\begin{pmatrix}10&0&0\\0&10&0\\0&0&10\end{pmatrix}-\begin{pmatrix}12&6&0\\6&12&6\\0&6&12\end{pmatrix}+\begin{pmatrix}5&4&1\\4&6&4\\1&4&5\end{pmatrix}\right)=\left(\frac{1}{4}\right)\begin{pmatrix}3&-2&1\\-2&4&-2\\1&-2&3\end{pmatrix}$$

(3.9.17)

The Hamilton–Cayley theorem cannot be used to find inverses for all matrices. The matrix

$$\begin{pmatrix}1&-1\\-1&1\end{pmatrix} \tag{3.9.18}$$

cannot have an inverse since its determinant is zero, i.e., it is singular. If the Hamilton–Cayley theorem were used, the characteristic polynomial in **A**,

$$-2\mathbf{A}+\mathbf{A}^2=0 \tag{3.9.19}$$

might be multiplied by $\mathbf{A}^{-2}$ to give

$$\mathbf{A}^{-1}=\frac{1}{2}\mathbf{I}=\frac{1}{2} \tag{3.9.20}$$

A $\lambda=0$ eigenvector leaves a matrix of rank 1, and $\frac{1}{2}$ is the inverse of the remaining eigenvalue $\lambda=2$.

The 3×3 matrix of Equation 3.9.12 and its characteristic polynomial (Equation 3.9.13) are closely associated with the eigenvalues of the matrix. Since the roots of the polynomial,

$$\lambda_1=2-\sqrt{2}, \qquad \lambda_2=2, \qquad \lambda_3=2+\sqrt{2} \tag{3.9.21}$$

are the eigenvalues of the matrix, the characteristic polynomial can also be written

$$P(\lambda)=(\lambda_1-\lambda)(\lambda_2-\lambda)(\lambda_3-\lambda)=0 \tag{3.9.22}$$

This is also the characteristic polynomial for a diagonal matrix with elements $\lambda_i-\lambda$,

$$\begin{pmatrix}\lambda_1-\lambda&0&0\\0&\lambda_2-\lambda&0\\0&0&\lambda_3-\lambda\end{pmatrix} \tag{3.9.23}$$

with characteristic determinant

$$\begin{vmatrix} \lambda_1-\lambda & 0 & 0 \\ 0 & \lambda_2-\lambda & 0 \\ 0 & 0 & \lambda_3-\lambda \end{vmatrix} \tag{3.9.24}$$

The matrices

$$\mathbf{A}=\begin{pmatrix} 2 & 1 & 0 \\ 1 & 2 & 1 \\ 0 & 1 & 2 \end{pmatrix} \qquad \mathbf{D}=\begin{pmatrix} 2-\sqrt{2} & 0 & 0 \\ 0 & 2 & 0 \\ 0 & 0 & 2+\sqrt{2} \end{pmatrix} \tag{3.9.25}$$

both generate the characteristic polynomial

$$P(\lambda)=0=4-10\lambda+6\lambda^2+\lambda^3 \tag{3.9.26}$$

The sum of the diagonal elements for each matrix is 6. This number equals the $a_{N-1}=a_2$ coefficient in the characteristic polynomial as well. The sum of the diagonal elements is called the trace of the matrix. All matrices which give the same characteristic polynomial will have the same trace and it is called an invariant of the matrix. The trace will always equal the a_{N-1} element of the characteristic polynomial.

The determinant of the matrix is a second invariant since the determinant of **A** is

$$\text{Det}=2\begin{vmatrix} 2 & 1 \\ 1 & 2 \end{vmatrix}-1\begin{vmatrix} 1 & 1 \\ 0 & 2 \end{vmatrix}=2(3)-1(2)=4 \tag{3.9.27}$$

and the determinant of **D** is just the product of its diagonal elements:

$$\text{Det}=(2-\sqrt{2})(2)(2+\sqrt{2})=(2)(4-2)=4 \tag{3.9.28}$$

The a_0 term in the characteristic polynomial is also 4.

The final coefficient in the characteristic equation, a_1, lies between the trace, which sums single elements along the diagonal, and the determinant, which selects all combinations of three elements from separate rows and columns in the matrix. The a_1 element must consider elements in pairs. For the diagonal matrix, three pairs are possible,

$$1\&2 \qquad 1\&3 \qquad 2\&3 \tag{3.9.29}$$

The final invariant for the 3×3 diagonal matrix is formed by summing these three pair products:

$$(2-\sqrt{2})(2)+(2-\sqrt{2})(2+\sqrt{2})+(2)(2+\sqrt{2})=10 \tag{3.9.30}$$

To obtain this invariant from **A**, all 2×2 determinants formed by eliminating the row and column for each diagonal element are evaluated and summed. The three 2×2 determinants are

$$\begin{vmatrix} 2 & 1 \\ 1 & 2 \end{vmatrix} \quad \begin{vmatrix} 2 & 1 \\ 1 & 2 \end{vmatrix} \quad \begin{vmatrix} 2 & 0 \\ 0 & 2 \end{vmatrix} \tag{3.9.31}$$

and their sum is

$$3+3+4=10 \tag{3.9.32}$$

the third invariant for the 3×3 matrix. The trace is formed by selecting and summing all 1×1 elements along the diagonal, a_1 is formed by selecting all 2×2 determinants along the diagonal, and a_0 is the single 3×3 determinant possible for the 3×3 matrix.

The 2×2 determinant sum gives the value of a_1 but not its sign. However, the signs do alternate for matrices with positive eigenvalues. A matrix with all negative eigenvalues will produce a characteristic equation whose coefficients have the same sign.

For an $N \times N$ matrix, there are N invariants beginning with the trace and continuing through 2×2, 3×3, etc. determinants about the diagonal. The final invariant is the $N \times N$ determinant for the full matrix. For example, the matrix

$$\begin{pmatrix} 1 & 0 & 0 & 0 \\ -1 & 2 & 0 & 0 \\ 0 & -2 & 3 & 0 \\ 0 & 0 & -3 & 4 \end{pmatrix} \tag{3.9.33}$$

has a characteristic determinant which can be evaluated directly to give the characteristic polynomial,

$$24-50\lambda+35\lambda^2-10\lambda^3+\lambda^4 \tag{3.9.34}$$

The trace of the matrix gives $a_3=10$. There are four 2×2 determinants generated by eliminating the rows and columns for two diagonal elements. The 2×2 determinants and the rows and columns which generate them are

$$\begin{vmatrix} 1 & 0 \\ -1 & 2 \end{vmatrix} \quad \begin{vmatrix} 1 & 0 \\ 0 & 3 \end{vmatrix} \quad \begin{vmatrix} 1 & 0 \\ 0 & 4 \end{vmatrix} \quad \begin{vmatrix} 2 & 0 \\ -2 & 3 \end{vmatrix} \quad \begin{vmatrix} 2 & 0 \\ 0 & 4 \end{vmatrix} \quad \begin{vmatrix} 3 & 0 \\ -3 & 4 \end{vmatrix}$$

$$1\&2 \qquad 1\&3 \qquad 1\&4 \qquad 2\&3 \qquad 2\&4 \qquad 3\&4$$

These determinants have the sum

$$2+3+4+6+8+12=35 \tag{3.9.35}$$

The 3×3 determinants are formed by eliminating the row and column associated with each diagonal element:

$$\begin{vmatrix} 1 & 0 & 0 \\ -1 & 2 & 0 \\ 0 & -2 & 3 \end{vmatrix} \quad \begin{vmatrix} 2 & 0 & 0 \\ -2 & 3 & 0 \\ 0 & -3 & 4 \end{vmatrix} \quad \begin{vmatrix} 1 & 0 & 0 \\ 0 & 3 & 0 \\ 0 & -3 & 4 \end{vmatrix} \quad \begin{vmatrix} 1 & 0 & 0 \\ -1 & 2 & 0 \\ 0 & 0 & 4 \end{vmatrix}$$

$$1\&2\&3 \qquad 2\&3\&4 \qquad 1\&3\&4 \qquad 1\&2\&4$$

The 3×3 determinants have the sum

$$6 + 24 + 12 + 8 = 50 \tag{3.9.36}$$

In general, the number of $i \times i$ matrices for an $N \times N$ matrix will be

$$N!/i!(N-i)! \tag{3.9.37}$$

Matrices which have no negative eigenvalues are called positive-definite matrices. This property can be established by examining the determinants formed from the single element in the first row and column, a_{11}, the determinant formed from the four elements in the first two rows and columns, etc. The final determinant is the $N \times N$ determinant using all the elements of the matrix. If all these determinants are positive, then the matrix is positive definite. The matrix

$$\begin{pmatrix} 2 & 1 & 0 \\ 1 & 2 & 1 \\ 0 & 1 & 2 \end{pmatrix} \tag{3.9.38}$$

has three determinants,

$$|2| = 2 \qquad \begin{vmatrix} 2 & 1 \\ 1 & 2 \end{vmatrix} = 3 \qquad \begin{vmatrix} 2 & 1 & 0 \\ 1 & 2 & 1 \\ 0 & 1 & 2 \end{vmatrix} = 4 \tag{3.9.39}$$

and is positive definite.

3.10. Alternate Techniques for Eigenvalue and Eigenvector Determination

As the matrix order increases, determination of the characteristic polynomial and its roots becomes more difficult. For such matrices, it is often convenient to estimate the eigenvalues. The 2×2 matrix

$$\begin{pmatrix} 6 & 1 \\ 1 & 1 \end{pmatrix} \tag{3.10.1}$$

has eigenvalues $\lambda = 0.8$ and $\lambda = 6.2$, which are close to the values of the diagonal elements. Gersgorin's theorem states that the eigenvalues are all located within circles formed using the diagonal elements a_{ii} (6 and 1 in Equation 3.10.1) as centers. The radii of each of these circles are determined by summing the absolute values of all the remaining elements in the row (or column). For the 2×2 matrix, the remaining element is 1 in each row. Thus, there are two circles of radius 1 centered at 1 and 6 as illustrated in Figure 3.19. These circles are drawn in the complex plane. Since the matrix is symmetric, the two eigenvalues will be real and are located on the real axis in the intervals $0 < \lambda < 2$ and $5 < \lambda < 7$. The exact eigenvalues, 0.8 and 6.2, fall in these intervals.

For larger matrices, one circle may contain several eigenvalues. A given circle will contain only one eigenvalue if the absolute values of the differences for its center (a_{ii}) and the other diagonal elements (a_{jj}) are all greater than the sum of the absolute values of nondiagonal elements (a_{ik}, a_{jk}) for each of these rows. In the 2×2 matrix, the difference in diagonal elements is

$$|6 - 1| = 5 \tag{3.10.2}$$

while the sum of nondiagonal elements for each row is

$$|1| + |1| = 2 \tag{3.10.3}$$

Since the difference of diagonals is larger than the sum of nondiagonal elements, the eigenvalues must appear in separate circles.

Gersgorin's theorem is most useful for larger matrices where it is more difficult to calculate the exact eigenvalues as roots of the characteristic polynomial. The symmetric 4×4 matrix

$$\begin{pmatrix} 5 & 0.1 & 0 & -0.1 \\ 0.1 & 7 & 0.5 & 0 \\ 0 & 0.5 & 3 & -0.2 \\ -0.1 & 0 & -0.2 & 9 \end{pmatrix} \tag{3.10.4}$$

generates circles with the following (center, radius) combinations: (5, 0.2), (7, 0.6), (3, 0.7) and (9, 0.3). These are illustrated in Figure 3.20. No circles overlap but it is still necessary to determine which of these circles contain only

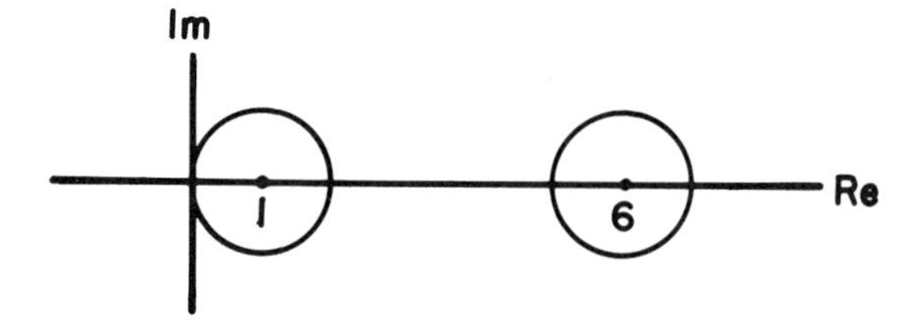

Figure 3.19. Circles containing matrix (Equation 3.10.1) eigenvalues as defined by Gersgorin's theorem.

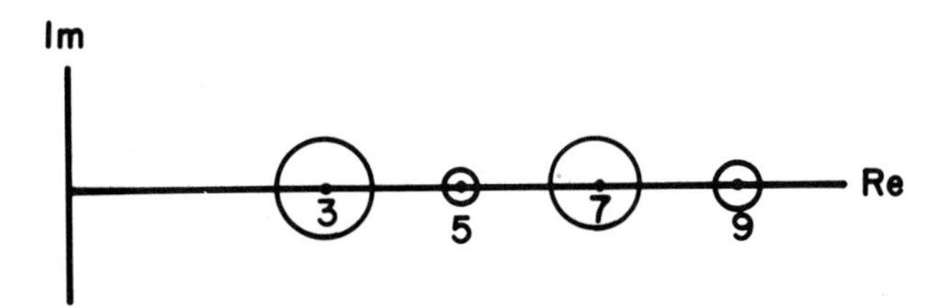

Figure 3.20. Overlapping Gersgorin circles containing the eigenvalues for the matrix of Equation 3.10.4.

one eigenvalue. The circle centered at 5 will be examined; the remainder are left as a problem (3.6). The elements of the first row must then be compared with those of each of the remaining rows via three equations:

$$\begin{aligned} |5-7| &= 2 > \{|0.1| + |-0.1|\} + \{|0.1| + |0.5|\} = 0.8 \\ |5-3| &= 2 > \{|0.1| + |-0.1|\} + \{|0.5| + |-0.2|\} = 0.9 \\ |5-9| &= 4 > \{|0.1| + |-0.1|\} + \{|-0.1| + |-0.2|\} = 0.5 \end{aligned} \tag{3.10.5}$$

There is a single eigenvalue in the interval $4.8 < \lambda < 5.2$.

The largest eigenvalue and its eigenvector can be determined via repeated application of the matrix to an arbitrary vector. If **A** operates on an eigenvector M times, the eigenvalue will be generated M times, i.e.,

$$\mathbf{A}^M |i\rangle = \lambda_i^M |i\rangle \tag{3.10.6}$$

If the same operation is performed on an arbitrary vector, this vector can be resolved into components on the eigenvectors of the system, i.e.,

$$|v\rangle = \sum c_i |i\rangle \tag{3.10.7}$$

When $\mathbf{A}^M$ operates on $|v\rangle$, it operates on the sum of eigenvectors. Each term in the sum will now be multiplied by its eigenvalue raised to the Mth power:

$$\mathbf{A}^M |v\rangle = \sum c_i \lambda_i^M |i\rangle \tag{3.10.8}$$

If M is large enough, the term containing the largest λ_i will dominate the sum.

The largest eigenvalue of the matrix

$$\begin{pmatrix} 2 & 1 & 1 \\ 1 & 2 & 1 \\ 1 & 1 & 2 \end{pmatrix} \tag{3.10.9}$$

is found by applying it to the arbitrary vector

$$\begin{pmatrix} 1 \\ 0 \\ 0 \end{pmatrix} \tag{3.10.10}$$

since the operations are simple. The first, second, and third operations give the vectors

$$\begin{pmatrix} 2 \\ 1 \\ 1 \end{pmatrix} \quad \begin{pmatrix} 6 \\ 5 \\ 5 \end{pmatrix} \quad \begin{pmatrix} 22 \\ 21 \\ 21 \end{pmatrix} \tag{3.10.11}$$

Since the components of the vectors are relative values, all three components of the third vector can be divided by 21 to give the vector

$$\begin{pmatrix} 1.0476 \\ 1 \\ 1 \end{pmatrix} \tag{3.10.12}$$

An additional **A** operation on this vector gives both the approximate eigenvalue and eigenvector:

$$\begin{pmatrix} 1.0476 \\ 1 \\ 1 \end{pmatrix} \quad \begin{pmatrix} 4.0952 \\ 4.0476 \\ 4.0476 \end{pmatrix} = 4 \begin{pmatrix} 1.0118 \\ 1.0000 \\ 1.0000 \end{pmatrix} \tag{3.10.13}$$

This result predicts an eigenvalue of 4 and an eigenvector with equal components. This can be verified by operating on this limit with **A**.

The largest eigenvalue and its eigenvector are eliminated from the matrix by deflation, i.e., by subtracting the product of the eigenvalue and its projection operator. For the vector $|1\rangle$, the projection operator (Chapter 4) is

$$|1\rangle\langle 1| = \left(\frac{1}{3}\right) \begin{pmatrix} 1 \\ 1 \\ 1 \end{pmatrix} (1 \quad 1 \quad 1) \tag{3.10.14}$$

The first component of the column vector is multiplied by each component of the row vector to produce the first row of a matrix; the second column component multiplied by each row vector component generates the second row, and the third column vector component generates the third row, i.e.,

$$|1\rangle\langle 1| = \left(\frac{1}{3}\right) \begin{pmatrix} 1 & 1 & 1 \\ 1 & 1 & 1 \\ 1 & 1 & 1 \end{pmatrix} \tag{3.10.15}$$

This matrix is multiplied by the eigenvalue (4) and subtracted from the original matrix to produce a matrix which no longer contains the maximal eigenvalue (its rank is reduced by 1):

$$\begin{pmatrix} 2 & 1 & 1 \\ 1 & 2 & 1 \\ 1 & 1 & 2 \end{pmatrix} - \left(\frac{4}{3}\right) \begin{pmatrix} 1 & 1 & 1 \\ 1 & 1 & 1 \\ 1 & 1 & 1 \end{pmatrix} = \left(\frac{1}{3}\right) \begin{pmatrix} 2 & -1 & -1 \\ -1 & 2 & -1 \\ -1 & -1 & 2 \end{pmatrix} \tag{3.10.16}$$

If the vector $|1\rangle$ now operates on the new matrix, $\mathbf{A}'$, it will give an eigenvalue of 0. Eigenvalue 4 has been eliminated.

The new matrix can now operate on an arbitrary vector to produce the next largest eigenvalue and eigenvector. $\mathbf{A}'$ operates on a vector with components (1, 0, 0) to produce the following sequence of vectors:

$$\begin{pmatrix} 1 \\ 0 \\ 0 \end{pmatrix} \quad \begin{pmatrix} 2 \\ -1 \\ -1 \end{pmatrix} \quad \begin{pmatrix} 6 \\ -3 \\ -3 \end{pmatrix} \tag{3.10.17}$$

The first and second operations produce the same relative components. The eigenvector

$$\begin{pmatrix} 2 \\ -1 \\ -1 \end{pmatrix} \tag{3.10.18}$$

is verified by operating with the original matrix $\mathbf{A}$:

$$\begin{pmatrix} 2 & 1 & 1 \\ 1 & 2 & 1 \\ 1 & 1 & 2 \end{pmatrix} \begin{pmatrix} 2 \\ -1 \\ -1 \end{pmatrix} = (1) \begin{pmatrix} 2 \\ -1 \\ -1 \end{pmatrix} \tag{3.10.19}$$

The second eigenvalue is 1. Since the trace of all three eigenvalues is 6, the final eigenvalue must be

$$6 - 4 - 1 = 1 \tag{3.10.20}$$

The matrix will have two degenerate, i.e., identical, eigenvalues.

Problems

3.1. Determine the cofactors and matrix inverse of the matrix

$$\begin{pmatrix} 2 & 1 & 0 \\ 1 & 2 & 1 \\ 0 & 1 & 2 \end{pmatrix}$$

3.2. Find the inverse of a 2×2 matrix

$$\begin{pmatrix} a_{11} & a_{12} \\ a_{21} & a_{22} \end{pmatrix}$$

using cofactors and show that

$$\mathbf{A}^{-1}\mathbf{A} = \mathbf{A}\mathbf{A}^{-1}$$

3.3. Determine the inverse of the nonsymmetric matrix

$$\begin{pmatrix} 2 & 2 & 0 \\ 1 & 2 & 2 \\ 0 & 1 & 2 \end{pmatrix}$$

using cofactors and using the Hamilton–Cayley theorem.

3.4. Show that the normalized eigenvectors

$$\begin{pmatrix} 1 \\ 0 \\ -1 \end{pmatrix} \quad \begin{pmatrix} 1 \\ -\sqrt{2} \\ 1 \end{pmatrix} \quad \begin{pmatrix} 1 \\ \sqrt{2} \\ 1 \end{pmatrix}$$

are orthogonal.

3.5. The matrix

$$\begin{pmatrix} 2 & 1 & 0 \\ 1 & 2 & 1 \\ 0 & 1 & 2 \end{pmatrix}$$

can be deflated by subtracting the product of the eigenvalue $\lambda_i = 2$ and its eigenvector projector $|i\rangle\langle i|$.

a. Evaluate the deflated matrix.
b. Show that the deflated matrix has rank 2.

3.6. Show that each of the eigenvalues for the matrix of Equation 3.10.4 lies within a distinct Gersgorin circle.

3.7. Determine the final vector which results when the column vector with components $(1, 0, 0)$ (the x axis) is rotated through the Eulerian angles $\varphi = 90°$, $\theta = 180°$, and $\psi = 45°$.

3.8. Evaluate the determinant of the matrix

$$\begin{vmatrix} \psi_1(1) & \psi_2(1) & \psi_3(1) \\ \psi_1(2) & \psi_2(2) & \psi_3(2) \\ \psi_1(3) & \psi_2(3) & \psi_3(3) \end{vmatrix}$$

and show that the final result will change sign if two of the electrons are interchanged.

3.9. Show that

$$\begin{pmatrix} \cos\theta & \sin\theta & 0 \\ -\sin\theta & \cos\theta & 0 \\ 0 & 0 & 1 \end{pmatrix} \quad \text{and} \quad \begin{pmatrix} \cos\theta & -\sin\theta & 0 \\ \sin\theta & \cos\theta & 0 \\ 0 & 0 & 1 \end{pmatrix}$$

are inverse to each other.

3.10. Determine the eigenvalues and eigenvectors of the matrix

$$\begin{pmatrix} 2 & 1 & 0 \\ 3 & -1 & 1 \\ 0 & 1 & 2 \end{pmatrix}$$

Determine the characteristic equation for this matrix using its invariants.

3.11. Show that a matrix in the Frobenius normal form,

$$\begin{pmatrix} a_{N-1} & a_{N-2} & \cdots & a_1 & a_0 \\ 1 & 0 & \cdots & 0 & 0 \\ 0 & 1 & \cdots & 0 & 0 \\ & & \vdots & & \\ 0 & 0 & \cdots & 1 & 0 \end{pmatrix}$$

gives the characteristic polynomial

$$P(\lambda) = \lambda^N + a_{N-1}\lambda^{N-1} + \cdots + a_1\lambda^1 + a_0 = 0$$

Use a 4×4 matrix.

3.12. Two matrices have elements which are functions of a variable x:

$$\begin{pmatrix} x & 2x^2 \\ 2x^2 & 2x \end{pmatrix} \quad \begin{pmatrix} x^3 & x \\ x & 3x \end{pmatrix}$$

Show that the two matrices satisfy the equality

$$d(\mathbf{AB})/dx = (d\mathbf{A}/dx)\,\mathbf{B} + \mathbf{A}(d\mathbf{B}/dx)$$

3.13. Solve the two simultaneous equations

$$2x + 3y = 16$$

$$x - 2y = 10$$

using the matrix inverse.

3.14. The inertia tensor is often expressed in the form

$$\mathscr{I} = r^2\mathbf{I} - |r\rangle\langle r|$$

where $|r\rangle$ has the components x and y and $r^2 = x^2 + y^2$. Show that this is equivalent to

$$\begin{pmatrix} xx & -xy \\ -yx & yy \end{pmatrix}$$

in two dimensions. Use the same vector equation to determine $\mathscr{I}$ for three dimensions.

3.15. Schur's inequality states that the sum of the squares of all the eigenvalues of a matrix is less than or equal to the sum of the squares of the absolute values of all elements in the matrix, i.e.,

$$\sum |\lambda_i|^2 \leqslant \sum |a_{jk}|^2$$

The inequality becomes an equality for real symmetric, unitary, and Hermitian matrices. Matrices which satisfy the equality are normal matrices. Show that the matrix

$$\begin{pmatrix} 2 & 1 & 0 \\ 1 & 2 & 1 \\ 0 & 1 & 2 \end{pmatrix}$$

is normal using the known eigenvalues for this matrix.

3.16. Show $\mathbf{R}^{-1}$ for the $\mathbf{R}$ rotation of Equation 3.5.11 is equal to the transpose of $\mathbf{R}$.

4

Similarity Transforms and Projections

4.1. The Similarity Transform

When a matrix operates on a vector, it normally generates a new vector in the space. However, if the vector chosen is an eigenvector, operation with the matrix will regenerate the original eigenvector and a scalar factor, the eigenvalue:

$$\mathbf{A}\,|i\rangle = \lambda_i\,|i\rangle \tag{4.1.1}$$

The spatial direction of an eigenvector is not altered by operations with the matrix **A**. The operation increases the length of the vector by the factor λ_i. The selection of a single eigenvector confines operations to one direction in the N-dimensional space.

The spatial properties of eigenvectors are illustrated for the 2×2 matrix

$$\begin{pmatrix} -1 & 1 \\ 1 & -1 \end{pmatrix} \tag{4.1.2}$$

with eigenvectors

$$|0\rangle = \begin{pmatrix} \dfrac{1}{\sqrt{2}} \\ \dfrac{1}{\sqrt{2}} \end{pmatrix} \qquad |-2\rangle = \begin{pmatrix} \dfrac{-1}{\sqrt{2}} \\ \dfrac{1}{\sqrt{2}} \end{pmatrix} \tag{4.1.3}$$

Since the x and y components for $|0\rangle$ are equal, the eigenvector is oriented $+45^\circ$ to the x axis. The second vector is rotated 135° from x or 45° from y. The rotation of the x and y components of any vector $|v\rangle$ through -45° requires a rotation matrix:

$$\begin{pmatrix} \cos 45^\circ & \sin 45^\circ \\ -\sin 45^\circ & \cos 45^\circ \end{pmatrix} = \begin{pmatrix} \dfrac{1}{\sqrt{2}} & \dfrac{1}{\sqrt{2}} \\ \dfrac{-1}{\sqrt{2}} & \dfrac{1}{\sqrt{2}} \end{pmatrix} \tag{4.1.4}$$

The elements in the first row of the rotation matrix are the components of $|0\rangle$ while the second-row elements are the components of $|-2\rangle$.

If the matrix **A** operates on $|v\rangle$ to give a new vector $|u\rangle$,

$$|u\rangle = \mathbf{A}\,|v\rangle \tag{4.1.5}$$

a rotation of $|v\rangle$ through $-45°$ requires a concomitant rotation of $|u\rangle$. If the rotation matrix of Equation 4.1.4 is designated $\mathbf{S}^{-1}$, the new vectors $|v'\rangle$ and $|u'\rangle$ are

$$\begin{aligned} |v'\rangle &= \mathbf{S}^{-1}\,|v\rangle \\ |u'\rangle &= \mathbf{S}^{-1}\,|u\rangle \end{aligned} \tag{4.1.6}$$

The rotated vectors will now require a new rotation matrix $\mathbf{A}'$ for consistency. This matrix is deduced from **A** by defining a second matrix **S** (an inverse) such that

$$\mathbf{S}\mathbf{S}^{-1} = \mathbf{I} \tag{4.1.7}$$

$\mathbf{I} = \mathbf{S}\mathbf{S}^{-1}$ is inserted between **A** and $|v\rangle$ without altering the result of the operation:

$$|u\rangle = \mathbf{A}\mathbf{I}\,|v\rangle = \mathbf{A}\mathbf{S}\mathbf{S}^{-1}\,|v\rangle \tag{4.1.8}$$

The equation can be converted to an equation for $|u'\rangle$ by operating on both sides with $\mathbf{S}^{-1}$ from the left:

$$|u'\rangle = \mathbf{S}^{-1}\,|u\rangle = \mathbf{S}^{-1}\mathbf{A}\mathbf{S}\mathbf{S}^{-1}\,|v\rangle = (\mathbf{S}^{-1}\mathbf{A}\mathbf{S})(\mathbf{S}^{-1}\,|v\rangle) = \mathbf{A}'\,|v'\rangle \tag{4.1.9}$$

The operation

$$\mathbf{A}' = \mathbf{S}^{-1}\mathbf{A}\mathbf{S} \tag{4.1.10}$$

is called a similarity transform.

For the 2×2 matrices, a rotation of $45°$ gave an $\mathbf{S}^{-1}$ with the components of each eigenvector in successive rows. The **S** matrix must have these components in successive columns,

$$\mathbf{S} = (|0\rangle \quad |-2\rangle) = \left(\frac{1}{\sqrt{2}}\right)\begin{pmatrix} 1 & -1 \\ 1 & 1 \end{pmatrix} \tag{4.1.11}$$

to satisfy $\mathbf{S}\mathbf{S}^{-1} = \mathbf{I}$:

$$\left(\frac{1}{2}\right)\begin{pmatrix} 1 & -1 \\ 1 & 1 \end{pmatrix}\begin{pmatrix} 1 & 1 \\ -1 & 1 \end{pmatrix} = \begin{pmatrix} 1 & 0 \\ 0 & 1 \end{pmatrix} \tag{4.1.12}$$

When the eigenvectors generate the rotation matrices **S** and $\mathbf{S}^{-1}$, the similarity transform of **A** gives a diagonal matrix:

$$\mathbf{A}' = \mathbf{S}^{-1}\mathbf{A}\mathbf{S}$$

$$= \begin{pmatrix}1\\ \overline{2}\end{pmatrix}\begin{pmatrix}1 & 1\\ -1 & 1\end{pmatrix}\begin{pmatrix}-1 & 1\\ 1 & -1\end{pmatrix}\begin{pmatrix}1 & -1\\ 1 & 1\end{pmatrix} = \begin{pmatrix}0 & 0\\ 0 & -2\end{pmatrix} \tag{4.1.13}$$

The diagonal matrix created in this manner is a manifestation of the "one-dimensional" nature of each eigenvector.

The similarity transform matrices are more general than the rotation matrices since they are applicable to a space of dimension N. The N-component eigenvectors for an $N \times N$ matrix **A** are placed in successive columns to generate **S**:

$$(|1\rangle \quad |2\rangle \quad \ldots \quad |N\rangle) \tag{4.1.14}$$

The $\mathbf{S}^{-1}$ matrix has bra eigenvectors in successive rows:

$$\begin{pmatrix}\langle 1|\\ \langle 2|\\ \vdots\\ \langle N|\end{pmatrix} \tag{4.1.15}$$

The row and column notation clearly shows the relationship between diagonalization and the eigenvector property (Equation 4.1.1). The **A** matrix will operate on each column of the **S** vector to produce a new set of components for that column. The operation

$$\mathbf{A}(|1\rangle \quad |2\rangle \quad \ldots \quad |N\rangle) \tag{4.1.16}$$

gives

$$(\mathbf{A}\,|1\rangle \quad \mathbf{A}\,|2\rangle \quad \ldots \quad |\mathbf{A}\,|N\rangle) \tag{4.1.17}$$

for the components of the matrix. Application of the $\mathbf{S}^{-1}$ matrix produces a new matrix with components (Problem 4.1)

$$\begin{pmatrix}\langle 1|\,\mathbf{A}\,|1\rangle & \langle 1|\,\mathbf{A}\,|2\rangle & \ldots & \langle 1|\,\mathbf{A}\,|N\rangle\\ \langle 2|\,\mathbf{A}\,|1\rangle & \langle 2|\,\mathbf{A}\,|2\rangle & \ldots & \langle 2|\,\mathbf{A}\,|N\rangle\\ \vdots & & \ddots & \\ \langle N|\,\mathbf{A}\,|1\rangle & \langle N|\,\mathbf{A}\,|2\rangle & \ldots & \langle N|\,\mathbf{A}\,|N\rangle\end{pmatrix} \tag{4.1.18}$$

Row and column eigenvectors with the same index appear only on the diagonal of the final matrix. Consider the fate of any off-diagonal element, e.g., $\langle 1|\,\mathbf{A}\,|2\rangle$. The eigenvector is regenerated after operation with the matrix **A**, i.e.,

$$\mathbf{A}\,|2\rangle = \lambda_2\,|2\rangle \tag{4.1.19}$$

The 12 element is

$$\langle 1|\, \mathbf{A}\, |2\rangle = \langle 1|\, \lambda_2\, |2\rangle = \lambda_2 \langle 1\,|\,2\rangle \tag{4.1.20}$$

For a complete set of orthonormal eigenvectors,

$$\langle 1\,|\,2\rangle = 0 \tag{4.1.21}$$

All off-diagonal elements vanish in the same way. The diagonal elements are the eigenvalues

$$\langle i|\, \mathbf{A}\, |i\rangle = \lambda_i \langle i\,|\,i\rangle = \lambda_i \tag{4.1.22}$$

The eigenvalues appear on the diagonal in the order in which the eigenvectors are arranged in the **S** matrix.

The eigenvectors are associated with principal axes. For example, the quadratic form

$$(x \quad y)\begin{pmatrix} 2 & 1 \\ 1 & 2 \end{pmatrix}\begin{pmatrix} x \\ y \end{pmatrix} = 1 \tag{4.1.23}$$

gives the equation for an ellipse with a mixed (xy) term (Figure 4.1):

$$2x^2 + 2xy + 2y^2 = 1 \tag{4.1.24}$$

Although the direction and magnitude of the two principal axes is not apparent from the equation, the **S** and $\mathbf{S}^{-1}$ matrices for this matrix,

$$\mathbf{S} = \begin{pmatrix} 1 & 1 \\ 1 & -1 \end{pmatrix} \qquad \mathbf{S}^{-1} = \left(\frac{1}{2}\right)\begin{pmatrix} 1 & 1 \\ 1 & -1 \end{pmatrix} \tag{4.1.25}$$

allow the coordinate transform

$$\langle x|\, \mathbf{SS}^{-1}\, |\mathbf{A}|\, \mathbf{SS}^{-1}\, |x\rangle = (\langle x|\, \mathbf{S})(\mathbf{S}^{-1}\mathbf{AS})(\mathbf{S}^{-1}\, |x\rangle) \tag{4.1.26}$$

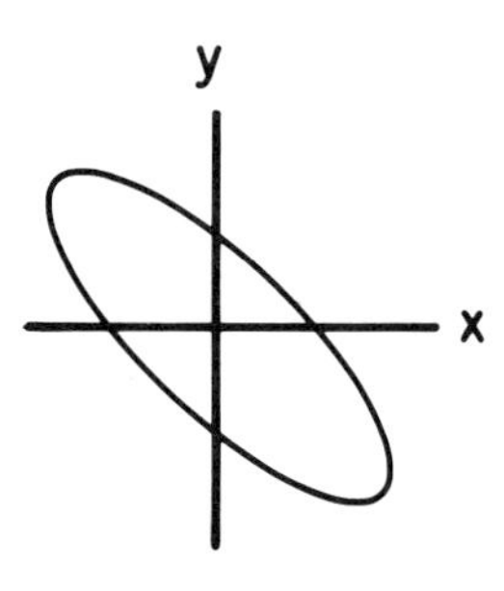

Figure 4.1. Original x and y coordinates and the eigenvector coordinate directions associated with the equation for an ellipse.

The transformed matrix is

$$\left(\frac{1}{2}\right)\begin{pmatrix}1 & 1\\ 1 & -1\end{pmatrix}\begin{pmatrix}2 & 1\\ 1 & 2\end{pmatrix}\begin{pmatrix}1 & 1\\ 1 & -1\end{pmatrix}=\begin{pmatrix}3 & 0\\ 0 & 1\end{pmatrix} \tag{4.1.27}$$

The original column vector $|x\rangle$ becomes

$$|x'\rangle = \mathbf{S}^{-1}\,|x\rangle$$

$$\begin{pmatrix}x'\\ y'\end{pmatrix}=\left(\frac{1}{\sqrt{2}}\right)\begin{pmatrix}1 & 1\\ 1 & -1\end{pmatrix}\begin{pmatrix}x\\ y\end{pmatrix}=\left(\frac{1}{\sqrt{2}}\right)\begin{pmatrix}x+y\\ x-y\end{pmatrix} \tag{4.1.28}$$

The row vector $(x' \quad y')$ has identical components for this symmetric matrix.

The transformed vectors involve components which are a sum and a difference of x and y values. If coordinates with $x = y$ were selected, i.e., a $+45^\circ$ axis, the second component would vanish. For an axis with $x = -y$, the first component vanishes. In this coordinate system, the equation for the ellipse is

$$(x' \quad y')\begin{pmatrix}3 & 0\\ 0 & 1\end{pmatrix}\begin{pmatrix}x'\\ y'\end{pmatrix}=3x'^2+y'^2=1$$

$$[x'/(1/\sqrt{3})]^2+y'^2=1 \tag{4.1.29}$$

This is the standard equation for an ellipse. The major and minor axes are oriented at -45° and 45°, respectively. The semimajor axis extends 1 unit from the origin. The semiminor axis extends a distance $\dfrac{1}{\sqrt{3}}$ (Figure 4.2). The component x' is selecting only those points for which $x = y$. The two points $x' = \dfrac{1}{\sqrt{3}}$ and $x' = \dfrac{-1}{\sqrt{3}}$ define the intersection of the ellipse and its semiminor axis.

For a nonsymmetric matrix, a complete set of biorthonormal eigenvectors generates the $\mathbf{S}$ and $\mathbf{S}^{-1}$ transform matrices. The nonsymmetric matrix

$$\begin{pmatrix}-1 & 2\\ 1 & -2\end{pmatrix} \tag{4.1.30}$$

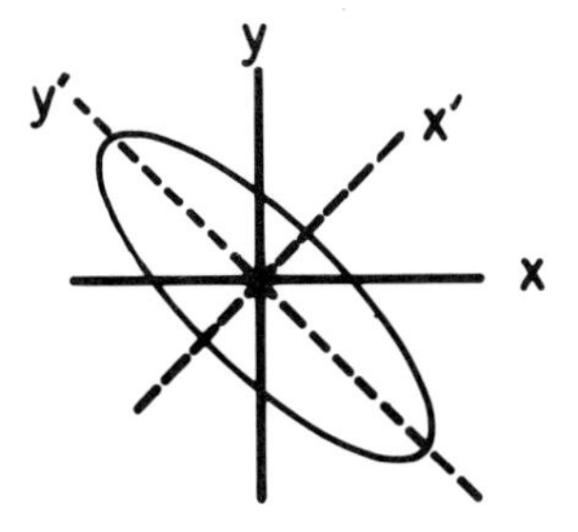

Figure 4.2. Orientation of the major and minor axes of the ellipse on the eigenvectors for the system.

has eigenvalues 0 and -3 with column eigenvectors

$$|0\rangle = \left(\frac{1}{\sqrt{3}}\right)\begin{pmatrix}2\\1\end{pmatrix} \qquad |-3\rangle = \left(\frac{1}{\sqrt{3}}\right)\begin{pmatrix}1\\-1\end{pmatrix} \tag{4.1.31}$$

and row eigenvectors

$$\langle 0| = \left(\frac{1}{\sqrt{3}}\right)(1 \quad 1); \qquad \langle -3| = \left(\frac{1}{\sqrt{3}}\right)(1 \quad -2) \tag{4.1.32}$$

These vectors are biorthonormal:

$$\begin{aligned}
\langle 0|0\rangle &= \left(\frac{1}{3}\right)(1 \quad 1)\begin{pmatrix}2\\1\end{pmatrix} = 1\\
\langle -3|-3\rangle &= \left(\frac{1}{3}\right)(1 \quad -2)\begin{pmatrix}1\\-1\end{pmatrix} = 1\\
\langle 0|-3\rangle &= \left(\frac{1}{3}\right)(1 \quad 1)\begin{pmatrix}1\\-1\end{pmatrix} = 0\\
\langle -3|0\rangle &= \left(\frac{1}{3}\right)(1 \quad -2)\begin{pmatrix}2\\1\end{pmatrix} = 0
\end{aligned} \tag{4.1.33}$$

The $\mathbf{S}$ matrix is formed from the column eigenvectors,

$$\mathbf{S} = \left(\frac{1}{\sqrt{3}}\right)\begin{pmatrix}2 & 1\\1 & -1\end{pmatrix} = (|0\rangle \quad |-3\rangle) \tag{4.1.34}$$

while the $\mathbf{S}^{-1}$ matrix is formed from the row vectors,

$$\mathbf{S}^{-1} = \left(\frac{1}{\sqrt{3}}\right)\begin{pmatrix}1 & 1\\1 & -2\end{pmatrix} = \begin{pmatrix}\langle 0|\\ \langle -3|\end{pmatrix} \tag{4.1.35}$$

The similarity transform of $\mathbf{A}$ is

$$\begin{aligned}
\mathbf{S}^{-1}\mathbf{A}\mathbf{S} &= \left(\frac{1}{3}\right)\begin{pmatrix}1 & 1\\1 & -2\end{pmatrix}\begin{pmatrix}-1 & 2\\1 & -2\end{pmatrix}\begin{pmatrix}2 & 1\\1 & -1\end{pmatrix}\\
&= \left(\frac{1}{3}\right)\begin{pmatrix}0 & 0\\0 & -9\end{pmatrix} = \begin{pmatrix}0 & 0\\0 & -3\end{pmatrix}
\end{aligned} \tag{4.1.36}$$

The normalization factors are combined as a factor for $\mathbf{S}^{-1}$ for convenience.

4.2. Simultaneous Diagonalization

Although a large number of symmetric 2×2 matrices can be created by varying the elements of the matrix, the eigenvectors for different matrices are often identical. The eigenvector pair

$$\frac{1}{\sqrt{2}}\begin{pmatrix}1\\1\end{pmatrix} \quad \frac{1}{\sqrt{2}}\begin{pmatrix}1\\-1\end{pmatrix} \tag{4.2.1}$$

is present for a variety of different 2×2 matrices. Such consistency continues for $N\times N$ matrices. The commutation property, introduced in this section, establishes whether matrices have a common set of eigenvectors. Since eigenvectors are often conveniently determined for some matrices, this commutation property provides the bridge to determine eigenvectors for more complicated matrices.

The symmetric matrix

$$\mathbf{A}=\begin{pmatrix}2 & 1\\1 & 2\end{pmatrix} \tag{4.2.2}$$

has eigenvalues 1 and 3 with corresponding eigenvectors

$$|1\rangle=\left(\frac{1}{\sqrt{2}}\right)(1 \quad -1) \qquad |3\rangle=\left(\frac{1}{\sqrt{2}}\right)(1 \quad -1) \tag{4.2.3}$$

which generate the similarity transform matrices

$$\mathbf{S}=\begin{pmatrix}1 & 1\\-1 & 1\end{pmatrix} \qquad \mathbf{S}^{-1}=\left(\frac{1}{2}\right)\begin{pmatrix}1 & -1\\1 & 1\end{pmatrix} \tag{4.2.4}$$

The similarity transform of the matrix $\mathbf{A}$ gives the diagonal matrix

$$\mathbf{A}_d=\begin{pmatrix}1 & 0\\0 & 3\end{pmatrix} \tag{4.2.5}$$

The similarity transform matrices will also diagonalize the matrix

$$\mathbf{B}=\begin{pmatrix}1 & -1\\-1 & 1\end{pmatrix} \tag{4.2.6}$$

to give

$$\mathbf{B}_d=\begin{pmatrix}0 & 0\\0 & 2\end{pmatrix} \tag{4.2.7}$$

The multiplication of $\mathbf{A}_d$ and $\mathbf{B}_d$ is independent of the order of the matrices since the diagonal element of one matrix multiplies only the corresponding diagonal element in the second matrix:

$$\mathbf{B}_d\mathbf{A}_d=\begin{pmatrix}1 & 0\\0 & 3\end{pmatrix}\begin{pmatrix}0 & 0\\0 & 2\end{pmatrix}=\begin{pmatrix}0 & 0\\0 & 6\end{pmatrix} \tag{4.2.8}$$

and

$$\mathbf{A}_d\mathbf{B}_d=\begin{pmatrix}0 & 0\\0 & 2\end{pmatrix}\begin{pmatrix}1 & 0\\0 & 3\end{pmatrix}=\begin{pmatrix}0 & 0\\0 & 6\end{pmatrix} \tag{4.2.9}$$

When the matrix product is independent of the order of multiplication, the two matrices are commuting matrices. The equality of the two products is often written as a difference, i.e.,

$$\mathbf{A}_d\mathbf{B}_d-\mathbf{B}_d\mathbf{A}_d=0 \tag{4.2.10}$$

This commutation relationship is written in a special square bracket notation,

$$[\mathbf{A}_d, \mathbf{B}_d]=0 \tag{4.2.11}$$

where the first matrix in the bracket is the matrix on the left in the first product pair. The full bracket is called a commutator.

The commutation property is not obeyed by all matrix pairs. For example, the 2×2 matrices

$$\mathbf{C}=\begin{pmatrix}1 & 0\\0 & 2\end{pmatrix} \qquad \mathbf{D}=\begin{pmatrix}2 & 1\\1 & 2\end{pmatrix}$$

produce a commutator

$$\begin{aligned}[\mathbf{C}, \mathbf{D}]&=\mathbf{CD}-\mathbf{DC}\\&=\begin{pmatrix}1 & 0\\0 & 2\end{pmatrix}\begin{pmatrix}2 & 1\\1 & 2\end{pmatrix}-\begin{pmatrix}2 & 1\\1 & 2\end{pmatrix}\begin{pmatrix}1 & 0\\0 & 2\end{pmatrix}\\&=\begin{pmatrix}2 & 1\\2 & 4\end{pmatrix}-\begin{pmatrix}2 & 2\\1 & 4\end{pmatrix}=\begin{pmatrix}0 & -1\\1 & 0\end{pmatrix}\end{aligned} \tag{4.2.12}$$

The presence of one diagonal matrix in the product does not guarantee commutation.

The original matrices **A** and **B**

$$\mathbf{A}=\begin{pmatrix}2 & 1\\1 & 2\end{pmatrix} \qquad \mathbf{B}=\begin{pmatrix}1 & -1\\-1 & 1\end{pmatrix} \tag{4.2.13}$$

do commute:

$$\begin{pmatrix} 2 & 1 \\ 1 & 2 \end{pmatrix}\begin{pmatrix} 1 & -1 \\ -1 & 1 \end{pmatrix} - \begin{pmatrix} 1 & -1 \\ -1 & 1 \end{pmatrix}\begin{pmatrix} 2 & 1 \\ 1 & 2 \end{pmatrix}$$

$$= \begin{pmatrix} 1 & -1 \\ -1 & 1 \end{pmatrix} - \begin{pmatrix} 1 & -1 \\ -1 & 1 \end{pmatrix} = \begin{pmatrix} 0 & 0 \\ 0 & 0 \end{pmatrix} \quad (4.2.14)$$

A is diagonalized by a similarity matrix **S** while **B** is diagonalized by a similarity matrix **T**:

$$\begin{aligned} \mathbf{A}_d &= \mathbf{S}^{-1}\mathbf{A}\mathbf{S} \\ \mathbf{B}_d &= \mathbf{T}^{-1}\mathbf{B}\mathbf{T} \end{aligned} \quad (4.2.15)$$

These diagonal matrices must commute:

$$[\mathbf{A}_d, \mathbf{B}_d] = 0 = \mathbf{A}_d\mathbf{B}_d - \mathbf{B}_d\mathbf{A}_d \quad (4.2.16)$$

Substituting the transforms of Equation 4.2.15 gives

$$(\mathbf{S}^{-1}\mathbf{A}\mathbf{S})(\mathbf{T}^{-1}\mathbf{B}\mathbf{T}) - (\mathbf{T}^{-1}\mathbf{B}\mathbf{T})(\mathbf{S}^{-1}\mathbf{A}\mathbf{S}) \quad (4.2.17)$$

If $\mathbf{T} = \mathbf{S}$, i.e., both matrices have the same eigenvectors, this equation is

$$\mathbf{S}^{-1}(\mathbf{A}\mathbf{B} - \mathbf{B}\mathbf{A})\mathbf{S} = \mathbf{A}_d\mathbf{B}_d - \mathbf{B}_d\mathbf{A}_d = 0 \quad (4.2.18)$$

The commutation property is useful for the determination of eigenvectors and eigenvalues for a complicated matrix. If such a matrix commutes with a second matrix with known eigenvectors, these eigenvectors are also the eigenvectors for the complicated matrix. For example, the matrix

$$\begin{pmatrix} -1 & 1 & 0 \\ 1 & -2 & 1 \\ 0 & 1 & -1 \end{pmatrix} \quad (4.2.19)$$

has the eigenvectors

$$|0\rangle = \left(\frac{1}{\sqrt{3}}\right)\begin{pmatrix} 1 \\ 1 \\ 1 \end{pmatrix} \qquad |-1\rangle = \left(\frac{1}{\sqrt{2}}\right)\begin{pmatrix} 1 \\ 0 \\ -1 \end{pmatrix} \qquad |-3\rangle = \left(\frac{1}{2}\right)\begin{pmatrix} 1 \\ -2 \\ 1 \end{pmatrix} \quad (4.2.20)$$

A second matrix

$$\begin{pmatrix} -2 & 1 & 1 \\ 1 & -2 & 1 \\ 1 & 1 & -2 \end{pmatrix} \quad (4.2.21)$$

is not tridiagonal but it does commute with the first matrix:

$$\begin{pmatrix} -1 & 1 & 0 \\ 1 & -2 & 1 \\ 0 & 1 & -1 \end{pmatrix} \begin{pmatrix} -2 & 1 & 1 \\ 1 & -2 & 1 \\ 1 & 1 & -2 \end{pmatrix} - \begin{pmatrix} -2 & 1 & 1 \\ 1 & -2 & 1 \\ 1 & 1 & -2 \end{pmatrix} \begin{pmatrix} -1 & 1 & 0 \\ 1 & -2 & 1 \\ 0 & 1 & -1 \end{pmatrix}$$

$$= \begin{pmatrix} 3 & -3 & 0 \\ -3 & 6 & -3 \\ 0 & -3 & 3 \end{pmatrix} - \begin{pmatrix} 3 & -3 & 0 \\ -3 & 6 & -3 \\ 0 & -3 & 3 \end{pmatrix} = 0 \tag{4.2.22}$$

Because the matrices commute, the three eigenvectors (Equation 4.2.20) are eigenvectors for both matrices. For example,

$$\begin{pmatrix} -2 & 1 & 1 \\ 1 & -2 & 1 \\ 1 & 1 & -2 \end{pmatrix} \begin{pmatrix} 1 \\ -2 \\ 1 \end{pmatrix} = (-3) \begin{pmatrix} 1 \\ -2 \\ 1 \end{pmatrix} \tag{4.2.23}$$

and

$$\begin{pmatrix} -2 & 1 & 1 \\ 1 & -2 & 1 \\ 1 & 1 & -2 \end{pmatrix} \begin{pmatrix} 1 \\ 0 \\ -1 \end{pmatrix} = (-3) \begin{pmatrix} 1 \\ 0 \\ -1 \end{pmatrix} \tag{4.2.24}$$

The matrix is doubly degenerate in the eigenvalue -3. The orthonormal basis generated by the tridiagonal matrix serves as the eigenvector set which completely diagonalizes the matrix with degenerate eigenvalues.

Commuting matrices play a major role in systems which are symmetric in space. A quantum mechanical system is illustrated in Figure 4.3. Three wavefunctions of the system are located by vectors at 0°, 120°, and 240°. The three functions obey the eigenvalue–eigenvector equations

$$\begin{aligned} \mathbf{H}f_1 &= E_1 f_1 \\ \mathbf{H}f_2 &= E_2 f_2 \\ \mathbf{H}f_3 &= E_3 f_3 \end{aligned} \tag{4.2.25}$$

The Hamiltonian operator, $\mathbf{H}$, operates on f_i (an eigenvector) to regenerate the eigenvector and the energy E_i (the eigenvalue) for this operator.

The vector f_1 moves to the location of f_2 via an $\mathbf{R}(120°)$ operation in two dimensions:

$$\begin{pmatrix} \cos 120° & -\sin 120° \\ \sin 120° & \cos 120° \end{pmatrix} = \begin{pmatrix} \dfrac{-1}{2} & -\dfrac{\sqrt{3}}{2} \\ \dfrac{\sqrt{3}}{2} & \dfrac{-1}{2} \end{pmatrix} \tag{4.2.26}$$

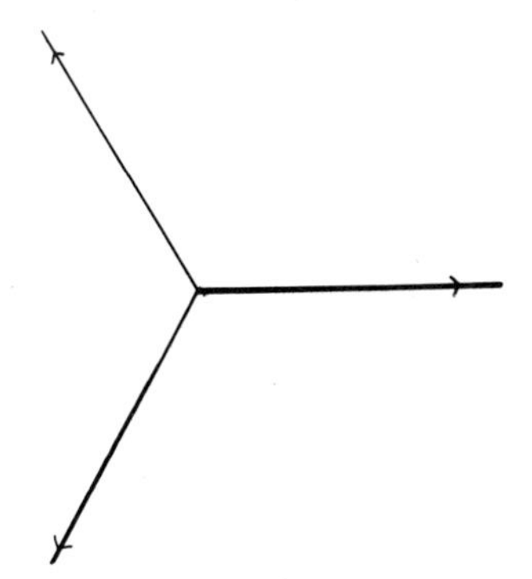

Figure 4.3. Vector analogue of a quantum system having orbitals oriented at 0°, 120°, and 240°.

Similarly, f_1 can be moved to the location of f_3 with the $\mathbf{R}_{240^\circ}$ matrix:

$$\begin{pmatrix} \cos 240^\circ & -\sin 240^\circ \\ \sin 240^\circ & \cos 240^\circ \end{pmatrix} = \begin{pmatrix} \dfrac{-1}{2} & \dfrac{\sqrt{3}}{2} \\ -\dfrac{\sqrt{3}}{2} & \dfrac{-1}{2} \end{pmatrix} \tag{4.2.27}$$

The order of rotation will make no difference so these two matrices commute with each other.

If the Hamiltonian operator depends on orientation, each of the three energies in Equation 4.2.25 can be different and the order of operation is important. The operation

$$\mathbf{R}_{120^\circ}\mathbf{H}f_1 = \mathbf{R}_{120^\circ}E_1 f_1 = E_1\mathbf{R}_{120^\circ}f_1 = E_1 f_2 \tag{4.2.28}$$

gives the energy eigenvalue E_1 for f_1. If the order of operations is reversed, the result is

$$\mathbf{H}\mathbf{R}_{120^\circ}f_1 = \mathbf{H}f_2 = E_2 f_2 \tag{4.2.29}$$

The final eigenvector is unchanged but the energy eigenvalue is E_2. The difference of the two above results,

$$[\mathbf{R}_{120^\circ}, \mathbf{H}]\, f_1 = (\mathbf{R}_{120^\circ}\mathbf{H} - \mathbf{H}\mathbf{R}_{120^\circ})\, f_1 = (E_1 - E_2)\, f_2 \tag{4.2.30}$$

shows clearly that the two operations will not commute when $E_1 \neq E_2$. If the Hamiltonian is not dependent on the actual angular orientation of the wavefunction, then each f_i will have the same energy. In this case, the $\mathbf{H}$ operator will commute with the rotation operator since

$$E_1 - E_2 = 0 \tag{4.2.31}$$

The commutation of a rotation (symmetry) operator is important for the generation of wavefunctions. If a Hamiltonian commutes with rotation operators $\mathbf{R}_i$,

$$[\mathbf{R}_i, \mathbf{H}] = 0 \tag{4.2.32}$$

this effectively states that a wavefunction can be rotated using the rotation operators $\mathbf{R}_i$ and the new wavefunction will have the same energy as the original wavefunction. Thus, the rotation operations generate all the wavefunctions having the same energy. For example, if f_1 is given, the intrinsic symmetry of the system would dictate rotations of 120° and 240° to generate f_2 and f_3. All three orbitals have the same energy.

Commuting operators enter quantum mechanics in a second fundamental way. The uncertainty principle states that the product of uncertainties for momentum p_x and position x must exceed $\hbar$:

$$\Delta p_x \, \Delta x > \hbar \tag{4.2.33}$$

If p_x and x are to be operators in quantum mechanics, the uncertainty principle means that the result of measurements of p_x and x depends on the order of operations. Thus, p_x and x cannot commute without violating the uncertainty principle. The uncertainty principle becomes

$$[x, p_x] = i\hbar \tag{4.2.34}$$

This result must hold for every value of x and the p_x at that position. For a single value of x, this condition is satisfied only if p_x is assumed proportional to the derivative with respect to x at that point, i.e.,

$$p_x = -i\hbar \, d/dx \tag{4.2.35}$$

(Problem 4.7). This momentum operator is normally presented as a postulate of quantum mechanics.

4.3. Generalized Characteristic Equations

The eigenvalue–eigenvector equation for a matrix $\mathbf{A}$,

$$\mathbf{A} \, |i\rangle = \lambda_i \, |i\rangle \tag{4.3.1}$$

must include the identity matrix $\mathbf{I}$ so that both sides of the equation have a matrix operating on the eigenvector,

$$\mathbf{A} \, |i\rangle = \lambda_i \mathbf{I} \, |i\rangle \tag{4.3.2}$$

or

$$(\mathbf{A} - \lambda_i \mathbf{I}) \, |i\rangle = 0 \tag{4.3.3}$$

The equation for a 2×2 matrix is then

$$\left[\begin{pmatrix} 2 & 1 \\ 1 & 2 \end{pmatrix} - \begin{pmatrix} \lambda & 0 \\ 0 & \lambda \end{pmatrix}\right]\begin{pmatrix} a \\ b \end{pmatrix} = 0 \quad (4.3.4)$$

The eigenvalues which are subtracted from the diagonal elements of **A** are identical, i.e., they have equal weight.

Under some circumstances, there are advantages to replacing the **I** matrix in the equation by a general matrix **B**. The eigenvalue–eigenvector equation is

$$(\mathbf{A} - \lambda_i \mathbf{B})\,|i\rangle = 0 \quad (4.3.5)$$

The eigenvectors for such cases will have special normalization properties.

The matrix

$$\begin{pmatrix} 2 & 1 & 0 \\ \frac{1}{2} & 1 & \frac{1}{2} \\ 0 & \frac{1}{3} & \frac{2}{3} \end{pmatrix} \quad (4.3.6)$$

is nonsymmetric and includes fractional elements. The matrix does contain an important pattern, however. The matrix would become symmetric if the center row was multiplied by 2 and the bottom row was multiplied by 3. This is accomplished by left-multiplying the matrix by the diagonal matrix

$$\mathbf{B} = \begin{pmatrix} 1 & 0 & 0 \\ 0 & 2 & 0 \\ 0 & 0 & 3 \end{pmatrix} \quad (4.3.7)$$

to give

$$\mathbf{BA} = \begin{pmatrix} 1 & 0 & 0 \\ 0 & 2 & 0 \\ 0 & 0 & 3 \end{pmatrix}\begin{pmatrix} 2 & 1 & 0 \\ \frac{1}{2} & 1 & \frac{1}{2} \\ 0 & \frac{1}{3} & \frac{2}{3} \end{pmatrix} = \begin{pmatrix} 2 & 1 & 0 \\ 1 & 2 & 1 \\ 0 & 1 & 2 \end{pmatrix} \quad (4.3.8)$$

The entire characteristic matrix equation can be multiplied by **B** to give

$$\mathbf{B}(\mathbf{A} - \lambda_i \mathbf{I})\,|i\rangle = (\mathbf{C} - \lambda_i \mathbf{B})\,|i\rangle \quad (4.3.9)$$

where

$$\mathbf{C} = \mathbf{BA} \quad (4.3.10)$$

is the symmetric matrix of Equation 4.3.8. The diagonal **B** matrix is also symmetric.

The characteristic determinant for Equation 4.3.9 is

$$\begin{vmatrix} 2-1\lambda & 1 & 0 \\ 1 & 2-2\lambda & 1 \\ 0 & 1 & 2-3\lambda \end{vmatrix} = 0 \tag{4.3.11}$$

The characteristic polynomial is complicated by the factors 2 and 3 associated with λ. However, since the entire equation was multiplied by the same matrix (**B**), the eigenvalues for the characteristic equation will be identical to those for the **A** matrix.

To illustrate the nature of eigenvalues in the two formats, the matrix

$$\begin{pmatrix} 2 & 2 \\ 1 & 1 \end{pmatrix} \tag{4.3.12}$$

is used. The eigenvalues are 0 and 3 and the eigenvectors are biorthogonal for this nonsymmetric matrix:

$$\langle 0| = \left(\frac{1}{3}\right)(1 \quad -2); \qquad \langle 3| = \left(\frac{1}{3}\right)(1 \quad 1) \tag{4.3.13}$$

$$|0\rangle = \begin{pmatrix} 1 \\ -1 \end{pmatrix} \qquad |3\rangle = \begin{pmatrix} 2 \\ 1 \end{pmatrix} \tag{4.3.14}$$

The characteristic equation for **A** can be converted to an equation with a symmetric matrix **C** by multiplying from the left with the diagonal matrix

$$\mathbf{B} = \begin{pmatrix} 1 & 0 \\ 0 & 2 \end{pmatrix} \tag{4.3.15}$$

to give

$$\mathbf{BA}\,|i\rangle - \lambda_i \mathbf{B}\,|i\rangle = 0$$
$$\left[\begin{pmatrix} 2 & 2 \\ 2 & 2 \end{pmatrix} - \begin{pmatrix} 1\lambda & 0 \\ 0 & 2\lambda \end{pmatrix}\right] |i\rangle = 0 \tag{4.3.16}$$

The characteristic determinant is

$$\begin{vmatrix} 2-2\lambda & 2 \\ 2 & 2-1\lambda \end{vmatrix} = 0 \tag{4.3.17}$$

which yields the same eigenvalues (0 and 3) as **A**.

For the symmetric matrix **C**, $\lambda = 0$ has the eigenvectors

$$\langle 0| = \left(\frac{1}{2}\right)(1 \quad -1) \qquad |0\rangle = \begin{pmatrix} 1 \\ -1 \end{pmatrix} \tag{4.3.18}$$

The $\lambda = 3$ eigenvectors are

$$\langle 3| = \left(\frac{1}{5}\right)(2 \quad 1) \qquad |3\rangle = \begin{pmatrix} 2 \\ 1 \end{pmatrix} \tag{4.3.19}$$

The symmetric **C** matrix produces bra and ket eigenvectors with equal components. However, the eigenvectors are not orthogonal:

$$\langle 0|3\rangle = \left(\frac{1}{2}\right)(1 \quad -1)\begin{pmatrix} 2 \\ 1 \end{pmatrix} = \left(\frac{1}{2}\right)(1) \neq 0 \tag{4.3.20}$$

The **B** matrix has altered the orthogonality condition.

The eigenvectors for the symmetric matrix become orthogonal if **B** is included in the orthogonality condition:

$$\langle 0|\,\mathbf{B}\,|3\rangle = (1 \quad -1)\begin{pmatrix} 1 & 0 \\ 0 & 2 \end{pmatrix}\begin{pmatrix} 2 \\ 1 \end{pmatrix} = 0 \tag{4.3.21}$$

and

$$\langle 3|\,\mathbf{B}\,|0\rangle = (2 \quad 1)\begin{pmatrix} 1 & 0 \\ 0 & 2 \end{pmatrix}\begin{pmatrix} 1 \\ -1 \end{pmatrix} = 0 \tag{4.3.22}$$

The normalization of the vectors must also include **B**:

$$\begin{aligned} \langle 0|\,\mathbf{B}\,|0\rangle &= (1 \quad -1)\begin{pmatrix} 1 & 0 \\ 0 & 2 \end{pmatrix}\begin{pmatrix} 1 \\ -1 \end{pmatrix} = 3 \\ \langle 3|\,\mathbf{B}\,|3\rangle &= (2 \quad 1)\begin{pmatrix} 1 & 0 \\ 0 & 2 \end{pmatrix}\begin{pmatrix} 2 \\ 1 \end{pmatrix} = 6 \end{aligned} \tag{4.3.23}$$

In general, the $|i\rangle$ are the eigenvectors for

$$\mathbf{C} = \mathbf{BA} \tag{4.3.24}$$

which gives

$$\langle i|\,\mathbf{C}\,|i\rangle = \lambda_i; \qquad \langle j|\,\mathbf{C}\,|i\rangle = 0 \tag{4.3.25}$$

The operator equation is

$$\mathbf{C}\,|i\rangle = \lambda_i \mathbf{B}\,|i\rangle \tag{4.3.26}$$

for the column eigenvectors and

$$\langle j|\,\mathbf{C} = \lambda_j \langle j|\,\mathbf{B} \tag{4.3.27}$$

for the row eigenvectors. The quadratic form $\langle j|\,\mathbf{C}\,|i\rangle$ may then be either

$$\langle j|\,\mathbf{C}\,|i\rangle = \lambda_i \langle j|\,\mathbf{B}\,|i\rangle \tag{4.3.28}$$

or

$$\langle j|\, \mathbf{C}\, |i\rangle = \lambda_j \langle j|\, \mathbf{B}\, |i\rangle \tag{4.3.29}$$

Subtracting this equation gives

$$(\lambda_i - \lambda_j)\langle j|\, \mathbf{B}\, |i\rangle = 0 \tag{4.3.30}$$

If $|i\rangle$ and $\langle j|$ have different eigenvalues, $\lambda_j - \lambda_i \neq 0$, and

$$\langle j|\, \mathbf{B}\, |i\rangle = 0 \tag{4.3.31}$$

In the $\mathbf{C} - \lambda \mathbf{B}$ format, the eigenvectors are those of the **C** matrix. However, **C** was generated from the original matrix **A**. The eigenvectors for the **A** matrix are related to those for **C**. The eigenvector equation for $|i\rangle$ of **C** is

$$\mathbf{C}\, |i\rangle = \mathbf{BA}\, |i\rangle = \lambda_i \mathbf{B}\, |i\rangle \tag{4.3.32}$$

Operating from the left with $\langle j|$ for **C** gives

$$\langle j|\, \mathbf{C}\, |i\rangle = \langle j|\, \mathbf{BA}\, |i\rangle = \lambda_i \langle j|\, \mathbf{B}\, |i\rangle \tag{4.3.33}$$

The right-hand side is the orthogonality relation for **C**. However, it is also the orthogonality relation for **BA**. Therefore, the valid eigenvectors for **A** must be

$$\langle j|\, \mathbf{B} \quad \text{and} \quad |i\rangle \tag{4.3.34}$$

The 2×2 **A** matrix must have row eigenvectors

$$\begin{aligned} \langle 0_{\mathbf{A}}| &= \langle 0_{\mathbf{C}}|\, \mathbf{B} = (1 \quad -1)\begin{pmatrix} 1 & 0 \\ 0 & 2 \end{pmatrix} = (1 \quad -2) \\ \langle 3_{\mathbf{A}}| &= \langle 3_{\mathbf{C}}|\, \mathbf{B} = (2 \quad 1)\begin{pmatrix} 1 & 0 \\ 0 & 2 \end{pmatrix} = (2 \quad 2) \end{aligned} \tag{4.3.35}$$

These eigenvectors were derived by direct analysis of **A** (Equation 4.3.13).

The general characteristic equation will appear in subsequent chapters. In Chapter 5, the equation of motion for a coupled set of oscillating atoms with different masses will be described by a characteristic equation

$$\mathbf{K} - \mathbf{M}\lambda = \mathbf{K} - \mathbf{M}\omega^2 \tag{4.3.36}$$

where **K** is a matrix of force constants and **M** is a diagonal matrix with the masses of each atom i in the system.

In Chapter 8, the matrix Hamiltonian generated by a set of nonorthogonal atomic orbitals will be

$$\mathbf{H} - \lambda \mathbf{S} = \mathbf{H} - E\mathbf{S} \tag{4.3.37}$$

where E is the energy eigenvalue and $\mathbf{S}$ will be a nondiagonal symmetric matrix with ij elements reflecting the overlap of the nonorthogonal i and j eigenvectors (atomic orbitals).

4.4. Matrix Decomposition Using Eigenvectors

An $N \times N$ Hermitian matrix with N distinct eigenvalues always has N orthogonal eigenvectors. When this $\mathbf{A}$ matrix operates on an eigenvector, it retains the orientation of that vector in the N-dimensional space multiplied by the eigenvalue for that vector:

$$\mathbf{A}\,|i\rangle = \lambda_i\,|i\rangle \tag{4.4.1}$$

Application of $\mathbf{A}$ a second time will produce the square of the eigenvalue while the eigenvector remains unchanged:

$$\mathbf{A}^2\,|i\rangle = \mathbf{A}(\mathbf{A}\,|i\rangle) = \lambda_i^2\,|i\rangle \tag{4.4.2}$$

Each eigenvalue–eigenvector equation describes one of the N dimensions. Some combination of all N dimensions will then describe the action of the matrix in the N-dimensional space. In fact, the matrix itself can be described in terms of its eigenvalues and eigenvectors.

For a matrix $\mathbf{A}$ with eigenvectors $|i\rangle$, an arbitrary vector $|v\rangle$ can be written as a linear combination of projections on the eigenvectors $|i\rangle$. The projection on the eigenvector $|i\rangle$ is

$$\langle i|v\rangle \tag{4.4.3}$$

The vector $|v\rangle$ is then a linear combination of the complete set of eigenvectors:

$$|v\rangle = \sum_i |i\rangle\langle i|v\rangle \tag{4.4.4}$$

When the matrix $\mathbf{A}$ operates on the vector $|v\rangle$, it operates on the linear combination of eigenvectors. In each case, it generates the eigenvalue associated with that eigenvector. The operation is

$$\mathbf{A}\,|v\rangle = \sum_i \mathbf{A}\,|i\rangle\langle i|v\rangle = \sum_i \lambda_i\,|i\rangle\langle i|v\rangle \tag{4.4.5}$$

The eigenvector operation has replaced the matrix with the ith eigenvector while leaving the ith projection operator intact.

The decomposition of the operation $\mathbf{A}\,|v\rangle$ in terms of the $|i\rangle$ produces the identity

$$\mathbf{AI}\,|v\rangle = \mathbf{A}\left(\sum |i\rangle\langle i|\right)|v\rangle \tag{4.4.6}$$

so

$$\mathbf{I} = \sum |i\rangle\langle i| = \sum \mathbf{P}_i \tag{4.4.7}$$

Because the $|i\rangle$ constitute a complete set, the sum of all projections must be unity.

Both sides of Equation 4.4.5 ultimately operate on the vector $|v\rangle$. This vector can be eliminated from each side of the equation, leaving an expression for the matrix **A** in terms of both the eigenvalues and the eigenvectors:

$$\mathbf{A} = \sum_i^N \lambda_i |i\rangle\langle i| \tag{4.4.8}$$

Although it may appear that the matrix has been reduced to a sum of vectors, this is so only if the matrix is applied to a vector. The projection operators

$$|i\rangle\langle i| \tag{4.4.9}$$

are actually matrices associated with each eigenvector. This projection operator will operate on a vector to give the scalar magnitude of the projection and the direction of the eigenvector. The vector combination $|j\rangle\langle i|$ has a column vector which precedes the row vector. The combination follows the normal rules of matrix multiplication. The element of the first row of the left matrix (a 2×1 matrix) is multiplied by the corresponding element of the right matrix (a 1×2 matrix), etc. This yields a 2×2 projection matrix:

$$\begin{pmatrix} c \\ d \end{pmatrix} (a \quad b) = \begin{pmatrix} ca & cb \\ da & db \end{pmatrix} \tag{4.4.10}$$

This is an example of a direct product of the 2×1 and 1×2 matrices. Such direct products are important in group theory and will be considered in Chapter 10.

The matrix **A** is

$$\mathbf{A} = \sum_i^N \lambda_i |i\rangle\langle i| = \sum_i \lambda_i \mathbf{P}_i \tag{4.4.11}$$

The equation relates the matrix and its eigenvalues and eigenvectors. However, the matrix is really only a mathematical device used to change the magnitude and/or direction of a vector in the space. Each eigenvalue–projection operator term in the series selects a specific direction in the space for some vector $|v\rangle$; then the eigenvalue simply tells the factor by which the matrix changes the vector in this specific direction. The decomposition occurred for matrices with the eigenvectors

$$(1 \quad 1) \qquad (1 \quad -1) \tag{4.4.12}$$

Although the eigenvectors may be valid for many 2×2 matrices, their eigenvalues will differ and these scalar factors will control how the matrix alters a vector $|v\rangle$ in the space.

This is illustrated with the matrix

$$\begin{pmatrix} 2 & 1 \\ 1 & 2 \end{pmatrix} \tag{4.4.13}$$

with eigenvalues $\lambda = 1$ and $\lambda = 3$ and normalized eigenvectors

$$|1\rangle = \left(\frac{1}{\sqrt{2}}\right)(1 \quad -1) \tag{4.4.14}$$

$$|3\rangle = \left(\frac{1}{\sqrt{2}}\right)(1 \quad 1) \tag{4.4.15}$$

The eigenvalue–eigenvector decomposition of the matrix is

$$\mathbf{A} = (+1)\left(\frac{1}{2}\right)\begin{pmatrix} 1 \\ -1 \end{pmatrix}(1 \quad -1) + (3)\left(\frac{1}{2}\right)\begin{pmatrix} 1 \\ 1 \end{pmatrix}(1 \quad 1) \tag{4.4.16}$$

and, in matrix format,

$$\left(\frac{1}{2}\right)\begin{pmatrix} 1 & -1 \\ -1 & 1 \end{pmatrix} + \left(\frac{3}{2}\right)\begin{pmatrix} 1 & 1 \\ 1 & 1 \end{pmatrix} \tag{4.4.17}$$

The result is verified by summing all elements:

$$\begin{pmatrix} \frac{1}{2} + \frac{3}{2} & \frac{-1}{2} + \frac{3}{2} \\ \frac{-1}{2} + \frac{3}{2} & \frac{1}{2} + \frac{3}{2} \end{pmatrix} = \begin{pmatrix} 2 & 1 \\ 1 & 2 \end{pmatrix} \tag{4.4.18}$$

If the decomposed matrix operates on the eigenvector $|i\rangle$, only the ith projection operator gives a finite projection,

$$\sum_i^N \lambda_i \, |i\rangle\langle i|i\rangle = \lambda_i \, |i\rangle \tag{4.4.19}$$

This corroborates the earlier observation that a matrix operating on an eigenvector must regenerate that eigenvector times its eigenvalue.

The decomposition technique is also useful for powers of the matrix $\mathbf{A}$. For the product $\mathbf{A}^2$, the right-hand matrix is decomposed to

$$\mathbf{A}^2 = \sum_i^N \mathbf{A}(\lambda_i \, |i\rangle\langle i|) \tag{4.4.20}$$

Application of the second **A** matrix regenerates the eigenvector and a second eigenvalue:

$$\mathbf{A}^2 = \sum \lambda_i^2 \, |i\rangle\langle i| \tag{4.4.21}$$

The projection operator portion of the decomposition maintains the proper direction. The magnitude of this component of the squared matrix is proportional to the square of the eigenvalue.

This result is generalized for an arbitrary power N of the matrix **A** (Problem 4.10):

$$\mathbf{A}^N = \sum \lambda^N \, |i\rangle\langle i| \tag{4.4.22}$$

Only the scalar eigenvalue is raised to the Nth power in each term of the summation. The projection operator technique permits the determination of matrix powers of **A** which would be quite tedious to evaluate by direct matrix multiplication. For cxamplc, thc tcnth power of the matrix **A**,

$$\mathbf{A} = \begin{pmatrix} 2 & 1 \\ 1 & 2 \end{pmatrix} \tag{4.4.23}$$

is

$$\begin{aligned} \mathbf{A}^{10} &= (1)^{10} \left(\frac{1}{2}\right) \begin{pmatrix} 1 \\ -1 \end{pmatrix} (1 \quad -1) + (3)^{10} \left(\frac{1}{2}\right) \begin{pmatrix} 1 \\ 1 \end{pmatrix} (1 \quad 1) \\ &= \left(\frac{1}{2}\right) \begin{pmatrix} 1 & -1 \\ -1 & 1 \end{pmatrix} + (3)^{10} \left(\frac{1}{2}\right) \begin{pmatrix} 1 & 1 \\ 1 & 1 \end{pmatrix} \end{aligned} \tag{4.4.24}$$

The factor $3^{10} = 59{,}049$ is significantly larger than $1^{10} = 1$. The larger eigenvalue and its eigenvector will dominate the summation as the power increases. Repeated applications of the matrix **A** to an arbitrary vector $|v\rangle$ must ultimately produce the eigenvector associated with the maximal eigenvalue.

The method is checked by comparing $\mathbf{A}^2$ from direct multiplication to $\mathbf{A}^2$ determined using projection operators. The product is

$$\mathbf{A}^2 = \begin{pmatrix} 2 & 1 \\ 1 & 2 \end{pmatrix} \begin{pmatrix} 2 & 1 \\ 1 & 2 \end{pmatrix} = \begin{pmatrix} 5 & 4 \\ 4 & 5 \end{pmatrix} \tag{4.4.25}$$

The projection operator decomposition gives

$$\mathbf{A}^2 = (1)^2/2 \begin{pmatrix} 1 & -1 \\ -1 & 1 \end{pmatrix} + (3)^2/2 \begin{pmatrix} 1 & 1 \\ 1 & 1 \end{pmatrix} = \begin{pmatrix} 5 & 4 \\ 4 & 5 \end{pmatrix} \tag{4.4.26}$$

as well.

The projection operator technique is useful for matrix inverses:

$$\begin{aligned}\mathbf{A}^{-1} &= \sum_{i=1}^{N} \lambda_i^{-1} |i\rangle\langle i| \\ &= \left(\frac{1}{1}\right)\left(\frac{1}{2}\right)\begin{pmatrix} 1 & -1 \\ -1 & 1 \end{pmatrix} + \left(\frac{1}{3}\right)\left(\frac{1}{2}\right)\begin{pmatrix} 1 & 1 \\ 1 & 1 \end{pmatrix} \\ &= \begin{pmatrix} \frac{2}{3} & -\frac{1}{3} \\ -\frac{1}{3} & \frac{2}{3} \end{pmatrix}\end{aligned} \tag{4.4.27}$$

The above result is easily verified:

$$\mathbf{A}\mathbf{A}^{-1} = \mathbf{I}\begin{pmatrix} 2 & 1 \\ 1 & 2 \end{pmatrix}\begin{pmatrix} \frac{2}{3} & -\frac{1}{3} \\ -\frac{1}{3} & \frac{2}{3} \end{pmatrix} = \begin{pmatrix} 1 & 0 \\ 0 & 1 \end{pmatrix} \tag{4.4.28}$$

Projection matrices for nonsymmetric matrices exist but must be generated from biorthonormal eigenvectors. For example, the matrix

$$\begin{pmatrix} -1 & 2 \\ 1 & -2 \end{pmatrix} \tag{4.4.29}$$

has eigenvalues 0 and -3 with associated eigenvectors,

$$\begin{aligned} |0\rangle &= \left(\frac{1}{\sqrt{3}}\right)\begin{pmatrix} 2 \\ 1 \end{pmatrix} & |-3\rangle &= \left(\frac{1}{\sqrt{3}}\right)\begin{pmatrix} 1 \\ -1 \end{pmatrix} \\ \langle 0| &= \left(\frac{1}{\sqrt{3}}\right)(1 \quad 1) & \langle -3| &= \left(\frac{1}{\sqrt{3}}\right)(1 \quad -2) \end{aligned} \tag{4.4.30}$$

The matrix decomposition is

$$\begin{aligned}\mathbf{A} &= (0)\left(\frac{1}{3}\right)\begin{pmatrix} 2 \\ 1 \end{pmatrix}(1 \quad 1) + (-3)\left(\frac{1}{3}\right)\begin{pmatrix} 1 \\ -1 \end{pmatrix}(1 \quad -2) \\ &= (-1)\begin{pmatrix} 1 \\ -1 \end{pmatrix}(1 \quad -2)\end{aligned} \tag{4.4.31}$$

Projection operator decompositions are not possible when the scalar function formed with the eigenvalue is not defined. For Equation 4.4.29, the presence of $\lambda = 0$ makes the definition of $\mathbf{A}^{-1}$ or $\mathbf{A}^{-N}$ impossible since $1/\lambda = 1/0$ is undefined.

Since a matrix **A** can be raised to any power, it is possible to generate matrix polynomials such as

$$\mathbf{A}^N + c_{N-1}\mathbf{A}^{N-1} + \cdots + c_1\mathbf{A} + c_0\mathbf{I} = 0 \tag{4.4.32}$$

Since each term in the summation can be decomposed using eigenvalues and projection operators, the matrix polynomial can be rewritten as a sum of scalar polynomials with variable λ_i. Each of these scalar polynomials must be multiplied by the appropriate projection matrix:

$$(\lambda_i^N + c_{N-1}\lambda_i^{N-1} + \cdots + c_0)\,\mathbf{P}_i \tag{4.4.33}$$

This decomposition illustrates the origin of the Hamilton–Cayley theorem. If the polynomial is the characteristic polynomial of the matrix **A**, then each term in the summation contains this characteristic equation. Since each eigenvalue is a root of the characteristic polynomial, each term is identically zero when all eigenvalues are distinct. The Hamilton–Cayley theorem results when the projections are reformed into the original matrix.

4.5. The Lagrange–Sylvester Formula

The decomposition of a matrix into eigenvalues and projection operators is a powerful method for solving matrix problems. However, the method does require a determination of eigenvalues and bra and ket eigenvectors. The efficiency increases if the projection operators are determined directly from both symmetric and nonsymmetric matrices if the eigenvalues are known. The Lagrange–Sylvester formula defines projection matrices from the discrete eigenvalues of the matrix.

The Hamilton–Cayley theorem states that any matrix is a solution of its own characteristic equation,

$$\mathbf{A}^N + c_{N-1}\mathbf{A}^{N-1} + \cdots + c_1\mathbf{A} + c_0\mathbf{I} = 0 \tag{4.5.1}$$

The characteristic equation may also be written as a product involving all the roots (eigenvalues) of the characteristic equation,

$$(\lambda - \lambda_1)(\lambda - \lambda_2)\cdots(\lambda - \lambda_N) = 0 \tag{4.5.2}$$

For this reason, the Hamilton–Cayley theorem is also

$$(\mathbf{A} - \lambda_1\mathbf{I})(\mathbf{A} - \lambda_2\mathbf{I})\cdots(\mathbf{A} - \lambda_N\mathbf{I}) = 0 \tag{4.5.3}$$

Since the characteristic equation of this polynomial is zero, there is no advantage to operating on a vector with the full equation; the result of the operation must be zero. However, if a single factor $(\mathbf{A} - \lambda_i\mathbf{I})$ is removed from the product, the product cannot become zero if the matrix polynomial operates on the eigenvector $|i\rangle$ since the factor $(\lambda-\lambda_i)$ is not present to null the result. This means that only vector components on $|i\rangle$ will give a nonzero result when the reduced polynomial operates on an arbitrary vector $|v\rangle$. The $|i\rangle$ component is projected from the remainder of the space.

A 3×3 matrix with eigenvalues λ_1, λ_2, and λ_3 will have a matrix characteristic equation

$$(\mathbf{A}-\lambda_1\mathbf{I})(\mathbf{A}-\lambda_2\mathbf{I})(\mathbf{A}-\lambda_3\mathbf{I})=0 \tag{4.5.4}$$

The projection on $|1\rangle$ is proportional to

$$(\mathbf{A}-\lambda_2\mathbf{I})(\mathbf{A}-\lambda_3\mathbf{I}) \tag{4.5.5}$$

The exact expression for the projection operator for $|1\rangle$ is determined by noting that

$$\mathbf{P}_i\,|i\rangle=|i\rangle \tag{4.5.6}$$

The projection operator matrix produces a factor of unity when it is applied to $|i\rangle$. Equation 4.5.5 applied to $|1\rangle$ gives

$$(\mathbf{A}-\lambda_2\mathbf{I})(\mathbf{A}-\lambda_3\mathbf{I})\,|1\rangle=(\lambda_1-\lambda_2)(\lambda_1-\lambda_3)\,|1\rangle \tag{4.5.7}$$

The result is not zero since $(\mathbf{A}-\lambda_1\mathbf{I})$ is absent. Since a projection matrix for $|1\rangle$ must regenerate $|1\rangle$, Equation 4.5.7 is divided by

$$(\lambda_1-\lambda_2)(\lambda_1-\lambda_3) \tag{4.5.8}$$

to give

$$\mathbf{P}_1=[(\mathbf{A}-\lambda_2\mathbf{I})(\mathbf{A}-\lambda_3\mathbf{I})]/[(\lambda_1-\lambda_2)(\lambda_1-\lambda_3)] \tag{4.5.9}$$

This is the Lagrange–Sylvester formula for a 3×3 matrix. The formula gives projection operators for symmetric or nonsymmetric matrices with nondegenerate eigenvalues.

The projection operators for $|2\rangle$ and $|3\rangle$ are

$$\mathbf{P}_2=[(\mathbf{A}-\lambda_1\mathbf{I})(\mathbf{A}-\lambda_3\mathbf{I})]/[(\lambda_2-\lambda_1)(\lambda_2-\lambda_3)] \tag{4.5.10}$$

$$\mathbf{P}_3=[(\mathbf{A}-\lambda_1\mathbf{I})(\mathbf{A}-\lambda_2\mathbf{I})]/[(\lambda_3-\lambda_1)(\lambda_3-\lambda_2)] \tag{4.5.11}$$

The Lagrange–Sylvester formula can be generalized for the N projection operators of an $N\times N$ matrix. Projection operator $\mathbf{P}_i$ is

$$\mathbf{P}_i=|i\rangle\langle i|=\prod_{\substack{j\\ j\neq i}}^{N}(\mathbf{A}-\lambda_j\mathbf{I})\Big/\prod_{\substack{j\\ j\neq i}}^{N}(\lambda_i-\lambda_j) \tag{4.5.12}$$

The product factors in the numerator $(\mathbf{A}-\lambda_j\mathbf{I})$ do not include $(\mathbf{A}-\lambda_i\mathbf{I})$. In the denominator, the factor $\lambda_i-\lambda_i$ is deleted.

Since the projection operators for the matrix

$$\begin{pmatrix} 2 & 1 \\ 1 & 2 \end{pmatrix} \tag{4.5.13}$$

were determined in Section 4.4, they are determined here using the Lagrange–Sylvester formula for comparison. For the 2×2 matrix, each "product" involves only one matrix factor. The projection matrices are

$$\mathbf{P}(\lambda = 1) = (\mathbf{A} - 3\mathbf{I})/(1-3)$$

$$= \left[\begin{pmatrix} 2 & 1 \\ 1 & 2 \end{pmatrix} - \begin{pmatrix} 3 & 0 \\ 0 & 3 \end{pmatrix}\right](-2)^{-1} \tag{4.5.14}$$

$$= \left(\frac{1}{2}\right)\begin{pmatrix} 1 & -1 \\ -1 & 1 \end{pmatrix} \tag{4.5.15}$$

and

$$\mathbf{P}(\lambda = 3) = (\mathbf{A} - 1\mathbf{I})/(3-1) \tag{4.5.16}$$

$$= \left(\frac{1}{2}\right)\begin{pmatrix} 1 & 1 \\ 1 & 1 \end{pmatrix} \tag{4.5.17}$$

in agreement with the previous result.

The Lagrange–Sylvester formula generates the projection operators directly. The eigenvectors are determined by applying the projection matrix to a vector with a finite projection on the eigenvector. The projection $\mathbf{P}(\lambda = 1)$ for the vector

$$|v\rangle = \begin{pmatrix} 2 \\ 1 \end{pmatrix} \tag{4.5.18}$$

is

$$\mathbf{P}(\lambda = 1)\,|v\rangle = \left(\frac{1}{2}\right)\begin{pmatrix} 1 & -1 \\ -1 & 1 \end{pmatrix}\begin{pmatrix} 2 \\ 1 \end{pmatrix} = \begin{pmatrix} \frac{1}{2} \\ \frac{-1}{2} \end{pmatrix} = \left(\frac{1}{2}\right)\begin{pmatrix} 1 \\ -1 \end{pmatrix} \tag{4.5.19}$$

The projection operators have some special properties. Since each projection operator spans one dimension of the space, the sum of all the projection matrices must equal the identity. The projection matrices for the 2×2 matrix sum as

$$\mathbf{P}_1 + \mathbf{P}_2 = \mathbf{I} \tag{4.5.20}$$

$$\left(\frac{1}{2}\right)\begin{pmatrix} 1 & -1 \\ -1 & 1 \end{pmatrix} + \left(\frac{1}{2}\right)\begin{pmatrix} 1 & 1 \\ 1 & 1 \end{pmatrix} = \begin{pmatrix} 1 & 0 \\ 0 & 1 \end{pmatrix} \tag{4.5.21}$$

For an $N \times N$ matrix with N distinct eigenvalues, this completeness condition is

$$\sum_{i=1}^{N} \mathbf{P}_i = \mathbf{I} \tag{4.5.22}$$

If $N-1$ projection matrices are known, the final projection matrix can be determined by subtracting the sum of the other projection matrices from the unit matrix.

A projection operator $\mathbf{P}_i$ operating on a vector $|v\rangle$ will isolate only the $|i\rangle$ projection,

$$|i\rangle\langle i|v\rangle \tag{4.5.23}$$

Once the projection is complete, subsequent operations with $\mathbf{P}_i$ will reproduce $|i\rangle$ with no additional change. In other words, a single projection $\mathbf{P}_i$ functions exactly as two or more $\mathbf{P}_i$ in sequence:

$$\mathbf{P}_i\mathbf{P}_i\,|v\rangle = \mathbf{P}_i\,|i\rangle\langle i|v\rangle = |i\rangle\langle i|v\rangle = \mathbf{P}_i\,|v\rangle \tag{4.5.24}$$

The $\mathbf{P}_i$ are idempotent matrices defined by

$$\mathbf{P}_i = \mathbf{P}_i\mathbf{P}_i \tag{4.5.25}$$

If a projection $\mathbf{P}_i$ which generates the $|i\rangle$ eigenvector is followed by a second projection $\mathbf{P}_j$ for a different eigenvalue, λ_j, the $\mathbf{P}_j$ projection gives a zero projection on $|i\rangle$, i.e.,

$$\mathbf{P}_j\mathbf{P}_i = 0 \tag{4.5.26}$$

The projection operator matrices are orthogonal under matrix multiplication.

The orthogonality properties can be tested with the 2×2 matrix projection operators:

$$\begin{aligned}\mathbf{P}(\lambda=1) = \mathbf{P}(\lambda=1)^2 &= \left(\frac{1}{4}\right)\begin{pmatrix} 1 & -1 \\ -1 & 1 \end{pmatrix}\begin{pmatrix} 1 & -1 \\ -1 & 1 \end{pmatrix} \\ &= \left(\frac{1}{4}\right)\begin{pmatrix} 2 & -2 \\ -2 & 2 \end{pmatrix} = \left(\frac{1}{2}\right)\begin{pmatrix} 1 & -1 \\ -1 & 1 \end{pmatrix}\end{aligned} \tag{4.5.27}$$

$$\begin{aligned}\mathbf{P}(\lambda=3) = \mathbf{P}(\lambda=3)^2 &= \left(\frac{1}{4}\right)\begin{pmatrix} 1 & 1 \\ 1 & 1 \end{pmatrix}\begin{pmatrix} 1 & 1 \\ 1 & 1 \end{pmatrix} \\ &= \left(\frac{1}{4}\right)\begin{pmatrix} 2 & 2 \\ 2 & 2 \end{pmatrix} = \left(\frac{1}{2}\right)\begin{pmatrix} 1 & 1 \\ 1 & 1 \end{pmatrix}\end{aligned} \tag{4.5.28}$$

The orthogonality property is

$$\mathbf{P}(\lambda=1)\,\mathbf{P}(\lambda=3) = \left(\frac{1}{4}\right)\begin{pmatrix} 1 & -1 \\ -1 & 1 \end{pmatrix}\begin{pmatrix} 1 & 1 \\ 1 & 1 \end{pmatrix} = 0 \tag{4.5.29}$$

The determination of the projection operators becomes more difficult as the order of the matrix increases. The nonsymmetric matrix

$$\begin{pmatrix} 2 & 1 & 0 \\ 3 & 2 & 1 \\ 0 & 3 & 2 \end{pmatrix} \tag{4.5.30}$$

has the characteristic equation

$$(2-\lambda)(\lambda^2-4\lambda-2)=0 \tag{4.5.31}$$

and eigenvalues

$$2, \qquad 2+\sqrt{6}, \qquad 2-\sqrt{6} \tag{4.5.32}$$

For the 3×3 matrix, the projection operators are

$$\mathbf{P}(\lambda=2)=[\mathbf{A}-(2+\sqrt{6})\mathbf{I}][\mathbf{A}-(2-\sqrt{6})\mathbf{I}]/\{[2-(2+\sqrt{6})][2-(2-\sqrt{6})]\}$$

$$=\left\{\begin{pmatrix} 2 & 1 & 0 \\ 3 & 2 & 1 \\ 0 & 3 & 2 \end{pmatrix}-\begin{pmatrix} 2+\sqrt{6} & 0 & 0 \\ 0 & 2+\sqrt{6} & 0 \\ 0 & 0 & 2+\sqrt{6} \end{pmatrix}\right\}\left\{\begin{pmatrix} 2 & 1 & 0 \\ 3 & 2 & 1 \\ 0 & 3 & 2 \end{pmatrix}\right.$$

$$\left.-\begin{pmatrix} 2-\sqrt{6} & 0 & 0 \\ 0 & 2-\sqrt{6} & 0 \\ 0 & 0 & 2-\sqrt{6} \end{pmatrix}\right\}\{[2-(2+\sqrt{6})][2-(2-\sqrt{6})]\}^{-1}$$

$$=\left(\frac{-1}{6}\right)\begin{pmatrix} -\sqrt{6} & 1 & 0 \\ 3 & -\sqrt{6} & 1 \\ 0 & 3 & -\sqrt{6} \end{pmatrix}\begin{pmatrix} \sqrt{6} & 1 & 0 \\ 3 & \sqrt{6} & 1 \\ 0 & 3 & \sqrt{6} \end{pmatrix}$$

$$=\left(\frac{-1}{6}\right)\begin{pmatrix} -3 & 0 & 1 \\ 0 & 0 & 0 \\ 9 & 0 & -3 \end{pmatrix} \tag{4.5.33}$$

$$\mathbf{P}(\lambda=2+\sqrt{6})=(\mathbf{A}-2\mathbf{I})[\mathbf{A}-(2-\sqrt{6})\mathbf{I}]/\{(2+\sqrt{6}-2)[2+\sqrt{6}-(2-\sqrt{6})]\}$$

$$=\left(\frac{1}{12}\right)\begin{pmatrix} 0 & 1 & 0 & \sqrt{6} \\ 3 & 0 & 1 & 3 \\ 0 & 3 & 0 & 0 \end{pmatrix}\begin{pmatrix} 1 & 0 \\ \sqrt{6} & 1 \\ 3 & \sqrt{6} \end{pmatrix}$$

$$=\left(\frac{1}{12}\right)\begin{pmatrix} 3 & \sqrt{6} & 1 \\ 3\sqrt{6} & 6 & \sqrt{6} \\ 1 & 3\sqrt{6} & 3 \end{pmatrix} \tag{4.5.34}$$

$$\mathbf{P}(\lambda=2-\sqrt{6})=(\mathbf{A}-2\mathbf{I})\{\mathbf{A}-(2+\sqrt{6})\mathbf{I}\}/\{(2-\sqrt{6}-2)[2-\sqrt{6}-(2+\sqrt{6})]\}$$

$$=\left(\frac{1}{12}\right)\begin{pmatrix}0 & 1 & 0\\ 3 & 0 & 1\\ 0 & 3 & 0\end{pmatrix}\begin{pmatrix}-\sqrt{6} & 1 & 0\\ 3 & -\sqrt{6} & 1\\ 0 & 3 & -\sqrt{6}\end{pmatrix}$$

$$=\left(\frac{1}{12}\right)\begin{pmatrix}3 & -\sqrt{6} & 1\\ -3\sqrt{6} & 6 & -\sqrt{6}\\ 9 & -3\sqrt{6} & 3\end{pmatrix} \quad (4.5.35)$$

Two sets of vectors are required to develop **P** from the bra and ket vectors so the Lagrange–Sylvester formula is quite efficient. These matrices are verified using the projection operator properties (Problem 4.12).

It is often convenient to leave the projection operators in the (unmultiplied) Lagrange–Sylvester form. For example, the matrix

$$\mathbf{A}=\begin{pmatrix}2 & 1\\ 1 & 2\end{pmatrix} \quad (4.5.36)$$

has projection operators

$$\mathbf{P}(\lambda=1)=(\mathbf{A}-3\mathbf{I})/(1-3) \qquad \mathbf{P}(\lambda=3)=(\mathbf{A}-1\mathbf{I})/(3-1) \quad (4.5.37)$$

The matrices **A** and **I** appear in the final expression and permit the development of polynomial expressions for many matrix functions.

The squared inverse matrix, $\mathbf{A}^{-2}$, is expanded to

$$\begin{aligned}\mathbf{A}^{-2}&=\lambda_1^{-2}\mathbf{P}(1)+\lambda_3^{-2}\mathbf{P}(3)\\ &=(1)^{-2}(\mathbf{A}-3\mathbf{I})/(-2)+(3)^{-2}(\mathbf{A}-\mathbf{I})/2\end{aligned} \quad (4.5.38)$$

Terms in **A** and **I** are collected to give

$$\mathbf{A}^{-2}=\frac{1}{9}(-4\mathbf{A}+|3|) \quad (4.5.39)$$

4.6. Degenerate Eigenvalues

The matrices selected as illustrative examples for similarity transforms and projection operators in previous sections have special properties which are not common to all matrices. For example, each $N \times N$ matrix had a complete set of N eigenvectors and could be completely diagonalized by a similarity transform. Such matrices are simple matrices. Hermitian and symmetric matrices are examples of simple matrices which give a complete set of orthogonal eigenvectors.

The matrices selected were also characterized by nondegenerate eigenvalues. A simple $N\times N$ matrix will still have a rank N, but there are $N\times N$ matrices with degenerate eigenvalues which are not simple. Because their rank is less than N, it will be impossible to find a complete set of eigenvectors. In this section, simple and nonsimple matrices with degenerate eigenvalues are compared.

The matrix

$$\begin{pmatrix} -2 & 1 & 0 & 1 \\ 1 & -2 & 1 & 0 \\ 0 & 1 & -2 & 1 \\ 1 & 0 & 1 & -2 \end{pmatrix} \tag{4.6.1}$$

is an example of a symmetric matrix with a degenerate eigenvalue. The characteristic determinant $|\mathbf{A}-\lambda\mathbf{I}|$ is expanded in its first row to give

$$-(2-\lambda)\begin{vmatrix} -2-\lambda & 1 & 0 \\ 1 & -2-\lambda & 1 \\ 0 & 1 & -2-\lambda \end{vmatrix} + (-1)\begin{vmatrix} 1 & 1 & 0 \\ 0 & -2-\lambda & 1 \\ 1 & 1 & -2-\lambda \end{vmatrix}$$

$$+(-1)\begin{vmatrix} 1 & -2-\lambda & 1 \\ 0 & 1 & -2-\lambda \\ 1 & 0 & 1 \end{vmatrix}$$

$$= -(2+\lambda)[-(2+\lambda)^3+2(2+\lambda)]-2(2+\lambda)^2=0 \tag{4.6.2}$$

which factors as

$$-(2+\lambda)^2\,(\lambda)(\lambda+4)=0 \tag{4.6.3}$$

There are two nondegenerate eigenvalues, 0 and -4, and the doubly degenerate eigenvalue $\lambda=-2$ arising from the square factor in the characteristic equation. All eigenvalues are real for this symmetric matrix.

The projection operators for the nondegenerate eigenvalues must be calculated using the degenerate eigenvalue twice. The $\mathbf{P}(\lambda=0)$ projection matrix is

$$\mathbf{P}(\lambda=0)=[\mathbf{A}-(-2)\mathbf{I}][\mathbf{A}-(-2)\mathbf{I}][\mathbf{A}+4\mathbf{I}]/\{[0-(-2)][0-(-4)]\}$$

$$=\begin{pmatrix} 0 & 1 & 0 & 1 \\ 1 & 0 & 1 & 0 \\ 0 & 1 & 0 & 1 \\ 1 & 0 & 1 & 0 \end{pmatrix}\begin{pmatrix} 0 & 1 & 0 & 1 \\ 1 & 0 & 1 & 0 \\ 0 & 1 & 0 & 1 \\ 1 & 0 & 1 & 0 \end{pmatrix}\begin{pmatrix} 2 & 1 & 0 & 1 \\ 1 & 2 & 1 & 0 \\ 0 & 1 & 2 & 1 \\ 1 & 0 & 1 & 2 \end{pmatrix}(2^4)^{-1}$$

$$=\left(\frac{1}{4}\right)\begin{pmatrix} 1 & 1 & 1 & 1 \\ 1 & 1 & 1 & 1 \\ 1 & 1 & 1 & 1 \\ 1 & 1 & 1 & 1 \end{pmatrix} \tag{4.6.4}$$

The projection operator $\mathbf{P}(\lambda = -4)$ is

$$\mathbf{P}(\lambda = -4) = [\mathbf{A} - (-2)\mathbf{I}]^2 [\mathbf{A} - 0\mathbf{I}]/\{[4 - (-2)]^2 (4 - 0)\}$$
$$= \left(\frac{1}{4}\right) \begin{pmatrix} 1 & -1 & 1 & -1 \\ -1 & 1 & -1 & 1 \\ 1 & -1 & 1 & -1 \\ -1 & 1 & -1 & 1 \end{pmatrix} \tag{4.6.5}$$

The corresponding eigenvectors for the symmetric matrix are (Problem 4.17)

$$\left(\frac{1}{2}\right)(1 \quad 1 \quad 1 \quad 1) \tag{4.6.6}$$

and

$$\left(\frac{1}{2}\right)(1 \quad -1 \quad 1 \quad -1) \tag{4.6.7}$$

The eigenvectors for the two nondegenerate eigenvalues are orthogonal.

Since the symmetric matrix has a complete set of eigenvectors, the projection operator for the two-dimensional subspace of $\lambda = -2$ is found using the completeness condition for projection matrices:

$$\mathbf{P}(\lambda = 0) + \mathbf{P}(\lambda = -4) + \mathbf{P}(\lambda = -2) = \mathbf{I} \tag{4.6.8}$$

so that

$$\mathbf{P}(\lambda = -2) = \mathbf{I} - \mathbf{P}(\lambda = 0) - \mathbf{P}(\lambda = -4) \tag{4.6.9}$$

$\mathbf{P}(\lambda = -2)$ must span two dimensions in the four-dimensional space. The matrix projection operator is

$$\begin{pmatrix} 1 & 0 & 0 & 0 \\ 0 & 1 & 0 & 0 \\ 0 & 0 & 1 & 0 \\ 0 & 0 & 0 & 1 \end{pmatrix} - \left(\frac{1}{4}\right) \begin{pmatrix} 1 & 1 & 1 & 1 \\ 1 & 1 & 1 & 1 \\ 1 & 1 & 1 & 1 \\ 1 & 1 & 1 & 1 \end{pmatrix} - \left(\frac{1}{4}\right) \begin{pmatrix} 1 & -1 & 1 & -1 \\ -1 & 1 & -1 & 1 \\ 1 & -1 & 1 & -1 \\ -1 & 1 & -1 & 1 \end{pmatrix}$$
$$= \left(\frac{1}{2}\right) \begin{pmatrix} 1 & 0 & -1 & 0 \\ 0 & 1 & 0 & -1 \\ -1 & 0 & 1 & 0 \\ 0 & -1 & 0 & 1 \end{pmatrix} \tag{4.6.10}$$

Since $\mathbf{P}(\lambda = -2)$ spans two dimensions in the space, there are two eigenvectors. However, these two eigenvectors are not unique. The characteristic matrix for $\lambda = -2$ is

$$(\mathbf{A} - \lambda\mathbf{I}) = \begin{pmatrix} 0 & 1 & 0 & 1 \\ 1 & 0 & 1 & 0 \\ 0 & 1 & 0 & 1 \\ 1 & 0 & 1 & 0 \end{pmatrix} \tag{4.6.11}$$

The vectors

$$\begin{aligned} &(1 \quad 1 \quad -1 \quad -1) \\ &(1 \quad 0 \quad -1 \quad 0) \\ &(1 \quad -1 \quad -1 \quad 1) \end{aligned} \tag{4.6.12}$$

all satisfy the equation

$$(\mathbf{A} + 2\mathbf{I})\,|-2\rangle = 0 \tag{4.6.13}$$

The eigenvectors are not proportional to each other, and each is orthogonal to $|0\rangle$ and $|-4\rangle$. However, they are not necessarily orthogonal to each other. Among the three eigenvectors of Equation 4.6.12, only the first and third are orthogonal.

Although the symmetric matrix must have four orthogonal eigenvectors, these eigenvectors are not unique for the degenerate eigenvalue. A number of eigenvectors for the subspace will be orthogonal to the eigenvectors for the remaining eigenvalues. Two mutually orthogonal eigenvectors for the subspace must be developed using, for example, the Gram–Schmidt orthogonalization procedure.

The projection operator $\mathbf{P}(\lambda = -2)$ is also generated from eigenvectors of the subspace. The vector

$$\langle -2| = (1 \quad 0 \quad -1 \quad 0) \tag{4.6.14}$$

is an eigenvector for $\lambda = -2$. Its projection operator is

$$|2\rangle\langle 2| = \left(\frac{1}{2}\right) \begin{pmatrix} 1 & 0 & -1 & 0 \\ 0 & 0 & 0 & 0 \\ -1 & 0 & 1 & 0 \\ 0 & 0 & 0 & 0 \end{pmatrix} \tag{4.6.15}$$

Columns 2 and 4 have only zero elements.

The full projection for the subspace requires a second eigenvector. For a mutually orthogonal eigenvector, Gram–Schmidt orthogonalization is applied to any vector $|v\rangle$ with an eigenvalue of -2. The eigenvector

$$|v\rangle = \begin{pmatrix} 1 \\ 1 \\ -1 \\ -1 \end{pmatrix} \tag{4.6.16}$$

gives an eigenvector $|2'\rangle$ orthogonal to $|2\rangle$,

$$|2'\rangle = |v\rangle - |2\rangle\langle 2|v\rangle$$

$$|2'\rangle = \begin{pmatrix} 1 \\ 1 \\ -1 \\ -1 \end{pmatrix} - \left(\frac{1}{2}\right)\begin{pmatrix} 1 \\ 0 \\ -1 \\ 0 \end{pmatrix}(1 \quad 0 \quad -1 \quad 0)\begin{pmatrix} 1 \\ 1 \\ -1 \\ 0 \end{pmatrix} = \begin{pmatrix} 0 \\ 1 \\ 0 \\ -1 \end{pmatrix} \tag{4.6.17}$$

with nonzero second and fourth components. Its projection operator is

$$\left(\frac{1}{2}\right)\begin{pmatrix} 0 \\ 1 \\ 0 \\ -1 \end{pmatrix}(0 \quad 1 \quad 0 \quad -1) = \left(\frac{1}{2}\right)\begin{pmatrix} 0 & 0 & 0 & 0 \\ 0 & 1 & 0 & -1 \\ 0 & 0 & 0 & 0 \\ 0 & -1 & 0 & 1 \end{pmatrix} \tag{4.6.18}$$

The sum of the two $\lambda = -2$ projection operators gives the full projection operator for the subspace:

$$\begin{aligned} &\mathbf{P}(\lambda = -2) + \mathbf{P}'(\lambda = -2) \\ &= \left(\frac{1}{2}\right)\begin{pmatrix} 1 & 0 & -1 & 0 \\ 0 & 0 & 0 & 0 \\ -1 & 0 & 1 & 0 \\ 0 & 0 & 0 & 0 \end{pmatrix} + \left(\frac{1}{2}\right)\begin{pmatrix} 0 & 0 & 0 & 0 \\ 0 & 1 & 0 & -1 \\ 0 & 0 & 0 & 0 \\ 0 & -1 & 0 & 1 \end{pmatrix} \\ &= \left(\frac{1}{2}\right)\begin{pmatrix} 1 & 0 & -1 & 0 \\ 0 & 1 & 0 & -1 \\ -1 & 0 & 1 & 0 \\ 0 & -1 & 0 & 1 \end{pmatrix} \end{aligned} \tag{4.6.19}$$

The selection of an initial eigenvector in the $\lambda = -2$ subspace was completely arbitrary. A second choice of eigenvectors for the subspace gives entirely different projection operators. For example, the eigenvectors

$$|-2\rangle = \left(\frac{1}{2}\right)\begin{pmatrix} 1 \\ -1 \\ -1 \\ 1 \end{pmatrix} \qquad |-2'\rangle = \left(\frac{1}{2}\right)\begin{pmatrix} 1 \\ 1 \\ -1 \\ -1 \end{pmatrix} \tag{4.6.20}$$

give the equally valid projection operators (Problem 4.27)

$$\mathbf{P}(\lambda=-2)=\left(\frac{1}{4}\right)\begin{pmatrix} 1 & -1 & -1 & 1 \\ -1 & 1 & 1 & -1 \\ -1 & 1 & 1 & -1 \\ 1 & -1 & -1 & 1 \end{pmatrix} \tag{4.6.21}$$

and

$$\mathbf{P}(\lambda=-2')=\left(\frac{1}{4}\right)\begin{pmatrix} 1 & 1 & -1 & -1 \\ 1 & 1 & -1 & -1 \\ -1 & -1 & 1 & 1 \\ -1 & -1 & 1 & 1 \end{pmatrix} \tag{4.6.22}$$

Although the $\mathbf{P}_i$ are different, their sum gives the proper subspace projection matrix:

$$\begin{aligned} &\mathbf{P}(\lambda=-2)+\mathbf{P}'(\lambda=-2) \\ &=\left(\frac{1}{4}\right)\begin{pmatrix} 1 & -1 & -1 & 1 \\ -1 & 1 & 1 & -1 \\ -1 & 1 & 1 & -1 \\ 1 & -1 & -1 & 1 \end{pmatrix}+\left(\frac{1}{4}\right)\begin{pmatrix} 1 & 1 & -1 & -1 \\ 1 & 1 & -1 & -1 \\ -1 & -1 & 1 & 1 \\ -1 & -1 & 1 & 1 \end{pmatrix} \\ &=\left(\frac{1}{2}\right)\begin{pmatrix} 1 & 0 & -1 & 0 \\ 0 & 1 & 0 & -1 \\ -1 & 0 & 1 & 0 \\ 0 & -1 & 0 & 1 \end{pmatrix} \end{aligned} \tag{4.6.23}$$

These results are general for simple matrices with degenerate eigenvalues. A mutually orthogonal set of eigenvectors must be developed. In many cases, mutually orthogonal eigenvectors within the subspace are found as sums or differences of other valid subspace eigenvectors. For example, the two orthogonal eigenvectors

$$\left(\frac{1}{2}\right)(1 \quad -1 \quad -1 \quad 1) \quad \text{and} \quad \left(\frac{1}{2}\right)(1 \quad 1 \quad -1 \quad -1) \tag{4.6.24}$$

can be added to give

$$(1 \quad 0 \quad -1 \quad 0) \tag{4.6.25}$$

and subtracted to give

$$(0 \quad -1 \quad 0 \quad 1) \tag{4.6.26}$$

This type of linear combination occurs frequently with eigenvectors in quantum mechanics.

A nonsymmetric matrix may give a complete set of eigenvectors only if it is simple. The matrix

$$\mathbf{A} = \begin{pmatrix} 1 & 0 & 0 \\ -1 & 1 & 0 \\ 0 & -1 & 0 \end{pmatrix} \tag{4.6.27}$$

has a doubly degenerate ($\lambda = 1$) eigenvalue. The $\lambda = 0$ eigenvalue has eigenvectors

$$\langle 0| = (1 \quad 1 \quad 1) \qquad |0\rangle = \begin{pmatrix} 0 \\ 0 \\ 1 \end{pmatrix} \tag{4.6.28}$$

and projection operator

$$\begin{pmatrix} 0 & 0 & 0 \\ 0 & 0 & 0 \\ 1 & 1 & 1 \end{pmatrix} \tag{4.6.29}$$

Since the doubly degenerate subspace constitutes the remainder of the space, the projection matrix for this subspace is determined as

$$\mathbf{P}(\lambda = 1) = \mathbf{I} - \mathbf{P}(\lambda = 0) \tag{4.6.30}$$

$$= \begin{pmatrix} 1 & 0 & 0 \\ 0 & 1 & 0 \\ -1 & -1 & 0 \end{pmatrix} \tag{4.6.31}$$

The projection matrices obey the equations for projection matrices (Problem 4.18):

$$\mathbf{P}_i \mathbf{P}_j = 0 \qquad \mathbf{P}_i \mathbf{P}_i = \mathbf{P}_i \tag{4.6.32}$$

Although the analysis appears consistent, other matrices with the same eigenvalues can have exactly the same projection operator. For example, the matrix

$$\begin{pmatrix} 1 & 0 & 0 \\ 0 & 1 & 0 \\ -1 & -1 & 0 \end{pmatrix} \tag{4.6.33}$$

has the same eigenvalues and gives exactly the same projection operators (Problem 4.28).

This ambiguity is resolved by examining the characteristic equation for $\lambda = 1$,

$$(\mathbf{A} - \lambda \mathbf{I})\,|1\rangle = \begin{pmatrix} 0 & 0 & 0 \\ -1 & 0 & 0 \\ 0 & -1 & -1 \end{pmatrix} \begin{pmatrix} a \\ b \\ c \end{pmatrix} = 0 \tag{4.6.34}$$

The equations are

$$\begin{aligned} -a &= 0 \\ -b - c &= 0 \end{aligned} \tag{4.6.35}$$

The final equation establishes a relationship between b and c but this produces only a single eigenvector. The second matrix (Equation 4.6.33) with the same projection operators also produces a single eigenvector. The matrix is not simple and does not have a complete set of eigenvectors, and the projection matrices and eigenvalues may not reproduce the initial **A** matrix (Problem 4.28).

The presence of degenerate eigenvalues for a symmetric matrix poses a problem for the determination of projection operators using the Lagrange–Sylvester formula. The formula has a denominator involving eigenvalue differences,

$$(\lambda_i - \lambda_1)(\lambda_i - \lambda_2) \cdots (\lambda_i - \lambda_N) \tag{4.6.36}$$

where λ_i cannot appear as the second eigenvalue in any factor. If there is a degenerate eigenvalue, one of the factors will be zero and the Lagrange–Sylvester formula fails. This problem can be circumvented by a simple technique, which is also applicable to functions of matrices (Section 6.7).

A simple matrix **A** with two degenerate eigenvalues, λ_1 and λ_2, can be expanded in its eigenvalues and projection matrices as

$$\mathbf{A} = \lambda_1 \mathbf{P}(1) + \lambda_2 \mathbf{P}(2) + \cdots \tag{4.6.37}$$

where only the first two terms for the degenerate eigenvalues concern us. The projection matrices are

$$\mathbf{P}(1) = (\mathbf{A} - \lambda_2 \mathbf{I}) D / (\lambda_1 - \lambda_2) \tag{4.6.38}$$

and

$$\mathbf{P}(2) = (\mathbf{A} - \lambda_1 \mathbf{I}) D / (\lambda_2 - \lambda_1) \tag{4.6.39}$$

with

$$D = (\mathbf{A} - \lambda_3 \mathbf{I}) \cdots (\mathbf{A} - \lambda_N \mathbf{I}) / [(\lambda_1 - \lambda_3) \cdots (\lambda_1 - \lambda_N)] \tag{4.6.40}$$

i.e., the remainder of each projection matrix.

The first two terms of the expansion for **A** are

$$\mathbf{A} = \lambda_1(\mathbf{A} - \lambda_2\mathbf{I})D/(\lambda_1 - \lambda_2) + \lambda_2(\mathbf{A} - \lambda_1\mathbf{I})D/(\lambda_2 - \lambda_1) \tag{4.6.41}$$

The terms can be combined to give

$$\begin{aligned} \mathbf{A} &= [(\lambda_1 - \lambda_2)\mathbf{A} + (-\lambda_1\lambda_2 + \lambda_1\lambda_2)]D/(\lambda_1 - \lambda_2) \\ &= \mathbf{A}D \end{aligned} \tag{4.6.42}$$

The two terms for the degenerate eigenvalues reduce to the matrix **A** times the factor D.

The technique is demonstrated for the matrix

$$\mathbf{A} = \begin{pmatrix} 2 & -1 & -1 \\ -1 & 2 & -1 \\ -1 & -1 & 2 \end{pmatrix} \tag{4.6.43}$$

with a single eigenvalue 0 and a doubly degenerate eigenvalue $+3$. $\mathbf{P}(3)$ is

$$\begin{aligned} \mathbf{P}(3) &= \mathbf{AD} = \mathbf{A}(\mathbf{A} - 0\mathbf{I})/(3 - 0) \\ &= \frac{1}{3}\mathbf{A}^2 \\ &= \left(\frac{1}{3}\right)\begin{pmatrix} 2 & -1 & -1 \\ -1 & 2 & -1 \\ -1 & -1 & 2 \end{pmatrix}\begin{pmatrix} 2 & -1 & -1 \\ -1 & 2 & -1 \\ -1 & -1 & 2 \end{pmatrix} = \left(\frac{1}{3}\right)\begin{pmatrix} 2 & -1 & -1 \\ -1 & 2 & -1 \\ -1 & -1 & 2 \end{pmatrix} \end{aligned} \tag{4.6.44}$$

which can be verified using the completeness relation.

4.7. Matrix Functions and Equations

The resolution of a matrix into eigenvalues and projection operators effectively resolves an N-dimensional space into N one-dimensional spaces. The eigenvalue for a specific eigenvector is then used as a scalar variable.

The advantages of these decompositions are demonstrated using a 3×3 diagonal matrix which operates on $|v\rangle$ to produce a new vector $|u\rangle$:

$$|u\rangle = \mathbf{A}\,|v\rangle \tag{4.7.1}$$

$$\begin{pmatrix} u_1 \\ u_2 \\ u_3 \end{pmatrix} = \begin{pmatrix} a_{11} & 0 & 0 \\ 0 & a_{22} & 0 \\ 0 & 0 & a_{33} \end{pmatrix}\begin{pmatrix} v_1 \\ v_2 \\ v_3 \end{pmatrix}$$

$$\begin{aligned} u_1 &= a_{11}v_1 \\ u_2 &= a_{22}v_2 \\ u_3 &= a_{33}v_3 \end{aligned} \tag{4.7.2}$$

The matrix equation resolves into three separate equations. The same situation holds if the variables u_i are replaced by the derivatives

$$u_i = dv_i/dt \tag{4.7.3}$$

The three scalar equations are now first-order differential equations,

$$\begin{aligned} dv_1/dt &= a_{11}v_1 \\ dv_2/dt &= a_{22}v_2 \\ dv_3/dt &= a_{33}v_3 \end{aligned} \tag{4.7.4}$$

The three equations together constitute a matrix differential equation,

$$\begin{pmatrix} dv_1/dt \\ dv_2/dt \\ dv_3/dt \end{pmatrix} = \begin{pmatrix} a_{11} & 0 & 0 \\ 0 & a_{22} & 0 \\ 0 & 0 & a_{33} \end{pmatrix} \begin{pmatrix} v_1 \\ v_2 \\ v_3 \end{pmatrix} \tag{4.7.5}$$

which is written formally as

$$d\,|v\rangle/dt = \mathbf{A}\,|v\rangle \tag{4.7.6}$$

For the diagonal $\mathbf{A}$ matrix, there are three independent first-order differential equations. If the initial value of v_i is v_i^0, the differential equation has a solution

$$v_i(t) = v_i^0 \exp(a_{ii}t) \tag{4.7.7}$$

This scalar solution predicts the full matrix solution of Equation 4.7.6.

The definition of

$$\exp(\mathbf{A}) \tag{4.7.8}$$

is ambiguous. The notation might suggest, for example, that each element in the matrix must be exponentiated. However, since the matrix function is determined by its action on vectors, it is convenient to expand the exponential as a polynomial series. An exponential of the scalar x is

$$\exp(x) = 1 + x/1! + x^2/2! + \cdots + x^n/n! + \cdots \tag{4.7.9}$$

The exponential of $\mathbf{A}$ is defined as a polynomial in powers of $\mathbf{A}$:

$$\exp(\mathbf{A}) = \mathbf{I} + \mathbf{A}/1! + \mathbf{A}^2/2! + \cdots + \mathbf{A}^n/n! + \cdots \tag{4.7.10}$$

This is merely an alternate functional statement for the exponential. However, each term in the Taylor series can now be resolved into N distinct terms through a projection operator decomposition. The full series definition of $\exp(\mathbf{A})$ is

$$\sum_n^\infty \sum_i^N (\lambda_i^n/n!)\,|i\rangle\langle i| \tag{4.7.11}$$

where the index i sums the set of N projection operators and the index n sums terms in the infinite Taylor series which defines the exponential.

The infinite sum is equal to the exponential function. If the summation is replaced by the exponential function, the infinite sum becomes the exponential of the eigenvalue:

$$\exp(\mathbf{A}) = \sum \exp(\lambda_i)\, |i\rangle\langle i| \tag{4.7.12}$$

This equation highlights the major advantage of the projection operator method. The matrix exponential is now a sum of scalar exponentials weighted by the projection matrices. In the previous section, the matrix $\mathbf{A}$ was decomposed as

$$\mathbf{A} = \sum_i \lambda_i\, |i\rangle\langle i| \tag{4.7.13}$$

This result can now be generalized for any function, $f(\mathbf{A})$, of $\mathbf{A}$ as

$$f(\mathbf{A}) = \sum_i^N f(\lambda_i)\, |i\rangle\langle i| \tag{4.7.14}$$

This decomposition works for each valid scalar function. For example, the matrix inverse, $\mathbf{A}^{-1}$, cannot be defined for a matrix with an eigenvalue of 0 since the function

$$1/\lambda \tag{4.7.15}$$

is not defined when $\lambda = 0$.

Since the matrix will resolve into N one-dimensional scalar problems, the matrix equation can often be solved exactly as if it were a scalar problem. However, this is true only if the system can be resolved in this manner. All matrices in a given equation must be diagonalizable by the same similarity transform, i.e., the matrices in the equation must commute. The product of matrix exponentials of matrices $\mathbf{A}$ and $\mathbf{B}$ is

$$\exp(\mathbf{A})\exp(\mathbf{B}) \tag{4.7.16}$$

and is resolved by expanding each exponential:

$$\begin{aligned} \exp(\mathbf{A})\exp(\mathbf{B}) &= (\mathbf{I} + \mathbf{A}/1! + \mathbf{A}^2/2! + \cdots)(\mathbf{I} + \mathbf{B}/1! + \mathbf{B}^2/2! + \cdots) \\ &= \mathbf{I} + (\mathbf{A} + \mathbf{B})/1! + \frac{1}{2}(\mathbf{A}^2 + 2\mathbf{AB} + \mathbf{B}^2) + \cdots \end{aligned} \tag{4.7.17}$$

This expansion appears to be the expansion for

$$\exp[(\mathbf{A} + \mathbf{B})t] \tag{4.7.18}$$

but this is not true if **A** and **B** do not commute. The quadratic term of Equation 4.7.18 must include **AB** and **BA**:

$$(\mathbf{A}+\mathbf{B})^2=\mathbf{A}^2+\mathbf{AB}+\mathbf{BA}+\mathbf{B}^2 \tag{4.7.19}$$

The difference

$$\exp(\mathbf{A}+\mathbf{B})-\exp(\mathbf{A})\exp(\mathbf{B}) \tag{4.7.20}$$

is

$$\begin{aligned}&\frac{1}{2}(\mathbf{A}^2+\mathbf{AB}+\mathbf{BA}+\mathbf{B}^2-\mathbf{A}^2-2\mathbf{AB}-\mathbf{B}^2)+\cdots\\&=\frac{1}{2}(\mathbf{BA}-\mathbf{AB})=\frac{1}{2}[\mathbf{B},\mathbf{A}]\end{aligned} \tag{4.7.21}$$

The difference terms vanish only if **A** and **B** commute,

$$[\mathbf{B},\mathbf{A}]=0 \tag{4.7.22}$$

Most of the equations considered here will involve a single matrix and the relevant functions of that matrix. In general,

$$[\mathbf{A},f(\mathbf{A})]=0 \tag{4.7.23}$$

The matrix approach to sets of coupled differential equations is illustrated with the differential equations,

$$\begin{aligned}dx/dt&=2x+y\\dy/dt&=x+2y\end{aligned} \tag{4.7.24}$$

expressed in matrix format as

$$\begin{gathered}d\,|x\rangle/dt=\mathbf{A}\,|x\rangle\\\begin{pmatrix}dx/dt\\dy/dt\end{pmatrix}=\begin{pmatrix}2&1\\1&2\end{pmatrix}\begin{pmatrix}x\\y\end{pmatrix}\end{gathered} \tag{4.7.25}$$

The formal solution to this equation with an initial condition vector

$$|x^0\rangle=\begin{pmatrix}x^0\\y^0\end{pmatrix} \tag{4.7.26}$$

is

$$|x\rangle=\exp(\mathbf{A}t)\,|x^0\rangle \tag{4.7.27}$$

where the column vector of initial values must appear to the right of the matrix function.

The equation is expanded in its eigenvalues and projection operators:

$$|x(t)\rangle = \sum \exp(\lambda_i t)\, \mathbf{P}(\lambda_i)\, |x^0\rangle \tag{4.7.28}$$

The projection operators for the eigenvalues 1 and 3 of matrix **A** have been given previously (Section 4.5). The solution is

$$\begin{aligned}\begin{pmatrix} x \\ y \end{pmatrix} &= \exp(3t)/2 \begin{pmatrix} 1 & 1 \\ 1 & 1 \end{pmatrix}\begin{pmatrix} x^0 \\ y^0 \end{pmatrix} + \exp(1t)/2 \begin{pmatrix} 1 & -1 \\ -1 & 1 \end{pmatrix}\begin{pmatrix} x^0 \\ y^0 \end{pmatrix} \\ &= \exp(3t)/2 \begin{pmatrix} x^0 + y^0 \\ x^0 + y^0 \end{pmatrix} + \exp(1t)/2 \begin{pmatrix} x^0 - y^0 \\ -x^0 + y^0 \end{pmatrix}\end{aligned} \tag{4.7.29}$$

The solution contains two exponential terms whose coefficients contain combinations of x^0 and y^0. The original x^0 and y^0 were not eigenvectors of the system. The projection operators project the initial condition vector onto each of the eigenvectors and the exponentials give the temporal change of the vector magnitude.

The matrix exponential is used for two types of time-dependent equations. In Chapter 6, coupled sequences of reactions will produce the matrix equation

$$d\,|c\rangle/dt = \mathbf{K}\,|c\rangle \tag{4.6.30}$$

with formal solution

$$|c\rangle = \exp(\mathbf{K}t)\,|c^0\rangle \tag{4.7.31}$$

where **K** is a matrix of rate constants for the reactions and $|c\rangle$ is a vector for the concentrations of each species in the reaction scheme. In Chapter 8, the time-dependent Schrödinger equation,

$$-i\hbar\, d\psi(t)/dt = \mathbf{H}\psi(t) \tag{4.7.32}$$

where **H** is the Hamiltonian and $\psi(t)$ is the time-dependent wavefunction is formally solved as

$$|\psi(t)\rangle = \exp(-i\mathbf{H}t/\hbar)\,|\psi^0\rangle \tag{4.7.33}$$

The time evolution of the wavefunction in the Schrödinger picture is observed on every eigencomponent of the wavefunction. If the wavefunction is a single eigenstate obeying

$$\mathbf{H}\,|\psi^0\rangle = E\,|\psi^0\rangle \tag{4.7.34}$$

the equation resolves into a single term,

$$|\psi(t)\rangle = \exp(-iEt/\hbar)\,|\psi^0\rangle \tag{4.7.35}$$

Although the wavefunction evolves in time, the probability density,

$$\begin{aligned}\langle \psi(t) | \psi(t) \rangle &= \langle \psi^0 | \exp(iEt/\hbar) \exp(-iEt/\hbar) | \psi^0 \rangle \\ &= \langle \psi^0 | \psi^0 \rangle\end{aligned} \tag{4.7.36}$$

is time independent.

The differential equation traces the time dependence of the vector components. However, in some cases, the time dependence of a matrix or operator can be monitored. The Heisenberg equation of motion for an operator $\mathbf{A}$ is related to the commutator of this operator with the time-independent Hamiltonian as

$$i\hbar\, d\mathbf{A}(t)/dt = [\mathbf{A}(t), \mathbf{H}] \tag{4.7.37}$$

$\mathbf{A}(t)$, not ψ, is time dependent in the Heisenberg format. To determine the eigenfunction at any time t, $\mathbf{A}(t)$ is determined from Equation 4.7.37 and used in the equation

$$\mathbf{A}(t)\psi = a(t)\psi \tag{4.7.38}$$

For a matrix with degenerate eigenvalues, the expansion for a function $f(\mathbf{A})$ must be modified. The expansion for $f(\mathbf{A})$ is

$$\sum f(\lambda_i)\, \mathbf{P}(\lambda_i) \tag{4.7.39}$$

The exponential for the matrix

$$\mathbf{A} = \begin{pmatrix} 2 & -1 & -1 \\ -1 & 2 & -1 \\ -1 & -1 & 2 \end{pmatrix} \tag{4.7.40}$$

with eigenvalues 0, 3, and 3 introduced in Section 4.6 is expanded as

$$\begin{aligned}\exp(\mathbf{A}) &= \sum_i \exp(\lambda_i)\, \mathbf{P}_i \\ &= \exp(0)\, \mathbf{P}(0) + \frac{\exp(\lambda_1)(\mathbf{A} - \lambda_2 \mathbf{I})(\mathbf{A} - 0\mathbf{I})}{(\lambda_1 - \lambda_2)(\lambda_1)} \\ &\quad + \frac{\exp(\lambda_2)(\mathbf{A} - \lambda_1 \mathbf{I})(\mathbf{A} - 0\mathbf{I})}{(\lambda_2 - \lambda_1)(\lambda_2)}\end{aligned} \tag{4.7.41}$$

The two degenerate terms can be combined to give

$$\exp(0)\, \mathbf{P}(0) + \exp(\lambda)\{\mathbf{A} - (\lambda - 1)\mathbf{T}\}\mathbf{A}/\lambda \tag{4.7.42}$$

The derivation is left as an exercise (Problem 4.29).

4.8. Diagonalization of Tridiagonal Matrices

Although matrix diagonalization becomes more difficult as the order of the matrix increases, the elements of many such matrices follow regular patterns because of the physical nature of the system. For example, particles may interact only with their nearest neighbors or intramolecular transitions may occur only between adjacent energy levels. In such cases, the ith component will only couple with the $(i+1)$th or $(i-1)$th component and the only nonzero matrix elements are $a_{i,i-1}$, $a_{i,i}$, or $a_{i,i+1}$.

The matrices will have elements on the main diagonal and on "diagonals" above and below the main diagonal:

$$\begin{pmatrix} a & b & 0 & \cdots & & \\ c & a & b & 0 & \cdots & \\ 0 & c & a & b & 0 & \cdots \\ 0 & 0 & c & a & b & \cdots \\ \vdots & & & & & \end{pmatrix} = [c, a, b] \tag{4.8.1}$$

This is a tridiagonal matrix. Eigenvalues and eigenvectors for a matrix of arbitrary order can be found if all the elements on each of the diagonals are identical (Equation 4.8.1).

The 2×2 matrix

$$\begin{pmatrix} 2 & 1 \\ 1 & 2 \end{pmatrix} \tag{4.8.2}$$

is tridiagonal. A symmetric 3×3 matrix with the same diagonal elements illustrates the tridiagonal pattern:

$$\begin{pmatrix} 2 & 1 & 0 \\ 1 & 2 & 1 \\ 0 & 1 & 2 \end{pmatrix} \tag{4.8.3}$$

This pattern continues for an arbitrary order N. Although the main diagonal has elements 2, a review of other symmetric 2×2 matrices of this type reveals an interesting regularity. If the upper and lower diagonals contain 1's, the main diagonal may have $-2 \leqslant z \leqslant 2$ to produce the same eigenvectors. The eigenvectors for 2×2 matrices of this type are

$$(1 \quad 1) \quad \text{and} \quad (1 \quad -1) \tag{4.8.4}$$

and the eigenvalues for the matrix

$$\begin{pmatrix} z & 1 \\ 1 & z \end{pmatrix} \tag{4.8.5}$$

with $-2 \leqslant z \leqslant 2$ are

$$\lambda = +z+1 \qquad \lambda = +z-1 \tag{4.8.6}$$

This result for the 2×2 matrix can be generalized to determine the eigenvalues and eigenvectors of an $N \times N$ tridiagonal matrix,

$$\begin{pmatrix} z & 1 & 0 & 0 & \dots & \\ 1 & z & 1 & 0 & \dots & \\ 0 & 1 & z & 1 & 0 & \dots \\ 0 & 0 & 1 & z & 1 & \dots \\ \vdots & & & & & \end{pmatrix} \tag{4.8.7}$$

where $-2 \leqslant z \leqslant 2$.

The eigenvector equation for a symmetric 4×4 tridiagonal matrix is

$$\begin{pmatrix} z & 1 & 0 & 0 \\ 1 & z & 1 & 0 \\ 0 & 1 & z & 1 \\ 0 & 0 & 1 & z \end{pmatrix} \begin{pmatrix} x_1 \\ x_2 \\ x_3 \\ x_4 \end{pmatrix} = \lambda \begin{pmatrix} x_1 \\ x_2 \\ x_3 \\ x_4 \end{pmatrix} \tag{4.8.8}$$

or

$$\begin{pmatrix} z-\lambda & 1 & 0 & 0 \\ 1 & z-\lambda & 1 & 0 \\ 0 & 1 & z-\lambda & 1 \\ 0 & 0 & 1 & z-\lambda \end{pmatrix} \begin{pmatrix} x_1 \\ x_2 \\ x_3 \\ x_4 \end{pmatrix} = 0 \tag{4.8.9}$$

The successive equations generated from the matrix equation are

$$\begin{aligned} (z-\lambda)\, x_1 + 1x_2 &= 0 \\ x_1 + (z-\lambda)\, x_2 + x_3 &= 0 \\ x_2 + (z-\lambda)\, x_3 + x_4 &= 0 \\ x_3 + (z-\lambda)\, x_4 &= 0 \end{aligned} \tag{4.8.10}$$

If the variables x_0 and x_5 are introduced into the first and last equations, respectively, they will become identical to the remaining equations. Although this does change the full characteristic equation, the system can be manipulated so that both x_0 and x_5 will become zero during the analysis. The equations with x_0 and x_5,

$$\begin{aligned} x_0 + (z-\lambda)\, x_1 + x_2 &= 0 \\ x_1 + (z-\lambda)\, x_2 + x_3 &= 0 \\ x_2 + (z-\lambda)\, x_3 + x_4 &= 0 \\ x_3 + (z-\lambda)\, x_4 + x_5 &= 0 \end{aligned} \tag{4.8.11}$$

produce equations which are identical. For a general index j, the equation is

$$x_{j-1} + (z-\lambda)\, x_j + x_{j+1} = 0 \tag{4.8.12}$$

A solution to this finite-difference equation gives both λ and the eigenvector components, x_j.

Equation 4.8.12 is the finite-difference equivalent of a second-order differential equation. The definitions of first and second derivatives for finite difference equations are illustrated in Figure 4.4. The first finite-difference derivative is the difference between two points on a one-dimensional grid with equal spacing:

$$\Delta_1 x = x_{j+1} - x_j \tag{4.8.13}$$

The second derivative is the difference of two such adjacent differences:

$$\begin{aligned} \Delta_2 x &= (x_{j+1} - x_j) - (x_j - x_{j-1}) \\ &= x_{j+1} - 2x_j + x_{j-1} \end{aligned} \tag{4.8.14}$$

Although the general characteristic equation contains a factor $z-\lambda$ rather than 2 for the central term, the similarity of the equations suggests that this solution is also related to the solution of a second-order, finite-difference equation. Since regular second-order equations of this type have sine and cosine solutions, a trial solution

$$x_j = c_1 \cos j\theta + c_2 \sin j\theta \tag{4.8.15}$$

is selected for the finite-difference equation.

The trial solution is substituted into the difference equation. Only $\cos j\theta$ is considered here; $\sin j\theta$ is left as an exercise (Problem 4.22). The row equation for index j is

$$x_{j-1} + (z-\lambda)\, x_j + x_{j+1} = 0 = \cos(j-1)\theta + (z-\lambda)\cos j\theta + \cos(j+1)\theta \tag{4.8.16}$$

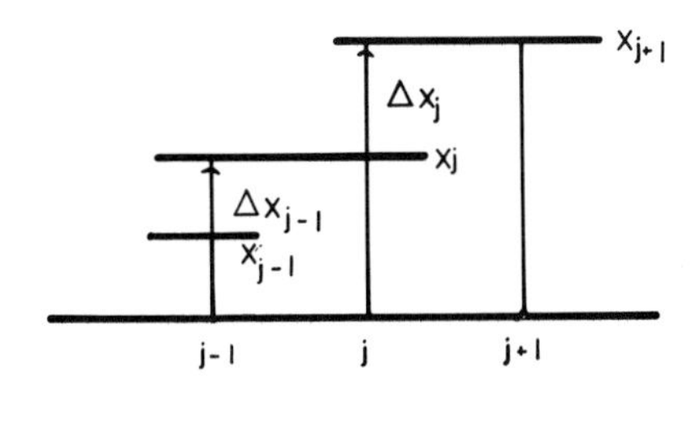

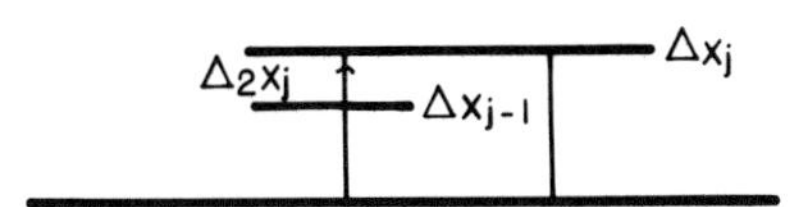

Figure 4.4. Definitions of finite-difference first and second derivatives for equally spaced x_i.

which is simplified by using the identity

$$\cos(\theta + \varphi) = \cos\theta\cos\varphi - \sin\theta\sin\varphi \tag{4.8.17}$$

The equation becomes

$$\begin{aligned}\cos j\theta\cos\theta + \sin j\theta\sin\theta + (z - \lambda)\cos j\theta\\ + \cos j\theta\cos\theta - \sin j\theta\sin\theta = 0\end{aligned} \tag{4.8.18}$$

The two sine products produced by the indices $j - 1$ and $j + 1$, respectively, cancel to leave

$$\cos j\theta\cos\theta + (z - \lambda)\cos j\theta + \cos j\theta\cos\theta = 0 \tag{4.8.19}$$

Cancellation of $\cos j\theta$ from all three terms yields

$$\cos\theta + (z - \lambda) + \cos\theta = 0 \tag{4.8.20}$$

to give solutions for λ and $z - \lambda$ in terms of $\cos\theta$:

$$\begin{aligned} z - \lambda &= -2\cos\theta \\ \lambda &= z + 2\cos\theta \end{aligned} \tag{4.8.21}$$

The index j is eliminated. To determine θ, the boundary conditions are applied to the original trial solution for x_j:

$$x_j = c_1\cos j\theta + c_2\sin j\theta \tag{4.8.22}$$

The two boundary conditions give the eigenvalue and a relative value c_1/c_2.

The parameters x_0 and x_5 were introduced to make the first and last equations for the 4×4 characteristic equation the same as the central equations. For an N-row system, the parameters x_0 and x_{N+1} are used. The boundary conditions are

$$x_0 = x_5 = 0 \tag{4.8.23}$$

$$x_0 = x_{N+1} = 0 \tag{4.8.24}$$

Substituting x_0 into the equation for $j = 0$ gives

$$c_1\cos 0\theta + c_2\sin 0\theta = c_1 = 0 \tag{4.8.25}$$

which is satisfied only if c_1 is identically equal to zero. The consistent equation reduces to

$$x_j = c_2\sin j\theta = 0 \tag{4.8.26}$$

with unknowns c_2 and θ. However, c_2 is common to each eigenvector component and is the normalization coefficient. The final boundary condition,

$$x_{N+1} = 0 \tag{4.8.27}$$

is substituted in Equation 4.8.26 to give

$$x_{N+1} = 0 = c_2 \sin [(N+1)\theta] \tag{4.8.28}$$

This equation can only be satisfied when the argument of the sine is an integer multiple of π, i.e.,

$$(N+1)\theta = n\pi \qquad \text{with} \quad n = 1, 2, 3, \ldots, N \tag{4.8.29}$$

$$\theta = n\pi/(N+1) \tag{4.8.30}$$

The index stops at N since there are only N distinct eigenvalues for this $N \times N$ matrix.

Now θ is substituted into the expressions for λ_j and x_j. First,

$$\lambda = z + 2 \cos \theta \tag{4.8.31}$$

becomes

$$\lambda_n = z + 2 \cos (n\pi/N+1) \qquad n = 1, 2, \ldots, N \tag{4.8.32}$$

The equation for x_j is

$$x_{jn} = c \sin [jn\pi/(N+1)] \tag{4.8.33}$$

for the jth component of the nth eigenvector. Both indices range from 1 to N. The 2×2 matrix

$$\begin{pmatrix} z & 1 \\ 1 & z \end{pmatrix} \tag{4.8.34}$$

will have the two eigenvalues

$$\begin{aligned} \lambda_1 &= z + 2 \cos (1\pi/3) = z + 2 \cos (60°) = z + 1 \\ \lambda_2 &= z + 2 \cos (2\pi/3) = z + 2 \cos (120°) = z - 1 \end{aligned} \tag{4.8.35}$$

The eigenvector $|1\rangle$ is

$$\begin{pmatrix} \sin [(1)(1)\pi/3] \\ \sin [(2)(1)\pi/3] \end{pmatrix} = \begin{pmatrix} \sin 60° \\ \sin 120° \end{pmatrix} = \begin{pmatrix} \dfrac{\sqrt{3}}{2} \\ -\dfrac{\sqrt{3}}{2} \end{pmatrix} \tag{4.8.36}$$

and $|2\rangle$ is

$$\begin{pmatrix} \sin[(1)(2)\pi/3] \\ \sin[(2)(2)\pi/3] \end{pmatrix} = \begin{pmatrix} \sin 120^\circ \\ \sin 240^\circ \end{pmatrix} = \begin{pmatrix} -\dfrac{\sqrt{3}}{2} \\ -\dfrac{\sqrt{3}}{2} \end{pmatrix} \tag{4.8.37}$$

The relative components are $(1, -1)$ and $(1, 1)$, respectively.

For the 3×3 matrix

$$\begin{pmatrix} 2 & 1 & 0 \\ 1 & 2 & 1 \\ 0 & 1 & 2 \end{pmatrix} \tag{4.8.38}$$

the eigenvalues are

$$\begin{aligned} \lambda_1 &= 2 + 2\cos(1\pi/4) = 2 + 2\left(\frac{1}{\sqrt{2}}\right) = 2 + \sqrt{2} \\ \lambda_2 &= 2 + 2\cos(2\pi/4) = 2 + 0 = 2 \\ \lambda_3 &= 2 + 2\cos(3\pi/4) = 2 - 2\left(\frac{1}{\sqrt{2}}\right) = 2 - \sqrt{2} \end{aligned} \tag{4.8.39}$$

with eigenvectors

$$|1\rangle = \begin{pmatrix} \sin[(1)(1)\pi/4] \\ \sin[(2)(1)\pi/4] \\ \sin[(3)(1)\pi/4] \end{pmatrix} = \begin{pmatrix} \dfrac{1}{\sqrt{2}} \\ 1 \\ \dfrac{1}{\sqrt{2}} \end{pmatrix} = \begin{pmatrix} 1 \\ \sqrt{2} \\ 1 \end{pmatrix} \tag{4.8.40}$$

$$|2\rangle = \begin{pmatrix} \sin[(1)(2)\pi/4] \\ \sin[(2)(2)\pi/4] \\ \sin[(3)(2)\pi/4] \end{pmatrix} = \begin{pmatrix} 1 \\ 0 \\ -1 \end{pmatrix} \tag{4.8.41}$$

$$|3\rangle = \begin{pmatrix} \sin[(1)(3)\pi/4] \\ \sin[(2)(3)\pi/4] \\ \sin[(3)(3)\pi/4] \end{pmatrix} = \begin{pmatrix} \dfrac{1}{\sqrt{2}} \\ -1 \\ \dfrac{1}{\sqrt{2}} \end{pmatrix} = \begin{pmatrix} 1 \\ -\sqrt{2} \\ 1 \end{pmatrix} \tag{4.8.42}$$

The sine functions impart periodicity to the components. The first eigenvector has only positive components, i.e., it never crosses the zero line. The second

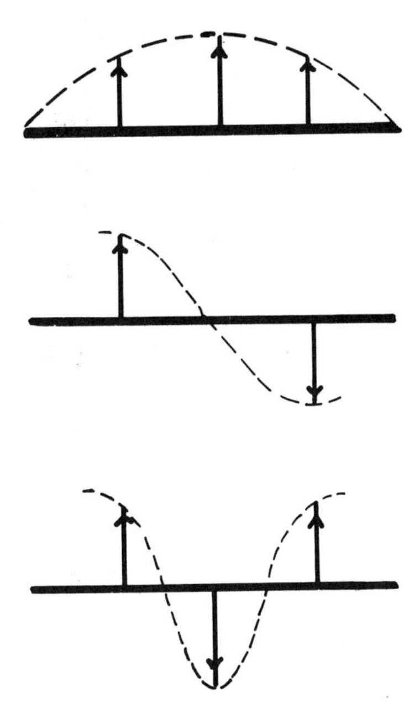

Figure 4.5. The magnitudes of eigenvector components for a 3×3 tridiagonal matrix and their determination with the sine function.

eigenvector crosses once, from $+1$ to -1, while the third eigenvector crosses the line twice, $+$ to $-$ and then $-$ to $+$. The eigenvector components for each eigenvalue are plotted with respect to a zero axis in Figure 4.5.

4.9. Other Tridiagonal Matrices

The tridiagonal matrices of Section 4.8 had a common factor z on the main diagonal. For a consistent row equation, the two terminal elements required parameters x_0 and x_{N+1}, which became the boundary conditions for determination of the eigenvectors and eigenvalues. These boundary conditions can be modified to determine eigenvalues for other types of tridiagonal matrices.

The symmetric tridiagonal matrix

$$\begin{pmatrix} -1 & 1 & 0 & 0 \\ 1 & -2 & 1 & 0 \\ 0 & 1 & -2 & 1 \\ 0 & 0 & 1 & -1 \end{pmatrix} \tag{4.9.1}$$

differs from the matrices of the last section only in its two terminal elements. Since these elements dictate the boundary conditions, the eigenvalues and eigenvectors can be determined by applying these new boundary conditions to the general finite-difference equation for x_j.

The row equations from the characteristic equation,

$$(\mathbf{A} - \lambda \mathbf{I})\,|x\rangle = 0 \tag{4.9.2}$$

are

$$\begin{aligned} (-1-\lambda)\,x_1 + 1x_2 &= 0 \\ x_1 + (-2-\lambda)\,x_2 + x_3 &= 0 \\ x_2 + (-2-\lambda)\,x_3 + x_4 &= 0 \\ x_3 + (-1-\lambda)\,x_4 &= 0 \end{aligned} \tag{4.9.3}$$

and have the general solution

$$x_j = c_1 \cos j\theta + c_2 \sin j\theta \tag{4.9.4}$$

which can only bc solved if

$$-2 - \lambda = 2 \cos \theta \tag{4.9.5}$$

The boundary conditions must give -1 at each terminal element. The boundary conditions must be

$$x_0 = x_1 \quad \text{and} \quad x_{N+1} = x_N \tag{4.9.6}$$

The result is verified by using these results with the modified terminal row equations

$$\begin{aligned} x_0 - (2+\lambda)\,x_1 + x_2 &= 0 \\ x_{N-1} - (2+\lambda)\,x_N + x_{N+1} &= 0 \end{aligned} \tag{4.9.7}$$

to give

$$x_1 + (-2-\lambda)\,x_1 + x_2 = (-1-\lambda)\,x_1 + x_2 = 0 \tag{4.9.8}$$

and

$$x_{N-1} + (-2-\lambda)\,x_N + x_N = x_{N-1} + (-1-\lambda)\,x_N = 0 \tag{4.9.9}$$

These new boundary conditions are used to evaluate the unknowns in the component x_j:

$$x_j = c_1 \cos j\theta + c_2 \sin j\theta \tag{4.9.10}$$

The solutions for x_1 and x_0 are equated to satisfy the first boundary condition:

$$c_1 \cos 0\theta + c_2 \sin 0\theta = c_1 \cos 1\theta + c_2 \sin 1\theta \tag{4.9.11}$$

Since the sine function is always zero when $j=0$ while the function for $j=1$ is nonzero, the sine function must be eliminated, leaving

$$\cos 0\theta = \cos 1\theta \tag{4.9.12}$$

and the general solution for x_j,

$$x_j = c \cos j\theta \tag{4.9.13}$$

The second boundary condition gives

$$\cos N\theta = \cos [(N+1)\theta] \tag{4.9.14}$$

This boundary condition cannot be satisfied by an integral number of π since the one additional θ due to $N+1$ on the right-hand side will change the sign of the function. Both sides become equal if the index j carries a phase shift of $\frac{\pi}{2}$, i.e.,

$$\theta_j = n\pi\left(j-\frac{1}{2}\right) = (n\pi/2)(2j-1) \tag{4.9.15}$$

Then x_j becomes

$$x_j = \cos\left[\left(j-\frac{1}{2}\right) n\pi/N\right] = \cos [(2j-1)\, n\pi/2N] \tag{4.9.16}$$

This solution satisfies both boundary conditions:

$$\begin{gathered} x_0 = x_1 \\ \cos \{[2(0)-1]\, n\pi/2N\} = \cos \{2(1)-1]\, n\pi/2N\} \\ \cos(-n\pi/2N) = \cos(n\pi/2N) \end{gathered} \tag{4.9.17}$$

$$x_N = x_{N+1}$$

$$\cos \{[2(N)-1]\, n\pi/2N\} = \cos \{[2(N+1)-1]\, n\pi/2N\} = \cos [(2N+1)\, n\pi/2N] \tag{4.9.18}$$

With the identity,

$$\cos(\theta+\varphi) = \cos\theta \cos\varphi - \sin\theta \sin\varphi \tag{4.9.19}$$

the equality becomes

$$\begin{aligned} &\cos n\pi \cos(n\pi/2N) + \sin n\pi \sin(n\pi/2N) \\ &\quad = \cos n\pi \cos(n\pi/2N) - \sin n\pi \sin(n\pi/2N) \end{aligned} \tag{4.9.20}$$

The sine terms are zero because of $\sin n\pi$; the remaining cosine functions are identical and the boundary condition is satisfied.

The factor

$$n\pi/N \tag{4.9.21}$$

multiplies the index $j-\frac{1}{2}$ and determines the eigenvalue, λ_n, for this matrix:

$$\begin{aligned} -2-\lambda_n &= 2\cos\theta = 2\cos(n\pi/N) \\ \lambda_n &= -2-2\cos(n\pi/N) \end{aligned} \tag{4.9.22}$$

The identity

$$\sin^2 x = \frac{1}{2}(1+\cos 2x) \tag{4.9.23}$$

permits a reduction to

$$\lambda_n = -4\sin^2(\pi n/2N) \tag{4.9.24}$$

For the 3×3 tridiagonal matrix,

$$\begin{pmatrix} -2 & 1 & 0 \\ 1 & -2 & 1 \\ 0 & 1 & -2 \end{pmatrix} \tag{4.9.25}$$

the eigenvalues are

$$\begin{aligned} \lambda_0 &= -4\sin^2[(0\pi/2\times 3)] = 0 \\ \lambda_1 &= -4\sin^2(1\pi/6) = -4\left(\frac{1}{4}\right) = -1 \\ \lambda_2 &= -4\sin^2(2\pi/6) = -4\left(\frac{3}{4}\right) = -3 \end{aligned} \tag{4.9.26}$$

The corresponding eigenvectors are

$$|-1\rangle = \begin{pmatrix} \cos[(2\times 1-1)(1)\pi/(2\times 3)] \\ \cos[(2\times 2-1)(1)\pi/(2\times 3)] \\ \cos[(2\times 3-1)(1)\pi/(2\times 3)] \end{pmatrix} = \begin{pmatrix} \frac{\sqrt{3}}{2} \\ 0 \\ -\frac{\sqrt{3}}{2} \end{pmatrix} = \begin{pmatrix} 1 \\ 0 \\ -1 \end{pmatrix} \tag{4.9.27}$$

$$|-3\rangle = \begin{pmatrix} \cos[(1)(2)\pi/6] \\ \cos[(3)(2)\pi/6] \\ \cos[(5)(2)\pi/6] \end{pmatrix} = \begin{pmatrix} \frac{1}{2} \\ -1 \\ \frac{1}{2} \end{pmatrix} \propto \begin{pmatrix} 1 \\ -2 \\ 1 \end{pmatrix} \tag{4.9.28}$$

$$|0\rangle = \begin{pmatrix} \cos[(1)(0)\pi/6] \\ \cos[(3)(0)\pi/6] \\ \cos[(5)(0)\pi/6] \end{pmatrix} = \begin{pmatrix} 1 \\ 1 \\ 1 \end{pmatrix} \tag{4.9.29}$$

These results can verified by operating with the matrix **A** (Problem 4.25).

When all the elements on the main diagonal of a tridiagonal matrix equaled z, a common set of eigenvectors was available for $-2 \leqslant z \leqslant 2$. When -1's appear in the terminal positions, a set of matrices with the same eigenvectors is found by adding the same integer to each diagonal element. For example, if $+3$ is added to each diagonal element of the 3×3 matrix, the new matrix

$$\begin{pmatrix} 2 & 1 & 0 \\ 1 & 1 & 1 \\ 0 & 1 & 2 \end{pmatrix} \tag{4.9.30}$$

gives eigenvalues of 3, 2, and 0 with the eigenvectors of the matrix in Equation 4.9.25.

Other triangular matrices require different boundary conditions. Although matrices like

$$\begin{pmatrix} 2 & 1 & 0 \\ 1 & 2 & 1 \\ 0 & 1 & 2 \end{pmatrix} \tag{4.9.31}$$

differ only in the sign of the diagonal elements, they require entirely different boundary conditions and their eigenvectors will be different. In this case, the general equation for the jth row is

$$x_{j-1} + (2 - \lambda)\, x_j + x_{j+1} = 0 \tag{4.9.32}$$

The first-row equation is

$$x_0 + (2 - \lambda)\, x_1 + x_2 = 0 \tag{4.9.33}$$

To match this equation with the actual first-row equation,

$$(1 - \lambda)\, x_1 + x_2 = 0 \tag{4.9.34}$$

the boundary condition must be

$$x_0 = -x_1 \tag{4.9.35}$$

This boundary condition requires solutions in sine rather than cosine functions.

The final-row equation is

$$x_{N-1} + (2 - \lambda)\, x_N = 0 \tag{4.9.36}$$

and the boundary condition is

$$x_{N+1} = -x_N \tag{4.9.37}$$

The eigenvalues and eigenvectors for an $N \times N$ matrix of this type can be determined using the methods of this section (Problem 4.31). The eigenvector components are

$$x_{jn} = \sin\left[(2j-1)\, n\pi/2N\right] \tag{4.9.38}$$

and the eigenvalues are

$$\lambda_n = +4\sin^2(n\pi/2N) \tag{4.9.39}$$

4.10. Asymmetric Tridiagonal Matrices

The symmetric tridiagonal matrices require a single set of eigenvector components for each eigenvalue. Other nonsymmetric tridiagonal matrices can be similarity transformed to such symmetric matrices. The eigenvectors for this symmetric matrix are determined and then transformed back to determine the bra and ket eigenvectors for the nonsymmetric matrix.

The Nth order nonsymmetric tridiagonal matrix

$$\begin{pmatrix} a & b & 0 & 0 & 0 & \cdots \\ c & a & b & 0 & 0 & \\ 0 & c & a & b & 0 & \\ 0 & 0 & c & a & b & \\ \vdots & & & & & \end{pmatrix} \tag{4.10.1}$$

can be written in a row format,

$$\mathbf{A} = [c, a, b] \tag{4.10.2}$$

for convenience. The elements within each diagonal are the same but the elements of different diagonals are different. This matrix can be transformed to a symmetric matrix.

The 2×2 matrix

$$\begin{pmatrix} a & b \\ c & a \end{pmatrix} \tag{4.10.3}$$

serves as an illustrative example. The transform must distribute b and c between the upper and lower diagonals to create a symmetric matrix. The geometric means $\sqrt{b}$ and $\sqrt{c}$ will appear in the similarity transform. Since $\sqrt{b}$ must be eliminated from a_{12} and $\sqrt{c}$ must be added, the transform factor is

$$c^{1/2}b^{-1/2} \tag{4.10.4}$$

Since this factor must change the a_{12} element, it appears as the second diagonal element in a diagonal transform matrix:

$$\mathbf{D}=\begin{pmatrix}1 & 0\\ 0 & \sqrt{c/b}\end{pmatrix} \tag{4.10.5}$$

with the symbol $\mathbf{D}$ selected to distinguish this similarity transform from the eigenvector similarity transforms. $\mathbf{D}^{-1}$ is

$$\mathbf{D}^{-1}=\begin{pmatrix}1 & 0\\ 0 & \sqrt{b/c}\end{pmatrix} \tag{4.10.6}$$

and the similarity transform gives

$$\begin{aligned}\mathbf{D}^{-1}\mathbf{A}\mathbf{D} &= \mathbf{D}^{-1}(\mathbf{A}\mathbf{D})\\ &= \begin{pmatrix} a & \sqrt{cb}\\ \sqrt{cb} & a\end{pmatrix}\end{aligned} \tag{4.10.7}$$

The transformed matrix is symmetric.

The 2×2 matrix suggests a general pattern in which each additional diagonal element must include an additional factor of $\sqrt{c/b}$. For a 3×3 matrix, the transform matrix must be

$$\mathbf{D}=\begin{pmatrix}1 & 0 & 0\\ 0 & \sqrt{c/b} & 0\\ 0 & 0 & c/b\end{pmatrix} \tag{4.10.8}$$

and the inverse matrix is

$$\mathbf{D}^{-1}=\begin{pmatrix}1 & 0 & 0\\ 0 & \sqrt{b/c} & 0\\ 0 & 0 & b/c\end{pmatrix} \tag{4.10.9}$$

The transform of the 3×3 tridiagonal matrix is

$$\begin{pmatrix}1 & 0 & 0\\ 0 & \sqrt{b/c} & 0\\ 0 & 0 & b/c\end{pmatrix}\begin{pmatrix}a & b & 0\\ c & a & b\\ 0 & c & a\end{pmatrix}\begin{pmatrix}1 & 0 & 0\\ 0 & \sqrt{c/b} & 0\\ 0 & 0 & c/b\end{pmatrix}=\begin{pmatrix}a & \sqrt{cb} & 0\\ \sqrt{bc} & a & \sqrt{bc}\\ 0 & \sqrt{bc} & a\end{pmatrix} \tag{4.10.10}$$

For an $N\times N$ matrix, each diagonal element of $\mathbf{D}$ will increase by a factor $\sqrt{c/b}$. The diagonal matrix elements are

$$\begin{aligned}[\mathbf{D}]_{ii} &= (c/b)^{[(i-1)/2]}\\ [\mathbf{D}]_{ii}^{-1} &= (b/c)^{[(i-1)/2]}\end{aligned} \tag{4.10.11}$$

The **D** matrix is

$$\begin{pmatrix} 1 & 0 & \cdots & & \\ 0 & \sqrt{c/b} & 0 & \cdots & \\ 0 & 0 & c/b & \cdots & \\ \vdots & & & \ddots & \\ 0 & 0 & 0 & (c/b)^{(N-1)/2} \end{pmatrix} \tag{4.10.12}$$

The eigenvalues and eigenvectors for the nonsymmetric matrix are determined from the eigenvalues and eigenvectors of the symmetric matrix. The symmetric 3×3 matrix is

$$\begin{pmatrix} a & \sqrt{bc} & 0 \\ \sqrt{bc} & a & \sqrt{bc} \\ 0 & \sqrt{bc} & a \end{pmatrix} \tag{4.10.13}$$

Since the eigenvalues and eigenvectors for a tridiagonal matrix

$$[1, z, 1] \tag{4.10.14}$$

are known, the symmetric matrix (Equation 4.10.13) is converted to this format by factoring $\sqrt{bc}$ from each element of the matrix to give

$$(\sqrt{bc})\begin{pmatrix} a/\sqrt{bc} & 1 & 0 \\ 1 & a/\sqrt{bc} & 1 \\ 0 & 1 & a/\sqrt{bc} \end{pmatrix} \tag{4.10.15}$$

If

$$z = a/\sqrt{bc} \tag{4.10.16}$$

then the eigenvalues for the matrix are identical to those derived in Section 4.8. These eigenvalues are

$$\lambda' = z - 2\cos(n\pi/N + 1) \tag{4.10.17}$$

Substituting the value for z gives

$$\lambda' = a/\sqrt{bc} - 2\cos(n\pi/N + 1) \tag{4.10.18}$$

Each element in the $[1, z, 1]$ matrix must be multiplied by the factor $\sqrt{bc}$ to recover the original matrix eigenvalues. The eigenvalues for the original matrix are determined by multiplying each $[1, z, 1]$ eigenvalue by this factor:

$$\begin{aligned} \lambda &= (\sqrt{bc})[a/\sqrt{bc} - 2\cos(n\pi/N + 1)] \\ &= a - 2\sqrt{bc}\cos(n\pi/N + 1) \qquad n = 1, 2, \ldots, N \end{aligned} \tag{4.10.19}$$

The eigenvectors for a $[1, z, 1]$ matrix are known and can be used to determine the biorthogonal eigenvectors for the nonsymmetric matrix. The characteristic equation for an eigenvector $|i'\rangle$ of the symmetric matrix $\mathbf{A}'$ is

$$\mathbf{A}' \, |i'\rangle = \lambda' \, |i'\rangle \tag{4.10.20}$$

Since the symmetric matrix was generated by the similarity transform

$$\mathbf{A}' = \mathbf{D}^{-1}\mathbf{A}\mathbf{D} \tag{4.10.21}$$

the transformation to determine $\mathbf{A}$ from $\mathbf{A}'$ is

$$\mathbf{A} = \mathbf{D}\mathbf{A}'\mathbf{D}^{-1} \tag{4.10.22}$$

The characteristic equation is converted to a characteristic equation for $\mathbf{A}$ by inserting $\mathbf{D}^{-1}\mathbf{D} = \mathbf{I}$ between $\mathbf{A}'$ and $|i'\rangle$ and left-multiplying each side of the equation by $\mathbf{D}$:

$$(\mathbf{D}\mathbf{A}'\mathbf{D}^{-1})(\mathbf{D} \, |i'\rangle) = \lambda'\mathbf{D} \, |i'\rangle \tag{4.10.23}$$

The eigenvector for $\mathbf{A}$,

$$|i\rangle = \mathbf{D} \, |i'\rangle \tag{4.10.24}$$

has the same eigenvalue, λ_i'. The eigenvectors for $\mathbf{A}$ are determined from the $|i'\rangle$ using Equation 4.10.24.

The transforms are illustrated with the 3×3 matrix

$$\begin{pmatrix} \sqrt{2} & \frac{1}{2} & 0 \\ 1 & \sqrt{2} & \frac{1}{2} \\ 0 & 1 & \sqrt{2} \end{pmatrix} \tag{4.10.25}$$

The similarity transform must contain the factor

$$\sqrt{c/b} = \sqrt{2} \tag{4.10.26}$$

and the $\mathbf{D}$ matrix is

$$\begin{pmatrix} 1 & 0 & 0 \\ 0 & \sqrt{2} & 0 \\ 0 & 0 & 2 \end{pmatrix} \tag{4.10.27}$$

$$\mathbf{D}^{-1}\mathbf{AD} = \begin{pmatrix} 1 & 0 & 0 \\ 0 & \frac{1}{\sqrt{2}} & 0 \\ 0 & 0 & \frac{1}{2} \end{pmatrix} \begin{pmatrix} \sqrt{2} & \frac{1}{2} & 0 \\ 1 & \sqrt{2} & \frac{1}{2} \\ 0 & 1 & \sqrt{2} \end{pmatrix} \begin{pmatrix} 1 & 0 & 0 \\ 0 & \sqrt{2} & 0 \\ 0 & 0 & 2 \end{pmatrix}$$

$$= (\sqrt{2}/2) \begin{pmatrix} 2 & 1 & 1 \\ 1 & 2 & 1 \\ 0 & 1 & 2 \end{pmatrix} \tag{4.10.28}$$

The symmetric matrix has eigenvalues

$$2-\sqrt{2}, \qquad 2, \qquad 2+\sqrt{2} \tag{4.10.29}$$

Since every element in the matrix is multiplied by the scalar factor $\sqrt{2}/2$, each eigenvalue must be multiplied by this factor. Since the eigenvalues of a matrix are invariant under a similarity transform, these are then the eigenvalues of the original matrix:

$$\sqrt{2}-1, \qquad \sqrt{2}, \qquad \sqrt{2}+1 \tag{4.10.30}$$

This can be verified by direct evaluation of the characteristic equation.

The eigenvectors of the symmetric matrix are

$$|2-\sqrt{2}\rangle = \begin{pmatrix} 1 \\ \sqrt{2} \\ 1 \end{pmatrix} \qquad |2\rangle = \begin{pmatrix} 1 \\ 0 \\ -1 \end{pmatrix} \qquad |2+\sqrt{2}\rangle = \begin{pmatrix} 1 \\ -\sqrt{2} \\ 1 \end{pmatrix} \tag{4.10.31}$$

The column eigenvectors for **A** are determined using

$$|i\rangle = \mathbf{D}\,|i'\rangle \tag{4.10.32}$$

where the $|i'\rangle$ are the eigenvectors of the symmetric matrices (Equations 4.10.31). The eigenvectors are

$$|\sqrt{2}-1\rangle = \begin{pmatrix} 1 & 0 & 0 \\ 0 & \sqrt{2} & 0 \\ 0 & 0 & 2 \end{pmatrix} \begin{pmatrix} 1 \\ \sqrt{2} \\ 1 \end{pmatrix} = \begin{pmatrix} 1 \\ 2 \\ 2 \end{pmatrix} \tag{4.10.33}$$

$$|\sqrt{2}\rangle = \begin{pmatrix} 1 & 0 & 0 \\ 0 & \sqrt{2} & 0 \\ 0 & 0 & 2 \end{pmatrix} \begin{pmatrix} 1 \\ 0 \\ -1 \end{pmatrix} = \begin{pmatrix} 1 \\ 0 \\ -2 \end{pmatrix} \tag{4.10.34}$$

$$|\sqrt{2}+1\rangle = \begin{pmatrix} 1 & 0 & 0 \\ 0 & \sqrt{2} & 0 \\ 0 & 0 & 2 \end{pmatrix} \begin{pmatrix} 1 \\ -\sqrt{2} \\ 1 \end{pmatrix} = \begin{pmatrix} 1 \\ -2 \\ 2 \end{pmatrix} \tag{4.10.35}$$

The row eigenvectors for **A** are generated using the transform

$$\langle i| = \langle i'|\, \mathbf{D}^{-1} \tag{4.10.36}$$

where the components of $\langle i'|$ are identical to those of $|i'\rangle$ since the $\mathbf{A}'$ matrix is symmetric. The eigenvectors $\langle i|$ are

$$\langle \sqrt{2}-1| = \begin{pmatrix} 1 & 1 & \frac{1}{2} \end{pmatrix} \tag{4.10.37}$$

$$\langle \sqrt{2}| = \begin{pmatrix} 1 & 0 & -\frac{1}{2} \end{pmatrix} \tag{4.10.38}$$

$$\langle \sqrt{2}+1| = \begin{pmatrix} 1 & -1 & \frac{1}{2} \end{pmatrix} \tag{4.10.39}$$

The biorthogonality properties of the **A** matrix eigenvectors are left as an exercise.

Although this symmetrization technique is effective for tridiagonal matrix systems, it can be used with some other special matrices. The technique will be used to symmetrize rate constant matrices using the principle of detailed balance at equilibrium in Chapter 6.

Problems

4.1. Show that the similarity transform

$$\mathbf{S}^{-1}\mathbf{AS}$$

produces a matrix with ij element $\langle i|\, \mathbf{A}\, |j\rangle$ (Equation 4.1.18).

4.2. An ellipse has the equation

$$(x/1)^2 + (y/2)^2 = 1$$

a. Determine the equation for this ellipse in terms of new x' and y' axes rotated $+45°$ (counterclockwise) relative to the original axes.
b. Determine the equation when the coordinate system is oriented 30° relative to the original coordinate axes.

4.3. Show that the selection of two orthogonal vectors for the degenerate subspace for matrix **A** (Equation 4.6.1) will produce a matrix which is completely diagonalized.

4.4. Show that the matrices

$$\begin{pmatrix} -1 & 1 & 0 \\ 1 & -2 & 1 \\ 0 & 1 & -1 \end{pmatrix} \qquad \begin{pmatrix} -2 & 1 & 1 \\ 1 & -2 & 1 \\ 1 & 1 & -2 \end{pmatrix}$$

commute. Find the common set of eigenvectors for both matrices.

4.5. Find the similarity transform for the matrix

$$\begin{pmatrix} 2 & \frac{\sqrt{2}}{2} & 0 \\ \frac{\sqrt{2}}{2} & 2 & \frac{\sqrt{2}}{2} \\ 0 & \frac{\sqrt{2}}{2} & 2 \end{pmatrix}$$

and determine its projection operators.

4.6. Perform the matrix multiplications to prove the rotation matrices $\mathbf{R}_{120}$ and $\mathbf{R}_{240}$ commute. Do these matrices commute with the $\mathbf{R}_{180}$ matrix?

4.7. Prove that $[\mathbf{x}, \mathbf{p_x}] = i\hbar$ if the $\mathbf{p_x}$ operator is selected as $\mathbf{p} = -i\hbar\, d/dx$.

4.8. Find the eigenvalues and eigenvectors for the characteristic equation

$$\left[\begin{pmatrix} 1 & 1 \\ 1 & 1 \end{pmatrix} - \lambda \begin{pmatrix} 2 & 1 \\ 1 & 2 \end{pmatrix}\right] |i\rangle = 0$$

Demonstrate the orthogonality properties for the eigenvectors.

4.9. Demonstrate the diagonal form of the characteristic equation of Problem 4.8 using a similarity transform on both matrices.

4.10. Prove that

$$\mathbf{A}^N = \sum \lambda^N |i\rangle\langle i|$$

4.11. Determine the projection operators for the matrix

$$\begin{pmatrix} 2 & -1 & -1 \\ -1 & 2 & -1 \\ -1 & -1 & 2 \end{pmatrix}$$

and demonstrate that

$$\mathbf{I} = \sum \mathbf{P}_i$$

$$\mathbf{P}_i \mathbf{P}_j = 0$$

4.12. Prove that Equations 4.5.33, 4.5.34, and 4.5.35 are valid projection operators.

4.13. Use the identity

$$\mathbf{A}^{-2}\mathbf{A}^{2}=\mathbf{I}$$

to verify the expression for $\mathbf{A}^{-2}$ in Equation 4.5.39.

4.14. Show that the matrices

$$\begin{pmatrix} 1 & 0 & 0 \\ -1 & 1 & 0 \\ 0 & -1 & 0 \end{pmatrix} \qquad \begin{pmatrix} 1 & 0 & 0 \\ 0 & 1 & 0 \\ -1 & -1 & 0 \end{pmatrix}$$

will produce the same projection matrix for the doubly degenerate eigenvalue $\lambda=1$.

4.15. Prove that

$$[\mathbf{A}, f(\mathbf{A})]=0$$

for any well-behaved function $f(\mathbf{A})$.

4.16. Demonstrate that the eigenvectors for the matrix

$$\begin{pmatrix} z & 1 \\ 1 & z \end{pmatrix}$$

are the same for all values of z between -2 and $+2$.

4.17. Develop a set of orthogonal eigenvectors for the matrix of Equation 4.6.1.

4.18. Show that the projection matrices in Equations 4.6.29 and 4.6.31 obey the relations

$$\mathbf{P}_i\mathbf{P}_j=0 \qquad \mathbf{I}=\sum \mathbf{P}_i$$

4.19. Show that it is impossible to find three orthogonal eigenvectors for the matrix of Equation 4.6.27.

4.20. Verify that the matrix of Equation 4.6.44 is a projection matrix.

4.21. Derive Equation 4.7.42.

4.22. Show that $\sin j\theta$ is a solution of the equation

$$x_{j-1}+(z-\lambda)\,x_j+x_{j+1}=0$$

4.23. Show that $\sin j\theta$ is eliminated as a viable solution for the boundary conditions $x_0=0$, $x_{N+1}=0$.

4.24. Prove that the eigenvalues for a tridiagonal matrix $[1, +2, 1]$ are given by the equation

$$\lambda_n = +4\sin^2(n\pi/N+1)$$

4.25. Demonstrate that the eigenvectors in Equations 4.9.27–4.9.29 will generate the eigenvalues when the matrix $\mathbf{A}$ operates on them.

4.26. Determine the eigenvalues and eigenvectors for the matrix

$$\begin{pmatrix} \sqrt{2} & \frac{1}{2} & 0 \\ 1 & \sqrt{2} & 1 \\ 0 & 1 & \sqrt{2} \end{pmatrix}$$

using the characteristic equation. Show that the eigenvectors are biorthogonal.

4.27. Verify the projection operators of Equations 4.6.21 and 4.6.22.

4.28. Show that the matrices of Equations 4.6.33 and 4.6.27 will generate the same projection matrices. Show that these projection operators do not sum to **I**.

4.29. Derive Equation 4.7.42 for the expansion of two degenerate eigenvalues. How could this analysis be extended for a triply degenerate eigenvalue?

4.30. Show that an $N \times N$ tridiagonal matrix

$$\begin{pmatrix} 1 & 1 & . & . & . & 0 \\ 1 & 2 & 1 & . & . & 0 \\ 0 & 1 & 2 & 1 & . & 0 \\ . & . & . & . & . & . \\ . & . & . & . & . & . \\ . & . & . & 1 & 2 & 1 \\ . & . & . & 0 & 1 & 1 \end{pmatrix}$$

will have eigenvalues

$$\lambda_n = 4 \sin^2 (n\pi/2N)$$

4.31. Determine the eigenvalues and eigenvectors for the matrices of Equations 4.9.30 and 4.9.31.

4.32. Evaluate the eigenvalues of the matrix of Equation 4.10.27 using the characteristic polynomial.

4.33. Determine the eigenvectors for the matrix of Equation 4.5.30.

4.34. For the matrix

$$\begin{pmatrix} -2 & 1 & 0 \\ 1 & -2 & 1 \\ 0 & 1 & -1 \end{pmatrix}$$

determine the boundary conditions. Show that

$$\lambda_n = 4 \sin^2 [(2n - 1/2N + 1)\pi/2]$$

and

$$x_{nj} = \sin [(2n - 1/2N + 1)\pi j]$$

5

Vibrations and Normal Modes

5.1. Normal Modes

Any molecule, even a homonuclear diatomic molecule, has a variety of distinct motions in three-dimensional space. It can translate in three different spatial directions, rotate about two orthogonal axes associated with significant moments of inertia, and vibrate along the internuclear axis. In general, however, a molecule moving in space will display all such motions. The observer really observes the motion of the two atoms of the molecule. These atomic motions must be resolved into molecular motions such as translation, rotation, and vibration.

The distinction between molecular and atomic motions can be clarified by restricting the diatomic to a single spatial dimension. The x axis is assumed coincident with the internuclear axis. The motions of atom 1 and atom 2 of the molecule are then expressed in terms of their x-coordinate values at any time t. The situation is illustrated in Figure 5.1. In one second, atom 1 moves from position -1 to position 4 while the second atom moves from position 1 to position 6. Under such a motion, the internuclear distance is unchanged, and the motion represents a translation along x. Since the position of each atom is known relative to the center of mass, the entire translation could be described by the motion of this center of mass from position 0 to position 5. This mode of motion is established only when both atoms move exactly the same distance. The translation must be described with two-dimensional vectors whose components are the positions of the two atoms:

$$\begin{pmatrix} -1 \\ 1 \end{pmatrix} + 5\begin{pmatrix} 1 \\ 1 \end{pmatrix} = \begin{pmatrix} 4 \\ 6 \end{pmatrix} \tag{5.1.1}$$

The vector

$$\begin{pmatrix} 1 \\ 1 \end{pmatrix} \tag{5.1.2}$$

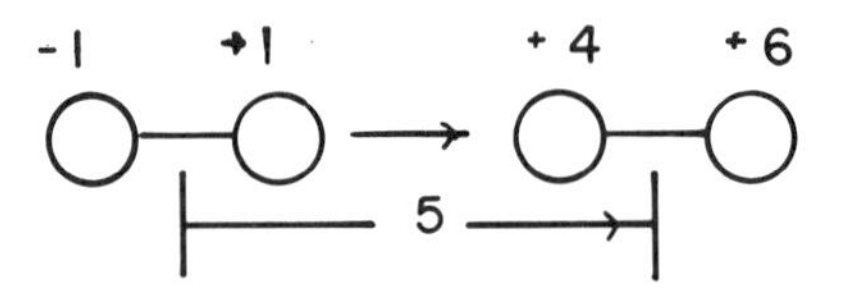

Figure 5.1. The translation of a diatomic molecule along the x axis. Each atom moves 5 units.

is a normal mode vector for translation. Both components are equal since both atoms move the same distance.

Equation 5.1.1 can be generalized to an expression for the location of each atom when these atoms at located at the positions x_1 and x_2 at time zero and move along the x axis with velocity v for a time t. This vector equation

$$\begin{pmatrix} x_1 \\ x_2 \end{pmatrix} = \begin{pmatrix} x_1 \\ x_2 \end{pmatrix} + vt \begin{pmatrix} 1 \\ 1 \end{pmatrix} \tag{5.1.3}$$

requires the same normal mode vector.

If the diatomic molecule is free to move in the x–y plane, each atom must be described by its motions along the orthogonal x and y spatial directions. The vector for a normal mode must now contain four components, x_1, y_1, x_2, and y_2. The molecule can have any orientation in space but a translation of the molecule in the x direction still requires equal displacement of atoms 1 and 2 (Figure 5.2). If the components are arranged in the order x_1, y_1, x_2, and y_2, this translation is

$$\begin{pmatrix} -0.7 \\ 0.7 \\ 0.7 \\ -0.7 \end{pmatrix} + 5 \begin{pmatrix} 1 \\ 0 \\ 1 \\ 0 \end{pmatrix} = \begin{pmatrix} 4.3 \\ 0.7 \\ 5.7 \\ -0.7 \end{pmatrix} \tag{5.1.4}$$

The normal mode translation on x permits no change in the two y components. A normal mode translation vector for y,

$$\begin{pmatrix} 0 \\ 1 \\ 0 \\ 1 \end{pmatrix} \tag{5.1.5}$$

has no atomic motions along x and is orthogonal to the x translation vector. These vectors describe a motion of the entire molecule in phase and are $|X\rangle$ and $|Y\rangle$ for x and y translation, respectively.

The extension to a system in two spatial directions increased the size of the coordinate space to four dimensions. There is no contradiction here. Four spatial coordinates must be specified to describe the motions of the diatomic molecule. These four numbers, not the three-dimensional space, describe the atomic

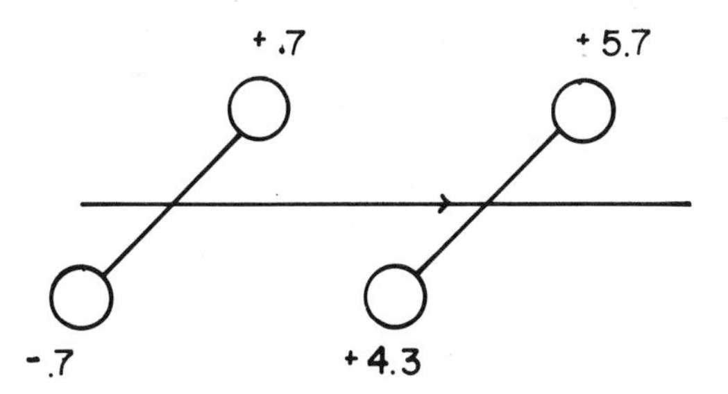

Figure 5.2. The x-axis translation of a diatomic molecule oriented 45° to the x axis. Both atoms move the same distance along the x axis.

motions. A diatomic molecule in three-dimensional space would require vectors of six components. In a phase space, both the six coordinate components and six velocity components are specified. The phase space "motion" of the diatomic molecule requires a 12-component vector.

The diatomic molecule constrained to the x axis had a translation vector characterized by equal components (1, 1). In a two-dimensional space, the vector orthogonal to this vector is

$$\begin{pmatrix} 1 \\ -1 \end{pmatrix} \tag{5.1.6}$$

which describes a motion where the two atoms always move in opposite directions. There can be no net translation—the motion is orthogonal to translation. The motion describes vibration about the internuclear axis as illustrated in Figure 5.3. The motion is periodic and the actual motion of each atom is described by the vector equation

$$\begin{pmatrix} x_1 \\ x_2 \end{pmatrix} = \begin{pmatrix} -1 \\ 1 \end{pmatrix} + A \sin(\omega t - \varphi) \begin{pmatrix} 1 \\ -1 \end{pmatrix} \tag{5.1.7}$$

where A describes the maximal displacement of either atom about its equilibrium position, i.e., with no vibration or translation. The angular frequency

$$\omega = 2\pi\nu \tag{5.1.8}$$

is dictated by the Hooke's law constant for the bond and the masses of the two atoms. The phase angle, φ, is used to establish the actual displacements of the two atoms at time zero.

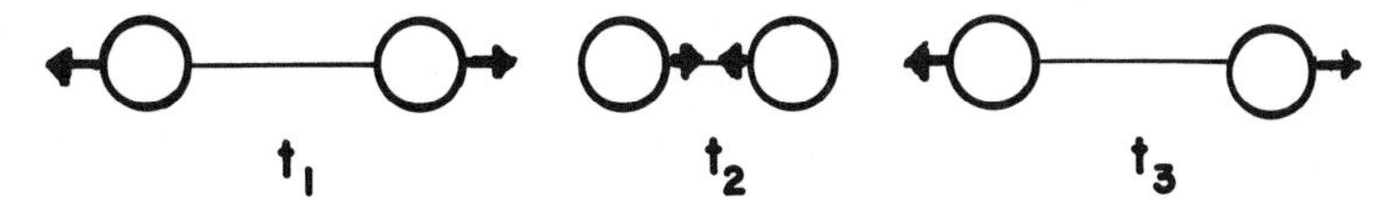

Figure 5.3. The motions of the atoms of a diatomic molecule as a function of time.

If both x and y coordinate directions are allowed, the molecule can also rotate in the space. The vector components which can be used to describe a counterclockwise rotation are illustrated in Figure 5.4. The vectors required for horizontal, vertical, and 45° orientations are

$$\begin{pmatrix} 0 \\ 1 \\ 0 \\ -1 \end{pmatrix} \quad \begin{pmatrix} -1 \\ 0 \\ 1 \\ 0 \end{pmatrix} \quad \begin{pmatrix} \frac{-1}{\sqrt{2}} \\ \frac{1}{\sqrt{2}} \\ \frac{1}{\sqrt{2}} \\ \frac{-1}{\sqrt{2}} \end{pmatrix} \tag{5.1.9}$$

As the molecule rotates, the components of the vector change. For an orientation at angle θ with respect to x (Figure 5.5), the vector is

$$\begin{pmatrix} -\sin\theta \\ \cos\theta \\ \sin\theta \\ -\cos\theta \end{pmatrix} \tag{5.1.10}$$

To describe rotation at some angular frequency ω, the angle θ is replaced by

$$\theta = \omega t \tag{5.1.11}$$

In its general form (Equation 5.1.10), the rotation vector is not orthogonal to the vibration vector defined earlier. The scalar product is

$$(1 \quad 0 \quad -1 \quad 0)\begin{pmatrix} -\sin\theta \\ \cos\theta \\ \sin\theta \\ -\cos\theta \end{pmatrix} = -2\sin\theta \tag{5.1.12}$$

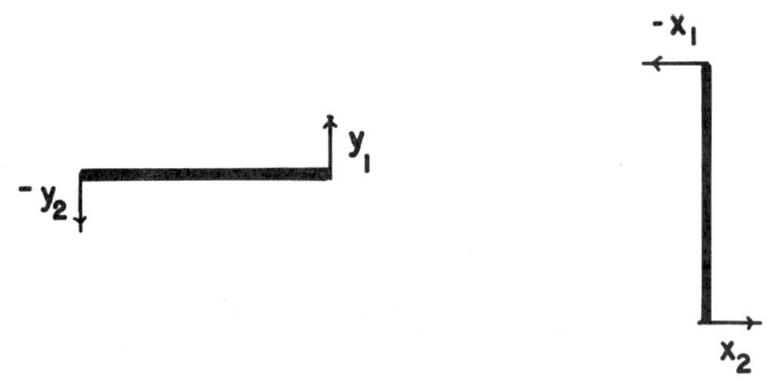

Figure 5.4. Counterclockwise rotation in two-dimensional space defined when the molecule is aligned with the x axis (a) and the y axis (b).

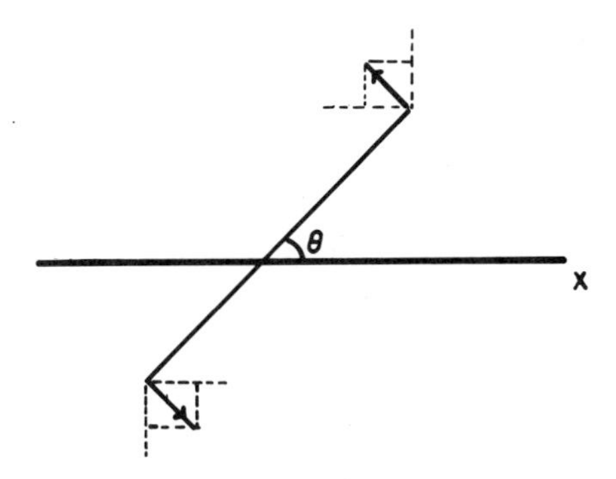

Figure 5.5. The vector components for rotation when the molecular axis is aligned at an angle θ with respect to the positive x axis.

The vectors are not orthogonal because the coordinates for vibration must also change as the molecule rotates. At any angle θ, the vibration must be described by components along the internuclear axis (Figure 5.6). The proper vector is

$$\begin{pmatrix} \cos\theta \\ \sin\theta \\ -\cos\theta \\ -\sin\theta \end{pmatrix} \tag{5.1.13}$$

which is orthogonal to the vector in Equation 5.1.10. Figures 5.5 and 5.6 show clearly that the rotation and vibration vectors on each atom are orthogonal in the three-dimensional space as well.

Since attention is normally focused on molecular vibrations rather than rotations, it is generally most convenient to "stop" the rotation at some convenient orientation and continue analysis without this rotation. Thus, when

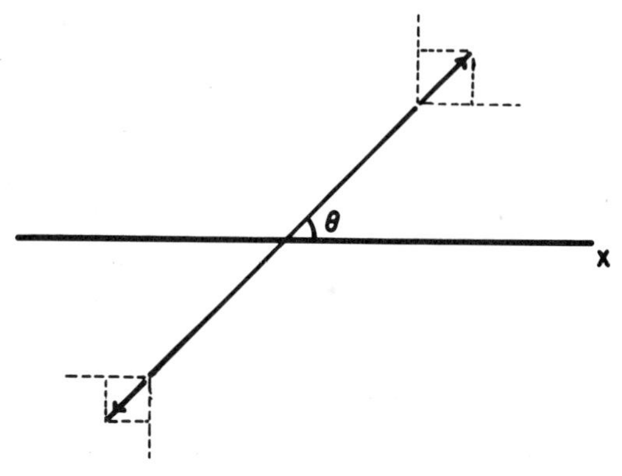

Figure 5.6. The x and y vector components for vibrations in a molecule aligned at an angle θ with respect to the positive x axis.

the internuclear axis is frozen on the x axis, the vibration and rotation normal modes are the vectors

$$|\text{vib}\rangle = \begin{pmatrix} 1 \\ 0 \\ -1 \\ 0 \end{pmatrix} \qquad |\text{rot}\rangle = \begin{pmatrix} 0 \\ 1 \\ 0 \\ -1 \end{pmatrix} \tag{5.1.14}$$

respectively.

The diatomic molecule has only a single vibrational mode. As the number of atoms in the molecule increases, an increasing number of modes will be vibrational. For any nonlinear molecule, three translation and three rotation modes are present and the remaining modes must be vibrations. For N atoms, the number of vibrational modes is

$$3N - 6 \tag{5.1.15}$$

and all modes must be described by a $3N$-component vector involving three spatial components for each atom.

Consider the homonuclear "ozone-like" molecule of Figure 5.7, where the bond angle is selected as 90°. In three-dimensional space for this nonlinear molecule, $3(3) - 6 = 3$ vibrations are possible. In the two-dimensional plane of the three atoms, there are $2(3) - 3 = 3$ vibrations as well since there are only two translations and one rotation in this plane. All three vibrations must lie in the plane. This poses some new questions concerning the choice of ideal coordinates for the three-dimensional space. The orthogonal x, y, and z spatial directions can be used to generate a coordinate set for the molecule. The components are ordered as $x_1, y_1, \ldots, y_3, z_3$. In this coordinate system, the two translations $|X\rangle$ and $|Y\rangle$ in the x–y plane are

$$|X\rangle = \begin{pmatrix} 1 \\ 0 \\ 0 \\ 1 \\ 0 \\ 0 \\ 1 \\ 0 \\ 0 \end{pmatrix} \qquad |Y\rangle = \begin{pmatrix} 0 \\ 1 \\ 0 \\ 0 \\ 1 \\ 0 \\ 0 \\ 1 \\ 0 \end{pmatrix} \tag{5.1.16}$$

The bond angle of 90° was chosen since the three atoms then lie on a semicircle centered midway between atoms 1 and 3. If the molecule is frozen in this position, then the rotation is described by $a - 1y_1$ motion of atom 1, a $-1x_2$

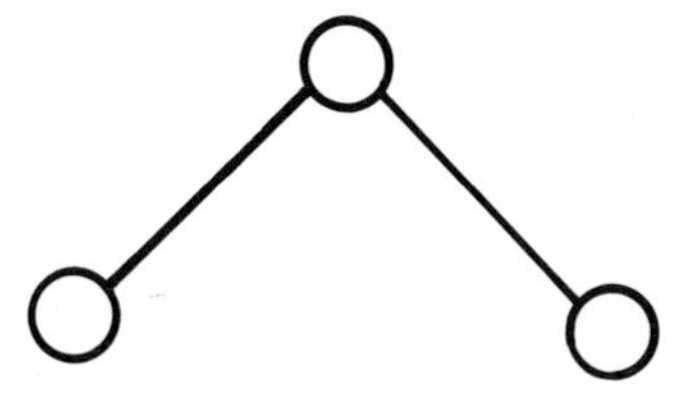

Figure 5.7. Homonuclear triatomic molecule with two bonds perpendicular to each other (90°).

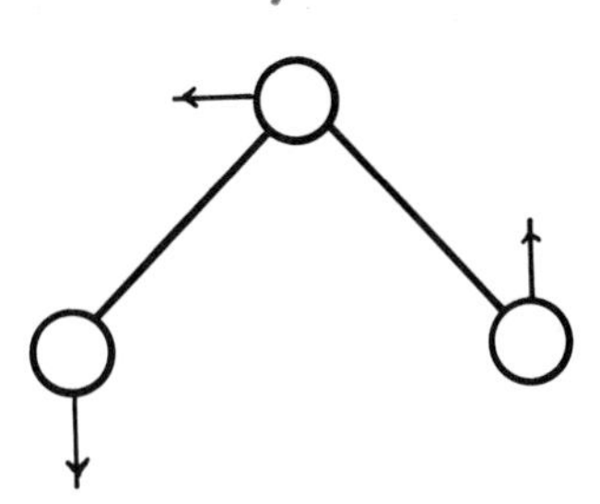

Figure 5.8. The rotation vector components for a triatomic molecule.

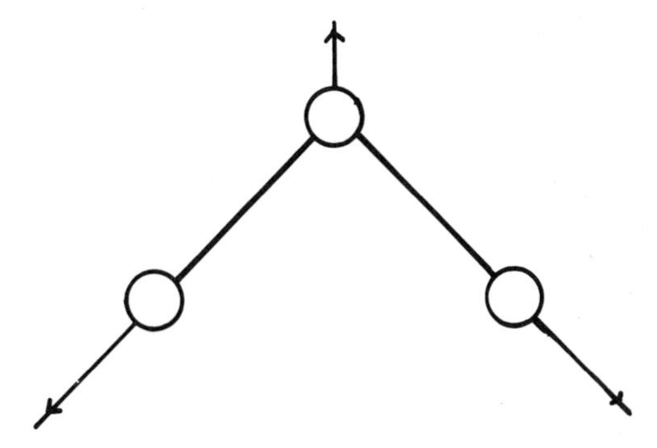

Figure 5.9. A symmetric stretch vibration in the homonuclear triatomic molecule. The displaced atoms all move away from the center of mass.

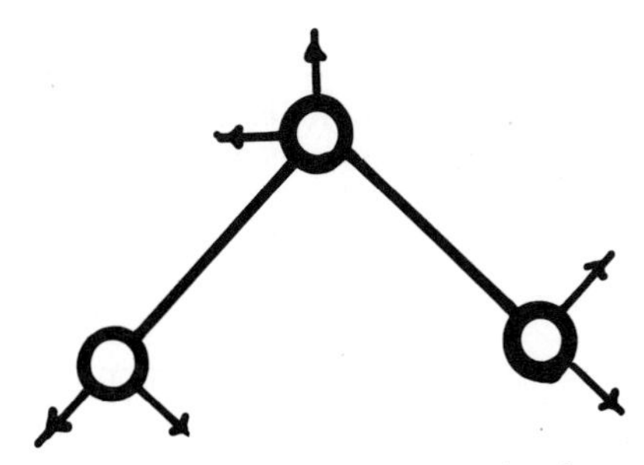

Figure 5.10. A set of six atomic coordinate vectors chosen using the symmetric stretch and bending vectors as a basis set.

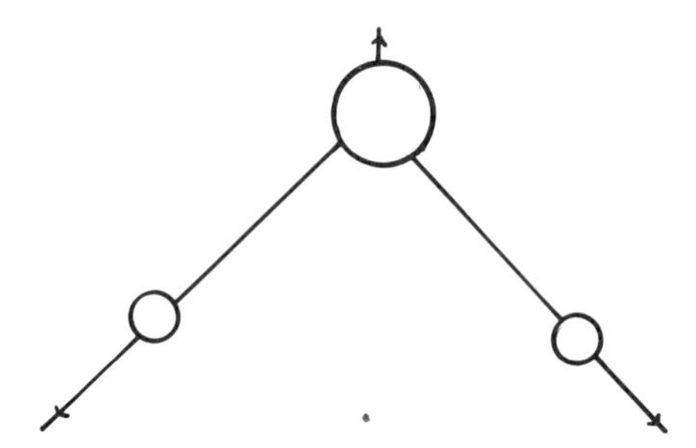

Figure 5.11. A symmetric stretch vibration for a triatomic molecule in which the central atom is much larger than the terminal atoms.

motion of atom 2, and a $+1y_3$ motion of atom 3, as illustrated in Figure 5.8. The nine ordered components (in row format) are then

$$(0 \ -1 \quad 0 \ -1 \quad 0 \quad 0 \quad 0 \quad 1 \quad 0) \tag{5.1.17}$$

Although translations are easily described using these nine coordinates, the normal modes of vibration for the molecule are not. A symmetric stretch vibration is illustrated in Figure 5.9. In the stretching phase, the two terminal atoms move along axes defined by the internuclear bonds as shown. The stretching vector for the central atom lies on an axis which bisects the molecule in the plane. The central atom moves a distance just sufficient to keep the center of mass of the molecule at a fixed location during the vibration.

If these three vectors are selected as part of the six-dimensional vector, the remaining three can be assigned to vectors on each atom perpendicular to these (Figure 5.10) in the molecular plane. If the two vectors on the terminal atoms move outward, the normal mode vibration is a bending vibration. The central atom most move downward along the bisection axis to maintain the center of mass (Problem 5.23). In three dimensions, the remaining three vectors will be perpendicular to the plane and centered on each atom.

If atom 2 is replaced by a heavier atom, the symmetric stretch components must be modified to maintain the center of mass of the system. The three symmetric stretch directions remain the same, but the actual magnitude of the heavier atom stretch must be reduced relative to the stretches of the lighter atoms. The modified symmetric stretch is shown in Figure 5.11.

A detailed description of normal modes requires solutions of the equations of motion for the atoms of the molecule. The techniques required are developed in the remainder of this chapter.

5.2. Equations of Motion for a Diatomic Molecule

The normal coordinates for a diatomic molecule in one and two dimensions were illustrated in Section 5.1. These normal modes appear naturally when the equations of motion for the two atoms in the system are solved. The two atoms of the molecule will give a two-dimensional system when the molecule is constrained to one spatial direction. This example will illustrate the matrix equations required for such systems and will suggest the format for systems requiring more coordinate directions.

A homonuclear diatomic molecule has atoms with spatial coordinates x_1 and x_2. Equations of motion for each atom are based on Hooke's law forces, i.e., the restoring force on each atom will be proportional to its displacement from some equilibrium position:

$$F = -k(x_1 - x_1^0) \tag{5.2.1}$$

If this atom has a mass m, then this force equals

$$F = ma = m d^2x_1/dt^2 \tag{5.2.2}$$

If a single atom is attached to a fixed surface (or very large atom) via a bond with force constant k, the dynamics of the system are described by a single force equation. Dropping subscripts and defining the equilibrium position as $x^0 = 0$ gives the equation

$$F = m d^2x/dt^2 = -kx \tag{5.2.3}$$

Sine or cosine functions provide the necessary sign change and are solutions to this equation. The general solution is

$$x(t) = A\cos(\omega t) + B\sin(\omega t) \tag{5.2.4}$$

where the constant

$$\omega^2 = k/m, \qquad \omega = \sqrt{(k/m)} \tag{5.2.5}$$

is the frequency at which the atom will oscillate. The constants A and B are determined from the position and velocity of the atom at $t = 0$. For example, if $x(0)$ and $v(0)$ are x^0 and v^0, respectively, A and B are solutions of the equations

$$x^0 = A\cos(\omega 0) + B\sin(\omega 0) = A$$

$$A = x_0 \tag{5.2.6}$$

$$v^0 = dx/dt\,|_{t=0} = -A\sin(\omega 0) + B\omega\cos(\omega 0) = B\omega$$

$$B = v^0/\omega \tag{5.2.7}$$

The full solution with these conditions,

$$x(t) = x^0\cos(\omega t) + (v^0/\omega)\sin(\omega t) \tag{5.2.8}$$

can be written as a sine function with a phase delay (Problem 5.19):

$$x(t) = A\ \sin(\omega t - \varphi) = [(x^0)^2 + (v^0/\omega)^2]^{-1}\ \sin(\omega t - \varphi) \tag{5.2.9}$$

The phase is

$$\tan\varphi = v^0/\omega x^0 \tag{5.2.10}$$

The constants A and φ in Equation 5.2.9 are based only on initial conditions. For more complicated systems with multicomponent normal modes, these constants will reappear as scalars in the full vector equation.

For the diatomic molecule, equations of motion are required for each of the atoms. These motions are coupled. If atom 2 moves from its equilibrium position, this will change the force on atom 1 as well. The molecule with its position vectors x_1 and x_2 is shown in Figure 5.12. If the atoms are stretched some distances x_1 and x_2 from their equilibrium positions, atom 2 will experience a restoring force if atom 2 moves a greater distance to the right than atom 1 does. The difference between the two variables determines the net restoring force on atom 2. The force equation is

$$md^2x_2/dt = -k(x_2 - x_1) \tag{5.2.11}$$

The force on atom 2 depends on both variables.

Equation 5.2.11 implies that the force is zero only when $x_1 = x_2$. Since each atomic coordinate is independent, it is convenient to define zero positions for x_1 and x_2 coordinates at the equilibrium positions of the atoms. For a translating, nonvibrating molecule, both atoms would start from (their) coordinate zero. A finite difference in these two coordinates would then mean a net stretch or compression of the bond.

The net force on atom 2 in one direction must be balanced by an equal and opposite force in the opposite direction on atom 1, i.e.,

$$m\ d^2x_1/dt^2 = +k(x_2 - x_1) \tag{5.2.12}$$

If x_1 and x_2 are selected as the first and second components of a two-component column vector, the two equations can be combined into a single matrix equation:

$$\begin{pmatrix} md^2x_1/dt^2 \\ m\ d^2x_1/dt^2 \end{pmatrix} = m\ d^2/dt^2 \begin{pmatrix} x_1 \\ x_2 \end{pmatrix} = \begin{pmatrix} -k & k \\ k & -k \end{pmatrix} \begin{pmatrix} x_1 \\ x_2 \end{pmatrix}$$

$$d^2|q\rangle/dt^2 = \mathbf{K}|q\rangle \tag{5.2.13}$$

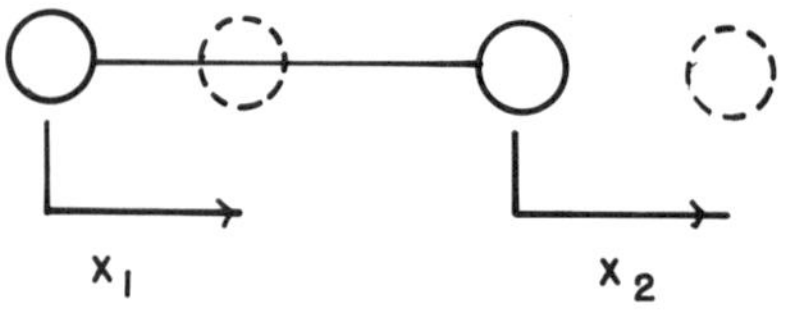

Figure 5.12. The displacement vectors x_1 and x_2 for a diatomic molecule confined to the x axis.

When the two atoms have different masses, the scalar m is replaced with a diagonal matrix, **M**, of masses.

If the **K** matrix can be transformed to a diagonal matrix, the coupled system can be replaced by two uncoupled equations. Factoring k from each element leaves the matrix

$$\begin{pmatrix} -1 & 1 \\ 1 & -1 \end{pmatrix} \tag{5.2.14}$$

with eigenvalues 0 and -2 and column eigenvectors

$$|0\rangle = \left(\frac{1}{\sqrt{2}}\right)\begin{pmatrix} 1 \\ 1 \end{pmatrix} \qquad |2\rangle = \left(\frac{1}{\sqrt{2}}\right)\begin{pmatrix} 1 \\ -1 \end{pmatrix} \tag{5.2.15}$$

Since the matrix is symmetric, the row vectors will have the same components. The similarity transform matrices are

$$\mathbf{S} = (|0\rangle \quad |2\rangle) = \begin{pmatrix} 1 & 1 \\ 1 & -1 \end{pmatrix} \tag{5.2.16}$$

$$\mathbf{S}^{-1} = \begin{pmatrix} \langle 0| \\ \langle 2| \end{pmatrix} = \left(\frac{1}{2}\right)\begin{pmatrix} 1 & 1 \\ 1 & -1 \end{pmatrix} \tag{5.2.17}$$

The full normalization is assigned to $\mathbf{S}^{-1}$ for convenience.

Equation 5.2.13 is recast in diagonal form by inserting the identity,

$$\mathbf{S}\mathbf{S}^{-1} = \mathbf{I} \tag{5.2.18}$$

between **K** and $|q\rangle$. Since $\mathbf{S}^{-1}$ now transforms $|q\rangle$ on the right-hand side, both sides of the equation are left multiplied by $\mathbf{S}^{-1}$ to give an equation in the new coordinates:

$$|q'\rangle = \mathbf{S}^{-1}|q\rangle = \left(\frac{1}{2}\right)\begin{pmatrix} 1 & 1 \\ 1 & -1 \end{pmatrix}\begin{pmatrix} x_1 \\ x_2 \end{pmatrix} = \begin{pmatrix} (x_1 + x_2)/2 \\ (x_1 - x_2)/2 \end{pmatrix} \tag{5.2.19}$$

This equation is

$$m\, d^2|q'\rangle/dt^2 = \mathbf{S}^{-1}\mathbf{K}\mathbf{S}|q'\rangle \tag{5.2.20}$$

with transformed matrix

$$\mathbf{K}' = \mathbf{S}^{-1}\mathbf{K}\mathbf{S} = \begin{pmatrix} 0 & 0 \\ 0 & -2 \end{pmatrix} \tag{5.2.21}$$

The new diagonal matrix equation,

$$\begin{aligned} d^2|q'\rangle/dt^2 &= \mathbf{K}'|q'\rangle \\ m\begin{pmatrix} d^2x_1'/dt^2 \\ d^2x_2'/dt^2 \end{pmatrix} &= \begin{pmatrix} 0 & 0 \\ 0 & -2 \end{pmatrix}\begin{pmatrix} x_1' \\ x_2' \end{pmatrix} \end{aligned} \tag{5.2.22}$$

immediately resolves into two uncoupled equations:

$$\begin{aligned} m\, d^2x_1'/dt^2 &= 0 \\ m\, d^2x_2'/dt^2 &= -(2k/m)x_2 \end{aligned} \tag{5.2.23}$$

delete

The first equation integrates to a linear equation with two constants of integration:

$$x_1' = c't + c \tag{5.2.24}$$

If $x_1'(0)$ and $v_1'(0)$ are the initial conditions, the constants are evaluated to give the equation

$$x_1' = v_1'(0)\, t + x_1'(0) \tag{5.2.25}$$

The vector moves linearly with time and has the form

$$(x_1 + x_2)/2 = \{[v_1(0) + v_2(0)]/2\}\, t + [x_1(0) + x_2(0)]/2 \tag{5.2.26}$$

The combination of atomic coordinates defines the center of mass of the molecule. The first equation with zero eigenvalue generates the equation of motion for the center of mass:

$$X = V^0 t + X^0 \tag{5.2.27}$$

The second equation,

$$d^2x_2'/dt^2 = -(2k/m)\, x_2 \tag{5.2.28}$$

parallels that for a single particle bound to a fixed point. The solution will oscillate with frequency

$$\omega = \sqrt{2k/m} \tag{5.2.29}$$

In this case, the oscillation will be dictated by the initial displacement of $x_1(0)$

and $x_2(0)$ from equilibrium. If maximal displacement occurs at $t=0$, the appropriate solution is

$$x_2'(t)=\{x_1(0)-x_2(0)\}\cos(\omega t) \tag{5.2.30}$$

The stretched bond will reach its equilibrium position and continue to maximal bond contraction $[\cos(\omega t)=-1]$ before returning to its initial coordinates. The entire system may be translating in space but the oscillation will proceed independently of this motion.

The motions of the atoms now involve two parts. The motion of the full molecule can be described by $X(t)$, which includes both initial position and molecular motion in time,

$$X(t)=X^0+V^0 t \tag{5.2.31}$$

and the vibrational component. $x_1(t)$ and $x_2(t)$ are recovered from $|q'\rangle$ by applying $\mathbf{S}$, i.e.,

$$\mathbf{S}\,|q'\rangle=\mathbf{S}\mathbf{S}^{-1}\,|q\rangle=|q\rangle \tag{5.2.32}$$

The result is

$$\begin{pmatrix}1 & 1\\ 1 & -1\end{pmatrix}\begin{pmatrix}X(t)\\ [(x_1-x_2)/2][\sin(\omega t-\varphi)]\end{pmatrix}=\begin{pmatrix}X(t)+\{(x_1(0)-x_2(0)\}\cos(\omega t\\ X(t)-\{x_1(0)-x_2(0)\}\cos(\omega t)\end{pmatrix} \tag{5.2.33}$$

The only difference between the two equations is the sign of the second term. The atoms must move in exactly opposite directions for the vibration. The vibration can be modified for any initial displacements and velocities (Problem 5.20).

When the diatomic molecule is heteroatomic, the two equations of motion become

$$\begin{aligned} m_1 d^2x_1/dt^2 &= -k(x_1-x_2)\\ m_2 d^2x_2/dt^2 &= k(x_1-x_2)\end{aligned} \tag{5.2.34}$$

Since m_1 and m_2 are different, the matrix format for these two equations requires a diagonal mass matrix $\mathbf{M}$,

$$\mathbf{M}=\begin{pmatrix}m_1 & 0\\ 0 & m_2\end{pmatrix} \tag{5.2.35}$$

The matrix equation is

$$\mathbf{M}\,d^2\,|q\rangle/dt^2=\mathbf{K}\,|q\rangle \tag{5.2.36}$$

Both sides can be multiplied by the inverse of **M**,

$$\mathbf{M}^{-1} = \begin{pmatrix} 1/m_1 & 0 \\ 0 & 1/m_2 \end{pmatrix} \tag{5.2.37}$$

to give the matrix equation

$$d^2|q\rangle/dt^2 = \mathbf{M}^{-1}\mathbf{K}\,|q\rangle \tag{5.2.38}$$

The entire problem parallels the homonuclear case. However, the non-symmetric matrix

$$\mathbf{M}^{-1}\mathbf{K} = \begin{pmatrix} -k/m_1 & k/m_1 \\ k/m_2 & -k/m_2 \end{pmatrix} \tag{5.2.39}$$

must now be diagonalized.

The characteristic polynomial for the matrix,

$$(k/m_1 - \lambda)(k/m_2 - \lambda) - k^2/m_1 m_2 = 0 \tag{5.2.40}$$

yields eigenvalues

$$\lambda = 0 \qquad \lambda = k/m_1 + k/m_2 = k/\mu \tag{5.2.41}$$

where μ is the reduced mass of the diatomic molecule. The similarity matrices which are determined by separate solutions for the row and column eigenvectors are (Problem 5.4):

$$\mathbf{S} = \begin{pmatrix} 1 & 1/m_1 \\ 1 & -1/m_2 \end{pmatrix} \tag{5.2.42}$$

$$\mathbf{S}^{-1} = \begin{pmatrix} 1/m_2 & 1/m_1 \\ 1 & -1 \end{pmatrix} \tag{5.2.43}$$

The transformed matrix, $\mathbf{S}^{-1}\mathbf{KS}$, is

$$\begin{pmatrix} 0 & 0 \\ 0 & k/\mu \end{pmatrix} \tag{5.2.44}$$

The system will again separate into two systems. The first equation will describe a translation in space while the nonzero eigenvalue gives the characteristic frequency ω for the oscillation:

$$\omega = \sqrt{k/\mu} \tag{5.2.45}$$

The reduced mass, μ, replaces $m/2$, the reduced mass for the homonuclear case.

The normal mode coordinates for the heteronuclear atom are

$$\mathbf{S}^{-1}|q\rangle = \begin{pmatrix} 1/m_1 & 1/m_2 \\ 1 & -1 \end{pmatrix} (\mu) \begin{pmatrix} x_1 \\ x_2 \end{pmatrix}$$

$$= \begin{pmatrix} x_1\mu/m_1 + x_2\mu/m_2 \\ \mu(x_1 - x_2) \end{pmatrix} \tag{5.2.46}$$

The vibrational component still depends on the difference between atoms 1 and 2. The first component locates the center of mass. The center of mass motion constitutes an independent motion in the diagonal (normal mode) system. Since

$$1/\mu = (m_1 + m_2)/m_1 m_2 \tag{5.2.47}$$

the first component of Equation 5.2.46 is

$$x_1 m_2/(m_1 + m_2) + x_2 m_1/(m_1 + m_2) = (m_2 x_1 + m_1 x_2)/(m_1 + m_2) \tag{5.2.48}$$

This is the standard equation for the location of the center of mass of a diatomic molecule. The first (translation) equation with eigenvalue $\lambda = 0$ describes the motion of this point with time.

In normal mode coordinates, the vibration depends only on the separation of atoms 1 and 2. Does each move the same distance during the vibration? The normal mode is transformed back to $|q\rangle$ as

$$|q\rangle = \mathbf{S}\,|q'\rangle = \begin{pmatrix} 1 & 1/m_1 \\ 1 & -1/m_2 \end{pmatrix} \begin{pmatrix} \mu X \\ \mu(x_1 - x_2) \end{pmatrix}$$

$$= \begin{pmatrix} X + (\mu/m_1)(x_1 - x_2) \\ X - (\mu/m_2)(x_1 - x_2) \end{pmatrix}$$

$$= \begin{pmatrix} X + [m_2/(m_1 + m_2)](x_1 - x_2) \\ X + -[m_1/(m_1 + m_2)](x_1 - x_2) \end{pmatrix} \tag{5.2.49}$$

The motions of the individual atoms are weighted by a mass fraction. If m_1 is much lighter than m_2, the x_1 coordinate will have the largest weight and will involve a greater fraction of the total motion. The relative motions of the two atoms have the ratio

$$x_2/x_1 = m_1/m_2 \tag{5.2.50}$$

The lighter atom will move more than the larger atom. This appears in the $|q\rangle$ vector for individual atom motions. The $|q'\rangle$ vector focuses on the net separation and gives the directions of motion without resolving the fraction of the motion attributed to each atom. In a normal vibration mode, $|q'\rangle$, all atoms will maintain consistent phase relations with respect to each other. For example, in the diatomic molecule, atom 2 always moves in a direction opposite to atom 1. The q' motion is equivalent to a single atom with reduced mass μ vibrating on a fixed surface. Diagonalization has produced the one-dimensional vibration.

5.3. Normal Modes for Nontranslating Systems

The diatomic molecule of Section 5.2 had one translational and one vibrational normal mode. If a chain of atoms can be constrained in space, the normal modes will be vibrations. In Figure 5.13, a diatomic molecule is fixed to the walls via two additional bonds. For convenience, the force constants for all three bonds are equal and the two masses are identical. The translation of the free diatomic molecule is now constrained by the outer bonds and will force the translation into a new vibrational motion. This model is the precursor of a one-dimensional linear crystal, which will be introduced in Section 5.6.

Although the normal modes for this system will arise naturally in the matrix analysis of the system, their atomic motions are presented first because they are similar to the vibration and translation of the free diatomic molecule. Both atoms move in the same direction in the mode of Figure 5.14a. The stretched bond on the left will eventually pull the two atoms in that direction. The two outer bonds have changed the translational motion of the molecule into a "vibration" of the whole molecule between the outer bonds.

The vibration of Figure 5.14b has atomic motions identical to those observed for the free diatomic. The atoms move in opposite directions to alternately expand and compress the central bond. When the central bond is compressed, both outer bonds will be expanded and vice versa.

The three bond–two atom system can be solved using the force equations for each atom and the techniques of Section 8.2. This same result is also found using the conserved total energy of the system.

The variables x_1 and x_2 are selected as the displacements of atoms 1 and 2 from equilibrium. The kinetic energies of each atom depend on the atomic velocities,

$$\begin{aligned} v_1 &= dx_1/dt \\ v_2 &= dx_2/dt \end{aligned} \tag{5.3.1}$$

and are

$$\begin{aligned} T_1 &= mv_1^2/2 \\ T_2 &= mv_2^2/2 \end{aligned} \tag{5.3.2}$$

The atomic kinetic energies are not coupled.

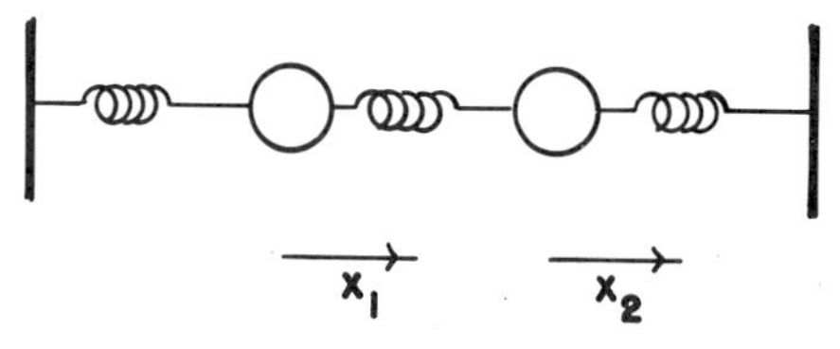

Figure 5.13. A diatomic molecule bonded between two fixed surfaces. The force constants for each bond are equal. The atomic displacements are x_1 and x_2.

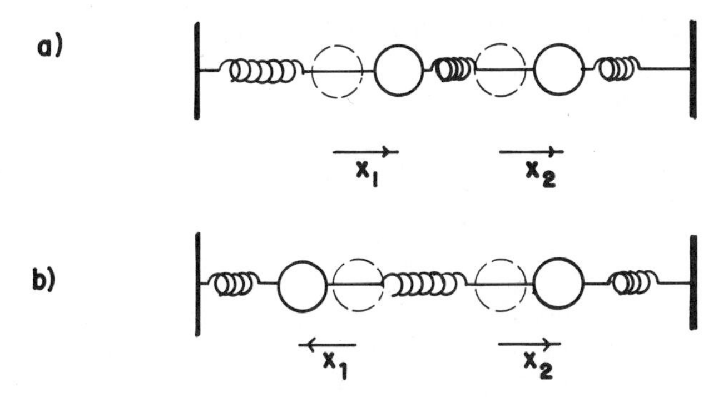

Figure 5.14. The two vibrational modes for a diatomic molecule bonded to two fixed surfaces: (a) vibration of both atoms in phase; (b) vibration of both atoms 180° out of phase, i.e., in exactly opposite directions.

The potential energy of the system requires three terms, one corresponding to each of the bonds in the system. The vectors of Figure 5.13 show that expansion or contraction of the bond on the left depends only on x_1 since only atom 1 is connected to this bond. The potential energy for this bond is

$$V_1 = kx_1^2/2 \tag{5.3.3}$$

The bond on the right will acquire potential energy only if atom 2 is displaced from equilibrium by an amount x_2. This potential energy is

$$V_3 = kx_2^2/2 \tag{5.3.4}$$

The center bond is affected by the displacements of both x_1 and x_2. However, the potential energy will change only if x_1 and x_2 experience different displacements. The potential for this bond will depend on the difference of these displacements, i.e.,

$$V_2 = (k/2)(x_1 - x_2)^2 \tag{5.3.5}$$

The order of variables is not crucial since the difference is squared.

The total energy is a scalar sum of all these terms:

$$\begin{aligned} E &= T_1 + T_2 + V_1 + V_2 + V_3 \\ &= (m/2)(v_1^2 + v_2^2) + (k/2)[x_1^2 + (x_1 - x_2)^2 + x_2^2] \end{aligned} \tag{5.3.6}$$

Since all terms are quadratic in the velocity and position variables, the energy equation can be written as a sum of quadratic forms. The kinetic energy terms generate a diagonal quadratic form since there is no coupling:

$$\frac{1}{2}(v_1 \quad v_2)\begin{pmatrix} m & 0 \\ 0 & m \end{pmatrix}\begin{pmatrix} v_1 \\ v_2 \end{pmatrix} = \frac{1}{2}\langle v|\mathbf{M}|v\rangle = (m/2)\langle v|\mathbf{I}|v\rangle \tag{5.3.7}$$

with $m_1 = m_2 = m$.

Because of the coupling of coordinates in the potential energy, its matrix has off-diagonal elements. When the force constants, k, are identical, the total potential energy becomes

$$\begin{aligned} V &= (k/2)(x_1^2 + x_1^2 - 2x_1x_2 + x_2^2 + x_2^2) \\ &= (k/2)(2x_1^2 - 2x_1x_2 + 2x_2^2) \end{aligned} \tag{5.3.8}$$

The squares of a single coordinate x_i result from diagonal elements of the matrix so each diagonal element is 2. The factor of -2 in the cross term will be generated from the two off-diagonal terms of the matrix so each of these elements must be -1. The appropriate quadratic form is

$$(k/2)(x_1, x_2)\begin{pmatrix} 2 & -1 \\ -1 & 2 \end{pmatrix}\begin{pmatrix} x_1 \\ x_2 \end{pmatrix} \tag{5.3.9}$$

The diagonal elements for this matrix are 2 while those for the free diatomic molecule were 1. The terminal bonds to the walls provide the two additional factors which appeared in the terminal diagonal terms. For an N-atom chain, the a_{11} and a_{NN} elements will be 1 if the atom is free and 2 if atoms 1 and N are bonded to fixed surfaces.

The total energy of the coupled system is now a sum of quadratic terms for the kinetic and potential energies:

$$\begin{aligned} E &= \frac{1}{2}\langle v|\mathbf{M}|v\rangle + \langle q|\mathbf{K}|q\rangle \\ &= \frac{1}{2}(v_1, v_2)\begin{pmatrix} m & 0 \\ 0 & m \end{pmatrix}\begin{pmatrix} v_1 \\ v_2 \end{pmatrix} + \frac{1}{2}(x_1, x_2)\begin{pmatrix} 2k & -1k \\ -1k & 2k \end{pmatrix}\begin{pmatrix} x_1 \\ x_2 \end{pmatrix} \end{aligned} \tag{5.3.10}$$

With a proper similarity transform, this energy equation can be stated in diagonal matrix form. The energy will then break into two energy components corresponding to the two distinct normal modes of the system. The eigenvalues $\lambda = 1$ and $\lambda = 3$ for the symmetric $\mathbf{K}$ matrix give the now familiar orthogonal eigenvectors,

$$|1\rangle = \left(\frac{1}{\sqrt{2}}\right)\begin{pmatrix} 1 \\ 1 \end{pmatrix} \qquad |3\rangle = \left(\frac{1}{\sqrt{2}}\right)\begin{pmatrix} 1 \\ -1 \end{pmatrix} \tag{5.3.11}$$

and similarity transform matrices,

$$\mathbf{S} = \begin{pmatrix} 1 & 1 \\ 1 & -1 \end{pmatrix} \qquad \mathbf{S}^{-1} = \left(\frac{1}{2}\right)\begin{pmatrix} 1 & 1 \\ 1 & -1 \end{pmatrix} \tag{5.3.12}$$

The identity matrix, $\mathbf{I}=\mathbf{SS}^{-1}$, is now inserted between the matrices and the vectors to give

$$E=\frac{1}{2}\langle v|\,\mathbf{SS}^{-1}\mathbf{MSS}^{-1}\,|v\rangle+\frac{1}{2}\langle q|\,\mathbf{SS}^{-1}\mathbf{KSS}^{-1}\,|q\rangle$$

$$=\frac{1}{2}\,\langle v'|\,\mathbf{M}\,|v'\rangle+\frac{1}{2}\langle q'|\,\mathbf{K}'\,|q'\rangle \tag{5.3.13}$$

$\mathbf{K}'$ is diagonal in the eigenvalues,

$$\mathbf{K}=\begin{pmatrix}\frac{1}{2}\end{pmatrix}\begin{pmatrix}1 & 1\\ 1 & -1\end{pmatrix}\begin{pmatrix}2k & -1k\\ -1k & 2k\end{pmatrix}\begin{pmatrix}1 & 1\\ 1 & -1\end{pmatrix}=\begin{pmatrix}1k & 0\\ 0 & 3k\end{pmatrix} \tag{5.3.14}$$

and the energy is

$$E=\frac{1}{2}\langle v'|\,\mathbf{M}\,|v'\rangle+\frac{1}{2}\langle q'|\,\mathbf{K}'\,|q'\rangle$$

$$=(v_1' \quad v_2')\begin{pmatrix}m & 0\\ 0 & m\end{pmatrix}\begin{pmatrix}v_1\\ v_2\end{pmatrix}+(x_1' x_2')\begin{pmatrix}1k & 0\\ 0 & 3k\end{pmatrix}\begin{pmatrix}x_1'\\ x_2'\end{pmatrix} \tag{5.3.15}$$

The diagonal matrices now separate into two independent energy expressions, one for each normal mode:

$$\begin{aligned}E_1'&=(m/2)\,v_1'^2+(k/2)\,x_1'^2\\ E_2'&=(m/2)\,v_2'^2+(3k/2)\,x_2'^2\end{aligned} \tag{5.3.16}$$

The x_2' normal mode has three times the potential energy of the x_1' normal mode. These normal mode motions are determined from the independent atomic displacements as (Figure 5.14)

$$|q'\rangle=S^{-1}|q\rangle$$

$$\begin{pmatrix}x_1'\\ x_2'\end{pmatrix}=\begin{pmatrix}\frac{1}{2}\end{pmatrix}\begin{pmatrix}1 & 1\\ 1 & -1\end{pmatrix}\begin{pmatrix}x_1\\ x_2\end{pmatrix}$$

$$=\begin{pmatrix}\frac{1}{2}\end{pmatrix}\begin{pmatrix}x_1+x_2\\ x_1-x_2\end{pmatrix} \tag{5.3.17}$$

For mode 1, the x' coordinate is the center-of-mass coordinate. This center of mass, i.e., the two-atom "molecule," oscillates about the midpoint of the system. Both atoms will always move in the same direction in this mode.

The second mode describes opposed motions of the two atoms (Figure 5.14b). The atoms vibrate like the atoms in the free molecule, but the potential energy is three times as large because this motion involves the simultaneous expansion or compression of all three springs.

The equations for the energies of the normal modes do not give the vibrational frequencies directly. However, both normal modes will vibrate and the displacement coordinates are proportional to

$$x_i' \propto \sin(\omega_i t) \tag{5.3.18}$$

The velocities are proportional to

$$v_i' \propto \omega_i \cos(\omega_i t) \tag{5.3.19}$$

These expressions can be substituted into the energy expressions for each normal mode to determine the frequencies. For mode 1, the energy is

$$E_1' = (m/2)\,\omega_1^2 \cos^2(\omega_1 t) + (k/2)\sin^2(\omega_1 t) \tag{5.3.20}$$

Since the left-hand side is time independent, the right-hand side must also be time independent. Since

$$\cos^2(\omega_1 t) + \sin^2(\omega_1 t) = 1 \tag{5.3.21}$$

the right-hand side will become time-independent only if the coefficients in the kinetic and potential energy terms are equal, i.e.,

$$m\omega_1^2/2 = k/2 \tag{5.3.22}$$

Solving for ω_1 gives

$$\omega_1 = \sqrt{k/m} \tag{5.3.23}$$

The frequency of the second mode is determined via the same procedure as

$$\omega_2 = \sqrt{3k/m} \tag{5.3.24}$$

Since this mode involves more compression and expansion of the bonds, the additional energy is manifested as a higher vibrational frequency.

For polyatomic systems, careful selection of the coordinates can simplify the problem. Once the coordinates are selected, they are used to define the kinetic and potential energies. These can then define a Lagrangian or Hamiltonian from which all the force equations in these coordinates are generated. The Lagrangian is the difference of the kinetic and potential energies:

$$\mathscr{L} = T - V \tag{5.3.25}$$

For the coordinates of the bound diatomic system, the Lagrangian is

$$\mathscr{L} = (m/2)(v_1^2 + v_2^2) - (k/2)(2x_1^2 - 2x_1x_2 + 2x_2^2) \tag{5.3.26}$$

The force equations are generated by selecting conjugate position and velocity components. For x_1, v_1 is the conjugate velocity since they describe position and velocity for a common atomic coordinate. Each conjugate pair satisfies an equation

$$d/dt(\partial \mathscr{L}/\partial v_i) = \partial \mathscr{L}/\partial x_i \tag{5.3.27}$$

Partial derivatives are used since v_i and x_i are independent variables in the equation. The force equation for atom 1 is

$$\begin{aligned} d/dt[(m/2)\,2v_1] &= -(k/2)(4x_1 - 2x_2) \\ m\,d^2x_1/dt^2 &= -k(2x_1 - x_2) \end{aligned} \tag{5.3.28}$$

The force equation for atom 2 is

$$d/dt[(m/2)\,2v_2] = -(k/2)(-2x_1 + 4x_2)$$

$$m\,d^2x_2/dt^2 = -k(-x_1 + 2x_2) \tag{5.3.29}$$

as determined in Section 5.2. In more complicated systems, the Lagrangian provides a convenient technique for generating the force equations with proper signs.

The Lagrangian technique is applicable to any coordinate system. The equation is generally written in terms of generalized position variables, q_i, and their conjugate velocities,

$$\dot{q}_i = dq_i/dt \tag{5.3.30}$$

as

$$d/dt(\partial \mathscr{L}/\partial \dot{q}_i) = \partial \mathscr{L}/\partial q_i \tag{5.3.31}$$

For example, the conjugate variables in this equation might be a polar angle, θ, with conjugate velocity

$$\omega = \dot{\theta} = d\theta/dt \tag{5.3.32}$$

The kinetic and potential energies for the system in these variables define the Lagrangian and the force equations.

Hamilton's equations, like Lagrange's equation, provide a convenient technique for translating the kinetic and potential energies in conjugate variables into force equations. The conjugate variables for the Hamiltonian are a position coordinate, q_i, and a generalized momentum,

$$p_i = m\,dx_i/dt \tag{5.3.33}$$

The Hamiltonian is defined in terms of the Lagrangian as

$$H = \sum p_i q_i - \mathscr{L} \tag{5.3.34}$$

However, in most cases, the Hamiltonian reduces to a sum of the kinetic and potential energies written in the variables q_i and p_i,

$$H(q_i, p_i) = T(q_i, p_i) + V(q_i, p_i) \tag{5.3.35}$$

Velocity and force equations are then determined directly from the equations

$$\dot{q}_i = \partial H/\partial p_i \qquad p_i = -\partial H/\partial q_i \tag{5.3.36}$$

For example, the Hamiltonian for a single atom bound to a wall is

$$H = p^2/2m + kx^2/2 \tag{5.3.37}$$

where x and p are the conjugate position and momentum for the particle in one spatial direction. Hamilton's equations become

$$\dot{x} = \partial H/\partial p = 2p/2m + 0 = p/m$$

$$p = m\dot{x} \tag{5.3.38}$$

$$\dot{p} = -\partial H/\partial x = 0 - 2kx/2 = -kx = F \tag{5.3.39}$$

since the time derivative of the momentum is the force:

$$dp/dt = m\, dv/dt = m\, d^2x/dt^2 = F \tag{5.3.40}$$

The utility of both the Lagrangian and Hamiltonian techniques for generating force equations will be demonstrated later in this chapter for polyatomic molecules.

5.4. Normal Modes Using Projection Operators

The diatomic molecule has been the focus of the past three sections because it clearly illustrates the connection between normal modes and the individual atomic motions. In previous sections, force and energy equations were resolved into separate scalar equations using matrix diagonalization. In this section, relevant matrix equations will be solved first as if they were scalar equations. Projection operator techniques will then resolve the formal matrix solution of the equation into distinct scalar solutions.

For variety, the diatomic molecule is fixed to a single surface with a force constant k equal to the "molecular" force constant (Figure 5.15). This model describes a diatomic molecule adsorbed at a surface. With the atomic displacement variables x_1 and x_2 for atoms 1 and 2, respectively, the total energy of the system for a homonuclear molecule is

$$E = (m/2)(v_1^2 + v_2^2) + (k/2)[x_1^2 + (x_1 - x_2)^2] \tag{5.4.1}$$

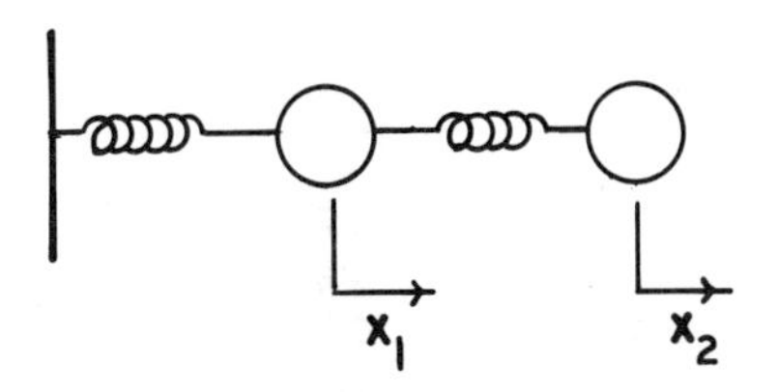

Figure 5.15. A diatomic molecule confined to the x direction and bound to a single surface via a bond with force constant k.

There are only two potential energy terms since there is no right-hand wall. The forces for this system are generated using the Lagrangian,

$$\begin{aligned}\mathscr{L} &= T - V \\ &= (m/2)(v_1^2 + v_2^2) - (k/2)[x_1^2 + (x_1 - x_2)^2]\end{aligned} \tag{5.4.2}$$

to give

$$\begin{aligned} m\, d^2x_1/dt^2 &= -(k/2)[2x_1 + 2(x_1 - x_2)] \\ &= -k(2x_1 - x_2) \\ m\, d^2x_2/dt^2 &= -(k/2)[2(x_1 - x_2)(-1)] \\ &= -k[-x_1 + x_2] \end{aligned} \tag{5.4.3}$$

The matrix equation for the two atomic displacements is

$$d^2/dt^2 \begin{pmatrix} x_1 \\ x_2 \end{pmatrix} = -(k/m) \begin{pmatrix} 2 & -1 \\ -1 & 1 \end{pmatrix} \begin{pmatrix} x_1 \\ x_2 \end{pmatrix} \tag{5.4.4}$$

The diagonal element in $\mathbf{K}$ for the bound end of the molecule is 2 while the unbound end has an element 1. The model is intermediate between the free and fully bound diatomic molecule systems.

Although the matrix is symmetric and the eigenvalues are real, these eigenvalues are more complicated than those encountered for the other two systems. They are

$$\lambda_- = (3 - \sqrt{5})/2 \qquad \lambda_+ = (3 + \sqrt{5})/2 \tag{5.4.5}$$

Since it is rather tedious to develop eigenvectors and similarity matrices for this system, it is better to move directly to the projection operator matrices for each eigenvalue. These are generated directly using the Lagrange–Sylvester formula,

$$\mathbf{P}_i = \prod_{\substack{j \\ j \neq i}}^{N} (K - \lambda_j) \Big/ \prod_{j}^{N} (\lambda_i - \lambda_j) \tag{5.4.6}$$

to give

$$\mathbf{P}(-) = -(1/2\sqrt{5})\begin{pmatrix} (1-\sqrt{5}) & -2 \\ -2 & (-1-\sqrt{5}) \end{pmatrix} \tag{5.4.7}$$

$$\mathbf{P}(+) = (1/2\sqrt{5})\begin{pmatrix} (1+\sqrt{5}) & -2 \\ -2 & (-1+\sqrt{5}) \end{pmatrix} \tag{5.4.8}$$

These matrices are idempotent and span the two-dimensional space.

The matrix problem can utilize the projection operators while avoiding similarity transforms if the matrix equation

$$\mathbf{M}d^2|q\rangle/dt^2 = -\mathbf{K}|q\rangle \tag{5.4.9}$$

is solved like the corresponding scalar equation,

$$m\, d^2x/dt^2 = -kx \tag{5.4.10}$$

The scalar solution was determined in Section 5.2 as

$$x(t) = (v^0/\omega)\sin(\omega t) + x^0\cos(\omega t) \tag{5.4.11}$$

In the matrix problem with two components, $x(t)$, x^0, and v^0 are all replaced by two-component vectors giving, respectively, the positions, initial positions, and initial velocities of the two atoms. If column vectors are selected, all vectors will be placed to the right.

The scalar solution determined the vibrational frequency, ω, as

$$\omega = \sqrt{k/m} \tag{5.4.12}$$

For the matrix problem, this quantity must be replaced by a frequency matrix defined as

$$\boldsymbol{\omega} = \mathbf{M}^{-1/2}\mathbf{K}^{1/2} = \sqrt{(1/m)}\,K^{1/2} \tag{5.4.13}$$

The full matrix solution for the displacements is

$$|q(t)\rangle = \boldsymbol{\omega}^{-1}\sin(\boldsymbol{\omega}t)|v^0\rangle + \cos(\boldsymbol{\omega}t)|x^0\rangle \tag{5.4.14}$$

The sine and cosine functions are matrix functions which operate on the initial vectors $|v^0\rangle$ and $|x^0\rangle$.

Both the solution $|q\rangle$ and the expression for vibrational frequencies are matrix equations. However, they can be resolved into two projections which recast most of the equations in scalar form. For example, the solution for $|q\rangle$ is

$$\begin{aligned} |q(t)\rangle = {} & [\sin(\omega_- t)/\omega_-]\,\mathbf{P}(-)|v^0\rangle + \cos(\omega_- t)\,\mathbf{P}(-)|x^0\rangle \\ & + [\sin(\omega_+ t)/\omega_+]\,\mathbf{P}(+)|v^0\rangle + \cos(\omega_+ t)\,\mathbf{P}(+)|x^0\rangle \end{aligned} \tag{5.4.15}$$

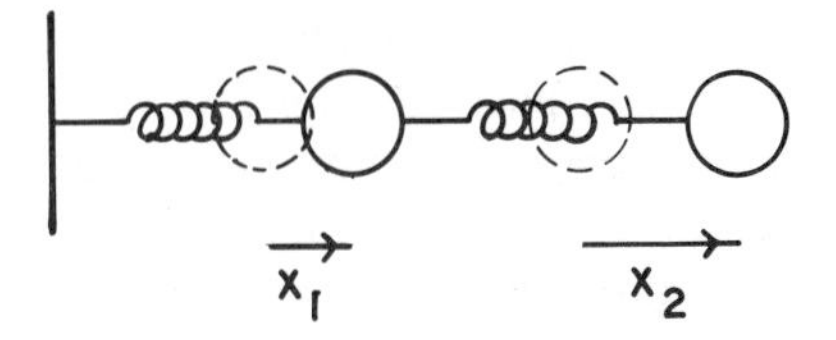

Figure 5.16. The relative displacements of two atoms of different mass bound to a single surface. The in-phase vibration.

The $\mathbf{P}_i$ matrices operate only on $|x_0\rangle$ and $|v_0\rangle$. The scalar factors ω_+ and ω_- are defined from the eigenvalues λ_+ and λ_-:

$$\omega_- = \sqrt{k(3-\sqrt{5})/2m}$$
$$\omega_+ = \sqrt{k(3+\sqrt{5})/2m} \tag{5.4.16}$$

These are the scalar components of Equation 5.4.13. The atomic motions x_1 and x_2 are now sums of their motions from each normal mode. For example, the low-energy-vibration projection matrix $\mathbf{P}(-)$ operates on $|x_0\rangle$ to give

$$-(1/2\sqrt{5})\begin{pmatrix} 1-\sqrt{5} & -2 \\ -2 & -1-\sqrt{5} \end{pmatrix}\begin{pmatrix} x_1^0 \\ x_2^0 \end{pmatrix}$$
$$= \begin{pmatrix} (1-\sqrt{5})\,x_1^0 & -2x_2^0 \\ -2x_1^0 & -(1+\sqrt{5})\,x_2^0 \end{pmatrix}$$
$$= \begin{pmatrix} -1.23x_1^0 & -2x_2^0 \\ -2x_1^0 & -3.23x_2^0 \end{pmatrix} \tag{5.4.17}$$

Both components describe the same relative displacements, which are illustrated in Figure 5.16. When both atoms are moving to the right, atom 2 will move through a larger distance than atom 1 since it benefits by the motion of atom 1 to the right. When the atoms both move in the opposite direction, atom 2 will again move further from its equilibrium position.

The $\mathbf{P}(+)$ projection operator gives the relative displacements of the two atoms in this normal mode as

$$(1+\sqrt{5})\,x_1^0 - 2x_2^0 = 3.236x_1^0 - 2x_2^0 \tag{5.4.18}$$

The two atoms now move in opposite directions (Figure 5.17) and atom 1 undergoes the larger displacement.

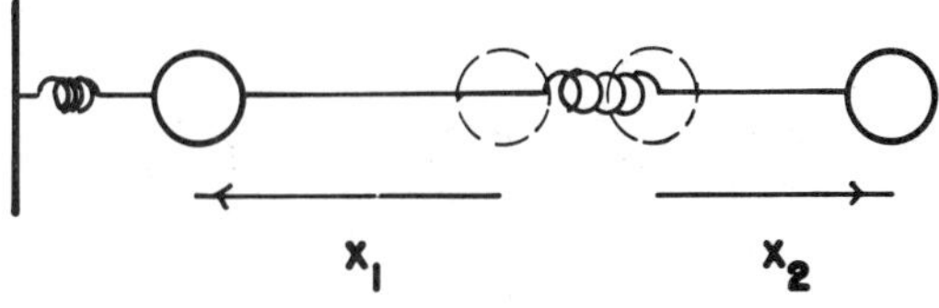

Figure 5.17. The relative displacements of two atoms of different mass bound to a single surface. The out-of-phase vibration.

The projection operator method emphasizes the difference between the initial conditions and the actual functional form of the solution. The projection matrix determines the vector projection of initial conditions which contributes to each mode. The eigenvalue determines the time-dependent characteristics of each mode. The normal modes of the system appear as a summation which yields the actual motions of each atom in the system.

The adsorbed molecule had complicated eigenvalues and eigenvectors because opposite ends of the molecule are different. By contrast, three atoms bound between two fixed surfaces with common masses and force constants (Figure 5.18) generate a 3×3 **K** matrix for the three displacements x_1, x_2, and x_3 (Problem 5.18):

$$\mathbf{K} = k \begin{pmatrix} 2 & -1 & 0 \\ -1 & 2 & -1 \\ 0 & -1 & 2 \end{pmatrix} \tag{5.4.19}$$

This matrix has eigenvalues

$$\begin{aligned} \omega_- &= (2 - \sqrt{2})\, k \\ \lambda_0 &= 2k \\ \lambda_+ &= (2 + \sqrt{2})\, k \end{aligned} \tag{5.4.20}$$

Since the normal modes for each eigenvalue will have solutions

$$\sin \omega_i t \tag{5.4.21}$$

this equation can be substituted directly for the x_i in the force equation to give

$$\begin{aligned} m\, d^2(\sin \omega_i t)/dt^2 &= -\lambda_i \sin \omega_i t \\ -m\omega_i^2 \sin \omega_i t &= -\lambda_i \sin \omega_i t \end{aligned} \tag{5.4.22}$$

The sine function is common to both sides and is eliminated to give the identity

$$m\omega_i^2 = +\lambda_i \tag{5.4.23}$$

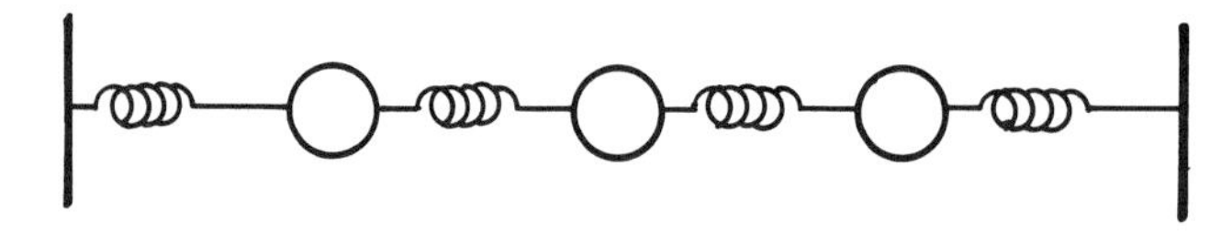

Figure 5.18. Three homonuclear atoms bound linearly between two fixed surfaces. All force constants equal k.

Each eigenvalue can be introduced to determine the corresponding vibration frequency. For Equation 5.4.19, these are

$$\begin{aligned}\lambda_- &= \sqrt{(2-\sqrt{2})\,k/m}\\ \lambda_0 &= \sqrt{2k/m}\\ \lambda_+ &= \sqrt{(2+\sqrt{2})\,k/m}\end{aligned} \tag{5.4.24}$$

The normal modes of the three-atom system are described by the three eigenvectors for the matrix:

$$|-\rangle = \begin{pmatrix} 1 \\ \sqrt{2} \\ 1 \end{pmatrix} \qquad |0\rangle = \begin{pmatrix} 1 \\ 0 \\ -1 \end{pmatrix} \qquad |+\rangle = \begin{pmatrix} 1 \\ -\sqrt{2} \\ 1 \end{pmatrix} \tag{5.4.25}$$

The normal modes are illustrated in Figure 5.19. The $|+\rangle$ vector moves all three atoms in the same direction. However, even though all three atoms have the same mass, the central atom moves a greater distance during each vibration phase. In the $|0\rangle$ eigenvector, the central atom remains motionless during the vibration. Both outer atoms move toward this central atom symmetrically. The high-frequency vibration moves two adjacent atoms toward each other while the remaining atom moves in the opposite direction. All adjacent atoms are moving in opposite directions. This pattern continues for larger linear systems. As the vibrational frequency increases, the displacement vectors change direction an increasing number of times as one moves sequentially from atom to atom in the chain.

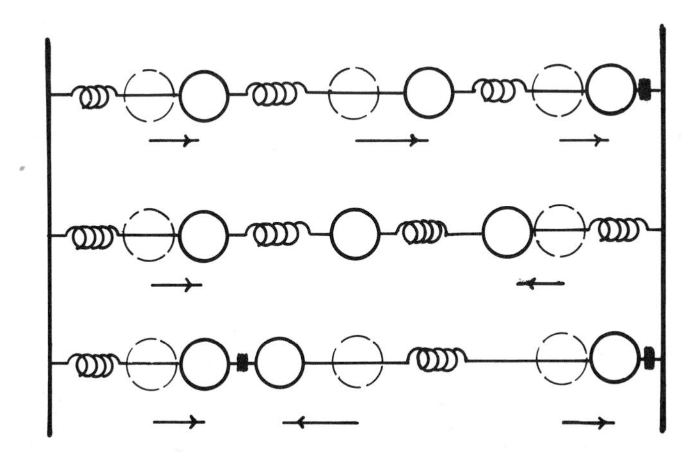

Figure 5.19. The three vibrational normal modes for the linear triatomic molecule fixed between two surfaces. (a) All atoms vibrate in phase. (b) The outer atoms vibrate equally 180° out of phase. The central atom is stationary. (c) The outer atoms move in the same direction while the central atom moves in the opposite direction.

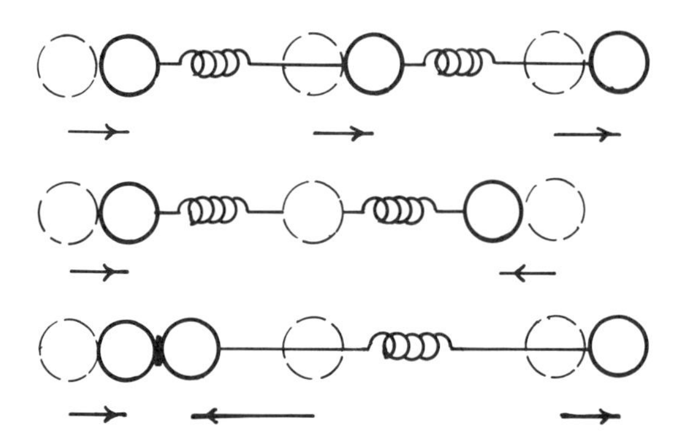

Figure 5.20. The normal modes for a free linear triatomic homonuclear molecule. (a) Translation along x. (b) The outer atoms move in phase while the central atom is stationary (symmetric stretch). (c) The outer atoms move in phase while the central atom moves in the opposite direction. The relevant distances can be compared with those of Figure 5.19.

The three normal modes for a free, three-atom linear molecule are similar to those of the fixed molecule (Figure 5.20). The corresponding eigenvectors are

$$|0\rangle = \begin{pmatrix} 1 \\ 1 \\ 1 \end{pmatrix} \qquad |1k\rangle = \begin{pmatrix} 1 \\ 0 \\ -1 \end{pmatrix} \qquad |3k\rangle = \begin{pmatrix} 1 \\ -2 \\ 1 \end{pmatrix} \tag{5.4.26}$$

The details of this derivation are left as a problem (5.7).

5.5. Normal Modes for Heteroatomic Systems

When a set of atoms are described by common masses and force constants, the common k and m can be factored from the **K** and **M** matrices, leaving integer matrix elements which can be analyzed readily. The introduction of different masses or force constants produces some new effects which are not apparent with these regular systems. Some of the effects are considered for systems with small numbers of components to provide background for the larger systems which will be examined in subsequent sections.

The matrix equation for a bound diatomic molecule with masses m_1 and m_2 is

$$M\, d^2|q\rangle/dt^2 = -K|q\rangle$$

$$\begin{pmatrix} m_1 & 0 \\ 0 & m_2 \end{pmatrix} d^2/dt^2 \begin{pmatrix} x_1 \\ x_2 \end{pmatrix} = -k \begin{pmatrix} 2 & -1 \\ -1 & 2 \end{pmatrix} \begin{pmatrix} x_1 \\ x_2 \end{pmatrix}$$

As k for the two wall-binding force constants decreases to zero, the system evolves to a free diatomic molecule. The matrix

$$\begin{pmatrix} a & -1 \\ -1 & a \end{pmatrix} \tag{5.5.1}$$

with

$$-2 \leqslant a \leqslant +2 \tag{5.5.2}$$

has a characteristic polynomial

$$\lambda^2 + 2a\lambda + (a^2 - 1) = 0 \tag{5.5.3}$$

with eigenvalues

$$\lambda_+ = a + 1 \qquad \lambda_- = a - 1 \tag{5.5.4}$$

The eigenvalues are distributed above and below the diagonal element a.

The eigenvectors for this system are

$$|-\rangle = \begin{pmatrix} 1 \\ 1 \end{pmatrix} \qquad |+\rangle = \begin{pmatrix} 1 \\ -1 \end{pmatrix} \tag{5.5.5}$$

and independent of a for this 2×2 matrix.

For a heteroatomic molecule, the normal modes include the effects of different masses which change the magnitudes of the individual atomic motions. This is demonstrated with a free, three-atom system. The $\mathbf{M}^{-1}\mathbf{K}$ matrix for a three-atom system in which the central atomic mass, m_2, differs from the two, equal outer masses, m_1, is

$$\begin{pmatrix} 1/m_1 & -1/m_1 & 0 \\ -1/m_2 & 2/m_2 & -1/m_2 \\ 0 & -1/m_1 & 2/m_1 \end{pmatrix} \tag{5.5.6}$$

The matrix is tridiagonal, but nonsymmetric. The matrix is symmetrized using the transform matrices (Chapter 4)

$$\mathbf{D} = \begin{pmatrix} 1 & 0 & 0 \\ 0 & \sqrt{m_1/m_2} & 0 \\ 0 & 0 & 1 \end{pmatrix} \tag{5.5.7}$$

$$\mathbf{D}^{-1} = \begin{pmatrix} 1 & 0 & 0 \\ 0 & \sqrt{m_2/m_1} & 0 \\ 0 & 0 & 1 \end{pmatrix} \tag{5.5.8}$$

These transform matrices are different from those for tridiagonal matrices,

$[b, a, c]$, because the elements along each diagonal differ. The proper transform matrices for symmetrization must be determined for each individual case.

The transform produces a symmetric matrix,

$$\begin{pmatrix} 1 & 0 & 0 \\ 0 & \sqrt{m_2/m_1} & 0 \\ 0 & 0 & 1 \end{pmatrix} \begin{pmatrix} 1/m_1 & -1/m_1 & 0 \\ -1/m_2 & 2/m_2 & -1/m_2 \\ 0 & -1/m_1 & 1/m_1 \end{pmatrix} \begin{pmatrix} 1 & 0 & 0 \\ 0 & \sqrt{m_1/m_2} & 0 \\ 0 & 0 & 1 \end{pmatrix}$$

$$= \begin{pmatrix} 1/m_1 & -1/\sqrt{m_1 m_2} & 0 \\ -1/\sqrt{m_1 m_2} & 2/m_2 & -1/\sqrt{m_1 m_2} \\ 0 & -1/\sqrt{m_1 m_2} & 1/m_1 \end{pmatrix} \tag{5.5.9}$$

so the eigenvalues for this matrix are real. For this 3×3 case, the eigenvalues and eigenvectors can be determined directly from the characteristic equation

$$(\mathbf{M}^{-1}\mathbf{K} - \lambda_i \mathbf{I})|i\rangle = 0 \tag{5.5.10}$$

The characteristic polynomial for the matrix is

$$(1/m_1 - \lambda)[\lambda^2 - \lambda(2/m_2 + 1/m_1)] = 0 \tag{5.5.11}$$

which factors easily to give the eigenvalues

$$\lambda = 0 \qquad \lambda = 1/m_1 \qquad \lambda = 2/m_2 + 1/m_1 \tag{5.5.12}$$

The eigenvectors for the symmetric matrix are

$$|0\rangle = \begin{pmatrix} 1 \\ \sqrt{m_1/m_2} \\ 1 \end{pmatrix} \qquad |1/m\rangle = \begin{pmatrix} 1 \\ 0 \\ -1 \end{pmatrix} \qquad |1/m_1 + 2/m_2\rangle = \begin{pmatrix} 1 \\ -2\sqrt{m_2/m_1} \\ 1 \end{pmatrix} \tag{5.5.13}$$

These normal modes are very similar to those found when all three atoms are identical. However, $|0\rangle$, which should be a translation with equal components, includes a ratio of masses as its second component. This eigenvector must be transformed to the column eigenvector for the nonsymmetric matrix to show the actual translation vector for the atomic components, i.e.,

$$|0'\rangle = \mathbf{D}^{-1}|0\rangle$$

$$= \begin{pmatrix} 1 & 0 & 0 \\ 0 & \sqrt{m_2/m_1} & 0 \\ 0 & 0 & 1 \end{pmatrix} \begin{pmatrix} 1 \\ \sqrt{m_2/m_1} \\ 1 \end{pmatrix} = \begin{pmatrix} 1 \\ 1 \\ 1 \end{pmatrix} \tag{5.5.14}$$

The normal vibrational modes for the symmetric matrix must be transformed to establish the remaining normal mode vectors. The eigenvector $|1/m\rangle$ becomes

$$\mathbf{D}^{-1}|1/m\rangle = \begin{pmatrix} 1 & 0 & 0 \\ 0 & \sqrt{m_2/m_1} & 0 \\ 0 & 0 & 1 \end{pmatrix} \begin{pmatrix} 1 \\ 0 \\ -1 \end{pmatrix} = \begin{pmatrix} 1 \\ 0 \\ -1 \end{pmatrix} \tag{5.5.15}$$

Since only the outer atoms of equal mass vibrate in this mode while the central atom remains stationary, the eigenvector remains the same. However, the second vibrational mode becomes

$$\mathbf{D}^{-1}|2/m_2 + 1/m_1\rangle = \begin{pmatrix} 1 & 0 & 0 \\ 0 & \sqrt{m_2/m_1} & 0 \\ 0 & 0 & 1 \end{pmatrix} \begin{pmatrix} 1 \\ -2\sqrt{m_2/m_1} \\ 1 \end{pmatrix} = \begin{pmatrix} 1 \\ -2m_2/m_1 \\ 1 \end{pmatrix}$$

$$= \begin{pmatrix} m_1 \\ -2m_2 \\ m_1 \end{pmatrix} \tag{5.5.16}$$

When the mass m_2 for the central atom is larger than m_1, the vector displacements of the outer atoms are less than the displacement of the central atom. The ratio of outer (d_0) and central displacements (d_c) is

$$d_0/d_c = m_1/2m_2 \tag{5.5.17}$$

The displacements when the central atom is one tenth the mass of the outer atoms are shown in Figure 5.21. If the masses are reversed so that the outer atoms are the heavier, the central atom will undergo the lesser displacement.

The vibrational frequencies vary with respect to each other as a function of the masses in the three-atom molecule. The $|1/m_1\rangle$ mode illustrates this clearly. The two outer atoms vibrate in this mode and their mass then dictates the vibration frequency,

$$\omega = \sqrt{k/\mu} \tag{5.5.18}$$

As the mass of the outer atoms increases relative to the central mass, the vibrational frequency will fall and will approach zero in the limit of very large m_1 (Figure 5.22).

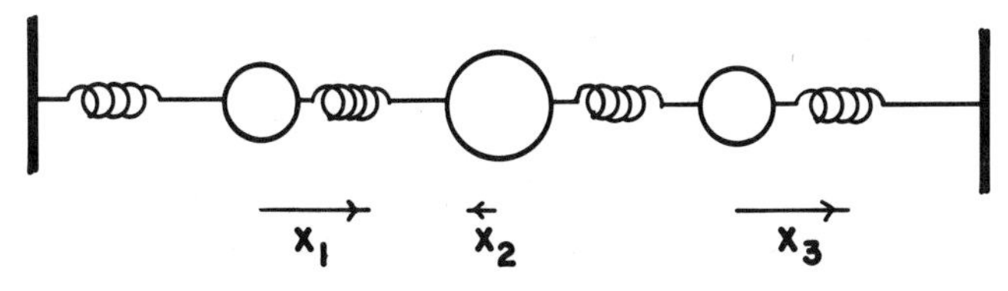

Figure 5.21. The relative displacements of the central and outer atoms of a triatomic heteronuclear molecule when the central mass is ten times each outer mass.

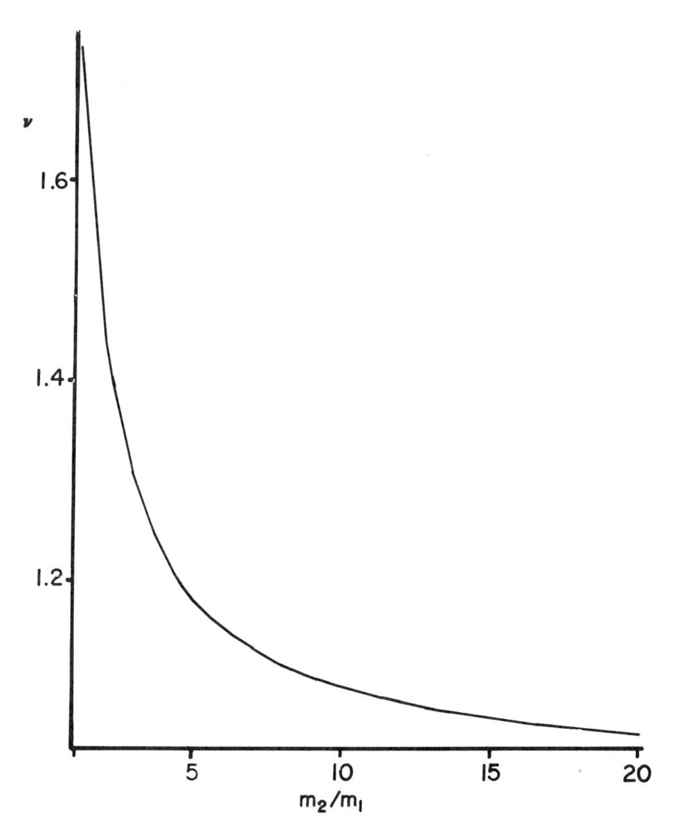

Figure 5.22. The variation in the vibrational frequency (Equation 5.5.18) for a linear triatomic molecule as the mass ratio, m_2/m_1, increases.

The vibrational frequency of the second vibrational mode,

$$\omega^2 = (2/m_2 + 1/m_1)\, k = [(k/m_1)(1 + 2m_1/m_2)] \tag{5.5.19}$$

also decreases as m_1 increases relative to m_2 but approaches a limiting frequency of $(2k/m_2)^{1/2}$. The mass of the central atom limits the minimum frequency of this mode. For large m_1, the central atom will vibrate between two nearly immobile atoms. The problem then approaches that of a single atom bound to two fixed walls (Problem 5.9). The second vibration becomes impossible as the atomic "walls" become more fixed.

The relative separation of the two normal mode frequencies remains essentially constant as the ratio of masses changes (Figure 5.22). For systems with alternating heavy and light atoms, the frequency spectrum separates into two regions; one region involves the motions of the heavier atoms while the second involves motions of lighter atoms.

An alternating atom system with two m_1 atoms and two m_2 atoms is fixed between two walls in Figure 5.23. The $\mathbf{M}^{-1}\mathbf{K}$ matrix for the system is

$$k\begin{pmatrix} 2/m_1 & -1/m_1 & 0 & 0 \\ -1/m_2 & 2/m_2 & -1/m_2 & 0 \\ 0 & -1/m_1 & 2/m_1 & -1/m_1 \\ 0 & 0 & -1/m_2 & 2/m_2 \end{pmatrix} \tag{5.5.20}$$

2×2 arrays are repeated along the diagonal of the system. A more general approach to the solutions of such systems is introduced in Section 5.8. For

$$m_2 = 10m_1, \qquad m_1 = 1, \qquad k = 1 \tag{5.5.21}$$

the matrix is

$$\begin{pmatrix} 2 & -1 & 0 & 0 \\ -0.1 & 0.2 & -0.1 & 0 \\ 0 & -1 & 2 & -1 \\ 0 & 0 & -0.1 & 0.2 \end{pmatrix} \tag{5.5.22}$$

An examination of the rows in conjunction with Gersgorin's theorem suggests that the eigenvalues of the matrix might be clustered about 0.2 and 2. The characteristic polynomial determined with matrix invariants is

$$\lambda^4 - 4.4\lambda^3 + 5.34\lambda^2 - 1.1\lambda + 0.05 = 0 \tag{5.5.23}$$

Two of the eigenvalues are found by locating a sign change in the polynomial for ranges of values near 0.2 and 2. The remaining two eigenvalues are then determined by dividing out these roots and using the quadratic formula on the polynomial which remains. The eigenvalues are

$$.0666, 0.177, 2.025, \text{ and } 2.1314 \tag{5.5.24}$$

The eigenvalues have separated into two pairs of eigenvalues. The two low values are ranged below 0.2 and they correspond to motions involving the larger mass, m_2. The second pair both lie above 2 and correspond to modes where the lighter atoms are the dominant vibrating species.

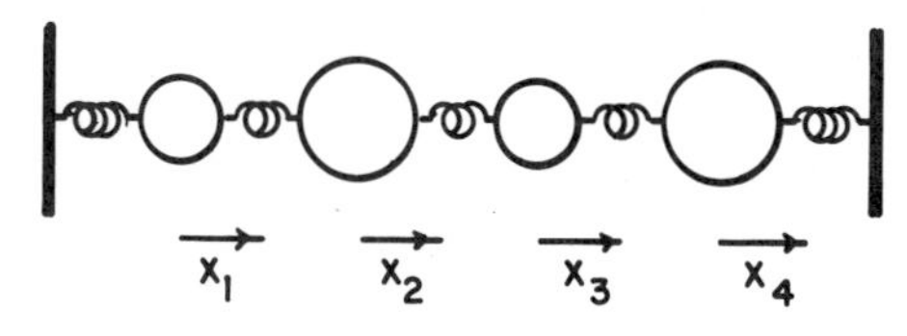

Figure 5.23. A heteronuclear four-atom system in which the mass of the heavy atoms, m_2, is ten times the mass of the light atoms, m_1

5.6. A Homogeneous One-Dimensional Crystal

The basic vibrational properties of linear chains with small numbers of atoms have been examined for chains with free ends and ends bound to fixed surfaces. The matrices have been tridiagonal and will continue to be so as the total number of atoms in the chain is increased to some arbitrary number N. Methods for the determination of eigenvalues and eigenvectors for N-dimensional tridiagonal systems are now applied to a study of the vibrational properties of these long chains. When the terminal atoms of such chains are bound to fixed surfaces, the chain becomes a model of a one-dimensional crystal.

A one-dimensional crystal of N identical atoms is illustrated in Figure 5.24. Although the chain now contains N atoms, each atom has vibrations which depend only on its displacements with respect to its two nearest neighbors. The **K** matrix with nearest-neighbor interactions will be tridiagonal. The ith atom in the chain will then experience net displacements

$$\begin{aligned} &x_i - x_{i+1} \\ &x_i - x_{i-1} \end{aligned} \tag{5.6.1}$$

where x_i is the displacement of atom i from its equilibrium position. The force equation for the ith atom is

$$\begin{aligned} m\,d^2x_i/dt^2 &= -k(x_i - x_{i-1}) - k(x_i - x_{i+1}) \\ d^2x_i/dt^2 &= -(k/m)(-x_{i-1} + 2x_i - x_{i-1}) \end{aligned} \tag{5.6.2}$$

This equation is valid for all i including $i = 1$ and $i = N$ since these terminal atoms are bound to the fixed surfaces via two additional bonds.

Because the vibrations involve only nearest neighbors, the matrix is tridiagonal, i.e.,

$$\mathbf{K} = (k/m)[-1, 2, -1] \tag{5.6.3}$$

The $N \times N$ matrix equation is

$$M\,d^2|q\rangle/dt^2 = -K\,|q\rangle$$

$$m\,d^2/dt^2 \begin{pmatrix} x_1 \\ x_2 \\ x_i \\ \cdot \\ x_N \end{pmatrix} = \begin{pmatrix} 2 & -1 & 0 & 0 & \cdots & 0 \\ -1 & 2 & -1 & 0 & \cdots & 0 \\ \cdots & -1 & 2 & & -1 & 0 \\ \cdot & \cdot & \cdot & & \cdot & \cdot \\ 0 & 0 & \cdots & & -1 & 2 \end{pmatrix} \begin{pmatrix} x_1 \\ x_2 \\ x_i \\ \cdot \\ x_N \end{pmatrix} \tag{5.6.4}$$

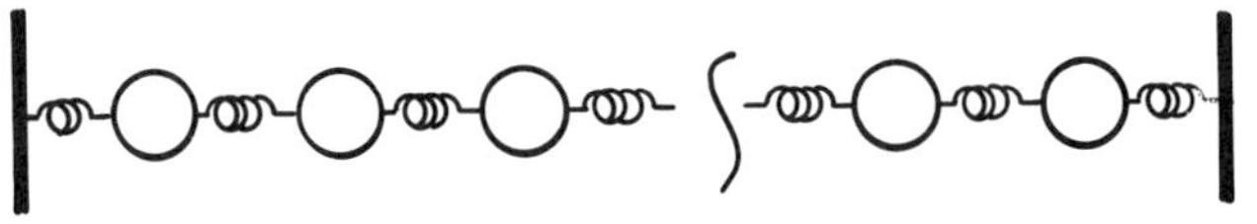

Figure 5.24. A one-dimensional crystal of N identical atoms bound to two fixed surfaces.

The eigenvalues and eigenvectors for this tridiagonal matrix are found using the techniques introduced in Chapter 4. The eigenvalues of this $N \times N$ matrix are

$$\lambda_n = 4\sin^2[n\pi/2(N+1)] \qquad n = 1, 2, ..., N \tag{5.6.5}$$

The vibrational frequency, ω_n, for the nth vibrational mode of the crystal is

$$\begin{aligned}\omega_n &= \{(k/m)\,4\sin^2[n\pi/2(N+1)]\}^{1/2}\\ &= \sqrt{k/m}\{2\sin[n\pi/2(N+1)]\}\end{aligned} \tag{5.6.6}$$

For the N normal modes, no vibrational frequency can exceed

$$\omega = 2\sqrt{k/m} = \sqrt{4k/m} \tag{5.6.7}$$

This could have been deduced directly from Gersgorin's theorem since the diagonal element of each row is 2 and the sum of the absolute values of the remaining elements in the row is

$$|-1| + |-1| = 2 \tag{5.6.8}$$

so the Gersgorin circle is centered at 2 with a radius of 2. As the value of N increases, the maximal eigenfrequency will approach the value given by Equation 5.6.7 while the minimal eigenfrequency will approach zero. All eigenvalues lie on the real axis because the matrix is symmetric.

The spectrum of all crystal frequencies is shown in Figure 5.25. The allowed spectral frequencies are more closely spaced near the central frequency

$$\omega_m = \sqrt{2k/m} \tag{5.6.9}$$

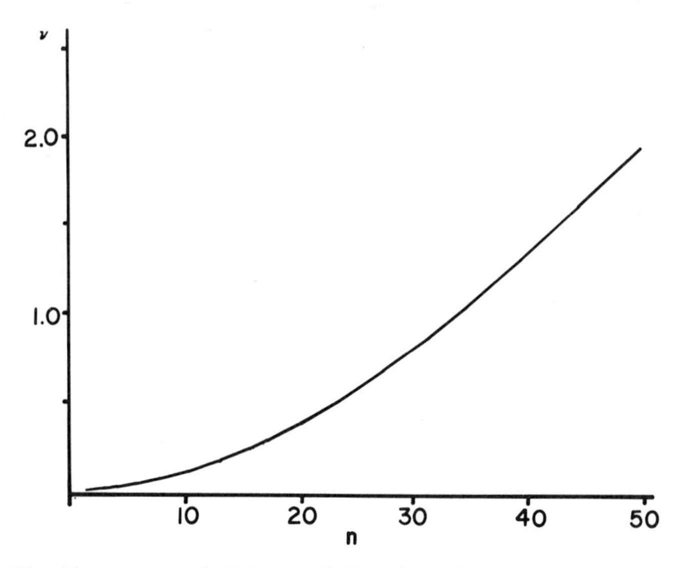

Figure 5.25. The spectrum of all frequencies for a linear homonuclear chain of 50 atoms.

Since each eigenvalue will produce an eigenvector with N components, the notation

$$|n\rangle_k \tag{5.6.10}$$

is used for the kth component of the eigenvector for λ_n.

The eigenvector components are (Chapter 4)

$$|n\rangle_k = \sin[n\pi k/(N+1)] \tag{5.6.11}$$

The appearance of n in the sine corroborates an observation made previously for two- and three-atom systems. The high-frequency vibrations characterized by a large value of n will contain atomic displacements which change direction more frequently along the chain. For example, when $n=1$, the first and last vector components are

$$\begin{aligned} |1\rangle_1 &= \sin[\pi/(N+1)] \\ |1\rangle_N &= \sin[N\pi/(N+1)] \end{aligned} \tag{5.6.12}$$

The maximal component never reaches the limit

$$\sin\pi = 0 \tag{5.6.13}$$

so there are no sign changes for any intermediate component in the vector, i.e., all atoms move in the same direction in this mode. By contrast, the signs of components, i.e., displacements, change signs with each atom for the $|N\rangle$ eigenvector. The first two components are

$$\begin{aligned} |N\rangle_1 &= \sin[N\pi/(N+1)] \\ |N\rangle_2 &= \sin[N\pi 2/(N+1)] \end{aligned} \tag{5.6.14}$$

The sine argument for $|N\rangle_1$ is less than π and the component is positive. The $|N\rangle_2$ component argument is between π and 2π and must be negative. If atom 1 moves right, atom 2 must move left in this mode. This alternation continues for the entire crystal chain.

Although the atoms all move in the same direction for $|1\rangle$, the displacement amplitudes are not equal. The displacement amplitudes are illustrated in Figure 5.26. The central atoms have the largest displacements, and a plot of displacement amplitude as a function of atom in the chain is a sinusoid. Although the displacement amplitudes are often shown perpendicular to the line of atoms,

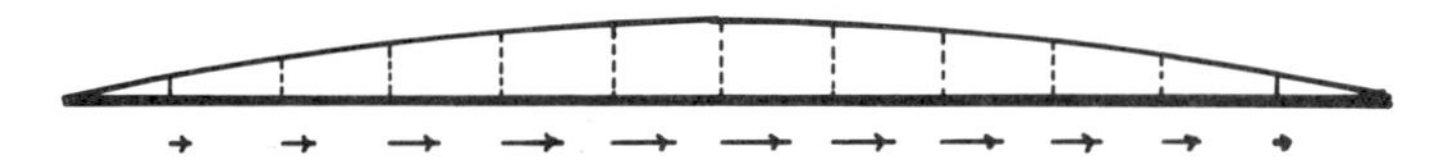

Figure 5.26. The displacement amplitudes for individual atoms in the lowest-frequency vibrational mode in the one-dimensional crystal.

the actual displacements are displacements along the one-dimensional axis containing all the atoms in the one-dimensional crystal. This is the normal mode associated with a single vibration frequency.

The $n=2$ vibrational mode is a low-energy mode with a single node. The eigenvector components,

$$|2\rangle_k = \sin[2\pi k/(N+1)] \tag{5.6.15}$$

fall to zero halfway through the chain of atoms so that all atoms in the left half of the crystal have positive motions when the atoms on the right half have negative motions (Figure 5.27). The amplitudes of the displacements lie on a sine curve with a single central node. The entire crystal is undergoing a sinusoidal motion in time. The amplitudes for the displacements are now maximal at fractional distances 0.25 and 0.75. These displacements are in opposite directions for $|2\rangle$.

The normal modes for the linear crystal are standing waves. Even though the direction and amplitude of each component of the mode will change sinusoidally with time, each component will still retain the same relative displacement with respect to the remaining components. For example, when $\sin \omega_i t$ becomes zero for the ith normal mode, all the atoms are at their equilibrium positions. Each atom then changes direction. These regular vibrations of the entire crystal store its energy.

A standing wave is a linear combination of propagating waves. For the nth normal mode, each component k of the eigenvector can be resolved into a sum of exponential functions:

$$\sin[n\pi k/(N+1)] \propto \exp[in\pi k/(N+1)] - \exp[-in\pi k/(N+1)] \tag{5.6.16}$$

The exponential with positive argument describes a wave which propagates through the atoms from left to right. The exponential with negative argument propagates in the opposite direction. The index k describes the motion of the wave through the atoms of the chain while the index n determines the direction each atom will be displaced by the wave, i.e., it is a measure of the wavelength of the propagating wave in the crystal. For example, the $n=1$ propagating wave displaces each atom in the positive direction as it propagates through the crystal. If the second traveling wave from Equation 5.6.16 is propagating at the same time, it is also producing positive displacements. As the waves passed in the chain center, they would both provide maximal positive displacement of the atoms in this region.

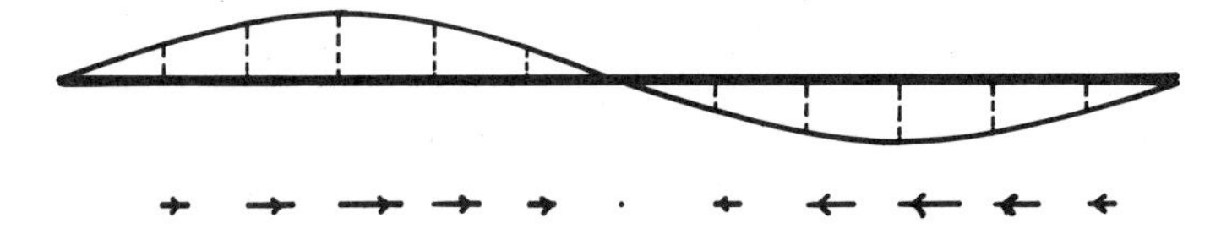

Figure 5.27. The displacement amplitudes for individual atoms for the $n=2$ normal mode in a one-dimensional crystal.

Although the normal modes for an N-atom linear crystal are a straightforward extension of studies of normal modes for two- and three-atom chains, such vibrational mode studies need not be restricted to one dimension. Figure 5.28 shows a 2×2 array of atoms which constitutes the simplest example of a two-dimensional crystal. All atoms have the same mass and the force constants, k, are all equal. Each atom must now be described with two spatial coordinates, x_i and y_i, and the system has eight dimensions. The system can actually be separated into x_i and y_i motions so the four equations involving x_i displacements can be examined in detail.

Since the atoms are arranged in a square array, the x displacements of the atoms are defined with a two-index system, x_{ij}. For instance, x_{12} is the displacement from equilibrium along x for the atom in the first row and the second column. Figure 5.28 illustrates the displacements which will produce a force on each atom. The atom at (1, 1) will experience a restoring force as

$$x_{11} - x_{12} \tag{5.6.17}$$

increases. The atom is also tied to a wall so the force equation will contain an additional $x_{11} - 0$ (since the wall cannot move). If atom 21 also moves, the bond between 11 and 21 will also be stretched along x and this will be proportional to the difference

$$x_{11} - x_{21} \tag{5.6.18}$$

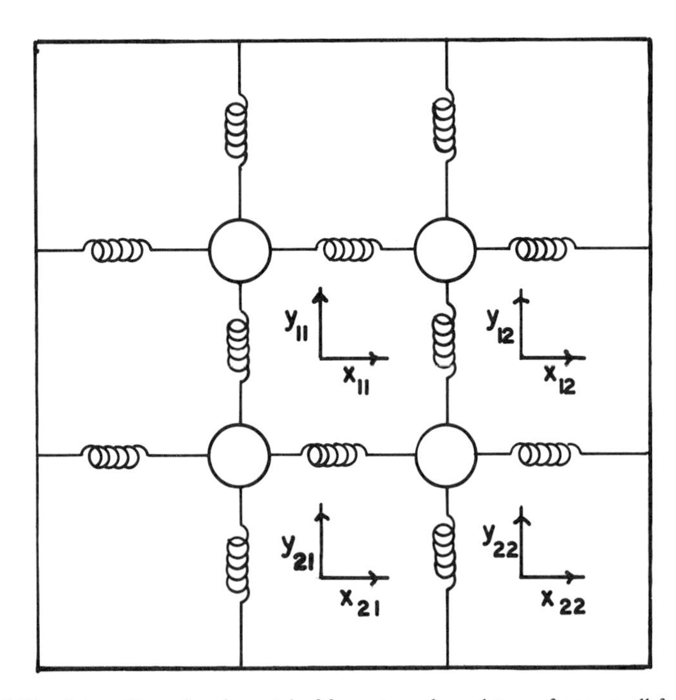

Figure 5.28. A two-dimensional crystal of four atoms bound to surfaces on all four sides.

Since x_{11} is still tied to the wall, this vibration also requires an additional x_{11}. The total force equation for the 11 atom is

$$m\, d^2x_{11}/dt^2 = -k[(2x_{11} - x_{12}) + (2x_{11} - x_{21})] \tag{5.6.19}$$

The equation includes four distinct displacements, the atom displaced against x_{12} and the wall and the atom displaced against x_{21} and the wall.

The remaining force equations are

$$\begin{aligned} m\, d^2x_{12}/dt^2 &= -k[(2x_{12} - x_{11}) + (2x_{12} - x_{22})] \\ m\, d^2x_{21}/dt^2 &= -k[(2x_{21} - x_{22}) + (2x_{21} - x_{11})] \\ m\, d^2x_{22}/dt^2 &= -k[(2x_{22} - x_{21}) + (2x_{22} - x_{12})] \end{aligned} \tag{5.6.20}$$

If the displacements are ordered x_{11}, x_{12}, x_{21}, and x_{22}, the $\mathbf{K}$ matrix for this vibrational system in the x spatial direction is

$$\mathbf{K} = k \begin{pmatrix} 4 & -1 & -1 & 0 \\ -1 & 4 & 0 & -1 \\ -1 & 0 & 4 & -1 \\ 0 & -1 & -1 & 4 \end{pmatrix} \tag{5.6.21}$$

The eigenvalues and eigenvectors for this matrix are easily found using direct products of matrices of lower order (Chapter 10). The eigenvalues for this symmetric matrix are determined from the orthogonal set of eigenvectors

$$\begin{pmatrix} 1 \\ 1 \\ 1 \\ 1 \end{pmatrix} \quad \begin{pmatrix} 1 \\ 1 \\ -1 \\ -1 \end{pmatrix} \quad \begin{pmatrix} 1 \\ -1 \\ 1 \\ -1 \end{pmatrix} \quad \begin{pmatrix} 1 \\ -1 \\ -1 \\ 1 \end{pmatrix} \tag{5.6.22}$$

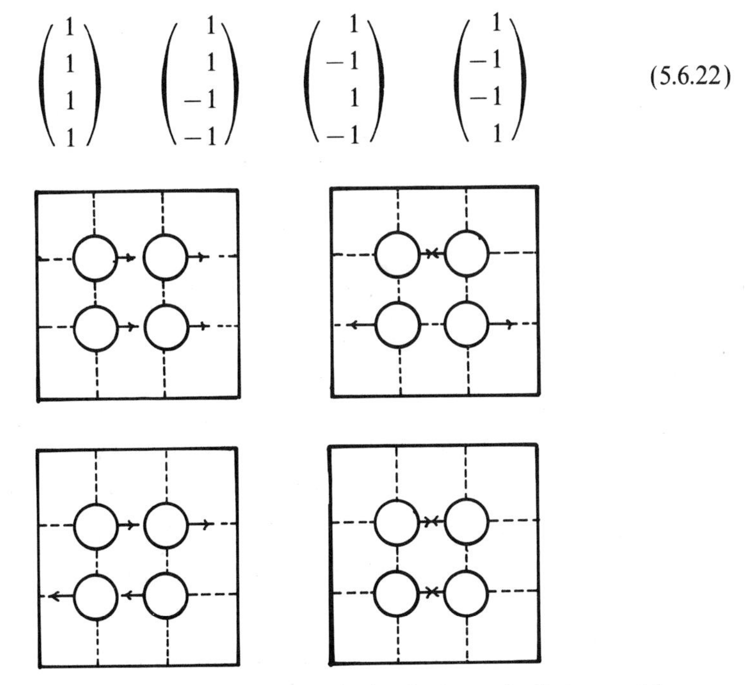

Figure 5.29. The four x-direction normal modes for the 2-atom by 2-atom crystal.

as

$$\lambda=2 \qquad \lambda=4 \qquad \lambda=4 \qquad \lambda=6 \tag{5.6.23}$$

respectively. The four frequencies of vibration for the x spatial direction are

$$\omega=\sqrt{2k/m} \qquad \omega=\sqrt{4k/m} \qquad \omega=\sqrt{4k/m} \qquad \omega=\sqrt{6k/m} \tag{5.6.24}$$

The normal modes for the four-atom crystal are illustrated in Figure 5.29. Direct product techniques can be used to extend such analyses to two-dimensional crystals of arbitrary length and width.

5.7. Cyclic Boundary Conditions

The vibrational frequencies and normal modes of a linear chain of atoms served as a basic model for a linear crystal. The problem is easily solved when each atom has the same mass and is bound with the same force constants because the **K** matrix has rows with the same arrangement of elements. The regularity of such chains can be exploited when atoms of different mass are present, e.g., a linear chain with alternating atoms of masses m_1 and m_2, respectively. In this section, some additional techniques for determining eigenvalues and eigenvectors using propagating waves are introduced and applied to heteroatomic, regular linear chains.

A chain of four atoms of mass m fixed between two walls is illustrated in Figure 5.30a. Its **K** matrix,

$$\mathbf{K}=k\begin{pmatrix} 2 & -1 & 0 & 0 \\ -1 & 2 & -1 & 0 \\ 0 & -1 & 2 & -1 \\ 0 & 0 & -1 & 2 \end{pmatrix} \tag{5.7.1}$$

has bonds to both the walls and adjacent atoms in the chain. However, the first and last rows of the matrix differ from the other rows since they include a single off-diagonal element, i.e., atom 1 changes only one interatomic distance. If the rows can be altered to act like the remaining rows, the eigenvalues and

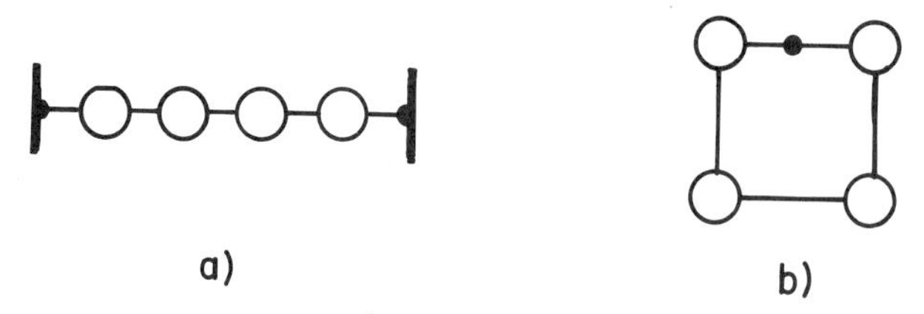

Figure 5.30. Introduction of cyclic boundary conditions for a linear chain of four homonuclear atoms: (a) the linear chain; (b) the linear chain with cyclic boundary conditions.

eigenvectors can be determined in a more general way. As the size of the chain increases, the "end effects" will become less important and a modified matrix will serve as a good approximation to the matrix of Equation 5.7.1.

The cyclic or Born–von Karman boundary conditions are illustrated for the four-atom system in Figure 5.30b. The last atom of the chain is bound to the first atom and the walls are eliminated. The displacements of atom 1 are now

$$\begin{aligned} &x_1 - x_2 \\ &x_1 - x_4 \end{aligned} \tag{5.7.2}$$

and the force equations for each of the atoms are

$$\begin{aligned} F_1 &= -k[(x_1 - x_2) + (x_1 - x_4)] \\ F_2 &= -k[(x_2 - x_3) + (x_2 - x_1)] \\ F_3 &= -k[(x_3 - x_4) + (x_3 - x_2)] \\ F_4 &= -k[(x_4 - x_1) + (x_4 - x_3)] \end{aligned} \tag{5.7.3}$$

where

$$F_i = m\, d^2x_i/dt^2 \tag{5.7.4}$$

Each force equation now contains four terms and the **K** matrix is

$$k\begin{pmatrix} 2 & -1 & 0 & -1 \\ -1 & 2 & -1 & 0 \\ 0 & -1 & 2 & -1 \\ -1 & 0 & -1 & 2 \end{pmatrix} \tag{5.7.5}$$

The cyclic boundary conditions "insert" additional -1 elements at positions 14 and 41 so the matrix is still symmetric. Although the system cycles back on itself, it is still one-dimensional since each atom moves only in the x spatial direction. The cyclic boundary conditions are only a construct since the two termini could never intersect in a one-dimensional system.

For an N-atom system, the two walls are replaced by atoms 0 and $N+1$ with displacements x_0 and x_{N+1}. When cyclic boundary conditions are introduced, atom $N+1$ is actually the first atom (1) in the chain:

$$x_{N+1} = x_1 \tag{5.7.6}$$

Atom 0 is actually atom N for the cyclic conditions:

$$x_0 = x_N \tag{5.7.7}$$

These two conditions ensure that the **K** matrix will have

$$a_{1N} = a_{N1} = -k \tag{5.7.8}$$

The cyclic boundary conditions effectively eliminate the starting and termination points of the linear crystal. An observer moving along the chain would continue to meet atoms without ever encountering a boundary. A propagating wave would continue to move around the chain indefinitely.

The matrix with cyclic boundary conditions approximates the real **K** matrix for large N. Since the eigenvalues and eigenvectors for the **K** matrix are known, analysis of the system with cyclic boundary conditions may seem superfluous. However, the cyclic matrix serves to introduce a new technique for the determination of eigenvalues and eigenvectors. If **K** commutes with a second matrix, they have common eigenvectors. A **T** matrix, which translates the atomic coordinates of the chain, does commute with **K** and has eigenvectors which are easily determined.

A translation matrix moves the displacement x_1 of atom 1 to atom 2, x_2 is moved to atom 3, and x_3 is moved to atom 4. Because of the cyclic boundary conditions, x_4 is transferred to atom 1. The translation matrix then describes the propagation of displacements by a single atom. The initial and final displacements dictate the **T**, matrix. For four atoms, the **T** matrix which moves the displacement for atom i to atom $i+1$ is

$$|q'\rangle = \mathbf{T}\,|q\rangle$$

$$\begin{pmatrix} x_2 \\ x_3 \\ x_4 \\ x_1 \end{pmatrix} = \begin{pmatrix} 0 & 1 & 0 & 0 \\ 0 & 0 & 1 & 0 \\ 0 & 0 & 0 & 1 \\ 1 & 0 & 0 & 0 \end{pmatrix} \begin{pmatrix} x_1 \\ x_2 \\ x_3 \\ x_4 \end{pmatrix} \tag{5.7.9}$$

The elements of an identity matrix have been moved one position to the right.

The translation matrix **T** is much simpler than the **K** matrix. If this matrix commutes with **K**, it is convenient to determine its eigenvalues and eigenvectors. The **K** matrix can then operate on these eigenvectors to give the eigenvalues of the **K** matrix. The commutator [**K**, **T**] is zero when the products of the matrices in either order are identical. **KT** is

$$\begin{pmatrix} 2 & -1 & 0 & -1 \\ -1 & 2 & -1 & 0 \\ 0 & -1 & 2 & -1 \\ -1 & 0 & -1 & 2 \end{pmatrix} \begin{pmatrix} 0 & 1 & 0 & 0 \\ 0 & 0 & 1 & 0 \\ 0 & 0 & 0 & 1 \\ 1 & 0 & 0 & 0 \end{pmatrix} = \begin{pmatrix} -1 & 2 & -1 & 0 \\ 0 & -1 & 2 & -1 \\ -1 & 0 & -1 & 2 \\ 2 & -1 & 0 & -1 \end{pmatrix} \tag{5.7.10}$$

while **TK** is

$$\begin{pmatrix} 0 & 1 & 0 & 0 \\ 0 & 0 & 1 & 0 \\ 0 & 0 & 0 & 1 \\ 1 & 0 & 0 & 0 \end{pmatrix} \begin{pmatrix} 2 & -1 & 0 & -1 \\ -1 & 2 & -1 & 0 \\ 0 & -1 & 2 & -1 \\ -1 & 0 & -1 & 2 \end{pmatrix} = \begin{pmatrix} -1 & 2 & -1 & 0 \\ 0 & -1 & 2 & -1 \\ -1 & 0 & -1 & 2 \\ 2 & -1 & 0 & -1 \end{pmatrix} \tag{5.7.11}$$

The two product matrices are identical and **T** and **K** commute.

The characteristic determinant for the **T** matrix,

$$\begin{vmatrix} 0-\lambda & 1 & 0 & 0 \\ 0 & 0-\lambda & 1 & 0 \\ 0 & 0 & 0-\lambda & 1 \\ 1 & 0 & 0 & 0-\lambda \end{vmatrix} = 0 \tag{5.7.12}$$

gives the characteristic polynomial

$$\lambda^4 - 1 = 0 \tag{5.7.13}$$

The eigenvalue roots of this equation are functions which give 1 when they are raised to the fourth power. $\lambda = 1$ is one eigenvalue. The remaining three roots of Equation 5.7.13 are

$$\begin{aligned} \lambda_2 &= \exp(i2\pi/4) \\ \lambda_3 &= \exp(i4\pi/4) \\ \lambda_4 &= \exp(i6\pi/4) \end{aligned} \tag{5.7.14}$$

since each root has the form

$$\exp(i2\pi n/4) \qquad n = 0, 1, 2, 3 \tag{5.7.15}$$

and

$$[\exp(i2\pi n/4)]^4 = \exp(i2\pi n) = \cos 2\pi n + i \sin 2\pi n = +1 \tag{5.7.16}$$

The eigenvectors for each eigenvalue of **T** are found using the **T** matrix. For $\lambda_1 = 1$, the components of the eigenvector are determined from the matrix equation,

$$\mathbf{T}\,|1\rangle = 1\,|1\rangle$$

$$\begin{pmatrix} -1 & 1 & 0 & 0 \\ 0 & -1 & 1 & 0 \\ 0 & 0 & -1 & 1 \\ 1 & 0 & 0 & -1 \end{pmatrix} \begin{pmatrix} x_1 \\ x_2 \\ x_3 \\ x_4 \end{pmatrix} = 0 \tag{5.7.17}$$

The solution of this equation is a translation eigenvector,

$$\begin{pmatrix} 1 \\ 1 \\ 1 \\ 1 \end{pmatrix} \tag{5.7.18}$$

which moves each atomic displacement to the next atom without changing its direction.

This eigenvector is also an eigenvector for $\mathbf{K}\cdot\mathbf{K}$ operates on this vector to give

$$\begin{pmatrix} 2 & -1 & 0 & -1 \\ -1 & 2 & -1 & 0 \\ 0 & -1 & 2 & -1 \\ -1 & 0 & -1 & 2 \end{pmatrix}\begin{pmatrix} 1 \\ 1 \\ 1 \\ 1 \end{pmatrix} = 0\begin{pmatrix} 1 \\ 1 \\ 1 \\ 1 \end{pmatrix} \tag{5.7.19}$$

For this matrix, the mode with identical components is a translation normal mode.

The remaining eigenvectors are more complicated. The general eigenvalue, $\exp(i2\pi n/4)$, is substituted in the characteristic matrix to determine the eigenvector for $\mathbf{T}$:

$$\begin{pmatrix} -\exp(i2\pi n/4) & 1 & 0 & 0 \\ 0 & -\exp(i2\pi n/4) & 1 & 0 \\ 0 & 0 & -\exp(i2\pi n/4) & 1 \\ 1 & 0 & 0 & -\exp(i2\pi n/4) \end{pmatrix}\begin{pmatrix} x_1 \\ x_2 \\ x_3 \\ x_4 \end{pmatrix} = 0 \tag{5.7.10}$$

Selecting $x_1 = 1$, the remaining components are determined from each row of the matrix. For x_2,

$$-\exp(i2\pi n/4)(1) + (1)\,x_2 = 0$$

$$x_2 = \exp(i2\pi n/4) \tag{5.7.21}$$

The equation for the next row gives

$$-\exp(i2\pi n/2\; x_2 + (1)\,x_3 = 0$$

$$x_3 = +\exp(i2\pi n/4)\,x_2 = \exp(i2\pi n/4)\exp(i2\pi n/4) = \exp(i4\pi n/4) = \exp(i\pi n) \tag{5.7.22}$$

The final component is

$$x_4 = \exp(i\pi n)\,x_3 = \exp(i6\pi n/4) = \exp(i3\pi n/2) \tag{5.7.23}$$

The three eigenvectors are found by substituting $n = 1$, 2, or 3 for these components:

$$|2\rangle = \begin{pmatrix} 1 \\ \exp(i\pi/2) \\ \exp(i\pi) \\ \exp(i3\pi/2) \end{pmatrix} \quad |3\rangle = \begin{pmatrix} 1 \\ \exp(i\pi) \\ \exp(i2\pi) \\ \exp(i3\pi) \end{pmatrix} \quad |4\rangle = \begin{pmatrix} 1 \\ \exp(3i\pi/2) \\ \exp(i3\pi) \\ \exp(i9\pi/2) \end{pmatrix} \tag{5.7.24}$$

These eigenvector components contain imaginary parts, e.g.,

$$\exp(i\pi/2) = \cos \pi/2 + i \sin \pi/2 = i(1) \tag{5.7.25}$$

The symmetric **K** matrix has real eigenvalues and displacement components. The imaginary components of Equation 5.7.24 are more general eigenvectors for both the **K** and nonsymmetric **T** matrix. If the alternate valid components

$$\exp(-i2\pi n/4) \tag{5.7.26}$$

for the characteristic equation are used, the eigenvector components for **T** will contain this negative sign as well. Linear combinations of exponentials with positive and negative arguments, i.e., sine and cosine functions, are also eigenvector components for the **K** matrix.

The three eigenvectors of Equation 5.7.24 have the general components

$$|n\rangle_k = \exp(i2\pi nk/4) \qquad n = 1, 2, 3; \quad k = 0, 1, 2, 3 \tag{5.7.27}$$

K will operate on these eigenvectors to give the eigenvalue and the eigenvector:

$$\begin{pmatrix} 2 & -1 & 0 & -1 \\ -1 & 2 & -1 & 0 \\ 0 & -1 & 2 & -1 \\ -1 & 0 & -1 & 2 \end{pmatrix} \begin{pmatrix} 1 \\ \exp(i2\pi n 1/4) \\ \exp(i2\pi n 2/4) \\ \exp(i2\pi n 3/4) \end{pmatrix}$$

$$= \begin{pmatrix} 2 - \exp(i\pi n/2) - \exp(i3\pi n/2) \\ -1 + 2\exp(i\pi n/2) - \exp(in\pi) \\ -\exp(i\pi n/2) + 2\exp(i\pi n) - \exp(i3\pi n/2) \\ -1 - \exp(i\pi n) + 2\exp(i3\pi n/2) \end{pmatrix} \tag{5.7.28}$$

The original eigenvector components are now factored from these terms. Since the function repeats itself each 2π, a motion of 270° $(3\pi/2)$ is equivalent to a motion of $-90°$ $(-\pi/2)$:

$$\exp(-in\pi/2) = \exp(+i3\pi n/2) \tag{5.7.29}$$

The vector factors as

$$\begin{aligned} &2 - [\exp(i\pi n/2) + \exp(-i\pi n/2)] \\ &\{2 - [\exp(i\pi n/2) + \exp(-i\pi n/2)]\} \exp(i\pi n/2) \\ &\{2 - [\exp(i\pi n/2) + \exp(-i\pi n/2)]\} \exp(i\pi n) \\ &\{2 - [\exp(i\pi n/2) + \exp(-i\pi n/2)\}] \exp(i3\pi n/2) \end{aligned} \tag{5.7.30}$$

The original components are regenerated and the common eigenvalue is

$$2 - [\exp(i\pi n/2) + \exp(-i\pi n/2)] \tag{5.7.31}$$

Since

$$\cos(n\pi/2) = \frac{1}{2}[\exp(i\pi n/2) + \exp(-i\pi n/2)] \tag{5.7.32}$$

these eigenvalues become

$$\lambda_n = 2 - 2\cos(n\pi/2) \tag{5.7.33}$$

For an $N \times N$ matrix with cyclic boundary conditions, this result becomes

$$\lambda_n = 2[1 - \cos(2\pi n/N)] = 4\sin^2(\pi n/N) \tag{5.7.34}$$

The N (not $N+1$) in the argument permits a zero eigenvalue since a wave can translate indefinitely along this cyclic system.

The eigenvalues λ_n lack a factor of 2 in the denominator. As the index n ranges from 1 to N, the factor $\sin(\pi n/N)$ can be either positive or negative although its square will always be positive. Degenerate eigenvalues appear. For example, the four-atom chain has eigenvalues,

$$\begin{aligned} 4\sin^2(0\pi/4) &= 0 \\ 4\sin^2(1\pi/4) &= 2 \\ 4\sin^2(2\pi/4) &= 4 \\ 4\sin^2(3\pi/4) &= 2 \end{aligned} \tag{5.7.35}$$

The exponential components are all real for the two nondegenerate eigenvalues

$$|0\rangle = \begin{pmatrix} 1 \\ 1 \\ 1 \\ 1 \end{pmatrix} \qquad |4\rangle = \begin{pmatrix} 1 \\ -1 \\ 1 \\ -1 \end{pmatrix} \tag{5.7.36}$$

The degenerate eigenvectors have imaginary components

$$\begin{pmatrix} 1 \\ i \\ -1 \\ -i \end{pmatrix} \quad \begin{pmatrix} 1 \\ -i \\ -1 \\ i \end{pmatrix} \tag{5.7.37}$$

and constitute a subspace for $\lambda = 2$. Subspace eigenvectors with real components are found as sums and differences. The sum is

$$\begin{pmatrix} 1 \\ i \\ -1 \\ -i \end{pmatrix} + \begin{pmatrix} 1 \\ -i \\ -1 \\ i \end{pmatrix} = \begin{pmatrix} 2 \\ 0 \\ -2 \\ 0 \end{pmatrix} \propto \begin{pmatrix} 1 \\ 0 \\ -1 \\ 0 \end{pmatrix} \tag{5.7.38}$$

while the difference is

$$\begin{pmatrix} 1 \\ i \\ -1 \\ -i \end{pmatrix} - \begin{pmatrix} 1 \\ -i \\ -1 \\ i \end{pmatrix} = \begin{pmatrix} 0 \\ 2i \\ 0 \\ -2i \end{pmatrix} = \begin{pmatrix} 0 \\ 1 \\ 0 \\ -1 \end{pmatrix} \tag{5.7.39}$$

5.8. Heteroatomic Linear Crystals

Homogeneous chains of atoms which function as a simple model of a one-dimensional crystal can be analyzed with or without the cyclic boundary conditions. The analyses provide a range of vibrational frequencies for the crystal and the nature of the normal mode vibrations associated with these frequencies. These results are a natural extension of the analyses for chains containing two or three atoms. When the atoms in such chains had different masses, an inverse mass matrix, $\mathbf{M}^{-1}$, mass was required. The resultant matrix for analysis, $\mathbf{M}^{-1}\mathbf{K}$, was not symmetric and gave complicated eigenvalues and eigenvectors for the two and three-atom systems.

The difficulties of the two- and three-atom systems should be magnified for the system of Figure 5.31, in which an atom of low mass, m_1, alternates with an atom of high mass, m_h, along the length of the chain. The $\mathbf{M}^{-1}\mathbf{K}$ matrix has m_1^{-1} in each element of the odd rows and m_h^{-1} in each element of the even rows.

The two- and three-atom heteroatomic chains did exhibit some pattern which might continue for the longer one-dimensional crystals. The high-frequency normal modes were dominated by the motions of the atoms of low mass. The low-frequency normal modes were dominated by the total mass of the two atoms since pairs of different atoms moved as a unit. Since its mass is larger, the low-frequency vibrations would be dominated by the larger atom's motion. This suggests that the motions of the light atoms and the heavy atoms could be described separately and then coupled in some way.

The cyclic boundary conditions are most convenient for the coupling process. The light atoms are periodic on the chain. If there are N heavy atoms and N light atoms on the chain, a propagating wave for the light atoms has the form

$$\exp(i2\pi nk/N) \qquad n, k = 0, 1, 2, \ldots, N-1 \tag{5.8.1}$$

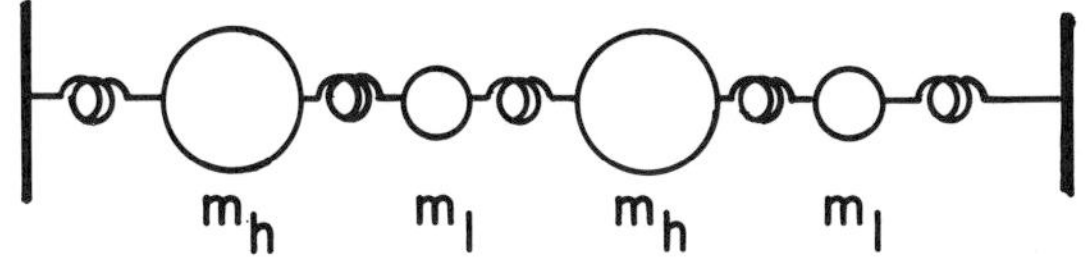

Figure 5.31. A linear chain in which heavy atoms alternate with light atoms.

This expression is often written using a wavevector, $\mathbf{k}'$, and the separation between atoms in the chain, a. The total length of a chain of N atoms is Na. For the index n which appears in the eigenvalue, a propagating wave will pass through zero n times. Thus, the wavelength is

$$\lambda' = Na/n \tag{5.8.2}$$

The wavevector is

$$\mathbf{k}' = 2\pi/\lambda' = 2\pi n/aN \tag{5.8.3}$$

The actual distance traveled by the wave is determined by the index k and the distance per atom, a:

$$x = ka \tag{5.8.4}$$

The propagating wave is rearranged as

$$\exp(i2\pi kna/Na) = \exp[i(2\pi n/aN)(ak)] = \exp(i\mathbf{k}'x) \tag{5.8.5}$$

which is often used in place of the indexed form.

The wave equation (Equation 5.8.1) for the light atoms describes only these atoms. The heavy atoms exist for the same length of chain and their propagating wave equation will be exactly the same. If the heavy and light atoms were not coupled, each of these waves would move independently around the chain. Since the bonds link heavy and light atoms, the two waves will couple and the actual wave for the full chain will be some linear combination of the two waves.

The actual chain is $2N$ atoms long with N light and N heavy atoms. Since every bond joins a heavy and a light atom, the same constant, k, can be used for every bond in the chain. The eigenvector component for the jth light atom in the system will be

$$|n\rangle_j = \exp(i2\pi nj/2N) \tag{5.8.6}$$

where $2N$ is used for the total "length" of the chain which contains the N light atoms. The eigenvector for the jth component of the heavy atom wave is identical. However, it will always be displaced from a light atom by one unit. If j is even for a light atom, it must be odd for the heavy atoms. To ensure consistency, the index $2j$ is used for a light atom at position $2j$ in the chain. The heavy atoms on either side are then indexed as $2j-1$ and $2j+1$, respectively.

The eigenvector components are now

$$\begin{aligned} &\text{Light: } \exp[i2\pi n(2j)/2N] \\ &\text{Heavy: } \exp[i2\pi n(2j+1)/2N] \end{aligned} \tag{5.7.7}$$

Wave propagation along the chain with frequency ω_n requires the additional exponential factor

$$\exp(i\omega t) \tag{5.8.8}$$

and the entire propagating function is

$$|n(t)\rangle_{2j} = x_{2j} = \exp[i2\pi n(2j)/2N + i\omega_n t] \tag{5.8.9}$$

for the wave at index $2j$.

The force equation for the light atom located at $2j$ depends on the displacements of the heavy atoms on either side of it:

$$m_1 d^2 x_{2j}/dt^2 = -k(-x_{2j-1} + 2x_{2j} - x_{2j+1}) \tag{5.8.10}$$

A heavy atom appears at $2j+1$ and its force equation is

$$m_h\, d^2 x_{2j+1}/dt^2 = -k(-x_{2j} + 2x_{2j+1} - x_{2j+2}) \tag{5.8.11}$$

The equations are similar but the amplitudes of the waves for the light atoms and the heavy atoms may be different. The eigenvalue which describes the entire coupled mode will be the same for both waves. The waves are assigned amplitudes A and B for the light and heavy atom systems, respectively; i.e.,

$$\begin{aligned} &\text{Light: } x_{2j} = A \exp[i2\pi n(2j)/2N + i\omega t] \\ &\text{Heavy: } x_{2j+1} = B \exp[i2\pi n(2j+1)/2N + i\omega t] \end{aligned} \tag{5.8.12}$$

These functions are now substituted into the force equations to develop some relationship between the amplitudes A and B.

The time derivatives act only on $\exp(-i\omega t)$ to give a factor of ω^2. The time-dependent exponential is canceled from each side to give the time-independent equations

$$\begin{aligned} &m_1 \omega^2 A \exp[i2\pi n(2j)/2N] \\ &\quad = -k\{-B \exp[i2\pi n(2j-1)/2N] + 2A \exp[i2\pi n(2j)/2N] \\ &\quad - B \exp[i2\pi n(2j+1)/2N]\} \end{aligned} \tag{5.8.13}$$

$$\begin{aligned} &m_h \omega^2 B \exp[i2\pi n(2j+1)/2N] \\ &\quad = -k\{-A \exp[i\pi n(2j)/2N] + 2B \exp[i2\pi n(2j+1)/2N] \\ &\quad - A \exp[i2\pi n(2j+2)/2N]\} \end{aligned} \tag{5.8.14}$$

The common exponentials cancel to give two equations in A and B:

$$\begin{aligned} m_1 \omega^2 A &= -k[-B \exp(-i2\pi n/2N) + 2A - B \exp(+i2\pi n/2N)] \\ m_h \omega^2 B &= -k[-A \exp(-i2\pi n/2N) + 2B - A \exp(+i2\pi n/2N)] \end{aligned} \tag{5.8.15}$$

This equation is an eigenvalue (ω_n)–eigenvector (A, B) equation:

$$\begin{pmatrix} 2k - m_1\omega^2 & -2k\cos(\pi n/N) \\ -2\cos(n\pi/N) & 2k - m_h\omega^2 \end{pmatrix} \begin{pmatrix} A \\ B \end{pmatrix} = 0 \tag{5.8.16}$$

Each index n has two eigenfrequencies which are determined from the characteristic polynomial for the matrix:

$$(2k - m_1\omega^2)(2k - m_h\omega^2) - k^2\cos^2(n\pi/N) = 0$$
$$m_1 m_h \omega^4 - 2k(m_1 + m_h)\,\omega^2 + (4k^2/m_1 m_h)[1 - \cos^2(n\pi/N)] = 0 \qquad (5.8.17)$$

The frequencies for n are

$$\omega_+^2(n) = k[m_1 + m_h)/m_1 m_h] + D = k/\mu + D$$
$$\omega_-^2(n) = k/\mu - D \qquad (5.8.18)$$

where

$$\mu = (m_1 + m_h)/m_1 m_h \qquad (5.8.19)$$

and

$$D - \{(k/\mu)^2 - (4k^2/m_l m_h)[1 - \cos^2(n\pi/N)]\}^{1/2}$$
$$= [(k/\mu)^2 - (4k^2/m_1 m_h)\sin^2(n\pi/N)]^{1/2} \qquad (5.8.20)$$

There are N pairs of frequencies ω_+ and ω_-. Since the indexing of N can start anywhere, n is allowed to range from $n = -N/2$ to $n = +N/2$. The $n = 0$ eigenfrequencies are then located in the center of the spectrum. Because the sine function in D is squared, this frequency spectrum is symmetric about $n = 0$.

Some limits establish the nature of the full frequency spectrum. When $n = 0$, Equations 5.8.18 become

$$\omega_-^2 = (k/\mu) - [(k/\mu)^2 + 0]^{1/2} = 0$$
$$\omega_+^2 = (k/\mu) + (k/\mu) = 2k/\mu \qquad (5.8.21)$$

The frequencies are maximally separated at $n = 0$.

The two frequencies at $n = N/2$ or $-N/2$ are identical since the spectrum is symmetric. For $n = N/2$, the discriminant D is

$$D = k[(1/\mu)^2 - 4/(m_1 m_h)]$$
$$= (k/m_1 m_h)[(m_1 + m_h)^2 - 4m_1 m_h]^{1/2}$$
$$= (k/m_1 m_h)(m_h - m_1) \qquad (5.8.22)$$

The eigenfrequency ω_-^2 is

$$(k/\mu) - D = k[(m_1 + m_h) - (m_h - m_1)]/(m_1 m_h)$$
$$= 2k_1/m_h m_1 = 2k/m_h \qquad (5.8.23)$$

The eigenfrequency depends only on m_h and represents the largest possible frequency of the ω_- spectrum. By contrast, ω_+^2 is

$$(k/\mu) + D = 2k/m_1 \tag{5.8.24}$$

Since $m_1 < m_h$, this frequency will always be higher than the ω_- frequencies even though it is the minimal ω_+. There is a gap in the frequency spectrum. The diatomic chain has two sets of vibrational frequencies or modes which are separated by at least

$$(2k)(m_1^{-1/2} - m_h^{-1/2}) \tag{5.8.25}$$

(Figure 5.32). The frequency spectrum generated by the positive discriminant is called the optical branch while the negative discriminant is called the acoustic branch. The high-energy modes are called the optical branch since simple diatomic crystals, for example, the alkali halides, will absorb infrared radiation to activate these vibrational modes.

The actual vibrational modes depend on the relative amplitudes A and B; i.e., each eigenvalue will give an eigenvector with components A and B. However, some eigenvector properties can be determined by selecting "simpler" eigenvalues. Since $\lambda = 0$ for the $n = 0$ acoustic mode, the eigenvector equation for this eigenvalue is

$$\begin{pmatrix} 2 - m_1(0) & -2\cos 0 \\ -2\cos 0 & 2 - m_h(0) \end{pmatrix}\begin{pmatrix} A \\ B \end{pmatrix} = 0$$

$$= \begin{pmatrix} 2 & -2 \\ -2 & 2 \end{pmatrix}\begin{pmatrix} A \\ B \end{pmatrix} = 0 \tag{5.8.26}$$

with eigenvector

$$\begin{pmatrix} 1 \\ 1 \end{pmatrix} \tag{5.8.27}$$

The adjacent h and l atoms move in the same direction and the same distance. The diatomics of the chain are vibrating as units. When $n = 0$, the amplitudes of diatomic motions at different points on the chain may be different but the two atoms of each diatomic remain in phase relative to each other.

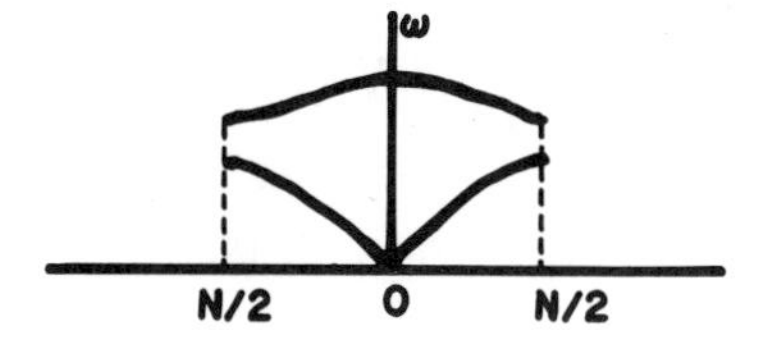

Figure 5.32. The optical and acoustic mode spectra for a linear chain with alternating heavy and light atoms.

The optical mode eigenvector for $n=0$ is determined via the same matrix equation with $\omega_+^2 = 2k/\mu$. The a_{11} element is

$$2-2m_1[(m_1+m_h)/m_1m_h] = = 2-2-2(m_1/m_h) = -2(m_1/m_h) \quad (5.8.28)$$

The a_{22} element is

$$-2(m_h/m_1) \quad (5.8.29)$$

and the matrix equation is

$$\begin{pmatrix} -2(m_1/m_h) & -2 \\ -2 & -2(m_1/m_h) \end{pmatrix}\begin{pmatrix} A \\ B \end{pmatrix} = 0 \quad (5.8.30)$$

The eigenvector is

$$\begin{pmatrix} A \\ B \end{pmatrix} = \begin{pmatrix} -m_h \\ m_1 \end{pmatrix} \quad (5.8.31)$$

Since m_h is larger, the lighter atom with relative amplitude A will move the larger distance during the vibration. The atoms of the diatomic will always move against each other. This is consistent with the higher frequencies of the optical mode.

5.9. Normal Modes for Molecules in Two Dimensions

A diatomic molecule which was confined to a single direction in space gave motions which were described with two normal modes. When both atoms moved in the same direction at the same rate, the molecule was translating. When they moved in opposite directions, it was undergoing a symmetric vibration. By confining the molecule to a single spatial axis, the entire system was described by two basis vectors. One basis vector described the motions of atom 1 while the second described the motion of atom 2. The translation and vibration eigenvectors gave the relative motions of these two vectors on the axis. Thc basis vectors themselves are orthogonal vectors in a phase space, and the eigenvectors give the projections on each of these coordinate axes.

The diatomic molecule can also be confined to a two-dimensional plane. The four eigenvectors for this system were discussed in Section 5.2. Three eigenvectors were associated with $\lambda = 0$ eigenvalues and correspond to the translations and one rotation in the plane. There is only one vibration for the diatomic molecule in both the one- and two-dimensional cases. This is true in the three-dimensional case as well since vibrations take place only along the single internuclear axis.

As the number of atoms in the molecule increases, the number of possible vibrations also increases. A set of coordinates must be selected for each atom,

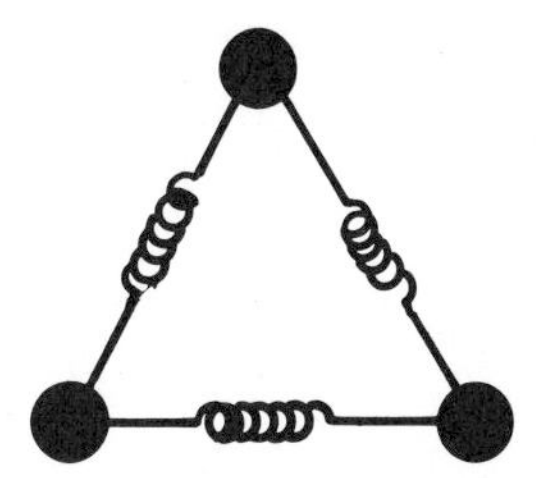

Figure 5.33. A homonuclear triangular molecule.

and net separations of any atomic pair must be incorporated into the force equations for the system. The Lagrangian and Hamiltonian techniques are extremely useful for generating the force equations once a coordinate system has been selected. The procedure is illustrated using the hypothetical triangular molecule of Figure 5.33. The three atoms have the same mass and are joined with bonds of the same force constant k to form an equilateral triangle with bond angles of 60°. Figure 5.34 illustrates possible coordinate systems for the molecule when it is constrained to a surface. In case (a), the coordinates for each atom lie on the standard x and y coordinate axes. The system is ideal to describe translations in the x and y directions but only the bottom internuclear bond is aligned with one of the molecular axes. In case (b), each atom has one coordinate aligned with the internuclear axis. However, the extension at the opposite end of each bond must be described by a sum of components.

The system selected for analysis is illustrated in Figure 5.35. The x_1, x_2, and x_3 coordinates leave the atoms along axes defined by the medians of the triangle as illustrated. The y_1, y_2, and y_3 vectors are constructed at right angles to each of the x_i vectors, as shown. The vectors are selected to match two major normal modes of the triangular molecule. The x_i vectors will be the only active components in a symmetric stretch vibration while the y_i vectors will be the only active components for rotation of this molecule.

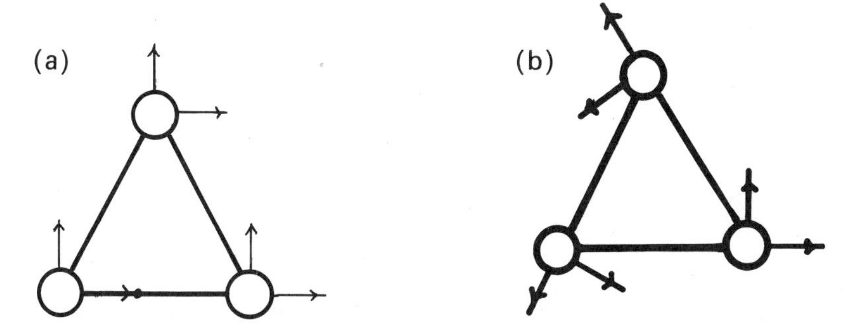

Figure 5.34. Two possible coordinate systems to describe the atomic motions of the triangular molecule: (a) atomic coordinates aligned on x and y spatial axes; (b) three atomic coordinates aligned along the internuclear axes, with the remaining three aligned perpendicular to the first three in the plane.

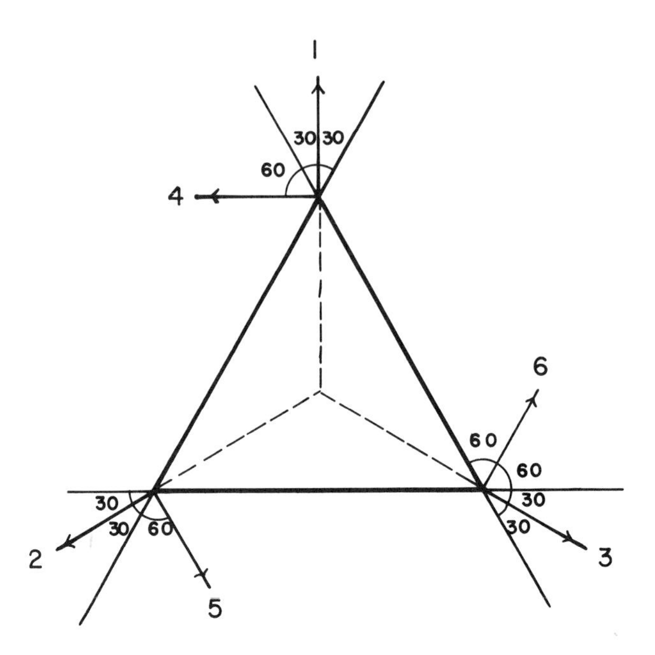

Figure 5.35. A coordinate system for the homonuclear triangular molecule which emphasizes molecular rotation and the symmetric stretch vibration.

The six coordinates for this system require six force equations. The force equations are generated using Hamilton's equations. The Hamiltonian was defined as the sum of kinetic and potential energies for this conservative system,

$$H = T + V \tag{5.9.1}$$

where the energy terms must be functions of position and momentum rather than position and velocity. The six coordinates selected for the triangular molecule are described using a generalized coordinate q_i, e.g., $q_1 = x_1$, $q_5 = y_2$. There are then six Hamiltonian equations for the six q_i:

$$\partial H/\partial q_i = -\partial p/\partial t = -F \tag{5.9.2}$$

Since the kinetic energy will depend on the velocities dq_i/dt, which are considered separate variables, Hamilton's equations are simply a definition of the force in terms of the potential:

$$\partial H/\partial q_i = \partial V/\partial q_i = -F_{q_i} \tag{5.9.3}$$

The equations express the forces with respect to the set of coordinates which were selected. It is much easier to determine the total potential energy for the molecule and develop the six forces using Hamilton's equation in the form of Equation 5.9.3.

The total potential energy of the triangular, homonuclear molecule is found by determining the potential energy for each bond as a function of the system vectors. The relevant components are shown in Figure 5.35. The coordinates x_i and y_i are used in place of the q_i to separate the perpendicular spatial directions. The potential energies for the three bonds are

$$V_{12} = (k/2)(\sqrt{3}x_1/2 - y_1/2 + \sqrt{3}x_2/2 + y_2/2)^2 \tag{5.9.4}$$

$$V_{23} = (k/2)(\sqrt{3}x_3/2 + y_3/2 + \sqrt{3}x_2/2 - y_2/2)^2 \tag{5.9.5}$$

$$V_{31} = (k/2)(\sqrt{3}x_3/2 - y_3/2 + \sqrt{3}x_1/2 + y_1/2)^2 \tag{5.9.6}$$

The total potential is the sum of these three bond potentials:

$$V = V_{12} + V_{23} + V_{31} \tag{5.9.7}$$

The six forces are determined via derivatives with respect to the six variables:

$$\begin{aligned} F_{x1} &= -\partial V/\partial x_1 = -(\sqrt{3}k/2)(\sqrt{3}x_1 + 0 + \sqrt{3}x_2/2 + y_2/2 + \sqrt{3}x_3/2 - y_3/2) \\ &= (-k/4)(6x_1 + 3x_2 + 3x_3 + \sqrt{3}\,y_2 - \sqrt{3}\,y_3) \end{aligned} \tag{5.9.8}$$

$$\begin{aligned} F_{x2} &= -\partial V/\partial x_2 \\ &= (-k/4)(3x_1 + 6x_2 + 3x_3 - \sqrt{3}y_1 + \sqrt{3}y_3) \end{aligned} \tag{5.9.9}$$

$$\begin{aligned} F_{x3} &= -\partial V/\partial x_3 \\ &= (-k/4)(3x_1 + 3x_2 + 6x_3 + \sqrt{3}\,y_1 - \sqrt{3}y_2) \end{aligned} \tag{5.9.10}$$

$$\begin{aligned} F_{y1} &= -\partial V/\partial y_1 \\ &= (-k/4)(-\sqrt{3}x_2 + \sqrt{3}x_3 + 2y_1 - y_2 - y_3) \end{aligned} \tag{5.9.11}$$

$$\begin{aligned} F_{y2} &= -\partial V/\partial y_2 \\ &= (-k/4)(\sqrt{3}x_1 - \sqrt{3}x_3 - y_1 + 2y_2 - y_3) \end{aligned} \tag{5.9.12}$$

$$\begin{aligned} F_{y3} &= -\partial V/\partial y_3 \\ &= (-k/4)(-\sqrt{3}x_1 + \sqrt{3}x_2 - y_1 - y_2 + 2y_3) \end{aligned} \tag{5.9.13}$$

The force constant matrix for the molecule is

$$(-k/4)\begin{pmatrix} 6 & 3 & 3 & 0 & \sqrt{3} & -\sqrt{3} \\ 3 & 6 & 3 & -\sqrt{3} & 0 & \sqrt{3} \\ 3 & 3 & 6 & \sqrt{3} & -\sqrt{3} & 0 \\ 0 & -\sqrt{3} & \sqrt{3} & 2 & -1 & -1 \\ \sqrt{3} & 0 & -\sqrt{3} & -1 & 2 & -1 \\ -\sqrt{3} & \sqrt{3} & 0 & -1 & -1 & 2 \end{pmatrix} \tag{5.9.14}$$

Since some of the normal modes of the system are known, it is easier to examine their properties. A symmetric stretch involves only equal displacements of the three vectors x_1, x_2, and x_3. The vector is

$$\begin{pmatrix} 1 \\ 1 \\ 1 \\ 0 \\ 0 \\ 0 \end{pmatrix} \tag{5.9.15}$$

The **K** matrix operates on this vector to give

$$(-k/4)\begin{pmatrix} 6 & 3 & 3 & 0 & \sqrt{3} & -\sqrt{3} \\ 3 & 6 & 3 & -\sqrt{3} & 0 & \sqrt{3} \\ 3 & 3 & 6 & \sqrt{3} & -\sqrt{3} & 0 \\ 0 & -\sqrt{3} & \sqrt{3} & 2 & -1 & -1 \\ \sqrt{3} & 0 & -\sqrt{3} & -1 & 2 & -1 \\ -\sqrt{3} & \sqrt{3} & 0 & -1 & -1 & 2 \end{pmatrix}\begin{pmatrix} 1 \\ 1 \\ 1 \\ 0 \\ 0 \\ 0 \end{pmatrix} = (-3k)\begin{pmatrix} 1 \\ 1 \\ 1 \\ 0 \\ 0 \\ 0 \end{pmatrix} \tag{5.9.16}$$

Since the eigenvalue is

$$\lambda = m\omega^2 \tag{5.9.17}$$

the vibrational frequency is

$$\omega = (3k/m)^{1/2} \tag{5.9.18}$$

The rotation eigenvector involves only the y_i vectors:

$$\begin{pmatrix} 0 \\ 0 \\ 0 \\ 1 \\ 1 \\ 1 \end{pmatrix} \tag{5.9.19}$$

When **K** operates on this vector, the eigenvalue is zero:

$$\begin{pmatrix} 6 & 3 & 3 & 0 & \sqrt{3} & -\sqrt{3} \\ 3 & 6 & 3 & -\sqrt{3} & 0 & \sqrt{3} \\ 3 & 3 & 6 & \sqrt{3} & -\sqrt{3} & 0 \\ 0 & -\sqrt{3} & \sqrt{3} & 2 & -1 & -1 \\ \sqrt{3} & 0 & -\sqrt{3} & -1 & 2 & -1 \\ -\sqrt{3} & \sqrt{3} & 0 & -1 & -1 & 2 \end{pmatrix} \begin{pmatrix} 0 \\ 0 \\ 0 \\ 1 \\ 1 \\ 1 \end{pmatrix} = 0 \qquad (5.9.20)$$

The x and y translations are determined by finding the projections of three parallel unit vectors on the basis coordinates. The x translation projections are shown in Figure 5.36 while the y translation projections are shown in Figure 5.37. The vectors are

$$|y\rangle = \begin{pmatrix} 3 \\ -1 \\ -1 \\ 0 \\ -\sqrt{3} \\ \sqrt{3} \end{pmatrix} \qquad |x\rangle = \begin{pmatrix} 0 \\ \sqrt{3} \\ -\sqrt{3} \\ 2 \\ -1 \\ -1 \end{pmatrix} \qquad (5.9.21)$$

The final two vectors must correspond to vibrations and must be orthogonal to the four known eigenvectors. The eigenvectors for symmetric stretch and rotation have equal positive elements in the first and last three positions, respectively. For orthogonality, the final two vectors must be partitioned into two regions. The sum of the first three elements must be zero and the sum of the second three elements must be zero. In addition, the vectors orthogonal to $|x\rangle$ and $|y\rangle$ are found by reversing the $+\sqrt{3}$ and $-\sqrt{3}$ components in these vectors. The final two orthogonal vectors must then be (Problem 5.15)

$$\begin{pmatrix} 2 \\ -1 \\ -1 \\ 0 \\ \sqrt{3} \\ -\sqrt{3} \end{pmatrix} \begin{pmatrix} 0 \\ -\sqrt{3} \\ \sqrt{3} \\ 2 \\ -1 \\ -1 \end{pmatrix} \qquad (5.9.22)$$

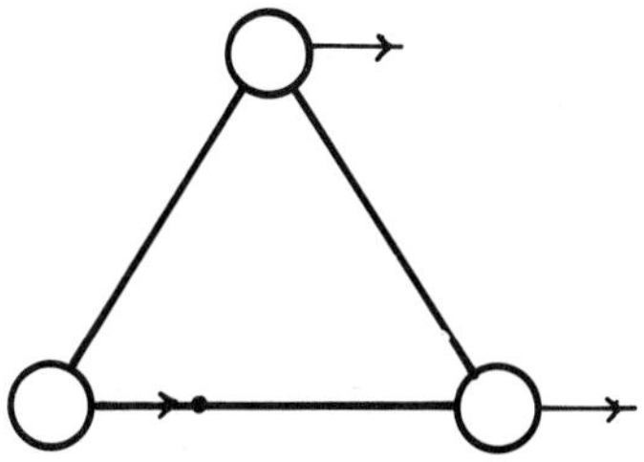

Figure 5.36. An x translation of the triangular molecule using the coordinates of Figure 5.35.

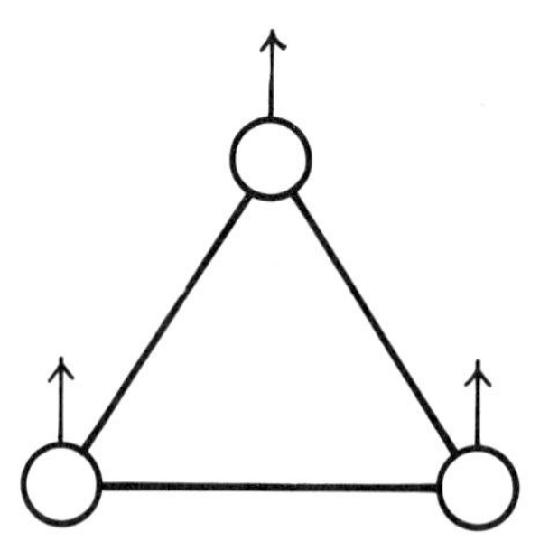

Figure 5.37. A y translation of the triangular molecule using the coordinate system of Figure 5.35.

The eigenvectors are degenerate with eigenvalues 3/2. This is illustrated with the first of these eigenvectors:

$$(k/4)\begin{pmatrix} 6 & 3 & 3 & 0 & \sqrt{3} & -\sqrt{3} \\ 3 & 6 & 3 & -\sqrt{3} & 0 & \sqrt{3} \\ 3 & 3 & 6 & \sqrt{3} & -\sqrt{3} & 0 \\ 0 & -\sqrt{3} & \sqrt{3} & 2 & -1 & -1 \\ \sqrt{3} & 0 & -\sqrt{3} & -1 & 3 & -1 \\ -\sqrt{3} & \sqrt{3} & 0 & -1 & -1 & 2 \end{pmatrix}\begin{pmatrix} 2 \\ -1 \\ -1 \\ 0 \\ 3 \\ -3 \end{pmatrix} = k\left(\frac{3}{2}\right)\begin{pmatrix} 2 \\ -1 \\ -1 \\ 0 \\ \sqrt{3} \\ -\sqrt{3} \end{pmatrix} \tag{5.9.23}$$

The six normal modes for the triangular molecule are shown in Figure 5.38. The rotation and symmetric stretch vibration have an intrinsic rotational symmetry. If the molecule is rotated through 120° or 240°, the unlabeled vectors look exactly the same. The vector positions will coincide with the original positions if the symmetric stretch vectors are reflected in a plane which intersects any atom and bisects the bond opposite this atom, as shown in Figure 5.39. The symmetric stretch vibration has high symmetry. Even though the rotation is symmetric with respect to rotations, it is not so with respect to the reflections. In Figure 5.40, each of these reflections produces a new vector which is colinear but directed in exactly the opposite direction. For coincidence, each vector must be multiplied

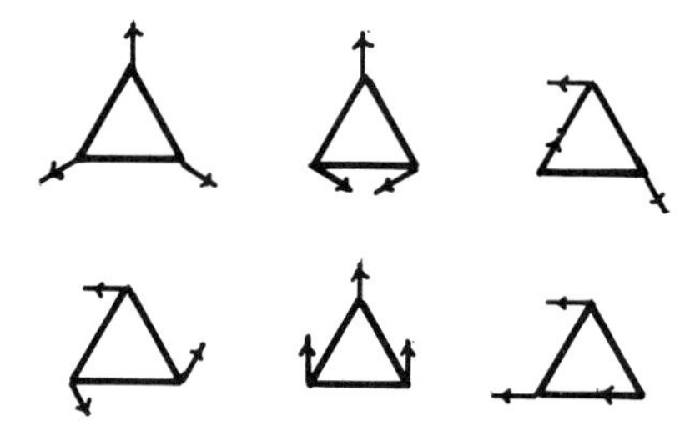

Figure 5.38. The six normal modes for a homonuclear triangular molecule.

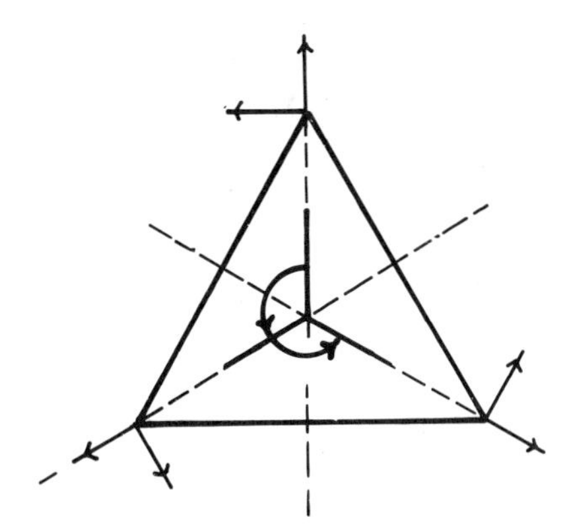

Figure 5.39. The rotation of symmetric stretch and rotation normal mode vector components by 120° and 240°. The rotation moves each component to the location of a second component.

by the scalar -1. Both the symmetric stretch and the rotation follow specific symmetry rules which are considered in Chapter 11.

The methods illustrated for the triangular molecule can be used for any molecule. In general, however, the masses of the atoms in the molecule will be different. Heteroatomic systems have been more complicated because $\mathbf{M}^{-1}\mathbf{K}$ is not symmetric. The matrix can be symmetrized, and this may permit a less complicated analysis of the heteroatomic systems.

The general force equation for an N-dimensional system is

$$d^2\,|q\rangle/dt^2 = -\mathbf{M}^{-1}\mathbf{K}\,|q\rangle \tag{5.9.24}$$

where $\mathbf{M}^{-1}\mathbf{K}$ is generally nonsymmetric. However, if $\mathbf{M}^{-1/2}$ is applied to each side and $\mathbf{I}=\mathbf{M}^{1/2}\mathbf{M}^{-1/2}$ is inserted between $\mathbf{K}$ and $|q\rangle$, the equation becomes

$$\begin{aligned} d^2\mathbf{M}^{1/2}\,|q\rangle/dt^2 &= \mathbf{M}^{1/2}\mathbf{M}^{-1}\mathbf{K}\mathbf{M}^{-1/2}\mathbf{M}^{1/2}\,|q\rangle \\ &= \mathbf{M}^{-1/2}\mathbf{K}\mathbf{M}^{-1/2}\mathbf{M}^{1/2}\,|q\rangle \end{aligned} \tag{5.9.25}$$

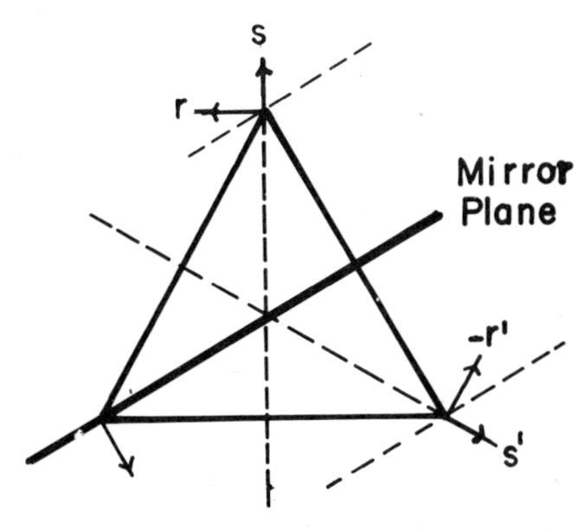

Figure 5.40. The reflections of symmetric stretch and rotation normal mode vector components in mirror planes which pass through one atom and the center of the bond opposite.

The coordinates have been transformed to

$$|q'\rangle = \mathbf{M}^{1/2}\,|q\rangle \tag{5.9.26}$$

and the new ($\mathbf{K}'$) matrix is symmetric:

$$\mathbf{K}' = \mathbf{M}^{-1/2}\mathbf{K}\mathbf{M}^{-1/2} \tag{5.9.27}$$

$\mathbf{K}'$ is not generated by a similarity transform. The transform matrices are identical. However, since the resultant matrix is symmetric, its eigenvectors define a similarity transform:

$$\mathbf{K}'' = \mathbf{S}^{-1}\mathbf{K}'\mathbf{S} = \mathbf{S}^{-1}\mathbf{M}^{-1/2}\mathbf{K}\mathbf{M}^{-1/2}\mathbf{S} \tag{5.9.28}$$

Since $\mathbf{K}''$ is diagonal, the transform matrices

$$\mathbf{M}^{-1/2}\mathbf{S} \quad \text{and} \quad \mathbf{S}^{-1}\mathbf{M}^{-1/2} \tag{5.9.29}$$

do diagonalize the $\mathbf{K}$ matrix when the masses of the atoms are different.

The symmetrization technique is illustrated with a linear triatomic molecule confined to a single spatial direction. The two outer masses are each equal to m_1 while the center atom has mass m_2. CO_2 is an example of such a molecule.

The $\mathbf{K}$ matrix for the molecule is

$$k\begin{pmatrix} 1 & -1 & 0 \\ -1 & 2 & -1 \\ 0 & -1 & 1 \end{pmatrix} \tag{5.9.30}$$

The $\mathbf{K}'$ matrix is

$$\begin{pmatrix} 1/\sqrt{m_1} & 0 & 0 \\ 0 & 1/\sqrt{m_2} & 0 \\ 0 & 0 & 1/\sqrt{m_1} \end{pmatrix}\begin{pmatrix} k & -k & 0 \\ -k & 2k & -k \\ 0 & -k & k \end{pmatrix}\begin{pmatrix} 1/\sqrt{m_1} & 0 & 0 \\ 0 & 1/\sqrt{m_2} & \\ 0 & 0 & 1/\sqrt{m_1} \end{pmatrix}$$

$$= \begin{pmatrix} m/m_1 & -k/m' & 0 \\ -k/m' & 2k/m_2 & -k/m' \\ 0 & -k/m' & k/m_1 \end{pmatrix} \tag{5.9.31}$$

where

$$m' = \sqrt{m_1 m_2} \tag{5.9.32}$$

The common off-diagonal elements can be factored from the matrix to give

$$(k/m')\begin{pmatrix} \sqrt{m_2/m_1} & -1 & 0 \\ -1 & 2\sqrt{m_1/m_2} & -1 \\ 0 & -1 & \sqrt{m_2/m_1} \end{pmatrix} \tag{5.9.33}$$

The matrix is symmetric and has a standard tridiagonal form. The characteristic polynomial,

$$(\sqrt{m_2/m_1} - \lambda)[-(2\sqrt{m_1/m_2} + \lambda\sqrt{m_2/m_1}) + \lambda^2] = 0 \qquad (5.9.34)$$

gives eigenvalues

$$\lambda = 0; \qquad \lambda = (k/m')\sqrt{m_2/m_1}; \qquad \lambda = (k/m')(2\sqrt{m_1/m_2} + \sqrt{m_2/m_1} \qquad (5.9.35)$$

The eigenvalues differ from those for the **K** matrix since **K**′ was not generated with a similarity transform. The eigenvectors $|q'\rangle$ are

$$\begin{pmatrix} 1 \\ \sqrt{m_2/m_1} \\ 1 \end{pmatrix} \quad \begin{pmatrix} 1 \\ 0 \\ -1 \end{pmatrix} \quad \begin{pmatrix} 1 \\ -2\sqrt{m_1/m_2} \\ 1 \end{pmatrix} \qquad (5.9.36)$$

for the sequence of eigenvalues in Equation 5.9.35. The eigenvectors are similar to those observed when the central element was 2 and the terminal diagonal elements were 1. Ratios of masses appear in the vector components.

The actual normal modes are developed from these symmetric matrix normal modes using the equation

$$|q'\rangle = \mathbf{M}^{1/2}\,|q\rangle; \qquad |q\rangle = \mathbf{M}^{-1/2}\,|q'\rangle \qquad (5.9.37)$$

The $\lambda = 0$ eigenvector is

$$\begin{pmatrix} x_1 \\ x_2 \\ x_3 \end{pmatrix} = \begin{pmatrix} 1/\sqrt{m_1} & 0 & 0 \\ 0 & 1/\sqrt{m_2} & 0 \\ 0 & 0 & 1/\sqrt{m_1} \end{pmatrix} \begin{pmatrix} 1 \\ \sqrt{m_2 m_1} \\ 1 \end{pmatrix} \propto 1/\sqrt{m_1} \begin{pmatrix} 1 \\ 1 \\ 1 \end{pmatrix} \qquad (5.9.38)$$

The remaining vibrational eigenvectors are

$$\begin{pmatrix} 1 \\ 0 \\ -1 \end{pmatrix} \quad \begin{pmatrix} m_2 \\ -2m_1 \\ m_2 \end{pmatrix} \qquad (5.9.39)$$

The first vibration involves only the motion of the outer atoms of equal mass and their displacements are the same. This is a symmetric stretch vibration of the linear molecule. The second vibration has displacements which depend on the mass. If $m_2 \gg m_1$, the outer atoms will move the larger relative distance. The larger atom will move a smaller distance in the opposite direction. During each of these vibrations, the central of mass of the molecule is unchanged.

Problems

5.1. For a diatomic molecule in two dimensions, determine the four-component vectors required to describe rotation and vibration when the molecule is aligned with the y axis.

5.2. A diatomic molecule in three dimensions will have three translations, two rotations, and a single vibration. Determine these normal modes as a six-component vector.

5.3. Develop the solution for the displacement of a single atom bound to a fixed surface with force constant k when the atom is fully extended at $t = 0$ and its initial velocity is zero. Determine the solution when the initial $x = 0$ and the velocity is v_0.

5.4. Determine the similarity transform matrices for the heteronuclear diatomic molecule with masses m_1 and m_2 and a force constant k (Equations 5.2.42 and 5.2.43).

5.5. Determine the diagonal **K** matrix for the heteroatomic diatomic molecule of Problem 5.4 (Equation 5.2.44).

5.6. If two disks are mounted on a shaft, the rotation of the first disk (1) through an angle θ_1 will produce a rotation in the second disk. The angular momenta of the two disks are $\mathscr{I}_1 \omega_1$ and $\mathscr{I}_2 \omega_2$, respectively. I_1 and I_2 are the moments of inertia and ω_i are the angular velocities. The restoring force constants on the three sections of shaft which join the disks all equal k. Determine the Lagrangian and Hamiltonian for this system. Use these to show that the force equations for the two disks must be

$$\mathscr{I}_1 d\theta_1/dt = -k(\theta_1 - \theta_2) - k\theta_1$$

$$\mathscr{I}_2 d\theta_2/dt = k(\theta_1 - \theta_2) - k\theta_2$$

Solve the matrix equation for this system.

5.7. For the free linear triatomic molecule, use the vibrational frequencies to determine the potential energies for each of the normal modes. Explain the difference between the normal mode energies.

5.8. Determine the eigenvectors for the heteronuclear triatomic molecule (Equation 5.5.13).

5.9. A single atom can be bound between two walls using bonds with force constants k. Develop the equation for vibration and compare the vibrational frequency with the frequency of an atom bound to a single wall with force constant k.

5.10. Develop the normal modes for the heteronuclear system with $m_2 = 10m_1$ One vibrational mode (Equation 5.5.16) is given.

5.11. Develop the eigenvectors and eigenvalues for a six-atom chain using translation matrix **T** for a system with cyclic boundary conditions.

5.12. For a three-atom chain with cyclic boundary conditions, demonstrate that the eigenvectors with components $\exp(i2\pi nk/N)$ are eigenvectors for the **K** matrix.

5.13. Generate the eigenvectors of Equation 5.7.37 as a linear combination of exponential eigenfunctions for the system, noting that the linear combinations can be generated by multiplication with a phase function, e.g., $\exp(i\pi n/2)$.

5.14. A linear triatomic molecule with two adjacent atoms of mass m_1 and the final atom of mass m_2 can be analyzed using the symmetrization technique of Section 5.9. Develop the symmetric matrix, its similarity transform matrices, and its eigenvectors.

5.15. Determine the $|x\rangle$ and $|y\rangle$ translation vectors for the homonuclear triangular molecule (Equation 5.9.21).

5.16. Many triatomic molecules have only two bonds which connect the central atom to the outer atoms and create a bond angle θ. Develop the Lagrangian for a triatomic molecule with a bond angle of 90° and three atoms of equal mass. Since there are only two bonds, the potential energies are determined by stretching along the internuclear axes with force constants k and a bending vibration in which the two atoms move outward to increase the bond angle with force constant k'.

5.17. Two pendulums of length l and mass m are connected to a surface. They are then connected by a spring with force constant k. The kinetic energy is

$$T = ml^2\omega 2/2 + ml^2\theta^2/2$$

and the potential energy is

$$V = mgl^2\theta_1^2 + mgl^2\theta_2^2/2 + kl^2(\theta_1 - \theta_2)/2$$

a. Develop a matrix equation for the system.
b. Determine the eigenvalues and vibration frequencies.
c. Determine the eigenvectors and symmetrize the equation for the total energy.

5.18. Use the Lagrangian technique to develop the force equations and **K** matrix for three identical particles confined between two fixed walls. All force constants are k.

5.19. Show that the force equation of Equation 5.2.8 in sines and cosines can be converted to a single sinusoid with a phase delay (Equation 5.2.9).

5.20. For the vibrations of Equation 5.2.33, develop a solution when there is no displacement at $t = 0$ and the velocity at $t = 0$ is v_0.

5.21. Set up the potential energy equations for a square molecule containing four identical atoms.

5.22. Use the traveling wave solution $\exp(-i2\pi n/4)$ with the four-atom chain with cyclic boundary conditions and show that it gives the same eigenvalues as $\exp(+i2\pi n/4)$.

5.23. Determine the vectors in a 90° triatomic molecule which give a bending normal mode without altering the center of mass of the total molecule.

6

Kinetics

6.1. Isomerization Reactions

In Chapter 5, the normal modes of vibration in multiparticle systems were determined using the matrix of force constants and the equations of motion. Each row of the matrix system was generated by the equation of motion for a single particle. Diagonalizing this total system of equations gave a new coordinate system in which there was no coupling between the different coordinate directions and the system resolved into a set of N independent scalar equations. In this chapter, similar behavior is observed with systems of kinetic equations. Although distinct chemical species are normally monitored during kinetic experiments, the concentrations of these species will be "blended" as linear combinations which correspond to the normal mode linear combinations in vibrating systems. The time evolution of distinct species is then easily separated from these combinations.

The basic isomerization reaction,

$$\mathrm{A} \underset{k_r}{\overset{k_f}{\rightleftharpoons}} \mathrm{B}$$

where k_f and k_r are the forward and reverse rate constants, respectively, illustrates the methods of analysis for kinetic systems. The rate equations for this two-component system are

$$d[\mathrm{A}]/dt = -k_f[\mathrm{A}] + k_r[\mathrm{B}] \tag{6.1.1}$$

$$d[\mathrm{B}]/dt = k_f[\mathrm{A}] - k_r[\mathrm{B}] \tag{6.1.2}$$

where the square brackets represent molar concentrations. The two equations are related since

$$d[\mathrm{A}]/dt = -d[\mathrm{B}]/dt \tag{6.1.3}$$

The identity indicates a linear dependence between the concentrations of A and B and a reduction in matrix rank. One eigenvalue is zero.

The linear dependence between [A] and [B] appears since the total concentration of particles must remain constant. If the total concentration of particles is $[A_0]$, then

$$[A(t)] + [B(t)] = [A_0] \tag{6.1.4}$$

This total remains constant at all times t although the relative amounts of the A and B isomers will change until the system reaches equilibrium.

Substitution of

$$[B(t)] = [A_0] - [A(t)] \tag{6.1.5}$$

into the first kinetic equation gives

$$\begin{aligned} d[A]/dt &= -k_f[A] + k_r\{[A_0] - [A]\} \\ &= k_r[A_0] - (k_f + k_r)[A] \end{aligned} \tag{6.1.6}$$

This equation can be rewritten as

$$d[A]/dt + (k_f + k_r)[A] = k_r[A_0] \tag{6.1.7}$$

and solved by multiplying both sides of the equation by the integrating factor

$$\exp[(k_f + k_r)t] = E = \exp(kt) \tag{6.1.8}$$

where the definitions

$$k = k_f + k_r \tag{6.1.9}$$

and the exponential E simplify the expressions for further manipulation. The equation becomes

$$Ed[A]/dt + kE[A] = k_r[A_0]E \tag{6.1.10}$$

This equation can be simplified using the identity

$$d\{E[A]\}/dt = Ed[A]/dt + kE[A] \tag{6.1.11}$$

The left-hand side is a perfect differential and is integrated from $[A_0]$ to $[A(t)]$. The right-hand side is integrated from $t = 0$ to the time t at which $[A] = [A(t)]$:

$$\int_0^t d\{E[A]\} = k_r[A_0] \int_0^t E\,dt \tag{6.1.12}$$

with a solution

$$E[A(t)] - [A_0] = k_r[A_0]\,k^{-1}(E - 1) \tag{6.1.13}$$

Multiplying the entire equation by E^{-1} gives

$$[A(t)] = [A_0] \exp(-kt) + (k_r/k)[A_0][1 - \exp(-kt)] \qquad (6.1.14)$$

This expression is simplified by placing the identity

$$(k_f + k_r)/(k_f + k_r) = 1 \qquad (6.1.15)$$

in the first term on the left to give

$$[A(t)] = (k_r/k)[A_0] + (k_f/k)[A_0] \exp(-kt) \qquad (6.1.16)$$

The solution for [B] is found using conservation of mass (Equation 6.1.4):

$$\begin{aligned} [B(t)] &= [A_0] - [A(t)] \qquad (6.1.17) \\ &= [A_0](k_f + k_r)/k - (k_r/k)[A_0] - (k_f/k)[A_0] \exp(-kt) \\ &= [A_0](k_f/k) - (k_f/k)[A_0] \exp(-kt) \qquad (6.1.18) \end{aligned}$$

The kinetic evolution of [A(t)] and [B(t)] is shown in Figure 6.1.

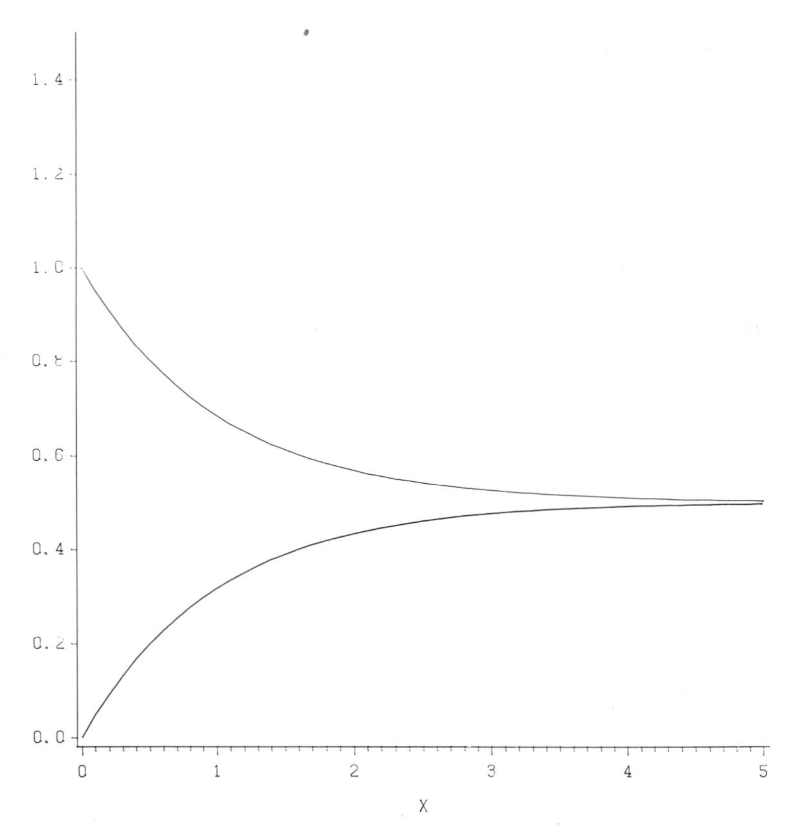

Figure 6.1. Evolution of [A(t)] and [B(t)] for a first-order isomerization reaction.

There are some common characteristics in the solutions for the concentrations of A and B. The time-dependent term is exactly the same for both species. The signs differ because a decrease in one must produce an equal increase in the other to conserve mass. Although an experimentalist might follow the evolution of [A] or [B], he will observe the same decay constant in either case. The eigenvector which describes the simultaneous decrease of [A] and increase of [B] is characterized by a single decay constant, the eigenvalue associated with this eigenvector.

Only the numerators of the time-independent terms in each expression differ. The terms give the concentrations at equilibrium:

$$[\mathrm{A}] = [\mathrm{A}_0](k_r/k) \tag{6.1.19}$$

$$[\mathrm{B}] = [\mathrm{A}_0](k_f/k) \tag{6.1.20}$$

The relative fractions of each species are determined by the rate constants leading to that species; e.g., the rate defined by k_r leads to species A.

The problem is restated as a matrix differential equation by assigning the two concentrations, $[\mathrm{A}(t)]$ and $[\mathrm{B}(t)]$, as the components of a vector. The two differential equations are

$$d[\mathrm{A}]/dt = -k_f[\mathrm{A}] + k_r[\mathrm{B}] \tag{6.1.21}$$

$$d[\mathrm{B}]/dt = k_f[\mathrm{A}] - k_r[\mathrm{B}] \tag{6.1.22}$$

with initial conditions

$$[\mathrm{A}(0)] = [\mathrm{A}_0] \qquad [\mathrm{B}(0)] = [\mathrm{B}_0]$$

This is recast as the matrix equation

$$d\,|c\rangle/dt = \mathbf{K}\,|c\rangle \tag{6.1.23}$$

where

$$\mathbf{K} = \begin{pmatrix} -k_f & k_r \\ k_f & -k_r \end{pmatrix}$$

and

$$|c\rangle = \begin{pmatrix} [\mathrm{A}(t)] \\ [\mathrm{B}(t)] \end{pmatrix} \qquad |c_0\rangle = \begin{pmatrix} [\mathrm{A}_0] \\ [\mathrm{B}_0] \end{pmatrix} \tag{6.1.24}$$

If such a system can be resolved into a sum of projection operators, the matrix system can be solved exactly as if it were a scalar equation. A first-order equation

$$dc/dt = kc \qquad c_0 = c(0) \tag{6.1.25}$$

has an exponential solution

$$c(t) = c_0 \exp(+kt) \tag{6.1.26}$$

The corresponding solution of the matrix differential equation is

$$|c\rangle = \exp(\mathbf{K}t)\,|c_0\rangle \tag{6.1.27}$$

where the column vector of initial concentrations, $|c_0\rangle$, must be placed to the right of the matrix function. The solution of this matrix equation is written as a sum of scalar functions by introducing the eigenvalues and projection operators to give

$$|c\rangle = \sum_i^N \exp(\lambda_i t)\,\mathbf{P}_i\,|c_0\rangle \tag{6.1.28}$$

Each exponential in the sum contains one of the decay constants for the reaction. This result is general; for a set of coupled kinetic equations of this type, the solution will always be a sum of exponential functions.

The two-component isomerization reaction has the 2×2 **K** matrix

$$\mathbf{K} = \begin{pmatrix} -k_f & k_r \\ k_f & -k_r \end{pmatrix} \tag{6.1.29}$$

The eigenvalues are determined from the characteristic equation

$$\begin{aligned} (k_f + \lambda)(k_r + \lambda) - k_r k_f = 0 \\ (k_f + k_r)\lambda + \lambda^2 = 0 \end{aligned} \tag{6.1.30}$$

as

$$\lambda = 0; \qquad \lambda = -(k_f + k_r) \tag{6.1.31}$$

The projection operators for each eigenvalue are

$$\begin{aligned} \mathbf{P}(\lambda = 0) &= [\mathbf{K} - (-\mathbf{I})(k_f + k_r)]/(0 + k_f + k_r) \\ &= k^{-1}\begin{pmatrix} -k_f + k_f + k_r & k_r \\ k_f & -k_r + k_f + k_r \end{pmatrix} \\ &= k^{-1}\begin{pmatrix} k_r & k_r \\ k_f & k_f \end{pmatrix} \end{aligned} \tag{6.1.32}$$

where

$$k = k_f + k_r \tag{6.1.33}$$

The second projection operator is

$$\mathbf{P}[\lambda = -(k_f + k_r)] = [\mathbf{K} - \mathbf{I}(0)]/[-(k_f + k_r) - 0]$$
$$= -\begin{pmatrix} -k_f & k_r \\ k_f & -k_r \end{pmatrix} k^{-1} \quad (6.1.34)$$

The solution to the matrix equation is

$$|c\rangle = \exp(0t)\,\mathbf{P}(\lambda = 0)\,|c_0\rangle + \exp(-kt)\,\mathbf{P}(\lambda = -k)\,|c_0\rangle$$
$$= (1/k)\begin{pmatrix} k_r & k_r \\ k_f & k_f \end{pmatrix}\begin{pmatrix} [\mathrm{A}_0] \\ [\mathrm{B}_0] \end{pmatrix} + \exp(-kt)/k\begin{pmatrix} k_f & -k_r \\ -k_f & k_r \end{pmatrix}\begin{pmatrix} [\mathrm{A}_0] \\ [\mathrm{B}_0] \end{pmatrix} \quad (6.1.35)$$

After multiplication, the full solution for $[\mathrm{A}(t)]$ appears as the first component of the vectors while $[\mathrm{B}(t)]$ is the second component:

$$[\mathrm{A}(t)] = k_r\{[\mathrm{A}_0] + [\mathrm{B}_0]\}/k + (\{k_f[\mathrm{A}_0] - k_r[\mathrm{B}_0]\}/k)\exp(-kt) \quad (6.1.36)$$

$$[\mathrm{B}(t)] = k_f\{[\mathrm{A}_0] + [\mathrm{B}_0]\}/k - (\{k_f[\mathrm{A}_0] - k_r[\mathrm{B}_0]\}/k)\exp(-kt) \quad (6.1.37)$$

This solution is more complicated than the previous one because the initial concentration vector included both molecular species. The original solution is regenerated if $[\mathrm{B}_0] = 0$:

$$[\mathrm{A}(t)] = (k_r/k)[\mathrm{A}_0] + (k_f/k)[\mathrm{A}_0]\exp(-kt) \quad (6.1.38)$$

$$[\mathrm{B}(t)] = (k_f/k)[\mathrm{A}_0] - (k_f/k)[\mathrm{A}_0]\exp(-kt) \quad (6.1.39)$$

Some general properties of such matrix solutions are present in these kinetic equations. For example, there is always one set of components, the eigenvector for $\lambda = 0$, which is time independent. This vector gives the concentrations of each species at equilibrium. The remaining eigenvector for this example with components proportional to $(1, -1)$ describes the equal losses and gains of the two concentrations. The magnitude of the term is proportional to

$$k_f[\mathrm{A}_0] - k_r[\mathrm{B}_0] \quad (6.1.40)$$

Since the forward and reverse rates will be equal at equilibrium, this difference measures the deviation from the equilibrium rate at the start of the reaction.

6.2. Properties of Matrix Solutions of Kinetic Equations

The projection operator technique for coupled kinetic equations introduced in Section 6.1 produced time-dependent solutions for a vector with components $[\mathrm{A}(t)]$ and $[\mathrm{B}(t)]$. Because the concentrations are the primary experimental observables, this is the preferred emphasis. However, the eigenvalue–eigenvector approach to these coupled kinetic equations places the emphasis on principal axes for these systems. These principal axes provide new insights into the nature of kinetic relaxation processes.

The matrix kinetic equation assigned the first vector component to the concentration [A] and the second vector component to the concentration [B]. [A] and [B] play roles analogous to x and y in a Cartesian coordinate system. The eigenvectors determined for a specific **K** matrix are always defined relative to the [A] and [B] component axes. This is particularly important for these kinetic studies since the kinetics of the individual chemical species are the more important experimental observables.

The nature of the eigenvectors for the **K** matrix can be clarified by selecting the special case where

$$k_f = k_r = 1 \tag{6.2.1}$$

and the initial condition is

$$[\mathrm{A}(0)] = A_0; \qquad [\mathrm{B}(0)] = 0$$

The solution in this case is (Equations 6.1.38 and 6.1.39)

$$\begin{pmatrix} [\mathrm{A}(t)] \\ [\mathrm{B}(t)] \end{pmatrix} = (A_0/2)\begin{pmatrix} 1 \\ 1 \end{pmatrix} + (A_0/2)\begin{pmatrix} 1 \\ -1 \end{pmatrix} \exp(-2t) \tag{6.2.2}$$

or

$$|c(t)\rangle = (A_0/\sqrt{2})\,|0\rangle + (A_0/\sqrt{2})\,|-2\rangle \exp(-2t) \tag{6.2.3}$$

with the normalized eigenvectors

$$|0\rangle = \begin{pmatrix} \dfrac{1}{\sqrt{2}} \\ \dfrac{1}{\sqrt{2}} \end{pmatrix} \qquad |-2\rangle = \begin{pmatrix} \dfrac{1}{\sqrt{2}} \\ \dfrac{-1}{\sqrt{2}} \end{pmatrix} \tag{6.2.4}$$

The equilibrium vector for this special case has equal concentrations of [A] and [B], i.e.,

$$[\mathrm{A}^{\mathrm{eq}}] = [\mathrm{B}^{\mathrm{eq}}] = A_0/2 \tag{6.2.5}$$

The equilibrium eigenvector would then lie at a 45° angle to the [A] and [B] coordinates as shown in Figure 6.2. The projections on each of the axes are $A_0/2$. The equilibrium eigenvector has a norm

$$(A_0^2/4 + A_0^2/4)^{1/2} = A_0/\sqrt{2} \tag{6.2.6}$$

which gives the proper projections on the [A] and [B] axes.

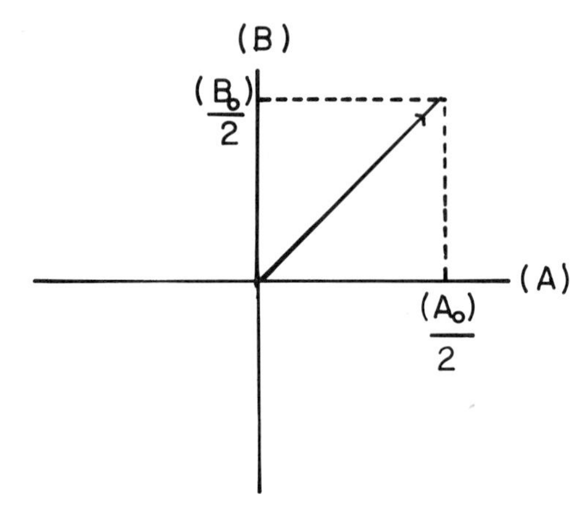

Figure 6.2. The 45° equilibrium vector in the [A]–[B] plane when $k_f = k_r = 1$.

The equilibrium vector and its projections are time independent. This gives some insight into the properties of matrix kinetics. With an eigenvector formalism, a single coordinate direction, the equilibrium vector direction, remains constant during the entire kinetic evolution. The nonequilibrium principal axes decay to zero with time, leaving the equilibrium vector in the concentration space.

The two-dimensional isomerization demonstrates these properties. The non-equilibrium eigenvector in the space is characterized by the components (1, −1) as shown in Figure 6.3. The direction of this vector is also fixed in the space, and projections on the [A] and [B] axes at time zero are

$$[A] = A_0/2; \qquad [B] = -A_0/2 \tag{6.2.7}$$

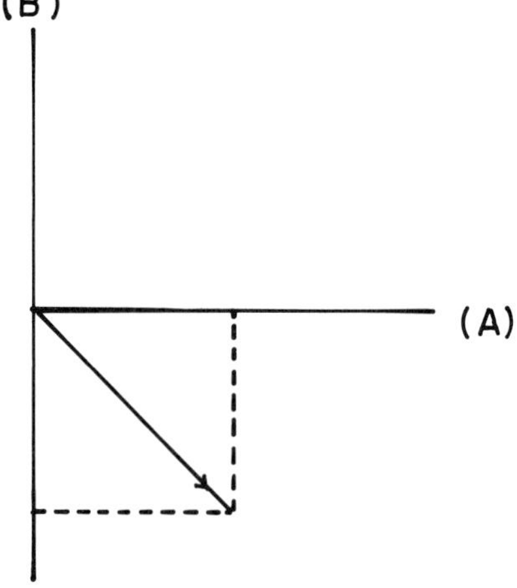

Figure 6.3. The −45° vector which contains the time dependence for the simple first-order reaction. Net concentration projections at time zero are shown.

The components might suggest that the concentration of B can be negative. However, to obtain [B], all [B] projections must be added. The concentrations of both components on the [B] axis sum to zero at time zero (Figure 6.3). In addition, each of the two eigenvectors contributes $A_0/2$ to the [A] coordinate. Their total on this axis gives A_0, the initial condition which was chosen for this example. The sum of all components is essential.

Although the direction of the $|-2\rangle$ vector is fixed in the two-dimensional space, its magnitude decays exponentially in time as

$$\exp(-2t) \tag{6.2.8}$$

The same relative projections on the two axes remain, but both projections become smaller at the same rate. This is illustrated in Figure 6.4. The net projection on [A] at time zero was A_0. This component must decay to $A_0/2$ at equilibrium. This decay is shown as a series of dots on the axis for equal time increments. As the system approaches equilibrium, the dots become more closely spaced, as expected for the exponential function.

On the [B] axis, the equilibrium component remains positive but the $|-2\rangle$ vector cancels progressively less of this vector with increasing time. Eventually, the net component approaches the equilibrium component. The two sets of components can be combined to produce a continuous trajectory in the two-dimensional space, as shown in Figure 6.5. The trajectory follows a straight line from $(A_0, 0)$ to $(A_0/2, A_0/2)$ in this two-dimensional case since all kinetic change takes place along the single coordinate direction $(1, -1)$.

For reaction schemes involving more than two species, the graphical description is more difficult. However, the eigenvalue–eigenvector formalism predicts similar properties. Each nonequilibrium eigenvector in the N-dimensional space is multiplied by an exponential decay function,

$$\exp(-\lambda_i t) \tag{6.2.9}$$

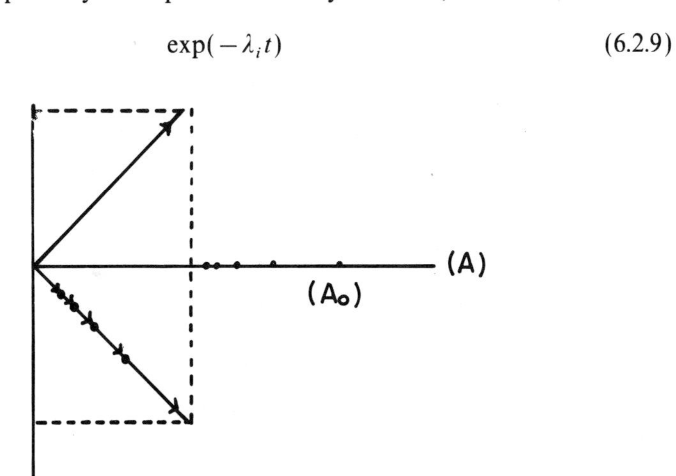

Figure 6.4. The decay of the $|-2\rangle$ vector with time. The decay projections on the [A] and [B] axes are shown.

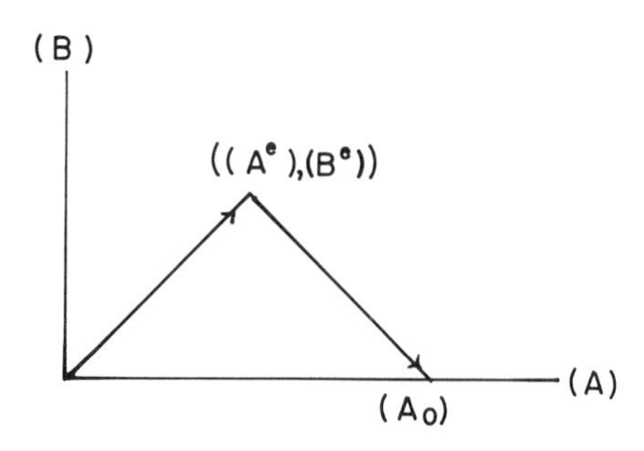

Figure 6.5. The net trajectory of the A–B isomerization in the [A]–[B] plane.

and their projections on the component axes (species concentrations) decay in time. At long times, only the equilibrium vector remains. The concentrations of the individual species at equilibrium are found by projecting the vector onto each "species" coordinate axis. The sum of such projections gives the actual species concentration.

The N-component vector which describes the approach to equilibrium can describe a complicated motion in the N-dimensional space. However, the resolution into a summation of eigenvector terms does permit a partial simplification. The terms containing the largest eigenvalues will decay more rapidly. The smallest, nonzero eigenvalue will produce the slowest decay. The complicated motion in the space is approximated as a two-dimensional subspace containing the equilibrium eigenvector and the eigenvector associated with the smallest nonzero eigenvalue. This approximation is quite effective for many systems containing large numbers of components.

The choice of equal rate constants for the forward and reverse rate constants produced two eigenvectors which were mutually orthogonal. In general, these rate constants will not be equal. Under such conditions, the two column vectors in the two-dimensional space need not be orthogonal. For example, when $k_f = 1$ and $k_r = 3$, the concentration vector is (Problem 6.16)

$$\begin{pmatrix} [\mathrm{A}] \\ [\mathrm{B}] \end{pmatrix} = (A_0/4)\begin{pmatrix} 3 \\ 1 \end{pmatrix} + (A_0/4)\begin{pmatrix} 1 \\ -1 \end{pmatrix} \exp(-4t) \tag{6.2.10}$$

The concentration of A is favored and the slope of the equilibrium vector will be small (Figure 6.6). The nonequilibrium eigenvector has a fixed orientation in the space. Its magnitude is dictated by the row eigenvector

$$\langle -4 | c_0 \rangle \tag{6.2.11}$$

The time evolution is a linear trajectory parallel to $|-4\rangle$ in the two-dimensional space.

Several trends for these two-dimensional systems are present in systems with more components. The equilibrium eigenvector gave the concentrations of each

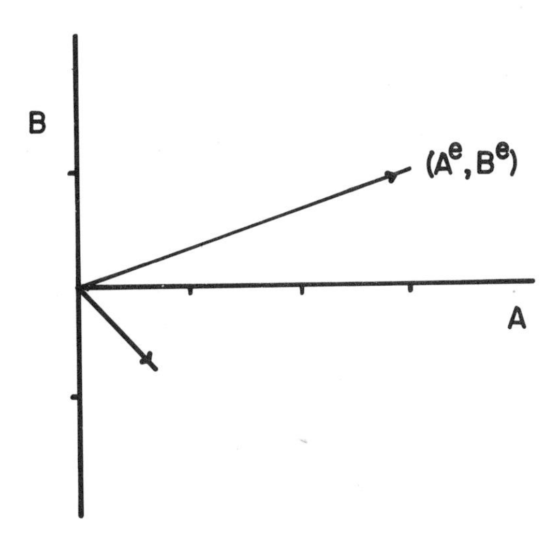

Figure 6.6. Eigenvectors for the A–B isomerization when $k_f = 1$ and $k_r = 3$. The trajectory to equilibrium parallels the nonequilibrium vector.

species as the appropriate component in the equilibrium eigenvector. These components depend on the actual rate constants for the system. However, in each case, the equilibrium row eigenvectors were identical with components

$$\langle 0| = (1, 1) \tag{6.2.12}$$

For an N-component system, each component will be unity. This is related to the fact that the equilibrium eigenvectors must give a zero eigenvalue:

$$\langle 0| \mathbf{K} = \langle 0| 0 \tag{6.2.13}$$

Because each column of the matrix must sum to zero if the mass of the system is conserved, the $\langle 0|$ eigenvector acts as a column summation vector and has identical components for this reason. The zero eigenvector also sums the concentrations of each component in the equilibrium ket eigenvector. The normalization for the two equilibrium vectors is the total concentration of all species in the system.

The $\langle 0|$ vector is orthogonal or biorthogonal to each nonequilibrium eigenvector. Since $\langle 0|$ is a summation vector, this means that each nonequilibrium ket eigenvector has components that sum to zero. Since the total concentration of the system must reside in the equilibrium column eigenvector, there is no net concentration in any of the remaining eigenvectors. The nonequilibrium eigenvectors simply redistribute the mass of the system among the different component species at each time t during the kinetic evolution.

The presence of a zero eigenvalue is easily established for the two-component systems since the characteristic equation can be solved for the two eigenvalues analytically. However, the conservation of mass condition always requires a zero eigenvalue. Consider component 1. The differential equation for component 1 can include rates of loss to all the other component species of the system. However, each of these loss rates appears as a rate of increase for one of the other components. The rate of loss for species 1 will appear as the 11 element of the **K** matrix. Each rate of gain produced by the population of component 1 must appear in the first column of the **K** matrix. The total sum of rates of loss and gain in the first column must be zero. This is the case for each other column in the **K** matrix.

The eigenvalues of the **K** matrix are determined from the characteristic determinant

$$[\mathbf{K}-\lambda\mathbf{I}]=0 \tag{6.2.14}$$

The value of the determinant is unchanged if columns or rows are added or subtracted component by component. The operation replaces a column or row of the determinant. It is most convenient in this case to form a new row by adding all the rows in the matrix to form the final row. By conservation of mass, all the rate constants in the column must sum to zero. Thus, the final row would contain only the eigenvalue $-\lambda$. Since this determinant must equal zero, at least one of the eigenvalues must be zero. A row of zeros for the determinant gives a determinant equal to zero.

The procedure can be illustrated for a three-component **K** matrix:

$$\mathbf{K}=\begin{pmatrix} -k_{21}-k_{31} & k_{12} & k_{13} \\ k_{21} & -k_{12}-k_{32} & k_{23} \\ k_{31} & k_{32} & -k_{13}-k_{23} \end{pmatrix} \tag{6.2.15}$$

The column sum of the characteristic determinant gives

$$\begin{vmatrix} -k_{21}-k_{31}-\lambda & k_{12} & k_{13} \\ k_{21} & -k_{12}-k_{32}-\lambda & k_{23} \\ -\lambda & -\lambda & -\lambda \end{vmatrix} \tag{6.2.16}$$

Although two other eigenvalues will also make this determinant equal to zero, $\lambda=0$ is definitely an eigenvalue since expansion in the final row will give a determinant of zero if $\lambda=0$.

6.3. Kinetics with Degenerate Eigenvalues

The sequence of consecutive kinetic reactions, e.g.,

$$\mathrm{A} \rightleftharpoons \mathrm{B} \rightleftharpoons \mathrm{C}$$

occurs frequently in chemical kinetics. For example, the irreversible case with consecutive forward rate constants k_1 and k_2, respectively, gives a tractable analytical solution. When reverse rate constants are included, the equations are conveniently solved using eigenvalue–eigenvector techniques. However, the technique is not generally applicable. The technique is first used for a sequential reaction scheme in which the eigenvalues of the matrix are nondegenerate. The failure of the technique then occurs when two eigenvalues become degenerate.

Two irreversible sequential reactions with rate constants k_1 and k_2 have rate equations

$$d[\mathrm{A}]/dt = -k_1[\mathrm{A}] \qquad [\mathrm{A}(0)] = A_0 \tag{6.3.1}$$

$$d[\mathrm{B}]/dt = k_1[\mathrm{A}] - k_2[\mathrm{B}] \qquad [\mathrm{B}(0)] = 0 \tag{6.3.2}$$

$$d[\mathrm{C}]/dt = k_2[\mathrm{B}] \qquad [\mathrm{C}(0)] = 0 \tag{6.3.3}$$

The solution for the irreversible reaction of A is

$$[\mathrm{A}(t)] = A_0 \exp(-k_1 t) \tag{6.3.4}$$

This decay is independent of subsequent rate processes. However, the time-dependent A concentration plays a role in subsequent kinetic equations. Substituting the solution for [A] into the differential equation for [B] gives

$$d[\mathrm{B}]/dt + k_2[\mathrm{B}] = k_1 A_0 \exp(-k_1 t) \tag{6.3.5}$$

The rate term in [B] is moved to the left-hand side to permit introduction of the integrating factor

$$\exp(k_2 t) \tag{6.3.6}$$

This gives

$$d\{[\mathrm{B}]\exp(k_2 t)\}/dt = k_1 A_0 \exp[(k_2 - k_1)t] \tag{6.3.7}$$

which is integrated from an initial $[\mathrm{B}(0)] = 0$ to some $[\mathrm{B}(t)]$ to give

$$[\mathrm{B}(t)]\exp(k_2 t) = [k_1 A_0/(k_2 - k_1)]\{\exp[(k_2 - k_1)t] - 1\} \tag{6.3.8}$$

$$[\mathrm{B}(t)] = [A_0 k_1/(k_2 - k_1)][\exp(-k_1 t) - \exp(-k_2 t)] \tag{6.3.9}$$

$[\mathrm{C}(t)]$ is found using conservation of total mass:

$$[\mathrm{C}(t)] = A_0 - [\mathrm{A}(t)] - [\mathrm{B}(t)] \tag{6.3.10}$$

$$\begin{aligned} &= A_0\{1 - [(k_2 - k_1)/(k_2 - k_1)]\exp(-k_1 t) \\ &\quad - [k_1/(k_2 - k_1)][\exp(-k_1 t) - \exp(-k_2 t)]\} \\ &= A_0\{1 - [k_2/(k_2 - k_1)]\exp(-k_1 t) \\ &\quad + [k_1/(k_2 - k_1)]\exp(-k_2 t)\} \end{aligned} \tag{6.3.11}$$

The variation of [A], [B], and [C] with time is shown in Figure 6.7.

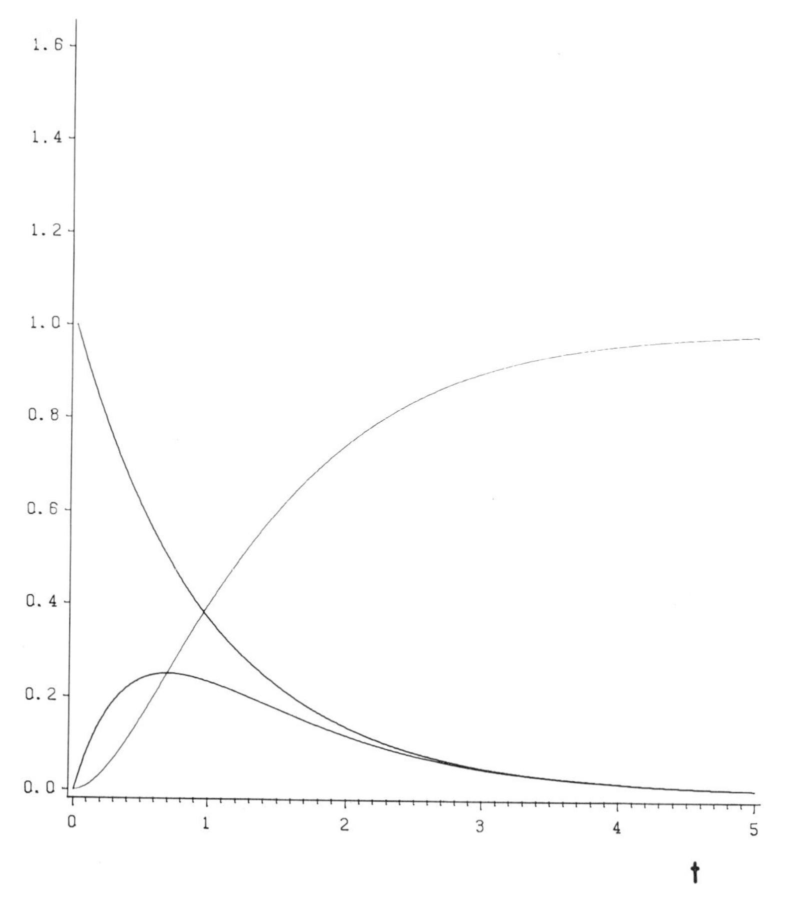

Figure 6.7. The time evolution of the concentrations [A], [B], and [C] for the sequential reaction A → B → C.

The situation changes for the special case where $k_1 = k_2 = k$. The solution for $[A(t)]$ is unchanged. However, the second equation requires an integrating factor,

$$\exp(kt) \tag{6.3.12}$$

to give

$$d\{[B]\exp(kt)\}/dt = kA_0 \exp[(k-k)t] = kA_0 \tag{6.3.13}$$

The integrated equation gives a single term on the right which includes a linear time dependence,

$$[B(t)] = kA_0 t \exp(-kt) \tag{6.3.14}$$

In this case, the concentration of B rises linearly until the exponential decay from B to C finally eliminates B molecules, as shown in Figure 6.8.

The appearance of a new functional dependence for [B] suggests that the form of the general matrix solution must be different for degenerate eigenvalues.

For the irreversible case with forward rate constants k_1 and k_2, the rate constant matrix for the vector with components [A] = A, [B] = B and [C] = C,

$$\begin{pmatrix} [A] \\ [B] \\ [C] \end{pmatrix} = \begin{pmatrix} A \\ B \\ C \end{pmatrix} \tag{6.3.15}$$

is

$$\begin{pmatrix} -k_1 & 0 & 0 \\ k_1 & -k_2 & 0 \\ 0 & k_2 & 0 \end{pmatrix} \tag{6.3.16}$$

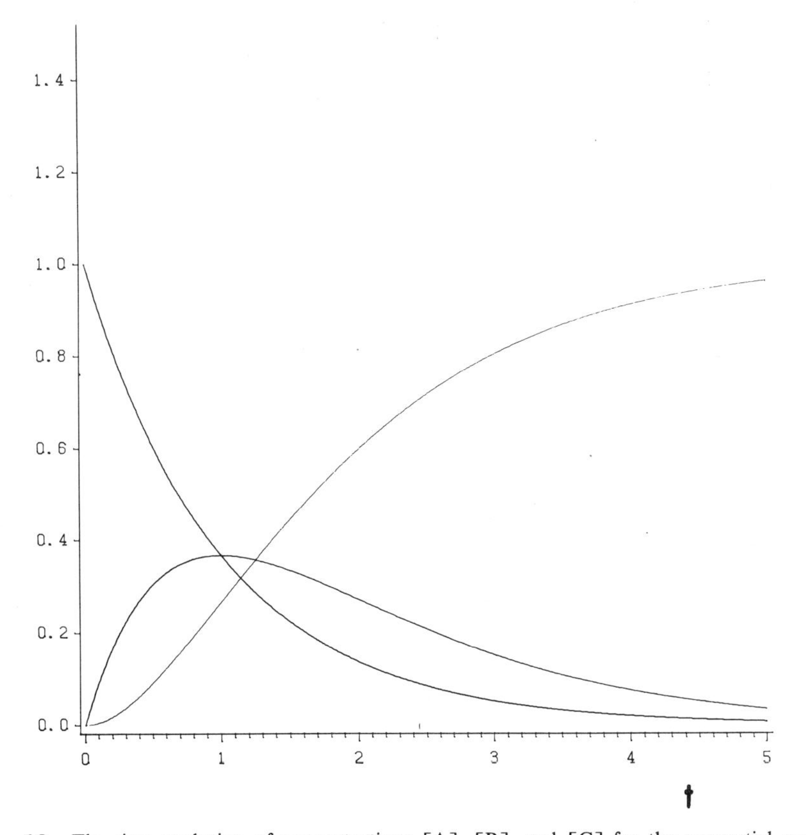

Figure 6.8. The time evolution of concentrations [A], [B], and [C] for the sequential reaction scheme when $k_1 = k_2$.

Since the matrix is nonsymmetric, the projection operator solutions are preferable. The characteristic equation is

$$(-k_1-\lambda)(-k_2-\lambda)(-\lambda)=0 \tag{6.3.17}$$

with eigenvalues

$$\lambda=-k_1 \qquad \lambda=-k_2 \qquad \lambda=0 \tag{6.3.18}$$

The eigenvalues are the elements which appear on the diagonal of the original matrix. The matrix is lower triangular. All the elements above the main diagonal are zero. An upper triangular matrix would have all zero elements below the main diagonal. In either of these cases, the eigenvalues will always be identical to the diagonal elements (Problem 6.17).

The equilibrium projection operator is

$$\mathbf{P}(\lambda=0)=(\mathbf{K}+k_1\mathbf{I})(\mathbf{K}+k_2\mathbf{I})/(0+k_1)(0+k_2)$$

$$=(k_1k_2)^{-1}\begin{pmatrix} 0 & 0 & 0 \\ +k_1 & -k_2+k_1 & 0 \\ 0 & k_2 & k_1 \end{pmatrix}\begin{pmatrix} -k_1+k_2 & 0 & 0 \\ +k_1 & 0 & 0 \\ 0 & +k_2 & k_2 \end{pmatrix}$$

$$=\begin{pmatrix} 0 & 0 & 0 \\ 0 & 0 & 0 \\ 1 & 1 & 1 \end{pmatrix} \tag{6.3.19}$$

This simple projection operator results because of the simple equilibrium configuration for this system. All the molecules must remain as species C since there is no way for them to reisomerize to species A or B. State C is a "trapping" state. There is no diagonal element in the third row of the **K** matrix. Molecules can enter but cannot leave. There are elements in each column of the bottom row since this projection operator must sum all the initial concentrations. For initial concentrations of A_0, B_0, and C_0, the projection operator gives an equilibrium vector

$$\begin{pmatrix} 0 & 0 & 0 \\ 0 & 0 & 0 \\ 1 & 1 & 1 \end{pmatrix}\begin{pmatrix} A_0 \\ B_0 \\ C_0 \end{pmatrix}=\begin{pmatrix} 0 \\ 0 \\ A_0+B_0+C_0 \end{pmatrix} \tag{6.3.20}$$

i.e., at equilibrium, all the mass is present as C.

The other two projection operators are

$$\mathbf{P}(\lambda=-k_1)=[\mathbf{K}-(-k_2)\mathbf{I}](\mathbf{K}-0\mathbf{I})/(-k_1+k_2)(-k_1-0)$$

$$=[(k_1-k_2)k_1]^{-1}\begin{pmatrix} k_2-k_1 & 0 & 0 \\ k_1 & 0 & 0 \\ 0 & k_2 & k_2 \end{pmatrix}\begin{pmatrix} -k_1 & 0 & 0 \\ k_1 & -k_2 & 0 \\ 0 & k_2 & 0 \end{pmatrix}$$

$$=\begin{pmatrix} +1 & 0 & 0 \\ -k_1/(k_1-k_2) & 0 & 0 \\ +k_2/(k_1-k_2) & 0 & 0 \end{pmatrix} \tag{6.3.21}$$

$$\mathbf{P}(\lambda=-k_2)=[\mathbf{K}-(-k_1)\mathbf{I}](\mathbf{K}-0\mathbf{I})/(k_2-k_1)(k_2)$$

$$=\begin{pmatrix} 0 & 0 & 0 \\ k_1/(k_1-k_2) & 1 & 0 \\ -k_1/(k_1-k_2) & -1 & 0 \end{pmatrix} \tag{6.3.22}$$

Both these projection matrices have many zero elements while the nonzero elements are more complicated than those for $\mathbf{P}(\lambda=0)$. They contain all the information necessary for arbitrary initial conditions. The initial conditions $[\mathrm{A}(0)]=A_0$, $[\mathrm{B}(0)]=[\mathrm{C}(0)]=0$ regenerate the original result. With general initial conditions, the solution is

$$|c(t)\rangle=\mathbf{P}(\lambda=0)\,|c_0\rangle+\exp(-k_1t)\,\mathbf{P}(\lambda=-k_1)\,|c_0\rangle$$
$$+\exp(-k_2t)\,\mathbf{P}(\lambda=-k_2)\,|c_0\rangle \tag{6.3.23}$$

which gives the following concentration vector:

$$\begin{pmatrix} A \\ B \\ C \end{pmatrix}=\begin{pmatrix} 0 \\ 0 \\ T \end{pmatrix}+A_0\exp(-k_1t)\begin{pmatrix} 1 \\ -k_1/(k_2-k_1) \\ k_2/(k_1-k_2) \end{pmatrix}$$
$$+\exp(-k_2t)\begin{pmatrix} 0 \\ -k_1A_0/(k_2-k_1)+B_0\exp(-k_2t) \\ k_1A_0/(k_2-k_1)-B_0\exp(-k_2t) \end{pmatrix} \tag{6.3.24}$$

with $T=A_0+B_0+C_0$.

The individual solutions for [A], [B], and [C] are

$$\begin{aligned} A&=A_0\exp(-k_1t) \\ B&=A_0[k_1/(k_2-k_1)][\exp(-k_1t)-\exp(-k_2t)]+B_0\exp(-k_2t) \\ C&=A_0\{1-[k_2/(k_2-k_1)]\exp(-k_1t) \\ &\quad+[k_1/(k_2-k_1)]\exp(-k_2t)\}-B_0\exp(-k_2t) \end{aligned} \tag{6.3.25}$$

Although the eigenvalue–eigenvector technique is quite effective for most sequential reaction schemes, it does fail for the special case where $k=k_1=k_2$. The **K** matrix becomes

$$\begin{pmatrix} -k & 0 & 0 \\ k & -k & 0 \\ 0 & k & 0 \end{pmatrix} \tag{6.3.26}$$

which is lower triangular with eigenvalues

$$0,\ -k,\ -k \tag{6.3.27}$$

i.e., two eigenvalues are now degenerate.

The projection operator for $\lambda=0$ is well behaved and can be found using the Lagrange–Sylvester projection operator equation,

$$\mathbf{P}(\lambda=0)=(\mathbf{K}+k\mathbf{I})^2/(0+k)^2$$

$$=(0+k)^{-2}\begin{pmatrix}0 & 0 & 0\\ k & 0 & 0\\ 0 & k & k\end{pmatrix}\begin{pmatrix}0 & 0 & 0\\ k & 0 & 0\\ 0 & k & k\end{pmatrix}$$

$$=\begin{pmatrix}0 & 0 & 0\\ 0 & 0 & 0\\ 1 & 1 & 1\end{pmatrix} \tag{6.3.28}$$

This projection operator is identical to that for the nondegenerate case since C is still a trapping state.

An equation for the exponential of a symmetric matrix with a doubly degenerate eigenvalue was given in Chapter 4 (Equation 4.7.42). This expression is applied to the nonsymmetric k matrix to give

$$\mathbf{P}(\lambda=-k)=[\mathbf{I}-\mathbf{K}t+kt\mathbf{I}]\exp(-kt)\frac{\mathbf{K}}{k} \tag{6.3.29}$$

$$=\exp(-kt)\begin{pmatrix}1 & 0 & 0\\ +kt & 1 & 0\\ 0 & +kt & 1+kt\end{pmatrix}\begin{pmatrix}1 & 0 & 0\\ -1 & 1 & 0\\ 0 & -1 & 0\end{pmatrix}$$

$$=\exp(-kt)\begin{pmatrix}+1 & 0 & 0\\ -kt-1 & +1 & 0\\ -kt & -1 & 0\end{pmatrix} \tag{6.3.30}$$

The final projection on the initial vector is now

$$\exp(-kt)\begin{pmatrix}1 & 0 & 0\\ -kt-1 & 1 & 0\\ kt & -1 & 0\end{pmatrix}\begin{pmatrix}A_0\\ 0\\ 0\end{pmatrix} \tag{6.3.31}$$

The final concentrations of A, B, and C are determined by adding the two projections:

$$\begin{aligned}A(t)&=A_0\exp(-kt)\\ B(t)&=A_0(-kt-1)\exp(-kt)\\ C(t)&=A_0-A_0(kt)\exp(-kt)\end{aligned} \tag{6.3.32}$$

Although the solutions do have a linear time dependence, the solutions differ from those found earlier by sequential integration of the differential equations. Why? Although the projection matrices for $\lambda=0$ and $\lambda=-k$ are orthogonal, they do not span the space since their sum is not equal to **I**.

The nonsymmetric K matrix simply does not provide enough information to determine eigenvectors and projection operators, the matrix equation with $k=1$ is

$$(\mathbf{K}-\lambda\mathbf{I})\,|c\rangle=(\mathbf{K}+\lambda\mathbf{I})\,|c\rangle=0$$

$$\begin{pmatrix} 0 & 0 & 0 \\ k & 0 & 0 \\ 0 & k & k \end{pmatrix}\begin{pmatrix} A \\ B \\ C \end{pmatrix}=0 \tag{6.3.33}$$

The three equations which must be solved to determine the eigenvector are

$$\begin{aligned} 0A+0B+0C&=0 \\ 1A+0B+0C&=0 \\ 0A+1B+1C&=0 \end{aligned} \tag{6.3.34}$$

It is now necessary to find consistent A, B, and C. The first equation gives no information on the relative magnitudes of these components. The second equation requires that $A=0$. Thus, the third equation provides the total information on the relative magnitudes of B and C. The eigenvector components are

$$A=0 \qquad B=1 \qquad C=-1 \tag{6.3.35}$$

The process must be repeated for the row eigenvector for $\lambda=-k$. The equations are

$$\begin{aligned} 0A+1B+0C&=0 \\ 0A+0B+1C&=0 \\ 0A+0B+1C&=0 \end{aligned} \tag{6.3.36}$$

C and B must be zero since each appears alone in the equations. A might be any arbitrary number, e.g., 1, but there is no opportunity to find working relations between the components. There is simply not enough information in the equations. A maximum of one dependent variable can be related to one independent variable. Under such circumstances, the projection operator technique fails. For example, the biorthonormalization $\langle -1|-1\rangle$ gives a zero norm:

$$(1 \quad 0 \quad 0)\begin{pmatrix} 0 \\ 1 \\ -1 \end{pmatrix}=0 \tag{6.3.37}$$

This example serves to illustrate the potential difficulties of matrices which are neither symmetric nor Hermitian. If the matrix has degenerate eigenvalues but is symmetric, an orthogonal set of eigenvectors can be found.

If reverse rate constants k_{-1} and k_{-2} are included in the sequential reaction scheme,

$$\mathrm{A} \underset{k_{-1}}{\overset{k_1}{\rightleftharpoons}} \mathrm{B} \underset{k_{-2}}{\overset{k_2}{\rightleftharpoons}} \mathrm{C}$$

the reaction matrix **K** is

$$\begin{pmatrix} -k_1 & k_{-1} & 0 \\ k_1 & -(k_{-1}+k_2) & k_{-2} \\ 0 & k_2 & -k_{-2} \end{pmatrix} \tag{6.3.38}$$

The columns all sum to zero as they must for conservation of mass to give a determinant of zero and a zero eigenvalue. The remaining two eigenvalues are determined from the characteristic equation,

$$\begin{aligned} \lambda^3 + d\lambda^2 + \lambda = 0 \\ \lambda^2 + d\lambda + s = 0 \end{aligned} \tag{6.3.39}$$

where d is the trace

$$d = -(k_1 + k_{-1} + k_2 + k_{-2}) \tag{6.3.40}$$

Since the sum of columns is zero, the determinant, i.e., the λ^0 term, is zero. The remaining λ^1 coefficient, p, is found using 2×2 determinants:

$$\begin{vmatrix} -k_1 & k_{-1} \\ k_1 & -(k_{-1}+k_2) \end{vmatrix} + \begin{vmatrix} -(k_{-1}+k_2) & k_{-2} \\ k_2 & -k_{-2} \end{vmatrix} + \begin{vmatrix} -k_1 & 0 \\ 0 & -k_{-2} \end{vmatrix}$$

$$= (k_1 k_2 + k_1 k_{-1} - k_1 k_{-1}) + (k_2 k_{-2} + k_{-1} k_{-2} - k_2 k_{-2}) + k_1 k_{-2}$$

$$s = k_1 k_2 + k_{-1} k_{-2} + k_1 k_{-2} \tag{6.3.41}$$

The eigenvalues are

$$\lambda_{\pm} = -d/2 \pm \frac{1}{2}(d^2 - 4s)^{1/2} \tag{6.3.42}$$

Since the two nonzero eigenvalues are quite complicated, it is difficult to substitute them into either the characteristic equation or the Lagrange–Sylvester expression to determine eigenvectors or projection operators. However, it is possible to determine the eigenvectors in terms of the λ_i. For the case where $[\mathrm{A}(0)] = A_0$ and $[\mathrm{B}(0)] = [\mathrm{C}(0)] = 0$, the products $\mathbf{P}(\lambda_i)\,|c\rangle$ for each eigenvalue are

$$\mathbf{P}(0)\,|c\rangle = \begin{pmatrix} k_{-2}k_{-1} \\ k_1k_{-2} \\ k_1k_2 \end{pmatrix} (A_0/\lambda_+\lambda_-)$$

$$\mathbf{P}(\lambda_+)\,|c\rangle = \begin{pmatrix} k_1(\lambda_+ - k_2 - k_{-2}) \\ k_1(k_{-2} - \lambda_+) \\ k_1k_2 \end{pmatrix} [A_0/\lambda_+(\lambda_+ - \lambda_-)] \tag{6.3.43}$$

$$\mathbf{P}(\lambda_-)\,|c\rangle = \begin{pmatrix} k_1(k_2 + k_{-2} - \lambda_-) \\ k_1(\lambda_- - k_{-2}) \\ k_1k_2 \end{pmatrix} [A_0/\lambda_-(\lambda_+ - \lambda_-)]$$

In Section 6.7, a graphical technique is used to permit rapid determination of the equilibrium vector for such systems.

6.4. The Master Equation

Matrix techniques have been used to study the kinetic evolution of mixtures of chemical species. The concentrations used as components in the vector actually represent the average behavior of a large number of molecules. The rate of reaction for a given molecule is proportional to the number of such molecules in the system.

A single reactant molecule may exist in a large number of discrete states; it may acquire vibrational, rotational, or translational energy via collisions with other molecules. These transitions at the microscopic level will ultimately determine the macroscopic rate of reaction for an ensemble of such molecules.

The microscopic transitions are studied by inserting the molecules in a "heat bath." The gas phase heat bath is simply a large number of nonreactive molecules, e.g., Ar or N_2, which act to transfer energy to or from the reactive molecule. The concentration of heat bath molecules is much larger than the concentration of reactive molecules so that transitions are caused almost exclusively by collisions with the nonreactive molecules.

The reactive molecule exists in a large number of distinct states. For convenience, only different vibrational states are considered. A collision may transfer energy to the molecule as increased vibrational energy, but this does not happen during each collision. Some fraction of the collisions will result in an actual change of the vibrational level of the molecule. The transition must then be defined via a transition probability, p_{ij}, that the molecule moves from vibrational level j to level i as a result of the collision.

Although a transition probability per collision is a natural definition, experimental observations are ultimately based on transition probabilities per unit time. The conversion to this new transition probability is accomplished using the average collision frequency z as an internal clock. Since z has the units of collisions per second, the product

$$z(\text{collisions/s})\, P_{ij}(\text{transition probability/collision}) \tag{6.4.1}$$

has the units of transition probability per second, i.e.,

$$w_{ij} = zp_{ij} \tag{6.4.2}$$

The transition probability can predict changes within a single molecule. However, the transition probability is an average determined from the behavior of a large number of molecules. This is important because the introduction of such probabilities generates a bridge between reversible microscopic events and irreversible macroscopic processes. For example, a large number of the molecules must proceed irreversibly to an equilibrium distribution. At the microscopic level, each process is time reversible. The system is as likely to be moving away from equilibrium as it is to be moving toward it. However, after defining transition probabilities, the system will "acquire" an equilibrium zero eigenvalue and move toward equilibrium. The transition from reversible to irreversible behavior via the introduction of probabilities is illustrated for a simple example in Section 7.1.

The kinetic equations produced by the transition probabilities can be described with a molecule which contains only two states, as shown in Figure 6.9. The transition probabilities are used as rate constants for transitions within the molecule. The transition from the lower level (1) to the upper level (2) is proportional to the transition probability w_{21}. The initial state index is placed to the right (column index) for consistency when the matrix operates on column eigenvectors. The transition probability from state 2 to state 1 is w_{12}. Kinetic equations describe the populations of state 1 (f_1) and state 2 (f_2):

$$df_1/dt = -w_{21} f_1 + w_{12} f_2 \tag{6.4.3}$$

$$df_2/dt = +w_{21} f_1 - w_{12} f_2 \tag{6.4.4}$$

The functional form of the equations is identical to that for the A–B isomerization of Section 6.1. In this case, the "kinetic" changes take place within the molecule.

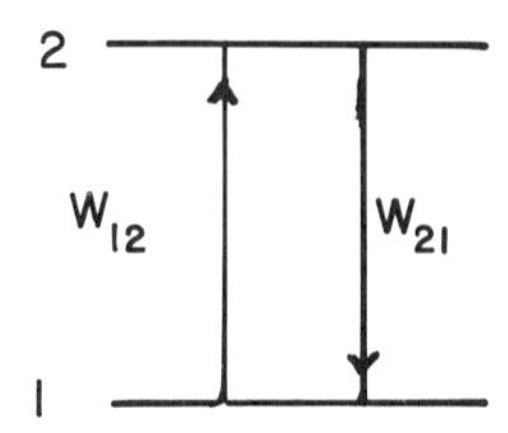

Figure 6.9. Transitions for a two-state molecule in a heat bath.

The transition probability matrix **W** is

$$\begin{pmatrix} -w_{21} & w_{12} \\ w_{21} & -w_{12} \end{pmatrix} \tag{6.4.5}$$

and the kinetic equation is

$$d\,|f\rangle/dt = \mathbf{W}\,|f\rangle \tag{6.4.6}$$

The kinetic equations retain properties observed for isomerization systems. For example, the sum of elements in each column is zero. The system conserves "mass." In this case, the conservation condition implies that the molecule must be in one of its microscopic states. No reaction to a new molecular species is allowed.

Although both the intermolecular and intramolecular kinetic schemes appear identical, there is one important difference. A real molecule has many distinct states, and each of these states may require a kinetic equation. Even if vibrational transitions alone are considered, the transition probability matrix is large. The set of intramolecular kinetic equations in matrix form is called a master equation. This equation for N states is

$$d/dt \begin{pmatrix} f_1 \\ f_2 \\ \vdots \\ f_N \end{pmatrix} = \begin{pmatrix} w_{11} & w_{12} & \cdots & w_{1N} \\ w_{21} & w_{22} & \cdots & w_{2N} \\ & & \vdots & \\ w_{N1} & w_{N2} & \cdots & w_{NN} \end{pmatrix} \begin{pmatrix} f_1 \\ f_2 \\ \vdots \\ f_N \end{pmatrix} \tag{6.4.7}$$

The diagonal elements, w_{ii}, must contain all rates of depopulation for state i. A diagonal element w_{ii} is the negative sum of all transition probabilities from the state i. The transition probabilities for this sum appear as the nondiagonal matrix elements of the ith column,

$$w_{ii} = -\sum_{i} w_{ji} \tag{6.4.8}$$

This, of course, is just the condition that the elements in a column of a **K** matrix must sum to zero if mass is conserved. The diagonal elements of the **W** matrix can always be found from the nondiagonal elements in this way.

The condition is illustrated with the two-state master equation. The diagonal elements are

$$w_{11} = -w_{21} \quad \text{and} \quad w_{22} = -w_{12} \tag{6.4.9}$$

A gain in state 2 is balanced by a loss in state 1.

Although most quantum mechanical systems will have discrete states and must be described by transition probabilities between these states, many such states may be present. Under such circumstances, it is convenient to approximate the levels as a continuum. The index i is replaced by a continuous variable, e.g., x, which locates the level in the continuum. The transition from x to x' is then a functional transition probability,

$$w(x' \mid x) \tag{6.4.10}$$

and the population of states at x is the function $f(x)$. Because the number of rows in a matrix for the continuous function would be infinite, it is easier to look at a single "row" for this function. The rate of change of $f(x)$ involves gains from each of the other x' states,

$$\int_{x'} w(x \mid x')\, f(x')\, dx' \tag{6.4.11}$$

where the integral is required to sum contributions from all other states. The loss from state x will also involve an integral, but it will sum all the states x' which are reached by depopulation of x:

$$-\int_{x} w(x' \mid x)\, f(x)\, dx \tag{6.4.12}$$

The master equation in x is then

$$df(x)/dt = -\int w(x' \mid x)\, f(x)\, dx + \int w(x \mid x')\, f(x')\, dx' \tag{6.4.13}$$

For vibrational transitions induced by collisions with heat bath molecules or atoms, there are parallels with the optically induced vibrational transitions. The heat bath molecule can be approximated as a plane matter wave; the resultant mathematics for determining the transition probabilities is then analogous to that of the optical transition probabilities. In particular, nonzero transition probabilities are observed for transitions from state i to either state $i+1$ or state $i-1$. A transition probability from state i to state $i+1$ can be related to the transition probability, p_{10}, the transition probability from level 0 to level 1 as

$$p_{i+1,i} = p_{10}(i+1) \tag{6.4.14}$$

Each up transition probability is proportional to the quantum number of the final level. For a down transition from a level i to a level $i-1$, the transition probability is

$$p_{i-1,i} = p_{01}\, i \tag{6.4.15}$$

and is proportional to the initial level quantum number i.

These two nearest-neighbor transition probabilities lead to a tridiagonal matrix. For example, with four vibrational levels, the transition probability matrix at high temperatures is

$$z\begin{pmatrix} p_{00} & p_{01} & p_{02} & p_{03} \\ p_{10} & p_{11} & p_{12} & p_{13} \\ p_{20} & p_{21} & p_{22} & p_{23} \\ p_{30} & p_{31} & p_{32} & p_{33} \end{pmatrix}$$

$$= z\begin{pmatrix} -p_{10} & p_{01} & 0 & 0 \\ p_{10} & -(p_{01}+2p_{10}) & 2p_{01} & 0 \\ 0 & 2p_{10} & -(2p_{01}+3p_{10}) & 3p_{01} \\ 0 & 0 & 3p_{10} & -3p_{01} \end{pmatrix} \tag{6.4.16}$$

The transition probabilities p_{01} and p_{10} are equal so the matrix becomes

$$zp_{10}\begin{pmatrix} -1 & 1 & 0 & 0 \\ 1 & -3 & 2 & 0 \\ 0 & 2 & -5 & 3 \\ 0 & 0 & 3 & -3 \end{pmatrix} \tag{6.4.17}$$

The pattern continues for an $N \times N$ $\mathbf{W}$ matrix. The diagonal above the main diagonal will increase regularly to a maximum of $N-1$, e.g., 3, in Equation 6.4.17. The diagonal below the main diagonal will begin at 1 and will also continue until $N-1$.

This tridiagonal matrix differs from normal mode matrices since the transition probabilities increase as the vibrational level increases. The ith row equation is

$$df_i/dt = (i-1)f_{i-1} - (2i-1)f_i + if_{i+1} \tag{6.4.18}$$

The factors $i-1$, $2i-1$, etc. complicate analysis of this system.

The nature of the transition probabilities plays a major role in determination of the macroscopic kinetics of a system. For example, if molecules react irreversibly, their reaction rates will be defined by the number of molecules which remain. In this case, f_i is then the probability of observing the system with i molecules.

The transition probability matrix is

$$z\begin{pmatrix} 0 & p_{01} & 0 & 0 & \cdots \\ 0 & -p_{01} & p_{12} & 0 & 0 \\ 0 & & -p_{12} & p_{23} & \\ \vdots & & & & \end{pmatrix} \tag{6.4.19}$$

The matrix is upper diagonal because only a single molecule irreversibly leaves the system. The Nth component f_N will be unity at time zero. The transition probabilities are dependent on the number, j, of molecules which remain at time t:

$$p_{ij} f_j \tag{6.4.20}$$

The equation for the ith row of the matrix equation is

$$df_i/dt = -p_{i-1,i} f_i + p_{i,i+1} f_{i+1} \tag{6.4.21}$$

Inserting the linear transition probabilities gives

$$df_i/d(pt) = -if_i + (i+1) f_{i+1} \tag{6.4.22}$$

The average number of molecules at time t is

$$\langle i \rangle = \sum if_i \tag{6.4.23}$$

Equation 6.4.22 can be multiplied by i and summed to determine the average rate of change of i:

$$d\left(\sum_0^\infty if_i\right) \Big/ dt = \sum_0^\infty -i^2 f_i + \sum_0^\infty i(i+1) f_{i+1} \tag{6.4.24}$$

The summations on the right can be simplified by noting the first several terms in the series

$$\begin{aligned} &-0f_0 + 0f_1 - 1f_1 + 2f_2 - 4f_2 + 6f_3 - \cdots \\ &= -1f_1 - 2f_2 - \cdots - if_i - \cdots \end{aligned} \tag{6.4.25}$$

The equation becomes

$$d\langle f \rangle / d(pt) = -\langle f \rangle \tag{6.4.26}$$

At the macroscopic level, the average number of particles will decay exponentially with time. The choice of linear transition probabilities led directly to this result.

If the transition probability contains nonlinear terms, the differential equation for the average decay will change. For example, the choice of a transition probability containing a term which is quadratic in the number of molecules,

$$p_{ij} = pj + qj^2 \tag{6.4.27}$$

gives an $\langle f \rangle$ equation which contains the linear dependence of Equation 6.4.20 and a second quadratic term. This term is

$$q[i^2(i+1) f_{i+1} - i^3 f_i] \tag{6.4.28}$$

The first several terms,

$$-0f_1 + 0f_0 - 4f_2 + 1f_1 - 18f_3 + 8f_2 - \cdots \tag{6.4.29}$$

give

$$f_1 + 4f_2 + \cdots + i^2 f_i + \cdots \tag{6.4.30}$$

The differential equation for the average rate of decay is

$$d\langle f \rangle / dt = -p\langle f \rangle + \sum_{i=0}^{\infty} i^2 f_i \tag{6.4.31}$$

The second term introduces nonlinearity into the differential equation for the average macroscopic kinetics.

6.5. Symmetrization of the Master Equation

The general expression for the master equation in the last section contains off-diagonal transition probabilities which are all independent. The diagonal elements are obtained by summing the off-diagonal elements in their row. The problem simplifies when the transition probabilities obey selection rules. For example, the selection rules observed for the Landau–Teller transition probabilities produced a tridiagonal matrix. Because the Landau–Teller probabilities were proportional to the molecular level index *i*, the tridiagonal matrix was still complicated albeit more tractable than a matrix with additional off-diagonal elements. For a two-level model with an up transition probability, p_{10}, and a down transition probability, p_{01}, the master equation transition probability matrix is

$$z\begin{pmatrix} -p_{10} & p_{01} \\ p_{10} & -p_{01} \end{pmatrix} \tag{6.5.1}$$

where the top row describes the kinetics of state 0 and the lower row describes those of state 1. For microscopic reversibility,

$$p_{01} = p_{10} \tag{6.5.2}$$

this system has an equilibrium eigenvector which predicts equal concentrations for each of the two states. This is actually a simple example of microscopic reversibility. The dynamics of the particle as it moves from state 0 to state 1 are reversed, i.e., the particle retraces its dynamic path, when negative time is substituted into the equations of motion.

The equality of up and down transition probabilities via microscopic reversibility is disturbing. If state 1 has a higher energy than state 0, Boltzmann statistics predict a lower population since there are fewer bath molecules which can supply this extra energy. The populations of states 0 and 1 are proportional to

$$(0)\quad 1; \qquad (1)\quad \exp(-E/kT) \tag{6.5.3}$$

where k is Boltzmann's constant and T is the temperature. At equilibrium, the vector has components

$$\begin{pmatrix} 1 \\ \exp(-E/kT) \end{pmatrix} \tag{6.5.4}$$

and not the equal populations predicted by the transition probability expression. The Boltzmann statistics of the system are required. The molecules undergo collisions with a heat bath of molecules whose distribution of energies will affect the rates of transition between the two states.

For an up transition, the molecule must accept an energy E from one of the bath molecules. For a down transition, it must transfer an energy E to the bath molecule. However, because of the Boltzmann distribution for the bath molecules, there will be a greater population of bath molecules without energy which are then available to accept the energy and permit the molecule to make a 1 to 0 transition. Molecules with energies 0 and E will have exactly the same transition probability because of microscopic reversibility, but the energetic molecules will deactivate at a greater rate because there are more bath molecules of low energy which can accept the A molecule energy.

The effects of the heat bath energy distribution are incorporated in the transition probabilities per unit time, w_{ij}. In Section 6.4, p was multiplied by a collision frequency z. Two collision frequencies are now required: a collision frequency z_0 for the number of collisions per second for bath molecules with no energy and z_1 for the number of collisions per second for bath molecules with an energy of E. If the bath molecules have the same energy differences as the reacting molecule, the modified collision frequencies are

$$z_0 = z(1)/[1+\exp(-E/kT)] = z/(1+F) \tag{6.5.5}$$

$$z_1 = zF/(1+F) \tag{6.5.6}$$

where the denominator $1+F = 1+\exp(-E/kT)$ is the partition function for this system and normalizes each change as a fraction of the total number of molecular collisions z.

The two z_i multiply the appropriate transition probabilities in the matrix, i.e., the 1 to 0 transition is multiplied by z_0 since this collision frequency includes all collisions which can produce a down transition. The collision frequency z_1 reflects collisions with bath molecules with energy E which can be transferred to the molecule for the transition from 0 to 1. The transition probability matrix is

$$\begin{pmatrix} -p_{10}z_1 & p_{01}z_0 \\ p_{10}z_1 & -p_{01}z_0 \end{pmatrix} \tag{6.5.7}$$

The transition probabilities are defined in Figure 6.10.

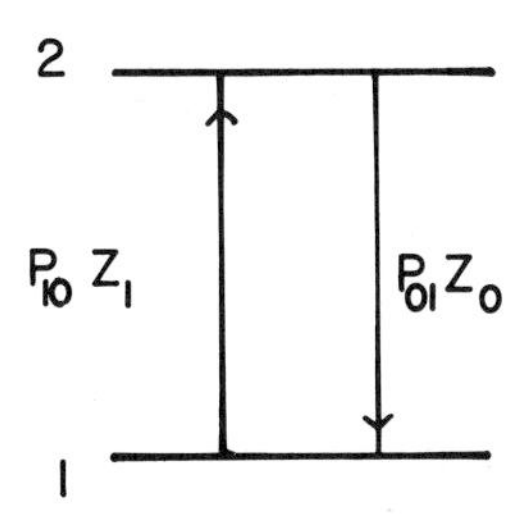

Figure 6.10. Transitions for a two-state molecule modified for effective collision frequencies.

The system still has the zero eigenvalue, and its eigenvector will have equilibrium components n_0 and n_1 which are determined with the matrix equation

$$\begin{pmatrix} -p_{10}z_1 & p_{01}z_0 \\ p_{10}z_1 & -p_{01}z_0 \end{pmatrix}\begin{pmatrix} n_0 \\ n_1 \end{pmatrix} = 0$$
$$zp_{10}\begin{pmatrix} -F & 1 \\ F & -1 \end{pmatrix}\begin{pmatrix} n_0 \\ n_1 \end{pmatrix} = 0 \tag{6.5.8}$$

The relative populations of state 0 and 1 are

$$-Fn_0 + n_1 = 0$$
$$n_1 = Fn_0 = F \tag{6.5.9}$$

With $n_0 = 1$, the eigenvector components are proportional to the Boltzmann factors for the equilibrium distribution,

$$\begin{pmatrix} 1 \\ \exp(-E/kT) \end{pmatrix} \tag{6.5.10}$$

Since the row eigenvector for this nonsymmetric matrix is

$$(1 \quad 1) \tag{6.5.11}$$

the normalization

$$(1 \quad 1)\begin{pmatrix} 1 \\ \exp(-E/kT) \end{pmatrix} = 1 + \exp(-E/kT) \tag{6.5.12}$$

gives the molecular partition function for this two-level system.

To produce a kinetic system with the proper Boltzmann distribution at equilibrium, Boltzmann factors were introduced into the transition probability matrix. However, will this inclusion be valid when the system is not at equilibrium? This is reasonable when the number of bath molecules is much greater than the number of A molecules. Transfer of energy between A and the bath molecules will have no effect on the Boltzmann distribution of the bath.

The $\lambda = 0$ eigenvalue for this two-level system with the collision frequencies included can be generalized as the condition of detailed balance at equilibrium. The system is at equilibrium when the rate of change is zero so that

$$\begin{aligned} 0 &= -zp_{10}Fn_0 + zp_{01}n_1 \\ (zp_{10}F)\, n_0 &= (zp_{01})\, n_1 \\ k_{10}n_0 &= k_{01}n_1 \end{aligned} \tag{6.5.13}$$

This is the condition of detailed balance at equilibrium. The forward and reverse rates of reaction between a pair of states must be equal at equilibrium. This holds even if factors within the rate constant expression are different, e.g., $n_i \neq n_j$. For multistate systems, this condition will be satisfied for every communicating pair of states.

For a multistate system, how are the probabilities modified to include a final equilibrium distribution? For the 3×3 transition probability matrix

$$z \begin{pmatrix} -(p_{10} + p_{20}) & p_{01} & p_{02} \\ p_{10} & -(p_{01} + p_{21}) & p_{12} \\ p_{20} & p_{21} & -(p_{02} + p_{12}) \end{pmatrix} \tag{6.5.14}$$

the bath molecules will have a distribution with components

$$1 \quad F \quad F^2 \tag{6.5.15}$$

These components must also occur in the collision frequencies z_i. A factor of F is multiplied by the transition probability for each level of rise. Downward transition probabilities are always multiplied by a factor of 1. For example, p_{20} will be multiplied by the factor F^2 since these are the bath molecules with sufficient energy for such an excitation. The matrix is

$$z \begin{pmatrix} -(p_{10}F + p_{20}F^2) & p_{01} & p_{02} \\ p_{10}F & -(p_{01} + p_{21}F) & p_{12} \\ p_{20}F^2 & p_{21}F & -(p_{02} + p_{12}) \end{pmatrix} \tag{6.5.16}$$

The first diagonal below the main diagonal contains factors of F; the second diagonal below the main diagonal contains F^2, etc. The Boltzmann factors now appear with the transition probabilities.

When the matrix of modified transition probabilities operates on an equilibrium (Boltzmann) distribution of states,

$$\begin{pmatrix} n_0 \\ n_1 \\ n_2 \end{pmatrix} = \begin{pmatrix} 1 \\ F \\ F^2 \end{pmatrix} \tag{6.5.17}$$

it gives row equations like

$$-(p_{10}F+p_{20}F^2)(1)+p_{01}F+p_{02}F^2=0 \tag{6.5.18}$$

which is satisfied for systems which obey the condition of microscopic reversibility,

$$p_{10}=p_{01} \qquad p_{20}=p_{02} \tag{6.5.19}$$

Thus, both detailed balance and microscopic reversibility are required for consistent results.

The introduction of Boltzmann factors in the master equation produces a nonsymmetric matrix. All the factors containing F lie on and below the diagonal. However, the eigenvalues, i.e., the decay constants, are real. This is proved by showing that the nonsymmetric matrix can be similarity transformed to a symmetric matrix with real eigenvalues. These tridiagonal matrices can be transformed to symmetric matrices using techniques introduced in Chapter 4. The 2×2 matrix

$$zp_{01}\begin{pmatrix} -F & 1 \\ F & -1 \end{pmatrix} \tag{6.5.20}$$

is tridiagonal with $b=1$ and $c=F$. The similarity transform matrix $\mathbf{D}$ is diagonal with elements

$$1, \sqrt{c/b}, c/b, \ldots \tag{6.5.21}$$

The inverse matrix, $\mathbf{D}^{-1}$, is diagonal with elements

$$1, \sqrt{b/c}, b/c, \ldots \tag{6.5.22}$$

For the two-level transform, the matrices are then

$$\mathbf{D}=\begin{pmatrix} 1 & 0 \\ 0 & F^{1/2} \end{pmatrix} \tag{6.5.23}$$

and

$$\mathbf{D}^{-1}=\begin{pmatrix} 1 & 0 \\ 0 & F^{-1/2} \end{pmatrix} \tag{6.5.24}$$

The matrix transforms to

$$\begin{aligned} \mathbf{D}^{-1}\mathbf{Z}\mathbf{D} &= \begin{pmatrix} 1 & 0 \\ 0 & F \end{pmatrix}\begin{pmatrix} -F & 1 \\ F & 1 \end{pmatrix}\begin{pmatrix} 1 & 0 \\ 0 & F \end{pmatrix} \\ &= \begin{pmatrix} -F & F^{1/2} \\ F^{1/2} & 1 \end{pmatrix} \end{aligned} \tag{6.5.25}$$

The diagonal elements are unchanged and the two off-diagonal elements are equal. The eigenvalues for this symmetric matrix are real. The method works for any system which obeys detailed balance between pairs of states. The rate constant matrix for a reversible isomerization of A and B,

$$\begin{pmatrix} -k_f & k_r \\ +k_f & -k_r \end{pmatrix} \tag{6.5.26}$$

will have concentrations A^{eq} and B^{eq} at equilibrium. The **D** matrices are

$$\mathbf{D} = \begin{pmatrix} 1 & 0 \\ 0 & (B^{eq}/A^{eq})^{1/2} \end{pmatrix} \tag{6.5.27}$$

$$\mathbf{D}^{-1} = \begin{pmatrix} 1 & 0 \\ 0 & (A^{eq}/B^{eq})^{1/2} \end{pmatrix} \tag{6.5.28}$$

The ratio $(B^{eq}/A^{eq})^{1/2}$ is selected for row 2 in the **D** matrix since the rate constant k_r must multiply B for detailed balance.

The similarity transform is

$$\begin{pmatrix} 1 & 0 \\ 0 & (A^{eq}/B^{eq})^{1/2} \end{pmatrix} \begin{pmatrix} -k_f & k_r \\ k_f & -k_r \end{pmatrix} \begin{pmatrix} 1 & 0 \\ 0 & (B^{eq}/A^{eq})^{1/2} \end{pmatrix}$$
$$= \begin{pmatrix} -k_f & k_r(B^{eq}/A^{eq})^{1/2} \\ k_f(A^{eq}/B^{eq})^{1/2} & -k_r \end{pmatrix} \tag{6.5.29}$$

The two off-diagonal components are equal via the condition of detailed balance:

$$\begin{aligned} k_f(A^{eq}/B^{eq})^{1/2} &= k_r(B^{eq}/A^{eq})^{1/2} \\ k_f A^{eq} &= k_r B^{eq} \end{aligned} \tag{6.5.30}$$

The eigenvalues are real for both the symmetric matrix and the original nonsymmetric **K** matrix.

The **D** matrices verify that the original matrix has real eigenvalues. Of course, the similarity transform is useful in situations where analysis of the symmetric matrix is easier. The eigenvectors for the symmetric matrix could be found and then transformed to the row and column vectors for the nonsymmetric matrix using the transforms **D** and $\mathbf{D}^{-1}$.

The symmetrization technique works for any **W** matrix when detailed balance holds. The 3×3 matrix

$$\begin{pmatrix} -(p_{10}F + p_{20}F^2) & p_{01} & p_{02} \\ p_{10}F & -(p_{01} + p_{21}F) & p_{12} \\ p_{20}F^2 & p_{21}F & -(p_{02} + p_{12}) \end{pmatrix} \tag{6.5.31}$$

has a **D** matrix with the diagonal elements $(1, F^{1/2}, F)$. The $\mathbf{D}^{-1}$ matrix has the reciprocals of these elements. The transform

$$\mathbf{D}^{-1}\mathbf{W}\mathbf{D} = z\begin{pmatrix} 1 & 0 & 0 \\ 0 & F^{-1/2} & 0 \\ 0 & 0 & F^{-1} \end{pmatrix}\begin{pmatrix} -(p_{10}F + p_{20}F^2) & p_{01} & p_{02} \\ p_{10}F & -(p_{01} + p_{21}F) & p_{12} \\ p_{20}F^2 & p_{21}F & -(p_{02} + p_{12}) \end{pmatrix}$$

$$\times\begin{pmatrix} 1 & 0 & 0 \\ 0 & F^{1/2} & 0 \\ 0 & 0 & F \end{pmatrix}$$

$$= z\begin{pmatrix} -(p_{10}F + p_{20}F^2) & p_{01}F^{1/2} & p_{02}F \\ p_{10}F^{+1/2} & -(p_{01} + p_{21}F) & p_{12}F^{1/2} \\ p_{20}F & p_{21}F^{1/2} & -(p_{02} + p_{12}) \end{pmatrix} \tag{6.5.32}$$

generates a symmetric matrix for any transition probability matrix where the pairs of states, e.g., i and j, are related via detailed balanced at equilibrium.

6.6. The Wegscheider Conditions and Cyclic Reactions

Reaction kinetics generally proceed in a unique direction. For a given concentration of reactant, the system will evolve to a stable concentration vector, the equilibrium eigenvector. However, there are some interesting kinetic situations in which the ultimate product of the reaction regenerates the starting reactant.

The reaction scheme of Section 6.3,

$$A \rightleftharpoons B \rightleftharpoons C$$

can be converted to a cyclic reaction scheme by introducing an additional reaction pathway from product C to reactant A, as shown in Figure 6.11. Because complete eigenvalue–eigenvector analysis for this cyclic system is quite involved, analysis will begin with the irreversible case. Forward reactions from A to B, B

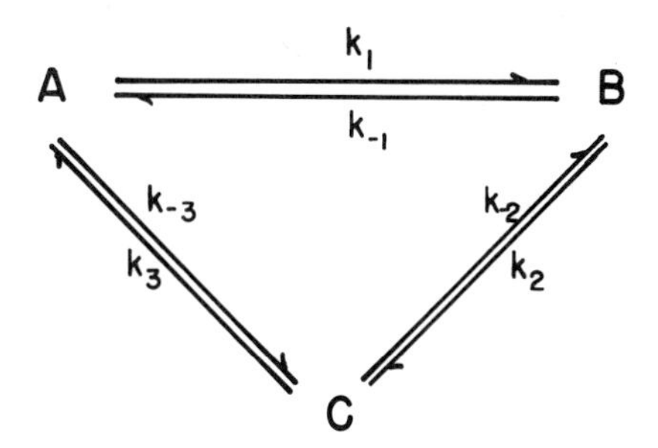

Figure 6.11. Cyclic reaction scheme involving forward (k_i) and reverse (k_{-i}) rate constants.

to C, and C to A with rate constants k_1, k_2, and k_3, respectively, are allowed (Figure 6.12a). The rate constant matrix **K** is

$$\begin{pmatrix} -k_1 & 0 & k_3 \\ k_1 & -k_2 & 0 \\ 0 & k_2 & -k_3 \end{pmatrix} \tag{6.6.1}$$

Conservation of total mass again yields the $\lambda = 0$ equilibrium eigenvector. This eigenvector is expressed in terms of the three rate constants, but this must be done with care. The first row generates an equilibrium relationship between [A] and [C]:

$$-k_1[\mathrm{A}] + k_3[\mathrm{C}] = 0 \tag{6.6.2}$$

The solution $[\mathrm{A}] = k_3$, $[\mathrm{C}] = k_1$ satisfies Equation 6.6.2:

$$-k_1(k_3) + k_3(k_1) = 0 \tag{6.6.3}$$

However, the equation generated by the third row of the matrix gives a different solution for [C]:

$$k_2[\mathrm{B}] - k_3[\mathrm{C}] = 0 \tag{6.6.4}$$

and

$$[\mathrm{B}] = k_3; \qquad [\mathrm{C}] = k_2 \tag{6.6.5}$$

The concentration of C is proportional to k_1 in one case and k_2 in the second.

The eigenvector components give the relative populations for each of the states of the system. The results generated by each of the two row equations are satisfied by a product of both solutions:

$$[\mathrm{C}] = k_1 k_2 \tag{6.6.6}$$

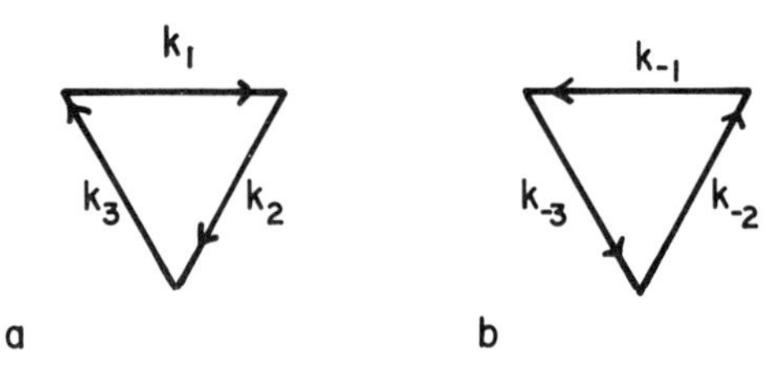

Figure 6.12. Irreversible cyclic reaction scheme with forward (a) and reverse (b) rate constants.

This result can be substituted into the first-row equation (Equation 6.6.2) to find the concentration [A]:

$$-k_1[A]+k_3[C]= -k_1[A]+k_3k_1k_2=0 \tag{6.6.7}$$

$$[A]=k_2k_3 \tag{6.6.8}$$

The third-row equation gives

$$[B]=k_1k_3 \tag{6.6.9}$$

The equilibrium eigenvector is

$$\begin{pmatrix} k_2k_3 \\ k_1k_3 \\ k_1k_2 \end{pmatrix} \tag{6.6.10}$$

The **K** matrix operates on this vector to produce the zero eigenvalue.

It may be noted that there are

$$3!/2!1!=3 \tag{6.6.11}$$

ways to arrange the rate constants in groups of two, and these are the components of the vector.

Equilibrium for a noncyclic reaction scheme is characterized by two properties: the concentrations of the different species remain constant in time and the rate of change between the species is zero. For cyclic systems, the concentration of each species also becomes constant at equilibrium. However, the rate of reaction does not have to become zero. The rate of change from A to B at equilibrium is

$$\text{Rate}(A \to B)=k_1[A] \tag{6.6.12}$$

Substituting the relative component for [A] (Equation 6.6.10) gives

$$\text{Rate}(A \to B)=k_1k_2k_3 \tag{6.6.13}$$

The rate for reaction from B to C is

$$\text{Rate}(B \to C)=k_2k_1k_3 \tag{6.6.14}$$

while the final rate for C is also

$$\text{Rate}(C \to A)=k_3k_1k_2 \tag{6.6.15}$$

Each rate is the same and is equal to the product of all the rate constants in the unidirectional cycle.

There is an important difference between these unidirectional rates and the net rate of change for any species in the system. The net rate of change for A includes the steady loss to B and a steady gain from C, i.e.,

$$d[\mathrm{A}]/dt = -k_1[\mathrm{A}] + k_3[\mathrm{C}] \tag{6.6.16}$$

The net rate of change of [A] is zero but this is maintained by the steady flow of molecules through A. The rate of which A is generated, $k_3[\mathrm{C}]$, is the same as the rate, $k_1[\mathrm{A}]$, at which it is lost to B.

This analysis is repeated for the irreversible kinetic scheme shown in Figure 6.12b. In this case, only reverse reaction rate constants are finite and reaction proceeds in a counterclockwise direction. The net rate from B to A in this case is

$$\mathrm{Rate}(\mathrm{B} \rightarrow \mathrm{A}) = k_{-1}k_{-2}k_{-3} \tag{6.6.17}$$

The two irreversible reaction schemes provide net rates from A to B and B to A, respectively. When both sets of rate constants are present in a reversible system, equilibrium is still defined as a zero net rate. If the rate from A to B is exactly equal to the rate from B to A, the net rate is zero since there are two cyclic flows in opposite directions. Equating equal and opposite rates at equilibrium gives

$$k_1k_2k_3 = k_{-1}k_{-2}k_{-3} \tag{6.6.18}$$

This is a Wegscheider condition; the product of all the forward rate constants must equal the product of all the reverse rate constants in a cyclic system.

The Wegscheider condition is more global than the principle of detailed balance. The Wegscheider condition results from unidirectional flows which involve all the species in the cycle. The principle of detailed balance at equilibrium requires equal rates between each pair of species. The Wegscheider condition for the cyclic system is

$$k_1k_2k_3/k_{-1}k_{-2}k_{-3} = 1 \tag{6.6.19}$$

Three separate conditions are required for detailed balance at equilibrium:

$$k_1[\mathrm{A}] = k_{-1}[\mathrm{B}] \tag{6.6.20}$$

$$k_2[\mathrm{B}] = k_{-2}[\mathrm{C}] \tag{6.6.21}$$

$$k_3[\mathrm{C}] = k_{-3}[\mathrm{A}] \tag{6.6.22}$$

These equalities define equilibrium concentration ratios for A, B, and C:

$$k_1/k_{-1} = B^{eq}/A^{eq}$$
$$k_2/k_{-2} = C^{eq}/B^{eq} \quad (6.6.23)$$
$$k_3/k_{-3} = A^{eq}/C^{eq}$$

The three independent detailed balance ratios are combined to generate the single Wegscheider condition for the reaction scheme:

$$(k_1/k_{-1})(k_2/k_{-2})(k_3/k_{-3}) = (B^{eq}/A^{eq})(C^{eq}/B^{eq})(A^{eq}/B^{eq}) = 1 \quad (6.6.24)$$

Detailed balance at equilibrium is a more restrictive condition than the Wegscheider condition. A and B will obey detailed balance even if C is not present. The Wegscheider condition requires the presence of all three species.

The Wegscheider and detailed balance criteria are used for equilibrium situations. To illustrate the time dependence of cyclic kinetics, the rate constants for the irreversible system are equated to unity to give the following **K** matrix:

$$\begin{pmatrix} -1 & 0 & 1 \\ 1 & -1 & 0 \\ 0 & 1 & -1 \end{pmatrix} \quad (6.6.25)$$

The system has a zero eigenvalue, and the characteristic equation

$$(\lambda)(\lambda^2 + 3\lambda + 3) = 0 \quad (6.6.26)$$

yields the eigenvalues

$$0, \quad \frac{-3}{2} + i\frac{\sqrt{3}}{2}, \quad \frac{-3}{2} - i\frac{\sqrt{3}}{2} \quad (6.6.27)$$

The eigenvalues have imaginary components for this nonsymmetric matrix.

The $\lambda_0 = 0$ eigenvector is

$$(1 \quad 1 \quad 1) \quad (6.6.28)$$

since the rate constants favor all states equally at equilibrium.

The projection operator approach is preferred for this nonsymmetric matrix. The projection operators are

$$\mathbf{P}\left(\frac{-3}{2}+i\frac{\sqrt{3}}{2}\right)$$

$$=\left[\mathbf{K}-\left(\frac{-3}{2}-i\frac{\sqrt{3}}{2}\right)\right](\mathbf{K}-0)\Big/\left(\frac{-3}{2}+i\frac{\sqrt{3}}{2}+\frac{3}{2}+i\frac{\sqrt{3}}{2}\right)$$

$$\times\left(\frac{-3}{2}+i\frac{\sqrt{3}}{2}\right)$$

$$=\left[(i\sqrt{3})\left(\frac{-3}{2}+i\frac{3}{2}\right)\right]^{-1/2}\begin{pmatrix}\frac{1}{2}+i\frac{\sqrt{3}}{2} & 0 & 1\\ 1 & \frac{1}{2}+i\frac{\sqrt{3}}{2} & 0\\ 0 & 1 & \frac{1}{2}+i\frac{\sqrt{3}}{2}\end{pmatrix}$$

$$\times\begin{pmatrix}-1 & 0 & 1\\ 1 & -1 & 0\\ 0 & 1 & -1\end{pmatrix}$$

$$=\left[(3)\left(\frac{-1}{2}-i\frac{\sqrt{3}}{2}\right)\right]^{-1/2}\begin{pmatrix}\frac{-1}{2}-i\frac{\sqrt{3}}{2} & 1 & \frac{-1}{2}+i\frac{\sqrt{3}}{2}\\ \frac{-1}{2}+i\frac{\sqrt{3}}{2} & \frac{-1}{2}-i\frac{\sqrt{3}}{2} & 1\\ 1 & \frac{-1}{2}+i\frac{\sqrt{3}}{2} & \frac{-1}{2}-i\frac{\sqrt{3}}{2}\end{pmatrix}$$

(6.6.29)

and

$$\mathbf{P}\left(\frac{-3}{2}-i\frac{\sqrt{3}}{2}\right)$$

$$=\left[\mathbf{K}-\left(\frac{-3}{2}+i\frac{\sqrt{3}}{2}\right)\right](\mathbf{K}-0)\Big/\left(\frac{-3}{2}-i\frac{\sqrt{3}}{2}+\frac{3}{2}-i\frac{\sqrt{3}}{2}\right)$$

$$\times\left(\frac{-3}{2}-i\frac{\sqrt{3}}{2}-0\right)$$

$$=\left[(-i\sqrt{3})\left(\frac{-3}{2}-i\frac{\sqrt{3}}{2}\right)\right]^{-1/2}\begin{pmatrix}\frac{1}{2}-i\frac{\sqrt{3}}{2} & 0 & 1\\ 1 & \frac{1}{2}-i\frac{\sqrt{3}}{2} & 0\\ 0 & 1 & \frac{1}{2}-i\frac{\sqrt{3}}{2}\end{pmatrix}$$

$$\times\begin{pmatrix}-1 & 0 & 1\\ 1 & -1 & 0\\ 0 & 1 & -1\end{pmatrix}$$

$$=\left[(3)\left(\frac{-1}{2}+\frac{\sqrt{3}}{2}\right)\right]^{-1/2}\begin{pmatrix}\frac{-1}{2}+i\frac{\sqrt{3}}{2} & 1 & \frac{-1}{2}-i\frac{\sqrt{3}}{2}\\ \frac{-1}{2}-i\frac{\sqrt{3}}{2} & \frac{-1}{2}+i\frac{\sqrt{3}}{2} & 1\\ 1 & \frac{-1}{2}-i\frac{\sqrt{3}}{2} & \frac{-1}{2}+i\frac{\sqrt{3}}{2}\end{pmatrix} \tag{6.6.30}$$

Since the eigenvalue–eigenvector solution is

$$|c\rangle=\sum \exp(-\lambda_i t)\,\mathbf{P}_i\,|c_0\rangle \tag{6.6.31}$$

the projection operator must act on the initial component vector. If all molecules are A isomers at $t=0$, the initial vector components are

$$[\mathrm{A}(0)]=A_0, \qquad [\mathrm{B}(0)]=[\mathrm{C}(0)]=0 \tag{6.6.32}$$

The projections $\mathbf{P}_+\,|c_0\rangle$ and $\mathbf{P}_-\,|c_0\rangle$ for the respective operators are

$$\mathbf{P}_+\,|c_0\rangle=\begin{pmatrix}\frac{-1}{2}-i\frac{\sqrt{3}}{2}\\ \frac{-1}{2}+i\frac{\sqrt{3}}{2}\\ 1\end{pmatrix}\left[3\left(\frac{-1}{2}-i\frac{\sqrt{3}}{2}\right)\right]^{-1} \tag{6.6.33}$$

and

$$\mathbf{P}_{-}\,|c_0\rangle = \begin{pmatrix} \dfrac{-1}{2} + i\dfrac{\sqrt{3}}{2} \\ \dfrac{-1}{2} - i\dfrac{\sqrt{3}}{2} \\ 1 \end{pmatrix} \left[3\left(\frac{-1}{2} + i\frac{\sqrt{3}}{2}\right)\right]^{-1} \tag{6.6.34}$$

The vector components

$$\frac{-1}{2} + i\frac{\sqrt{3}}{2} \quad \text{or} \quad \frac{-1}{2} - i\frac{\sqrt{3}}{2} \tag{6.6.35}$$

define angles in the complex plane (Figure 6.13). DeMoirve's theorem,

$$\exp(i\theta) = \cos\theta + i\sin\theta \tag{6.6.36}$$

gives the equalities

$$\exp(i120°) = \cos 120° + i\sin 120° = \frac{-1}{2} + i\frac{\sqrt{3}}{2} \tag{6.6.37}$$

$$\exp(i240°) = \cos 240° + i\sin 240° = \frac{-1}{2} - i\frac{\sqrt{3}}{2} \tag{6.6.38}$$

$$\exp(i0) = \cos 0 + i\sin 0 = 1 \tag{6.6.39}$$

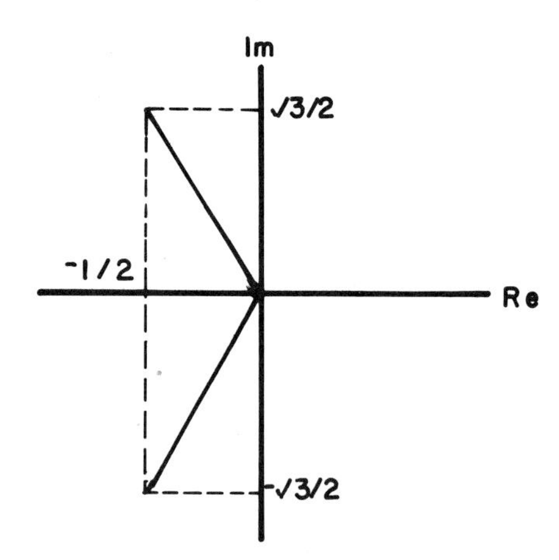

Figure 6.13. Geometry of the vectors $\frac{-1}{2} - i\frac{\sqrt{3}}{2}$ and $\frac{-1}{2} + i\frac{\sqrt{3}}{2}$ in the complex plane, showing associated angles of 240° and 120°.

The vector projections $\mathbf{P}_+ |c_0\rangle$ and $\mathbf{P}_- |c_0\rangle$ are

$$\mathbf{P}_+ |c\rangle = \exp(-i240°)/3 \begin{pmatrix} \exp(i240°) \\ \exp(i120°) \\ \exp(i0) \end{pmatrix} = \left(\frac{1}{3}\right) \begin{pmatrix} \exp(-i0) \\ \exp(-i120°) \\ \exp(-i240°) \end{pmatrix} \tag{6.6.40}$$

and

$$\mathbf{P}_- |c\rangle = \exp(-i120°)/3 \begin{pmatrix} \exp(i120°) \\ \exp(i240°) \\ \exp(i0) \end{pmatrix} = \left(\frac{1}{3}\right) \begin{pmatrix} \exp(+i0) \\ \exp(i120°) \\ \exp(i240°) \end{pmatrix} \tag{6.6.41}$$

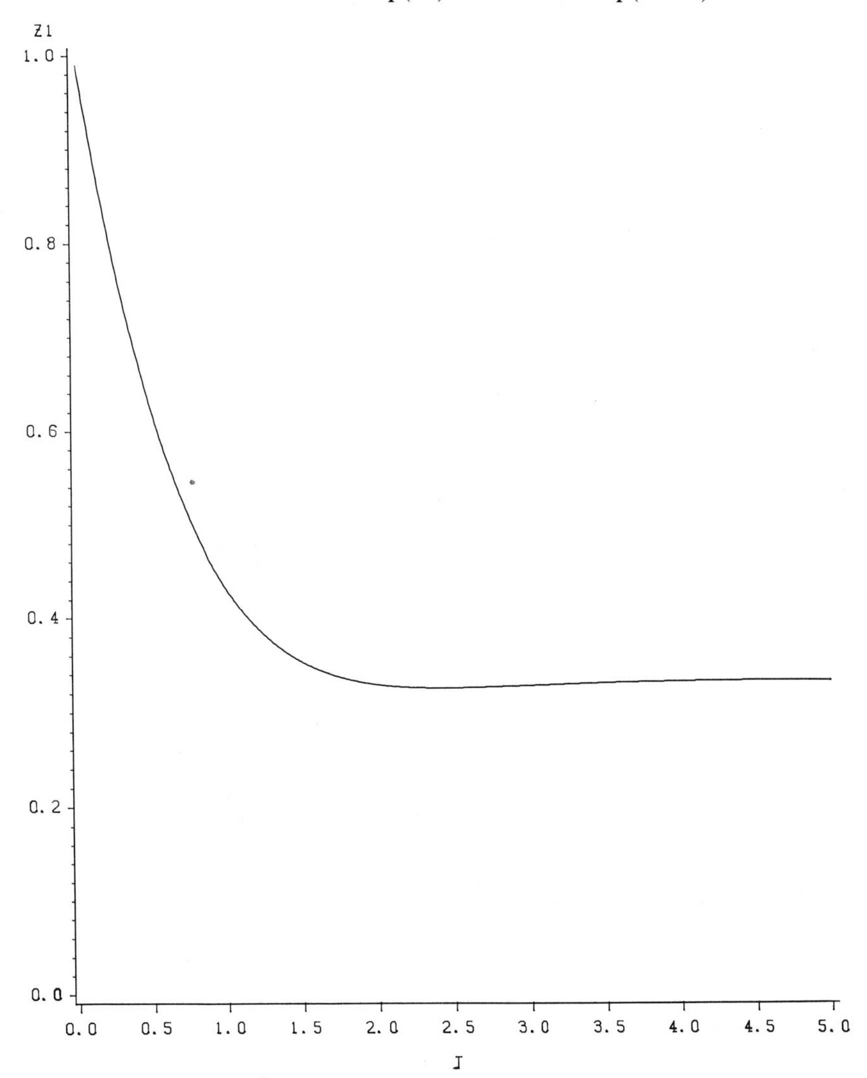

Figure 6.14. Evolution of the oscillating A concentration component with time.

The leading exponentials in each case are generated from the denominators of Equations 6.6.33 and 6.6.34. For $\mathbf{P}_+ \, |c_0\rangle$, the phase increases sequentially by 120° for each component. The eigenvectors describe a phase relation between the components just as (1, −1) describes two components which are 180° out of phase. For $\mathbf{P}_- \, |c_0\rangle$, the phase decreases sequentially by 120°.

The projection vectors are combined with the appropriate exponential decays to generate the kinetic evolution of the components. The exponential decay will contain real and imaginary time-dependent components:

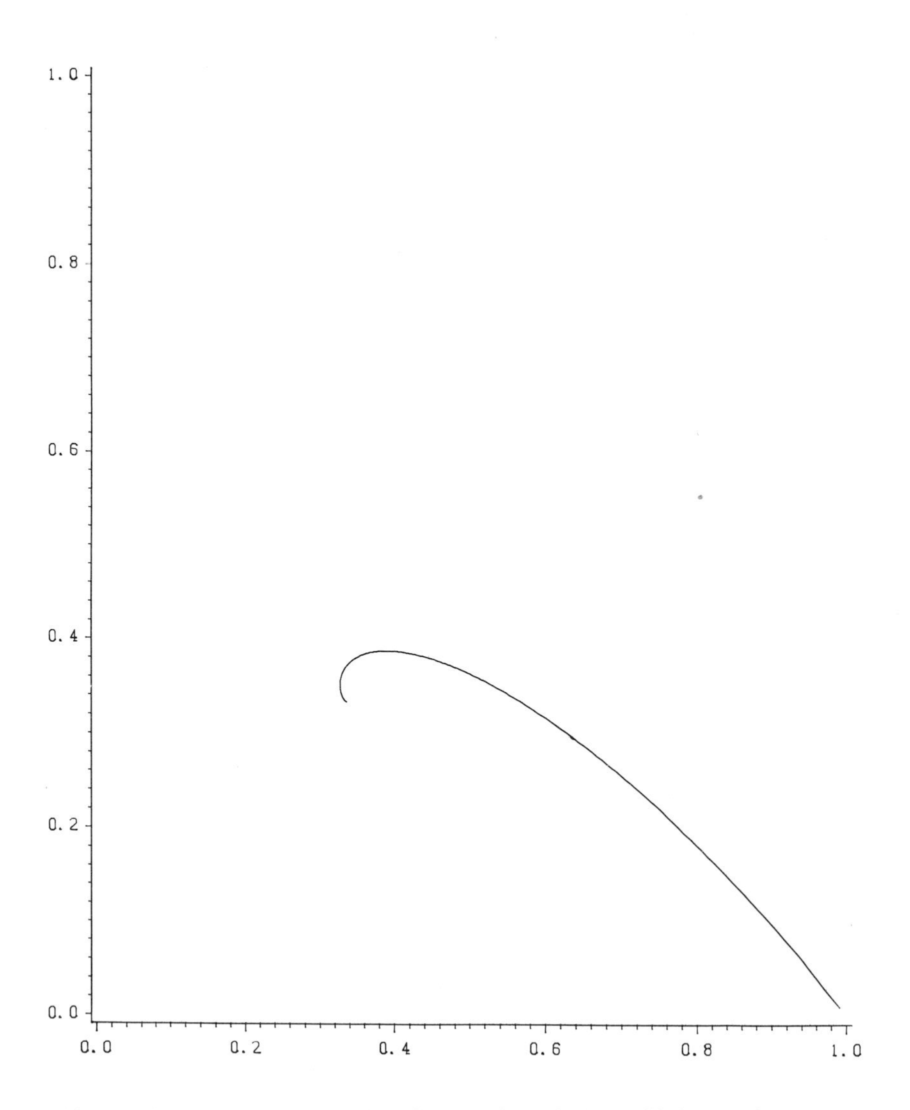

Figure 6.15. The temporal evolution of [A] and [B] in the oscillating reaction system.

$$\begin{pmatrix} A \\ B \\ C \end{pmatrix} = \begin{pmatrix} \frac{1}{3} \\ \frac{1}{3} \\ \frac{1}{3} \end{pmatrix} + \exp(-3t/2)\exp(i\sqrt{3}t/2)\begin{pmatrix} \exp(-i0) \\ \exp(-i120^\circ) \\ \exp(-i240^\circ) \end{pmatrix}$$

$$+\exp(-i\sqrt{3}t/2)\begin{pmatrix} \exp(i0) \\ \exp(i120^\circ) \\ \exp(i240^\circ) \end{pmatrix} \tag{6.6.42}$$

The imaginary exponentials for each component can now be combined and reduced. The factors in each row become

$$\exp(+i\sqrt{3}t/2) + \exp(-i\sqrt{3}t/2) = 2\cos(\sqrt{3}t/2) \tag{6.6.43}$$

$$\exp[i(\sqrt{3}t/2 - 120^\circ)] + \exp[-i(\sqrt{3}t/2 - 120^\circ)] = 2\cos(\sqrt{3}t/2 - 120^\circ) \tag{6.6.44}$$

$$\exp[i(\sqrt{3}t/2 - 240^\circ)] + \exp[-i(\sqrt{3}t/2 - 240^\circ)] = 2\cos(\sqrt{3}t/2 - 240^\circ) \tag{6.6.45}$$

The complete solution for the unidirectional kinetic system is

$$\begin{pmatrix} A \\ B \\ C \end{pmatrix} = \begin{pmatrix} \frac{1}{3} \\ \frac{1}{3} \\ \frac{1}{3} \end{pmatrix} + \frac{2}{3}\exp(-3t/2)\begin{pmatrix} \cos(\sqrt{3}t/2) \\ \cos(\sqrt{3}t/2 - 120^\circ) \\ \cos(\sqrt{3}t/2 - 240^\circ) \end{pmatrix}$$

$$= \begin{pmatrix} \frac{1}{3} \\ \frac{1}{3} \\ \frac{1}{3} \end{pmatrix} + \frac{2}{3}\exp(-3t/2)\begin{pmatrix} \cos\omega t \\ \cos(\omega t - 120^\circ) \\ \cos(\omega t - 240^\circ) \end{pmatrix} \tag{6.6.46}$$

The time-dependent term for this equation contains two factors. The common exponential decay will decay to zero, leaving only the equilibrium eigenvector. However, during this decay, the system will move cyclically from A to B to C to A. For example, if the concentration of A is maximal at $t = 0$, the concentration of B will reach its maximum when $\omega t = 120^\circ$, that of C will reach its maximum when $\omega t = 240°$, etc. Because of the exponential decay, the size of these concentration oscillations will decrease with each successive cycle. The decay of A with time is shown in Figure 6.14 while the oscillatory evolution of the system to equilibrium with time is shown on a [B]–[A] concentration plot in Figure 6.15.

6.7. Graph Theory in Kinetics

Although the solutions to sets of coupled kinetic equations are found using eigenvalue–eigenvector techniques, analytical solutions are limited to matrices which give tractable eigenvalues and eigenvectors. Such systems yield a complete analytical solution for arbitrary initial conditions. For the more complicated systems, it can be advantageous to sacrifice some information and use a simpler technique. In this section, equilibrium eigenvectors for complicated systems are determined using a graph theory approach.

The graph technique for equilibrium eigenvectors is first developed for simple systems such as the irreversible isomerization

$$A \xrightarrow{k_1} B$$

with rate constant matrix

$$\begin{pmatrix} -k_1 & 0 \\ k_1 & 0 \end{pmatrix} \tag{6.7.1}$$

The projection operator for the $\lambda = 0$ eigenvalue is

$$\mathbf{P}(\lambda = 0) = \begin{pmatrix} 0 & 0 \\ k_1 & k_1 \end{pmatrix} (1/k_1) \tag{6.7.2}$$

The matrix format demonstrates two characteristics which are present in more complicated systems. Only the elements of the bottom row are nonzero because B is a trapping state. The equilibrium bra vector has components of 1, and this produces the two identical columns for **P**. Any initial concentrations are summed by the projection matrix to give the final concentration of C at equilibrium. For an arbitrary initial vector with components A_0 and B_0, the equilibrium vector is

$$\mathbf{P}(\lambda = 0) \begin{pmatrix} A_0 \\ B_0 \end{pmatrix} = \begin{pmatrix} 0 & 0 \\ 1 & 1 \end{pmatrix} \begin{pmatrix} A_0 \\ B_0 \end{pmatrix} = \begin{pmatrix} 0 \\ A_0 + B_0 \end{pmatrix} \tag{6.7.3}$$

The rate constant k_1 which appears in the bottom row is associated with the reaction which leads to the state B. The rate constant leading to A is zero and the equilibrium concentration of A is zero. The rate constants dictate the final concentrations of A and B.

For two sequential irreversible reactions,

$$A \xrightarrow{k_1} B \xrightarrow{k_2} C$$

the **K** matrix is

$$\begin{pmatrix} -k_1 & 0 & 0 \\ k_1 & -k_2 & 0 \\ 0 & k_2 & 0 \end{pmatrix} \tag{6.7.4}$$

The projection operator for $\lambda = 0$ is determined as the product of two matrices so the final matrix contains products of rate constants:

$$\begin{aligned} \mathbf{P}(\lambda=0) &= (\mathbf{K}+k_1\mathbf{I})(\mathbf{K}+k_2\mathbf{I})/(0+k_1)(0+k_2) \\ &= \begin{pmatrix} 0 & 0 & 0 \\ k_1 & -k_2+k_1 & 0 \\ 0 & k_2 & k_1 \end{pmatrix} \begin{pmatrix} -k_1+k_2 & 0 & 0 \\ k_1 & 0 & 0 \\ 0 & k_2 & k_2 \end{pmatrix} (k_1k_2)^{-1} \\ &= \begin{pmatrix} 0 & 0 & 0 \\ 0 & 0 & 0 \\ k_1k_2 & k_1k_2 & k_1k_2 \end{pmatrix} (k_1k_2)^{-1} = \begin{pmatrix} 0 & 0 & 0 \\ 0 & 0 & 0 \\ 1 & 1 & 1 \end{pmatrix} \end{aligned} \tag{6.7.5}$$

The nonzero elements occupy the bottom row for this irreversible process. In this case, the elements are products of the rate constants for the two reactions required to place all molecules in the trapping state. The "1"s in the bottom row sum all the initial concentrations to give the final total concentration of C.

$$f_C = f_A(0) + f_B(0) + f_C(0) = 1 \tag{6.7.6}$$

One new rate constant is added per irreversible reaction since each new reaction will add one matrix to the Lagrange–Sylvester expression for the equilibrium projection operator (Problem 6.6).

The pattern is more complicated for reversible reactions. For the one-step reversible isomerization

$$\mathrm{A} \underset{k_{-1}}{\overset{k_1}{\rightleftharpoons}} \mathrm{B}$$

the rate constant matrix is

$$\begin{pmatrix} -k_1 & k_{-1} \\ k_1 & -k_{-1} \end{pmatrix} \tag{6.7.7}$$

with eigenvalues 0 and $-(k_1 + k_{-1})$. The projection operator for $\lambda = 0$ involves a single matrix and is

$$\begin{pmatrix} k_{-1} & k_{-1} \\ k_1 & k_1 \end{pmatrix} (k_1 + k_{-1})^{-1} \tag{6.7.8}$$

The elements in a given row are equal to permit summation of the initial concentrations. However, each row contains different elements. The first row contains k_{-1}, which is the rate constant directed to species A. The second row contains k_1, which is the rate constant directed to species B. Since B is no longer a trapping state, the denominator, which equals the sum in all column elements, produces fractional values for each row:

$$f_{\mathrm{A}} = k_{-1}/(k_1 + k_{-1}) \tag{6.7.9}$$

$$f_{\mathrm{B}} = k_1/(k_1 + k_{-1}) \tag{6.7.10}$$

These are the fractional concentrations determined in Section 6.1. They now reveal a pattern. If a reaction leads to a species, the rate constant for the reaction dictates the equilibrium fraction of that species.

The pattern is extended for two sequential reversible reactions,

$$\mathrm{A} \underset{k_{-1}}{\overset{k_1}{\rightleftharpoons}} \mathrm{B} \underset{k_{-2}}{\overset{k_2}{\rightleftharpoons}} \mathrm{C}$$

The rate constant matrix is

$$\begin{pmatrix} -k_1 & k_{-1} & 0 \\ k_1 & -k_{-1}-k_2 & k_{-2} \\ 0 & k_2 & -k_{-2} \end{pmatrix} \tag{6.7.11}$$

The $\mathbf{P}(\lambda = 0)$ operator requires the three eigenvalues for the matrix. These were determined in Section 6.3 as

$$\lambda = 0 \tag{6.7.12}$$

$$\lambda_+ = -d/2 + (d^2 - 4s)^{1/2} \tag{6.7.13}$$

$$\lambda_- = -d/2 - (d^2 - 4s)^{1/2} \tag{6.7.14}$$

where

$$d = -(k_1 + k_{-1} + k_2 + k_{-2}) \tag{6.7.15}$$

$$s = k_1 k_2 + k_{-1} k_{-2} + k_1 k_{-2} \tag{6.7.16}$$

Although the projection operator for $\lambda = 0$ will involve these complicated eigenvalues, the entire problem can be simplified before the matrices are multiplied. Define

$$A = d/2 \quad \text{and} \quad B = (d^2 - 4s)^{1/2}/2 \tag{6.7.17}$$

for the Lagrange–Sylvester formula for $\mathbf{P}(\lambda = 0)$:

$$(\mathbf{K} - \mathbf{I}\lambda_+)(\mathbf{K} - \mathbf{I}\lambda_-)/(0 - \lambda_+)(0 - \lambda_-) \tag{6.7.18}$$

The denominator is the product

$$\begin{aligned} \lambda_+ \lambda_- &= (-A + B)(-A - B) = A^2 - B^2 \\ &= (d^2/4) - (d^2/4 - s) = s \end{aligned} \tag{6.7.19}$$

The terms in the numerator can be formally multiplied to give

$$\mathbf{K}^2-(\lambda_+ +\lambda_-)\mathbf{K}+\lambda_+\lambda_-\mathbf{I} \tag{6.7.20}$$

$$=\mathbf{K}^2-2A\mathbf{K}+(\mathbf{A}^2-\mathbf{B}^2)\mathbf{I} \tag{6.7.21}$$

The projection operator is

$$\mathbf{P}(\lambda=0)=(\mathbf{K}^2-d\mathbf{K}+s\mathbf{I})/s \tag{6.7.22}$$

Evaluation of this projection operator is still complicated by the $\mathbf{K}^2$ term. However, in each previous sequential reaction example, the elements of each row of the $\mathbf{P}(\lambda=0)$ projection operator were identical. Determination of a single column of the projection matrix gives the rows for the full projection operator. To determine a single column, the projection operator operates on an initial concentration vector with components

$$(1, 0, 0)$$

to give

$$[(\mathbf{K}^2-d\mathbf{K}+s\mathbf{I})/s]\;|c\rangle=\begin{pmatrix} k_{-1}k_{-2}/s \\ k_1k_{-2}/s \\ k_1k_2/s \end{pmatrix} \tag{6.7.23}$$

The projection matrix is

$$\begin{pmatrix} k_{-1}k_{-2} & k_{-1}k_{-2} & k_{-1}k_{-2} \\ k_1k_{-2} & k_1k_{-2} & k_1k_{-2} \\ k_1k_2 & k_1k_2 & k_1k_2 \end{pmatrix}(k_{-1}k_{-2}+k_1k_{-2}+k_1k_2)^{-1} \tag{6.7.24}$$

The denominator for the operator is the sum of all the elements in a column. The elements for a given row are products of rate constants for the reaction steps which lead to the component for that row. For example, the product $k_{-1}k_{-2}$ involves the two reaction steps which would lead all species to the component A.

These simple sequential reaction schemes have produced products proportional to the fractions of each species. For example, factor k_1k_{-2} is proportional to the fraction of B. In general, however, the fractional concentration of a component can involve several such terms. The reversible cyclic reaction scheme of Figure 6.16 was analyzed as forward and reverse cycles. Its equilibrium properties are now analyzed completely.

The rate constant matrix for this system has no zero elements:

$$\begin{pmatrix} -k_1-k_{-3} & k_{-1} & k_3 \\ k_1 & -k_{-1}-k_2 & k_{-2} \\ k_{-3} & k_2 & -k_3-k_{-2} \end{pmatrix} \tag{6.7.25}$$

The two nonzero eigenvalues for this matrix can be determined using the invariants of the matrix (Problem 6.2). The constants d and s for the characteristic equation

$$\lambda^2+d\lambda+s=0 \tag{6.7.26}$$

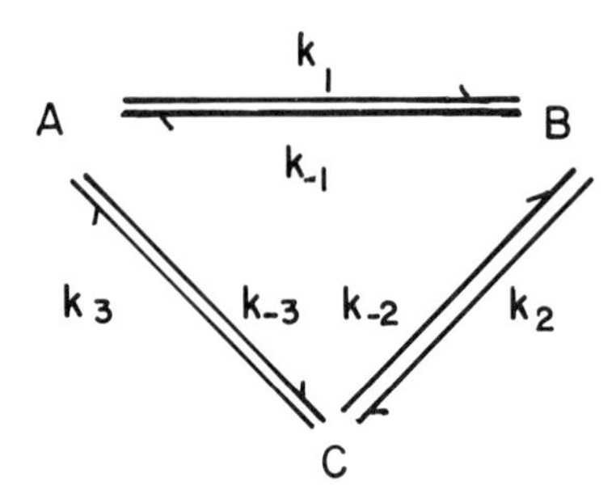

Figure 6.16. The cyclic reversible reaction scheme.

are

$$d = -(k_1 + k_{-1} + k_2 + k_{-2} + k_3 + k_{-3}) \tag{6.7.27}$$

$$s = k_1 k_2 + k_{-3} k_{-1} + k_2 k_{-3} + k_{-1} k_{-2} \\ + k_2 k_3 + k_1 k_3 + k_1 k_{-2} + k_{-3} k_{-2} \tag{6.7.28}$$

The formal projection operator for $\lambda = 0$ again has the form of a characteristic equation with a factor of $\mathbf{K}$ eliminated:

$$\mathbf{P}(\lambda = 0) = (\mathbf{K}^2 - d\mathbf{K} + s\mathbf{I})/s \tag{6.7.29}$$

The column vector generated from the initial concentration vector with components (1, 0, 0) is

$$\begin{pmatrix} k_3 k_2 + k_{-1} k_3 + k_{-1} k_{-2} \\ k_{-2} k_{-3} + k_1 k_3 + k_1 k_{-2} \\ k_1 k_2 + k_{-1} k_{-3} + k_2 k_{-3} \end{pmatrix} s^{-1} \tag{6.7.30}$$

where s is the sum of the nine products which appear in the numerator. Each element in the projection operator will consist of a sum of three pairs of rate constants. The three elements in each row will be identical.

Are the equilibrium fractions of A, B, and C consistent with the earlier observation that the rate constants in each term are associated with reaction steps which lead to the component of interest? The equilibrium concentrations were proportional to the products of two rate constants so the line graphs must involve two lines. Cycles, consisting of three connected lines, are forbidden. There are three distinct ways to arrange two lines around the triangle. This will produce three products for each species in the cycle. The vector graphs for each species are illustrated in Figure 6.17. Vectors lead to the species of interest. The denominator is the sum of all nine product pairs and is equivalent to the summation s determined from the characteristic equation.

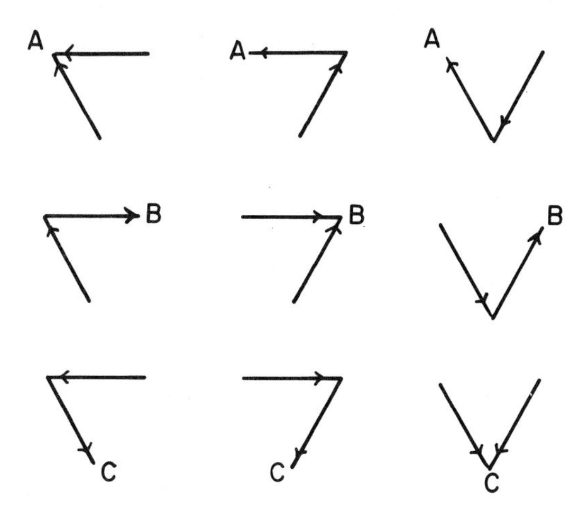

Figure 6.17. The vector graphs associated with the rate constant products which lead to components A, B, and C for the cyclic reaction.

6.8. Graphs for Kinetics

The basic graph theory of Section 6.7 can be generalized for any reaction system. The procedure is outlined in this section and used for a determination of equilibrium concentrations in a variety of multistate reaction systems.

The fractional concentrations of a two-step sequential reversible reaction are already known. The reaction sequence

$$\mathrm{A} \underset{k_{-1}}{\overset{k_1}{\rightleftharpoons}} \mathrm{B} \underset{k_{-2}}{\overset{k_2}{\rightleftharpoons}} \mathrm{C}$$

is converted to a line graph by assigning each distinct species in the system a vertex, i.e., a point on the graph. The location of these vertices is not crucial although it is often convenient to place them in their positions in the reaction scheme since communicating species will be connected. For the sequential system, the three vertices for A, B, and C are then sequentially ordered on a line:

A . . B . C

If the species are kinetically coupled, the vertices are connected by a line without arrows. Since A couples to B and B couples to C, there are two lines:

A .——— . B ——— . C

Since A and C are not coupled directly, no line connects these two vertices.

This single-line graph is converted into vector graphs for each of the three species. In each case, arrows are added to direct the lines between vertices to the molecule of interest. The directed graphs for C, B, and A, respectively, are

$$\mathrm{A}\,.\xrightarrow{k_1}.\,\mathrm{B}\xrightarrow{k_2}.\,\mathrm{C}$$

$$\mathrm{A}\,.\xrightarrow{k_1}.\,\mathrm{B}\xleftarrow{k_{-2}}.\,\mathrm{C}$$

$$\mathrm{A}\,.\xleftarrow{k_{-1}}.\,\mathrm{B}\xleftarrow{k_{-2}}.\,\mathrm{C}$$

Each vector is now labeled with the rate constant associated with that reaction step.

The directed vectors determine the equilibrium fractions of each species. The population of a species is directly proportional to the product of rate constants for the vectors which lead to that species. The three fractions are proportional to the products

$$f_{\mathrm{C}} \propto k_1 k_2 \tag{6.8.1}$$

$$f_{\mathrm{B}} \propto k_1 k_{-2} \tag{6.8.2}$$

$$f_{\mathrm{A}} \propto k_{-2} k_{-1} \tag{6.8.3}$$

Since the fractions of the three components sum to 1, the normalization is

$$\begin{aligned} f_{\mathrm{A}} + f_{\mathrm{B}} + f_{\mathrm{C}} = 1 = N(k_1 k_2 + k_1 k_{-2} + k_{-2} k_{-1}) \\ N = 1/s = (k_1 k_2 + k_1 k_{-2} + k_{-1} k_{-2})^{-1} \end{aligned} \tag{6.8.4}$$

where s is the sum of the products for all species. The fractions are

$$f_{\mathrm{A}} = k_{-1} k_{-2}/s \tag{6.8.5}$$

$$f_{\mathrm{B}} = k_1 k_{-2}/s \tag{6.8.6}$$

$$f_{\mathrm{C}} = k_1 k_2/s \tag{6.8.7}$$

The actual concentration of each species at equilibrium is found by multiplying the appropriate fraction by the total concentration of the system. For example, if the initial concentrations of A, B, and C are A_0, B_0, and C_0, respectively, the equilibrium concentration of A is

$$[\mathrm{A}] = f_{\mathrm{A}}(A_0 + B_0 + C_0) \tag{6.8.8}$$

Two additional aspects of sequential reactions occur for Michaelis–Menten enzyme kinetics. A substrate S reacts reversibly with an enzyme E to form an enzyme complex ES which then reacts to a product and the free enzyme:

$$\mathrm{E} + \mathrm{S} \underset{k_{-1}}{\overset{k_1'}{\rightleftharpoons}} \mathrm{ES} \xrightarrow{k_2} \mathrm{P} + \mathrm{E}$$

The first forward reaction step is bimolecular. Since first-order rates are required for graphical analysis, the bimolecular rate

$$k_1'[\mathrm{S}][\mathrm{E}] \tag{6.8.9}$$

is approximated as a pseudo-first-order reaction rate,

$$k_1[\mathrm{E}] = \{k_1'[\mathrm{S}]\}[\mathrm{E}] \tag{6.8.10}$$

when $[\mathrm{S}] \gg [\mathrm{E}]$. Although the reaction has two reaction steps, there are only two enzyme states: the free enzyme with concentration $[\mathrm{E}]$ and the enzyme in the $[\mathrm{ES}]$ complex. The line graph has only these two vertices:

$$\mathrm{E}\,.\!\!\frac{\qquad\quad}{}\!\!.\,\mathrm{ES}$$

The vector graph for the fraction of $[\mathrm{ES}]$ is

$$\mathrm{E}\,.\xrightarrow{\;\;k_1\;\;}.\,\mathrm{ES}$$

Two separate pathways lead from the ES complex to E. The reverse reaction has a rate constant k_{-1} while the reaction to form product has the rate constant k_2. Since each process connects the same enzyme vertices, the rate constants are added and the vector graph for E is

$$\mathrm{E}\,.\xleftarrow{\;k_{-1}+k_2\;}.\,\mathrm{ES}$$

The fractions of each state are

$$f_{\mathrm{E}} = (k_{-1} + k_2)/s \tag{6.8.11}$$

$$f_{\mathrm{ES}} = k_1/s \tag{6.8.12}$$

$$s = k_1 + (k_{-1} + k_2) \tag{6.8.13}$$

Although the fractional concentrations of E and ES remain constant in time, there will be a steady reaction of S to P. This is generally determined by the rate of formation of P,

$$\mathrm{Rate(P)} = k_2[\mathrm{ES}] \tag{6.8.14}$$

The loss of substrate S with time is assumed to have no effect on the forward pseudo-first-order rate.

Graph theory techniques are applicable to a wide variety of kinetic schemes. The membrane channel of Figure 6.18 contains two ion binding sites and has the

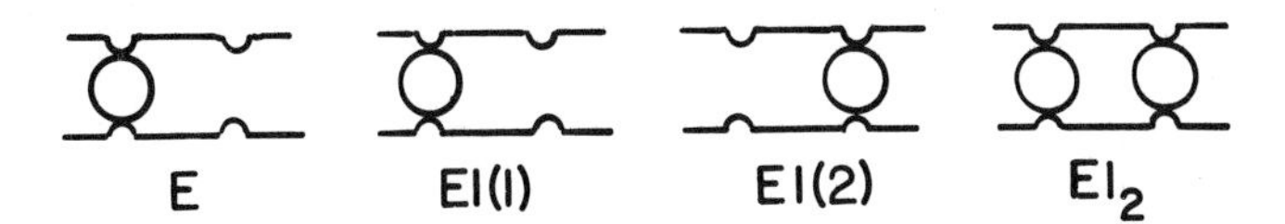

Figure 6.18. States of a membrane channel with two ion binding sites.

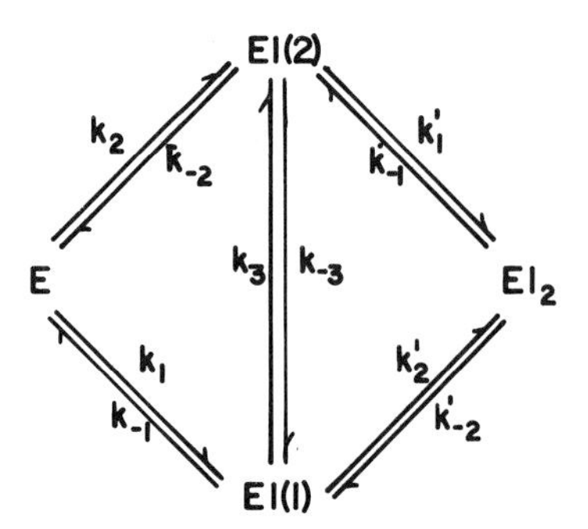

Figure 6.19. The kinetic scheme for transport in a membrane channel with two binding sites.

following possible channel states: (1) ion-free channel, E; (2) ion at channel site 1, EI(1); (3) ion at channel site 2, EI(2); (4) ions at both channel sites, EI_2. The system will have four vertices. The full kinetic scheme is shown in Figure 6.19.

The line graphs for the two-state channel kinetics (Figure 6.20) contain only three lines since one additional line must always form a cycle. The eight line graphs appear as vector graphs for each of the four state fractions. The normalization sum, s, then involves $8 \times 4 = 32$ separate terms; each is the product of three rate constants.

Great care is required to ensure that all posible graphs for the system have been found. Once the line graphs are known, the vector graphs and their associated rate constants are easily determined. Although the results can be quite complicated, they flow easily from the complete set of vector graphs.

6.9. Mean First Passage Times

In Section 6.4, the master equation was developed for kinetic transitions in a conservative system. The energy level could change as the result of a collision with a heat bath molecule but the structure of the molecule was unchanged, i.e.,

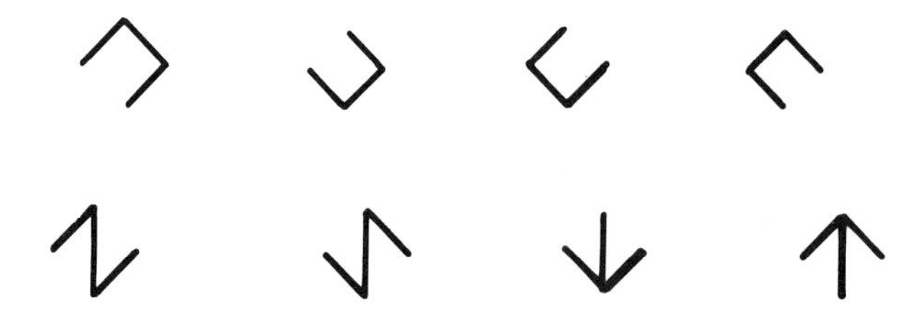

Figure 6.20. Line graphs for the two-state ion binding membrane channel.

there was no reaction to a new chemical species. The experimentalist will often be interested in some rate of reaction rather than the individual transitions within a nonreacting molecule. Therefore, it is advantageous to retain the master equation formalism but introduce additional rate processes which permit a change of molecule A to some new molecule B. Although conservation of mass for the ensemble of A molecules must be sacrificed, the approach leads to an average rate of reaction for A.

A unimolecular reaction has the rate expression

$$d[\mathrm{A}]/dt = -k_{\mathrm{uni}}[\mathrm{A}]$$

i.e., the rate of reaction depends only on the A molecule concentration. However, at the microscopic level, the A molecule collides with the heat bath molecules. Excess energy gained in a collision is normally lost on subsequent collisions. On rare occasions, the A molecule might experience a series of favorable collisions which leave it with an internal energy sufficient to rupture a molecular bond. The molecule would then undergo a "unimolecular" reaction to product. The unimolecular rate constant is dependent on all the energy transfer collisions which provided sufficient energy for reaction.

For a real molecule, the vibrational levels are anharmonic, and the energy spacing between these levels decreases with increasing quantum number. The levels eventually reach the continuum, which corresponds to dissociation. Since the energy gap between levels changes with the level, the transition probabilities are complicated for such a system.

The truncated harmonic oscillator of Figure 6.21 has a finite number of levels with well-defined transition probabilities. For example, Landau–Teller transition probabilities can be used. The truncated harmonic oscillator is terminated at some level N. However, B molecules are formed when a molecule in the Nth level makes an additional transition to an $(N+1)$th or continuum level. This nonconservative step removes the molecule from the A manifold. The total rate for such a reaction is determined by the average time required to reach the $(N+1)$th "level."

An analysis of the truncated harmonic oscillator must ultimately produce

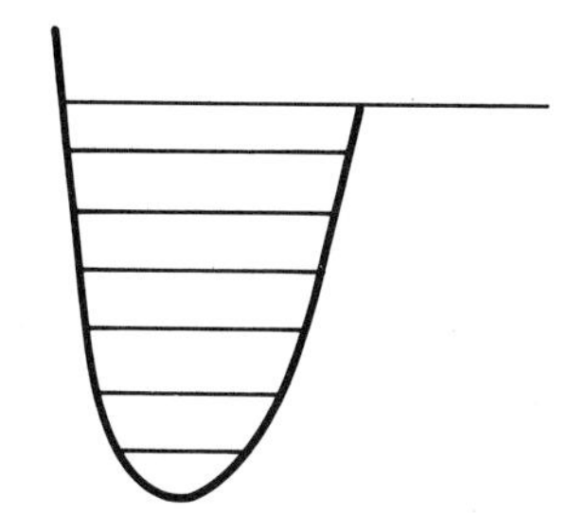

Figure 6.21. The truncated harmonic oscillator with equally spaced energy levels.

some parameter which can be related to the macroscopic rate of reaction. This is the mean first passage time, which is the average time required for a molecule to move through the vibrational levels and react from the Nth level.

The relationship between the mean passage time and the unimolecular rate constant k is illustrated by determining the dissociation time for an A molecule with a single level and a reaction rate constant w. The average time for this reaction should be some function of the rate constant w.

An ensemble of reactive molecules which decayed irreversibly with rate constant w would decay exponentially:

$$A = A_0 \exp(-wt) \tag{6.9.1}$$

The probability that a single particle reacts in some time t is also proportional to $\exp(-wt)$. If the particle has not reacted at time zero, the initial probability is

$$p(0) = 1 \tag{6.9.2}$$

and the probability that the molecule is present at time t is

$$p(t) = \exp(-kt) \tag{6.9.3}$$

The probability that it has reacted at time t is

$$1 - p(t) = 1 - \exp(-kt) \tag{6.9.4}$$

The average time a particle survives is determined from the probability that the particle has survived until some time t. This normalized probability is

$$\exp(-wt) \Big/ \int \exp(-wt)\, dt \tag{6.9.5}$$

The average survival time is

$$\int t \exp(-wt)\, dt \Big/ \int \exp(-wt)\, dt = t_{\text{av}} = \langle t \rangle \tag{6.9.6}$$

which can be written in differential form

$$\langle t \rangle = -\partial \ln S / \partial w \tag{6.9.7}$$

where

$$S = \int_0^\infty \exp(-wt)\, dt = 1/w \tag{6.9.8}$$

(Problem 6.8). The average time is

$$\langle t \rangle = -d \ln(1/w)/dw = d \ln(w)/dw = 1/w = \tau \tag{6.9.9}$$

The average survival time equals the natural lifetime τ of the molecule.

When a molecule has two levels and a transition to the continuum, the average time is dictated by two transition probabilities. If the transitions are irreversible with rate constants w (Figure 6.22), the average time for reaction will be the sum of the two average transition times:

$$\langle t \rangle = 1/w + 1/w = 2/w \tag{6.9.10}$$

The average molecule requires $\tau = 1/w$ to reach the upper molecular level and $\tau = 1/w$ to react.

For the irreversible two-level system, the **W** matrix is

$$\begin{pmatrix} -w & 0 \\ w & -w \end{pmatrix} \tag{6.9.11}$$

where the upper row describes the lower-energy state (state 1) and the lower row describes the upper (reactive) state.

The inverse of this matrix can be expanded as a sum of inverse rate constants (times) and projection operators. The inverse exists because the system does not conserve mass. The characteristic equation for the matrix is

$$\lambda^2 + 2w\lambda + w^2 = 0 \tag{6.9.12}$$

which is used with the Hamilton–Cayley theorem to find an inverse:

$$W^{-1} = -(1/w^2)(2w\mathbf{I} + \mathbf{W}) \tag{6.9.13}$$

$$= -\begin{pmatrix} 1/w & 0 \\ 1/w & 1/w \end{pmatrix} \tag{6.9.14}$$

The sum of the diagonal elements of the inverse **W** matrix is the sum of average times for transitions from each of the levels (two in this case). Since the sum of the diagonal elements is the trace of the matrix, the mean first passage time is

$$\langle t \rangle = -\text{tr}\,\mathbf{W}^{-1} \tag{6.9.15}$$

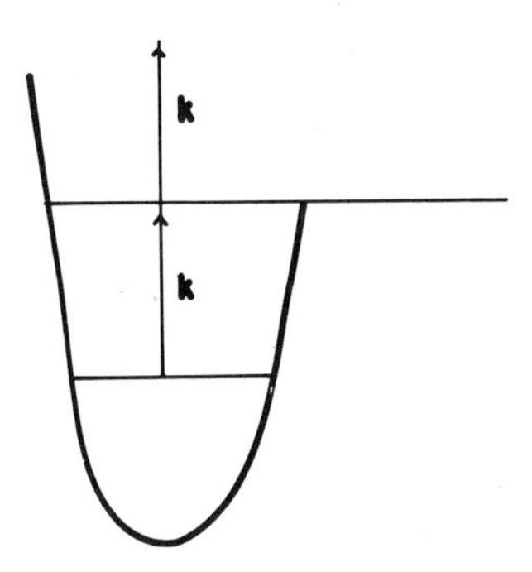

Figure 6.22. Two-level truncated harmonic oscillator with only allowed upward transitions. Each irreversible upward rate constant is k.

The multilevel model is similar to the one-level model derived earlier. In particular, the probability that the molecule is still in one level of the system at time t is proportional to

$$\exp(\mathbf{W}t) \tag{6.9.16}$$

The average time to reaction is

$$\langle t \rangle = \int t \exp(Wt)\, dt \Big/ \int \exp(Wt)\, dt = -\mathbf{W}^{-1} \tag{6.9.17}$$

If the **W** matrix is diagonal, the inverse matrix is

$$\begin{pmatrix} -1/w_1 & 0 & 0 & \cdots \\ 0 & -1/w_2 & 0 & \\ 0 & 0 & -1/w_3 & \\ \vdots & & & \end{pmatrix} \tag{6.9.18}$$

The mean first passage time sums the times in each of the levels. The trace of the inverse **W** matrix is required:

$$\langle t \rangle = -\mathrm{tr}\,(\mathbf{W}^{-1}) \tag{6.919}$$

Since the trace is a matrix invariant, Equation 6.9.19 gives the proper mean first passage time for any $\mathbf{W}^{-1}$ matrix.

When the transition between levels 1 and 2 is reversible (Figure 6.23), the up and down transition probabilities play a role in the magnitude of the mean first passage time. The mean passage time must increase since molecules are also removed from level 2 via deactivation.

If a simple model with equal transition probabilities p for each level is selected, the activation and deactivation probabilities will differ only because there are more deactivating collisions with the heat bath molecules. The upward transition is multiplied by the Boltzmann factor,

$$a = \exp(-E/kT) \tag{6.9.20}$$

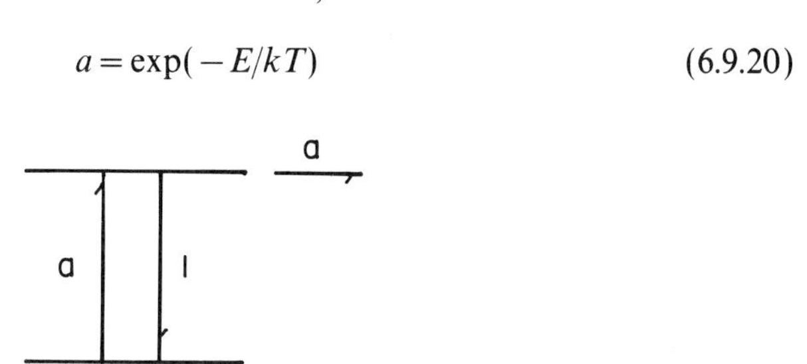

Figure 6.23. Two-level truncated harmonic oscillator with reversible intramolecular kinetics and irreversible reaction to products.

where E is the energy difference between the first and second levels. The **W** matrix is

$$w\begin{pmatrix} -1 & a \\ 1 & -1-a \end{pmatrix} \tag{6.9.21}$$

Note that the second level (lower row) can be depopulated by a return to level 1 with rate constant 1 or reaction via a rate constant a.

The Hamilton–Cayley theorem is again used to find $\mathbf{W}^{-1}$:

$$\begin{aligned}\mathbf{W}^{-1} &= (-1/a^2)[(1+2a)\mathbf{I}+\mathbf{W}] \\ &= -(1/a^2)\begin{pmatrix} 1+a & 1 \\ a & a \end{pmatrix}\end{aligned} \tag{6.9.22}$$

In each element of the trace of $\mathbf{W}^{-1}$, w^{-1} will replace w. The mean first passage time is

$$\begin{aligned}\langle t\rangle &= w^{-1}a^{-2}(1+a+a) \\ &= w^{-1}a^{-2}(1+2a)\end{aligned} \tag{6.9.23}$$

If $a=1$, the reversible model has a mean first passage time

$$\langle t\rangle = 3/w \tag{6.9.24}$$

which is longer than the mean first passage time for the nonreversible case,

$$\langle t\rangle = 2/w \tag{6.9.25}$$

A molecule in level 2 can react or deactivate in the reversible model, and this increases the mean first passage time.

For an N-level model with constant nearest-neighbor transition probabilities, each up transition has a transition probability wa while each down transition has a transition probability w. The **W** matrix is tridiagonal with elements

$$b_{ii} = -w(1+a); \qquad b_{i,i-1} = wa; \qquad b_{i,i+1} = w \tag{6.9.26}$$

For example, a three-level system has the matrix

$$\mathbf{W} = w\begin{pmatrix} -a & 1 & 0 \\ a & -(1+a) & 1 \\ 0 & a & -(1+a) \end{pmatrix} \tag{6.9.27}$$

Each upward transition probability is the same for the model of Equation 6.9.27. With Landau–Teller transition probabilities, the elements for the **W** matrix are

$$b_{ii} = -w(i-1+ia); \qquad b_{i,i-1} = w(i-1)a; \qquad b_{i,i+1} = wi \tag{6.9.28}$$

For example, a four-level system with reaction from level 3 has

$$\mathbf{W} = w \begin{pmatrix} -a & 1 & 0 & 0 \\ a & -(1+2a) & 2 & 0 \\ 0 & 2a & -(2+3a) & 3 \\ 0 & 0 & 3a & -(3+4a) \end{pmatrix} \tag{6.9.29}$$

Only the final column of each **W** matrix has elements which do not sum to zero. The system becomes nonconservative because of the steady loss from the uppermost level of the truncated harmonic oscillator. This rate process acts as a slight perturbation of the equilibrium system so that the kinetics can be evaluated using perturbation theory (Chapter 10).

The rate constant for reaction to B need not have the same type of transition probability as the remaining levels of the system. The NN element is

$$b_{NN} = -w - k \tag{6.9.30}$$

where k is the rate constant for the transition of an A molecule in the Nth level to a new molecule B. This perturbing rate constant appears only in the NN element of the matrix if the reaction is irreversible.

The **W** matrices for the mean first passage time are nonsymmetric. However, a similarity transform with the $\mathbf{D}^{1/2}$ and $\mathbf{D}^{-1/2}$ matrices of Section 6.5 generates a transform matrix which is symmetric through the principle of detailed balance. The symmetric matrix is

$$\begin{pmatrix} -a & d & 0 & 0 \\ d & -(1+2a) & 2d & 0 \\ 0 & 2d & -(2+3a) & 3d \\ 0 & 0 & 3d & -(3+4a) \end{pmatrix} \tag{6.9.31}$$

with

$$d = a^{1/2} \tag{6.9.32}$$

The mean first passage time is equal to the trace of the inverse of **W**. The systems with equal transition probabilities or Landau–Teller transition probabilities can be evaluated because their matrices are tridiagonal. A graph theory technique for evaluating such mean passage times is introduced in the next section.

6.10. Evaluation of Mean First Passage Times

The determination of the mean first passage time for the truncated harmonic oscillator models of Section 6.9 requires the determination of the matrix inverse $\mathbf{W}^{-1}$, which gives the mean first passage time as

$$\langle t \rangle = -\operatorname{tr} \mathbf{W}^{-1} \tag{6.10.1}$$

Since the **W** matrix is tridiagonal and the nonzero elements are related to each other in regular ways, the determination of mean first passage times for molecules with only a few levels will permit the determination of mean first passage times for a general N-level model. Matrices are inverted using cofactors and graph theory techniques.

The tridiagonal **W** matrices can be inverted using the method of cofactors (Chapter 3). The cofactor A_{ji} for the **W** matrix is proportional to the determinant formed by eliminating all the elements of the ith row and jth column. This cofactor determinant is multiplied by $(-1)^{i+j}$ and divided by the determinant of the full **W** matrix to form the ij element of the inverse **W** matrix.

The **W** matrix for a three-level truncated harmonic oscillator with constant transition probabilities is

$$\begin{pmatrix} -a & 1 & 0 \\ a & -1-a & 1 \\ 0 & a & -1-a \end{pmatrix} \tag{6.10.2}$$

Although the full inverse for the **W** matrix can be determined using the method of cofactors, the trace of this inverse matrix is all that is required for the mean first passage time. The diagonal elements $[\mathbf{W}^{-1}]_{ii}$ are determined from the cofactors of these diagonal elements:

$$c_{11} = \begin{vmatrix} -1-a & 1 \\ a & -1-a \end{vmatrix} = 1 + 2a + a^2 - a \tag{6.10.3}$$

$$c_{22} = \begin{vmatrix} -a & 0 \\ 0 & -1-a \end{vmatrix} = a(1+a) \tag{6.10.4}$$

$$c_{33} = \begin{vmatrix} -a & 1 \\ a & -1-a \end{vmatrix} = a^2 \tag{6.10.5}$$

Each of these cofactors is multiplied by

$$(-1)^{i+i} = +1 \tag{6.10.6}$$

and divided by the determinant of the **W** matrix,

$$\det \mathbf{W} = -a^3 \tag{6.10.7}$$

to generate the diagonal elements of the $\mathbf{W}^{-1}$ matrix:

$$\begin{pmatrix} 1+a+a^2 & & \\ & a+a^2 & \\ & & a^2 \end{pmatrix} (-a^{-3}) \tag{6.10.8}$$

The mean first passage time for the three-level system is

$$\begin{aligned}\langle t\rangle &= -\mathrm{tr}\,\mathbf{W}^{-1}\\ &= (1/a^3)(1+2a+3a^2)\end{aligned} \tag{6.10.9}$$

A pattern emerges from a comparison of mean first passage times for the two-level (Section 6.9), the three-level (Equation 6.10.9), and the four-level model (Problem 6.10). The mean first passage times are

$$\begin{aligned}\langle t\rangle(\text{two-level}) &= w^{-1}a^{-2}(1+2a)\\ \langle t\rangle(\text{three-level}) &= w^{-1}a^{-3}(1+2a+3a^2)\\ \langle t\rangle(\text{four-level}) &= w^{-1}a^{-4}(1+2a+3a^2+4a^3)\end{aligned} \tag{6.10.10}$$

This regular development continues and the mean first passage time for any N-level molecule is

$$\langle t\rangle = w^{-1}a^{-N}\sum (i+1)\,a^i \tag{6.10.11}$$

When the energy spacing of the levels is large and a is small, the leading term will dominate and the time for passage is

$$\langle t\rangle = w^{-1}a^{-N} \tag{6.10.12}$$

The physical significance of this result is clear since a system with small a will have a slow rate of reaction to B and will be close to equilibrium. In this case, the rate of reaction is proportional to the population of state N, i.e.,

$$a^{N-1} \tag{6.10.13}$$

and the rate of reaction from this level is

$$\text{Rate} = wa[f_N] = wa[a^{N-1}] = wa^N \tag{6.10.14}$$

For a two-level model with Landau–Teller transition probabilities, the $\mathbf{W}$ matrix is

$$w\begin{pmatrix} -a & 1\\ a & -(1+2a)\end{pmatrix} \tag{6.10.15}$$

with a mean first passage time

$$(1+3a)/2wa^2 \tag{6.10.16}$$

The three-level molecule with matrix

$$w\begin{pmatrix} -a & 1 & 0\\ a & -(1+2a) & 2\\ 0 & 2a & -(2+3a)\end{pmatrix} \tag{6.10.17}$$

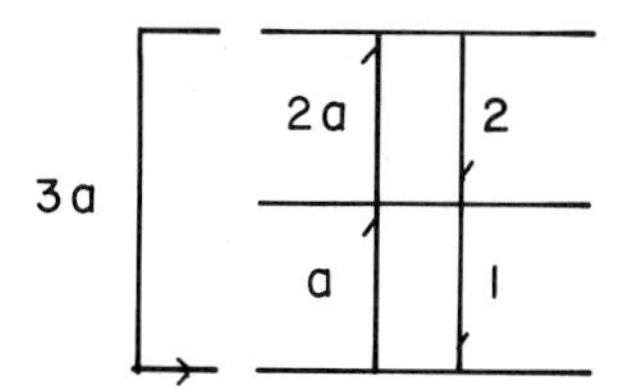

Figure 6.24. Kinetic scheme for a conservative mean first passage system which gives the proper mean first passage reaction times.

has diagonal elements

$$B_{11} = -2a^2/6a^3 \tag{6.10.18}$$

$$B_{22} = -(2a + 3a^2)/6a^3 \tag{6.10.19}$$

$$B_{33} = -(2 + 3a + 6a^2)/6a^3 \tag{6.10.20}$$

and the mean first passage time is

$$\langle t \rangle = (2 + 5a + 11a^2)/6wa^3 \tag{6.10.21}$$

The generalization to an N-level model is conveniently done with graph theory techniques. The mean first passage time averages the irreversible loss of A molecules and does not conserve mass. However, it can be adapted to graph theory techniques if the loss of A via the Nth level is compensated by the appearance of a new A molecule in the lowest ground state level. The total concentration of A remains constant with this conservative modification.

The distinct vector graphs for a three-level model are illustrated in Figure 6.25. The irreversible step produces a situation in which each of the states requires a different number of vector graphs. The upper level can only be reached

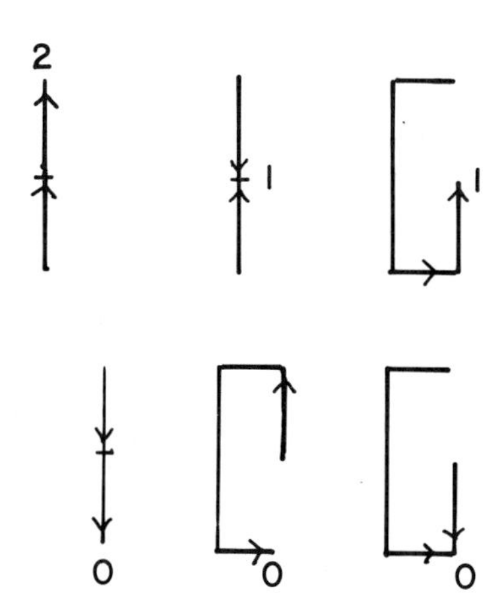

Figure 6.25. Vector graphs for the kinetic scheme of Figure 6.26.

by a single vector pathway containing two vectors. The intermediate (2) state can be reached via two different vector graphs which contain two vectors, and the ground state can be reached using three different vector graphs. For an N-level system, only a single vector graph reaches the N state, two vector graphs reach the $N-1$ state, and N vector graphs reach the ground state (Problem 6.11).

The rate constants associated with each vector in the graphs give the equilibrium fraction of each state:

$$f_3 = 2a^2/s \tag{6.10.22}$$

$$f_2 = (2a + 3a^2)/s \tag{6.10.23}$$

$$f_1 = (2 + 3a + 6a^2)/s \tag{6.10.24}$$

with

$$s = 2 + 5a + 11a^2 \tag{6.10.25}$$

The "irreversible" rate of reaction from level 3 is

$$R = 3af_3 = 6wa^3/s \tag{6.10.26}$$

and the mean first passage time is the reciprocal of this rate:

$$\langle t \rangle = 1/R = s/6wa^3 = (2 + 5a + 11a^2)/6wa^3 \tag{6.10.27}$$

This result is identical to the result obtained using the cofactor method. However, the graph technique provides a consistent way to generate expressions for the mean first passage time of an N-level system. For an N-level system, a total of

$$N(N+1)/2 \tag{6.10.28}$$

graphs are required.

Although the graphical technique gives the correct mean passage times, it does so by using a matrix which differs from the original matrix. The matrix for the three-level system of Figure 6.24 is

$$w \begin{pmatrix} -a & 1 & 3a \\ a & -(1+2a) & 2 \\ 0 & 2a & -(2+3a) \end{pmatrix} \tag{6.10.29}$$

Because this matrix is conservative, the fractional populations of each level are constant. However, because reaction from level 3 to level 1 is irreversible, the condition of detailed balance is violated. The system maintains constant fractions by cycling through the states. The unidirectional flux from level 3 to level 1 under these conditions is equivalent to the rate of formation of B in the nonconservative system.

6.11. Stepladder Models

Mean first passage times are calculated by using a master equation to describe reversible transitions within a manifold of states of the reactant molecule. An irreversible rate process permitted exit from this manifold, and this established the rate of unimolecular reaction. The model will become more realistic if the reverse reaction from the product molecule B to the A manifold is included. However, even though this reverse reaction may proceed via a single B state, all the other states for the B molecule will play a role in establishing the population of the reactive B state. To do a thorough study of the isomerization of A to B, the master equation must include all the relevant states of both the A and B molecules. Such models are important since they permit a careful definition of the macroscopic rate constants k_f and k_r for the isomerization.

To illustrate the models and minimize the total number of states, the molecules A and B are selected with the same ground state energy. The upper state is common to both molecules (Figure 6.26). Only the two ground state levels define A and B unambiguously. However, for a unidirectional rate from A to B, the upper level must be assigned to one molecule. If it is assigned to A, the transition from A to B involves reaction from the common upper level to B (Figure 6.27). It is clear that the distinction between A and B can be arbitrary for such models.

The downward transition probabilities (w) are equal for each molecule. The upward transition probabilities are weighted with

$$a = \exp(-E/kT) \tag{6.11.1}$$

where E is the energy difference between the levels. The **W** matrix is

$$w \begin{pmatrix} -a & 1 & 0 \\ a & -2 & a \\ 0 & 1 & -a \end{pmatrix} \tag{6.11.2}$$

Common elements do not appear on each diagonal, and matrix analysis is more difficult.

This **W** matrix might also describe transitions within a single molecule. Eigenvalue–eigenvector analysis does not distinguish the physical origin of the

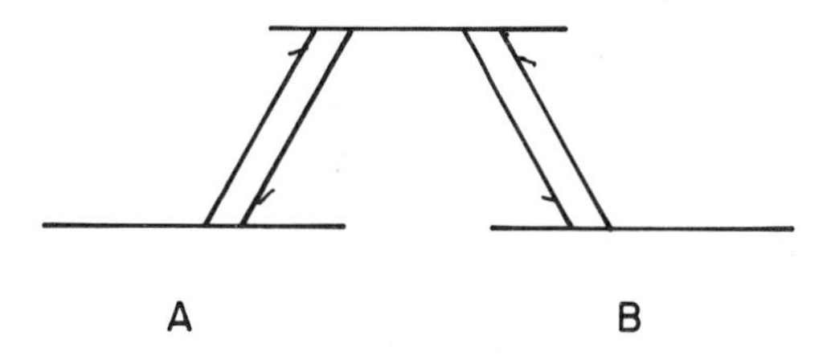

Figure 6.26. A stepladder model with three levels. The upper level is common to both A and B.

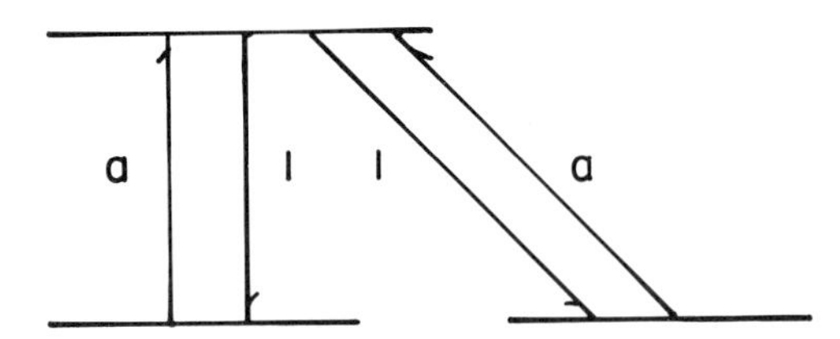

Figure 6.27. Separation of total three-state manifold into A and B molecules to establish kinetics between the two species.

transition probabilities. In fact, the decay times (eigenvalues) can include transition probabilities for B even if there is no B present when the reaction starts. This leads to some interesting definitions of reaction rates and rate constants.

The eigenvalues and eigenvectors for this three-level system are

$$\lambda = 0, \qquad \lambda = -wa, \qquad \lambda = -w(1+a) \tag{6.11.3}$$

The smallest nonzero eigenvalue $(-a)$ is well separated from the remaining eigenvalue $(-1-a)$ and will dominate the kinetic relaxation of the system.

The $\lambda = 0$ eigenvector is

$$|0\rangle = \begin{pmatrix} 1 \\ a \\ 1 \end{pmatrix} \tag{6.11.4}$$

This is the Boltzmann distribution for the system. The two remaining eigenvalues are

$$|-wa\rangle = \begin{pmatrix} 1 \\ 0 \\ -1 \end{pmatrix} \qquad |-w(1+a)\rangle = \begin{pmatrix} 1 \\ -2 \\ 1 \end{pmatrix} \tag{6.11.5}$$

The components of these two vectors sum to zero as they must since all concentration resides in the equilibrium vector. They do describe the time-dependent transfer of molecules between the two isomers. The vector $|-wa\rangle$ describes an increase in the ground level of one species coupled with an equal decrease from the ground level of the second species. The upper-level concentration remains unchanged in this mode. Although this suggests a direct flow from ground level A to ground level B, it actually describes a flow from A which is equal to the concomitant flow to B.

A four-state model with two levels of equal energy for A and B (Figure 6.28) provides a clear separation between A and B manifolds. Intermanifold trans-

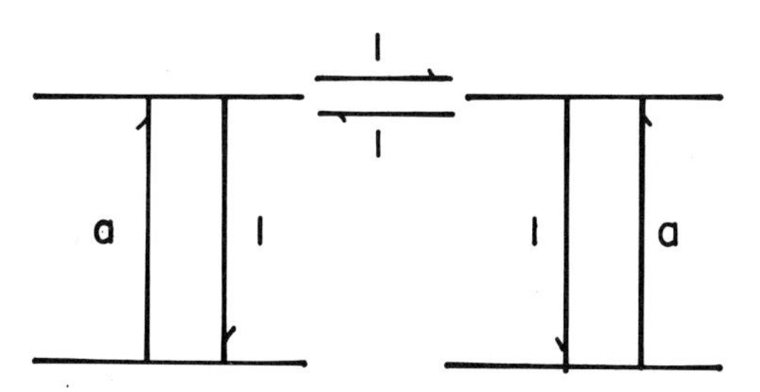

Figure 6.28. A four-level isomerization scheme. A and B each have two levels. Energies for the levels in each molecule are the same.

itions occur only via the two upper levels. For convenience, the A–B transition probabilities are equated to w (Figure 6.28). The **W** matrix is

$$w\begin{pmatrix} -a & 1 & 0 & 0 \\ a & -2 & 1 & 0 \\ 0 & 1 & -2 & a \\ 0 & 0 & 1 & -a \end{pmatrix} \tag{6.11.6}$$

The eigenvector $|0\rangle$ is the Boltzmann distribution:

$$|0\rangle = \begin{pmatrix} 1 \\ a \\ a \\ 1 \end{pmatrix} \tag{6.11.7}$$

The similarity transform of the **W** matrix to a symmetric matrix will prove that all eigenvalues must be real. In addition, the symmetric matrix is easier to analyze for eigenvalues and eigenvectors.

The **D** matrix which symmetrizes the **W** matrix is

$$\begin{pmatrix} 1 & 0 & 0 & 0 \\ 0 & \sqrt{a} & 0 & 0 \\ 0 & 0 & \sqrt{a} & 0 \\ 0 & 0 & 0 & 1 \end{pmatrix} \tag{6.11.8}$$

The symmetric **W**′ matrix is

$$w\begin{pmatrix} -a & \sqrt{a} & 0 & 0 \\ \sqrt{a} & -2 & \sqrt{a} & 0 \\ 0 & 1 & -2 & \sqrt{a} \\ 0 & 0 & \sqrt{a} & -a \end{pmatrix} \tag{6.11.9}$$

The symmetric $\mathbf{W}'$ matrix contains the A and B energy levels. If the two molecules were not coupled, the 4×4 matrix would reduce to a matrix with two 2×2 blocks along the diagonal. Since these molecules are identical, there is a correlation between the 1, 2 and 3, 4 eigenvector component pairs. Each of the four eigenvectors for the full 4×4 matrix is expressed in terms of two unknowns (u and v) as

$$\begin{array}{cccc} (+u & +v & +v & +u) \\ (+u & +v & -v & -u) \\ (+u & -v & -v & +u) \\ (+u & -v & v & -u) \end{array} \tag{6.11.10}$$

The first eigenvector is the equilibrium eigenvector with

$$u=1 \qquad v=\sqrt{a} \tag{6.11.11}$$

The eigenvector for the lowest nonzero eigenvalue will have a single node, i.e., $(u \quad v \quad -v \quad -u)$, and the characteristic equation is

$$\begin{pmatrix} -a-\lambda & \sqrt{a} & 0 & 0 \\ \sqrt{a} & -2-\lambda & 1 & 0 \\ 0 & 1 & -2-\lambda & \sqrt{a} \\ 0 & 0 & \sqrt{a} & -a-\lambda \end{pmatrix} \begin{pmatrix} u \\ v \\ -v \\ -u \end{pmatrix} = 0 \tag{6.11.12}$$

The system generates two pairs of identical equations for u and v:

$$\begin{aligned} (-a-\lambda)u+\sqrt{a}v&=0 \\ \sqrt{a}u-(3+\lambda)v&=0 \end{aligned} \tag{6.11.13}$$

The system reduces to a 2×2 matrix. Its eigenvalues are found using the characteristic determinant

$$\begin{vmatrix} -(a+\lambda) & \sqrt{a} \\ \sqrt{a} & -(3+\lambda) \end{vmatrix} = 0 \tag{6.11.14}$$

The characteristic polynomial

$$2a+(3+a)\lambda+\lambda^2=0 \tag{6.11.15}$$

gives eigenvalues

$$\begin{aligned} \lambda_+ &= -(3+a)/2+\frac{1}{2}\left[(3+a)^2-2a\right]^{1/2} \\ \lambda_- &= -(3+a)/2-\frac{1}{2}\left[(3+a)^2-2a\right]^{1/2} \end{aligned} \tag{6.11.16}$$

The smallest nonzero eigenvalue, λ_+, determines the evolution to equilibrium.

The lowest eigenvalue permits a definition of macroscopic forward and reverse rate constants. Since both molecules have exactly the same energy levels, the forward and reverse rate constants between the upper levels must be equal:

$$k_f + k_r = 2k \qquad (6.11.17)$$

The macroscopic decay constant for this isomerization is $2k$ since it is the sum of the forward and reverse rate constants. For the eigenvalue analysis, this decay to equilibrium is dominated by the smallest nonzero eigenvalue, $\lambda_+ = \lambda$. Equating the two decay constants with $k_f = k_r$ gives

$$k_f = k/2 = \lambda/2 \qquad k_r = k/2 = \lambda/2 \qquad (6.11.18)$$

The forward rate constant can be related to the molecular transition probabilities by substituting for λ_+,

$$k_f = -(3+a)/4 + [(3+a)^2 - 8a]^{1/2}/4 \qquad (6.11.19)$$

with $w = 1$.

Equation 6.11.19 for k_f includes transition probabilities for both the A and B molecules. This result contrasts sharply with approaches which relate k_f with the irreversible flow of molecules from A to B. In this case, the macroscopic forward rate is

$$R_f = k_f[\mathrm{A}] \qquad (6.11.20)$$

The irreversible microscopic rate from the upper level of A to the upper level of B dictates this rate as

$$R_f = (1)\,f_{\mathrm{A}2}[\mathrm{A}] \qquad (6.11.21)$$

The equilibrium fraction of excited A molecules is

$$[a/(1+a)]A \qquad (6.11.22)$$

where A is the total concentration of A molecules at time t. The irreversible rate is

$$R_f = (1w)[a/(1+a)][\mathrm{A}] \qquad (6.11.23)$$

Equating the rates of Equations 6.11.20 and 6.11.23 gives

$$k_f = wa/(1+a) = a/(1+a) \qquad (6.11.24)$$

This "equilibrium" approximation differs significantly from the proper lowest eigenvalue definition of k_f.

Problems

6.1. Find the eigenvalues and eigenvectors for the matrix

$$\mathbf{K} = \begin{pmatrix} -1 & 0 & 1 \\ 1 & -1 & 0 \\ 0 & 1 & -1 \end{pmatrix}$$

6.2. A cyclic reaction has all six rate constants (k_1, k_{-1}, k_2, k_{-2}, k_3, and k_{-3}) equal to unity.

a. Write the **K** matrix for this system.
b. Find the eigenvalues and eigenvectors for this system.
c. Write the solution when $A(0) = A_0$ and $B(0) = C(0) = 0$.

6.3. A membrane channel can bind ions at either of its two binding sites but it cannot bind ions at each site simultaneously because of ion–ion repulsion.

a. Set up the kinetic scheme for this system.
b. Develop the line graphs and vector graphs for each species and determine the steady-state fractions of each state.
c. Determine the net rate of ion flow through the channel if bath 1 has a concentration c_1 and bath 2 has a concentration c_2.

6.4. Write a **K** matrix for the system of Problem 6.3.

6.5. For the cyclic reaction of Problem 6.2 (all $k = 1$), develop the characteristic matrix of the equation using the invariants of the system.

6.6. (a) Show that an irreversible sequential reaction of N steps will have a concentration for the Nth species equal to

$$1 = k_1 k_2 \cdots k_N / k_1 k_2 \cdots k_N$$

using projection operators techniques. (b) Determine the other projection operators. Is there a pattern?

6.7. Show that Equation 6.6.46 of Section 6.6 is consistent with the initial conditions.

6.8. Derive the equation

$$\langle t \rangle = -d \ln S/dk$$

6.9. Use the $\mathbf{D}^{1/2}$ and $\mathbf{D}^{-1/2}$ transform matrices to symmetrize a four-level system obeying Landau–Teller transition probabilities.

6.10. Determine the mean first passage time for a four-level truncated harmonic oscillator when $a = 1$.

6.11. A truncated harmonic oscillator of four levels has downward transitions proportional to 1 and upward transitions proportional to a. Use graph theory with a conservative system to determine the mean first passage time for this system.

6.12. For a stepladder model with upward and downward transition probabilities proportional to 1, show that the smallest nonzero eigenvalue approaches 0 with increasing total number of levels N.

6.13. a. For the four-state reaction model of Section 6.11, determine u and v for the equilibrium vector.

b. Determine the equilibrium eigenvector for the four-level symmetric system and show that $\lambda = 0$.

c. Use the transform $|\lambda\rangle = \mathbf{D}\,|\lambda'\rangle$ to show the equilibrium eigenvector for the nonsymmetric **K** matrix.

6.14. Develop a four-level isomerization model when the two energy levels for molecule B are displaced from those of A even though the energy level difference in each molecule is the same.

6.15. The reaction rate between A and B manifolds was w. Consider the case where this reversible rate constant is $k \neq w$ and determine what happens as k goes to zero.

6.16. Develop a matrix solution for the reversible isomerization reaction of A and B when $k_f = 1$ and $k_r = 3$.

6.17. Show that the eigenvalues of a lower triangular matrix are always equal to the diagonal elements of the matrix.

7

Statistical Mechanics

7.1. The Wind–Tree Model

When a system is at equilibrium, it will always have the maximal possible entropy for a given total internal energy. Boltzmann's equation for the entropy S,

$$S = k \ln \Omega \tag{7.1.1}$$

where k is Boltzmann's constant and Ω is the total number of states for the system, illustrates the direct connection between randomness, characterized by a large number of states, and the entropy.

Boltzmann's equation must include every possible state of the system including states which are distinctly different from the expected equilibrium states of the system. Each state is weighted equally but equilibrium states are observed experimentally. Is there a contradiction?

The nature of the problem can be illustrated with a simple model developed by Ehrenfest which is called the "dog–flea" or urn model. N labeled fleas are allowed to move freely between two dogs or urns. Even if all the fleas are on dog 1 at time zero, the system will ultimately reach an equilibrium in which $N/2$ fleas are found on each dog. This is the macroscopic equilibrium. However, the entropy calculation includes "nonequilibrium" states, e.g., all the fleas on a single dog. Since each flea can select either dog, there are a total of

$$2^N \tag{7.1.2}$$

different ways to distribute the fleas between the two dogs, and this factor must be used in the entropy expression.

Since each flea can choose either dog, the situation with all fleas on a single dog is a definite state of the system. However, there is only a single way to construct this state. Each labeled flea must be on the one dog. The number of different states which produce $N/2$ fleas on each dog is significantly larger if N is large. There are a total of

$$\Omega = N!/[(N/2)!]^2 \tag{7.1.3}$$

different arrangements possible for the 50–50 situation. The case where the fleas labeled 1 to 50 are on dog 1 is one of these arrangements, and it is weighted equally with the state with all fleas on one dog. Each other labeled arrangement which gives $N/2$ fleas on dog 1 is equally probable. At the thermodynamic level, the observer is interested only in the total number of fleas on the dogs and not their labels. The 50–50 arrangement, i.e., distribution, will be observed most often because it can be made in so many distinct ways. The observer would then define the arrangement as the equilibrium distribution.

The appearance of dominant equilibrium states for the system introduces a second paradox into studies of statistical mechanics. If molecules replace the fleas of the dog–flea model and undergo classical motion, they must obey Newton's force equation,

$$F = m\, d^2r/dt^2 \tag{7.1.4}$$

Because time appears in a second derivative, the equations are reversible with respect to time. Substituting $(-t)$ gives

$$F = m\, d^2r/d(-t)^2 = m\, d^2r/dt^2 \tag{7.1.5}$$

Newton's equation makes no distinction between particles moving in "positive" time and those moving in "negative" time. This is the principle of microscopic reversibility. If time is reversed, the molecules will simply retrace their paths in space.

The microscopically reversible behavior of the molecules is in conflict with the evolution of systems of molecules to their maximal entropy. A system is started in a very "nonequilibrium" state. After some time, the system will have evolved to an equilibrium state where it will remain the major percentage of the time. The principle of microscopic reversibility states that negative time can replace positive time without altering the equations of motion. Thus, the

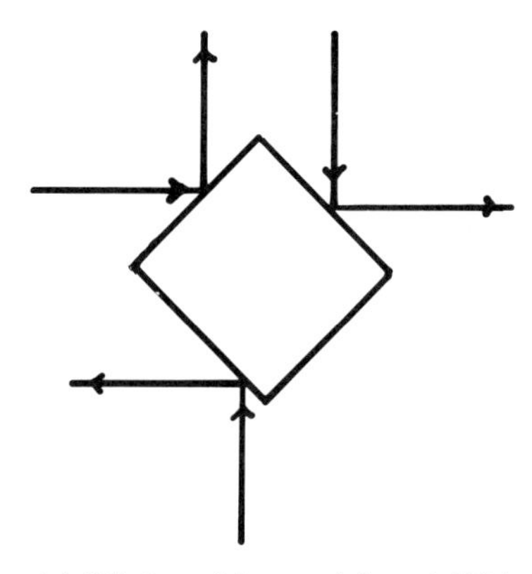

Figure 7.1. The wind–tree model. Wind particles are deflected 90° by collisions with the randomly placed, diamond-shaped trees.

equilibrium system has an equal probability of evolving back to its initial state, i.e., the system is as likely to order as it is to disorder in time.

The nature of the observed irreversible evolution to the maximal entropy is illustrated with a second Ehrenfest model, the wind–tree model. The model effectively reduces the number of states possible for each "wind" molecule using diamond-shaped trees (Figure 7.1). The trees are all oriented so that a wind particle will be deflected to a second direction. The total possible directions (or states) for each molecule is four. The observables are the numbers of wind particles moving in each of these four directions at any time t. The collisions are elastic so the particles always have equal energies.

The system will obey microscopic reversibility. Figure 7.2 shows the trajectory of a single wind particle with time. If t is replaced with $-t$, the particle will traverse this path in reverse. There is no preferred motion toward maximal entropy.

The system can be changed from a "time-reversible" to a "time-irreversible" system by introducing transition probabilities which describe the behavior of a large number of wind particles. Figure 7.3 shows that in some time t, a certain number of wind particles will strike the face of the tree, which changes their direction from "east" to "north." The probability of such a transition can be related to the cross section a (Figure 7.3) and the total number of trees, N, which the particles could strike in some interval t. The probability of an east–north transition is then determined as the ratio of this total cross section, Na, divided by the total length, L, of field accessible to the eastbound particles in time t:

$$p = (Na/L)/t \tag{7.1.6}$$

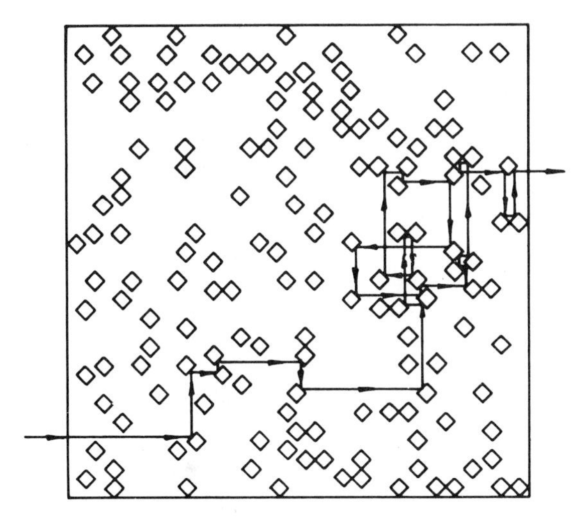

Figure 7.2. The trajectory of a single wind particle through a set of randomly placed trees.

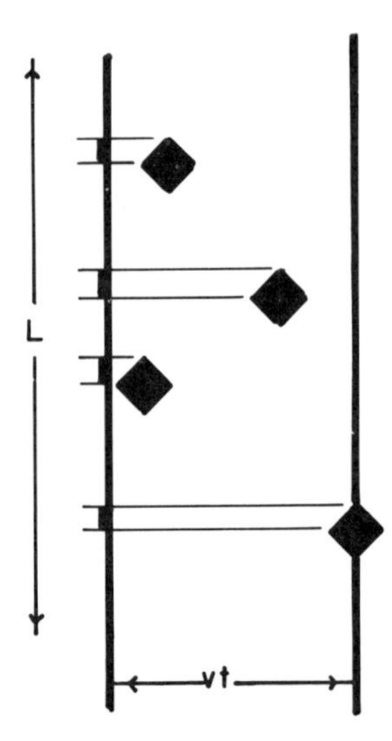

Figure 7.3. The relative cross section for direction-changing collisions relative to the total length accessible to the particles. The distance $d = vt$ is the distance traveled by a wind particle in time t.

Because of the symmetry of the trees, all changes of direction have the same transition probability.

The introduction of transition probabilities reduces the evolution of a large number of particles to a set of kinetic equations. The number of particles, N_i, or the fraction of particles, f_i, moving in the ith direction is dictated by transitions to and from two other directions. The kinetic equations for the f_i are

$$\begin{aligned} df_1/dt &= -2pf_1 + pf_2 + pf_4 \\ df_2/dt &= pf_1 - 2pf_2 + pf_3 \\ df_3/dt &= pf_2 - 2pf_3 + pf_4 \\ df_4/dt &= pf_1 + pf_3 - 2pf_4 \end{aligned} \tag{7.1.7}$$

which become the matrix equation

$$d\,|f\rangle/dt = \mathbf{K}\,|f\rangle \tag{7.1.8}$$

with

$$\mathbf{K} = p\begin{pmatrix} -2 & 1 & 0 & 1 \\ 1 & -2 & 1 & 0 \\ 0 & 1 & -2 & 1 \\ 1 & 0 & 1 & -2 \end{pmatrix} \tag{7.1.9}$$

The formal solution is

$$|f(t)\rangle = \exp(\mathbf{K}t)\,|f(0)\rangle \tag{7.1.10}$$

where $|f(0)\rangle$ can describe any nonequilibrium distribution of wind particles. The projection operator solution is

$$|f(t)\rangle = \sum \exp(\lambda_i pt)\,\mathbf{P}_i\,|f(0)\rangle \tag{7.1.11}$$

The eigenvalues of the matrix are 0, -2, -2, and -4. The full kinetic solution is left as a problem (7.1). The equilibrium fractions of the system associated with $\lambda = 0$ having a projection matrix

$$\mathbf{P}_0 = \begin{pmatrix} 1 & 1 & 1 & 1 \\ 1 & 1 & 1 & 1 \\ 1 & 1 & 1 & 1 \\ 1 & 1 & 1 & 1 \end{pmatrix} \left(\frac{1}{4}\right) \tag{7.1.12}$$

are

$$|f^{\text{eq}}\rangle = \exp(-0pt)\,\mathbf{P}_0\,|f(0)\rangle = \begin{bmatrix} \frac{1}{4} \\ \frac{1}{4} \\ \frac{1}{4} \\ \frac{1}{4} \end{bmatrix} \tag{7.1.13}$$

since the projection matrix sums all the initial fractions and these must sum to one for any initial distribution. The transition probabilities describe a system which evolves to the proper equilibrium distribution and is irreversible in time.

The wind–tree model can also describe the evolution of the system to equilibrium. Although this progress can be monitored by observing the evolution of each component toward its equilibrium value, this becomes impractical for systems with larger numbers of states. It is convenient to define a single parameter which includes the time-dependent fractions of each state and monitor its evolution in time.

Although Boltzmann originally defined an "H" function,

$$H(t) = \sum N_i(t) \ln[N_i(t)] \tag{7.1.14}$$

for the evolution of the system with time, the entropy is more convenient. The time dependence is included by using the information theory definition of entropy:

$$S(t) = -k \sum_{i=1}^{4} f_i(t) \ln[f_i(t)] \tag{7.1.15}$$

Since the f_i are fractions, $S(t)$ is always a positive function.

The rate of change of entropy is

$$dS(t)/dt = -k\, d/dt\left(\sum_{i=1}^{4} f_i \ln f_i\right)$$

$$= -k\left[\sum_{i=1}^{4} f_i(f_i^{-1})\, df_i/dt + \sum_{i=1}^{4} \ln(f_i)\, df_i/dt\right] \tag{7.1.16}$$

Since

$$\sum f_i(t) = 1 \tag{7.1.17}$$

the first summation is zero:

$$\sum df_i(t)/dt = 0 \tag{7.1.18}$$

The second summation is evaluated for the four-state wind–tree model. The time derivatives of the fractions are known and can be substituted into the summation to give

$$-d(S/k)/p\, dt = \ln(f_1)(f_4 + f_2 - 2f_1) + \ln(f_2)(f_1 - 2f_2 + f_3)$$
$$+ \ln(f_3)(f_2 - 2f_3 + f_4) + \ln(f_4)(f_1 + f_3 - 2f_4) \tag{7.1.19}$$

The total function is simplified by noting a pattern for each pair of indices. Terms in f_1 and f_2 are collected to give

$$(f_2 - f_1)\ln(f_1) + (f_1 - f_2)\ln(f_2)$$
$$= (f_2 - f_1)\ln(f_1/f_2) \tag{7.1.20}$$

The remaining pairs of indices, 13, 14, 23, 24, and 34, will each produce an expression like Equation 7.1.20 with the appropriate indices (Problem 7.2). For any pair of fractions, f_i and f_j, each product, e.g., Equation 7.1.20, is always negative. Since the sum of these factors is

$$d(-S/k)/dt \tag{7.1.21}$$

the rate of change of the entropy is always a positive quantity. Entropy will always increase with time in the transition probability formulation.

The actual rate of entropy increase becomes larger as the system is displaced from equilibrium. Both the fraction difference, $f_i - f_j$, and the logarithmic ratio, f_j/f_i, increase in absolute magnitude as the fractions deviate from their equilibrium fractions. A system with a large displacement from equilibrium will evolve back toward equilibrium more rapidly than one with a small displacement. The system will then tend to stay near equilibrium. This is an example of Liapunov stability and the entropy is a Liaponouv function.

The Kac ring model is similar to the wind–tree model but involves particles which can exist in only two states. The Kac ring can be modeled as a series of balls arranged in a circle. Each ball can be white or black. A set of m markers are placed randomly among the N balls in the circle (Figure 7.4). Each ball is then moved one unit for each time interval. If it passes a marker during the transition, it will change color. The system is completely reversible. If the balls are moved in the opposite direction, the system will retrace its path to the initial arrangement of balls even if this arrangement is very "nonequilibrium." Equilibrium in this case is an equal number of white and black balls on the ring.

Transition probabilities for the system are formed by noting that a transition from white to black or from black to white is dictated by the number of markers relative to the total number of atoms, i.e.,

$$p \propto m/N \tag{7.1.22}$$

In this case, there are only two fractional populations. The fractions of white and black balls are f_w and f_b, respectively. The kinetic equations are

$$\begin{aligned} df_w/dt &= -pf_w + pf_b \\ df_b/dt &= +pf_w - pf_b \end{aligned} \tag{7.1.23}$$

Since the system will reach equilibrium when the fractions of white and black balls are equal, the kinetic equations can be subtracted to develop a kinetic equation for the deviation from equilibrium:

$$\begin{aligned} d(f_w - f_b)/dt &= -2pf_w + 2pf_b \\ d\Delta/dt &= -2p\Delta \end{aligned} \tag{7.1.24}$$

with

$$\Delta = f_w - f_b \tag{7.1.25}$$

Figure 7.4. The Kac ring model. The balls change color when they pass a marker (×).

If the initial deviation is Δ_0, the deviation decays to zero as

$$\Delta = \Delta_0 \exp(-2pt) \tag{7.1.26}$$

The decay constant, $2p$, is the nonzero eigenvalue of the matrix. Subtraction of the two equations eliminates the equilibrium projection, leaving $|2p\rangle$ which will decay exponentially with time.

7.2. Statistical Mechanics of Linear Polymers

The entropy of the system of Section 7.1 was determined by counting all possible states of the system. For the dog–flea model, there were 2^N possible arrangements (states) and, for large N, the majority of these states produced the 50–50 distribution. The total states can be written as a sum of the numbers of states for each distribution using the binomial theorem,

$$\Omega = 2^N = (1+1)^N = \sum_j^N [N!/j!(N-j)!] \tag{7.2.1}$$

where the jth term in the summation is the distribution with j fleas on dog 1 and $N-j$ fleas on dog 2. The states for the distribution $(j, N-j)$ are associated with an entropy S_j and the equation becomes

$$\Omega_j = \exp(S_j/k) \tag{7.2.2}$$

This function peaks steeply when $j = N/2$ (Figure 7.5), and the peak value can be used as an approximation to the full summation. The fraction of 50–50 states is determined by dividing its total number of states by the total for all distributions (2^N):

$$f_{N/2} = N!/[(N/2)!]^2/2^N \tag{7.2.3}$$

These techniques can be applied to systems which have states of different energy, i.e., ε_j. The probability of finding a state with energy ε_j is proportional to the Boltzmann factor

$$\exp(-\varepsilon_j/kT) \tag{7.2.4}$$

where the temperature must be included for a dimensionless exponential argument. The probability of finding a molecule with energy ε_j is

$$\begin{aligned} p_j &= \exp(-\varepsilon_j/kT) \Big/ \sum \exp(-\varepsilon_i/kT) \\ &= \exp(-\varepsilon_j/kT)/q \end{aligned} \tag{7.2.5}$$

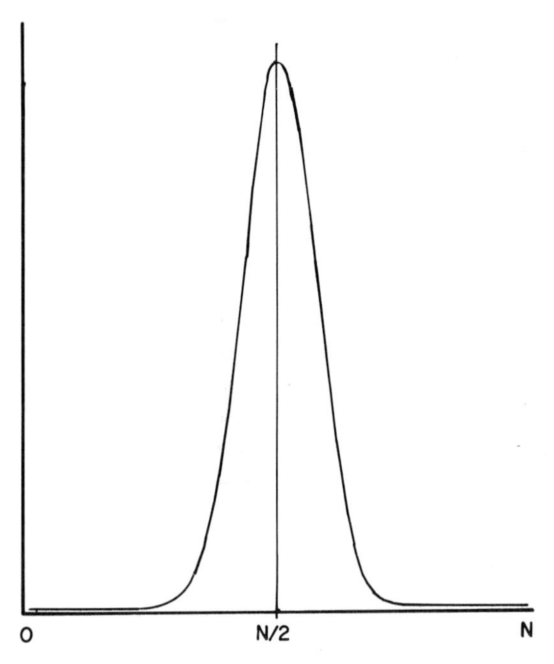

Figure 7.5. The increase in the number of distinct states in the vicinity of a 50–50 distribution of particles. $N = 100$ for this case.

where the summation over all possible energy states of the system is the partition function q.

In general, the probability of finding a given state is proportional to the exponential of all energies associated with that state. A state characterized by an energy ε_j and entropy S_j has a Boltzmann factor

$$\exp(S_j/k)\exp(-\varepsilon_j/kT) = \omega_j \exp(-\varepsilon_j/kT) \tag{7.2.6}$$

where ω_j is the degeneracy for states with energy ε_j. If a molecule with energy ε_j can bind a second molecule, the bound state will contain the additional factor

$$\exp(\mu/kT)\exp(-\varepsilon_j'/kT) \tag{7.2.7}$$

where μ is the chemical potential of the second molecule and ε_j' is the energy of the second molecule plus binding energies for the complex. Since

$$\mu = kT \ln c \tag{7.2.8}$$

where c is the molecular concentration of the solution,

$$c = \exp(\mu/kT) \tag{7.2.9}$$

the chemical potential, μ, determines the concentration for binding. Equation 7.2.7 is written thermodynamically as

$$\exp(\mu/kT)\exp(-\beta\varepsilon_j') = c/K \tag{7.2.10}$$

where K is the dissociation constant for the complex. The dissociation constant, like the Boltzmann factor, provides a quantitative measure of the extent of binding.

Although statistical mechanics can be used for any system, it is convenient to focus on the treatment of a simple one-dimensional system, which is easily expanded to more complex analyses. The model of choice is a long polymer containing identical residues. The energies of each residue are then identical. The orientation of each residue relative to its neighbors in the chain will have a distinct energy, and each of these energies should generate an exponential in the partition function. This total analysis is complicated, and it is convenient to characterize each residue by only two energies. If the residue has an orientation relative to its neighbors which is characteristic of a helical conformation in the polymer, it is assigned an energy ε_h. Any orientations which are not helical are assigned an energy ε_c.

This classification is unrealistic since several residues are required to form a helix. In this model, a residue could be classified as helical even if its neighbors are classified as random coil. However, this model does illustrate the use of Boltzmann statistics for a series of independent residues.

The probability that a residue is helical is

$$\begin{aligned} p_h &= \exp(-\varepsilon_h/kT)/[\exp(-\varepsilon_h/kT) + \exp(-\varepsilon_c/kT)] \\ &= \exp(-\varepsilon_h/kT)/q \end{aligned} \tag{7.2.11}$$

while the probability that it is random coil is

$$p_c = \exp(\varepsilon_c/kT)/q \tag{7.2.12}$$

Since each residue in the chain is independent and all have the same energies, the probabilities for each residue are equal. For a chain of N residues, the number of helical residues is

$$N_h = Np_h = N\exp(-\varepsilon_h/kT)/q \tag{7.2.13}$$

The probability p_h for independent residues then determines the probability that a specific residue is helical and the number of residues within a chain of N units which have the helical conformation.

The analysis gives the partition function for a single residue in the chain. However, because the residues are independent, the partition function for a full chain is also quite simple. A chain of three independent residues has eight distinct conformations:

$$\begin{array}{cccc} & hhc & cch & \\ hhh & hch & chc & ccc \\ & chh & hcc & \end{array} \tag{7.2.14}$$

The Boltzmann factors for each of these chain conformations are determined from the total energy for all residues. The Boltzmann factors for the successive columns in Equation 7.2.14 are

$$\begin{aligned} &\exp(-3\varepsilon_h/kT); \qquad \exp[-(2\varepsilon_h+\varepsilon_c)/kT]; \\ &\exp[-(\varepsilon_h+2\varepsilon_c)/kT]; \qquad \exp(-3\varepsilon_c/kT) \end{aligned} \tag{7.2.15}$$

Defining the single-residue Boltzmann factors,

$$h=\exp(-\varepsilon_h/kT); \qquad c=\exp(-\varepsilon_c/kT) \tag{7.2.16}$$

the partition function Q for the full three-unit chain is

$$Q=h^3+3h^2c+3hc^2+c^3 \tag{7.2.17}$$

This binomial form reduces to

$$Q=(h+c)^3=q^3 \tag{7.2.18}$$

For independent residues, the full chain partition function for an N-unit chain is determined directly from the partition function for a single residue,

$$Q=q^N \tag{7.2.19}$$

When the residues are independent, all analyses can be performed with a single residue. For example, consider a case where the residue has coil and helical conformers and the helix form can bind some S molecule. The partition function for the residue now contains three terms:

$$\begin{aligned} q &= c+h+h\exp(\mu/kT)\exp(-\varepsilon'_S/kT) \\ &= c+h+h\lambda_S q_S \end{aligned} \tag{7.2.20}$$

The final term contains the additional energy for bound S. The fraction of residues with bound S is

$$p_S=h\lambda_S q_S/q \tag{7.2.21}$$

For one S binding site per residue, this fraction also gives the total bound S for a chain of N units:

$$\langle S\rangle = Nh\lambda_S q_S/q \tag{7.2.22}$$

If each residue can bind S at M distinct sites, this partition function becomes (Problem 7.3)

$$q = c + h(1 + \lambda_S q_S)^M \tag{7.2.23}$$

and the average number bound per residue is found using the formula (Problem 7.4)

$$\langle S \rangle = \lambda_S d(\ln q)/d\lambda_S \tag{7.2.24}$$

The nature of $\langle S \rangle$ is illustrated by considering the case with two binding sites per residue. A single residue and its possible states are illustrated in Figure 7.6. The partition function is

$$q = c + h + 2h\lambda_S q_S + h(\lambda_S q_S)^2 \tag{7.2.25}$$

The average number of bound S is found by determining the probabilities for each of the five residue states, p_i. The average bound S is found by multiplying each probability by the number of S bound in that conformation:

$$\begin{aligned}\langle S \rangle &= \sum p_i n_i \\ &= [c(0) + h(0) + h_S q_S(1) + h\lambda_S q_S(1) + h(\lambda_S q_S)^2\,(2)]/q \\ &= [2h_S q_S + 2h(\lambda_S q_S)^2]/q\end{aligned} \tag{7.2.26}$$

The partition function can be used to determine thermodynamic parameters as well as probabilities and average binding. The average energy for a residue is found by multiplying each probability by its energy and summing:

$$\begin{aligned}\langle \varepsilon \rangle &= p_h \varepsilon_h + p_c \varepsilon_c \\ &= (h\varepsilon_h + c\varepsilon_c)/(h + c)\end{aligned} \tag{7.2.27}$$

The average energy can also be generated using the equation

$$\langle \varepsilon \rangle = -\partial \ln q/\partial \beta \tag{7.2.28}$$

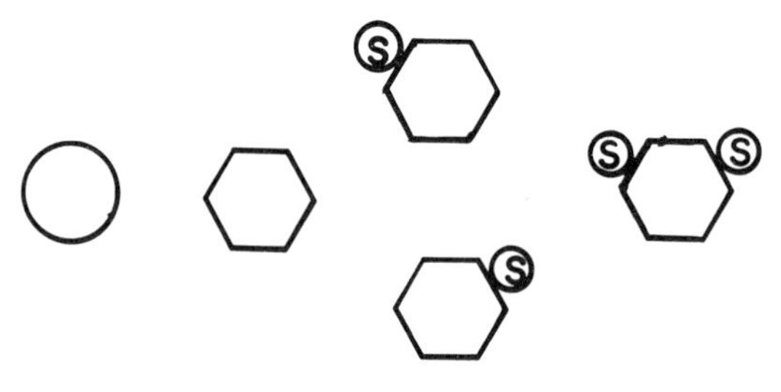

Figure 7.6. The conformations available to a single independent residue when it can exist in two conformations (h and c) and the h conformation can bind S at two independent, equivalent sites.

where

$$\beta = 1/kT \tag{7.2.29}$$

The entropy of the link is generated from the information theory formula for entropy:

$$\begin{aligned} \langle S \rangle &= -k[p_h \ln(p_h) + p_c \ln(p_c)] \\ &= -k[(h/q)\ln(h/q) + (c/q)\ln(c/q)] \end{aligned} \tag{7.2.30}$$

The $\ln q$ are separated and combined to give

$$-\langle S \rangle/k = (h/q)\ln(h) + (c/q)\ln(c) - \ln(q)(h/q + c/q) \tag{7.2.31}$$

Since

$$h + c = q \tag{7.2.32}$$

$$h/q + c/q = 1 \tag{7.2.33}$$

and

$$\begin{aligned} \ln h &= \ln[\exp(-\varepsilon_h/kT)] = -\varepsilon_h/kT \\ \ln c &= \ln[\exp(-\varepsilon_c/kT)] = -\varepsilon_c/kT \end{aligned} \tag{7.2.34}$$

the entropy expression becomes

$$-\langle S \rangle/k = [p_h(-\varepsilon_h/kT) + p_c(-\varepsilon_c/kT)] - \ln q \tag{7.2.35}$$

Since the first two terms define the average energy divided by kT, the equation is

$$\begin{aligned} -\langle S \rangle/k &= -\langle E \rangle/kT - \ln q \\ \langle E \rangle - \langle S \rangle T &= -kT \ln q \end{aligned} \tag{7.2.36}$$

If the average E and S are associated with the thermodynamic energy and entropy of a single residue, the thermodynamic equation

$$A = E - TS \tag{7.2.37}$$

defines an average Helmholtz free energy in terms of the single-residue partition function:

$$\langle A \rangle = -kT \ln q \tag{7.2.38}$$

The partition function plays a central role in defining state probabilities and thermodynamic parameters. For the independent residues, the partition function is easily determined. The problem becomes more complicated when the residues are not independent. For example, an α-helix does not complete a helix "loop" until the N–H group of the ith residue hydrogen bonds to the CO group of the $(i+4)$th residue, as shown in Figure 7.7. The helix encloses three residues but there is only one hydrogen bond. The simplest helix requires two different energies—an energy for the residues which have the helix conformation but no hydrogen bond and an energy for the residue with a hydrogen bond. The helix of n residues then requires two helix starter energies, ε_s, and $n-2$ hydrogen-bonded helix energies, ε_h. The partition function for this length of helix is then

$$\exp(-2\varepsilon_s/kT)\exp[(n-2)\,\varepsilon_h/kT] \tag{7.2.39}$$

The total partition function for a chain of N residues which contains only a single length of helix is

$$Q = c^N + \sum_{j=3}^{N} (N-j+1)\, c^{N-j} v^2 h^{j-2} \tag{7.2.40}$$

where

$$v = \mathrm{cxp}(-\varepsilon_s/kT) \tag{7.2.41}$$

and the summation includes the minimum helix of three units through the maximum helix of N units. The degeneracy of $N-j+1$ is present since a helix of j units can begin at $N-j+1$ positions on the chain. It is common practice to include two additional residues at the end of the chain which do not participate in the helix formation. The modified partition function has a total of $N+2$ residues (Problem 7.5).

Since only the relative energies of each state are important in most calculations, ε_c is often defined as zero energy. The starter and hydrogen-bonded residues then have the Boltzmann factors

$$\sigma = v^2/c^2 \tag{7.2.42}$$

$$s = h/c \tag{7.2.43}$$

The partition function of Equation 7.2.40 becomes

$$\begin{aligned} Q &= 1 + \sum_{j=3}^{N} (N-j+1)\,\sigma s^{j-2} \\ &= 1 + \sum_{1}^{N-2} (N-j-1)\,\sigma s^{j} \end{aligned} \tag{7.2.44}$$

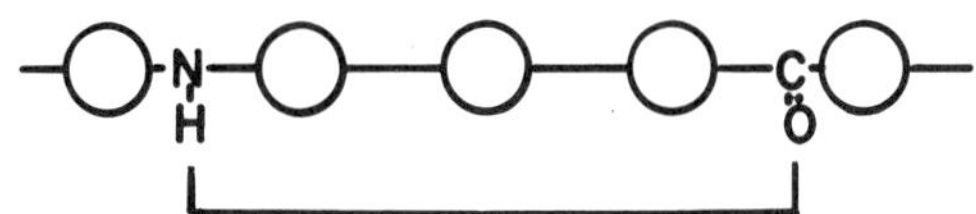

Figure 7.7. The hydrogen bonding scheme for an α-helix.

The starter residues highlight the difficulties which the independent residue models ignore. The energy of a given residue depends on the states of its neighbors. A single residue partition function must include residue–residue interaction energies as well as residue energies. This problem becomes tractable with matrix techniques.

7.3. Polymers with Nearest-Neighbor Interactions

The partition function of a system provides access to a set of thermodynamic and equilibrium parameters for the system. For polymer chains with independent residues, the partition function for a single residue yielded detailed information on the full chain. The problem becomes more complicated when interactions between residues are included. A helix includes binding between the ith and $(i+4)$th residues and a specific orientation of residues with the helical turn, and the associated energy must be included in the Boltzmann factors. The matrix methods introduced in this section permit the definition of a Boltzmann factor for each residue even though this residue has interactions with its neighbors.

Although it would be ideal to consider the i to $i+3$ interactions of helix, it is more convenient to start with a model which is a simple extension of the independent unit model of Section 7.1. Each residue is still restricted to two conformations, helix or random coil. In addition, the probabilities for these conformations will be modified by an energy of interaction with neighboring residues. The model will include only interactions with adjacent residues, i.e., nearest-neighbor interactions. Although this model is an approximation to real helical interactions, it does illustrate the techniques which must be used for these systems.

Although each independent residue had two energies for the two conformations, the addition of nearest-neighbor interactions requires a total of four energies; a helix surrounded by helix residues will have a different energy than a helix surrounded by random coil residues. Since the partition function must ultimately add energies for all residues in the chain, all interaction energies between two adjacent residues are assigned to the residue on the left. This "bookkeeping" procedure will include all $N-1$ interactions for an N-residue polymer. For the terminal residues, interactions with the solvent can also be included.

A given residue with interactions toward the right has four possible conformational energies. Residue i can be helix while the $(i+1)$th unit is either helix or random coil. If the ith residue is random coil, it can also be followed by a residue which is either helix or coil. These four possibilities will establish the Boltzmann factors associated with residue i. Boltzmann factors for the other identical residues in the chain are formed exactly the same way.

The different Boltzmann factors are defined using the energies ε_h and ε_c for the ith residue in the absence of interactions and four new interaction energies.

The interaction energy ε_{hh} is the interaction energy which must be added to the energy of the ith residue when this residue interacts with a helix to its right, and ε_{hc} is the interaction energy which must be added to ε_h for the ith residue when the $(i+1)$th residue is a random coil. The remaining two interaction energies, ε_{ch} and ε_{cc}, are defined in the same manner.

The energies and Boltzmann factors for each of the four possibilities are

i	$i+1$	ε	q_{ii}
h	h	$\varepsilon_h+\varepsilon_{hh}$	$\exp(-\varepsilon_h-\varepsilon_{hh})\beta=q_{hh}$
h	c	$\varepsilon_h+\varepsilon_{hc}$	$\exp(-\varepsilon_h-\varepsilon_{hc})\beta=q_{hc}$
c	h	$\varepsilon_c+\varepsilon_{ch}$	$\exp(-\varepsilon_c-\varepsilon_{ch})\beta=q_{ch}$
c	c	$\varepsilon_c+\varepsilon_{cc}$	$\exp(-\varepsilon_c-\varepsilon_{cc})\beta=q_{cc}$

(7.3.1)

The total Boltzmann factors for each conformation are q_{hh}, q_{hc}, q_{ch}, and q_{cc} with

$$\beta=1/kT \tag{7.3.2}$$

The four possible Boltzmann factors for a residue i require a matrix which is defined in tabular format. The first row contains Boltzmann factors for a helical residue when it is followed by helix or coil, respectively. The second row contains Boltzmann factors when a random coil is followed by a helix or coil. With this format, the columns define an $(i+1)$th residue which is either helix or coil. The matrix for the $(i+1)$th residue will have these helix or coil conformers as its rows. Matrix multiplication will couple the Boltzmann factors for successive residues to generate the full partition function.

The matrix of Boltzmann factors

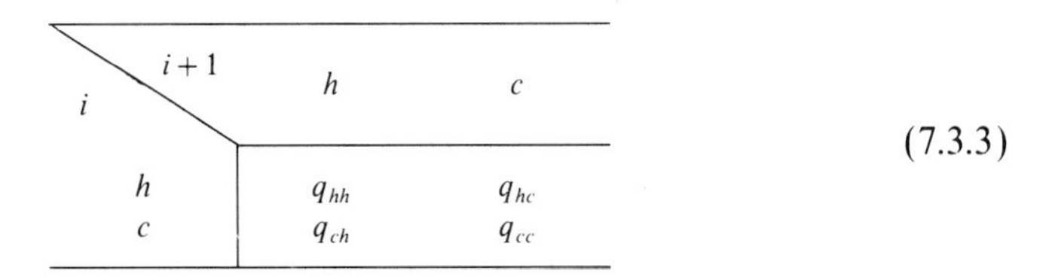

i \ $i+1$	h	c
h	q_{hh}	q_{hc}
c	q_{ch}	q_{cc}

(7.3.3)

generates the partition function for an N-residue chain. The process is best illustrated with a three-residue chain since the full partition function for this chain can be generated directly. For independent residues, the conformations *hch*, *chh*, and *hhc* all had equal energies and constituted a threefold degeneracy. With nearest-neighbor interactions, each of the eight possible products for the three residues is different. For example, the conformation with three helix units is now

$$q_{0h}q_{hh}q_{hh}q_{h0} \tag{7.3.4}$$

Because the terminal residues may interact with the solvent, Boltzmann factors

q_{0h} and q_{h0} are introduced to incorporate terminal residue–solvent interactions at the left and right, respectively. The product has four Boltzmann terms for the three residues and the solvent. Since a three-residue chain has only two nearest-neighbor interactions, q_{hh} appears twice in the product. The energy ε_h for the final residue appears in q_{h0}.

Each of the eight possible conformations is described by a product of four Boltzmann factors. For example, the conformation *hhc* gives

$$q_{0h}q_{hh}q_{hc}q_{c0} \tag{7.3.5}$$

while the conformation *hch* has the product

$$q_{0h}q_{hc}q_{ch}q_{h0} \tag{7.3.6}$$

Eight such products are summed to give the full partition function

$$\begin{aligned} Q = {} & q_{0h}q_{hh}q_{hh}q_{h0} + q_{0h}q_{hh}q_{hc}q_{c0} + q_{0h}q_{hc}q_{ch}q_{h0} + q_{0c}q_{ch}q_{hh}q_{h0} \\ & + q_{0h}q_{hc}q_{cc}q_{c0} + q_{0c}q_{ch}q_{hc}q_{c0} + q_{0c}q_{cc}q_{ch}q_{h0} + q_{0c}q_{cc}q_{cc}q_{c0} \end{aligned} \tag{7.3.7}$$

Although matrix multiplication is required to generate Boltzmann factors for several residues, the final partition function must be a simple summation of such factors. The terminal vectors produce the necessary scalar summation. A terminal helix or a terminal random coil interacts with a single neighbor, the solvent. The right terminal residue is described with a column vector rather than a matrix, i.e.,

$$\begin{pmatrix} q_{h0} \\ q_{c0} \end{pmatrix} \tag{7.3.8}$$

The left terminal residue requires a row vector:

$$\begin{pmatrix} q_{0h} & q_{0c} \end{pmatrix} \tag{7.3.9}$$

The row and column vectors will bracket a sequence of identical matrices. The entire quadratic form gives the partition function for the full chain.

The matrix format will generate the partition function for a three-residue chain. If the row and column vectors are defined as $\langle q|$ and $|q\rangle$, respectively, the partition function is

$$\begin{aligned} \langle q| \mathbf{QQ} |q\rangle &= \langle q| \mathbf{Q}^2 |q\rangle \\ &= \begin{pmatrix} q_{0h} & q_{0c} \end{pmatrix} \begin{pmatrix} q_{hh} & q_{hc} \\ q_{ch} & q_{cc} \end{pmatrix} \begin{pmatrix} q_{hh} & q_{hc} \\ q_{ch} & q_{cc} \end{pmatrix} \begin{pmatrix} q_{h0} \\ q_{c0} \end{pmatrix} \end{aligned} \tag{7.3.10}$$

For this case, it is most convenient to operate to the right with one of the **Q** matrices and to the left with the other. This produces the row vector

$$\begin{pmatrix} q_{0h}q_{hh} + q_{0c}q_{ch} & q_{0h}q_{hc} + q_{0c}q_{cc} \end{pmatrix} \tag{7.3.11}$$

and the column vector

$$\begin{pmatrix} q_{hh}q_{h0}+q_{hr}q_{r0} \\ q_{rh}q_{h0}+q_{rr}q_{r0} \end{pmatrix} \tag{7.3.12}$$

The product of these two vectors reproduces the partition function deduced directly from all the configurations:

$$\begin{aligned} &q_{0h}q_{hh}q_{hh}q_{h0}+q_{0h}q_{hh}q_{hc}q_{c0}+q_{0c}q_{ch}q_{hh}q_{h0}+q_{0c}q_{ch}q_{hc}q_{c0} \\ &\quad +q_{0h}q_{hc}q_{hh}q_{h0}+q_{0h}q_{hc}q_{cc}q_{c0}+q_{0c}q_{cc}q_{ch}q_{h0}+q_{0c}q_{cc}q_{cc}q_{c0} \end{aligned} \tag{7.3.13}$$

The partition function for a chain of N residues requires 2^N products and, for large N, generation of the entire partition function is tedious. However, the partition can be written simply in formal notation as

$$Q_t = \langle q| \mathbf{Q}^{N-1} |q\rangle \tag{7.3.14}$$

The matrix quadratic form can be converted to a more tractable form by expanding the function in its eigenvalues and projection operators. For the 2×2 matrix, the partition function becomes

$$\begin{aligned} \langle q| \mathbf{Q}^{N-1} |q\rangle &= \lambda_1^{N-1}\langle q| \mathbf{P}_1 |q\rangle + \lambda_2^{N-1}\langle q| \mathbf{P}_2 |q\rangle \\ Q &= c_1\lambda_1^{N-1}+c_2\lambda_2^{N-1} \end{aligned} \tag{7.3.15}$$

The extensive matrix product is reduced to a sum of two terms. The projection matrices operate only on the terminal vectors to give scalar factors.

The reduction technique is illustrated for the special case of independent units, i.e., no interactions. If ε_c is selected as the reference energy,

$$\begin{aligned} \exp(-\varepsilon_h/kT) &= h \\ \exp(-\varepsilon_c/kT) &= 1 \end{aligned} \tag{7.3.16}$$

The $\mathbf{Q}$ matrix is

$$\begin{pmatrix} h & h \\ 1 & 1 \end{pmatrix} \tag{7.3.17}$$

The right terminal residue has no solvent interactions; its helix form has Boltzmann factor h while its coil form has a factor 1:

$$|q\rangle = \begin{pmatrix} h \\ 1 \end{pmatrix} \tag{7.3.18}$$

The left-hand residue is generated by the first matrix in the sequence. The row vector contains the additional energy due to interactions with solvent. For convenience, both components of this vector are equated to 1, i.e., all solvation energy is placed in the column vector components:

$$\langle q| = (1 \quad 1) \tag{7.3.19}$$

The eigenvalues for the **Q** matrix are 0 and $1+h$. $\lambda = 0$ appears here because there are no interactions. Each unit is described by a single-residue partition function. The zero eigenvalue indicates the reduction of matrix order when no nearest-neighbor interactions are present.

The projection operator for $\lambda = 1+h$ is found using the Lagrange–Sylvester formula,

$$\begin{aligned} \mathbf{P}(1+h) &= \left[\begin{pmatrix} h & h \\ 1 & 1 \end{pmatrix} - \begin{pmatrix} 0 & 0 \\ 0 & 0 \end{pmatrix} \right] (1+h-0)^{-1} \\ &= \begin{pmatrix} h & h \\ 1 & 1 \end{pmatrix} (1+h)^{-1} \end{aligned} \tag{7.3.20}$$

The factor $\langle q| \, \mathbf{P} \, |q\rangle$ is

$$\begin{aligned} C &= \langle q| \, \mathbf{P} \, |q\rangle \\ &= (1 \quad 1) \begin{pmatrix} h & h \\ 1 & 1 \end{pmatrix} \begin{pmatrix} h \\ 1 \end{pmatrix} (1+h)^{-1} \\ &= 1+h \end{aligned} \tag{7.3.21}$$

The total partition function is

$$Q = C\lambda^{N-1} = (1+h)(1+h)^{N-1} = (1+h)^N \tag{7.3.22}$$

As expected, the partition function for this chain of independent units is the single residue partition function, $1+h$, raised to the Nth power.

With nearest-neighbor interactions, the analysis becomes slightly more difficult because there are two nonzero eigenvalues. The transfer matrix when helix units have an interaction energy is

$$\begin{pmatrix} ha & h \\ 1 & 1 \end{pmatrix} \tag{7.3.23}$$

where

$$a = \exp(-\varepsilon_{hh}/kT) \tag{7.3.24}$$

The matrix has two nonzero eigenvalues:

$$\begin{aligned} \lambda_+ &= \{(1+qa) + [(1-qa)^2 + 4q]^{1/2}\}/2 \\ \lambda_- &= \{(1+qa) - [(1-qa)^2 + 4q]^{1/2}\}/2 \end{aligned} \tag{7.3.25}$$

Although the eigenvalues are complicated, the two projection operators are still obtained using the Lagrange–Sylvester formula:

$$\mathbf{P}(+)=\begin{pmatrix}(ha-1)/2+D & h \\ 1 & (1-ha)/2+D\end{pmatrix}(2D)^{-1} \tag{7.3.26}$$

$$\mathbf{P}(-)=\begin{pmatrix}(ha-1)/2-D & h \\ 1 & (1-ha)/2-D\end{pmatrix}(-2D)^{-1} \tag{7.3.27}$$

with

$$D=[(1-ha)^2+4h]^{1/2} \tag{7.3.28}$$

These projection operators define the two scalar termination coefficients,

$$\begin{aligned} C_+ &= \langle q|\,\mathbf{P}(+)\,|q\rangle \\ C_- &= \langle q|\,\mathbf{P}(-)\,|q\rangle \end{aligned} \tag{7.3.29}$$

for the total partition function,

$$Q_t=C_+\lambda_+^{N-1}+C_-\lambda_-^{N-1} \tag{7.3.30}$$

Equation 7.3.30 is the complete partition function for the N-unit chain with nearest-neighbor interactions and is the sum of two separate products. In many cases, the partition function can be approximated very accurately with a single term. The factor D is subtracted from λ_- while it is added to λ_+. Then, λ_+ will be larger than λ_-. Since each eigenvalue must be raised to the $N-1$ power, the term containing λ_+ will dominate the sum. The partition function can be accurately approximated by deleting the $C_-\lambda_-^{(N-1)}$ term, i.e.,

$$Q_t\sim C_+\lambda_+^{N-1} \tag{7.3.31}$$

Many thermodynamic parameters depend on the logarithm of the partition function, and the logarithm of Equation 7.3.31 is a sum

$$\ln Q_t=\ln C_+ +\ln \lambda_+^{N-1} \tag{7.3.32}$$

Since C_+ involves only the terminal residues while λ_+^{N-1} includes all the inner residues, the second logarithm is the dominant term, and the partition function is accurately approximated as

$$\ln Q_t\sim (N-1)\ln \lambda_+ \tag{7.3.33}$$

The maximal eigenvalue of the matrix will dominate the partition function when N is large.

7.4. Other One-Dimensional Systems

Although long-chain homogeneous polymers with nearest-neighbor interactions are important, other systems also require interactions. These diverse physical systems can be studied using the transfer matrix–maximal eigenvalue techniques of Section 7.3. In this section, the technique is applied to a study of two such systems: (1) a line of ferromagnets which can be induced to align by an applied magnetic field or by the energy of interaction with their neighbors and (2) a lattice gas in which each "site" on a one-dimensional line can be empty or occupied by a particle. Interactions are possible between sites containing particles.

The line of identical, equidistant ferromagnets is illustrated in Figure 7.8. The ferromagnet is permitted to assume two configurations, in which the ferromagnet is aligned with the field (vector upward) or opposed to the field (vector downward). If there is no magnetic field, both states have the same energy and half the ferromagnets have the up or "+1" orientation. With an applied magnetic field, the fraction of "up" or "+1" vectors increases relative to the fraction of "down" or "−1" vectors.

The +1 and −1 units are distributed randomly along the chain if there are no nearest-neighbor interactions. This situation changes with the introduction of an interaction energy between nearest neighbors when they are aligned in the same direction; a +1 unit stabilizes +1 units on either side. Two adjacent −1 units will also stabilize each other, but adjacent ferromagnets of opposite orientation will be destabilizing. This interaction scheme will stabilize sequences of ferromagnets with the same orientation. There may be equal numbers of +1 and −1 units along the chain, but their location on the chain is no longer random. Like orientations will cluster as illustrated in Figure 7.9.

Since ferromagnets with the same sign, i.e., orientation, will have a net stabilization energy, it is convenient to define the interaction energy in terms of orientation "markers" +1 and −1. A stabilizing interaction energy is

$$\varepsilon_s = -\varepsilon(+1)(+1) = -(-1)(-1)\varepsilon = -\varepsilon \tag{7.4.1}$$

for adjacent ferromagnets of the same orientation and

$$\varepsilon_d = -\varepsilon(+1)(-1) = -\varepsilon(-1)(+1) = +\varepsilon \tag{7.4.2}$$

for two adjacent ferromagnets with opposed orientation. The Boltzmann factors are

$$q(+1, +1) = q(-1, -1) = q = \exp(+\beta\varepsilon) \tag{7.4.3}$$

$$q(+1, -1) = q(-1, +1) = q^{-1} = \exp(-\beta\varepsilon) \tag{7.4.4}$$

Figure 7.8. A line of identical ferromagnets with two possible orientations. Fifty percent have a +1 orientation and are distributed randomly along the line.

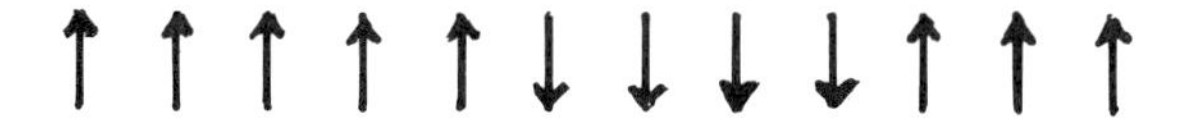

Figure 7.9. A 50–50 mixture of ferromagnets with clusters due to favorable interactions for ferromagnets with the same orientation.

where $(+1, +1)$ indicates the ith unit in a $+1$ state followed by an $(i+1)$th unit in the $+1$ state.

When nearest-neighbor interactions are allowed and no magnetic field is present, the fractions of $+1$ and -1 orientations are equal. However, the interactions will produce clusters of ferromagnets with the same orientation. The transfer matrix for this case is

$i \backslash i+1$	$+1$	-1
$+1$	q	q^{-1}
-1	q^{-1}	q

(7.4.5)

The characteristic equation for this matrix is

$$q^2 - q^{-2} - 2q\lambda + \lambda^2 = 0$$
$$q^4 - 1 - 2q^3\lambda + q^2\lambda^2 = 0 \tag{7.4.6}$$

with eigenvalues

$$\lambda_+ = q + q^{-1}$$
$$\lambda_- = q - q^{-1} \tag{7.4.7}$$

Once again, one of the two eigenvalues will dominate unless the interaction energy is extremely high. One eigenvalue will approach zero as the interaction energy approaches zero (Problem 7.7).

The system can be analyzed using the transfer matrix if there are boundary conditions. Terminal conditions for the long polymers appeared as row and column vectors. The resultant quadratic form gave the partition function. Cyclic boundary conditions provide an alternative termination technique.

With cyclic boundary conditions, the final ferromagnet is relocated adjacent to the first ferromagnet so all ferromagnets have the same interaction environment. The partition function for this system is simply $\mathbf{Q}$ raised to the Nth [not the $(N-1)$th] power,

$$\mathbf{Q}^N \tag{7.4.8}$$

The partition function is a scalar while Equation 7.4.8 is still a matrix. Some summation of $\mathbf{Q}^N$ is required. If $\mathbf{Q}$ were diagonal, then $\mathbf{Q}^N$ would also be diagonal:

$$\begin{pmatrix} \lambda^N & 0 \\ 0 & \lambda^N \end{pmatrix} \tag{7.4.9}$$

The partition function for the linear polymer was a sum of two terms. In this case, the partition function would be the sum of the two diagonal elements. There are no C_i because of the cyclic boundary conditions. The partition function is the trace of the diagonal matrix:

$$\text{tr}\, \mathbf{Q}_d^N = Q_t \tag{7.4.10}$$

Since the trace is an invariant of the matrix, the trace of the original, non-diagonal $\mathbf{Q}^N$ matrix will also give the partition function:

$$Q_t = \text{tr}\, \mathbf{Q}^N \tag{7.4.11}$$

Since $\mathbf{Q}^N$ is difficult to determine, it is easier to determine the eigenvalues for $\mathbf{Q}$. The partition function for this 2×2 system is then

$$Q_t = \sum \lambda_i^N \tag{7.4.12}$$

for the cyclic boundary conditions.

The partition function for the interacting ferromagnets in the absence of a magnetic field is

$$\begin{aligned} Q_t &= \lambda_+^N + \lambda_-^N \\ &= (q + q^{-1})^N + (q - q^{-1})^N \end{aligned} \tag{7.4.13}$$

This expression can be simplified using the exponential forms for q and q^{-1}. Defining

$$J = \varepsilon / kT \tag{7.4.14}$$

the exponential form of the partition function is

$$Q_t = [\exp(J) + \exp(-J)]^N + [\exp(J) - \exp(-J)]^N \tag{7.4.15}$$

The sum and difference of the exponentials define the cosh and sinh functions, respectively:

$$\begin{aligned} 2 \cosh J &= \exp(J) + \exp(-J) \\ 2 \sinh J &= \exp(J) - \exp(-J) \end{aligned} \tag{7.4.16}$$

Thus, the partition function is

$$Q_t = (2 \cosh J)^N + (2 \sinh J)^N \tag{7.4.17}$$

Because the λ_+ eigenvalue has the exponential sum, it is the larger eigenvalue, and for large N, the partition function is approximately

$$Q_t = (2 \cosh J)^N \tag{7.4.18}$$

Since $J = \varepsilon/kT$, the partition function is a function of temperature. If the interaction energy is larger than kT, ferromagnets of like orientation will cluster. As the temperature increases, the amount of clustering will decrease. This destabilization is monitored using the free energy of the system defined in Section 7.2,

$$A = -kT \ln Q_t \tag{7.4.19}$$

Since Q_t is the partition function for the entire chain, both sides of this equation are divided by N to give the free energy per ferromagnet,

$$-A'/kT = \ln(Q_t)/N \tag{7.4.20}$$

which provides a measure of like-orientation clustering with temperature.

A magnetic field applied to a chain of ferromagnets alters the 50–50 distribution. A greater fraction of the ferromagnets will align with the magnetic field, and the Boltzmann factors for each state must be changed accordingly. The energies of ferromagnets of either orientation will be proportional to the magnetic field H, and the ratio H/kT for the aligned ferromagnet is

$$B = -(+1)H/kT \tag{7.4.21}$$

The ratio B' for the field-opposed ferromagnets is

$$B' = -(-1)H/kT = +H/kT = -B \tag{7.4.22}$$

The Boltzmann factors are

$$\begin{aligned} q_+ &= \exp(B) \\ q_- &= \exp(-B) \end{aligned} \tag{7.4.23}$$

The transfer matrix must now include these additional Boltzmann factors. There are again four possibilities. If a $+1$ is followed by a $+1$, the ith term must have the factors $\exp(+B)$ and $\exp(+J)$. If the ith term is $+1$ followed by a -1, however, the ith unit still has the factor $\exp(+B)$ since it remains aligned but it has an interaction energy $\exp(-J)$ because of the opposition of the ith and $(i+1)$th units. The situation is similar for the ith unit in a -1 configuration. A -1 followed by $+1$ will have $\exp(-B)$ and $\exp(-J)$ but the $(-1, -1)$

sequence will have $\exp(-B)$ and a net stabilization indicated by $\exp(J)$. The transfer matrix

$i \backslash i+1$	$+1$	-1
$+1$	$\exp(B)\exp(J)$	$\exp(B)\exp(-J)$
-1	$\exp(-B)\exp(-J)$	$\exp(-B)\exp(J)$

(7.4.24)

determines the partition function of a cyclic system. For three units, the partition function can be evaluated directly (Problem 7.19).

The eigenvalues for the transfer matrix are determined from the characteristic equation for the matrix,

$$\exp(2J) - \exp(-2J) - \lambda\{\exp(J)[\exp(B) - \exp(-B)]\} + \lambda^2 = 0 \tag{7.4.25}$$

as

$$\begin{aligned} &\exp(J)[\exp(B) - \exp(-B)]/2 \\ &\quad \pm \left(\frac{1}{2}\right)\{\exp(2J)[\exp(B) - \exp(-B)]^2 - 4[\exp(2J) - \exp(-2J)]\}^{1/2} \\ &= \exp(J)\cosh(B) \pm [\exp(2J)\cosh^2(B) - 2\sinh(2J)]^{1/2} \end{aligned} \tag{7.4.26}$$

The partition function is now the sum of the Nth powers of these two eigenvalues:

$$Q_t = \lambda_+^N + \lambda_-^N \tag{7.4.27}$$

This partition function defines the thermodynamic properties for the ferromagnetic chain.

The same techniques define partition functions and free energies for a one-dimensional lattice gas or a one-dimensional alloy. For the one-dimensional gas, a line is broken into a series of "cells" of equal length. The number M of such cells can be increased so that the "end" effects can be ignored and the transfer matrix eigenvalues will determine the partition function. A set of N molecules ($N < M$) are permitted to occupy the cells although each cell will hold no more than one molecule. If there are interactions between the atoms, the molecules will cluster in adjacent cells.

The Boltzmann factor for occupation of a cell is simply

$$z = \exp(\mu/kT) \tag{7.4.28}$$

where μ is the chemical potential of the molecule in the cell relative to the unoccupied cell. The Boltzmann factor z must appear for all cells which contain a molecule. In addition, an occupied cell followed by a second occupied cell will have an additional interaction energy, ε, with a Boltzmann factor

$$a = \exp(-\varepsilon/kT) \tag{7.4.29}$$

The transfer matrix (o, occupied; u, unoccupied) is

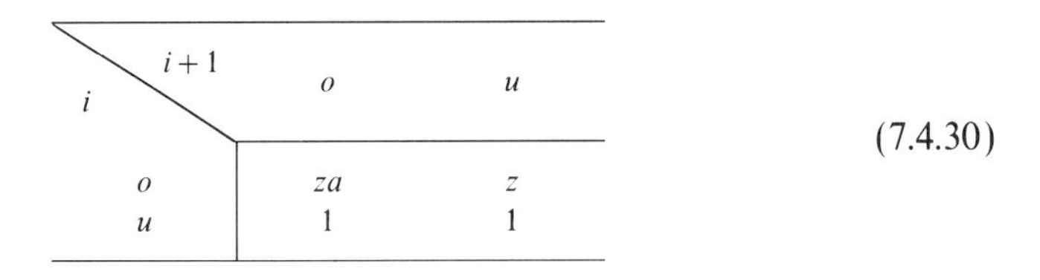

(7.4.30)

The characteristic equation

$$za - z - \lambda(za + 1) + \lambda^2 = 0 \tag{7.4.31}$$

parallels that determined for a polymer with helix–helix interactions in Section 7.3 and has the eigenvalues

$$\begin{aligned} &(1 + za)/2 + \frac{1}{2}(1 - 2za + z^2a^2 + 4z)^{1/2} \\ &(1 + za)/2 - \frac{1}{2}(1 - 2za + z^2a^2 + 4z)^{1/2} \end{aligned} \tag{7.4.32}$$

The sum of these eigenvalues raised to their Mth power gives the partition function for the lattice gas. If no interactions are present, $a = 1$ and the eigenvalues are

$$\lambda = 0; \qquad \lambda = (1 + z) \tag{7.4.33}$$

Each cell is independent and can be either occupied (z) or unoccupied (1). This suggests a simple definition for the chemical potential Boltzmann factor, z. The fraction of occupied cells when the cells are independent is

$$z/(1 + z) \tag{7.4.34}$$

For large M, the ratio N/M also equals the fraction of occupied cells and

$$\begin{aligned} z/(1 + z) &= N/M \\ z &= N/(M - N) \\ \mu/kT = \ln z &= \ln(N) - \ln(M - N) \end{aligned} \tag{7.4.35}$$

The chemical potential is proportional to the logarithm of the number (concentration) of molecules in the cells.

The one-dimensional binary alloy model is similar to the lattice gas. Each cell on the line can be occupied by species A or species B, and each cell must be occupied by one of these species (Problem 7.8).

7.5. Two-Dimensional Systems

The transfer matrix and its eigenvalues provide a convenient technique for the generation of partition functions for one-dimensional systems. The transfer matrix technique is applicable to two- and three-dimensional systems. Some examples for two-dimensional systems are illustrated in this section.

Two linear polymer chains are placed parallel to each other so that the corresponding units from each chain can bind as shown in Figure 7.10. For example, chain 1 might be poly-C with chain 2 poly-G, and the system is then a simple model for a double helix. If all residues from both chains are independent of the states of their neighbors, the total partition function for two chains of length N will be

$$Q_t = q^{2N} \tag{7.5.1}$$

This is equivalent to a partition function for a single chain of length $2N$ since the actual location of the residues is irrelevant when these residues are independent.

The two-dimensional model includes assumptions used for the one-dimensional systems. Each residue for the double helix has only two states, helix and random coil. Even though residues in each polymer are different, the helix and coil residue energies for each polymer are identical. The state of each residue, however, is now dictated by its neighbors along its chain, i.e., $i-1$, C and $i+1$, C for poly-C, and also by its nearest neighbor on the second chain, i.e., i, G on poly-G.

The transfer matrix is generated by considering matching residues on the two chains as a single unit. Each of these units has four distinct conformations. The conformations are listed with two indices for the conformations of the first and second residues of the pair, i.e., the conformations of the C and G residues. The conformations for the ith pair are

$$hh',\ hc',\ ch',\ cc' \tag{7.5.2}$$

Since the $(i+1)$th pair also has four conformations, the transfer matrix for the double helix is a 4×4 matrix.

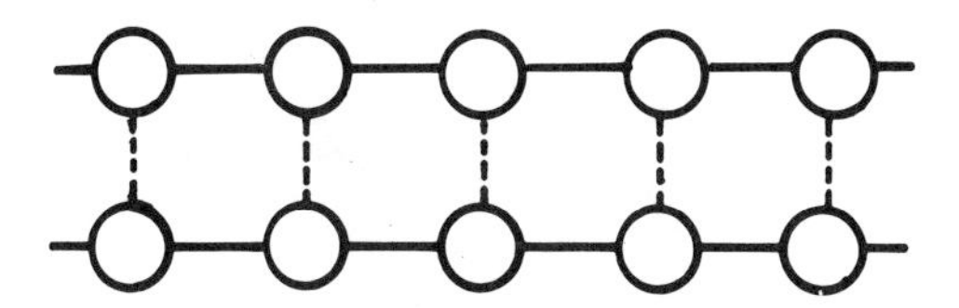

Figure 7.10. A simple double helix with intra- and interchain interactions.

The transfer matrix includes any interactions between the two paired residues on the chains. For example, the Boltzmann factor hh' includes the energy for creating each helical conformation plus the energy of interaction between these units. The (hh', hh') matrix element contains factors with these energies plus the energy of interaction if hh' is followed by an hh' at the $(i+1)$th position. Sixteen matrix elements are necessary to include all possible energy combinations. An hh' pair followed by an hc' pair has the h and h' helix energies and their interaction energy plus the interaction energy for h followed by h on chain 1 and h' followed by c' on chain 2.

The total transfer matrix is now

$$\begin{pmatrix} q_{hh',hh'} & q_{hh',hc'} & q_{hh',ch'} & q_{hh',cc'} \\ q_{hc',hh'} & q_{hc',hc'} & q_{hc',ch'} & q_{hc',cc'} \\ q_{ch',hh'} & q_{ch',hc'} & q_{ch',ch'} & q_{ch',cc'} \\ q_{cc',hh'} & q_{cc',hc'} & q_{cc',ch'} & q_{cc',cc'} \end{pmatrix} \tag{7.5.3}$$

Although the matrix is more complicated because there are more possibilities, the Nth power of its largest eigenvalue provides an excellent approximation for the chain partition function. The eigenvalues give the full solutions for cyclic boundary conditions when the final pair of units are placed adjacent to the initial pair (Figure 7.11). For shorter double helices, terminal boundary conditions must be used and determination of the projection matrices and terminal row and column vectors for the system is required.

To illustrate the form of this matrix, consider the case where the helix and random units have the Boltzmann factors

$$h = \exp(-\varepsilon_h/kT) \tag{7.5.4}$$

$$c = \exp(-\varepsilon_c/kT) = 1 \tag{7.5.5}$$

and any two adjacent helix units (on the same chain or adjacent units on the two

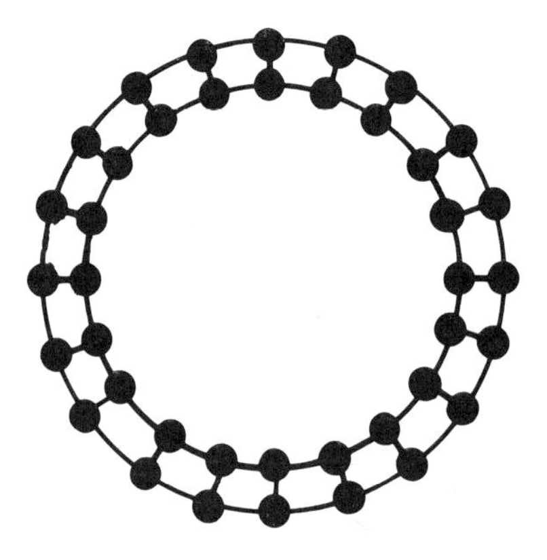

Figure 7.11. Cyclic boundary conditions for the double helix.

chains) are characterized by an interaction energy, ε, which gives a Boltzmann factor

$$a = \exp(-\varepsilon/kT) \tag{7.5.6}$$

The transfer matrix is

$$\begin{pmatrix} h^2a^3 & h^2a^2 & h^2a^2 & h^2a \\ hca^2 & hca & hc & hc \\ hca & hc & hca & hc \\ cc & cc & cc & cc \end{pmatrix} \tag{7.5.7}$$

$$= \begin{pmatrix} h^2a^3 & h^2a^2 & h^2a^2 & h^2a \\ ha^2 & ha & h & h \\ ha & h & ha & h \\ 1 & 1 & 1 & 1 \end{pmatrix} \tag{7.5.8}$$

Since even this matrix is complicated, it is valuable to consider some simpler cases. For example, with no interaction energy, i.e., all units independent, the matrix is

$$\begin{pmatrix} h^2 & h^2 & h^2 & h^2 \\ h & h & h & h \\ h & h & h & h \\ 1 & 1 & 1 & 1 \end{pmatrix} \tag{7.5.9}$$

The appearance of a transfer matrix with identical columns was also observed for single-chain systems when the matrix provided extraneous information. At least one eigenvalue must be zero. In fact, for this 4×4 matrix, three of the four eigenvalues are zero since all four columns are identical. The three zero eigenvalues are associated with the eigenvectors

$$\begin{pmatrix} 1 \\ -1 \\ 1 \\ -1 \end{pmatrix} \quad \begin{pmatrix} 1 \\ -1 \\ -1 \\ 1 \end{pmatrix} \quad \begin{pmatrix} 1 \\ 1 \\ -1 \\ -1 \end{pmatrix} \tag{7.5.10}$$

Since the trace of the matrix is invariant, the fourth, nonzero eigenvalue is the sum of the diagonal elements. This eigenvalue is

$$h^2 + 2h + 1 \tag{7.5.11}$$

The components of the eigenvector for this eigenvalue are (Problem 7.9)

$$(h^2, h, h, 1) \tag{7.5.12}$$

The form of partition function derived in this manner could have been anticipated. If each unit is independent, then the 4×4 matrix describes pairs of units from chain 1 and chain 2. The partition function for such a pair is

$$h^2 + 2h + 1 = (h + 1)^2 \tag{7.5.13}$$

For two chains of N units each, the total partition function is

$$(h + 1)^{2N} \tag{7.5.14}$$

for 2^N independent residues.

If $a \gg 1$, the interaction energy is large and repeated matrix multiplications to generate the full partition function will generate the eigenvalue in the highest power of a. In the matrix the 11 element is the only element which is raised to the third power. For this case, two matrix operations on the trial column eigenvector with components

$$(1, 0, 0, 0) \tag{7.5.15}$$

give an eigenvalue product and eigenvector

$$(a^3)^2 \begin{pmatrix} a^3 \\ a^2 \\ a \\ 1 \end{pmatrix} \tag{7.5.16}$$

For these limiting conditions, a^3 becomes the maximal eigenvalue. The strong helix–helix interaction energy favors helical conformations for each member of the pairs and propagates this helical character along each of the chains as well. The third power contains the energies of the three interactions for one residue pair.

This method is easily generalized to larger two-dimensional systems. For example, a sequence with three aligned polymer chains would define a single i state by all possible arrangements of the two configurations in the three units, i.e.,

$$2^3 = 8 \text{ conformations per triplet} \tag{7.5.17}$$

The conformers for the ith triplet are

$$\begin{array}{cccc} & hhc & hcc & \\ hhh & hch & chc & ccc \\ & chh & cch & \end{array} \tag{7.5.18}$$

Although only nearest-neighbor interactions are allowed, an 8×8 transfer matrix is required in order to include all arrangements of the interacting triplets. In general, a series of n aligned chains with nearest-neighbor interactions would require a total of

$$2^n \qquad (7.5.19)$$

conformations for each n-tuplet. The transfer matrix would be

$$2^n \times 2^n \qquad (7.5.20)$$

The model with two adjacent polymer chains was selected because the total number of conformations for each pair was limited and the transfer matrix was only a 4×4 matrix. The transfer matrix for a two-dimensional ferromagnet system with a length of N units and a width of M units with nearest-neighbor interactions is

$$2^M \times 2^M \qquad (7.5.21)$$

but the procedures for its generation are identical to those used for the smaller system. Cyclic boundary conditions are used to bring the final M-tuplet adjacent to the initial M-tuplet.

The development of the transfer matrix parallels the development of transfer matrices for simpler systems. The major problem is the determination of the maximal eigenvalue for the large transfer matrix. Onsager developed an expression for the largest eigenvalue of a two-dimensional array of interacting ferromagnets in the absence of a magnetic field. Although the one-dimensional Ising models developed in Section 7.4 do not exhibit phase transitions, the two-dimensional Ising model does.

7.6. Non-Nearest-Neighbor Interactions

The transfer matrices developed for one- and two-dimensional systems all involved residues with nearest-neighbor interactions. Linear systems such as the α-helix can include residue–residue interactions which are not nearest neighbor. The transfer matrices for linear systems of this type are presented in this section, starting with an analysis of nearest- and next-nearest-neighbor interactions.

Consider a polymer in which the ith unit can interact with both the $(i+1)$th and $(i+2)$th units. How many configurations are possible for this ith unit? With nearest-neighbor interactions and residues with only two conformations, the four sequential conformations *hh*, *hc*, *ch*, and *cc* required four different Boltzmann factors:

$$2 \times 2 = 2^2 = 4 \qquad (7.6.1)$$

For interactions with two nearest neighbors, the ith unit can have

$$2 \times 2 \times 2 = 2^3 = 8 \qquad (7.6.2)$$

separate Boltzmann factors; each of these can have a different net energy. The eight configurations are

$$\begin{matrix} & hhc & cch & \\ hhh & hch & chc & ccc \\ & chh & hcc & \end{matrix} \tag{7.6.3}$$

The eight distinct Boltzmann factors must be placed into a 4×4 matrix with 16 matrix elements. In general, a residue which interacts with k neighbors requires a

$$2^k \times 2^k \tag{7.6.4}$$

matrix. Although a 3×3 matrix has nine elements for the eight Boltzmann factors of a next-nearest-neighbor model, these factors cannot be introduced in a consistent way (Problem 7.11).

For the 2×2 nearest-neighbor model, the configurations of the ith residue were used as indices for the columns while the configurations of the $(i+1)$th residue served as row indices. The four indices for the 4×4 matrix require two sequential labeling indices, i.e., the column indices must be labeled *hh*, *hc*, *ch*, and *cc*. The first unit is i while the second is $i+1$. The row indices for the transfer matrix must begin with the $(i+1)$th unit. The pairs *hh*, *hc*, *ch*, and *cc* for the column indices label $i+1$ and $i+2$ residues in the three-residue sequence for i. The $i+1$ index must appear in both the row and column indices for the transfer matrix of the next-nearest-neighbor model.

The transfer matrix is generated with the table

$i, i+1$ \ $i+1, i+2$	*hh*	*hc*	*ch*	*cc*
hh				
hc				
ch				
cc				

(7.6.5)

Some elements in this transfer matrix have physical significance while others do not. For example, the (*hh*, *hh*) element has i, $i+1$, and $i+2$ all in the h configuration. This element for the ith residue includes the extra interaction energy generated by $i+1$ and $i+2$ as helices. The (*hc*, *ch*) element at (2, 3), *hch*, is a helical ith residue followed by c and h residues. There are only 8 distinct conformations but the transfer matrix has a total of 16 distinct elements. Some elements in the matrix are zero.

The (*hh*, *ch*) (1, 3) element has an i index *hh* but an $i+1$ index of *ch*. There is contradictory information here. One index states that the $(i+1)$th element must be h while the second states that it must be c. Because the element has no physical significance, it is equated to zero. Within the 16 elements of the matrix, there are 8 nonzero elements and 8 zero, or nonsense, elements.

With the nonsense zeros, the transfer matrix is

$$\begin{pmatrix} q_{hhh} & q_{hhc} & 0 & 0 \\ 0 & 0 & q_{hch} & q_{hcc} \\ q_{chh} & q_{chc} & 0 & 0 \\ 0 & 0 & q_{cch} & q_{ccc} \end{pmatrix} \tag{7.6.6}$$

Each row has only two nonzero elements.

A minimum of three helix units is required to generate the first hydrogen bond of this helix. In other words, an energy ε_h will appear for the ith unit only when $i+1$ and $i+2$ are also helix. A sequence hhc will have the two starter residues. The ith residue in this case will have the starter energy ε_s and the Boltzmann factor

$$v = \exp(-\varepsilon_s/kT) \tag{7.6.7}$$

Sequences hcc with a single helix residue will also have this starter energy. Any sequence which begins with c will have an energy ε_c since the coil orientation is not affected by the presence of a nearby helical residue.

When the starter factor v is used for the first two helix units in a section of helix, the Boltzmann factors h and c are redefined as

$$\begin{aligned} u &= \exp(-\varepsilon_c/kT) \\ w &= \exp(-\varepsilon_h/kT) \end{aligned} \tag{7.6.8}$$

The transfer matrix for the two nearest-neighbor interacting polymers is

$$\begin{pmatrix} w & v & 0 & 0 \\ 0 & 0 & v & v \\ u & u & 0 & 0 \\ 0 & 0 & u & u \end{pmatrix} \tag{7.6.9}$$

Each nonzero element contains only a single factor because this transfer matrix describes only the ith residue. The $(i+1)$th and $(i+2)$th residues simply determine the Boltzmann factors which must be placed in the transfer matrix. All nonzero elements in the lower two rows are u because the ith unit is a noninteracting random coil.

The presence of many zero elements in the transfer matrix results in a tractable characteristic equation for the matrix. The equation can be determined directly or via matrix invariants (Problem 7.14) as

$$\lambda^4 - (w+u)\,\lambda^3 + (wu - uv)\,\lambda^2 - (uv^2 - uvw)\,\lambda = 0 \tag{7.6.10}$$

λ is common to each term so that one eigenvalue is

$$\lambda = 0 \tag{7.6.11}$$

The appearance of a zero eigenvalue indicates that the transfer matrix has too much information and could have been reduced to a 3×3 matrix. The techniques for reducing the matrix order before generating the characteristic polynomial are introduced in the next section.

The remaining three eigenvalues are the roots of the third-order polynomial

$$\lambda^3 - (w+u)\,\lambda^2 + (wu - uv)\,\lambda^1 - (uv^2 - uvw) = 0 \tag{7.6.12}$$

An α-helix requires interaction with three sequential nearest neighbors. Determination of the energy and Boltzmann factor for the ith residue requires knowledge of four sequential conformations. The matrix in this case is

$$2^3 \times 2^3 = 8 \times 8 \tag{7.6.13}$$

The matrix will have 64 possible elements. For the sequence of four residues with two conformers for each residue, a total of

$$2^4 = 16 \tag{7.6.14}$$

Boltzmann factors are required. These must be incorporated into the 64 elements of the transfer matrix.

In order to sequence four residues in the eight columns and rows of the transfer matrix, each column and row must be indexed with a sequence of three residues. The indices are

$$hhh,\ hhc,\ hch,\ hcc,\ chh,\ chc,\ cch,\ ccc \tag{7.6.15}$$

The first index defines the ith residue for the rows and the $(i+1)$th residue for the columns. The $(i+1)$th and $(i+2)$th residues must match in both the row and column index for a nonzero matrix element. The table with acceptable sequences listed is

			$i+3$	*h*	*c*	*h*	*c*	*h*	*c*	*h*	*c*
			$i+2$	*h*	*h*	*c*	*c*	*h*	*h*	*c*	*c*
			$i+1$	*h*	*h*	*h*	*h*	*c*	*c*	*c*	*c*
i,	$i+1$,	$i+2$									
h	*h*	*h*		*hhhh*	*hhhc*	0	0	0	0	0	0
h	*h*	*c*		0	0	*hhch*	*hhcc*	0	0	0	0
h	*c*	*h*		0	0	0	0	*hchh*	*hchc*	0	0
h	*c*	*c*		0	0	0	0	0	0	*hcch*	*hccc*
c	*h*	*h*		*chhh*	*chhc*	0	0	0	0	0	0
c	*h*	*c*		0	0	*chch*	*chcc*	0	0	0	0
c	*c*	*h*		0	0	0	0	*cchh*	*cchc*	0	0
c	*c*	*c*		0	0	0	0	0	0	*ccch*	*cccc*

(7.6.16)

Each row contains only two nonzero elements.

7.7. Reduction of Matrix Order

In Section 7.6, the model for a linear polymer with nearest- and next-nearest-neighbor interactions required a 4×4 matrix. The characteristic polynomial for this transfer matrix contained a single $\lambda = 0$ eigenvalue. Since this eigenvalue makes no contribution to the total partition function, a 3×3 matrix gives exactly the same information. The 4×4 matrix is reduced by locating redundant rows and columns within the transfer matrix which contain exactly the same information. These rows and columns can be combined to reduce the rank of the matrix.

Zero eigenvalues have been encountered for simpler transfer matrix systems. For example, the nearest-neighbor model without interactions has a transfer matrix

$$\begin{pmatrix} q & q \\ 1 & 1 \end{pmatrix} \tag{7.7.1}$$

where q is the Boltzmann factor for the helix and 1 indicates the random coil configuration used as a reference ($\varepsilon_c = 0$). The characteristic equation

$$-(1+q)\lambda + \lambda^2 = 0 \tag{7.7.2}$$

gives eigenvalues of 0 and $1+q$. The latter is immediately recognizable as the unit partition function since 1 and q are the Boltzmann factors possible for this two-state model.

This result suggests that the 2×2 matrix should be reducible to a 1×1 matrix. The diagonalized transfer matrix shows this clearly:

$$\begin{pmatrix} 1+q & 0 \\ 0 & 0 \end{pmatrix} \tag{7.7.3}$$

This diagonalization normally requires a knowledge of eigenvalues and eigenvectors for the 2×2 matrix. This can be avoided if the matrix can be reduced to a 1×1 matrix before diagonalization. The 2×2 matrix assigns separate indices for the helix and random coil since these are distinct residue states. These two distinct states must be reduced to a single index when the residues are independent of their neighbors.

The table for the transfer matrix is

i \ $i+1$	h	c
h	q	q
c	1	1

(7.7.4)

The two columns are identical because it does not matter what the $(i+1)$th unit is for this independent unit case. This is the key to the reduction. Both columns are combined by noting that h or c $(h \cup c)$ was acceptable for the $(i+1)$th unit, i.e., each gave the same matrix element. This reduction generates a new transfer matrix table and a 2×1 transfer matrix:

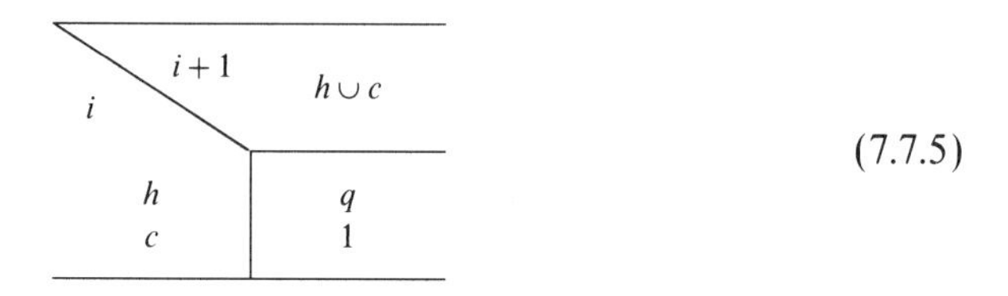

i \ $i+1$	$h \cup c$
h	q
c	1

(7.7.5)

The reduction to one column states that $i+1$ does not affect the ith unit. However, what about the $(i-1)$th unit? It will precede the 2×1 matrix above but it should also encounter a situation where $h \cup c$ is possible for i. In other words, the symmetry for successive residues in the chain requires a transfer matrix table

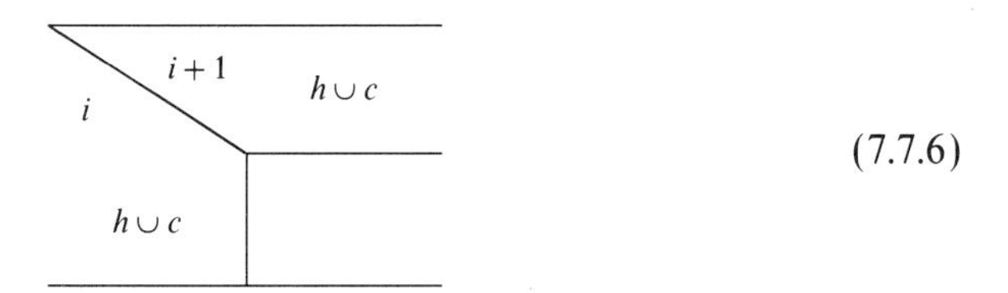

i \ $i+1$	$h \cup c$
$h \cup c$	

(7.7.6)

Eigenvalue analysis has established that the single element in the transfer matrix table must be $1+q$, the partition function for one residue. This sum and the reduction of matrix size both reflect the fact that the ith unit is an independent unit. Its full partition function is prepared without regard to the elements before or after it. Because helix (q) and coil (1) are possible, they are added to form the single element.

The 4×4 transfer matrix for next-nearest-neighbor interactions (Section 7.6) gave a characteristic polynomial which was third order in the nonzero eigenvalues. This suggests that an analysis similar to that for the 2×2 matrix will permit a reduction of matrix order from 4 to 3. The transfer matrix is

$i, i+1$ \ $i+1, i+2$	hh	hc	ch	cc
hh	w	v	0	0
hc	0	0	v	v
ch	u	u	0	0
cc	0	0	u	u

(7.7.7)

The appearance of the factor u in the lower two rows suggests that a random unit at the ith position is not affected by the configuration of the $(i+1)$th and $(i+2)$th units. When i is random, $i+1$ and $i+2$ can be either helix or random coil.

Since the ith residue as coil conformer has a Boltzmann factor u if it is followed by either helix or coil, consider the consequences of replacing the final two columns by a single entry,

$$c \quad c \cup h \tag{7.7.8}$$

since u is the result in each case. The row indices must also be reduced. The transfer matrix is

$i, i+1$ \ $i+1, i+2$	hh	hc	$c \quad c \cup h$
hh	w	v	0
hc	0	0	v
$c \quad c \cup h$	u	u	u

(7.7.9)

The bottom row is always u no matter what the subsequent states might be. Note also that before the reduction, two columns of the matrix were identical. Since a determinant is zero when any two columns (or rows) have identical elements, there had to be one $\lambda = 0$ eigenvalue. The reduction of the matrix has eliminated both the duplicate row and the duplicate column to give a matrix with no zero eigenvalue.

This result can be verified by determining the characteristic equation for the new matrix. The characteristic polynomial is

$$\begin{gathered} (w-\lambda)(-\lambda)(u-\lambda) + uv^2 - uv(w-\lambda) = 0 \\ -\lambda^3 + (w+u)\,\lambda^2 - (uw - uv)\,\lambda + uv^2 - uvw = 0 \end{gathered} \tag{7.7.10}$$

as obtained previously from the 4×4 transfer matrix.

The reduction of the transfer matrices is facilitated by observing certain row and column characteristics. For the 4×4 matrix, the appearance of the Boltzmann factor u in pairs of corresponding columns and rows indicated a characteristic determinant with two identical columns and a zero eigenvalue. Combining the pairs of rows and columns produced a 3×3 matrix with no zero eigenvalue.

The 8×8 matrix for three nearest-neighbor interactions,

		$i+3$	h	c	h	c	h	c	h	c
		$i+2$	h	h	c	c	h	h	c	c
		$i+1$	h	h	h	h	c	c	c	c
i	$i+1$	$i+2$								
h	h	h	w	w	0	0	0	0	0	0
h	h	c	0	0	v	v	0	0	0	0
h	c	h	0	0	0	0	v	v	0	0
h	c	c	0	0	0	0	0	0	u	u
c	h	h	u	u	0	0	0	0	0	0
c	h	c	0	0	u	u	0	0	0	0
c	c	h	0	0	0	0	u	u	0	0
c	c	c	0	0	0	0	0	0	u	u

(7.7.11)

has elements which were chosen for next-nearest-neighbor conditions. The pattern is apparent. Every pair of columns has exactly the same elements. The $i+3$ element makes absolutely no difference, and this system could be reduced quickly to the 4×4 system discussed above by replacing each successive *hc* pair for the $i+3$ index by $h \cup c$. The presence of identical pairs of columns signaled this reduction immediately.

The 8×8 transfer matrix is more complicated if configurations with only one or two helix units are excluded. In the transfer matrix above, if the ith unit was the first or second helical unit in any series of h units, it was assigned a Boltzmann factor v. Now, a v is only assigned to the ith unit if both the $(i+1)$th and $(i+2)$th units are also helical. In this case, configurations like *hchc* must be assigned a Boltzmann factor of 0 since they cannot exist. To be able to determine if residue i is part of a sequence of at least three helix residues, the conformations of residues on either side of this residue must be known. The transfer matrix indexing scheme is changed to begin with the $(i-2)$th residue. The ith unit for the matrix now appears in the center of the sequence. The transfer matrix is

		$i+1$	h	c	h	c	h	c	h	c
		i	h	h	c	c	h	h	c	c
		$i-1$	h	h	h	h	c	c	c	c
$i-2$	$i-1$	i								
h	h	h	w	v	0	0	0	0	0	0
h	h	c	0	0	u	u	0	0	0	0
h	c	h	0	0	0	0	v	0.	0	0
h	c	c	0	0	0	0	0	0	u	u
c	h	h	w	0.	0	0	0	0	0	0
c	h	c	0	0	u	u	0	0	0	0
c	c	h	0	0	0	0	v	0.	0	0
c	c	c	0	0	0	0	0	0	u	u

(7.7.12)

A decimal point has been added to indicate the zeros which result from chains which are definitely one or two units, e.g., *cchc* or *chhc*. In addition, since each helix must contain two v units, this configuration is used to mark an i which either ends or begins a helical chain. For example, the 12 element is v because an ith unit h is followed by an $(i+1)$th unit c.

The transfer matrix has pairs of columns with the same elements. When the columns are numbered sequentially from 1 to 8 moving left to right, columns 3 and 4 and columns 7 and 8 have the same elements and they should reduce to a single column. However, the elements for the corresponding rows are not all equal. For example, row 7 contains a v while row 8 does not. The same situation is observed for rows 3 and 4. However, there is no contradictory information. The v in row 7 results from an allowed configuration while the 0 in row 8 arises because there is no conformation. The transfer matrix is reduced via 7 and 8 and 3 and 4 to give

$$\begin{pmatrix} w & v & 0 & 0 & 0 & 0 \\ 0 & 0 & u & 0 & 0 & 0 \\ 0 & 0 & 0 & v & 0 & u \\ w & 0 & 0 & 0 & 0 & 0 \\ 0 & 0 & u & 0 & 0 & 0 \\ 0 & 0 & 0 & v & 0 & u \end{pmatrix} \tag{7.7.13}$$

When columns 3, 4 and 7, 8 are combined in this manner, the new matrix has row pairs (2, 5) and (3, 6) with identical elements. These rows are combined and rearranged to the order 1, (2, 5), 4, (3, 6) to give

$$\begin{pmatrix} w & v & 0 & 0 \\ 0 & 0 & 0 & u \\ w & 0 & 0 & 0 \\ 0 & 0 & v & u \end{pmatrix} \tag{7.7.14}$$

The columns could be combined because the elements were identical. The corresponding rows retained the nonzero elements when combined. The order of the rows and columns was then changed to demonstrate a logic in the choice of combinations. The first and third rows of the final matrix retain their original factors. In addition, the first column contains w, which indicates that $i-1$, i, and $i+1$ must be h. However, the two factors w could be obtained from configurations *hhhh* and *chhh* so these become the first- and third-row indices. The first column must have indices *hhh* to generate w. The table has the partial indices

		$i+1$	h
		i	h
		$i-1$	h
$i-2$	$i-1$	i	
h	h	h	w
—	—	—	0
c	h	h	w
—	—	—	0

(7.7.15)

Consider the second row which contains a single factor u. The ith residue must be c but the $(i-1)$th and $(i-2)$th residues can be either h or c. Since the first column is known and declares that $i-1$ must be h to generate a w, the second row has the indices $h \cup c$, h, c. The corresponding column must have consistent indices. This is true for row 3 and column 3 since each residue must follow the same indexing. The indices for the first three rows and columns are

		$i+1$	h	c	h	
		i	h	h	h	
		$i-1$	h	$h \cup c$	c	
$i-2$	$i-1$	i				
h	h	h	w	v	0	—
$h \cup c$	h	c	0	0	0	u
c	h	h	w	0	0	0
—	—	—	0	0	v	u

(7.7.16)

Note the role of $h \cup c$ in each position. Since the ith unit h is followed by c, it makes absolutely no difference whether $i-1$ is an h or a c; the result is v in either case.

The final row and column indices are more difficult. The $i-2$ unit will have no effect on the row elements v and u, which appear only for hc, ch, or cc type configurations. The $i-2$ row element and the $i-1$ column element must then be $h \cup c$. Since the third-column indices are chh and the element itself is v, the $(i-1)$th row index is c. Since $i-2$ is already $h \cup c$, an $h \cup c$ at $i-1$ leaves open the possibility of a helix string. The c prevents this possibility. The $(i+1)$th column index is also established because the u configuration depends on neither $i-1$ nor $i+1$. Thus, $i+1$ can be $h \cup c$. The final set of indices is

		$i+1$	h	c	h	$h \cup c$
		i	h	h	h	c
		$i-1$	h	$h \cup c$	c	$h \cup c$
$i-2$	$i-1$	i				
h	h	h	w	v	0	0
$h \cup c$	h	c	0	0	0	u
c	h	h	w	0	0	0
$h \cup c$	c	$h \cup c$	0	0	v	u

(7.7.17)

Since such reductions can be complicated, it is best to determine common rows and columns in the matrix, combine them, and then verify the result by determining a consistent set of indices for the transfer matrix.

7.8. The Kinetic Ising Model

Ising models are used to determine the partition function and the physical properties of systems where there are energies of interaction between the units of the system. The interaction energy will enhance certain configurations relative to others. However, these configurations will characterize a system at equilibrium. If the system is perturbed from equilibrium, the unit–unit interactions will influence the kinetics of relaxation to equilibrium. Units without interactions decay independently and produce the characteristic first-order kinetics. When such units interact, the kinetic process becomes more complicated.

The equilibrium properties of a line of interacting ferromagnets were developed in Section 7.5. Each ferromagnet can exist only in $+1$ (up) and -1 (down) configurations. The equilibrium fractions of these configurations depend on their Boltzmann factors. These equilibrium fractions, $p^{\mathrm{eq}}(+1)$ and $p^{\mathrm{eq}}(-1)$, are time independent. When this system is perturbed from equilibrium, these probabilities will change as the system returns to equilibrium. The probability that unit i is $+1$ at time t is

$$p(+1, t) \tag{7.8.1}$$

while the probability that it is -1 is

$$p(-1, t) \tag{7.8.2}$$

Since these are the only two states possible for this unit, these probabilities must sum to unity,

$$p(+1, t) + p(-1, t) = 1 \tag{7.8.3}$$

at all times.

In the absence of a magnetic field, the equilibrium probabilities are equal:

$$p(+1, t) = p(-1, t) = 0.5 \tag{7.8.4}$$

When independent units are perturbed from equilibrium, their evolution to equilibrium follows first-order kinetics,

$$\begin{aligned} dp(+1, t)/dt &= -(k/2)\, p(+1, t) + (k/2)\, p(-1, t) \\ dp(-1, t)/dt &= (k/2)\, p(+1, t) - (k/2)\, p(-1, t) \end{aligned} \tag{7.8.5}$$

where the rate constant $k/2$ will lead to a decay constant of k.

Equation 7.8.4 relates $p(-1, t)$ to $p(+1, t)$:

$$p(-1, t) = 1 - p(+1, t) \tag{7.8.6}$$

This is substituted into the first differential equation to give

$$dp(+1, t)/dt = (k/2) - kp(+1, t) \tag{7.8.7}$$

The equation can be solved with the integrating factor

$$\exp(kt) \tag{7.8.8}$$

to give

$$p(+1, t) = \frac{1}{2} + [p(+1, 0) - p(-1, 0)] \exp(-kt) \tag{7.8.9}$$

The difference between the up and down probabilities at $t = 0$ when the perturbation is applied will decay to zero leaving the equilibrium probabilities of $\frac{1}{2}$.

The probability $p(-1, t)$ for independent units is

$$p(-1, t) = \frac{1}{2} - [p(+1, 0) - p(-1, 0)] \exp(-kt) \tag{7.8.10}$$

since $p(+1, t)$ and $p(-1, t)$ must sum to unity.

The nature of the kinetics changes when nearest neighbors influence the kinetics of the ith residue. If the ith unit interacts with its nearest neighbors, eight possible arrangements exist for this unit:

$$\begin{array}{cccccccccccc} & & & +1 & +1 & -1 & -1 & -1 & +1 & & & \\ +1 & +1 & +1 & +1 & -1 & +1 & -1 & +1 & -1 & -1 & -1 & -1 \\ & & & -1 & +1 & +1 & +1 & -1 & -1 & & & \end{array} \tag{7.8.11}$$

The ith unit is more likely to react if it can attain a configuration which stabilizes it. The configuration (1, 1, 1) will have a small transition probability since it is difficult to leave the stabilized configuration. Similarly, the configuration $(-1, -1, -1)$ is equally stable due to stabilization by its neighbors in like configurations. On the other hand, the configuration

$$+1 \quad -1 \quad +1 \tag{7.8.12}$$

must have a larger transition probability since reaction of the ith unit stabilizes it. The same is true for the transition from $(-1, +1, -1)$ to $(-1, -1, -1)$.

The remaining four configurations have a central unit surrounded by units with opposed orientations:

$$+1 \quad +1 \quad -1; \qquad -1 \quad +1 \quad +1; \qquad -1 \quad -1 \quad +1; \qquad +1 \quad -1 \quad -1 \tag{7.8.13}$$

The nearest neighbors can produce no net stabilization because they are opposed.

The kinetics of this system differs from the system with independent units since transition probabilities for the ith unit will depend on the orientations of its neighbors. Since the rate expression will depend on both the ith residue and its neighbors, its order is greater than first order.

Since the ith residue can be located in three distinct environments, three transition probabilities are required in a rate expression for the change of the ith ferromagnet. If two neighboring states are oriented opposite to the ith unit, the transition probability is enhanced as

$$(k/2)(1+a) \tag{7.8.14}$$

where a is the fractional enhancement of the transition probability for this favorable configuration. If these neighbors have the same sign as the ith unit, the transition probability is decreased by the fraction a:

$$(k/2)(1-a) \tag{7.8.15}$$

Finally, if the two neighbors have opposite signs, there is no net interaction and the transition probability is

$$(k/2) \tag{7.8.16}$$

The three kinetic transitions are illustrated in Figure 7.12.

These expressions are generalized as a single transition probability using $s_i = +1$ or -1 as the index for the ith unit. The transition probabilities are now

$$k_i = (k/2)[1 - as_i(s_{i-1} + s_{i+1})] \tag{7.8.17}$$

This expression gives each of the three possible transition probabilities.

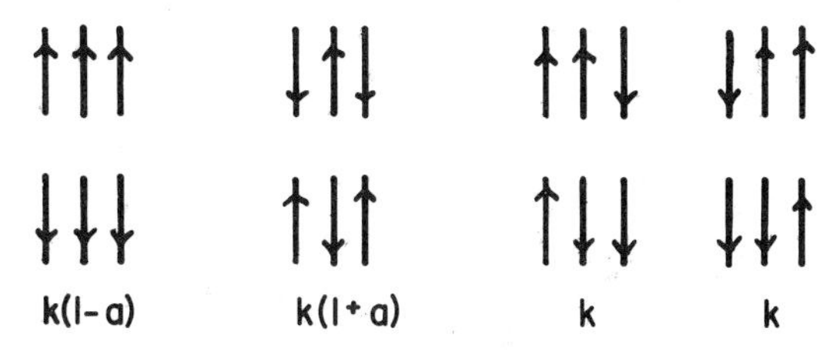

Figure 7.12. The interactive configurations for a residue undergoing kinetic interactions with its nearest neighbors.

The transition probabilities defined by Equation 7.8.17 differ from normal first-order transition probabilities which generally have the Arrhenius form,

$$k = A \exp(-E_a/kT) \tag{7.8.18}$$

However, the parameter a can be related to the interaction energy using the condition of detailed balance. For the transition,

$$p(+1, -1, +1) \rightleftharpoons p(+1, +1, +1) \tag{7.8.19}$$

detailed balance requires that the forward and reverse rates be equal at equilibrium. Using the transition probabilities for the forward and reverse kinetic steps gives

$$k(1+a)\, p(+1, -1, +1) = k(1-a)\, p(+1, +1, +1) \tag{7.8.20}$$

or

$$p(+1, +1, +1)/p(+1, -1, +1) = k(1+a)/k(1-a) \tag{7.8.21}$$

The ratio of the two probabilities will also depend on the interaction energies for each of the two configurations. If $p(+1, +1, +1)$ is stabilized by an energy $-\varepsilon/2$ while $p(+1, -1, +1)$ is destabilized by $+\varepsilon/2$, then the ratio of probabilities of the two conformations is

$$\begin{aligned} p(+1, +1, +1)/p(+1, -1, +1) &= \exp(+\varepsilon/2kT)/\exp(-\varepsilon/2kT) \\ &= \exp(\varepsilon/kT) \end{aligned} \tag{7.8.22}$$

Only the interaction energy appears since the $+1$ and -1 single units have the same energy.

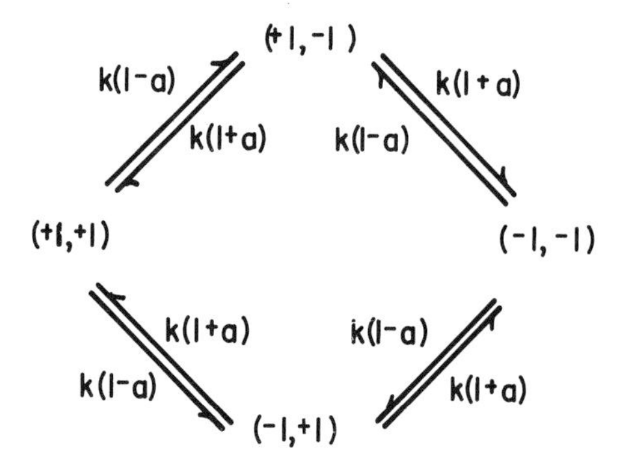

Figure 7.13. The kinetic scheme for transitions of pairs of interacting units—a two-unit scheme.

The probability ratios of Equations 7.8.21 and 7.8.22 can be equated and solved for the parameter a:

$$k(1+a)/k(1-a) = (1+a)/(1-a) = \exp(\varepsilon/kT) = A$$
$$A - aA = 1 + a \tag{7.8.23}$$
$$a = -(1-A)/(1+A) = -[1-\exp(\varepsilon/kT)]/[1+\exp(\varepsilon/kT)]$$

Multiplying numerator and denominator by $\exp(-\varepsilon/2kT)$ and using the identities

$$\sinh x = \frac{1}{2}[\exp(x) - \exp(-x)]$$
$$\cosh x = \frac{1}{2}[\exp(x) + \exp(x)] \tag{7.8.24}$$

gives

$$a = \sinh(\varepsilon/2kT)/\cosh(\varepsilon/2kT) = \tanh(\varepsilon/2kT) \tag{7.8.25}$$

which is valid for positive and negative interaction energies. If the interaction energy is zero, the hyperbolic tangent and a are zero and the transition probabilities are those for the independent unit model.

Although a has been determined for a single reacting pair of conformations, each other conformation pair, e.g.,

$$p(+1, -1, -1) \rightarrow p(+1, +1, -1) \tag{7.8.26}$$

gives Equation 7.8.25 for a. Verification is left as an excercise.

Even with these definitions for the transition probabilities, the evaluation of the kinetics for the general case can be difficult since it is impossible to focus on a single unit in the chain. A transition must be defined in terms of the whole chain. For example, a chain with four units with $+1$ conformation defines one state of the chain:

$$p(+1, +1, +1, +1) \tag{7.8.27}$$

This state can be lost to four different states since any one of the four units might change. The four possibilities are

$$p(-1, +1, +1, +1), \quad p(+1, -1, +1, +1),$$
$$p(+1, +1, -1, +1), \quad \text{and} \quad p(+1, +1, +1, -1) \tag{7.8.28}$$

Since each unit can have two conformations, even this short four-unit model will have a total of $2^4 = 16$ distinct states and a total of 16 different rate equations.

In order to develop tractable analytical solutions for such systems, the

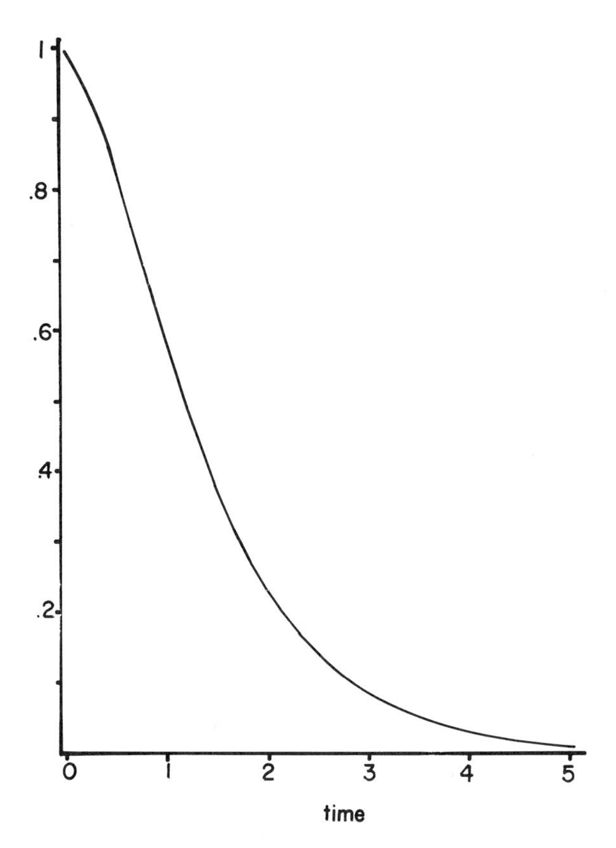

Figure 7.14. The sigmoidal kinetic decay of the fraction of proteins in the closed conformation.

temporal behavior of the average of s, $\langle s \rangle$, must be described in some rate expression. $\langle s \rangle$ must be defined using the probabilities for the full chain, e.g., Equation 7.8.28. For example, the rate expression for the state $p(+1, +1, +1, +1)$ is

$$-4(k/2)(1-a)\, p(+1, +1, +1, +1) \tag{7.8.29}$$

This state reacts via four different pathways which all have identical rate constants if cyclic boundary conditions are used. The state is created from four other states in which only one unit is -1. Since the -1 unit which must make the transition to create $p(+1, +1, +1, +1)$ is surrounded by $+1$ units, each rate expression is the same to give

$$+4(k/2)[p(-1, +1, +1, +1) + p(+1, -1, +1, +1) + p(+1, +1, -1, +1) + p(+1, +1, +1, -1)] \tag{7.8.30}$$

There are 16 such rate expressions.

The average $\langle s \rangle$ is formed by considering every possible conformation,

$$\langle s \rangle = \sum s_i \, p(\ldots s_i \ldots) \tag{7.8.31}$$

For example, when $p(+1, +1, +1, +1)$ is known at time t, it will contribute four terms to the total summation. The first unit will contribute

$$(+1)\, p(+1, +1, +1, +1) \tag{7.8.32}$$

to the average. The second, third, and fourth units contribute the same amount. The next probability, $p(+1, +1, +1, -1)$, will produce three factors,

$$(+1)\, p(+1, +1, +1, -1) \tag{7.8.33}$$

and a single

$$(-1)\, p(+1, +1, +1, -1) \tag{7.8.34}$$

The averaging procedure can be illustrated for changes involving a single unit in a single state. The two rate terms give the rate expression

$$\begin{aligned} &d[(+1)\, p(+1, +1, +1, +1)]/dt \\ &\quad = -(+1)\, p(+1, +1, +1, +1)(k/2)(1-a) \\ &\quad\quad + (-1)\, p(-1, +1, +1, +1)(k/2)(1+a) \end{aligned} \tag{7.8.35}$$

The summation over all probabilities will give

$$d\langle s \rangle/dt \tag{7.8.36}$$

on the left-hand side. The two terms on the right average the first unit in this probability. The -1 in the second term makes it the same sign as the first term. The right-hand side is

$$\begin{aligned} &-(k/2)[1p(+1, +1, +1, +1) + 1p(-1, +1, +1, +1)] \\ &\quad -[-(ka/2)\, p(+1, +1, +1, +1) + (ka/2)\, p(-1, +1, +1, +1)] \end{aligned} \tag{7.8.37}$$

The first bracketed term becomes

$$-k\langle s \rangle \tag{7.8.38}$$

when all probabilities are summed. The second term includes a, which depends on the nearest neighbors. The term averages as

$$k\langle a \rangle = k\langle \tanh(\varepsilon_i/kT) \rangle \tag{7.8.39}$$

The kinetic equation for $\langle s \rangle$ is

$$d\langle s \rangle/dt = -k\langle s \rangle + k\langle \tanh(\varepsilon_i/kT) \rangle \quad (7.8.40)$$

This single equation depends on a determination of $\langle a \rangle$.

The equation becomes tractable if the mean field approximation is used to introduce $\langle s \rangle$ into the second term. With this approximation, the average of the hyperbolic tangents of the energy is replaced with the hyperbolic tangent of the average energy,

$$\langle \tanh(\varepsilon_i/kT) \rangle = \tanh(b\langle s \rangle) \quad (7.8.41)$$

where

$$b = m\varepsilon/kT \quad (7.8.42)$$

with ε the interaction energy for a single unit and m the number of nearest neighbors. The hyperbolic tangent expands as

$$\tanh(b\langle s \rangle) = b\langle s \rangle + (b\langle s \rangle)^3/3 \quad (7.8.43)$$

and the equation for $\langle s \rangle$ is a nonlinear differential equation:

$$\begin{aligned} d\langle s \rangle/dt &= -k'\langle s \rangle + k''\langle s \rangle^3 \\ &= -k'\langle s \rangle(1 - c\langle s \rangle^2) \end{aligned} \quad (7.8.44)$$

This same equation can be obtained from simple probability arguments. The rate of change of some average unit $\langle s \rangle$ will depend on the value of $\langle s \rangle$. In addition, it will depend on the average value of its neighbors. The probability that at least one of the neighbors is not $\langle s \rangle$ is

$$1 - \langle s \rangle^2 \quad (7.8.45)$$

The joint probability of $\langle s \rangle$ and $1 - \langle s \rangle^2$ determines the rate of reaction:

$$d\langle s \rangle/dt = -k\langle s \rangle(1 - \langle s \rangle^2) \quad (7.8.46)$$

The solution of the equation is

$$\langle s \rangle = \langle s_0 \rangle \exp(-kt)/[1 - c\langle s_0 \rangle^2 + c\langle s_0 \rangle^2 \exp(-2kt)]^{1/2} \quad (7.8.47)$$

The factor in the denominator slows the initial rate of reaction. Since its exponential argument contains a factor of 2, this factor decays to zero more rapidly and the system evolves to a simple exponential decay when the exponential in the denominator decays to zero.

Problems

7.1. Develop the full kinetic solution for the four-state wind–tree model.

7.2. Show that $d(f \ln f)/dt$ breaks entirely into functions

$$(f_i - f_j) \ln (f_j/f_i)$$

7.3. Show that

$$q = c + h(1 + q_S \lambda_S)^M$$

for a system of independent residues which can bind M S molecules per residue.

7.4. Verify the equation

$$\langle S \rangle = \lambda_S d(\ln q)/d\lambda_S$$

7.5. Modify Equation 7.2.40 to include two additional c residues at each end of the polymer.

7.6. Determine the maximal eigenvalue and partition function for the transfer matrix

$$\begin{pmatrix} qa & q \\ 1 & 1 \end{pmatrix}$$

7.7. Show the eigenvalue $q - q^{-1}$ approaches zero as the interaction energy approaches zero.

7.8. Develop the partition function for a binary alloy. The metals, A and B, are present in mole fractions X_A and X_B.

7.9. Show that

$$\begin{pmatrix} h^2 \\ h \\ h \\ 1 \end{pmatrix}$$

is an eigenvector for the 4×4 transfer matrix of independent units (Equation 7.5.9).

7.10. Develop the transfer matrix for a double helix in which the intrachain interaction energy between nearest-neighbor units is different from the interchain interaction energy.

7.11. Show that there is no consistent way to place the eight Boltzmann factors for next-nearest-neighbor interactions into a 3×3 transfer matrix without using matrix reduction techniques.

7.12. a. Determine the transfer matrix for two rows of ferromagnets when there are no interactions and the magnetic field is B.

b. Show the transfer matrix with interactions is

$$\begin{pmatrix} \exp(3J) & 1 & 1 & \exp(-J) \\ 1 & \exp(J) & \exp(-3J) & 1 \\ 1 & \exp(-3J) & \exp(J) & 1 \\ \exp(-J) & 1 & 1 & \exp(-3J) \end{pmatrix}$$

when $B = 0$.

c. Develop the transfer matrix when both interactions and a magnetic field are present.

7.13. Show that

$$a = \tanh(\beta\varepsilon)$$

for a kinetic Ising model.

7.14. Develop the characteristic equation for the transfer matrix of Equation 7.6.9.

7.15. a. Develop the transition probability **K** matrix for interactive kinetics of a two-unit chain when transitions

$$(+1, -1) \rightarrow (-1, +1)$$
$$(-1, +1) \rightarrow (+1, -1)$$

with rate constants k are also allowed.

b. Determine the eigenvalues and eigenvectors for the kinetic system.

7.16. Use graph theory to determine the equilibrium vector for the two-unit chain of Figure 7.13.

7.17. The dog–flea model kinetics have a **K** matrix

$$\begin{pmatrix} -1 & \frac{1}{4} & 0 & 0 & 0 \\ 1 & -1 & \frac{1}{2} & 0 & 0 \\ 0 & \frac{3}{4} & -1 & \frac{3}{4} & 0 \\ 0 & 0 & \frac{1}{2} & -1 & 1 \\ 0 & 0 & 0 & \frac{1}{4} & -1 \end{pmatrix}$$

for a four-flea system.

a. Show that the $\lambda = 0$ eigenvalue has an eigenvector whose components obey a binomial distribution.

b. Develop the general **K** matrix for an N-flea system.

7.18. Write the dog–flea **K** matrix for a three-flea system.

a. Determine the equilibrium eigenvector.

b. Show that the eigenvalues for the **K** matrix are 0, -2, -4, and -6.

c. Show that all eigenvectors use the binomial coefficients 1, 3, 3, 1 in various orders.

7.19. Use the transfer matrix of Equation 7.4.24 to develop a partition function for a three-unit system. Derive the same partition function by considering all possible conformations and compare the two results.

8

Quantum Mechanics

8.1. Hybrid Atomic Orbitals

The sp^2 and sp^3 hybrid orbitals appear consistently in conjunction with the carbon atom. The resultant electronic configurations do provide maximal separation of the electrons about the carbon nucleus, but it is often difficult to see exactly how they can be constructed from the s and p electrons of the carbon atom. For example, does the spherically symmetric s orbital wavefunction play any role in the final geometric arrangement of the hybrid orbitals? A total of four electron orbitals are required to create the four hybrid orbitals, but how is the geometric information for these hybrid orbitals introduced?

The wavefunctions for the s, p_x, p_y, and p_z atomic orbitals must be examined geometrically. Their geometry suggests the geometric combination of orbitals necessary to create the hybrid orbitals. Although many hybrid geometries can be created in this manner, quantum analysis ultimately selects the hybrid orbitals which provide both the lowest total energy and the best reproduction of the molecular structure of the system.

If the hydrogenic wavefunctions of the electrons in the $n=2$ level are considered for hybridization, they are written as functions of r, θ, and φ:

$$p_z = f(r) \cos \theta \tag{8.1.1}$$

$$p_x = f(r) \sin \theta \cos \varphi \tag{8.1.2}$$

$$p_y = f(r) \sin \theta \sin \varphi \tag{8.1.3}$$

The sine and cosine functions indicate the values of the wavefunctions as a function of angular orientation. However, to consider these functions from a geometric point of view, it is better to convert to Cartesian coordinates, as illustrated in Figure 8.1. Since $r \cos \theta$ is the projection of a point at r on the z axis,

$$z = r \cos \theta \tag{8.1.4}$$

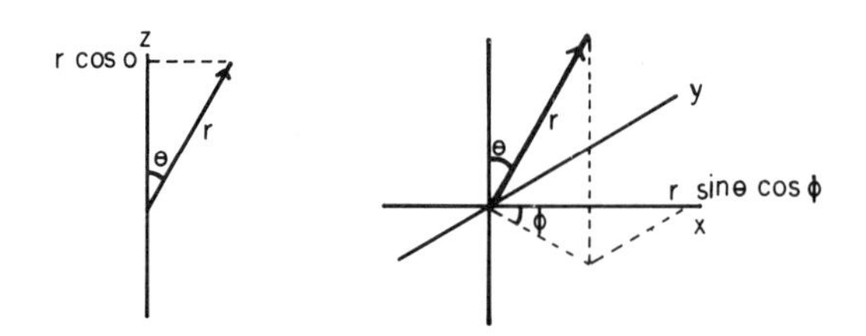

Figure 8.1. The Cartesian and spherical coordinates for orbitals.

$\cos\theta$ is

$$\cos\theta = z/r \tag{8.1.5}$$

In a similar manner, $\sin\theta$ projects the vector **r** onto the x–y plane while $\cos\phi$ projects onto the x axis in this plane, i.e.,

$$x = r\sin\theta\cos\varphi \tag{8.1.6}$$

or

$$x/r = \sin\theta\cos\varphi \tag{8.1.7}$$

The projection on y gives

$$y/r = \sin\theta\sin\varphi \tag{8.1.8}$$

The three p orbitals in Cartesian coordinates are

$$\begin{aligned} p_z &= f(r)z/r \\ p_x &= f(r)x/r \\ p_y &= f(r)\,y/r \end{aligned} \tag{8.1.9}$$

In Cartesian coordinates, the vector **r** has magnitude

$$r = (x^2 + y^2 + z^2)^{1/2} \tag{8.1.10}$$

and it has spherical symmetry. The s orbital in Cartesian coordinates is a function only of r, and it is completely symmetric with respect to orientation. The geometric information for the p orbitals lies in the factors x, y, and z for the p_x, p_y, and p_z orbitals, respectively.

The Cartesian and spherical coordinate formats describe the same wavefunctions. These functions have amplitude maxima and minima along the Cartesian coordinate axes. For example, p_z has a maximal amplitude when $\theta = 0$ for each value of r. In Cartesian coordinates, the maximal amplitude occurs for $r = z$. The radius is oriented along z and $z/r = 1$, a maximum for this function.

The major advantage of the Cartesian coordinates lies in their use to construct combinations of wavefunctions. The x, y, and z variables select the maximum amplitude for each wavefunction. These maximal amplitudes are selected as directional vectors in the three-dimensional space, and these vectors are then combined to generate a new wavefunction with a new geometry. Since the remainder of the factors for the wavefunction are spherically symmetric, only the factors x, y, and z are required for this geometric development. The three p orbitals are written as column vectors with components x, y, and z:

$$|x\rangle = \begin{pmatrix} 1 \\ 0 \\ 0 \end{pmatrix} \qquad |y\rangle = \begin{pmatrix} 0 \\ 1 \\ 0 \end{pmatrix} \qquad |z\rangle = \begin{pmatrix} 0 \\ 0 \\ 1 \end{pmatrix} \tag{8.1.11}$$

A set of hybrid orbitals is constructed by selecting the geometric directions in which the hybrid will have maximal amplitude and defining a three-component vector for this direction. This vector is then resolvable as a sum of projections on $|x\rangle$, $|y\rangle$, and $|z\rangle$. The three-dimensional vectors $|i\rangle$ are replaced by p_i wavefunctions to reproduce the hybrid wavefunction in the space. The technique is illustrated for the tetrahedral (sp^3) hybrid orbitals using the cubical arrangement of Figure 8.2. The C lies in the exact center of the cube. The cube dimensions are selected so that vertex projections on each axis are unity. For a tetrahedron, the orbitals have maxima on alternate vertices. The three-dimensional vectors which define these maxima are (Figure 8.2)

$$|1\rangle = \begin{pmatrix} 1 \\ 1 \\ 1 \end{pmatrix} \qquad |2\rangle = \begin{pmatrix} 1 \\ -1 \\ -1 \end{pmatrix} \qquad |3\rangle = \begin{pmatrix} -1 \\ -1 \\ 1 \end{pmatrix} \qquad |4\rangle = \begin{pmatrix} -1 \\ 1 \\ -1 \end{pmatrix} \tag{8.1.12}$$

These vectors are now resolved into components on the $|x\rangle$, $|y\rangle$, and $|z\rangle$ coordinate vectors, e.g.,

$$\begin{aligned} |1\rangle &= |x\rangle\langle x|1\rangle + |y\rangle\langle y|1\rangle + |z\rangle\langle z|1\rangle \\ &= 1\,|x\rangle + 1\,|y\rangle + 1\,|z\rangle \end{aligned} \tag{8.1.13}$$

The $|1\rangle$ vector has positive projections of unity on each spatial coordinate. The

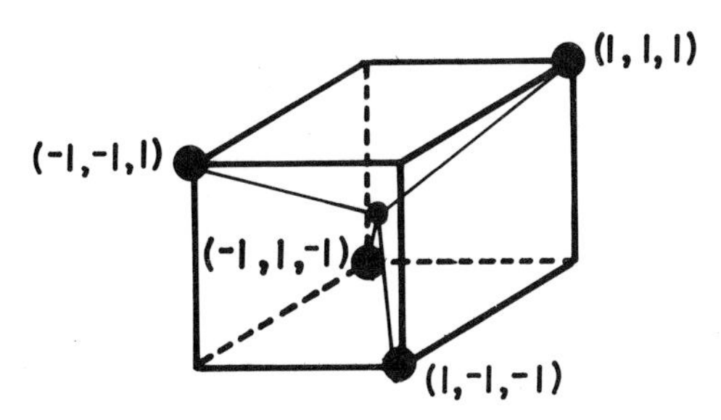

Figure 8.2. The use of cube vertices to define the location of sp^3 orbitals with respect to a carbon atom at the center of the cube.

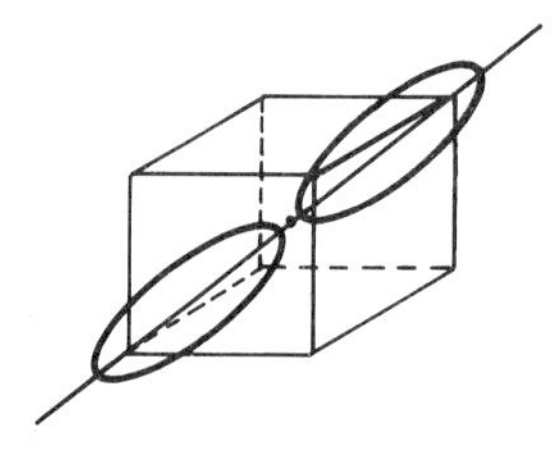

Figure 8.3. The linear combination $1p_x + 1p_y + 1p_z$.

hybrid wavefunction is produced by replacing the $|i\rangle$ vectors with the corresponding atomic orbital. The geometric portion of the hybrid wavefunction is

$$|1\rangle = 1p_x + 1p_y + 1p_z \tag{8.1.14}$$

However, the hybridization is not yet complete. The sum of the three p orbitals to give the hybrid orbital is shown in Figure 8.3. The orbital has the p orbital shape but is now directed along the spatial direction (1, 1, 1). The function minimum is located along the $(-1, -1, -1)$ spatial direction and the orbital amplitudes extend in two major directions.

A single orientation with maximal amplitude will dominate if the $2s$ orbital is added to the linear combinations of p orbitals. The $2s$ orbital will have the same amplitude in every spatial direction. If this amplitude is positive, it will supplement the electron amplitude in the (1, 1, 1) direction. The wavefunction along $(-1, -1, -1)$ has a negative amplitude, and the positive $2s$ amplitude will reduce the total electron amplitude in this direction. The linear combination of $2s$ and $2p$ orbitals creates a new hybrid function which concentrates the maximal amplitude, and hence the maximal electron density, along a single spatial direction. An amplitude map for a single radius r is shown in Figure 8.4. The total electron density map for the function is shown in Figure 8.5.

The remaining tetrahedral directions can be expressed in terms of the p orbitals using the projection operators. The hybrid orbitals including the $2s$ orbital are

$$\begin{aligned}|2\rangle &= 1s + 1p_x - 1p_y - 1p_z\\ |3\rangle &= 1s - 1p_x - 1p_y + 1p_z\\ |4\rangle &= 1s - 1p_x + 1p_y - 1p_z\end{aligned} \tag{8.1.15}$$

Figure 8.4. Regions of largest wave amplitude for an sp^3 orbital on the surface of a sphere.

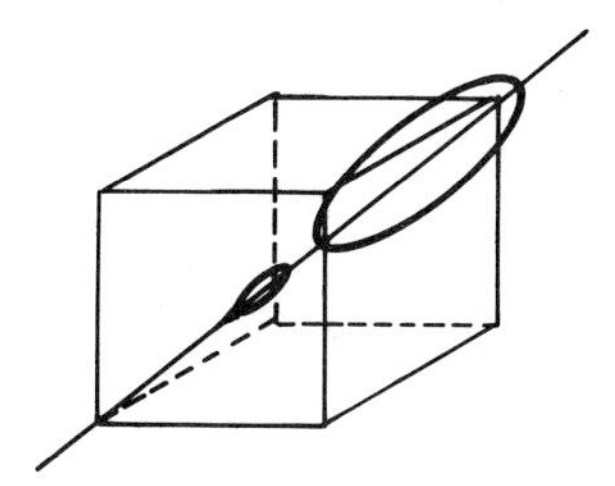

Figure 8.5. A probability density map of a hybrid sp^3 orbital.

The appropriate p_i orbitals have replaced the $|i\rangle$ vectors in each case. The result is a function in the three-dimensional space with a maximum along the appropriate geometric direction. For the sp^3 orbitals, these directions are illustrated in Figure 8.2.

The hybrid sp^2 orbitals are confined to a plane and are oriented at angles of $0°$, $120°$, and $240°$ in this plane. Selecting the x–y plane will ultimately yield a hybrid wavefunction which is a linear combination of p_x and p_y but not p_z. The planar orientations are illustrated in Figure 8.6. The first vector is located on the x coordinate axis; the remaining two vectors have projections on both the x and y axes. These three spatial orientations in two dimensions are described by the two-component vectors

$$|1\rangle = \begin{pmatrix} 1 \\ 0 \end{pmatrix} \qquad |2\rangle = \begin{pmatrix} \dfrac{-1}{2} \\ \dfrac{\sqrt{3}}{2} \end{pmatrix} \qquad |3\rangle = \begin{pmatrix} \dfrac{-1}{2} \\ \dfrac{-\sqrt{3}}{2} \end{pmatrix} \tag{8.1.16}$$

These vectors must be projected onto the two-component coordinate vectors

$$|x\rangle = \begin{pmatrix} 1 \\ 0 \end{pmatrix} \qquad |y\rangle = \begin{pmatrix} 0 \\ 1 \end{pmatrix} \tag{8.1.17}$$

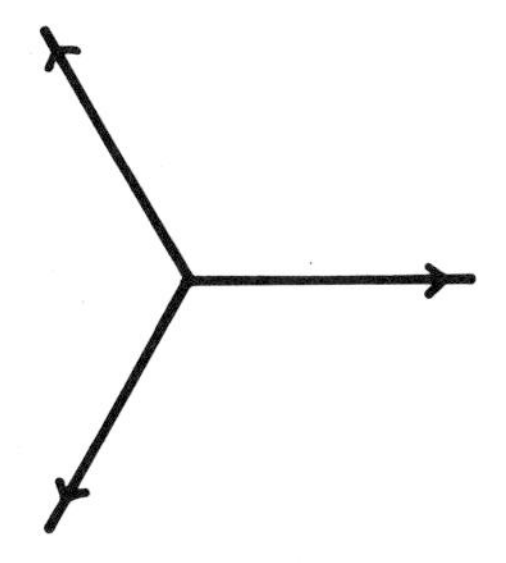

Figure 8.6. The planar orientations for the three sp^2 orbitals in space.

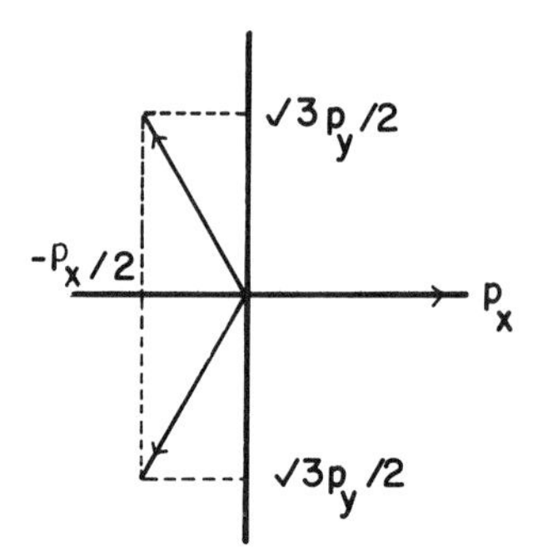

Figure 8.7. Regions of maximal wave amplitude for the three sp^2 orbitals on a ring of constant radius.

The projections on $|x\rangle$, $|y\rangle$, and $|z\rangle$ are

$$|1\rangle = 1\ |x\rangle \tag{8.1.18}$$

$$|2\rangle = -\frac{1}{2}\,|x\rangle + \frac{\sqrt{3}}{2}\,|y\rangle \tag{8.1.19}$$

$$|3\rangle = -\frac{1}{2}\,|x\rangle - \frac{\sqrt{3}}{2}\,|y\rangle \tag{8.1.20}$$

When the radial portions of each component are included and the 2s orbital is added to provide a net enhancement for one orientation of the orbital, the hybrid wavefunctions become

$$|1\rangle = 1s + 1p_x \tag{8.1.21}$$

$$|2\rangle = 1s - \frac{1}{2}\,p_x + \frac{\sqrt{3}}{2}\,p_y \tag{8.1.22}$$

$$|3\rangle = 1s - \frac{1}{2}\,p_x - \frac{\sqrt{3}}{2}\,p_y \tag{8.1.23}$$

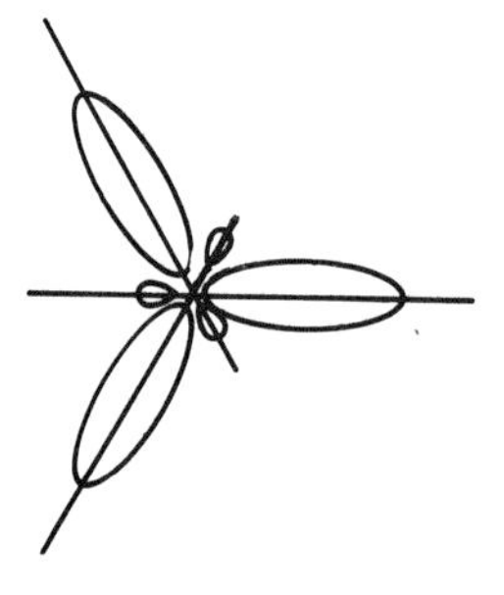

Figure 8.8. Probability density contours for the sp^2 hybrid orbitals.

The orientations of the wavefunctions for the p orbitals and the full hybrid are shown as boldface lines in Figure 8.7. The final result shown in Figure 8.8 for the square of the full radius-dependent function is the typical description of the sp^2 hybrid orbitals.

8.2. Matrix Quantum Mechanics

The matrices and eigenvectors of the previous three chapters have been finite-dimensional. Some modifications are required in quantum mechanics. The wavefunctions of quantum mechanics are Hilbert space vectors with infinite, i.e., continuous, components. In addition, an infinite number of these wavefunctions are required to span the space. This requires a modified point of view. If the wavefunction is a continuous function, it essetially defines every component of the eigenvector. Any value of the independent variable on the interval of interest defines the value of the function, i.e., its component, at that point.

A matrix served as an operator for finite-dimensional systems. The matrix operated on a vector to produce some change in the components of that vector. For example, a matrix **A** operates on its eigenvector to give a vector in which each component is multiplied by a scalar, the eigenvalue:

$$\mathbf{A}\,|\psi\rangle = \lambda\,|\psi\rangle \tag{8.2.1}$$

The operator acts on the function at every value of x. For example, the product

$$x\psi(x) \tag{8.2.2}$$

is the operator x acting on the function $\psi(x)$. Each component of the vector $\psi(x)$ has been multiplied by the value of x at that point. It is impossible to write each component but the functional product of Equation 8.2.2 does exactly the same job.

The operator may be a derivative which acts on the function. For example, the derivative d/dx acts on $\exp(-kx)$ to give

$$\mathbf{A}\psi(x) = d/dx\,[\exp(-kx)] = -k[\exp(-kx)] \tag{8.2.3}$$

The operation modifies the function by multiplying each component by $-k$. This is an example of an eigenvalue–eigenfunction equation in the Hilbert space. The operation returns the original function, i.e., vector, multiplied by a scalar, the eigenvalue. Only a finite number of vectors were eigenvectors for a specific matrix in finite-dimensional space. In Hilbert space, most operations will not be eigenvalue–eigenfunction operations. For example,

$$d/dx\,(\sin kx) = k(\cos kx) \tag{8.2.4}$$

is an operator equation but $\sin kx$ is not an eigenfunction for the operator since it generates $\cos kx$ rather than $\sin kx$.

Quantum mechanics requires operators for physically observable parameters like position and momentum. The actual scalar observables must appear when this operator operates on the function and returns a scalar and the function, i.e., an eigenfunction for the operator. An example of this is, of course, the operator form of the Schrödinger equation for the energy using a Hamiltonian operator:

$$\mathbf{H}\psi(x) = E\psi(x) \tag{8.2.5}$$

The scalar E multiplies every component of the wavefunction. The operation

$$\mathbf{x}\psi(x) = x\psi(x) \tag{8.2.6}$$

is not an eigenoperation since each component of the function is multiplied by a different value of x.

Although $\mathbf{x}$ is not an eigenoperator when it operates on a function of x, it can be used to define an operator for the momentum of the system. The starting point is Heisenberg's uncertainty principle. Although this is commonly written

$$\Delta x\, \Delta p \sim h \tag{8.2.7}$$

i.e., the position and momentum of a particle in one coordinate direction will always have an intrinsic uncertainty, the principle can be converted to operator format by stating that the order of operations on an eigenfunction to determine scalar observables can affect the results. Stated formally, a position operator $\mathbf{x}$ and momentum operator $\mathbf{p}_x$ will give one result when they operate in the order

$$\mathbf{x}\mathbf{p}_x\psi(x) \tag{8.2.8}$$

and a different result for the opposite order,

$$\mathbf{p}_x\mathbf{x}\psi(x) \tag{8.2.9}$$

The difference for these two operations will be approximately $i\hbar$:

$$(\mathbf{p}_x\mathbf{x} - \mathbf{x}\mathbf{p}_x)\,\psi(x) \approx i\hbar\psi(x) \tag{8.2.10}$$

The imaginary i is required for a consistent Hamiltonian. This result is often written as a commutator for the two operators:

$$[\mathbf{p}_x, \mathbf{x}] \approx i\hbar \tag{8.2.11}$$

The commutation relation is satisfied if the momentum operator, p_x, contains a derivative in x. The operator is

$$\mathbf{p}_x = -i\hbar\, d/dx \tag{8.2.12}$$

This can be verified by performing the operations for Equation 8.2.10. The first order (Equation 8.2.9) gives

$$(\mathbf{p_x})(\mathbf{x})\,\psi(x) = [-i\hbar\, d/dx]\, x\psi(x)$$
$$= -i\hbar[x\, d\psi(x)/dx + \psi(x)] \tag{8.2.13}$$

The d/dx of $\mathbf{p_x}$ must operate on both the function $\psi(x)$ and the second operator x, i.e., the momentum operator acts on the modified function created by the x operation.

The second term in Equation 8.2.10 is

$$(\mathbf{x})(\mathbf{p_x})\,\psi(x) = x[-i\hbar\, d\psi(x)/dx] \tag{8.2.14}$$

and the difference is

$$(\mathbf{p_x x} - \mathbf{x p_x})\,\psi(x) = -i\hbar[x\, d\psi(x)/dx + \psi(x) - x\, d\psi(x)/dx]$$
$$= -i\hbar\psi(x) \tag{8.2.15}$$

Defining $\mathbf{p_x}$ using Equation 8.2.12 "builds in" the intrinsic uncertainty of the Heisenberg principle.

The $\mathbf{x}$ and $\mathbf{p_x}$ operators are used to construct additional operators from classical equations in which these variables appear. For example, the Hamiltonian for a one-dimensional system is

$$H = p^2/2m + V(x) \tag{8.2.16}$$

While x remains a variable, each p is replaced with its operator. The Hamiltonian operator is

$$\mathbf{H} = (-i\hbar\, d/dx)(-i\hbar\, d/dx)/2m + V(x)$$
$$= (-\hbar^2/2m)\, d^2/dx^2 + V(x) \tag{8.2.17}$$

Despite its intrinsic x dependence, this operator can operate on certain eigenfunctions to give the eigenfunction and a scalar eigenvalue.

For the example given, the operator $\mathbf{p_x}$ is reexpressed as a function of x, and p is eliminated. The commutation relation can also be used to define the operator $\mathbf{x}$ in terms of the variable p. In this momentum space formulation, the relevant operators are

$$\mathbf{p_x}, \mathbf{x} = -i\hbar\, d/dp_x, x \tag{8.2.18}$$

Although each operator will have an infinite set of eigenvalues and eigenfunctions, it is often possible to truncate the basis set without a major loss in accuracy. This truncation reduces the system to a finite-matrix system. In general, the basis functions can be ordered so that the first functions are less complicated, i.e., fewer nodes and lower energies. A function might then be approximated with the first several eigenfunctions. The Fourier series introduced in Chapter 2 illustrates this approach. The index n in $\sin n\theta$ must continue from 1 to infinity. However, a well-behaved function can be approximated using only three or four terms in the series.

The general Schrödinger equation,

$$\mathbf{H}\psi(x) = E\psi(x) \tag{8.2.19}$$

for a one-dimensional system might require a very complicated $\psi(x)$. In such cases, the function is expanded in a well-behaved basis set,

$$|\psi\rangle = \sum |i\rangle\langle i|\psi\rangle \qquad i = 1, \ldots, \tag{8.2.20}$$

where the scalar product $\langle i|\psi\rangle$ defines an integration for the domain of the functions, i.e., $a \leqslant x \leqslant b$,

$$\langle i|\psi\rangle = \int_a^b \phi_i(x)\,\psi(x)\,dx \tag{8.2.21}$$

where $\varphi_i(x)$ is the basis function

$$|i\rangle = \varphi_i(x) \tag{8.2.22}$$

Since determination of the infinite scalar products for Equation 8.2.20 is impossible, a truncated set of these functions is defined as the approximate basis set for the system. The summation of Equation 8.2.20 might be replaced by a sum of only two basis functions:

$$|\psi\rangle = |1\rangle\langle 1|\psi\rangle + |2\rangle\langle 2|\psi\rangle \tag{8.2.23}$$

These functions are assumed to be the complete basis set so they obey the completeness relation,

$$|1\rangle\langle 1| + |2\rangle\langle 2| = 1 \tag{8.2.24}$$

This completeness relation is inserted into the original Schrödinger operator equation to give

$$\mathbf{H}\,|\psi\rangle = \mathbf{H}(|1\rangle\langle 1| + |2\rangle\langle 2|)\,|\psi\rangle = E(|1\rangle\langle 1| + |2\rangle\langle 2|)\,|\psi\rangle \tag{8.2.25}$$

The operator equation still contains functions $|1\rangle$ and $|2\rangle$ which depend on x. The equation can be changed to two scalar equations by applying the bra vectors $\langle 1|$ and $\langle 2|$ to Equation 8.2.25 to generate the equations

$$\begin{aligned} \langle 1|\,\mathbf{H}\,|1\rangle\langle 1|\psi\rangle + \langle 1|\,\mathbf{H}\,|2\rangle\langle 2|\psi\rangle &= E(\langle 1|1\rangle\langle 1|\psi\rangle + \langle 1|2\rangle\langle 2|\psi\rangle) \\ \langle 2|\,\mathbf{H}\,|1\rangle\langle 1|\psi\rangle + \langle 2|\,\mathbf{H}\,|2\rangle\langle 2|\psi\rangle &= E(\langle 2|1\rangle\langle 1|\psi\rangle + \langle 2|2\rangle\langle 2|\psi\rangle) \end{aligned} \tag{8.2.26}$$

The quadratic forms and scalar products generated in this manner are all scalars:

$$\begin{aligned}
H_{11} &= \int \varphi_1^*(x)\,\mathbf{H}\varphi_1(x)\,dx \\
H_{12} &= \int \varphi_1^*(x)\,\mathbf{H}\varphi_2(x)\,dx \\
H_{21} &= \int \varphi_2^*(x)\,\mathbf{H}\varphi_1(x)\,dx \\
H_{22} &= \int \varphi_2^*(x)\,\mathbf{H}\varphi_2(x)\,dx
\end{aligned} \tag{8.2.27}$$

If the basis vectors are not orthogonal, four scalar products must be defined for the basis vectors:

$$\begin{aligned}
S_{11} &= \langle 1 | 1 \rangle = \int \varphi_1^*(x)\,\varphi_1(x)\,dx \\
S_{12} &= \langle 1 | 2 \rangle = \int \varphi_1^*(x)\,\varphi_2(x)\,dx \\
S_{21} &= \langle 2 | 1 \rangle = \int \varphi_2^*(x)\,\varphi_1(x)\,dx \\
S_{22} &= \langle 2 | 2 \rangle = \int \varphi_2^*(x)\,\varphi_2(x)\,dx
\end{aligned} \tag{8.2.28}$$

Finally, the scalar projections of $|\psi\rangle$ on the basis functions $\langle 1|$ and $\langle 2|$ are

$$\begin{aligned}
c_1 &= \langle 1 | \psi \rangle = \int \varphi_1^*(x)\,\psi(x)\,dx \\
c_2 &= \langle 2 | \psi \rangle = \int \varphi_2^*(x)\,\psi(x)\,dx
\end{aligned} \tag{8.2.29}$$

With these defined scalars, Equations 8.2.26,

$$\begin{aligned}
H_{11}c_1 + H_{12}c_2 &= E(S_{11}c_1 + S_{12}c_2) \\
H_{21}c_1 + H_{22}c_2 &= E(S_{21}c_1 + S_{22}c_2)
\end{aligned} \tag{8.2.30}$$

become the matrix equation

$$\begin{gathered}
\mathbf{H}\,|c\rangle = E\mathbf{S}\,|c\rangle \\
\begin{pmatrix} H_{11} & H_{12} \\ H_{21} & H_{22} \end{pmatrix}\begin{pmatrix} c_1 \\ c_2 \end{pmatrix} = E\begin{pmatrix} S_{11} & S_{12} \\ S_{21} & S_{22} \end{pmatrix}\begin{pmatrix} c_1 \\ c_2 \end{pmatrix}
\end{gathered} \tag{8.2.31}$$

The matrix elements are quadratic forms,

$$\begin{pmatrix} \langle 1|\,\mathbf{H}\,|1\rangle & \langle 1|\,\mathbf{H}\,|2\rangle \\ \langle 2|\,\mathbf{H}\,|1\rangle & \langle 2|\,\mathbf{H}\,|2\rangle \end{pmatrix} \tag{8.2.32}$$

$$\begin{pmatrix} \langle 1\,|\,1\rangle & \langle 1\,|\,2\rangle \\ \langle 2\,|\,1\rangle & \langle 2\,|\,2\rangle \end{pmatrix} \tag{8.2.33}$$

The matrix elements may involve integration over one, two, or three spatial variables. For example, H_{11} for three-dimensional basis functions is

$$H_{11} = \iiint \varphi_1(|r\rangle)\,\mathbf{H}(|r\rangle, |p\rangle)\,\varphi_1(|r\rangle)\,dV \tag{8.2.34}$$

where $|r\rangle$ and $|p\rangle$ are three-dimensional vectors which can be expressed in an appropriate coordinate system, and dV is the volume element in this coordinate system. The remaining matrix elements are defined in the same manner.

A truncated basis set of N functions generates a set of N equations and an $N \times N$ matrix equation. The approach can also be extended for the infinite set of basis vectors. In this case, the matrix would have infinite rank and could not be solved unless there was some regularity in the matrices which permitted the infinite-dimensional solution to be obtained by induction.

The system with two basis vectors (Equation 8.2.31) contains the experimental information required from the operator equation. The eigenvalue E is the energy of the system determined using only two basis functions. The scalars c_1 and c_2 are the eigenvector components associated with the eigenvalue. These scalar components also are the projections of the unknown function $|\psi\rangle$ on each of the basis functions:

$$\begin{pmatrix} c_1 \\ c_2 \end{pmatrix} = \begin{pmatrix} \langle 1\,|\,\psi\rangle \\ \langle 2\,|\,\psi\rangle \end{pmatrix} \tag{8.2.35}$$

When these components are known, the actual wavefunction is generated as a sum of these projections:

$$|\psi\rangle = |1\rangle\langle 1\,|\,\psi\rangle + |2\rangle\langle 2\,|\,\psi\rangle = c_1\,|1\rangle + c_2\,|2\rangle \tag{8.2.36}$$

The solution of the eigenvalue–eigenvector equation is complicated by the presence of the **S** matrix. If the two basis vectors are orthonormal,

$$\begin{aligned} \langle 1\,|\,1\rangle &= \langle 2\,|\,2\rangle = 1 \\ \langle 1\,|\,2\rangle &= \langle 2\,|\,1\rangle = 0 \end{aligned} \tag{8.2.37}$$

the **S** matrix reduces to the identity matrix and the equation reduces to the "standard" matrix eigenvalue–eigenvector equation

$$\mathbf{H}\,|c\rangle = E\,|c\rangle \tag{8.2.38}$$

It is preferable to choose an orthonormal basis set because of this simplification. If a nonorthonormal set is selected, the equations can be modified in several ways to facilitate its solution. For example, each side can be multiplied from the left by $\mathbf{S}^{-1}$ to give

$$\mathbf{S}^{-1}\,\mathbf{H}\,|c\rangle = E\,|c\rangle \tag{8.2.39}$$

Since the $\mathbf{H}$ operator is Hermitian, the eigenvalues generated by Equation 8.2.38 must be real. The new operator

$$\mathbf{S}^{-1}\mathbf{H} \tag{8.2.40}$$

is not Hermitian and this can complicate the analysis.

The equation

$$\mathbf{H}\,|\psi\rangle = E\mathbf{S}\,|\psi\rangle \tag{8.2.41}$$

is a generalized eigenvalue–eigenvector equation. It can be solved in this format to give the eigenvalue E and eigenvalues defined using the $\mathbf{S}$ matrix, i.e.,

$$\langle i|\,\mathbf{S}\,|j\rangle = \begin{cases} 1 & i = j \\ 0 & i \neq j \end{cases} \tag{8.2.42}$$

Equation 8.2.41 can be converted into the standard format while retaining Hermitian character by transforming the vector $|c\rangle$ with $\mathbf{S}^{1/2}$:

$$\begin{aligned} |c'\rangle &= \mathbf{S}^{1/2}\,|c\rangle \\ |c\rangle &= \mathbf{S}^{-1/2}\,|c'\rangle \end{aligned} \tag{8.2.43}$$

The equation is then left multiplied by $\mathbf{S}^{-1/2}$ to give

$$\begin{gathered} \mathbf{S}^{-1/2}\mathbf{H}\mathbf{S}^{-1/2}\,|c'\rangle = E\mathbf{S}^{-1/2}\mathbf{S}\,|c\rangle = E\mathbf{S}^{1/2}\,|c\rangle = E\,|c'\rangle \\ \mathbf{S}^{-1/2}\mathbf{H}\mathbf{S}^{-1/2}\,|c'\rangle = E\,|c'\rangle \end{gathered} \tag{8.2.44}$$

The matrix

$$\mathbf{S}^{-1/2}\mathbf{H}\mathbf{S}^{-1/2} \tag{8.2.45}$$

is Hermitian (Problem 8.3). The analysis is similar to that required for a heteronuclear vibrating molecule.

8.3. Hückel Molecular Orbitals for Linear Molecules

Matrix eigenvalue–eigenvector equations describe molecular systems with limited basis functions. For molecular systems, certain atomic wavefunctions are selected as the basis set. The ethene molecule, C_2H_4, is made from carbon atoms which originally contained six electrons and hydrogen atoms which contained a single electron. Each of these electrons can be described by a hydrogenic atomic orbital. Since the electrons might also be in higher-energy states, the atomic orbitals associated with these states must also be considered as part of the basis set.

Hückel theory simplifies this problem by excluding many of these orbitals from the working basis set. The inner-shell $1s$ electrons of carbon are excluded. The basic molecular skeleton of ethene consists of bonds between the two carbons and the carbons and hydrogens. The electrons which constitute these bonds are also excluded from the basis set so that a total of two atomic orbitals remain. Each carbon atom has a final $2p_z$ electron, and these two electrons must ultimately become the second carbon–carbon bond in ethene. The two $2p_z$ atomic orbitals, which are labeled $|1\rangle$ and $|2\rangle$ for carbons 1 and 2, respectively, are the basis set for ethene in Hückel theory.

The two-orbital system will produce a 2×2 matrix eigenvalue–eigenvector equation which is identical to that of the two-dimensional system developed in the last section. The atomic orbitals are both $2p$ orbitals. Although these orbitals may have some overlap because the atoms are relatively close together, basic Hückel theory ignores these overlaps and assumes

$$\langle 1 | 2 \rangle = 0 \tag{8.3.1}$$

The atomic orbitals are normalized,

$$\langle 1 | 1 \rangle = \langle 2 | 2 \rangle \tag{8.3.2}$$

and the **S** matrix is equal to the identity matrix,

$$\mathbf{H} \, |c\rangle = E\mathbf{S} \, |c\rangle = E \, |c\rangle \tag{8.3.3}$$

The matrix equation for the p_z carbon orbitals in ethene is

$$\begin{pmatrix} H_{11} & H_{12} \\ H_{21} & H_{22} \end{pmatrix} \begin{pmatrix} c_1 \\ c_2 \end{pmatrix} = E \begin{pmatrix} c_1 \\ c_2 \end{pmatrix} \tag{8.3.4}$$

where c_1 and c_2 will give the relative contributions of the two atomic orbitals to the final molecular orbital.

There are two distinct types of matrix elements for ethene. The elements H_{11} and H_{22} require an integral where both atomic orbital functions in the integral have the same coordinates, i.e., they are on the same carbon. The elements H_{12} and H_{21} have integrals involving wavefunctions from different carbons. Because the carbons are identical, each pair of elements is equal:

$$H_{11} = H_{22} \tag{8.3.5}$$

$$H_{12} = H_{21} \tag{8.3.6}$$

Since the evaluation of the integrals is involved, the system is simplified by assigning scalar values to each pair of integrals. These scalar values can then be determined from experimental observations of the energies associated with the states of the ethene molecules. The scalars are

$$\alpha = H_{11} = H_{22} \tag{8.3.7}$$

$$\beta = H_{21} = H_{12} \tag{8.3.8}$$

and the matrix equation is

$$\begin{pmatrix} \alpha & \beta \\ \beta & \alpha \end{pmatrix} \begin{pmatrix} c_1 \\ c_2 \end{pmatrix} = E \begin{pmatrix} c_1 \\ c_2 \end{pmatrix} \tag{8.3.9}$$

The characteristic determinant is

$$\begin{vmatrix} \alpha - E & \beta \\ \beta & \alpha - E \end{vmatrix} = 0 \tag{8.3.10}$$

with characteristic polynomial

$$(\alpha - E)^2 - \beta^2 = E^2 - 2\alpha E + (\alpha^2 - \beta^2) = 0 \tag{8.3.11}$$

and eigenvalues

$$\begin{aligned} E_1 &= \alpha + \beta \\ E_2 &= \alpha - \beta \end{aligned} \tag{8.3.12}$$

The eigenvector for E_1 is obtained from

$$\begin{pmatrix} \alpha - \alpha - \beta & \beta \\ \beta & \alpha - \alpha - \beta \end{pmatrix} \begin{pmatrix} c_1 \\ c_2 \end{pmatrix} = 0$$

as

$$|c_1\rangle = \begin{pmatrix} 1 \\ 1 \end{pmatrix} \tag{8.3.13}$$

The eigenvector for E_2 is

$$|c_2\rangle = \begin{pmatrix} 1 \\ -1 \end{pmatrix} \tag{8.3.14}$$

Since the integral for β is negative, E_1 is the lower energy. The wavefunction for this molecular orbital is the linear combination of p_z atomic orbitals from each carbon:

$$|\psi\rangle = c_1\,|1\rangle + c_2\,|2\rangle = |1\rangle + |2\rangle \tag{8.3.15}$$

When c_1 and c_2 are given by the eigenvector, the orbitals can be combined in space to generate the "shape" of the molecular orbital. Since the wave amplitude of the p_z orbitals is positive on the positive z axis, the wave amplitudes for the two atomic orbitals will add in the region between the carbons to produce the familiar shape of a π orbital for ethene. Hückel theory has shown that the identical orbitals from each carbon make equal positive contributions to the final orbital. The energy for the orbital is

$$E_1 = \alpha + \beta \tag{8.3.16}$$

If the system had a single p orbital, the Hamiltonian would be the 1×1 matrix:

$$H_{11} c_1 = E c_1 = \alpha c_1 \tag{8.3.17}$$

α is the energy for the electron on a single carbon and is called the Coulomb integral. β is the stabilization energy which results when the atomic orbital becomes part of a molecular orbital and is called the resonance integral for the system.

The eigenvector for E_2 is

$$|\psi\rangle = |1\rangle - |2\rangle \tag{8.3.18}$$

The first p_z orbital still has a positive amplitude on the $+z$ axis but the negative sign reverses the amplitude of the electron on the second atom. In the region between the nuclei, the positive and negative amplitudes will cancel to produce a net zero amplitude, i.e., a node. The same cancellation takes place in the negative z direction to produce the antibonding π^* molecular orbital. Both orbitals are illustrated in Figure 8.9.

While the energy of the π orbital was lower than the atomic orbital energy by β, the energy, $\alpha - \beta$, of the π^* orbital is higher by β. The average energy of the two molecular orbitals is equal to the energy of the atomic orbitals but the molecular energies are stabilized and destabilized by equal amounts (Figure 8.10).

Although the atomic and molecular orbitals are functional descriptions of electrons when they function as waves, the molecular orbital approach suggests that the orbitals exist to be filled. In actuality, they describe the electron wave amplitudes when the electron is present. Two electrons with opposite electron spin can have the same electron probability. Since the bond contains two electrons, only the π orbital electrons are observed. There are no π^* orbitals unless

Figure 8.9. The π and π^* molecular orbitals for the ethene molecule.

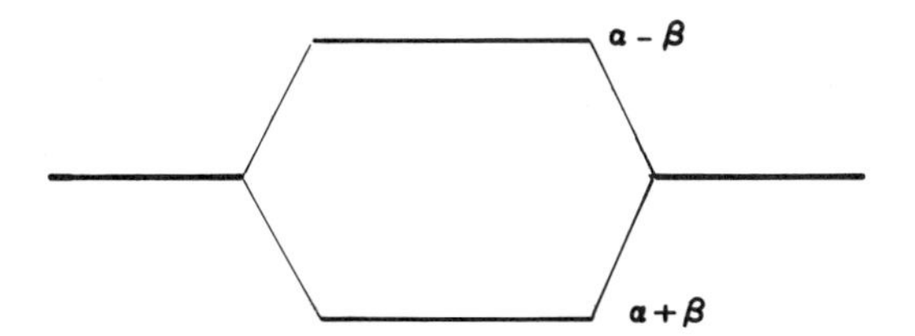

Figure 8.10. The energy levels for the π and π^* molecular orbitals for ethene.

there are electrons. If an electron was excited by the addition of energy, its probability density could change to that of the π^* orbital. The wavefunctions describe the delocalization of a single electron in the molecule.

The coefficients c_i give the amplitudes contributed by each atomic orbital in the molecule. In addition, they provide a measure of the location of the electron in the molecule. The wavefunction

$$|\psi\rangle = |1\rangle + |2\rangle \tag{8.3.19}$$

can be normalized by forming the scalar product, $\langle\psi|\psi\rangle$:

$$\langle\psi|\psi\rangle = (\langle 1| + \langle 2|)(|1\rangle + |2\rangle) = \langle 1|1\rangle + \langle 2|2\rangle + \langle 1|2\rangle + \langle 2|1\rangle = 2 \tag{8.3.20}$$

and

$$|\psi\rangle = \frac{1}{\sqrt{2}}(|1\rangle + |2\rangle) \tag{8.3.21}$$

The probability of finding the electron at carbon 1 will be equal to the square of the normalized coefficient for $|1\rangle$:

$$|c_1|^2 = \left(\frac{1}{\sqrt{2}}\right)^2 = \frac{1}{2} \tag{8.3.22}$$

The result is expected since the two carbon atoms are identical; the electron in the molecular orbital will be found half the time in the region of carbon atom 1.

The molecular orbitals for propene, C_3H_6, are generated using the same techniques. Although the formal bond structure for propene shows a single double bond, the molecular orbital is constructed with one atomic (p_z) orbital from each of the three carbons. The basis set of three atomic orbitals produces the 3×3 matrix eigenvector equation

$$\begin{pmatrix} H_{11} & H_{12} & H_{13} \\ H_{21} & H_{22} & H_{23} \\ H_{31} & H_{32} & H_{33} \end{pmatrix} \begin{pmatrix} c_1 \\ c_2 \\ c_3 \end{pmatrix} = E \begin{pmatrix} c_1 \\ c_2 \\ c_3 \end{pmatrix} \tag{8.3.23}$$

where $\mathbf{S}=\mathbf{I}$ once again. The matrix equation now contains three different types of integral. The diagonal elements H_{11} and H_{33} are integrals for the outer carbons and are defined as α like the corresponding integrals for ethene. Since α represents the electron localized on a single carbon, Hückel theory assigns this Coulomb integral for the central carbon a value of α as well. Each diagonal element of the matrix is α.

The integrals H_{12}, H_{21}, H_{23}, and H_{32} all describe a resonance between two orbitals on adjacent carbons. Since each carbon–carbon internuclear environment is identical, all these elements equal β:

$$\beta = H_{12} = H_{21} = H_{23} = H_{32} \tag{8.3.24}$$

The final two matrix elements, H_{13} and H_{31}, are resonance integrals for orbitals which are separated by an intermediate carbon atom. The overlap between these orbitals will be small, and the matrix elements are equated to zero:

$$H_{13} = H_{31} = 0 \tag{8.3.25}$$

Inserting the elements of the $\mathbf{H}$ matrix for propene in Equation 8.3.23 gives

$$\begin{pmatrix} \alpha & \beta & 0 \\ \beta & \alpha & \beta \\ 0 & \beta & \alpha \end{pmatrix} \begin{pmatrix} c_1 \\ c_2 \\ c_3 \end{pmatrix} = E \begin{pmatrix} c_1 \\ c_2 \\ c_3 \end{pmatrix} \tag{8.3.26}$$

The eigenvalues and eigenvectors for the matrix are found by factoring β from each term and defining

$$z = \alpha/\beta \tag{8.3.27}$$

to give the matrix

$$\beta \begin{pmatrix} z & 1 & 0 \\ 1 & z & 1 \\ 0 & 1 & z \end{pmatrix} \tag{8.3.28}$$

The tridiagonal matrix has appeared frequently and has the eigenvalues

$$\lambda = z; \qquad \lambda = z - \sqrt{2}; \qquad \lambda = z + \sqrt{2} \tag{8.3.29}$$

These eigenvalues must be multiplied by β to give the energies:

$$\begin{aligned} E_1 &= \alpha + \sqrt{2}\,\beta \\ E_2 &= \alpha \\ E_3 &= \alpha - \sqrt{2}\,\beta \end{aligned} \tag{8.3.30}$$

The energy level diagram for these orbitals is shown in Figure 8.11. E_2 is equal to

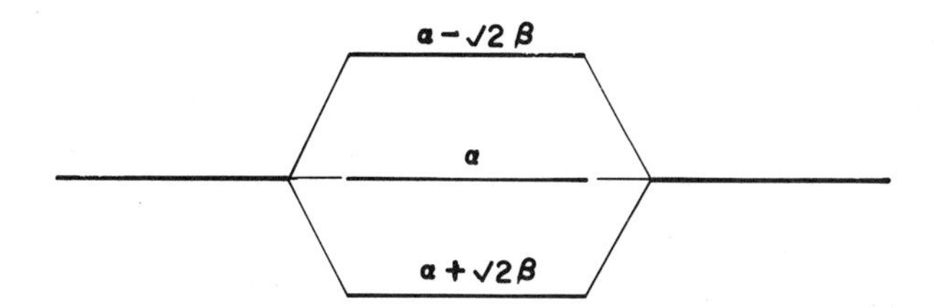

Figure 8.11. The energy levels for the propene molecule.

the energy for electrons localized on single carbon atoms because of a node at the central carbon. The electron probability is located about the two terminal carbon atoms.

The construction of the molecular orbitals requires determination of the coefficients c_i for each eigenvalue. Using the eigenvalues of Equation 8.3.29, the eigenvectors are

$$|1\rangle = \left(\frac{1}{2}\right)\begin{pmatrix} 1 \\ +\sqrt{2} \\ 1 \end{pmatrix} \qquad |2\rangle = \left(\frac{1}{\sqrt{2}}\right)\begin{pmatrix} 1 \\ 0 \\ -1 \end{pmatrix} \qquad |3\rangle = \left(\frac{1}{2}\right)\begin{pmatrix} 1 \\ -\sqrt{2} \\ 1 \end{pmatrix} \tag{8.3.31}$$

The linear combinations of atomic orbitals for each energy eigenvalue are

$$|\psi_1\rangle = \frac{1}{2}(|1\rangle + \sqrt{2}\,|2\rangle + |3\rangle) \tag{8.3.32}$$

$$|\psi_2\rangle = \frac{1}{\sqrt{2}}(|1\rangle - |3\rangle) \tag{8.3.33}$$

$$|\psi_3\rangle = \frac{1}{2}(|1\rangle - \sqrt{2}\,|2\rangle + |3\rangle) \tag{8.3.34}$$

The molecular orbitals are illustrated in Figure 8.12. $|\psi_1\rangle$ is similar to the π orbital for ethene but the central carbon has the larger wave amplitude since it receives contributions from both terminal carbons. The energy for $|\psi_2\rangle$ is identical to that of independent carbon atoms because of the node at the central carbon. There are two distinct regions of electron probability. The final

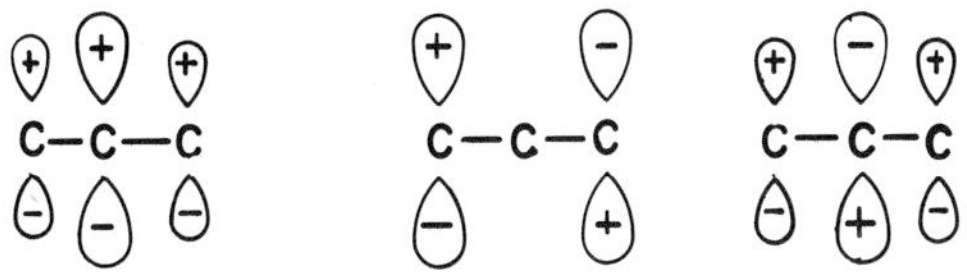

Figure 8.12. The probability densities for the three molecular orbitals of the propene molecule.

molecular orbital has nodes between each of the carbon atoms and there is electron amplitude for each of the three carbon atoms (Figure 8.12).

The coefficient vectors are normalized so the coefficients for each atomic orbital can be used directly to find the probability that the electron is located on each of the carbon atoms. The probabilities for the three molecular orbitals in the order 1, 2, 3 are

$$\begin{aligned} |\psi_1\rangle &: \left(\frac{1}{4}, \frac{1}{2}, \frac{1}{4}\right) \\ |\psi_2\rangle &: \left(\frac{1}{2}, 0, \frac{1}{2}\right) \\ |\psi_3\rangle &: \left(\frac{1}{4}, \frac{1}{2}, \frac{1}{4}\right) \end{aligned} \tag{8.3.35}$$

For the first and third molecular orbitals, the electron is twice as likely to be observed at the central carbon. The node for $|\psi_2\rangle$ at the central carbon gives zero electron density at this carbon.

Any linear molecule with carbons with atomic orbitals available for binding will have a tridiagonal **H** matrix if the simplifying assumptions of basic Hückel theory are used. The $N \times N$ tridiagonal matrix

$$\mathbf{H} = [1, z, 1] \tag{8.3.36}$$

has eigenvalues

$$\lambda_n = z + 2 \cos [n\pi/(N+1)] \qquad n = 1, 2, ..., N \tag{8.3.37}$$

and energy eigenvalues

$$E_n = \lambda_n = \alpha + 2\beta \cos [n\pi/(N+1)] \qquad n = 1, 2, ..., N \tag{8.3.38}$$

This can be verified for $N=3$. The development of energy levels with increasing chain length N is illustrated in Figure 8.13. The number of levels increases with N but the levels are ultimately bracketed between $\alpha + 2\beta$ and $\alpha - 2\beta$. The energy levels are more closely spaced for larger N and the transition energies are also smaller so the long molecules absorb energy of longer wavelength.

As the energy of the levels increases, the number of nodes in the molecular orbital increases, i.e., the coefficients change sign an increasing number of times. The jth component of the nth eigenvalue is (Chapter 4)

$$|n\rangle_j = \sin [jn\pi/(N+1)] \tag{8.3.39}$$

For $N=3$, $|2\rangle$ is

$$\begin{pmatrix} \sin [(1)(2)\pi/4] \\ \sin [(2)(2)\pi/4] \\ \sin [(3)(2)\pi/4] \end{pmatrix} = \begin{pmatrix} 1 \\ 0 \\ -1 \end{pmatrix} \tag{8.3.40}$$

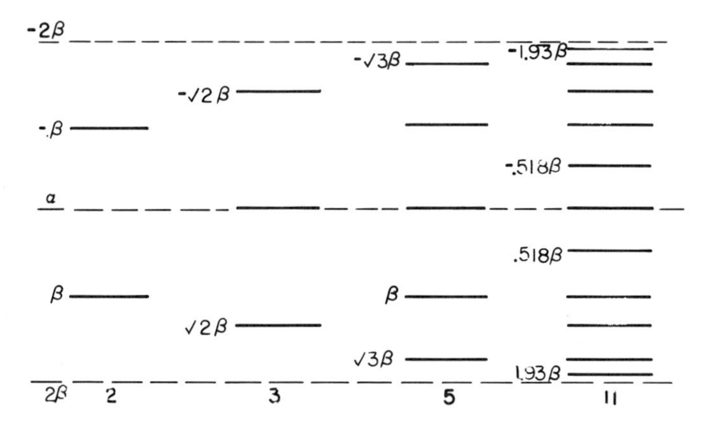

Figure 8.13. The variation in energy levels of long unsaturated chains as a function of the total number of carbons.

The molecular orbitals for a chain of five carbon atoms are shown in Figure 8.14 for the energies

$$\alpha + \sqrt{3}\,\beta; \qquad \alpha + \beta; \qquad \alpha; \qquad \alpha - \beta; \qquad \alpha - \sqrt{3}\,\beta \tag{8.3.41}$$

The electron has a positive amplitude in the positive z direction for every carbon in the lowest-energy wavefunction. The amplitude of the molecular wavefunction at each carbon is

$$\left(\frac{1}{2}, \frac{\sqrt{3}}{2}, 1, \frac{\sqrt{3}}{2}, \frac{1}{2}\right) \tag{8.3.42}$$

The amplitude is greatest at the central carbon, reflecting the shape of the sine function which generates the coefficients.

The vector can be normalized, and the square of the normalized com-

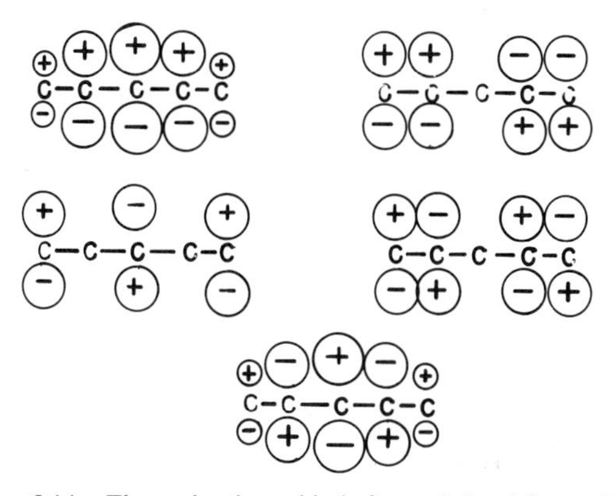

Figure 8.14. The molecular orbitals for a chain of five carbons.

ponents gives the probabilities for finding the electron at each of the carbons. These are

$$\left(\frac{1}{12}, \frac{3}{12}, \frac{4}{12}, \frac{3}{12}, \frac{1}{12}\right) \tag{8.3.43}$$

The eigenvector for the highest-energy eigenvalue has normalized components

$$\left(\frac{1}{2}, \frac{-\sqrt{3}}{2}, +1, \frac{-\sqrt{3}}{2}, \frac{1}{2}\right)\left(\frac{1}{\sqrt{3}}\right) \tag{8.3.44}$$

The electron densities at each carbon are

$$\left(\frac{1}{12}, \frac{3}{12}, \frac{4}{12}, \frac{3}{12}, \frac{1}{12}\right) \tag{8.3.45}$$

These densities are the same as those of the lowest eigenvalue. However, the high-energy eigenvalues has a wavefunction with four nodes.

8.4. Hückel Theory for Cyclic Molecules

Linear chains of carbon atoms which each contribute one p_z atomic orbital toward the creation of molecular orbitals require **H** matrices which are tridiagonal. Analysis for a cyclic molecule includes all the assumptions introduced for the linear chain. In addition, the final carbon atom in the chain must be σ-bonded to the initial carbon, i.e.,

$$\begin{aligned} C_{N+1} &= C_1 \\ C_0 &= C_N \end{aligned} \tag{8.4.1}$$

These conditions are equivalent to the cyclic boundary conditions for vibrating atom chains (Chapter 5).

When N carbon atoms form a ring, there are two additional nearest-neighbor resonance integrals, β,

$$\beta = H_{1N} = H_{N1} \tag{8.4.2}$$

which occupy the $1N$ and $N1$ elements of the **H** matrix. For example, the cyclobutadiene ring with four carbons has the following **H** matrix:

$$\begin{pmatrix} H_{11} & H_{12} & 0 & H_{14} \\ H_{21} & H_{22} & H_{23} & 0 \\ 0 & H_{32} & H_{33} & H_{34} \\ H_{41} & 0 & H_{43} & H_{44} \end{pmatrix} \begin{pmatrix} c_1 \\ c_2 \\ c_3 \\ c_4 \end{pmatrix} = E \begin{pmatrix} c_1 \\ c_2 \\ c_3 \\ c_4 \end{pmatrix} \tag{8.4.3}$$

Substituting α and β for the Coulomb (diagonal) and resonance (off-diagonal) matrix elements gives

$$\begin{pmatrix} \alpha & \beta & 0 & \beta \\ \beta & \alpha & \beta & 0 \\ 0 & \beta & \alpha & \beta \\ \beta & 0 & \beta & \alpha \end{pmatrix} \begin{pmatrix} c_1 \\ c_2 \\ c_3 \\ c_4 \end{pmatrix} = E \begin{pmatrix} c_1 \\ c_2 \\ c_3 \\ c_4 \end{pmatrix} \tag{8.4.4}$$

An N-carbon chain will have an $N \times N$ **H** matrix with tridiagonal elements and two additional elements at the corners of the matrix.

When the system is cyclic, the eigenvector components can be exponential functions,

$$|n\rangle_j = \exp(2\pi inj/N) \tag{8.4.5}$$

Since the actual location of a starting carbon is irrelevant in the cyclic chain, Equation 8.4.5 can be used for any row of the matrix. For $|n\rangle_j$, the single row equation

$$c_{j-1}\beta + c_j\alpha + \beta c_{j+1} = Ec_j \tag{8.4.6}$$

suffices to determine the E_n eigenvalue. Substituting the exponential wavefunction components gives

$$\begin{aligned} &\beta \exp[2\pi in(j-1)/N] + \alpha \exp[2\pi in(j)/N] + \beta \exp[2\pi in(j+1)/N] \\ &\quad = E_n \exp[2\pi in(j)/N] \\ &\qquad \beta \exp(-2\pi in/N) + \alpha + \beta(2\pi in/N) = E_n \end{aligned} \tag{8.4.7}$$

Since

$$\exp(2\pi in/N) + \exp(-2\pi in/N) = 2 \cos(2\pi n/N) \tag{8.4.8}$$

the nth energy eigenvalue is

$$E_n = \alpha + 2\beta \cos(2\pi n/N) \tag{8.4.9}$$

The energy eigenvalues for cyclobutadiene ($N = 4$) are

$$E_0 = \alpha + 2\beta \cos(2\pi 0/4) = \alpha + 2\beta \tag{8.4.10}$$

$$E_1 = \alpha + 2\beta \cos(2\pi 1/4) = \alpha \tag{8.4.11}$$

$$E_2 = \alpha + 2\beta \cos(2\pi 2/4) = \alpha - 2\beta \tag{8.4.12}$$

$$E_3 = \alpha + 2\beta \cos(2\pi 3/4) = \alpha \tag{8.4.13}$$

The factor of 2π in the cosine generates some doubly degenerate eigenvalues.

Such degeneracies are characteristic of cyclic systems. The eigenvalues repeat if n exceeds N. For $n = N + 1$, the eigenvalue is

$$\alpha + 2\beta \cos [2\pi(N+1)/N] = \alpha + 2\beta \cos [(2\pi) + (2\pi/N)] \qquad (8.4.14)$$

Since 2π is equivalent to $0°$, the eigenvalue is

$$\alpha + 2\beta \cos (2\pi/N) \qquad (8.4.15)$$

For $N = 4$, this is

$$E_{N+1} = \alpha \qquad (8.4.16)$$

which is equivalent to E_1.

The eigenvalues are regularly spaced around a circle every $2\pi/N$ degrees. This is illustrated for cyclobutadiene (Figure 8.15) with levels spaced every $2\pi/4 = 90°$ on a circle of radius 2β. Since the energy is ordered in the vertical direction, two of the levels with $E_n = \alpha$ are degenerate as predicted by the equations for these eigenvalues.

The procedure can be repeated for the benzene molecule with $N = 6$. The eigenvalues are determined from

$$E_n = \alpha + 2\beta \cos (2\pi n/6) \qquad (8.4.17)$$

as

$$\begin{aligned} E_0 &= \alpha + 2\beta \\ E_1 &= \alpha + \beta \\ E_2 &= \alpha - \beta \\ E_3 &= \alpha - 2\beta \\ E_4 &= \alpha - \beta \\ E_5 &= \alpha + \beta \end{aligned} \qquad (8.4.18)$$

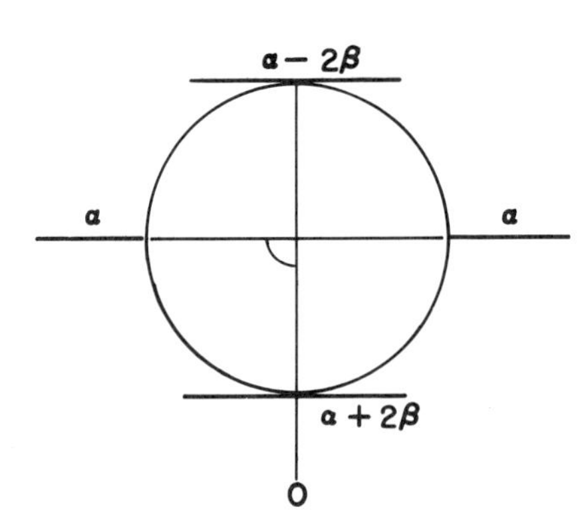

Figure 8.15. The energy levels of cyclobutadiene on a circle of radius 2β.

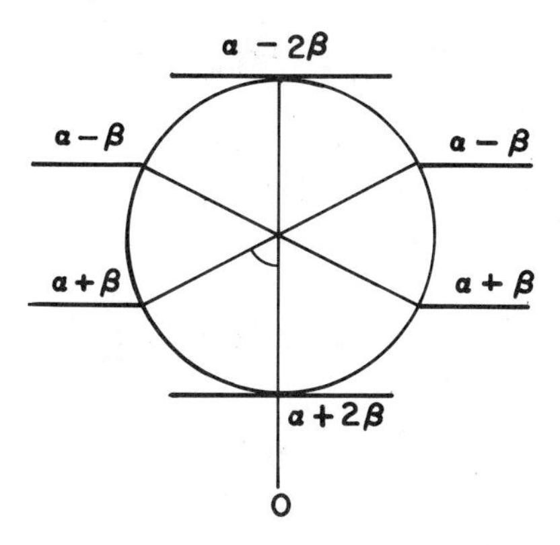

Figure 8.16. The energy levels of benzene on a circle of radius 2β.

Figure 8.16 shows the circle diagram for these eigenvalues. In this case, two pairs of eigenvalues are degenerate; the maximal and minimal eigenvalues are nondegenerate.

For an N-carbon system, the eigenvalues are located at angular intervals of $2\pi/N$ about the circle. The eigenvalue for $n=0$ is the lowest eigenvalue in each case since the integral β is negative. In addition, $2\pi 0/N$ generates a cosine argument of zero. In general, the intervals on the circle will begin with this argument; $\theta=0$ is defined on the negative vertical axis. For a ring with an even number of carbons, an argument of π is also possible, and this defines a maximal nondegenerate eigenvalue. The remainder of the eigenvalues must be doubly degenerate for rings with even numbers of carbons. For example, the cyclooctatriene molecule is a ring of eight carbons and the energy levels must be arranged every $2\pi/8=\pi/4$ on a circle of radius 2β as shown in Figure 8.17. There are three pairs of degenerate eigenvalues, and the projections of each line onto the vertical (energy) axis show that the pairs must have energies

$$\begin{aligned} \alpha+\beta\left(\frac{2}{\sqrt{2}}\right) &= \alpha+\sqrt{2}\,\beta \\ \alpha-\beta\left(\frac{2}{\sqrt{2}}\right) &= \alpha-\sqrt{2}\,\beta \\ \alpha-(0)\,\beta &= \alpha \end{aligned} \tag{8.4.19}$$

The maximal and minimal eigenvalues have energies of $\alpha-2\beta$ and $\alpha+2\beta$, respectively.

The orbitals exist only if there is an electron with that wave character. For an N-carbon ring, there are Np_z orbitals which assume the molecular orbital shapes. Because of electron spin, two electrons have the same spatial molecular orbital. This condition determines the stability of the ring system. For example, the benzene molecule with the energy levels of Figure 8.16 will have its six electrons in the lowest energy level $(\alpha+2\beta)$ and the two degenerate $(\alpha+\beta)$

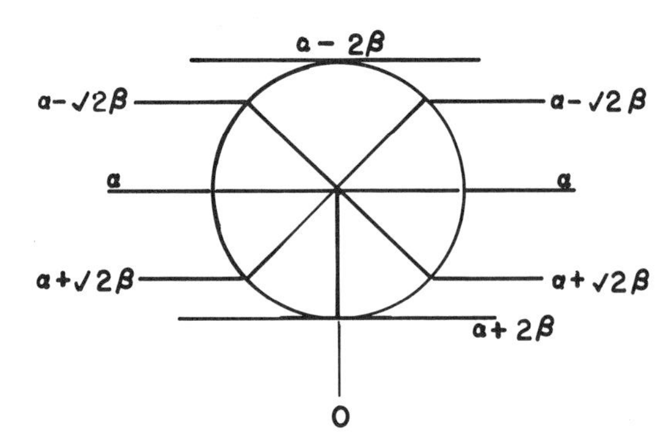

Figure 8.17. The energy levels of cyclooctatriene on a circle of radius 2β.

levels. Since each of these levels is more stable than the α level for an electron localized on a single carbon, the molecule will be stable. The cyclobutadiene system of Figure 8.15 will have only a single electron pair in a stabilizing orbital. The remaining two electrons are in "atomic-like" orbitals which do not stabilize the molecule.

Benzene and cyclobutadiene have even numbers of carbon atoms. Although rings with odd numbers of carbons have energy levels which can be calculated using the general expressions for the energy eigenvalues, they do have some different properties. The simplest odd-carbon system is cyclopropene. Its energy eigenvalues are

$$\begin{aligned} E_0 &= \alpha + 2\beta \cos(2\pi/3) = \alpha + 2\beta \\ E_1 &= \alpha + 2\beta \cos(2\pi/3) = \alpha - \beta \\ E_2 &= \alpha + 2\beta \cos(4\pi/3) = \alpha - \beta \end{aligned} \tag{8.4.20}$$

and their arrangement on the circle is shown in Figure 8.18. The first (lowest) energy remains nondegenerate. The two upper levels are now degenerate.

For rings with even numbers of carbon atoms, energy levels in the upper half of the ring are reflections of those in the lower half. Since β determines the

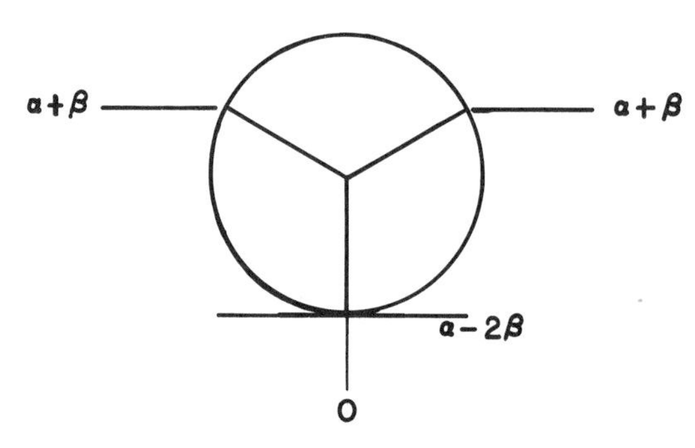

Figure 8.18. The energy levels of cyclopropene on a circle of radius 2β.

Figure 8.19. (a) The benzene molecule as an alternate carbon system. (b) The cyclopropene molecule as a non-alternate carbon system.

distance from the horizontal axis, positive values of β can be replaced with negative values to generate the upper energy levels from the lower energy levels. A molecule will have this symmetry if alternate carbons can be marked with an asterisk so that no two adjacent carbons have an asterisk. Benzene has this symmetry since asterisks can label carbons 1, 3, and 5 (Figure 8.19a). Cyclopropene cannot, since an asterisk on carbon 1 leaves two adjacent unmarked carbons. A second asterisk creates two adjacent marked carbons (Figure 8.19b).

The Hückel theory gives all the molecular orbital energies for the electron. The stability of the molecule depends on the number of electrons which are actually present in the orbitals. For cyclopropene with three p_z electrons, the nondegenerate lowest-energy state is filled with two electrons; the third electron must have the high energy, $\alpha - \beta$, and this will destabilize the cyclopropene molecule. However, the cyclopropenyl cation has lost one of the three electrons. The remaining electrons have the low energy, and the cation is predicted to be a stable species.

Hückel theory can be used to determine the energy levels for molecules which contain both ring and linear regions or fused rings. For example, the bicyclobutadiene molecule of Figure 8.20 has four carbons but carbons 2 and 3 each bond to three other carbons. The **H** matrix for the system is generated by considering each row. The first row has α on the diagonal (11). Since the atom is adjacent to carbons 2 and 3, β is added at the 12 and 13 elements. Element 14 is zero. The process is repeated with carbon 2 for row 2, etc., to give the matrix

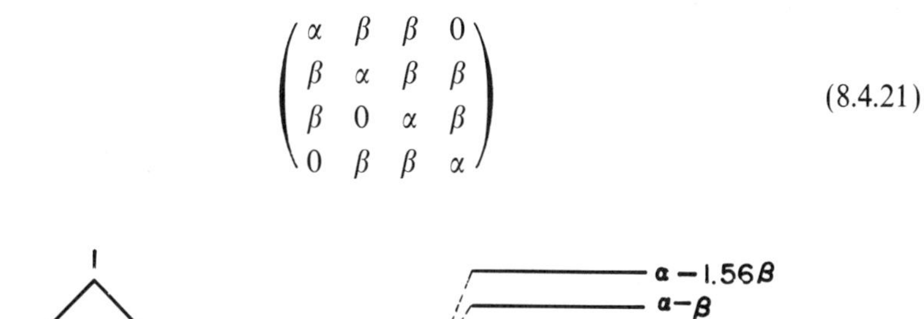

$$\begin{pmatrix} \alpha & \beta & \beta & 0 \\ \beta & \alpha & \beta & \beta \\ \beta & 0 & \alpha & \beta \\ 0 & \beta & \beta & \alpha \end{pmatrix} \tag{8.4.21}$$

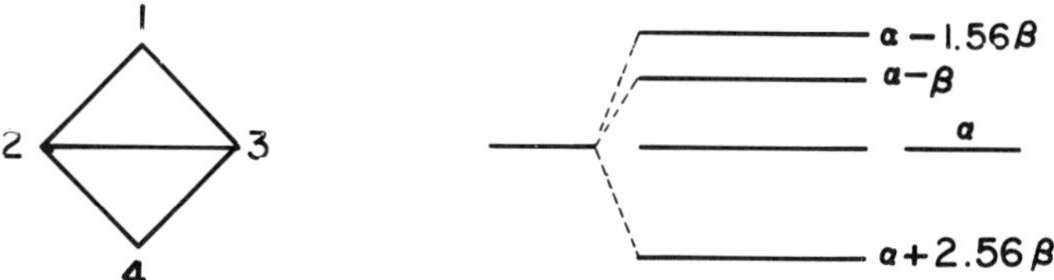

Figure 8.20. The energy level diagram for bicyclobutadiene.

Factoring β from each term gives the matrix

$$\beta\begin{pmatrix} z & 1 & 1 & 0 \\ 1 & z & 1 & 1 \\ 1 & 0 & z & 1 \\ 0 & 1 & 1 & z \end{pmatrix} \quad \text{with} \quad z=\alpha/\beta \tag{8.4.22}$$

The additional bond creates a matrix which has two lower and two upper subdiagonals on either side of the main matrix diagonal. The characteristic determinant is

$$\begin{vmatrix} x & 1 & 1 & 0 \\ 1 & x & 1 & 1 \\ 1 & 1 & x & 1 \\ 0 & 1 & 1 & x \end{vmatrix} \tag{8.4.23}$$

where the substitution

$$x=z-\lambda \tag{8.4.24}$$

has been made to simplify solution of this problem. The characteristic polynomial

$$x(x^3-5x+4)=0 \tag{8.4.25}$$

gives one root $x=0$ which becomes

$$\begin{gathered} \lambda=z-x=z=\alpha/\beta \\ E=\alpha \end{gathered} \tag{8.4.26}$$

One of the four energy levels is equal to the atomic energy levels.

The root $x=1$ is verified by substitution in Equation 8.4.25. The energy is

$$\begin{aligned} \lambda&=z-x=z-1=\alpha/\beta-1 \\ E_n&=\beta\lambda=\alpha-\beta \end{aligned} \tag{8.4.27}$$

Division by $x-1$ gives the quadratic equation for the remaining two roots:

$$x^2+x-4=0 \tag{8.4.28}$$

The roots are

$$\begin{aligned} x_+ &= -\frac{1}{2}+\frac{1}{2}\sqrt{17}=1.56 \\ x_- &= -\frac{1}{2}-\frac{1}{2}\sqrt{17}=-2.56 \end{aligned} \tag{8.4.29}$$

These are converted to the energy eigenvalues

$$E_+ = \alpha - 1.56\beta$$
$$E_- = \alpha + 2.56\beta \tag{8.4.30}$$

The energy diagram for cyclobutadiene is shown in Figure 8.20.

8.5. Degenerate Molecular Orbitals for Cyclic Molecules

While the energy levels for molecular orbitals constructed from p_z orbitals in long carbon chains were nondegenerate, the energy levels constructed from the atomic orbitals of carbon atoms in rings generally did have some degenerate levels. The cyclobutadiene molecule introduced in Section 8.4 has the two non-degenerate levels with energies $\alpha + 2\beta$ and $\alpha - 2\beta$ and two degenerate levels with energy α. For this cyclic system the eigenvector has components

$$|n\rangle_j = \exp(2\pi inj/N) \tag{8.5.1}$$

These components give the nondegenerate eigenvectors of the **H** matrix directly. For $n = 0$, the components are

$$\begin{pmatrix} |0\rangle_1 \\ |0\rangle_2 \\ |0\rangle_2 \\ |0\rangle_4 \end{pmatrix} = \begin{pmatrix} \exp[i2\pi(0)(1)/4] \\ \exp[i2\pi(0)(2)/4] \\ \exp[i2\pi(0)(3)/4] \\ \exp[i2\pi(0)(4)/4] \end{pmatrix} = \begin{pmatrix} 1 \\ 1 \\ 1 \\ 1 \end{pmatrix} \tag{8.5.2}$$

All arguments are multiples of 2π and all components are real. The wavefunction has the appearance of "electron clouds" above and below the plane of the molecule (Figure 8.21). The upper cloud is the linear combination of positive electron amplitudes for all four p_z electrons; the lower cloud is the linear combination of negative electron amplitudes.

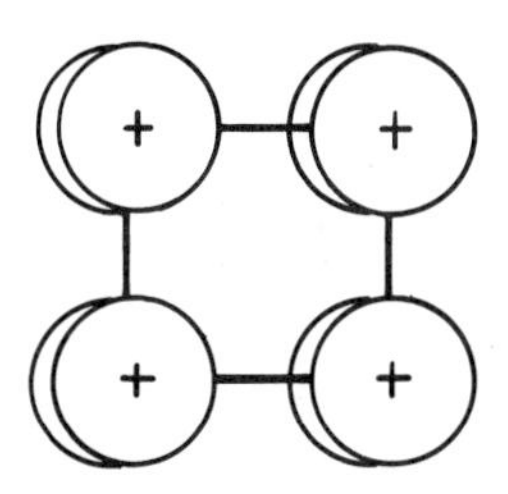

Figure 8.21. The lowest-energy molecular orbital for the butadiene molecule. All φ_i have a coefficient of $+1$.

The second nondegenerate level, for $n=2$, has an eigenvector

$$\begin{pmatrix} |2\rangle_1 \\ |2\rangle_2 \\ |2\rangle_3 \\ |2\rangle_4 \end{pmatrix} = \begin{pmatrix} \exp[i2\pi(2)(1)/4] \\ \exp[i2\pi(2)(2)/4] \\ \exp[i2\pi(2)(3)/4] \\ \exp[i2\pi(2)(4)/4] \end{pmatrix} = \begin{pmatrix} -1 \\ 1 \\ -1 \\ 1 \end{pmatrix} \tag{8.5.3}$$

Alternate p_z orbitals in the molecular orbital have opposite sign. The orbital has nodes between each pair of carbons and the orbital looks like two ethene π^* orbitals on each side of the molecule (Figure 8.22).

The eigenvectors of Equations 8.5.2 and 8.5.3 have appeared often as the eigenvectors of 4×4 matrices. The two eigenvectors which are orthogonal to these eigenvectors are

$$\begin{pmatrix} 1 \\ 1 \\ -1 \\ -1 \end{pmatrix} \quad \begin{pmatrix} 1 \\ -1 \\ -1 \\ 1 \end{pmatrix} \tag{8.5.4}$$

However, the two eigenvectors which are generated using Equation 8.5.1 are

$$\begin{pmatrix} |1\rangle_1 \\ |1\rangle_2 \\ |1\rangle_3 \\ |1\rangle_4 \end{pmatrix} = \begin{pmatrix} \exp[i2\pi(1)(1)/4] \\ \exp[i2\pi(1)(2)/4] \\ \exp[i2\pi(1)(3)/4] \\ \exp[i2\pi(1)(4)/4] \end{pmatrix} = \begin{pmatrix} i \\ -1 \\ -i \\ 1 \end{pmatrix} \tag{8.5.5}$$

$$\begin{pmatrix} |3\rangle_1 \\ |3\rangle_2 \\ |3\rangle_3 \\ |3\rangle_4 \end{pmatrix} = \begin{pmatrix} \exp[i2\pi(3)(1)/4] \\ \exp[i2\pi(3)(2)/4] \\ \exp[i2\pi(3)(3)/4] \\ \exp[i2\pi(3)(4)/4] \end{pmatrix} = \begin{pmatrix} -i \\ -1 \\ i \\ 1 \end{pmatrix} \tag{8.5.6}$$

These eigenvectors have imaginary components. Despite this, they are still orthogonal to the two nondegenerate eigenvectors. The change from 1 to i indicates a change of 90° rather than the 1 to -1 change of 180°. The single change of sign suggests a single node, intermediate between the zero nodes for

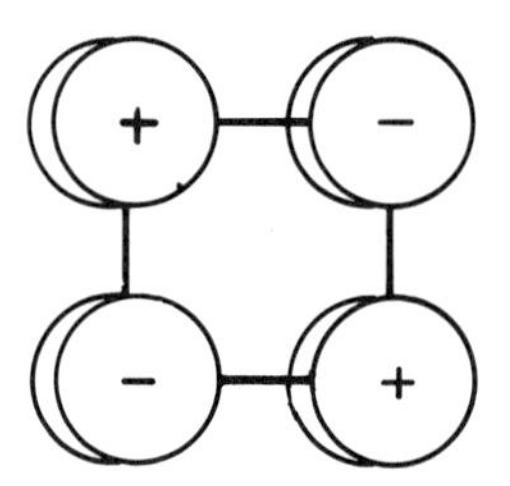

Figure 8.22. The highest-energy, antibonding molecular orbital for cyclobutadiene.

the lowest-energy eigenvector and the two nodes of the highest-energy eigenvector.

The eigenvectors for the degenerate eigenvalues are orthogonal to eigenvectors for other eigenvalues but they may not be mutually orthogonal. For example,

$$(i \quad -1 \quad -i \quad 1)\begin{pmatrix} -i \\ -1 \\ i \\ 1 \end{pmatrix} = 4 \tag{8.5.7}$$

Any linear combination of eigenvectors within the subspace will remain orthogonal to eigenvectors outside the subspace. Since this Hermitian matrix must have a complete set of orthogonal eigenvectors, the eigenvectors with imaginary components can be combined to give orthogonal eigenvectors with real components. For these two eigenvectors with imaginary components, the components can be added and subtracted to produce two eigenvectors with real coefficients:

$$\begin{pmatrix} i \\ -1 \\ -i \\ 1 \end{pmatrix} + \begin{pmatrix} -i \\ -1 \\ i \\ 1 \end{pmatrix} = \begin{pmatrix} 0 \\ -2 \\ 0 \\ 2 \end{pmatrix} \propto \begin{pmatrix} 0 \\ -1 \\ 0 \\ 1 \end{pmatrix} \tag{8.5.8}$$

$$\begin{pmatrix} i \\ -1 \\ -i \\ 1 \end{pmatrix} - \begin{pmatrix} -i \\ -1 \\ i \\ 1 \end{pmatrix} = \begin{pmatrix} 2i \\ 0 \\ -2i \\ 0 \end{pmatrix} \propto \begin{pmatrix} 1 \\ 0 \\ -1 \\ 0 \end{pmatrix} \tag{8.5.9}$$

These two eigenvectors would have to be added and subtracted again to generate the equally valid eigenvectors of Equation 8.5.4.

Both the eigenvectors of Equation 8.5.4 and those of Equations 8.5.8 and 8.5.9 are valid eigenvectors for the **H** matrix and eigenenergy α. This does not mean that the eigenvectors for the states are arbitrary; the phase of the eigenvalue, i.e., the position of the maximum and minimum amplitudes, is different for each case. This is shown in Figure 8.23. For Equations 8.5.8 and 8.5.9, the

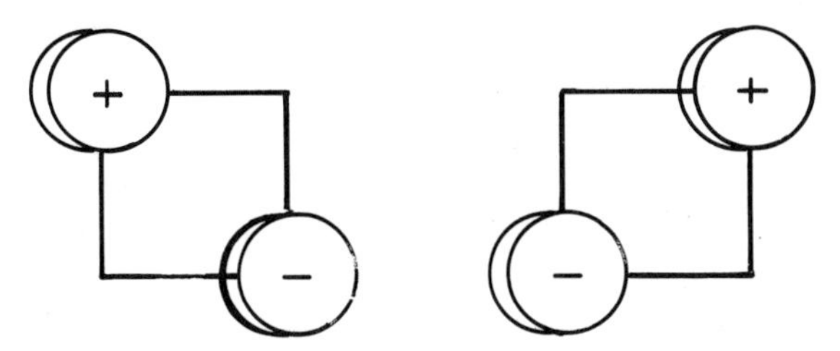

Figure 8.23. The high-energy (a) and low-energy (b) degenerate pairs of molecular orbitals for cyclobutadiene.

maximal amplitudes $\left(\frac{1}{\sqrt{2}}\right)$ are centered on atoms 1 and 3 and atoms 2 and 4, respectively (Figure 8.23a). The alternate carbons have zero electron density, i.e., the nodes lie on these carbons. The orbitals for Equation 8.5.4 (Figure 8.23b) have maximal amplitude between atoms 1 and 2 and atoms 3 and 4 in the first case and atoms 2 and 3 and atoms 4 and 1 in the second case. The electron densities of $\left(\frac{1}{2}\right)^2$ at each carbon for the normalized eigenvector are smaller because the maximal amplitude of the orbital lies midway between the carbons. The nodes in this case also lie between carbons.

This phase uncertainty is present for all the eigenvectors because their components are relative. For example, the nondegenerate eigenvector with components

$$(-1, 1, -1, 1) \tag{8.5.10}$$

is identical to

$$(1, -1, 1, -1) \tag{8.5.11}$$

The positive electron amplitude is transposed to the next carbon atom. Since all carbon atoms are identical, the new wavefunction is identical to the original. Either can be used.

The cyclopropene molecule has the nondegenerate eigenvector

$$\begin{pmatrix} 1 \\ 1 \\ 1 \end{pmatrix} \tag{8.5.12}$$

continuing a pattern for these systems. The lowest-energy eigenvector is generally a linear combination of all p_z orbitals with a $+1$ coefficient. This is the most energy-efficient way to create such orbitals. Once again, the electron density of the cyclic molecule will consist of a cloud of electron density generated by the positive electron wave amplitude above the molecule and a second cloud of electron density generated by the negative electron amplitude below the molecule. Of course, the entire vector can be multiplied by -1 so that the upper cloud is generated from negative amplitudes. This new phase will have exactly the same energy and electron densities as the original.

Although the eigenvector components for the degenerate eigenvectors can be generated using Equation 8.5.1, the eigenvectors for $\alpha + \beta$ can be determined directly from the matrix equation. For example, the eigenvectors

$$\begin{pmatrix} 1 \\ -1 \\ 0 \end{pmatrix} \quad \begin{pmatrix} -2 \\ 1 \\ 1 \end{pmatrix} \tag{8.5.13}$$

give the proper energy eigenvalue and are orthogonal to the third eigenvector (Equation 8.5.12). They are not mutually orthogonal but a Gram–Schmidt orthogonalization will rectify this. The normalized vector,

$$|2'\rangle = \left(\frac{1}{\sqrt{2}}\right)\begin{pmatrix}1\\-1\\0\end{pmatrix} \tag{8.5.14}$$

is selected as the starting vector. The second eigenvector is

$$|3'\rangle = |3\rangle - |2'\rangle\langle 2'|3\rangle$$

$$= \begin{pmatrix}2\\-1\\-1\end{pmatrix} - \left(\frac{1}{2}\right)\begin{pmatrix}1\\-1\\0\end{pmatrix}(1 \quad -1 \quad 0)\begin{pmatrix}2\\-1\\-1\end{pmatrix} = \begin{pmatrix}\frac{1}{2}\\\frac{1}{2}\\-1\end{pmatrix} = \begin{pmatrix}1\\1\\-2\end{pmatrix} \tag{8.5.15}$$

The orthogonalization simply shifted the three elements of the vector. The component 2 was shifted to coincide with the component 0 in the second vector. The combined wavefunctions distribute the electron evenly over the three carbons.

The wavefunctions for the degenerate eigenvalues are illustrated in Figure 8.24. There are two nodes in each case. For Equation 8.5.14, the nodes lie between carbons 1 and 2 and on carbon 3 so there is no electron density on carbon 3. For the second orbital, the density on carbon 3 is large and generated by a negative electron amplitude. The two nodes appear between carbons 1 and 3 and carbons 2 and 3. The amplitude on carbon 3 is larger because the maximal wave amplitude lies directly on this carbon. For the first orbital, this maximal amplitude lies between carbons 1 and 3 and carbons 2 and 3, i.e., at the positions of the nodes for the second orbital.

The benzene molecule has two nondegenerate and two pairs of degenerate orbitals. Since the nondegenerate orbitals are the maximal and minimal energy

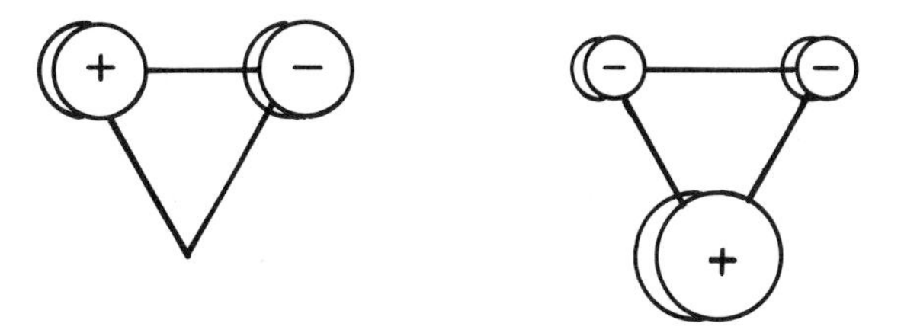

Figure 8.24. The degenerate molecular orbitals for the cyclopropene molecule.

orbitals, they are expected to have no nodes and the maximal number of nodes, respectively. The eigenvectors

$$\left(\frac{1}{\sqrt{6}}\right)\begin{pmatrix}1\\1\\1\\1\\1\\1\end{pmatrix} \quad \left(\frac{1}{\sqrt{6}}\right)\begin{pmatrix}1\\-1\\1\\-1\\1\\-1\end{pmatrix} \tag{8.5.16}$$

are valid for benzene and are illustrated in Figure 8.25. The high-energy orbital looks like a combination of three two-carbon antibonding orbitals. With this phase, the maximum positive and negative electron amplitudes lie directly on alternate carbons. Nodes lie midway between each carbon.

The two sets of degenerate eigenvalues lie between these two extremes. Using the smaller cyclic molecules as a guide, the lower-energy degenerate pair will have two nodes while the higher-energy degenerate pair will have four nodes, i.e., two nodal planes. These nodes will be used to predict the eigenvector components.

If the nodes for the $\alpha + \beta$ eigenvectors are placed at carbons 1 and 4, the maximal amplitudes must lie between carbons 2 and 3 and carbons 5 and 6.

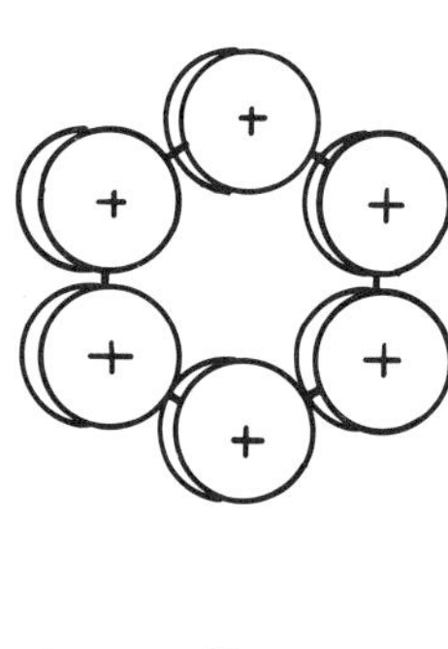

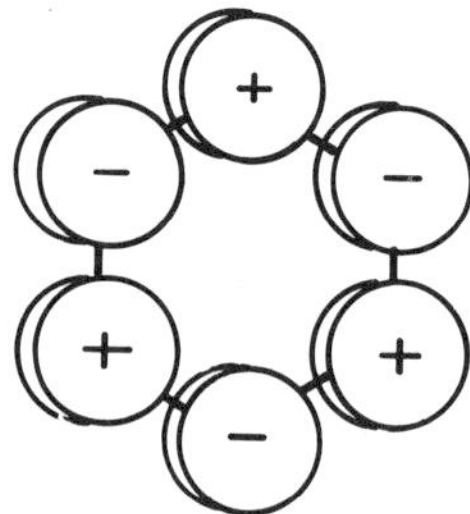

Figure 8.25. The lowest-energy (a) and highest-energy (b) molecular orbitals for the benzene molecule.

Since the carbons lie equidistant from these maxima, each pair must have the same amplitude. The predicted eigenvector has components

$$(0, 1, 1, 0, -1, -1) \tag{8.5.17}$$

and this is indeed valid as verified by operation with the **H** matrix.

The second degenerate eigenvector must be orthogonal to the first. Since the nodal plane of the first eigenvector passes through carbons 1 and 4, an orthogonal eigenvector might be generated by preparing a nodal plane perpendicular to the first, i.e., passing midway between carbons 2 and 3 and carbons 5 and 6 (Figure 8.26). This plane could also pass midway between carbons 1 and 2 and carbons 4 and 5 (Problem 8.8).

The maximal amplitudes of the second orbital appear at carbons 1 and 4. The eigenvector must have the normalized eigenvector components

$$\frac{1}{\sqrt{3}}\left(1, \frac{1}{2}, -\frac{1}{2}, -1, -\frac{1}{2}, \frac{1}{2}\right) \tag{8.5.18}$$

The two degenerate molecular orbitals are

$$|\psi\rangle = \frac{1}{\sqrt{2}}(\varphi_2 + \varphi_3 - \varphi_5 - \varphi_6) \tag{8.5.19}$$

$$|\psi\rangle = \frac{1}{\sqrt{3}}\left(\varphi_1 + \frac{1}{2}\varphi_2 - \frac{1}{2}\varphi_3 - \varphi_4 - \frac{1}{2}\varphi_5 + \frac{1}{2}\varphi_6\right) \tag{8.5.20}$$

The upper and lower degenerate levels are equidistant from the free atom energy levels. This horizontal symmetry again correlates the energy expressions in the upper and lower halves of the circle. There is also a correlation between the coefficients of Equations 8.5.19 and 8.5.20 and the coefficients of the higher energy level. Two nodal planes are needed for the higher-energy molecular orbitals. If carbons 1 and 4 define one plane while the midpoints (2–3) and (5–6) define the second (Figure 8.27b), the coefficients must be

$$(0, 1, -1, 0, 1, -1) \tag{8.5.21}$$

The coefficients are identical to those of Equation 8.5.19; only the signs are changed to permit the two nodal planes. The coefficients for the second eigenvec-

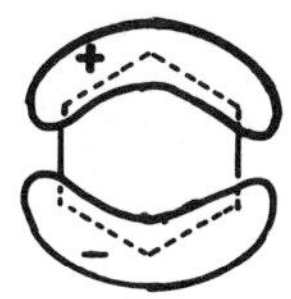

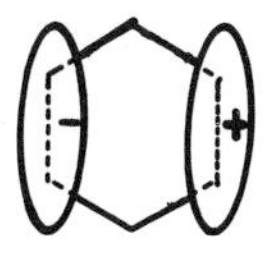

Figure 8.26. The nodal planes and wavefunctions for the degenerate, low-energy orbital pair in the benzene molecule.

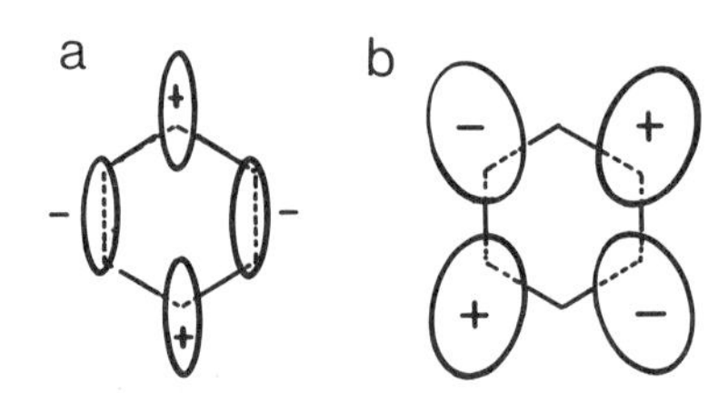

Figure 8.27. The nodal planes required for the high-energy degenerate orbital pair in the benzene molecule.

tor are found by assuming that the orbital has maximum amplitudes at carbons 1 and 4 and the midpoints (2–3) and (5–6) (Figure 8.27a):

$$\left(1, -\frac{1}{2}, -\frac{1}{2}, 1, -\frac{1}{2}, -\frac{1}{2}\right) \tag{8.5.22}$$

The two normalized molecular orbitals are

$$|\psi\rangle = \frac{1}{2}(\varphi_2 - \varphi_3 + \varphi_5 - \varphi_6) \tag{8.5.23}$$

$$|\psi\rangle = \frac{1}{\sqrt{3}}\left(\varphi_1 - \frac{1}{2}\varphi_2 - \frac{1}{2}\varphi_3 + \varphi_4 - \frac{1}{2}\varphi_5 - \frac{1}{2}\varphi_6\right) \tag{8.5.24}$$

Both higher-energy eigenvalues have four sign changes per circuit of the molecule. The lower-energy degenerate wavefunctions have only two sign changes.

8.6. The Pauli Spin Matrices

Hückel molecular orbital theory restricted the basis set of eigenfunctions which described the wave properties of the electrons. For example, only two electrons from ethene became its basis set. An electron with two possible spin components also generates 2×2 matrices which are called Pauli spin matrices.

The intrinsic spin of the electron gives rise to a spin angular momentum which has z components of $+\frac{1}{2}$ and $-\frac{1}{2}$ on the z axis in three-dimensional space since this spin will couple with parameters in the three-dimensional space. The two spin angular momentum z components are $\hbar/2$ and $-\hbar/2$ with

$$\hbar = h/2\pi \tag{8.6.1}$$

This may be expressed formally by defining a spin angular momentum operator for the z axis, σ_z. If an electron in the states with angular momentum components $\hbar/2$ and $-\hbar/2$ is defined as $\left|\frac{1}{2}\right\rangle$ and $\left|-\frac{1}{2}\right\rangle$, respectively, the formal operator equations for the system are

$$\sigma_z\left|\frac{1}{2}\right\rangle=(+\hbar/2)\left|\frac{1}{2}\right\rangle \tag{8.6.2}$$

$$\sigma_z\left|-\frac{1}{2}\right\rangle=(-\hbar/2)\left|-\frac{1}{2}\right\rangle \tag{8.6.3}$$

In general, the equation is

$$\boldsymbol{\sigma}_z\,|s\rangle=s_z\,|s\rangle \tag{8.6.4}$$

where s_z is the z component angular momentum eigenvalue and $|s\rangle$ is a wavefunction.

This basis set does not have to be truncated. The two electron spin states constitute a full basis set and the identity operator,

$$\left|\frac{1}{2}\right\rangle\left\langle\frac{1}{2}\right|+\left|-\frac{1}{2}\right\rangle\left\langle-\frac{1}{2}\right| \tag{8.6.5}$$

is used to convert the formal equation into a 2×2 matrix operator equation. Inserting the identity (Equation 8.6.5) between the operator and the vector and applying the bra eigenvectors $\left\langle\frac{1}{2}\right|$ and $\left\langle-\frac{1}{2}\right|$ from the left generates the two equations

$$\begin{aligned}\left\langle\frac{1}{2}\right|\boldsymbol{\sigma}_z\left|\frac{1}{2}\right\rangle\left\langle\frac{1}{2}\middle|s\right\rangle+\left\langle\frac{1}{2}\right|\boldsymbol{\sigma}_z\left|-\frac{1}{2}\right\rangle\left\langle-\frac{1}{2}\middle|s\right\rangle&=s_z\left\langle\frac{1}{2}\middle|s\right\rangle\\ \left\langle-\frac{1}{2}\right|\boldsymbol{\sigma}_z\left|\frac{1}{2}\right\rangle\left\langle\frac{1}{2}\middle|s\right\rangle+\left\langle-\frac{1}{2}\right|\boldsymbol{\sigma}_z\left|-\frac{1}{2}\right\rangle\left\langle-\frac{1}{2}\middle|s\right\rangle&=s_z\left\langle-\frac{1}{2}\middle|s\right\rangle\end{aligned} \tag{8.6.6}$$

since

$$\left\langle-\frac{1}{2}\middle|\frac{1}{2}\right\rangle=\left\langle\frac{1}{2}\middle|-\frac{1}{2}\right\rangle=0;\qquad \left\langle\frac{1}{2}\middle|\frac{1}{2}\right\rangle=\left\langle-\frac{1}{2}\middle|-\frac{1}{2}\right\rangle=1 \tag{8.6.7}$$

The matrix format is

$$\begin{pmatrix}\left\langle+\frac{1}{2}\right|\boldsymbol{\sigma}_z\left|+\frac{1}{2}\right\rangle & \left\langle+\frac{1}{2}\right|\boldsymbol{\sigma}_z\left|-\frac{1}{2}\right\rangle\\ \left\langle-\frac{1}{2}\right|\boldsymbol{\sigma}_z\left|+\frac{1}{2}\right\rangle & \left\langle-\frac{1}{2}\right|\boldsymbol{\sigma}_z\left|-\frac{1}{2}\right\rangle\end{pmatrix}\begin{pmatrix}\left\langle\frac{1}{2}\middle|s\right\rangle\\ \left\langle-\frac{1}{2}\middle|s\right\rangle\end{pmatrix}=s_z\begin{pmatrix}\left\langle\frac{1}{2}\middle|s\right\rangle\\ \left\langle-\frac{1}{2}\middle|s\right\rangle\end{pmatrix} \tag{8.6.8}$$

Since the angular momentum is defined by its observable projections on the z axis, the matrix has a very simple form. The matrix elements are

$$\begin{aligned}
\left\langle \frac{1}{2} \middle| \boldsymbol{\sigma}_z \middle| \frac{1}{2} \right\rangle &= (+\hbar/2) \left\langle \frac{1}{2} \middle| \frac{1}{2} \right\rangle = (+\hbar/2) \\
\left\langle -\frac{1}{2} \middle| \boldsymbol{\sigma}_z \middle| -\frac{1}{2} \right\rangle &= (-\hbar/2) \left\langle -\frac{1}{2} \middle| -\frac{1}{2} \right\rangle = (-\hbar/2) \\
\left\langle \frac{1}{2} \middle| \boldsymbol{\sigma}_z \middle| -\frac{1}{2} \right\rangle &= (-\hbar/2) \left\langle \frac{1}{2} \middle| -\frac{1}{2} \right\rangle = 0 \\
\left\langle -\frac{1}{2} \middle| \boldsymbol{\sigma}_z \middle| \frac{1}{2} \right\rangle &= (+\hbar/2) \left\langle -\frac{1}{2} \middle| \frac{1}{2} \right\rangle = 0
\end{aligned} \tag{8.6.9}$$

and the matrix is

$$(\hbar/2) \begin{pmatrix} 1 & 0 \\ 0 & -1 \end{pmatrix} \tag{8.6.10}$$

If the eigenvector $|s\rangle = \left| \frac{1}{2} \right\rangle$, its column vector is

$$\begin{pmatrix} \left\langle +\frac{1}{2} \middle| s \right\rangle \\ \left\langle -\frac{1}{2} \middle| s \right\rangle \end{pmatrix} = \begin{pmatrix} 1 \\ 0 \end{pmatrix} \tag{8.6.11}$$

The eigenvector $|s\rangle = \left| -\frac{1}{2} \right\rangle$ has the column vector

$$\begin{pmatrix} \left\langle +\frac{1}{2} \middle| s \right\rangle \\ \left\langle -\frac{1}{2} \middle| s \right\rangle \end{pmatrix} = \begin{pmatrix} 0 \\ 1 \end{pmatrix} \tag{8.6.12}$$

For example,

$$(\hbar/2) \begin{pmatrix} 1 & 0 \\ 0 & -1 \end{pmatrix} \begin{pmatrix} 1 \\ 0 \end{pmatrix} = (+\hbar/2) \begin{pmatrix} 1 \\ 0 \end{pmatrix} \tag{8.6.13}$$

Since the spin angular momentum component on the z axis is a component of the total spin angular momentum, there must be some total spin angular momentum which includes this component. In addition, there must be spin angular momentum components on the x and y axes as well. In this Cartesian coordinate system, the total spin angular momentum is the sum of the squares of the spin angular momentum components:

$$s^2 = s_z^2 + s_x^2 + s_y^2 \tag{8.6.14}$$

Since there is no preferred spatial direction for the electron, each of these squared components will make the same contribution to the total spin angular momentum. The square of the angular momentum for either z component is found by squaring the $\boldsymbol{\sigma}_z$ matrix

$$\begin{aligned}\boldsymbol{\sigma}_z^2 = (\hbar/2)^2 \begin{pmatrix} 1 & 0 \\ 0 & -1 \end{pmatrix}\begin{pmatrix} 1 & 0 \\ 0 & -1 \end{pmatrix} &= (\hbar/2)^2 \begin{pmatrix} 1 & 0 \\ 0 & 1 \end{pmatrix} \\ &= (\hbar^2/4)\mathbf{I}\end{aligned} \tag{8.6.15}$$

With the identity, the squared component is effectively a scalar. The total spin angular momentum must be the sum of three identical square components and is

$$\sigma^2 = 3(\hbar^2/4) \tag{8.6.16}$$

This total angular momentum follows the general equation for the total angular momentum squared with $s = \frac{1}{2}$:

$$\frac{1}{2}\left(\frac{1}{2} + 1\right)\hbar^2 \tag{8.6.17}$$

The x and y components of angular momentum are also represented with 2×2 matrices. Each squared component is proportional to the identity. In addition, the trace of the $\boldsymbol{\sigma}_z$ matrix is zero and $\boldsymbol{\sigma}_x$ and $\boldsymbol{\sigma}_y$ must also have a zero trace. The two matrices which satisfy these conditions for the x and y components are

$$\boldsymbol{\sigma}_x = (\hbar/2) \begin{pmatrix} 0 & 1 \\ 1 & 0 \end{pmatrix} \tag{8.6.18}$$

$$\boldsymbol{\sigma}_y = (\hbar/2) \begin{pmatrix} 0 & -i \\ i & 0 \end{pmatrix} \tag{8.6.19}$$

The squares of these Hermitian matrices are

$$(\hbar^2/4) \begin{pmatrix} 0 & 1 \\ 1 & 0 \end{pmatrix}\begin{pmatrix} 0 & 1 \\ 1 & 0 \end{pmatrix} = \begin{pmatrix} 1 & 0 \\ 0 & 1 \end{pmatrix} (\hbar^2/4) \tag{8.6.20}$$

$$(\hbar^2/4) \begin{pmatrix} 0 & -i \\ i & 0 \end{pmatrix}\begin{pmatrix} 0 & -i \\ i & 0 \end{pmatrix} = \begin{pmatrix} 1 & 0 \\ 0 & 1 \end{pmatrix} (\hbar^2/4) \tag{8.6.21}$$

Since the determinants of all three matrices are unity, they are unitary matrices, i.e., they will rotate a vector without changing its length. Since the total angular momentum and the z component of angular momentum are known, an exact determination of one of the two remaining components would permit an exact characterization of this system and violate Heisenberg's uncertainty principle. This does not occur since $\boldsymbol{\sigma}_x$ and $\boldsymbol{\sigma}_y$ do not give eigenvalues when they operate on the basis vectors developed using the z components:

$$(\hbar/2)\begin{pmatrix}0 & 1\\ 1 & 0\end{pmatrix}\begin{pmatrix}1\\ 0\end{pmatrix}=\begin{pmatrix}0\\ 1\end{pmatrix}(\hbar/2) \qquad (\hbar/2)\begin{pmatrix}0 & 1\\ 1 & 0\end{pmatrix}\begin{pmatrix}0\\ 1\end{pmatrix}=\begin{pmatrix}1\\ 0\end{pmatrix}(\hbar/2) \tag{8.6.22}$$

$$(\hbar/2)\begin{pmatrix}0 & -i\\ i & 0\end{pmatrix}\begin{pmatrix}1\\ 0\end{pmatrix}=\begin{pmatrix}0\\ i\end{pmatrix}(\hbar/2) \qquad (\hbar/2)\begin{pmatrix}0 & -i\\ i & 0\end{pmatrix}\begin{pmatrix}0\\ 1\end{pmatrix}=\begin{pmatrix}-i\\ 0\end{pmatrix}(\hbar/2) \tag{8.6.23}$$

The operations produce new vectors in each case.

The three spin angular momentum matrices anticommute:

$$\boldsymbol{\sigma}_i\boldsymbol{\sigma}_j+\boldsymbol{\sigma}_j\boldsymbol{\sigma}_i=0 \tag{8.6.24}$$

i.e., the two matrix products are added, not subtracted, to give zero. For $\boldsymbol{\sigma}_x$ and $\boldsymbol{\sigma}_y$,

$$\begin{aligned}(\hbar^2/4)\begin{pmatrix}0 & 1\\ 1 & 0\end{pmatrix}\begin{pmatrix}0 & -i\\ i & 0\end{pmatrix}+(\hbar^2/4)\begin{pmatrix}0 & -i\\ i & 0\end{pmatrix}\begin{pmatrix}0 & 1\\ 1 & 0\end{pmatrix}\\ =(\hbar^2/4)\begin{pmatrix}i & 0\\ 0 & -i\end{pmatrix}+(\hbar^2/4)\begin{pmatrix}-i & 0\\ 0 & i\end{pmatrix}=\begin{pmatrix}0 & 0\\ 0 & 0\end{pmatrix}\end{aligned} \tag{8.6.25}$$

Although the spin matrices do not commute, the commutation operation does establish relations between the three σ matrices. The commutator of matrices $\boldsymbol{\sigma}_x$ and $\boldsymbol{\sigma}_y$ found by subtracting the product matrices of Equation 8.6.25 is

$$\begin{aligned}(\hbar/2)^2\begin{pmatrix}i & 0\\ 0 & -i\end{pmatrix}-(\hbar/2)^2\begin{pmatrix}-i & 0\\ 0 & i\end{pmatrix}&=2i(\hbar/2)^2\begin{pmatrix}1 & 0\\ 0 & -1\end{pmatrix}\\ &=i\hbar(\hbar/2)\begin{pmatrix}1 & 0\\ 0 & -1\end{pmatrix}\end{aligned} \tag{8.6.26}$$

The commutator of $\boldsymbol{\sigma}_x$ and $\boldsymbol{\sigma}_y$ in that order generates the third component, $\boldsymbol{\sigma}_z$, multiplied by $i\hbar$. The general commutation relation is

$$[\boldsymbol{\sigma}_i, \boldsymbol{\sigma}_j]=i\hbar\boldsymbol{\sigma}_k \tag{8.6.27}$$

where the indices must follow the order $xyzx$, e.g.,

$$[\boldsymbol{\sigma}_z, \boldsymbol{\sigma}_x]=i\hbar\boldsymbol{\sigma}_y \tag{8.6.28}$$

If the order is reversed, e.g., yxz, the result is negative.

Although $\boldsymbol{\sigma}_x$ and $\boldsymbol{\sigma}_y$ are not eigenvalue operators for the electron system defined via z-axis components, these operators can still have some average values, $\langle \boldsymbol{\sigma}_x \rangle$ and $\langle \boldsymbol{\sigma}_y \rangle$, which are determined from the Pauli spin matrices for these components.

The $\boldsymbol{\sigma}_z$ matrix has the two z components of angular momentum as the diagonal elements. For an ensemble of these electrons, there should be equal numbers of $\left|\frac{1}{2}\right\rangle$ and $\left|-\frac{1}{2}\right\rangle$ spin states and the average z component should be zero. This result is determined directly as the trace of the $\boldsymbol{\sigma}_z$ matrix:

$$\text{tr}\, \boldsymbol{\sigma}_z = 0 \tag{8.6.29}$$

Even though $\boldsymbol{\sigma}_x$ and $\boldsymbol{\sigma}_y$ are not diagonal matrices, their average values are also determined via the invariant trace. Since the diagonal elements for these matrices are all zero,

$$\langle \boldsymbol{\sigma}_x \rangle = \langle \boldsymbol{\sigma}_y \rangle = \text{tr}\, \boldsymbol{\sigma}_x = \text{tr}\, \boldsymbol{\sigma}_y = 0 \tag{8.6.30}$$

Since there is no preferred spin orientation, the projections on $+x$ will appear as often as the projections on $-x$.

The trace of the square of any Pauli matrix gives the square of the angular momentum in that direction. For $\boldsymbol{\sigma}_x$,

$$(\hbar/2)^2 \begin{pmatrix} 0 & 1 \\ 1 & 0 \end{pmatrix} \begin{pmatrix} 0 & 1 \\ 1 & 0 \end{pmatrix} = (\hbar/2)^2 \begin{pmatrix} 1 & 0 \\ 0 & 1 \end{pmatrix} \tag{8.6.31}$$

and

$$\text{tr}\, \boldsymbol{\sigma}_x^2 = (\hbar/2)^2\, (2) = \hbar^2/2 \tag{8.6.32}$$

as determined previously. The traces of products with different indices, e.g.,

$$\boldsymbol{\sigma}_x \boldsymbol{\sigma}_y \tag{8.6.33}$$

are zero (Problem 8.9) since the projections on x and y are uncorrelated.

The Pauli spin matrices are only 2×2 matrices but there are three of them for the three components of a three-dimensional space. For this reason, transforms of the Pauli matrices in a two-dimensional space can be correlated with transforms in the three-dimensional space.

The Pauli spin matrices are selected as "vectors" for the three-dimensional space. A vector with components x, y, and z is then a sum of these components times the appropriate vector, i.e.,

$$\mathbf{r} = x\boldsymbol{\sigma}_x + y\boldsymbol{\sigma}_y + z\boldsymbol{\sigma}_z \tag{8.6.34}$$

The vector $\mathbf{r}$ is actually a 2×2 matrix but it will contain all the x-, y-, and

z-component information. The unitless Pauli spin matrices are substituted to determine the matrix $\mathbf{r}$:

$$\begin{pmatrix} 0 & x \\ x & 0 \end{pmatrix} + \begin{pmatrix} 0 & -iy \\ iy & 0 \end{pmatrix} + z\begin{pmatrix} 1 & 0 \\ 0 & -1 \end{pmatrix} = \begin{pmatrix} z & x-iy \\ x+iy & -z \end{pmatrix} \tag{8.6.35}$$

A transform operation in the three-dimensional space converts x, y, and z into new components x', y', and z'. Since these three components appear in the 2×2 $\mathbf{r}$ matrix, there will be a corresponding similarity transform which converts the matrix of Equation 8.6.35 into a new 2×2 matrix containing the new components x', y', and z':

$$\mathbf{r}' = \mathbf{S}^{-1}\mathbf{r}\mathbf{S} \tag{8.6.36}$$

For example, the unitary transform matrices,

$$\mathbf{S} = \begin{pmatrix} i & 0 \\ 0 & -i \end{pmatrix} \tag{8.6.37}$$

$$\mathbf{S}^{-1} = \begin{pmatrix} -i & 0 \\ 0 & i \end{pmatrix} \tag{8.6.38}$$

transform $\mathbf{r}$ as

$$\begin{pmatrix} -i & 0 \\ 0 & i \end{pmatrix}\begin{pmatrix} z & x-iy \\ x+iy & -z \end{pmatrix}\begin{pmatrix} i & 0 \\ 0 & -i \end{pmatrix} = \begin{pmatrix} z & -x-iy \\ -x-iy & z \end{pmatrix} \tag{8.6.39}$$

The z components are unchanged but both x and y are replaced by $-x$ and $-y$ in this transformation. The three-dimensional transform is

$$\begin{pmatrix} -x \\ -y \\ z \end{pmatrix} = \begin{pmatrix} -1 & 0 & 0 \\ 0 & -1 & 0 \\ 0 & 0 & 1 \end{pmatrix}\begin{pmatrix} x \\ y \\ z \end{pmatrix} \tag{8.6.40}$$

The $\mathbf{S}$ and $\mathbf{S}^{-1}$ similarity matrices for the two-dimensional matrix $\mathbf{r}$ gave the transformation of the three-dimensional components for a rotation of 180° about the z axis. Both operations are different. In three dimensions, the rotation operation involves a single matrix. In two dimensions, the two similarity transform matrices are required.

It would be convenient to have the 2×2 similarity transform matrices which give a rotation of θ about the z axis in the three-dimensional space. The $\mathbf{S}$ matrix for a rotation of 180° was diagonal with elements i and $-i$. These components are generated by the exponential functions

$$\begin{aligned}\exp(i\pi 180/2) &= +i \\ \exp(-i\pi 180/2) &= -i\end{aligned} \tag{8.6.41}$$

The 180° rotation in three-dimensional space must be divided by 2 to obtain the proper components, and this suggests that **S** must be

$$\begin{pmatrix}\exp(i\theta/2) & 0 \\ 0 & \exp(-i\theta/2)\end{pmatrix} \tag{8.6.42}$$

for a rotation of θ in three-dimensional space. This is verified by evaluating the similarity transform

$$\begin{aligned}\mathbf{S}^{-1}\mathbf{r}\mathbf{S} &= \begin{pmatrix}\exp(-i\theta/2) & 0 \\ 0 & \exp(i\theta/2)\end{pmatrix}\begin{pmatrix} z & x-iy \\ x+iy & z\end{pmatrix} \\ &\quad\times\begin{pmatrix}\exp(i\theta/2) & 0 \\ 0 & \exp(-i\theta/2)\end{pmatrix} \\ &= \begin{pmatrix} z & (x-iy)\exp(-i\theta) \\ (x+iy)\exp(i\theta) & z\end{pmatrix}\end{aligned} \tag{8.6.43}$$

z is unchanged in the transform so the rotation is about the z axis. Since the new off-diagonal components must be

$$\begin{aligned}&x' - iy' \\ &x' + iy'\end{aligned} \tag{8.6.44}$$

x' is determined as one-half of the sum of the two off-diagonal elements in Equation 8.6.43:

$$\begin{aligned}x' &= (x-iy)\exp(-i\theta) + (x+iy)\exp(i\theta) \\ &= x[\exp(i\theta)+\exp(-i\theta)]/2 + iy[\exp(i\theta)-\exp(-i\theta)]/2 \\ &= x\cos\theta - y\sin\theta\end{aligned} \tag{8.6.45}$$

The y' component is the difference of the 21 element and the 12 element divided by $2i$, i.e.,

$$\begin{aligned}y' &= (x+iy)\exp(i\theta)/2i - (x-iy)\exp(-i\theta)/2i \\ &= x[\exp(i\theta)-\exp(-i\theta)]/2i + y[\exp(i\theta)+\exp(-i\theta)]/2 \\ &= x\sin\theta + y\cos\theta\end{aligned} \tag{8.6.46}$$

The three-dimensional transform matrix is

$$\begin{pmatrix} \cos\theta & -\sin\theta & 0 \\ \sin\theta & \cos\theta & 0 \\ 0 & 0 & 1 \end{pmatrix} \tag{8.6.47}$$

The rotations about the z axis require diagonal similarity transform matrices. Rotations about the x and y axes must change the diagonal elements of the **r** matrix, i.e., z components, and the transform matrices will require off-diagonal elements (Problem 8.10).

8.7. Lowering and Raising Operators

The total squared spin angular momentum for a single electron is equal to the sum of the squares of the three spin angular momentum components:

$$s^2 = s_x^2 + s_y^2 + s_z^2 \tag{8.7.1}$$

For a quantized system, only the s_z component is known exactly. Since the total spin angular momentum is also known, the sum of the squares of the x and y components is

$$\begin{aligned} s_x^2 + s_y^2 &= s^2 - s_z^2 \\ &= 2(\hbar/2)^2 = \hbar^2/2 \end{aligned} \tag{8.7.2}$$

so this sum is known exactly.

The sum of squares of x and y spin components can be factored as

$$s_x^2 + s_y^2 = (s_x + is_y)(s_x - is_y) \tag{8.7.3}$$

This product of complex conjugates gives a spin component in planes perpendicular to the x–y plane.

Since the two-dimensional Pauli spin matrices describe the components of spin angular momentum in the 2×2 matrix representation, they will also satisfy Equation 8.7.3. The classical factors of Equation 8.7.3 which are defined as

$$s_+ = s_x + is_y \tag{8.7.4}$$

and

$$s_- = s_x - is_y \tag{8.7.5}$$

respectively, are replaced by the corresponding Pauli spin matrix sums:

$$\boldsymbol{\sigma}_+ = \boldsymbol{\sigma}_x + i\boldsymbol{\sigma}_y \tag{8.7.6}$$

$$\boldsymbol{\sigma}_- = \boldsymbol{\sigma}_x - i\boldsymbol{\sigma}_y \tag{8.7.7}$$

These matrices are

$$\boldsymbol{\sigma}_+ = (\hbar/2)\begin{pmatrix} 0 & 1 \\ 1 & 0 \end{pmatrix} + i(\hbar/2)\begin{pmatrix} 0 & -i \\ i & 0 \end{pmatrix}$$

$$= (\hbar/2)\begin{pmatrix} 0 & 2 \\ 0 & 0 \end{pmatrix} = \hbar\begin{pmatrix} 0 & 1 \\ 0 & 0 \end{pmatrix} \tag{8.7.8}$$

$$\boldsymbol{\sigma}_- = (\hbar/2)\begin{pmatrix} 0 & 1 \\ 1 & 0 \end{pmatrix} - i(\hbar/2)\begin{pmatrix} 0 & -i \\ i & 0 \end{pmatrix}$$

$$= (\hbar/2)\begin{pmatrix} 0 & 0 \\ 2 & 0 \end{pmatrix} = \hbar\begin{pmatrix} 0 & 0 \\ 1 & 0 \end{pmatrix} \tag{8.7.9}$$

The $\boldsymbol{\sigma}_+$ matrix operator will raise an eigenstate to the eigenstate with the next higher eigenvalue. For example, operation on $\left|-\frac{1}{2}\right\rangle$ gives

$$\boldsymbol{\sigma}_+\left|-\frac{1}{2}\right\rangle = \hbar\begin{pmatrix} 0 & 1 \\ 0 & 0 \end{pmatrix}\begin{pmatrix} 0 \\ 1 \end{pmatrix} = \hbar\begin{pmatrix} 1 \\ 0 \end{pmatrix} = \hbar\left|+\frac{1}{2}\right\rangle \tag{8.7.10}$$

The zero eigenvector is generated when $\boldsymbol{\sigma}_+$ operates on $\left|+\frac{1}{2}\right\rangle$:

$$\boldsymbol{\sigma}_+\left|\frac{1}{2}\right\rangle = \hbar\begin{pmatrix} 0 & 1 \\ 0 & 0 \end{pmatrix}\begin{pmatrix} 1 \\ 0 \end{pmatrix} = \hbar\begin{pmatrix} 0 \\ 0 \end{pmatrix} \tag{8.7.11}$$

The raising operator will raise the eigenvalue and eigenvector but will not generate eigenvectors outside the basis set.

The lowering operator generates the next lower eigenvector:

$$\begin{aligned} \boldsymbol{\sigma}_-\left|+\frac{1}{2}\right\rangle &= \hbar\left|-\frac{1}{2}\right\rangle \\ \boldsymbol{\sigma}_-\left|-\frac{1}{2}\right\rangle &= \hbar\,|0\rangle \end{aligned} \tag{8.7.12}$$

In each case, the factor $\hbar$, not $\hbar/2$ or $-\hbar/2$, multiplies the final eigenvector. The operation involves the x- and y-component operators, and these can be related to the total spin angular momentum and the z component of spin angular momentum. The total angular momentum must include the products $\boldsymbol{\sigma}_+\boldsymbol{\sigma}_-$ and $\boldsymbol{\sigma}_-\boldsymbol{\sigma}_+$:

$$\begin{aligned} \boldsymbol{\sigma}^2 &= \frac{1}{2}(\boldsymbol{\sigma}_+\boldsymbol{\sigma}_- + \boldsymbol{\sigma}_-\boldsymbol{\sigma}_+) + \boldsymbol{\sigma}_z^2 \\ \boldsymbol{\sigma}^2 - \boldsymbol{\sigma}_z^2 &= \frac{1}{2}(\boldsymbol{\sigma}_+\boldsymbol{\sigma}_- + \boldsymbol{\sigma}_-\boldsymbol{\sigma}_+) \end{aligned} \tag{8.7.13}$$

The sum of products of raising and lowering operators is proportional to the identity matrix:

$$(\hbar^2/2)\begin{pmatrix} 0 & 1 \\ 0 & 0 \end{pmatrix}\begin{pmatrix} 0 & 0 \\ 1 & 0 \end{pmatrix} + (\hbar^2/2)\begin{pmatrix} 0 & 0 \\ 1 & 0 \end{pmatrix}\begin{pmatrix} 0 & 1 \\ 0 & 0 \end{pmatrix} = (\hbar^2/2)\begin{pmatrix} 1 & 0 \\ 0 & 1 \end{pmatrix} \tag{8.7.14}$$

since $\sigma^2 - \sigma_z^2$ is known exactly. For more complicated systems, Equation 8.7.13 is generalized and used to determine the raising and lowering operators.

The raising and lowering matrices can be used for systems with electron spin angular momentum or spatial angular momentum. For example, a p electron has a total angular momentum characterized by an angular momentum quantum number, $l = 1$, which gives a total angular momentum

$$l^2 = l(l+1)\,\hbar^2 \tag{8.7.15}$$

The basis set is selected so that the angular momentum component on the z axis is also an eigenvalue with quantum number m. The eigenvector satisfies the equation

$$l_z\,|l, m\rangle = m\hbar\,|l, m\rangle \tag{8.7.16}$$

This information, with Equation 8.7.13, permits a determination of the raising and lowering operators for this three-dimensional system ($m = -1, 0, +1$). The matrix for

$$\frac{1}{2}(l_+ l_- + l_- l_+) \tag{8.7.17}$$

is equal to

$$l^2 - l_z^2 = (\hbar^2)\begin{pmatrix} 2 & 0 & 0 \\ 0 & 2 & 0 \\ 0 & 0 & 2 \end{pmatrix} - (\hbar^2)\begin{pmatrix} 1 & 0 & 0 \\ 0 & 0 & 0 \\ 0 & 0 & 1 \end{pmatrix}$$

$$= (\hbar^2)\begin{pmatrix} 1 & 0 & 0 \\ 0 & 2 & 0 \\ 0 & 0 & 1 \end{pmatrix} \tag{8.7.18}$$

The matrices are diagonal since the basis set is selected for the z component. The components of the total angular momentum matrix are determined from Equation 8.7.15 while the components of the second matrix are determined as the squares of Equation 8.7.16.

The raising and lowering matrices must have a single upper diagonal and lower diagonal, respectively, and the matrices which do this are

$$\mathbf{l}_+ = \hbar \begin{pmatrix} 0 & \sqrt{2} & 0 \\ 0 & 0 & \sqrt{2} \\ 0 & 0 & 0 \end{pmatrix} \tag{8.7.19}$$

$$\mathbf{l}_- = \hbar \begin{pmatrix} 0 & 0 & 0 \\ \sqrt{2} & 0 & 0 \\ 0 & \sqrt{2} & 0 \end{pmatrix} \tag{8.7.20}$$

which is verified by substitution:

$$\frac{1}{2}(\mathbf{l}_+\mathbf{l}_- + \mathbf{l}_-\mathbf{l}_+) = \left(\frac{1}{2}\right) \begin{pmatrix} 0 & \sqrt{2} & 0 \\ 0 & 0 & \sqrt{2} \\ 0 & 0 & 0 \end{pmatrix} \begin{pmatrix} 0 & 0 & 0 \\ \sqrt{2} & 0 & 0 \\ 0 & \sqrt{2} & 0 \end{pmatrix}$$

$$-\left(\frac{1}{2}\right) \begin{pmatrix} 0 & 0 & 0 \\ \sqrt{2} & 0 & 0 \\ 0 & \sqrt{2} & 0 \end{pmatrix} \begin{pmatrix} 0 & \sqrt{2} & 0 \\ 0 & 0 & \sqrt{2} \\ 0 & 0 & 0 \end{pmatrix}$$

$$= \begin{pmatrix} 1 & 0 & 0 \\ 0 & 2 & 0 \\ 0 & 0 & 1 \end{pmatrix} \tag{8.7.21}$$

The technique can be extended to systems with any number of eigenstates. For example, if the angular momentum and spin angular momentum are considered, the total angular momentum for Russell–Saunders coupling is defined by the j quantum number,

$$j = l + s \tag{8.7.22}$$

The maximum z component for the p electron is

$$j_z = l_z + s_z = 1 + \frac{1}{2} = \frac{3}{2} \tag{8.7.23}$$

The matrix

$$\frac{1}{2}(\mathbf{j}_+\mathbf{j}_- + \mathbf{j}_-\mathbf{j}_+) = \mathbf{j}^2 - \mathbf{j}_z^2 \tag{8.7.24}$$

is (Problem 8.11)

$$\begin{bmatrix} \frac{3}{2} & 0 & 0 & 0 \\ 0 & \frac{7}{2} & 0 & 0 \\ 0 & 0 & \frac{7}{2} & 0 \\ 0 & 0 & 0 & \frac{3}{2} \end{bmatrix} \tag{8.7.25}$$

The raising and lowering matrices which satisfy Equation 8.7.24 are

$$\mathbf{j}_+ = \hbar \begin{pmatrix} 0 & \sqrt{3} & 0 & 0 \\ 0 & 0 & 2 & 0 \\ 0 & 0 & 0 & \sqrt{3} \\ 0 & 0 & 0 & 0 \end{pmatrix} \tag{8.7.26}$$

$$\mathbf{j}_- = \hbar \begin{pmatrix} 0 & 0 & 0 & 0 \\ \sqrt{3} & 0 & 0 & 0 \\ 0 & 2 & 0 & 0 \\ 0 & 0 & \sqrt{3} & 0 \end{pmatrix} \tag{8.7.27}$$

The raising operation increases m by one unit. The new wavefunction $|j, m+1\rangle$ is multiplied by a constant c which appears as the $(m+1, m)$ element in the matrix:

$$\begin{gathered} \mathbf{j}_+ \, |j, m\rangle = c\,|j, m+1\rangle \\ \langle j, m+1|\,\mathbf{j}_+\,|j, m\rangle = c\langle j, m+1\,|\,j, m+1\rangle = c \end{gathered} \tag{8.7.28}$$

In general, for initial (j, m),

$$c = \hbar[(j-m)(j+m+1)]^{1/2} \tag{8.7.29}$$

For example, the 12 element $\left\langle \frac{3}{2}, \frac{3}{2} \middle| \frac{3}{2}, \frac{1}{2} \right\rangle$ is

$$c = \hbar\left[\left(\frac{3}{2}-\frac{1}{2}\right)\left(\frac{3}{2}+\frac{1}{2}+1\right)\right]^{1/2} = \sqrt{3} \tag{8.7.30}$$

The lowering operation,

$$\mathbf{j}_- \,|j, m\rangle = c\,|j, m-1\rangle \tag{8.7.31}$$

gives $|0\rangle$ for $m < -l$. For $m > -l$, the coefficient c is

$$c = h[(j+m)(j-m+1)]^{1/2} \tag{8.7.32}$$

These c appear as the appropriate elements in the matrix of Equation 8.7.27.

The equation for the lowering operations is

$$|j, m-1\rangle = h[(j+m)(j-m+1)]^{1/2}\,|j, m\rangle \tag{8.7.33}$$

The coefficient for the raising operation becomes zero when $m > l$; the coefficient for the lowering operation is zero when $m < -l$.

The raising and lowering operators do generate the eigenvectors of the system, but this has limited utility in the matrix format. The basis set of eigenvectors are column vectors which have a single nonzero component. Application of a raising matrix to

$$\begin{pmatrix} 0 \\ 1 \\ 0 \end{pmatrix} \tag{8.7.34}$$

gives

$$c\begin{pmatrix} 1 \\ 0 \\ 0 \end{pmatrix} \tag{8.7.35}$$

The real utility of the raising and lowering operations rests in their ability to generate wavefunctions for multielectron systems from single-electron wavefunctions. For a two-electron system, the z components of the spin for each electron must add to give the total angular momentum for the combination,

$$S_z = s_{1z} + s_{2z} \tag{8.7.36}$$

where the upper case S is used for total angular momenta and lower case s is used for the spin angular momenta of individual electrons.

When both electrons have a spin component of $+\frac{1}{2}$, the total spin component of the system must be $+1$ since the projections in one direction must add:

$$S_z = (+\hbar/2) + (+\hbar/2) = +1\hbar \tag{8.3.37}$$

The total wavefunction will be written as a product of the two electron spin wavefunctions:

$$|1, 1\rangle = \left|\frac{1}{2}, \frac{1}{2}\right\rangle \left|\frac{1}{2}, \frac{1}{2}\right\rangle \tag{8.7.38}$$

$S = 1$ is the total angular momentum quantum number, i.e.,

$$p_t = [S(S+1)]^{1/2} \hbar \tag{8.7.39}$$

and its maximal S_z equals 1 as well. The total wavefunction $|1, 1\rangle$ is the direct product (Chapter 10) of the two single-electron wavefunctions:

$$|1, 1\rangle = \left|\frac{1}{2}, \frac{1}{2}\right\rangle \left|\frac{1}{2}, \frac{1}{2}\right\rangle \tag{8.7.40}$$

For a direct product, operations for electron 1 will operate only on the the electron 1 wavefunction. The electron 2 operator operates only on the electron 2 wavefunction. The first single-electron wavefunction will always be that of electron 1.

Since each raising and lowering operator is constructed from x- and y-component spin operators, the total lowering operator must equal the sum of the lowering operators for each electron:

$$\mathbf{S}_- = \mathbf{s}_{1-} + \mathbf{s}_{2-} \tag{8.7.41}$$

The left-hand side will generate the lowered eigenvector for the full system. The right-hand side will generate the combination of single-electron wavefunctions for this lowered eigenvector.

The first lowering operation on the maximal eigenvector of Equation 8.7.38 is

$$\begin{aligned}
\mathbf{S}_- |1, 1\rangle &= (\mathbf{s}_{1-} + \mathbf{s}_{2-}) \left|\frac{1}{2}, \frac{1}{2}\right\rangle \left|\frac{1}{2}, \frac{1}{2}\right\rangle \\
\hbar[(1+1)(1-1+1)]^{1/2} |1, 0\rangle & \\
&= \sqrt{2}\,\hbar\, |1, 0\rangle \\
&= \hbar\left[\left(\frac{1}{2}+\frac{1}{2}\right)\left(\frac{1}{2}-\frac{1}{2}+1\right)\right]^{1/2} \left|\frac{1}{2}, -\frac{1}{2}\right\rangle \left|\frac{1}{2}, \frac{1}{2}\right\rangle \\
&\quad + \hbar\left[\left(\frac{1}{2}+\frac{1}{2}\right)\left(\frac{1}{2}-\frac{1}{2}+1\right)\right]^{1/2} \left|\frac{1}{2}, \frac{1}{2}\right\rangle \left|\frac{1}{2}, -\frac{1}{2}\right\rangle \\
&= \hbar\left(\left|\frac{1}{2}, -\frac{1}{2}\right\rangle \left|\frac{1}{2}, \frac{1}{2}\right\rangle + \left|\frac{1}{2}, \frac{1}{2}\right\rangle \left|\frac{1}{2}, -\frac{1}{2}\right\rangle\right)
\end{aligned} \tag{8.7.42}$$

Two terms are generated because each of the two single-electron operators operates on only one wavefunction. The total wavefunction which gives a total angular momentum of 1 and a net z component of 0 must have the two electrons with equal and opposite spin components. The wavefunction is often written as

$$\left|\frac{1}{2}, -\frac{1}{2}\right\rangle \left|\frac{1}{2}, -\frac{1}{2}\right\rangle + \left|\frac{1}{2}, -\frac{1}{2}\right\rangle \left|\frac{1}{2}, \frac{1}{2}\right\rangle = \alpha(1)\,\beta(2) + \alpha(2)\,\beta(1) \quad (8.7.43)$$

where α is the spin wavefunction with positive z component and β is the function with negative z component.

The factor of $\sqrt{2}$ which appears with $|1, 0\rangle$ can be moved to the right to give the normalized wavefunction

$$|1, 0\rangle = \frac{1}{\sqrt{2}}\left(\left|\frac{1}{2}, -\frac{1}{2}\right\rangle \left|\frac{1}{2}, \frac{1}{2}\right\rangle + \left|\frac{1}{2}, \frac{1}{2}\right\rangle \left|\frac{1}{2}, -\frac{1}{2}\right\rangle\right) \quad (8.7.44)$$

if normalized single-electron wavefunctions are used.

The final state $|1, -1\rangle$ is generated by performing the lowering operation on $|1, 0\rangle$ as given in Equation 8.7.44:

$$\begin{aligned} \mathbf{S}_-\,|1, 0\rangle &= \hbar[(1+0)(1-0+1)]^{1/2}\,|1, -1\rangle = \sqrt{2}\,\hbar\,|1, -1\rangle \\ &= (\mathbf{s}_{1-} + \mathbf{s}_{2-})\left(\frac{1}{\sqrt{2}}\right)\left(\left|\frac{1}{2}, -\frac{1}{2}\right\rangle \left|\frac{1}{2}, \frac{1}{2}\right\rangle + \left|\frac{1}{2}, \frac{1}{2}\right\rangle \left|\frac{1}{2}, -\frac{1}{2}\right\rangle\right) \\ &= \hbar\left(\frac{1}{\sqrt{2}}\right)\left[0 + (1)\left|\frac{1}{2}, -\frac{1}{2}\right\rangle \left|\frac{1}{2}, -\frac{1}{2}\right\rangle + (1)\left|\frac{1}{2}, -\frac{1}{2}\right\rangle \left|\frac{1}{2}, -\frac{1}{2}\right\rangle + 0\right] \\ &= \hbar\left(\frac{2}{\sqrt{2}}\right)\left|\frac{1}{2}, -\frac{1}{2}\right\rangle \left|\frac{1}{2}, -\frac{1}{2}\right\rangle \end{aligned} \quad (8.7.45)$$

The final wavefunction is

$$|1, -1\rangle = \left|\frac{1}{2}, -\frac{1}{2}\right\rangle \left|\frac{1}{2}, -\frac{1}{2}\right\rangle \quad (8.7.46)$$

The three wavefunctions constitute the triplet observed when the two electrons can have the same spin. The lowering operations do not generate the singlet state which is formed as the difference

$$\left|\frac{1}{2}, -\frac{1}{2}\right\rangle \left|\frac{1}{2}, \frac{1}{2}\right\rangle - \left|\frac{1}{2}, \frac{1}{2}\right\rangle \left|\frac{1}{2}, -\frac{1}{2}\right\rangle \quad (8.7.47)$$

with a total spin angular momentum of zero. Lowering or raising this eigenfunction gives only $|0\rangle$. The lowering operations generate all wavefunctions for each total angular momentum, e.g.,

$$S^2 = s(s+1)\,\hbar^2 = 2\hbar^2 \qquad s = 1 \quad (8.7.48)$$

A single p electron has both angular momentum and electron spin angular momentum. The wavefunctions including both wavefunctions are also generated with the lowering operator techniques. The largest z component for this Russell–Saunders coupling is

$$j_z = l_z + s_z = 1 + \frac{1}{2} \tag{8.7.49}$$

and the maximal eigenvector is associated with the total $j = \frac{3}{2}$ and a maximal $j_z = +\frac{3}{2}$:

$$\left|\frac{3}{2}\right\rangle = |1\rangle \left|\frac{1}{2}\right\rangle \tag{8.7.50}$$

where only z-component quantum numbers are listed in the brackets.

The lowering operator identity is

$$\mathbf{j}_- = \boldsymbol{l}_- + \mathbf{s}_- \tag{8.7.51}$$

and

$$\begin{aligned}
\mathbf{j}_- \left|\frac{3}{2}\right\rangle &= \hbar \left[\left(\frac{3}{2} + \frac{3}{2}\right)\left(\frac{3}{2} - \frac{3}{2} + 1\right)\right]^{1/2} \left|\frac{1}{2}\right\rangle = \sqrt{3}\,\hbar \left|\frac{1}{2}\right\rangle \\
&= (\boldsymbol{l}_- + \mathbf{s}_-)\,|1\rangle \left|\frac{1}{2}\right\rangle \\
&= \hbar \left[\sqrt{(1+1)(1-1+1)}\,|0\rangle \left|\frac{1}{2}\right\rangle \right. \\
&\qquad \left. + \sqrt{\left(\frac{1}{2} + \frac{1}{2}\right)\left(\frac{1}{2} - \frac{1}{2} + 1\right)}\,|1\rangle \left|-\frac{1}{2}\right\rangle \right] \\
&= \hbar \left(\sqrt{2}\,|0\rangle \left|\frac{1}{2}\right\rangle + |1\rangle \left|-\frac{1}{2}\right\rangle\right)
\end{aligned} \tag{8.7.52}$$

The total wavefunction is

$$\left|\frac{1}{2}\right\rangle = \frac{1}{\sqrt{3}} \left(\sqrt{2}\,|0\rangle \left|\frac{1}{2}\right\rangle + |1\rangle \left|-\frac{1}{2}\right\rangle\right) \tag{8.7.53}$$

The lowering operations can be continued to give the eigenvectors for $\left|-\frac{1}{2}\right\rangle$ and $\left|-\frac{3}{2}\right\rangle$:

$$\left|-\frac{1}{2}\right\rangle=\frac{1}{\sqrt{3}}\left(|-1\rangle\left|\frac{1}{2}\right\rangle+2\,|0\rangle\left|-\frac{1}{2}\right\rangle\right) \tag{8.7.54}$$

$$\left|-\frac{3}{2}\right\rangle=|-1\rangle\left|-\frac{1}{2}\right\rangle \tag{8.7.55}$$

The final operation should produce a wavefunction like the maximal wavefunction; only the signs of the z components are reversed. This serves as a check on the operations. Each wavefunction has terms with z components equal to those of the total wavefunction.

8.8. Projection Operators

Wavefunctions $|l, m\rangle$ for multielectron systems are generated using projection operator techniques. The full wavefunctions for a pair of electrons,

$$|0, 0\rangle, |1, 1\rangle, |1, 0\rangle, |1, -1\rangle \tag{8.8.1}$$

must be generated from their single-electron wavefunctions. The triplet and singlet wavefunctions are desired. Since the eigenvalues for total angular momentum are 0 and $3\hbar^2/2$ for the singlet and triplet manifolds, respectively, projections into each of these subspaces will separate the two manifolds. The equations for total angular momentum for the two systems are

$$\mathbf{S}^2\psi(1)=(1)(1+1)\,\hbar^2\psi=\lambda_1\psi \tag{8.8.2}$$

$$\mathbf{S}^2\psi(0)=(0)(0+1)\,\hbar^2\psi=\lambda_0\psi=0 \tag{8.8,3}$$

Both these possible total angular momenta must arise from the single-electron wavefunctions. The triplet component of the total system is determined using a projection operator defined by the Lagrange–Sylvester theorem:

$$\mathbf{P}_1=(\mathbf{S}^2-\lambda_0\mathbf{I})/(\lambda_1-\lambda_0) \tag{8.8.4}$$

The projection operator for the singlet state is

$$\mathbf{P}_0=(\mathbf{S}^2-\lambda_1\mathbf{I})/(\lambda_0-\lambda_1) \tag{8.8.5}$$

These projection operators will operate on any trial wavefunction to produce a new wavefunction which has either triplet or singlet character. For the two-electron system, the trial eigenvector must be a two-electron wavefunction. For example, the wavefunction $|1, 1\rangle$ which is already a wavefunction exclusively for the triplet system gives a triplet projection:

$$\begin{aligned}\mathbf{P}_1\,|1, 1\rangle&=[\mathbf{S}^2-0(0+1)\mathbf{I}]/[1(1+1)-0(0+1)]\,|1, 1\rangle\\&=\mathbf{S}^2\,|1, 1\rangle=(1)(1+1)\,1\hbar^2\,|1, 1\rangle\end{aligned} \tag{8.8.6}$$

The projection operator returns the original wavefunction because this was an eigenfunction for the triplet state.

The projection of the $|1, 1\rangle$ trial vector onto the singlet system is

$$\begin{aligned} \mathbf{P}_0 \,|1, 1\rangle &= [\mathbf{S}^2 - (1)(1+1)\mathbf{I}]/(0-2)\,|1, 1\rangle \\ &= -\frac{1}{2}[(\mathbf{S}^2\,|1, 1\rangle) - 2\,|1, 1\rangle] \\ &= -\frac{1}{2}(2\,|1, 1\rangle - 2\,|1, 1\rangle) = 0 \end{aligned} \tag{8.8.7}$$

This triplet wavefunction has no projection on the singlet manifold.

Although the projection operators do separate the singlet and triplet manifolds for the two-electron wavefunction, the ultimate goal is the development of such wavefunctions in terms of the single-electron wavefunctions. The total angular momentum, $\mathbf{S}$, requires a sum of squared components and is difficult to break into simple sums of single-electron functions. However, $\mathbf{S}^2$ can be written as a sum of components which are easily expressed as sums of one-electron operators. The total squared angular momentum

$$\mathbf{S}^2 = \frac{1}{2}(\mathbf{S}_+\mathbf{S}_- + \mathbf{S}_-\mathbf{S}_+) + \mathbf{S}_z^2 \tag{8.8.8}$$

is modified with the commutation relation

$$[\mathbf{S}_+, \mathbf{S}_-] = 2\hbar S_z \tag{8.8.9}$$

to give (Problem 8.12)

$$\mathbf{S}^2 = \mathbf{S}_+\mathbf{S}_- + \mathbf{S}_z^2 - \mathbf{S}_z \tag{8.8.10}$$

Each of the terms in the sum can be broken into sums of one-electron operators. For example,

$$\mathbf{S}_+ = \mathbf{s}_+(1) + \mathbf{s}_+(2); \qquad \mathbf{S}_z = \mathbf{s}_z(1) + \mathbf{s}_z(2) \tag{8.8.11}$$

This modified form of $\mathbf{S}^2$ will replace $\mathbf{S}^2$ in each of the projection operators. The projection operator can operate on a product of single-electron wavefunctions such as

$$\left|\frac{1}{2}\right\rangle\left|-\frac{1}{2}\right\rangle \tag{8.8.12}$$

for electrons 1 and 2 with quantum numbers $\left(\frac{1}{2}, \frac{1}{2}\right)$ and $\left(\frac{1}{2}, -\frac{1}{2}\right)$, respectively.

Each term in the expression for $\mathbf{S}^2$ is determined with this trial eigenvector. The first term from Equation 8.8.10 is

$$\mathbf{S}_+\mathbf{S}_-\left|\frac{1}{2}\right\rangle\left|-\frac{1}{2}\right\rangle=(\mathbf{s}_{1+}+\mathbf{s}_{2+})(\mathbf{s}_{1-}-\mathbf{s}_{2-})\left|\frac{1}{2}\right\rangle\left|-\frac{1}{2}\right\rangle$$
$$=(\mathbf{s}_{1+}+\mathbf{s}_{2+})\left|-\frac{1}{2}\right\rangle\left|-\frac{1}{2}\right\rangle$$
$$=\left|\frac{1}{2}\right\rangle\left|-\frac{1}{2}\right\rangle+\left|-\frac{1}{2}\right\rangle\left|\frac{1}{2}\right\rangle \tag{8.8.13}$$

The first pair of lowering operators produced a single term since the second lowering operation on $\left|-\frac{1}{2}\right\rangle$ gave $|0\rangle$. The sum of raising operators then acted on this single eigenfunction to generate a two-term wavefunction. The coefficients c for all these operations were 1.

The $\mathbf{S}_z$ operation is

$$\mathbf{S}_z\left|\frac{1}{2}\right\rangle\left|-\frac{1}{2}\right\rangle=(\mathbf{s}_{1z}+\mathbf{s}_{2z})\left|\frac{1}{2}\right\rangle\left|-\frac{1}{2}\right\rangle$$
$$=\left[\frac{1}{2}+\left(-\frac{1}{2}\right)\right]\left|\frac{1}{2}\right\rangle\left|-\frac{1}{2}\right\rangle=0\left|\frac{1}{2}\right\rangle\left|-\frac{1}{2}\right\rangle \tag{8.8.14}$$

Since $\mathbf{S}_z^2$ is formed by operating with a second $\mathbf{S}_z$ from the left, this term is also zero, and the total squared angular momentum operating on the trial wavefunction is expressed as the following sum of single-electron wavefunctions:

$$\mathbf{S}^2\left|\frac{1}{2}\right\rangle\left|-\frac{1}{2}\right\rangle=\left|\frac{1}{2}\right\rangle\left|-\frac{1}{2}\right\rangle+\left|-\frac{1}{2}\right\rangle\left|\frac{1}{2}\right\rangle \tag{8.8.15}$$

The operation $\mathbf{S}^2\left|\frac{1}{2}\right\rangle\left|-\frac{1}{2}\right\rangle$ is now introduced into the expression for $\mathbf{P}_1\left|\frac{1}{2}\right\rangle\left|-\frac{1}{2}\right\rangle$. $\mathbf{S}^2$ and the trial wavefunction are used together with this technique. The projection operation for the triplet state is

$$\mathbf{P}_1\left|\frac{1}{2}\right\rangle\left|-\frac{1}{2}\right\rangle=(\mathbf{S}^2-0)/(2-0)\left|\frac{1}{2}\right\rangle\left|-\frac{1}{2}\right\rangle$$
$$=\frac{1}{2}\left(\mathbf{S}^2\left|\frac{1}{2}\right\rangle\left|-\frac{1}{2}\right\rangle-0\left|\frac{1}{2}\right\rangle\left|-\frac{1}{2}\right\rangle\right)$$
$$=\frac{1}{2}\left(\left|\frac{1}{2}\right\rangle\left|-\frac{1}{2}\right\rangle+\left|-\frac{1}{2}\right\rangle\left|\frac{1}{2}\right\rangle\right) \tag{8.8.16}$$

The Lagrange–Sylvester projection operator has converted the trial wavefunction into the sum of single-electron product wavefunctions which was determined using the lowering operators in Section 8.7. Since the trial eigenvector had a total z-axis component of 0 $\left(+\frac{1}{2}+-\frac{1}{2}\right)$, the final projection was the eigenvector with a net z component of zero.

The Lagrange–Sylvester expression for the singlet projects a singlet wavefunction from this trial wavefunction as

$$\mathbf{P}_0\left|\frac{1}{2}\right\rangle\left|-\frac{1}{2}\right\rangle=(\mathbf{S}^2-2\mathbf{I})/(0-2)\left|\frac{1}{2}\right\rangle\left|-\frac{1}{2}\right\rangle \tag{8.8.17}$$

Since $\mathbf{S}^2\left|\frac{1}{2}\right\rangle\left|-\frac{1}{2}\right\rangle$ is already known, the projected wavefunction is

$$\begin{aligned}
&-\frac{1}{2}\left(\mathbf{S}^2\left|\frac{1}{2}\right\rangle\left|-\frac{1}{2}\right\rangle-2\left|\frac{1}{2}\right\rangle\left|-\frac{1}{2}\right\rangle\right)\\
&\quad=-\frac{1}{2}\left[\left(\left|\frac{1}{2}\right\rangle\left|-\frac{1}{2}\right\rangle+\left|-\frac{1}{2}\right\rangle\left|\frac{1}{2}\right\rangle\right)-2\left|\frac{1}{2}\right\rangle\left|-\frac{1}{2}\right\rangle\right]\\
&\quad=\frac{1}{2}\left(\left|\frac{1}{2}\right\rangle\left|-\frac{1}{2}\right\rangle-\left|-\frac{1}{2}\right\rangle\left|\frac{1}{2}\right\rangle\right)
\end{aligned} \tag{8.8.18}$$

The projection operator gives the difference of direct products expected for the singlet. The trial function had to have a net z component of 0 to give the finite projection on the singlet state.

A product state that definitely belongs in the triplet manifold will give no singlet projection. The trial wavefunction $\left|-\frac{1}{2}\right\rangle\left|-\frac{1}{2}\right\rangle$ is abbreviated with the notation

$$\left|-\frac{1}{2}\right\rangle\left|-\frac{1}{2}\right\rangle=|--\rangle \tag{8.8.19}$$

The operations with each term in the expression for $\mathbf{S}^2$ are examined. The first term is

$$\mathbf{S}_+\mathbf{S}_-\,|--\rangle=\mathbf{S}_+(\mathbf{s}_{1-}+\mathbf{s}_{2-})\,|--\rangle=|0\rangle \tag{8.8.20}$$

since the eigenvector cannot be lowered further. $\mathbf{S}_z$ gives

$$\begin{aligned}
\mathbf{S}_z\,|--\rangle&=(\mathbf{s}_{1z}+\mathbf{s}_{2z})\,|--\rangle\\
&=\left[-\frac{1}{2}+\left(-\frac{1}{2}\right)\right]|--\rangle=-1\,|--\rangle
\end{aligned} \tag{8.8.21}$$

Operation from the left with $\mathbf{S}_z$ again gives

$$\mathbf{S}_z^2\,|--\rangle=+1\,|--\rangle \tag{8.8.22}$$

The singlet projection is

$$\mathbf{P}_0 \,|--\rangle = -\frac{1}{2}\,[\mathbf{S}_+\mathbf{S}_- + \mathbf{S}_z^2 - \mathbf{S}_z - 2\mathbf{I}]\,|--\rangle$$

$$= -\frac{1}{2}\,(0+2-2)\,|--\rangle = |0\rangle \tag{8.8.23}$$

which is the expected result for this trial eigenvector.

The projection operator technique can be extended to N-electron systems. For a four-electron system, the maximum spin angular momentum has a quantum number of $4\left(\frac{1}{2}\right)=2$. There is a manifold of five eigenvectors for the case where the total spin angular momentum is 2. There is a manifold of three eigenvectors with total angular momentum quantum number 1 and a single eigenvector for the system with total angular quantum number 0. The projection operator technique can resolve these nine wavefunctions.

A trial eigenvector,

$$|+++-\rangle \tag{8.8.24}$$

describes a situation in which the electrons labeled 1 to 3 have $m = +\frac{1}{2}$ and the fourth electron has $m = -\frac{1}{2}$. The total z component for this trial direct product is

$$+\frac{1}{2}+\frac{1}{2}+\frac{1}{2}-\frac{1}{2} = +1 \tag{8.8.25}$$

Since this wavefunction can never give a total angular momentum of 0, there is no projection on the single "0" eigenvector.

Because there is no projection for the "0" or singlet manifold, this factor does not have to be included in the Lagrange–Sylvester formula and there are only two nonzero projection operators, $\mathbf{P}_2$ for the fivefold set and $\mathbf{P}_1$ for the triplet. For $S=2$, the projection operation on the trial function is

$$\mathbf{P}_2\,|+++-\rangle = [\mathbf{S}^2 - (1)(2)\mathbf{I}]/[(2)(3)-(1)(2)]\,|+++-\rangle$$

$$= \frac{1}{4}\,(\mathbf{S}_+\mathbf{S}_- + \mathbf{S}_z^2 - \mathbf{S}_z - 2\mathbf{I})\,|+++-\rangle \tag{8.8.26}$$

The terms in the projection operation are

$$\mathbf{S}_z\,|+++-\rangle = \left(\frac{1}{2}+\frac{1}{2}+\frac{1}{2}-\frac{1}{2}\right)|+++-\rangle = 1\,|+++-\rangle \tag{8.8.27}$$

$$\mathbf{S}_z^2\,|+++-\rangle = \mathbf{S}_z(1)\,|+++-\rangle = 1\,|+++-\rangle \tag{8.8.28}$$

The first operation in $\mathbf{S}_+\mathbf{S}_-$ is

$$\mathbf{S}_- \,|+++-\rangle = (\mathbf{s}_{1-} + \mathbf{s}_{2-} + \mathbf{s}_{3-} + \mathbf{s}_{4-})\,|+++-\rangle$$
$$= |-++-\rangle + |+-+-\rangle + |++--\rangle + 0 \quad (8.8.29)$$

since only the first three electron states can be lowered. Applying $\mathbf{S}_+$ to this result gives

$$\mathbf{S}_+(|-++-\rangle + |+-+-\rangle + |++--\rangle)$$
$$= (\mathbf{s}_{1+} + \mathbf{s}_{2+} + \mathbf{s}_{3+} + \mathbf{s}_{4+})(|-++-\rangle + |+-+-\rangle + |++--\rangle)$$
$$= |+++-\rangle + 0 + 0 + |-+++\rangle + 0 + |+++-\rangle + 0 + |+-++\rangle$$
$$+ 0 + 0 + |+++-\rangle + |++-+\rangle$$
$$= 3\,|+++-\rangle + |-+++\rangle + |+-++\rangle + |++-+\rangle \quad (8.8.30)$$

The total projection is now

$$\frac{1}{4}[3\,|+++-\rangle + |-+++\rangle + |+-++\rangle$$
$$+ |++-+\rangle + (1-1-2)\,|+++-\rangle]$$
$$= \frac{1}{4}(|+++-\rangle + |-+++\rangle + |+-++\rangle + |++-+\rangle) \quad (8.8.31)$$

Each product state with a single $-\frac{1}{2}$ electron is weighted equally in the final result. This is the total wavefunction for $S_z = +1$ and $S = 2$.

The linear combination of four-electron product functions for the triplet manifold is determined in the same way. The projection operation is

$$\mathbf{P}_1\,|+++-\rangle = [\mathbf{S}^2 - 2(3)\mathbf{I}]/[(1)(2) - (2)(3)]\,|+++-\rangle \quad (8.8.32)$$

The terms for the singlet state are excluded since this trial function cannot have a singlet projection. $\mathbf{S}^2\,|+++-\rangle$ was determined above and can be substituted into Equation 8.8.32 to give

$$\mathbf{P}_1\,|+++-\rangle = -\frac{1}{4}[3\,|+++-\rangle + |++-+\rangle$$
$$+ |+-++\rangle + |-+++\rangle + (1-1-6)\,|+++-\rangle]$$
$$= \frac{1}{4}(3\,|+++-\rangle - 1\,|++-+\rangle$$
$$- 1\,|+-++\rangle - 1\,|-+++\rangle) \quad (8.8.33)$$

If the four different single-electron product functions are labeled as the components of a four-component vector, the vectors for each of the states are

$$\begin{pmatrix} 1 \\ 1 \\ 1 \\ 1 \end{pmatrix} \quad \begin{pmatrix} 3 \\ -1 \\ -1 \\ -1 \end{pmatrix} \tag{8.8.34}$$

The two vectors are orthogonal. The projection operator technique generated the appropriate eigenvectors and included the orthogonality required by their presence in different manifolds. To generate the full set of nine orthogonal eigenvectors for the four-electron system, other trial vectors with different s_z must be selected and projected onto their appropriate manifolds. For example, the trial vector $|++--\rangle$ must have projections on all three manifolds (Problem 8.13).

Problems

8.1. Convert the classical Hamiltonian

$$H = p^2/2m + (k/2)\,x^2$$

to a Schrödinger equation in momentum space.

8.2. Develop the linear combinations of orbitals required to generate an *sp* orbital. Show how the function correlates with the geometry.

8.3. Show that the transform

$$\mathbf{S}^{-1/2}\mathbf{H}\mathbf{S}^{1/2}$$

where **H** is the Hamiltonian matrix and **S** is the orbital overlap matrix, is Hermitian.

8.4. Determine the energy levels and wavefunctions for the ethene molecule in terms of α and β when the four elements in the 2×2 **S** matrix all equal 0.3.

8.5. Determine the eigenvectors for the **H** matrix for propene (Equation 8.3.31).

8.6. Determine the eigenvalues for the matrix of Equation 8.3.28.

8.7. Determine the energy levels for cyclopentadiene. Show that these eigenvalues inscribe properly in a circle of radius 2π.

8.8. The highest molecular orbital of benzene has alternating coefficients of $+1$ and -1 when the maximum amplitude of the molecular orbital begins on carbon 1. Determine the wavefunction coefficients when the maximum amplitude begins on carbon 2 and when it begins halfway between carbons 1 and 2. Show that all these vectors are eigenvectors of the **H** matrix.

8.9. Show that Pauli matrix products

$$\boldsymbol{\sigma}_i \boldsymbol{\sigma}_j = 0$$

when $i \neq j$.

8.10. A rotation involving the three Euler angles in the three-dimensional case is equivalent to a similarity transform in the two-dimensional space defined by the 2×2 Pauli matrices. Show that the matrix product for rotation through Euler angles φ, θ, and ψ is

$$\begin{pmatrix} \exp(i\psi/2) & 0 \\ 0 & \exp(-i\psi/2) \end{pmatrix} \begin{pmatrix} \cos(\theta/2) & \sin(\theta/2) \\ -\sin(\theta/2) & \cos(\theta/2) \end{pmatrix} \begin{pmatrix} \exp(-i\varphi/2) & 0 \\ 0 & \exp(-i\varphi/2) \end{pmatrix}$$

Show that all elements in $\mathbf{S}^{-1}$ are complex conjugates of the elements in $\mathbf{S}$ for the similarity transform.

8.11. a. Develop the matrix for $\mathbf{S}^2 - \mathbf{S}_z^2$ for a system with $j = \frac{3}{2}$ (Equation 8.7.25).

b. Verify your result using the $\mathbf{j}_+$ and $\mathbf{j}_-$ matrices (Equations 8.7.26 and 8.7.27).

8.12. Prove that

$$\mathbf{S}^2 = \mathbf{S}_+ \mathbf{S}_- + \mathbf{S}_z^2 - \mathbf{S}_z$$

8.13. Using the four-electron product eigenvector $|++--\rangle$ as a trial vector, develop the wavefunctions consistent with this eigenvector for the singlet, triplet, and fivefold manifolds. Remember that each projection operator now contains two differences.

8.14. Determine the wavefunctions for two $2p$ electrons when electron spin is not included.

8.15. The coupling between two dipole states is described by the symmetric matrix

$$\begin{pmatrix} 0 & -\Delta \\ -\Delta & 0 \end{pmatrix}$$

Determine the changes in energy which will be produced by this interaction process.

9

Driven Systems and Fluctuations

9.1. Singlet–Singlet Kinetics

The kinetics of Chapter 6 dealt exclusively with systems which conserved mass. For such conservative systems, a transition always leads to a new state within the system. The sum of elements in any column of the rate constant matrix is always zero and the system must have a zero (equilibrium) eigenvalue. This chapter considers an alternate kinetic situation in which the state populations are generated by some input process.

The simple photochemical model of Figure 9.1 illustrates this nonconservative kinetics. The molecule has two electronic singlet states—the ground state, S_0, and S_1. S_1 lies sufficiently above the ground state so that it is unlikely to be populated by thermal collisions. At equilibrium, the fraction of molecules in the excited singlet will be extremely small, and this state can be eliminated from the equilibrium manifold.

The situation changes when light of sufficient energy is used to populate the S_1 levels of the molecules. The states are populated, but, simultaneously, there are a number of kinetic processes which immediately begin to depopulate S_1 levels. The transition from S_1 to S_0 which occurs spontaneously is fluorescence. The rate of depopulation of S_1 states is proportional to the number N_1 of the excited state molecules:

$$\text{Rate} = k_f N_1 \tag{9.1.1}$$

The excited singlet state can also be depopulated via a singlet–triplet transition when one of the electrons changes spin. Once again, this rate is proportional to the population N_1 of S_1 states:

$$\text{Rate} = k_i N_1 \tag{9.1.2}$$

The singlet may also be depopulated via collisions with other molecules Q. This bimolecular process has a rate

$$k_q[\text{Q}]\, N_1 \tag{9.1.3}$$

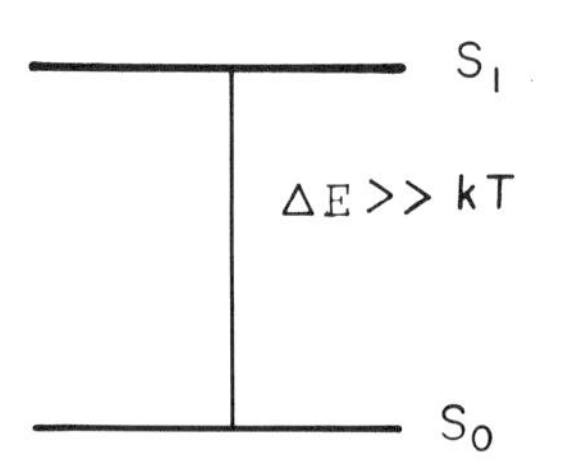

Figure 9.1. The ground and excited singlet states of a molecule.

When the population of Q is large, this rate is approximated as a pseudo-first-order rate process in N_1:

$$k'N_1 = \{k_q[\mathrm{Q}]\}\, N_1 \tag{9.1.4}$$

There are some major differences between this system and a conservative kinetic system. The pool of ground state molecules N_0 will be large and the net change of its population will be small. For this reason, it is excluded from the states to be included in the kinetic study. Only the population of the S_1 state is kinetically analyzed. Since the population of this state is zero at equilibrium, its population is maintained only by a photon flow. The population of S_1 states cannot increase indefinitely because kinetic processes such as fluorescence continually deplete this population. The actual population of S_1 represents a balance between the photon input and the various depopulating processes. With a steady photon input, the system will attain a nonequilibrium stationary state where the rates of population and depopulation are equal. This is a simple example of a flow system; the flow of photons through S_1 determines its nonequilibrium population.

The kinetics of S_1 are dictated by the combination of input and output rates. The state N_1 is depopulated by several distinct first-order rate processes. Although accessible depopulation pathways may differ from system to system, these depopulating first-order and pseudo-first-order rates are added to produce a single first-order rate in N_1:

$$-kN_1 = -(k_f + k_i + k' + \cdots)\, N_1 \tag{9.1.5}$$

The population rate for S_1 will depend on the population of S_0 molecules, which is essentially constant since only a small number reach the S_1 state. It is also proportional to the total number of photons, $I(t)$, entering the reaction vessel per unit time to give the "bimolecular" rate

$$R_i = k_e N_0 I(t) \tag{9.1.6}$$

Since N_0 is essentially constant, this rate is pseudo first order in the intensity $I(t)$. Although a variety of units can be selected for $I(t)$, Equation 9.1.6 simply describes the number of S_1 molecules created as a function of time and it is convenient to define this rate as a single input flux, $J(t)$:

$$J(t) = k_e N_0 I(t) \tag{9.1.7}$$

The rate equation for the population N_1 of state S_1 is

$$dN_1/dt = J(t) - kN_1 \tag{9.1.8}$$

The equation is rewritten as

$$dN_1/dt + kN_1 = J(t) \tag{9.1.9}$$

to permit a general solution using the integrating factor

$$\exp(+kt) \tag{9.1.10}$$

The equation becomes

$$\exp(kt)\, dN_1/dt + \exp(kt)\, kN_1 = d[N_1 \exp(kt)]/dt = J(t) \exp(kt) \tag{9.1.11}$$

Both sides may now be integrated. The population of N_1 is assumed to be zero when the excitation is started. The equation becomes

$$\exp(kt)\, N_1(t) = \int_0^t \exp(kt')\, J(t')\, dt' \tag{9.1.12}$$

where t' is used for the integration variable to distinguish it from the time t at which the population is observed.

The time-dependent solution for $N_1(t)$ is

$$N_1(t) = \exp(-kt) \int_0^t \exp(kt')\, J(t')\, dt' \tag{9.1.13}$$

This general equation is sometimes written in a Green's function formalism,

$$N_1(t) = \int_0^t G(t, t')\, J(t')\, dt' \tag{9.1.14}$$

where

$$G(t, t') = \exp[-k(t - t')] \tag{9.1.15}$$

The inputs at each time t' produce some response which "lingers" at time t. The integral collects the inputs at all times t' and presents their net effect on the system at time t.

The use of this general response equation can be illustrated using some special input fluxes, $J(t)$, and their effect on the population, N_1.

Delta Function Input

The photochemical kinetics of the excited state can be observed in time if the system is irradiated with a light pulse which is short relative to the time scale of the kinetic depopulation events. The limiting case of very short light pulses is the delta function, which introduces all the exciting photons in an infinitely short burst at $t=0$. The delta function is nonzero at a single time so it "selects" this time from an integration over a time regime. The integral of some function $f(t')$ and the delta function $\delta(t-t')$ over a time interval which includes t is

$$\int \delta(t'-t)\, f(t')\, dt' = f(t) \tag{9.1.16}$$

The input pulse, $J(t)$, of S_1 molecules is

$$J(t) = J\delta(t'-0) \tag{9.1.17}$$

where J is a constant. The population of state S_1 as a function of time is

$$\begin{aligned} N_1(t) &= \exp(-kt) \int_0^t \exp(kt')\, J\delta(t'-0)\, dt' \\ &= \exp(-kt)\, J \exp(0) = J \exp(-kt) \end{aligned} \tag{9.1.18}$$

The photon pulse creates an instantaneous population, J, of S_1 molecules; this population decays exponentially with rate constant k. S_1 is transiently populated.

To record this decay experimentally, some observable is required. The rate of fluorescence is determined from the instantaneous population, N_1, and the rate constant for fluorescence,

$$\text{Rate(fluorescence)} = k_f N_1 \tag{9.1.19}$$

The decay constant k is determined by observing the exponential decay of fluorescence. The analysis gives k, the total rate constant, and not just the fluorescence rate constant, k_f. Since all rate processes depopulate the system, the rate constant k dictates the total depopulation.

Constant Intensity

If a light beam of constant intensity is directed onto the sample starting at time zero, the input flux is

$$J(t) = J \tag{9.1.20}$$

The Green's function equation for N_1 is

$$N_1(t) = \exp(-kt) \int_0^t \exp(kt')\, J\, dt'$$
$$= J \exp(-kt)(1/k)[\exp(kt) - \exp(0)]$$
$$= (J/k)[1 - \exp(-kt)] \tag{9.1.21}$$

The time scale for the exponential is inversely related to the rate constant k which has values in the range 10^{-9}–10^{-6}. On a laboratory time scale (seconds), the equation reduces to the steady-state solution,

$$N_1(\mathrm{ss}) = J/k \tag{9.1.22}$$

and the population is the ratio of the input to the total decay rate constant. The rate or intensity of fluorescence is

$$\mathrm{Rate} = k_f N_1(\mathrm{ss}) = (k_f/k)\, J$$
$$= [k_f/(k_f + k_i + \cdots)]\, J \tag{9.1.23}$$

The intensity of fluorescence is proportional to the ratio of the fluorescence rate constant and the sum of all depopulating rate constants. This constitutes one definition of quantum yield. The fraction of excited S_1 molecules which decay via fluorescence is the rate constant for fluorescence divided by the sum of rate constants for all depopulation pathways from S_1.

Phase Fluorometry

There is an additional mode of time-dependent light intensity input which blends some of the properties of pulse and steady-state excitation of S_1. The intensity of the light is modulated at some frequency ω which is comparable to the total rate constant k (s^{-1}) for depopulation. The input flux is

$$J(t) = J \sin \omega t \tag{9.1.24}$$

The integral equation for N_1 is

$$N_1(t) = J \exp(-kt) \int_0^t \exp(kt') \sin(\omega t')\, dt' \tag{9.1.25}$$

which is solved for N_1 as

$$N_1(t) = J[(k \sin \omega t - \omega \cos \omega t)/(\omega^2 + k^2) - \exp(-kt)\omega/(\omega^2 + k^2)] \tag{9.1.26}$$

The exponential term will again decay to zero rapidly, leaving a steady-state solution

$$N_1(\mathrm{ss}) = J(k \sin \omega t - \omega \cos \omega t)/(k^2 + \omega^2) \tag{9.1.27}$$

This expression can be simplified significantly by defining a phase angle φ as illustrated in Figure 9.2. In this figure, k is the abscissa while ω is the ordinate. They relate to the phase angle as

$$\cos \varphi = k/(k^2 + \omega^2)^{1/2} \tag{9.1.28}$$

$$\sin \varphi = \omega/(k^2 + \omega^2)^{1/2} \tag{9.1.29}$$

The steady-state equation is

$$N_1(\text{ss}) = [J/(k^2 + \omega^2)^{1/2}](\cos \varphi \sin \omega t - \sin \varphi \cos \omega t) \tag{9.1.30}$$

The sine and cosine products reduce to

$$\sin \omega t \cos \varphi - \cos \omega t \sin \varphi = \sin (\omega t - \varphi) \tag{9.1.31}$$

and

$$N_1(\text{ss}) = [J/(k^2 + \omega^2)^{1/2}] \sin (\omega t - \varphi) \tag{9.1.32}$$

The population $N_1(\text{ss})$ varies sinusoidally at the same frequency as the modulated input intensity but it is phase delayed by the angle φ. Since the fluorescent intensity is

$$\text{Rate} = k_f N_1(\text{ss}) \tag{9.1.33}$$

this intensity will also vary sinusoidally. Both the amplitude and phase delay of this signal relative to the input signal J can be determined. The phase delay can be related to the rate constant k by noting the geometry of Figure 9.2. The tangent of φ is

$$\tan \varphi = \omega/k \tag{9.1.34}$$

Since

$$\tau = 1/k \tag{9.1.35}$$

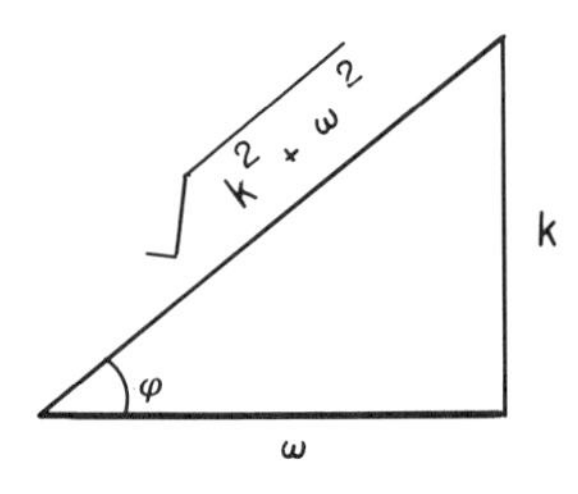

Figure 9.2. Trigonometric arrangement linking k and ω to the phase delay φ.

where τ is the natural lifetime of the S_1 state, this equation is also

$$\tan \varphi = \omega\tau \tag{9.1.36}$$

A measurement of the phase delay of fluorescence relative to excitation at frequency ω gives the rate constant k for the excited S_1 state.

9.2. Multilevel Driven Photochemical Systems

The differential equations of Section 9.1 were scalar equations since they monitored the input and output rates from a single electronic level. For multilevel systems, the scalar variable can be replaced by a vector, and a rate constant matrix will include rate constants which depopulate the system and transition probabilities which describe transitions between levels within the system. The solutions of these equations will parallel the solutions of the scalar differential equations of Section 9.1.

The extension to a multilevel system is accomplished by examining vibrational levels within the excited singlet S_1. For larger molecules, the number of such vibrational levels is extremely large and the **K** matrix order is high. Under such circumstances, it is convenient to describe the manifold of vibrational states using the strong-collision assumption. The manifold is approximated as a two-level system. The upper level in S_1 contains excess vibrational energy introduced by the exciting photons. However, each collision with a heat bath molecule removes this excess energy, leaving the molecule in the lowest vibrational level of the excited electronic singlet, S_1. The heat bath molecules do not quench the electronic energy of the molecule.

The two vibrational levels are labeled as upper (u) and lower (l), as illustrated in Figure 9.3. Each level can emit fluorescence with rate constant k_f, but the rate constant for intersystem crossing from the upper level is k_u and the rate constant for intersystem crossing from the lower level is k_l. For this basic model, only irreversible transitions from u to l are allowed since the molecule is unlikely to reacquire the excess vibrational energy before S_1 itself is depopulated. Since each collision is completely effective in removing vibrational energy, the

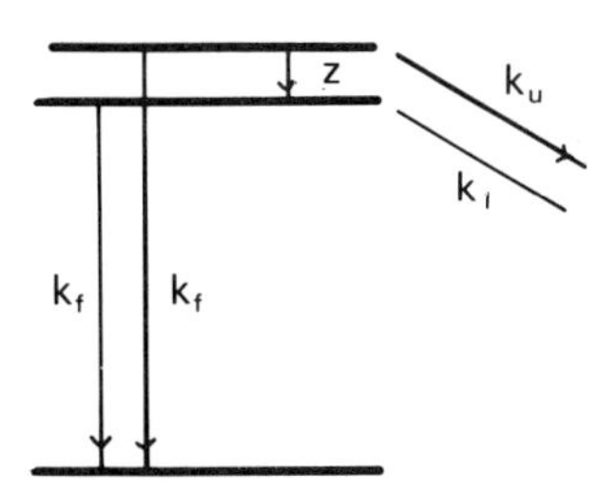

Figure 9.3. An excited singlet with two vibrational levels, showing depopulation rate constants for each level.

vibrational transition is proportional to the collision frequency z for a single molecule. The total rate of vibrational transitions is the product of z and the N_u population.

The rates of depopulation for each of these states are

$$\text{Rate}(u) = -(k_f + k_u + z)\, N_u \tag{9.2.1}$$

$$\text{Rate}(l) = -(k_f + k_l)\, N_l \tag{9.2.2}$$

In addition, the lower state is populated at a rate proportional to the population of the upper state,

$$+zN_u \tag{9.2.3}$$

Both of the levels can be populated by the exciting photons. However, by exciting with monochromatic light, it is also possible to excite only a single level. In this model, only the upper state is populated by some input flux $J_u(t)$. The equations for the rates of change of N_u and N_l are

$$\begin{aligned} dN_u/dt &= -(k_f + k_u + z)\, N_u + J_u(t) \\ dN_l/dt &= +zN_u - (k_f + k_l)\, N_l \end{aligned} \tag{9.2.4}$$

There is one major difference between this pair of coupled equations and the kinetic equations for conservative kinetic systems. All terms containing factors N_u or N_l can be condensed into a product of a matrix and a vector, i.e.,

$$\begin{pmatrix} -(k_f + k_u + z) & 0 \\ z & -(k_f + k_l) \end{pmatrix} \begin{pmatrix} N_u \\ N_l \end{pmatrix} \tag{9.2.5}$$

The input flux J is independent of the populations of the excited states. It acts on ground state molecules to produce the excited state and must be described as a vector. In this case, J_u is the vector component for the upper level, J_l, the flux into the lower level, is zero for this example. The excitation flux vector for this case is

$$\begin{pmatrix} J_u \\ 0 \end{pmatrix} \tag{9.2.6}$$

The entire two-level system of equations is expressed in matrix vector format as

$$d/dt \begin{pmatrix} N_u \\ N_l \end{pmatrix} = \begin{pmatrix} -(k_f + k_u + z) & 0 \\ w & -(k_f + k_l) \end{pmatrix} \begin{pmatrix} N_u \\ N_l \end{pmatrix} + \begin{pmatrix} J_u \\ 0 \end{pmatrix} \tag{9.2.7}$$

which is written formally as

$$d\,|N\rangle/dt = -\mathbf{K}\,|N\rangle + |J(t)\rangle \tag{9.2.8}$$

with

$$\mathbf{K} = \begin{pmatrix} +(k_f + k_u + z) & 0 \\ -z & +(k_f + k_l) \end{pmatrix} \tag{9.2.9}$$

and

$$|N\rangle = \begin{pmatrix} N_u \\ N_l \end{pmatrix} \qquad |J\rangle = \begin{pmatrix} J_u \\ J_l \end{pmatrix} \tag{9.2.10}$$

Since the one-level system was solved using the integrating factor, the formal solution for this multilevel system is solved in a parallel manner using a **K** matrix integrating factor

$$\exp(\mathbf{K}t) \tag{9.2.11}$$

The formal solution is

$$|N(t)\rangle = \exp(-\mathbf{K}t) \int_0^t \exp(\mathbf{K}t')\, |J(t')\rangle\, dt' \tag{9.2.12}$$

The order of parameters is now important. Since $|J\rangle$ and $|N(t)\rangle$ have been defined as column vectors, these vectors must always lie to the right of any matrices or matrix functions.

The solution of this equation for a specific $|J(t)\rangle$ will give the concentrations in each of the levels of the excited electronic state. To determine the total fluorescence from the two levels, each vector component, i.e., state population, must be multiplied by the fluorescence rate constant. This gives the fluorescence intensity for a single level. The intensities are added to give a total fluorescence intensity. A monochromator in the fluorescence detection system can be used to select intensity from a single level or a manifold of levels.

The fluorescence intensity is formally written by introducing a fluorescence rate constant matrix,

$$\mathbf{K}_f = \begin{pmatrix} k_f & 0 \\ 0 & k_f \end{pmatrix} \tag{9.2.13}$$

and a summation vector

$$\langle 1| = (1 \quad 1) \tag{9.2.14}$$

If the fluorescence were selected with a monochromator, then the summation vector would contain a 1 for level emissions detectable after the monochromator and a 0 for the levels rejected by the monochromator setting. If fluorescence from the lower level was monitored exclusively for the two-level model, the summation vector would be

$$\langle 1| = (0 \quad 1) \tag{9.2.15}$$

Since the intersystem crossing rate constants differ for each level, the intersystem rate constant matrix could have different rate constants for each level, i.e., for the two-level model,

$$\mathbf{K}_i = \begin{pmatrix} k_u & 0 \\ 0 & k_l \end{pmatrix} \tag{9.2.16}$$

The general expression for total observed fluorescence intensity is written formally using $\mathbf{K}_f$ and $\langle 1|$ as

$$I_f = \langle 1| \, \mathbf{K}_f \, |N(t)\rangle \tag{9.2.17}$$

where $|N(t)\rangle$ is determined by the nature of the forcing input, $|J(t)\rangle$.

To illustrate these techniques, the two-level system is subjected to a delta function input at time zero. The equation is

$$|N(t)\rangle = \exp(-\mathbf{K}t) \int_0^t \exp(\mathbf{K}t') \, |J\rangle \, \delta(t'-0) \, dt' \tag{9.2.18}$$

where $\mathbf{K}$ and $|J\rangle$ have been defined above. The equation is solved exactly like the one-level equation. The integration selects only $t'=0$ and the solution is

$$|N(t)\rangle = \exp(-\mathbf{K}t) \exp(\mathbf{K} \times 0) \, |J\rangle = \exp(-\mathbf{K}t) \, |J\rangle \tag{9.2.19}$$

The system can now be solved as a summation of the $\mathbf{K}$ matrix eigenvalues λ_i and projection operators $\mathbf{P}_i$:

$$|N(t)\rangle = \sum \exp(-\lambda_i t) \, \mathbf{P}_i \, |J\rangle \tag{9.2.20}$$

There is no zero eigenvalue since this would imply some equilibrium population of the excited electronic levels.

Evaluation of the eigenvalues and projection operators is simplified by defining the total rate constants,

$$\begin{aligned} k &= k_f + k_u + z \\ k' &= k_f + k_l \end{aligned} \tag{9.2.21}$$

The characteristic determinant and eigenvalues are

$$\begin{gathered} \begin{vmatrix} k-\lambda & 0 \\ z & k'-\lambda \end{vmatrix} = 0 \\ (k-\lambda)(k'-\lambda) = 0 \\ \lambda_1 = k \qquad \lambda_2 = k' \end{gathered} \tag{9.2.22}$$

The projection operator $\mathbf{P}(k)$ is

$$\mathbf{P}(k) = (\mathbf{K} - k'\mathbf{I})/(k - k')$$
$$= \begin{pmatrix} k-k' & 0 \\ -z & k'-k' \end{pmatrix} (k-k')^{-1}$$
$$= \begin{pmatrix} z+k_u-k_l & 0 \\ z & 0 \end{pmatrix} (z+k_u-k_l)^{-1}$$
$$= \begin{pmatrix} 1 & 0 \\ z/(z+k_u-k_l) & 0 \end{pmatrix} \tag{9.2.23}$$

and $\mathbf{P}(k')$ is

$$\mathbf{P}(k') = (\mathbf{K} - k\mathbf{I})/(k' - k)$$
$$= \begin{pmatrix} 0 & 0 \\ -z & -(z+k_u-k_l) \end{pmatrix} [-(z+k_u-k_l)]^{-1}$$
$$= \begin{pmatrix} 0 & 0 \\ z/(z+k_u-k_l) & 1 \end{pmatrix} \tag{9.2.24}$$

The concentration vector $|N(t)\rangle$ is the summation of the scalar projections:

$$|N(t)\rangle = \left[\exp(-kt) \begin{pmatrix} k-k' & 0 \\ -z & 0 \end{pmatrix} \begin{pmatrix} J \\ 0 \end{pmatrix} \right.$$
$$\left. + \exp(k't) \begin{pmatrix} 0 & 0 \\ +z & k-k' \end{pmatrix} \begin{pmatrix} J \\ 0 \end{pmatrix} \right] (k-k')^{-1}$$
$$= J\exp(-kt) \begin{pmatrix} 1 \\ -z/(k-k') \end{pmatrix} + \exp(-k't) \begin{pmatrix} 0 \\ +z/(k-k') \end{pmatrix} \tag{9.2.25}$$

The rows give the time dependence of each of the levels. The concentration of the upper level decays as

$$N_u(t) = J\exp[-(z+k_f+k_u)t] \tag{9.2.26}$$

There are three irreversible depopulating rate constants for this level, and the decay is identical to that observed for the single excited state in Section 9.1. The light flux J populates the upper level, and the depopulation rate is proportional to all the rate constants including z.

The population of the lower level depends on two decay constants:

$$N_l(t) = -\exp[-(z+k_f+k_u)t][+Jz/(z+k_u-k_l)]$$
$$+ \exp[-(k_f+k_l)t][Jz/(z+k_u-k_l)]$$
$$= [Jz/(z+k_u-k_l)]\{\exp[-(k_f-k_l)t] - \exp[-(z+k_f+k_u)t]\} \tag{9.2.27}$$

The population of the lower level is described by a difference of exponential decays. This parallels the kinetics observed for sequential reactions like

$$A \rightarrow B \rightarrow C$$

As the rate constant z populates the lower level at the expense of the upper level, N_l rises. As the upper level is depopulated, the transfer to the lower level decreases and its irreversible decay processes reduce its population to zero. The scheme is illustrated in Figure 9.4.

When J is a constant photon flux, the equation for the concentration vector becomes

$$|N(t)\rangle = \exp(-\mathbf{K}t) \int_0^t \exp(\mathbf{K}t')\, |J\rangle\, dt' \tag{9.2.28}$$

For this case, only the exponential has a time dependence. Integration for a single level produced a factor of $1/k$. For this matrix system, the solution requires $\mathbf{K}^{-1}$, the rate constant matrix inverse. The integration gives

$$|N(t)\rangle = \mathbf{K}^{-1}[\mathbf{I} - \exp(-\mathbf{K}t)]\, |J\rangle \tag{9.2.29}$$

Since there are no zero eigenvalues, a projection operator expansion of the exponential would give a sum of exponential decays. In the steady state, these would disappear to give

$$|N(\mathrm{ss})\rangle = \mathbf{K}^{-1}\, |J\rangle \tag{9.2.30}$$

The inverse of the $\mathbf{K}$ matrix gives the steady-state concentrations of all the levels.

The projection operator solution for this steady state for the two-state model is

$$|N\rangle = (J/k)\begin{pmatrix} 1 & 0 \\ -z/(k-k') & 0 \end{pmatrix}\begin{pmatrix} 1 \\ 0 \end{pmatrix} + (J/k')\begin{pmatrix} 0 & 0 \\ z/(k-k') & 0 \end{pmatrix}\begin{pmatrix} 1 \\ 0 \end{pmatrix} \tag{9.2.31}$$

The population of the upper level is

$$N_u = [J/(k_f + z + k_u)] \tag{9.2.32}$$

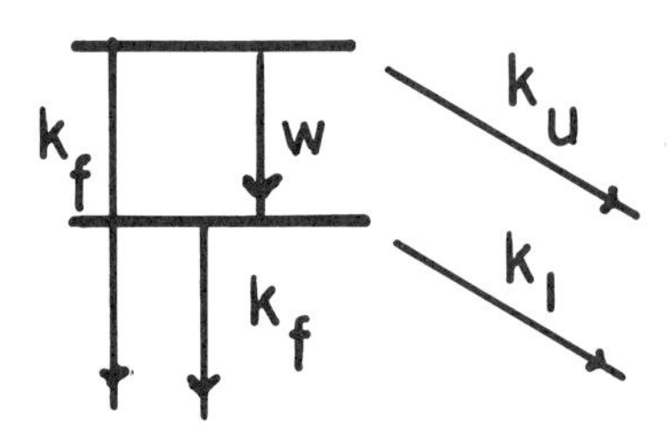

Figure 9.4. Temporal decay of the two excited singlet state levels after transient excitation to the upper level.

while the lower level has a population

$$N_l = [Jz/(z + k_u - k_l)][1/(k_f + k_l) - 1/(z + k_f + k_u)] \quad (9.2.33)$$

The total steady-state fluorescence is determined with the $\mathbf{K}_f$ matrix and summation vector

$$\langle 1| \mathbf{K}_f |N(\text{ss})\rangle \quad (9.2.34)$$

Both the pulse and steady input fluxes gave solutions which paralleled the one-level examples. The same holds true for the phase fluorometric solution. The input function is

$$|J(t)\rangle = |J\rangle \sin \omega t \quad (9.2.35)$$

which is incorporated into the equation for $|N(t)\rangle$:

$$|N(t)\rangle = \exp(-\mathbf{K}t) \int_0^t \exp(\mathbf{K}t') \sin(\omega t')\, dt' \,|J\rangle \quad (9.2.36)$$

The formal solution to this equation parallels that of the one-level system. The steady-state solution is

$$|N(\text{ss})\rangle = (\mathbf{K} \sin \omega t + \omega \mathbf{I} \cos \omega t)(\mathbf{K}^2 + \omega^2 \mathbf{I})^{-1} |J\rangle \quad (9.2.37)$$

where $\sin \omega t$ and $\cos \omega t$ are scalars. This equation can be rearranged to give both the amplitude and phase delay of the fluorescence which would be observed for this multilevel system.

The experimental parameters are calculated by determining the total fluorescence intensity using the $\mathbf{K}_f$ matrix and the summation vector $\langle u|$:

$$\langle 1| \mathbf{K}_f |N\rangle = \langle 1| \mathbf{K}_f \mathbf{K}(\mathbf{K}^2 + \omega^2 \mathbf{I})^{-1} |J\rangle \sin \omega t$$
$$+ \langle 1| \mathbf{K}_f \omega \mathbf{I}(\mathbf{K}^2 + \omega^2 \mathbf{I})^{-1} |J\rangle \cos \omega t \quad (9.2.38)$$

The quadratic forms which multiply $\sin \omega t$ and $\cos \omega t$ are scalars. The phase angle and amplitude cannot be defined by separating the inverse into a product of square roots by analogy with the one-level case (Section 9.1). The expression can be expanded in its projection operators to establish the phase angle, φ. For a two-level system, the equation for the intensity of fluorescence is

$$I = k_f(\lambda_1^2 + \omega^2)^{-1} \lambda_1 \langle 1| \mathbf{P}(1) |J\rangle \sin \omega t$$
$$+ k_f(\lambda_2^2 + \omega^2)^{-1} \omega \langle 1| \mathbf{P}(2) |J\rangle \cos \omega t \quad (9.2.39)$$

Each term in the summation is a scalar and defines a phase delay, φ_i. Each of these phase angles is defined using the equations

$$\cos \varphi_i = k_f \lambda_i (\lambda_i^2 + \omega^2)^{-1/2} \langle 1| \mathbf{P}(i) |J\rangle \quad (9.2.40)$$

$$\sin \varphi_i = k_f \omega (\lambda_i^2 + \omega^2)^{-1/2} \langle 1| \mathbf{P}(i) |J\rangle \quad (9.2.41)$$

φ_i is defined from $\tan \varphi_i$ for the ith term in the projection operator expansion:

$$\tan \varphi_i = \sin \varphi_i / \cos \varphi_i \qquad (9.2.42)$$

The intensity of fluorescence is

$$I = \sum k_f \langle 1 | \mathbf{P}(i) | J \rangle (\lambda_i^2 + \omega^2)^{-1/2} \sin(\omega t - \varphi_i) \qquad (9.2.43)$$

The final experimentally observed phase angle is a weighted sum of the phase delays associated with each eigenvalue.

9.3. Laser Systems

For photokinetic studies with conventional light sources, only the excited states were considered because the pool of molecules which remained in the ground state was always extremely large relative to the population of molecules in excited states. The few molecules which were excited from the ground state had a negligible effect on its population. The rate at which excited state molecules are created depends on the total number of photons entering the system, i.e., the light intensity I and the number of molecules in the system which are available to absorb these photons. The rate of absorption is "bimolecular" in the ground state population, N_0, and the intensity:

$$J = k_e' N_0 I \qquad (9.3.1)$$

where k_e' is a rate constant which gives an input flux J in units of excited molecules per second. When N_0 is large relative to N_1, the flux can be written as a pseudo-first-order rate in the intensity:

$$J = k_e I \qquad (9.3.2)$$

For laser systems, the population of the excited state molecules must become comparable to the population of ground state molecules. This requires a significant change in the population of the ground state population, and the kinetics of the ground state must be considered in the calculations. Under such circumstances, the bimolecular rate expression (9.3.1) must be used. The system conserves mass since any molecule which is lost by the ground state appears in an excited state. The photon flux establishes the excitation rate.

The excited state can still be depopulated by the first-order pathways such as fluorescence and intersystem crossing. The fluorescence is a spontaneous emission. Photons will be emitted by the excited molecules at a rate proportional to their population. The emission continues after the photon flux is terminated.

When the population of excited state molecules becomes comparable to the population of the ground state molecules, a new depopulation pathway becomes significant. Stimulated emission from the excited state is the reverse of excitation.

While spontaneous emission is a first-order process ($\propto N_1$), stimulated emission is proportional to both the excited state population and the intensity of photons with energy equal to that of the transition from 1 to 0. The molecule does not absorb these photons. Their presence dictates the rate of stimulated emission to the ground state.

The three rate processes are

1. excitation

$$\text{Rate} = k_e I N_0 = W N_0 \tag{9.3.3}$$

2. stimulated emission

$$\text{Rate} = k_e I N_1 = W N_1 \tag{9.3.4}$$

3. spontaneous emission

$$\text{Rate} = k N_1 = N_1/\tau \tag{9.3.5}$$

where τ is the natural lifetime, i.e., the average time required for the molecules to decay in the absence of a large photon background. The light intensity which is required for both absorption (1) and stimulated emission (2) is included in the rate constant W.

The rate constant k_e' is the same for transitions from both the ground and excited states. The only real difference between the excited and ground states in the presence of a large photon intensity is the population of each of these states. Under low-intensity conditions, the rate of excitation will always be much larger than the rate of stimulated emission because N_0 is so much larger than N_1. The kinetics will change as these two populations approach each other.

Since the absorption and stimulated emission rates depend on both the light intensity and the state populations, this system has no forcing function comparable to those used in previous sections. Since the ground state is included, the system conserves mass. However, it is maintained away from equilibrium by the light intensity I, which appears as a factor in two of the rate processes; without an intensity-induced absorption, only level N_0 is possible. The rate expressions are

$$dN_0/dt = -WN_0 + WN_1 + kN_1 \tag{9.3.6}$$

$$dN_1/dt = +WN_0 - WN_1 - kN_1 \tag{9.3.7}$$

The matrix equation for these two populations is

$$d/dt \begin{pmatrix} N_0 \\ N_1 \end{pmatrix} = \begin{pmatrix} -W & W+k \\ W & -W-k \end{pmatrix} \begin{pmatrix} N_0 \\ N_1 \end{pmatrix} \tag{9.3.8}$$

The columns sum to zero for this conservative matrix, so one eigenvalue is 0 while the second is $-2W-k$. The system is not a real equilibrium system since

W depends on the intensity I. For small I, $W \ll k$ and the system would evolve into the two differential equations

$$dN_0/dt = -J \tag{9.3.9}$$

$$dN_1/dt = +J - kN_1 \tag{9.3.10}$$

which were developed in Section 9.1.

Since the nonzero eigenvalue will appear as an exponential decay constant in the solution of Equation 9.3.8, it is eliminated from the stationary-state solution of this equation. The solution is

$$|N(\mathrm{ss})\rangle = \exp(-0t)\, \mathbf{P}(0)\, |N(0)\rangle \tag{9.3.11}$$

with

$$\mathbf{P}(0) = [\mathbf{K} - (-2W - k)\mathbf{I}]/(0 + 2w + k)$$

$$= \begin{pmatrix} W+k & W \\ W & W \end{pmatrix} (2W + k)^{-1} \tag{9.3.12}$$

Since all the molecules are in the ground state at time zero, the initial condition concentration vector has components

$$\begin{pmatrix} N_t \\ 0 \end{pmatrix} \tag{9.3.13}$$

where N_t is the total population of molecules in the system.

The steady-state populations are determined from Equation 9.3.11 as

$$(2W + k)^{-1} \begin{pmatrix} W+k & W \\ W & W \end{pmatrix} \begin{pmatrix} N_t \\ 0 \end{pmatrix} \tag{9.3.14}$$

to give

$$N_0 = [(W + k)/(2W + k)]\, N_t \tag{9.3.15}$$

$$N_1 = [W/(2W + k)]\, N_t \tag{9.3.16}$$

This result can also be obtained using graph theory (Problem 9.6). A single rate constant (W) leads to state 1 while the sum of rate constants ($W + k$) leads to the ground state.

This two-level system will function as a laser only if the population can be inverted, i.e., the population N_1 must exceed the population N_0. This will be impossible since the excited state is also depopulated via spontaneous emission with rate constant k. This pathway maintains N_1 below N_0 for any light intensity I and the system will not lase.

This kinetic restriction is illustrated by determining the difference

$$N_0 - N_1 \tag{9.3.17}$$

for the system. Substituting the stationary-state solutions gives

$$\begin{aligned} N_0 - N_1 &= [N_t/(2W+k)](W+k-W) \\ &= [k/(2W+k)]\, N_t \end{aligned} \tag{9.3.18}$$

The two populations could become equal only if $k \ll 2W$. There is no way that N_1 could ever exceed N_0.

What conditions are required to produce a situation where N_1 is greater than N_0? Under these conditions, the photons will induce stimulated emission from the excited molecules and the system will lase. This "inversion" can be created in systems containing additional absorption and emission states. A typical example using four levels is shown in Figure 9.5. Light pumps molecules from the ground state (0) to an excited level (3) as shown. The molecule may emit a photon and return to the ground state, but it may also decay to an intermediate level (2). The molecule undergoes a second transition to a second intermediate level (1) from which it may return to the ground state. A system of this type has distinct advantages over a two-level system. If level 2 is populated by transitions from level 3, then the population of level 2 molecules can increase. Moreover, level 1 is not the molecular ground state and its equilibrium population can be neglected. With the proper rate constants, the four-level system can produce a population inversion between levels 2 and 1. This population inversion can produce lasing with a photon energy equal to the energy difference between these two levels.

The lasing action of this system is illustrated with an idealized model in which all transitions except the $2 \rightarrow 1$ transition are irreversible, as illustrated in Figure 9.5. The $2 \rightarrow 1$ transition is reversible to demonstrate that the population inversion exists between two reversibly coupled levels. For a constant photon input for the transition from 0 to 3, the system will attain a stationary-state distribution of populations for the four levels which can be determined with graph theory.

The graphs for each of the four levels are illustrated in Figure 9.6. The population of level 2 requires two graphs of three vectors. The remaining levels

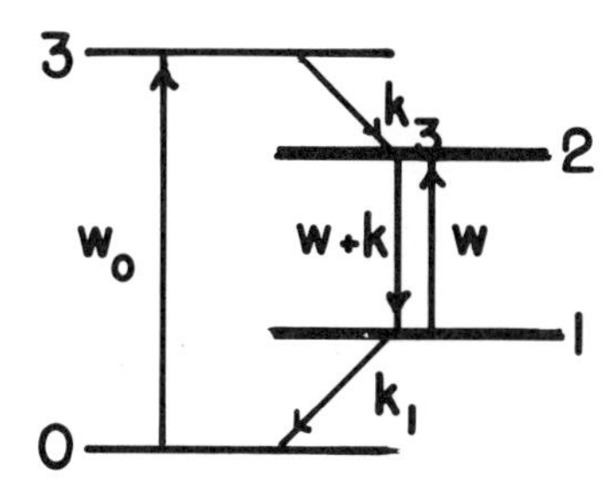

Figure 9.5. Scheme for a four-level laser system.

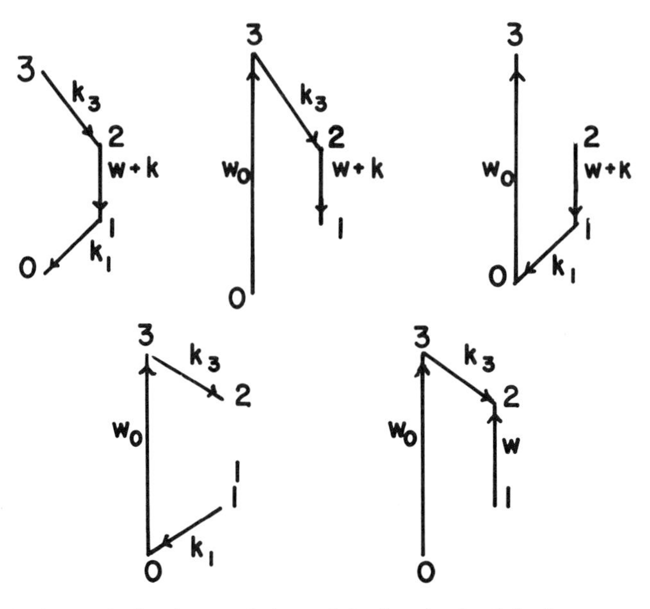

Figure 9.6. Kinetic graphs for the populations of the four levels of the laser system at stationary state.

require single graphs of three vectors. Using the rate constants listed in Figure 9.6, the fractional populations of each level are

$$N_3/N_t = (W+k)\,k_1\,W_0/S \tag{9.3.19}$$

$$N_2/N_t = (k_1\,W_0 k_3 + W_0 k_3\,W)/S \tag{9.3.20}$$

$$N_1/N_t = W_0 k_3(W+k)/S \tag{9.3.21}$$

$$N_0/N_t = k_3(W+k)\,k_1/S \tag{9.3.22}$$

where S is the sum of the four numerators in these equations:

$$S = (W+k)\,k_1\,W_0 + k_1\,W_0 k_3 + W_0 k_3\,W + W_0 k_3(W+k) + k_3(W+k)\,k_1 \tag{9.3.23}$$

If the difference between populations of levels 2 and 1 becomes positive, the system will lase. The difference is

$$\begin{aligned} N_2 - N_1 &= S^{-1}(k_1\,W_0 k_3 + W_0 k_3\,W - W_0 k_3\,W - W_0 k_3 k) \\ &= S^{-1}(W_0 k_3)(k_1 - k) \end{aligned} \tag{9.3.24}$$

The only difference term in the final expression is

$$(k_1 - k) \tag{9.3.25}$$

A large rate constant for depopulating level 1 will produce the required population inversion. This rate constant must be larger than the rate constant k for spontaneous depopulation of level 2 so that the population in level 2 exceeds the population of level 1.

9.4. Ionic Channels

Biological membranes have hydrophobic regions which make them excellent insulators for transverse ion flow. However, such membranes usually contain a wide variety of proteins as well, and some of these proteins have structures which produce a channel through the membrane. The interiors of such channels are compatible with polar moieties and will permit ions to traverse the membrane under applied gradients.

A typical channel might be arranged like the model in Figure 9.7. The polar interior of the channel contains two sites which permit ion binding. An ion enters and binds at one such site; it may then dissociate and return to the original solution or it may move to the second binding site from which it may escape into the other solution. The sequence of events is shown in Figure 9.8. In this case, only one ion can be in the channel at any time. Other models permit a state with ions at each binding site.

The ion flow through the channel can be described as a discrete state kinetic process. For example, transitions between the sites within the channel are assigned the rate constant k_m for the transition from state 1 to state 2 within the membrane and rate constant k_{-m} for the reverse transition (state 2 to state 1). The rate constants are defined as the product of a first-order rate constant, k', and a distance parameter, λ. The product $k'\lambda c = kc$, where c is the ion concentration, has the units of a flux.

The special character of this system lies in the interpretation of the bathing solutions on either side of the membrane which "feed" ions to these channel sites. To develop a closed kinetic system, baths 1 and 2 must be included as states. A transition from bath 1 to channel site 1 would then be proportional to the total concentration c_{b1} in bath 1

$$\text{Rate(in, 1)} = k_1 c_{b1} \tag{9.4.1}$$

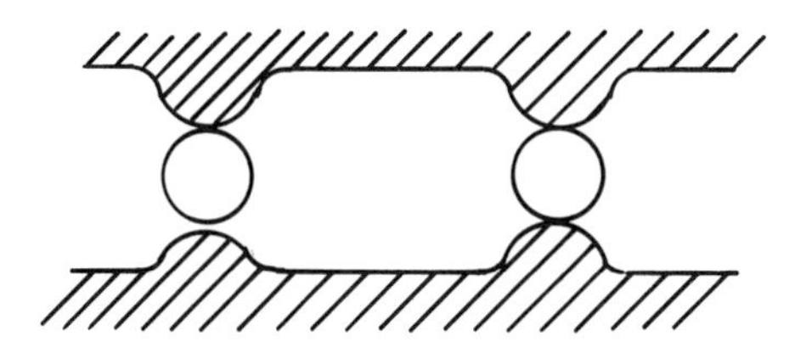

Figure 9.7. A membrane channel containing two internal ion binding sites.

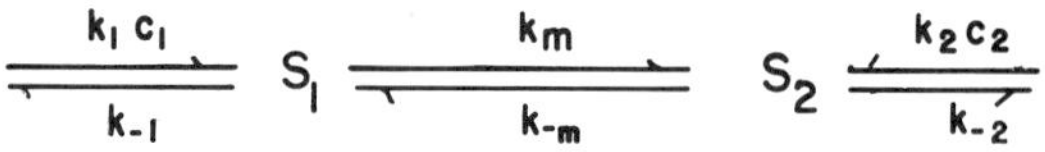

$$\underset{k_{-1}}{\overset{k_1 c_1}{\rightleftharpoons}} S_1 \underset{k_{-m}}{\overset{k_m}{\rightleftharpoons}} S_2 \underset{k_{-2}}{\overset{k_2 c_2}{\rightleftharpoons}}$$

Figure 9.8. Kinetic scheme for the transport of an ion through a two-state membrane channel molecule.

A particle which moved in the reverse direction (site 1 to bath 1) would have a rate proportional to the number of ions on membrane site 1,

$$k_{-1} c_1 \tag{9.4.2}$$

On the opposite side of the membrane, the rate from bath 2 into the membrane is

$$\text{Rate(in, 2)} = k_2 c_{b2} \tag{9.4.3}$$

where c_{b2} is the bath 2 concentration. The reverse rate is proportional to the concentration of channels with ions at site 2,

$$k_{-2} c_2 \tag{9.4.4}$$

Although the system can be prepared with each of the four states, the total number of ions in the baths will always be considerably larger than the number of ions within the membrane channels at any given time. On the time scale of the majority of such membrane channel systems, the bath concentrations will change very little. The first-order process in c_{bi} becomes a pseudo-zero-order process, i.e., a constant rate of ion input. By analogy with other kinetic systems, the constant input flux from bath 1 to channel site 1 is defined as

$$j_1 = k_1 c_{b1} = \text{constant} \tag{9.4.5}$$

while the constant input from bath 2 to site 2 is

$$j_2 = k_2 c_{b2} \tag{9.4.6}$$

The two constant fluxes generate a driven channel membrane system which parallels the excited state photochemical systems discussed earlier. The channel sites are populated by the two input fluxes. If these fluxes are eliminated, i.e., ions are removed from the bathing solutions, then the populations of channel ions will also decay to zero. The kinetics for the model of Figure 9.7 involve only the two membrane sites and the rate expressions for these sites are

$$dc_1/dt = -(k_m + k_{-1})\, c_1 + k_{-m} c_2 + j_1 \tag{9.4.7}$$

$$dc_2/dt = +k_m c_1 - (k_{-m} + k_{-2})\, c_2 + j_2 \tag{9.4.8}$$

The relevant rates are illustrated in Figure 9.8.

The matrix equation format parallels that observed for excited photochemical states:

$$d/dt\begin{pmatrix} c_1 \\ c_2 \end{pmatrix} = -\begin{pmatrix} +(k_m+k_{-1}) & -k_{-m} \\ -k_m & +(k_{-m}+k_{-2}) \end{pmatrix}\begin{pmatrix} c_1 \\ c_2 \end{pmatrix} + \begin{pmatrix} j_1 \\ j_2 \end{pmatrix}$$

$$d\,|c(t)\rangle/dt = -\mathbf{K}\,|c\rangle + |J\rangle \tag{9.4.9}$$

where

$$|c\rangle = \begin{pmatrix} c_1 \\ c_2 \end{pmatrix} \qquad |J\rangle = \begin{pmatrix} j_1 \\ j_2 \end{pmatrix} \tag{9.4.10}$$

$$\mathbf{K} = \begin{pmatrix} +(k_m+k_{-1}) & -k_{-m} \\ -k_m & +(k_{-m}+k_{-2}) \end{pmatrix} \tag{9.4.11}$$

If the two channel states have populations described as the vector

$$|c_0\rangle = \begin{pmatrix} c_1(0) \\ c_2(0) \end{pmatrix} \tag{9.4.12}$$

the formal solution to the equation will be

$$|c(t)\rangle = \exp(-\mathbf{K}t)\,|c_0\rangle + [\mathbf{I} - \exp(-\mathbf{K}t)]\,\mathbf{K}^{-1}\,|J\rangle \tag{9.4.13}$$

For constant bath concentrations, i.e., constant $|J\rangle$, this equation will evolve rapidly to the steady state:

$$|c(\text{ss})\rangle = \mathbf{K}^{-1}\,|J\rangle \tag{9.4.14}$$

To study the properties of this equation for a channel protein, consider the simple example in which all the rate constants are equal to 1, i.e.,

$$k_m = k_{-m} = k_{-1} = k_{-2} = 1 \tag{9.4.15}$$

The **K** matrix is

$$\mathbf{K} = \begin{pmatrix} 2 & -1 \\ -1 & 2 \end{pmatrix} \tag{9.4.16}$$

with characteristic equation

$$3 - 4\lambda + \lambda^2 = 0 \tag{9.4.17}$$

and eigenvalues

$$\lambda = 1 \qquad \lambda = 3 \tag{9.4.18}$$

The projection operators are

$$\mathbf{P}(1)=\left(\frac{1}{2}\right)\begin{pmatrix}1 & 1\\ 1 & 1\end{pmatrix} \tag{9.4.19}$$

$$\mathbf{P}(3)=\left(\frac{1}{2}\right)\begin{pmatrix}1 & -1\\ -1 & 1\end{pmatrix} \tag{9.4.20}$$

The steady-state solution is

$$\begin{pmatrix}c_1\\ c_2\end{pmatrix}=\left(\frac{1}{2}\right)\begin{pmatrix}1 & 1\\ 1 & 1\end{pmatrix}\begin{pmatrix}j_1\\ j_2\end{pmatrix}(1)^{-1}+\left(\frac{1}{2}\right)\begin{pmatrix}1 & -1\\ -1 & 1\end{pmatrix}\begin{pmatrix}j_1\\ j_2\end{pmatrix}(3)^{-1} \tag{9.4.21}$$

Since both matrices operate on the same vector, the terms can be combined to give a single matrix operation:

$$\begin{pmatrix}c_1\\ c_2\end{pmatrix}=\begin{pmatrix}\frac{1}{2}+\frac{1}{6} & \frac{1}{2}-\frac{1}{6}\\ \frac{1}{2}-\frac{1}{6} & \frac{1}{2}+\frac{1}{6}\end{pmatrix}\begin{pmatrix}j_1\\ j_2\end{pmatrix}$$

$$=\begin{pmatrix}\frac{2}{3} & \frac{1}{3}\\ \frac{1}{3} & \frac{2}{3}\end{pmatrix}\begin{pmatrix}j_1\\ j_2\end{pmatrix} \tag{9.4.22}$$

The stationary-state concentrations of channels with ions at sites 1 and 2 are, respectively,

$$c_1=\frac{2}{3}j_1+\frac{1}{3}j_2 \tag{9.4.23}$$

$$c_2=\frac{1}{3}j_1+\frac{2}{3}j_2 \tag{9.4.24}$$

The concentrations at steady state are apportioned in a consistent manner; a flux from bath 1 is twice as effective in populating site 1 as a flux from bath 2. An ion entering from bath 2 has only half the chance of populating site 1 because, after it reaches site 2, it may either move to site 1 or it may move back to bath 2. Only half the particles which reach site 2 actually reach site 1. Note also that contributions from each input flux add to give the total steady-state concentration of each state.

In the study of electronic excited states, it was necessary to find some property proportional to the steady-state concentration which could be observed experimentally. For excited states, the steady-state concentrations defined an observable fluorescence intensity. For channel transport, the observable is either the net flow of material through the channel, or, if the permeant species are charged, a net membrane current. The net permeant flow is related to the state concentrations.

The stationary-state concentrations of channel molecules with ions at sites 1 and 2 are known as functions of the rate constants and input fluxes. The net flux can be defined using these concentrations. Some ions will be entering channels while others are returning to the bath. For example, consider the flux from state 2 to bath 2. There are two unidirectional flows involved. Particles will leave state 2 for bath 2 with a rate

$$k_{-2}c_2 = (1)\, c_2 \tag{9.4.25}$$

using the k_i of Figure 9.8. However, while particles flow from membrane to the bath, other particles are leaving bath 2 and entering site 2 of the membrane. This rate is just j_2. The net rate from membrane site 2 to bath 2 is

$$j_{\text{net}} = k_{-2}c_2 - j_2 \tag{9.4.26}$$

Inserting the relevant rate constants and the steady-state concentration c_2 gives

$$\begin{aligned} j_{\text{net}} &= (1)(j_1/3 + 2j_2/3) - j_2 \\ &= j_1/3 - j_2/3 = \frac{1}{3}(j_1 - j_2) \end{aligned} \tag{9.4.27}$$

The net flux is proportional to the difference of the input fluxes,

$$j_1 - j_2 \tag{9.4.28}$$

The two sites within the membrane give a measure of the effectiveness of transport. In this case, they lead to a permeability coefficient of $\frac{1}{3}$.

The net flux can also be determined using bath 1 and site 1. In this case, the net flux is

$$j_{\text{net}} = j_1 - k_{-1}c_1 \tag{9.4.29}$$

Substituting the relevant parameters gives

$$\begin{aligned} j_{\text{net}} &= j_1 - (1)(2j_1/3 + j_2/3) \\ &= \frac{1}{3}(j_1 - j_2) \end{aligned} \tag{9.4.30}$$

The result is identical to that observed for net flow into bath 2. Since the system is at steady state, particles cannot build up on the protein sites with time. Every particle which enters from bath 1 must produce a net loss of one from the membrane so that the intramembrane concentrations remain constant.

The treatment of the special two-state membrane with rate constants of unity is easily extended to a general two-state channel. The **K** matrix in this case is

$$\begin{pmatrix} (k_m + k_{-1}) & -k_{-m} \\ -k_m & (k_{-m} + k_{-2}) \end{pmatrix} \tag{9.4.31}$$

The characteristic equation can be determined from the trace and determinant of this 2×2 matrix as

$$\lambda^2 - (k_m + k_{-1} + k_{-m} + k_{-2})\,\lambda + k_m k_{-2} + k_{-1} k_{-m} + k_{-1} k_{-2} = 0 \tag{9.4.32}$$

It is convenient to determine the matrix inverse using the Hamilton–Cayley theorem. For a characteristic equation,

$$\mathbf{K}^2 + a\mathbf{K} + b\mathbf{I} = 0 \tag{9.4.33}$$

the inverse is

$$\begin{aligned} \mathbf{K}^{-1} &= -(1/b)(\mathbf{K} + a\mathbf{I}) \\ &= -(1/b)\begin{pmatrix} -(k_{-m} + k_{-2}) & -k_{-m} \\ -k_m & -(k_m + k_{-1}) \end{pmatrix} \end{aligned} \tag{9.4.34}$$

and the steady-state concentrations are

$$\begin{aligned} |c\rangle &= \mathbf{K}^{-1}\,|J\rangle \\ &= (1/b)\begin{pmatrix} (k_{-m} + k_{-2})\,j_1 + k_{-m} j_2 \\ k_m j_1 + (k_m + k_{-1})\,j_2 \end{pmatrix} \end{aligned} \tag{9.4.35}$$

with

$$b = k_m k_{-2} + k_{-1} k_{-m} + k_{-1} k_{-2} \tag{9.4.36}$$

The net flux from channel site 2 to bath 2 is

$$\begin{aligned} j_{\text{net}} &= k_{-2} c_2 - j_2 \\ &= (1/b)(k_{-2} k_m j_1 + k_{-2} k_m j_2 + k_{-2} k_{-1} j_2 - j_1 b) \\ &= (k_{-2} k_m j_1 - k_{-1} k_{-m} j_2)/b \end{aligned} \tag{9.4.37}$$

In this case, the net flux cannot be reduced to a simple difference of the input fluxes.

The net flux equation can be rearranged by dividing both numerator and denominator by $k_{-2} k_m$. The equation becomes

$$j_{\text{net}} = \frac{j_1 - [(k_{-1}/k_{-2})(k_{-m}/k_m)]\,j_2}{1 + (k_{-1}/k_m) + (k_{-1}/k_{-2})(k_{-m}/k_m)} \tag{9.4.38}$$

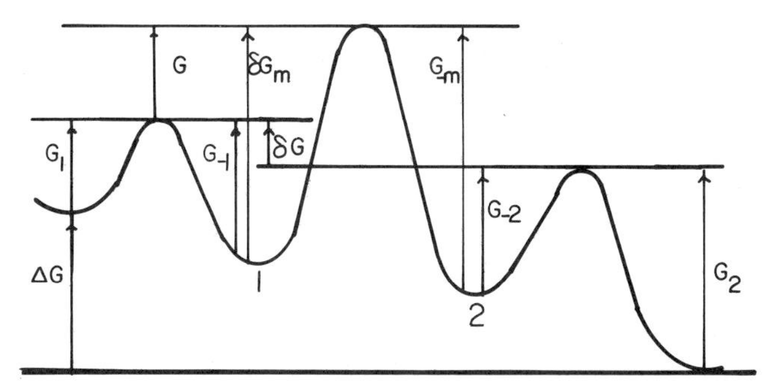

Figure 9.9. Energy barrier model for the two-state membrane transport model, showing the relations between the activation free energies and the transport rate constants.

Each of the rate constants in this expression can be related to a free energy of activation using Eyring rate theory,

$$k_i = (k_b T/h) \exp(-\Delta G_i^\ddagger/kT) \tag{9.4.39}$$

where $\Delta G_i^\ddagger$ is the free energy difference between the activated complex and the reactant state. The free energies for the two-site model are shown in Figure 9.9. The rate constant expression in the numerator becomes

$$(k_{-1}/k_{-2})(k_{-m}/k_m) = \exp(-\Delta G/kT) \tag{9.4.40}$$

where ΔG is the free energy difference between baths 1 and 2. The terms in the denominator become the free energy differences between the first energy maximum and the subsequent maxima as shown. The free energy expression for net flux is

$$j_{\text{net}} = [j_1 - \exp(-\Delta G/RT)\, j_2]/[1 + \exp(\delta G_1/RT) + \exp(\delta G_2/RT)] \tag{9.4.41}$$

9.5. Equilibrium and Stationary-State Properties

Because channel permeation models generally have a small number of states, they are ideal to use for a comparison of equilibrium and stationary-state properties. This section presents the differences between equilibrium and stationary-state distributions and the effects these differences will have on the thermodynamic properties observed for these systems.

The analysis utilizes an ensemble of membrane channels which contain two binding sites and which permit only a single ion in each channel. When the two bathing solutions have equal concentrations and the sites are identical, the number of channels with an ion at site 1 would be equal to the number of channels with an ion at site 2. If the channels were restricted to the two states with ions bound at site 1 or site 2, i.e., no empty or doubly occupied states, the equilibrium fractions of each of the states would form a two-component vector,

$$\begin{pmatrix} f_1 \\ f_2 \end{pmatrix} = \begin{pmatrix} \frac{1}{2} \\ \frac{1}{2} \end{pmatrix} \tag{9.5.1}$$

For a cation-permeable channel, an applied transmembrane potential difference will change the relative energies of the two sites. This, in turn, will alter the relative populations of channels with ions bound at the respective sites. The transmembrane potential will produce a potential difference between the two binding sites. Since the reference potential can be selected arbitrarily, it is convenient to assign site 1 zero potential. The relative potential of site 2 is then V.

In an equilibrium system, the probabilities p_1 and p_2 of finding the ion bound at site 1 or 2, respectively, would be proportional to the Boltzmann factors for each site:

$$p_1 \propto \exp(-zFV_1/RT) = \exp(0) = 1 \tag{9.5.2}$$

$$p_2 \propto \exp(-zFV_2/RT) = \exp(-zFV/RT) = a \tag{9.5.3}$$

The equilibrium fractions of each state are

$$\begin{pmatrix} 1 \\ a \end{pmatrix} (1+a)^{-1} \tag{9.5.4}$$

where the vector is normalized with the partition function

$$q = 1 + a \tag{9.5.5}$$

which is the sum of Boltzmann factors for the two possible states.

Although these results are developed in the framework of statistical mechanics, there is a physical difficulty with this approach. If both bath concentrations of ions are equal and the transmembrane potential is applied, this potential will drive ions through the membrane channels and the system will be nonequilibrium. Equilibrium is possible if the bath concentrations are altered to generate a chemical potential which opposes the electrical potential.

This equilibrium balance is clarified by forming the ratio of Equations 9.5.2 and 9.5.3 to give

$$c_2/c_1 = a/1 = \exp(-zFV/RT) \tag{9.5.6}$$

This result could be obtained directly by equating the electrochemical potentials μ_i of each site (Problem 9.9):

$$\mu_i = RT \ln c_i + z_i FV_i \tag{9.5.7}$$

These results must now be extended to the kinetics of the system. If the system had state populations proportional to 1 and a at equilibrium, this would imply that the rate constants for the transitions would also be different. The principle of detailed balance at equilibrium would require

$$k_{21} f_1 = k_{12} f_2 \tag{9.5.8}$$

where k_{21} is the rate constant for the transition from site 1 to site 2 and k_{12} is the rate constant for the reverse transition. Since the equilibrium fractions f_1 and f_2 are proportional to 1 and a at equilibrium, the condition of detailed balance gives

$$k_{21}(1) = k_{12}(a) \tag{9.5.9}$$

Since the sites are identical, transitions would be described by a common rate constant k if the populations of each site were equal. To compensate for the different populations and maintain detailed balance, the rate constants must be defined as

$$k_{21} = ka \tag{9.5.10}$$

$$k_{12} = k \tag{9.5.11}$$

as shown in Figure 9.10. The rate constant matrix $\mathbf{K}$ is

$$\mathbf{K} = k\begin{pmatrix} -a & 1 \\ a & -1 \end{pmatrix} \tag{9.5.12}$$

which has eigenvalues

$$\lambda = 0 \qquad \lambda = -k(a+1) \tag{9.5.13}$$

and normalized eigenvectors

$$|0\rangle = \begin{pmatrix} 1 \\ a \end{pmatrix}(1+a)^{-1} \qquad |-k(1+a)\rangle = \begin{pmatrix} 1 \\ -1 \end{pmatrix}(1+a)^{-1} \tag{9.5.14}$$

The normalization for each eigenvector is just the equilibrium partition function even though the system is not at equilibrium.

The kinetic results above are observed when there is some potential difference V between the two ion binding sites. These results simply reflect the kinetics between the two sites so that detailed balance is maintained. However,

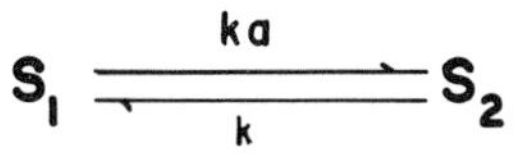

$$S_1 \underset{k}{\overset{ka}{\rightleftharpoons}} S_2$$

Figure 9.10. The rate constants for two intrachannel binding sites which obey the condition of detailed balance when the potentials at each site differ.

the application of a potential can also alter the rates at which the ions are transferred from the baths to these sites. If the transported ions are exclusively cations and the applied potential V' is negative, cations will flow from bath 1 to bath 2 through the channels. Since $a > 1$ in this case, the transition rate constant from state 1 to state 2 will be larger than the reverse rate constant. The full transmembrane potential is distributed between the transition from bath 1 to site 1, the transition from site 1 to site 2, and the final transition from site 2 to bath 2.

In Section 9.4, ions could enter the channel from either bath and the magnitude of these input fluxes determined the concentrations at each site. In this case, the concentrations were maintained by the flux, and a direct comparison of equilibrium and stationary-state systems was difficult. The comparison is possible for a stationary-state system which conserves mass. The same **K** matrix is used for both the equilibrium and the stationary states. To conserve mass in the driven system, the flux vector must introduce a net ion input of zero. A flux j populates site 1 from bath 1. At the same time, an equal flux of ions leaves site 2 for bath 2. This output flux is $-j$ and the driving vector is

$$\begin{pmatrix} j \\ -j \end{pmatrix} \tag{9.5.15}$$

This vector is orthogonal to the $|0\rangle$ eigenvector,

$$(1 \quad 1)\begin{pmatrix} j \\ -j \end{pmatrix} = 0 \tag{9.5.16}$$

refecting the fact that the flux vector can add no new ions to the membrane system, i.e., mass is conserved while ions flow through the channel. However, the flux will increase the fractional population of site 1 relative to site 2, i.e., the flux will drive the system from equilibrium to a steady-state distribution which can then be compared with equilibrium.

The stationary-state concentrations are determined by dividing the system into two parts. Thc $|0\rangle$ equilibrium vector is not perturbed by $|j\rangle$. The equilibrium component is always present, even when there is no flux:

$$|c^{\text{eq}}\rangle = \begin{pmatrix} 1 \\ a \end{pmatrix} (1 + a)^{-1} \tag{9.5.17}$$

The remaining projection operators (one in this case) will produce a finite concentration perturbation when j is finite.

The stationary-state perturbation is found by expanding the solution

$$|c(\text{ss})\rangle = \mathbf{K}^{-1}\,|j\rangle \tag{9.5.18}$$

in all its projection operators except $\mathbf{P}(\lambda=0)$. The two-site model gives

$$\mathbf{P}[-k(1+a)] = (\mathbf{K} - 0\mathbf{I})/(-1-a-0)$$
$$= \begin{pmatrix} a & -1 \\ -a & 1 \end{pmatrix} (1+a)^{-1} \tag{9.5.19}$$

and the perturbation is

$$\mathbf{K}^{-1}\,|J\rangle = (1/\lambda)\,\mathbf{P}\,|J\rangle$$
$$= [1/k(1+a)^2] \begin{pmatrix} ja+j \\ -ja-j \end{pmatrix}$$
$$= \begin{pmatrix} j \\ -j \end{pmatrix} [k(1+a)]^{-1} \tag{9.5.20}$$

The population of site 1 is increased by a factor proportional to j while the site 2 population is decreased by exactly the same factor. This proportionality factor contains the ratio of the net flux j to the rate constant k and the equilibrium partition function:

$$q = 1 + a \tag{9.5.21}$$

Since the flux j represents the number of ions entering a unit area of membrane or fixed number of channels per second and k has the units of s^{-1}, the ratio has units of ions per unit area. The perturbation from equilibrium depends on how rapidly ions are added and removed from the channels relative to their redistribution within the channel.

For a flux j, the fractional populations of the states are the sum of the equilibrium and perturbation eigenvectors:

$$\begin{pmatrix} f_1 \\ f_2 \end{pmatrix} = \left[\begin{pmatrix} 1 \\ a \end{pmatrix} + \begin{pmatrix} j/k \\ -j/k \end{pmatrix} \right] (1+a)^{-1} \tag{9.5.22}$$

The equilibrium partition function $(1+a)$ is common to both the equilibrium and perturbation terms since the perturbation retains mass conservation.

As the number of membrane states increases, the number of terms for the stationary-state perturbation increases accordingly. There are $N-1$ perturbation terms for an N-state system. Each term is proportional to a ratio

$$j/\lambda_i \tag{9.5.23}$$

Since the smaller eigenvalues, λ_i, will produce the largest ratios, j/λ_i, the perturbation terms can be ordered. The perturbation can often be effectively approximated by the term generated by the smallest nonzero eigenvalue.

For conservative stationary-state systems, the state populations can be used to define stationary-state thermodynamic properties. These thermodynamic properties will depend on the ratio j/k or j/λ_i. The approach is illustrated for the two-state model but is easily generalized (Problem 9.11).

The average energy of the system at equilibrium is calculated as the sum of Boltzmann probabilities for each site multiplied by the energy associated with that site:

$$E = f_1 E_1 + f_2 E_2 + \cdots \tag{9.5.24}$$

The stationary-state fractional populations or probabilities differ from the equilibrium probabilities, but they can be used in the same way. The energy of each membrane site can be expressed as the electrical potential V or the energy associated with this potential:

$$E = zeV \tag{9.5.25}$$

The potential at site 1 is 0 while the potential at site 2 is V so the average potential is

$$\begin{aligned}
&(1+a)^{-1}\{[1+(j/k)](0)+[a-(j/k)]V\}\\
&\quad = \frac{(1)(0)+a(V)}{1+a} + \frac{(j/k)(0)-(j/k)(V)}{1+a}\\
&\quad = aV/(1+a) - (j/k)V/(1+a) = V^{\text{eq}} - V^{s}
\end{aligned} \tag{9.5.26}$$

The first term is the average equilibrium potential while the second term is the change in potential due to the driving flux j.

The perturbation reduces the average potential since the population of the $V=0$ state (site 1) is increased relative to the $V=V$ (site 2) state. The change in energy is directly proportional to the stationary-state ratio (j/k).

The entropy of the system in steady state is determined using the information theory definition of entropy:

$$S = -k_b[f_1 \ln(f_1) + f_2 \ln(f_2)] \tag{9.5.27}$$

The stationary-state fractional populations define the entropy of the system:

$$\begin{aligned}
-S/k_b = {} & [(1+j/k)/(1+a)]\ln[(1+j/k)/(1+a)]\\
& + [(a-j/k)/(1+a)]\ln[(a-j/k)/(1+a)]
\end{aligned} \tag{9.5.28}$$

The entropy expression can be divided into two parts. Selecting the logarithm of the partition function, $q = 1+a$, from each term gives

$$\begin{aligned}
&-\ln(1+a)[(1+j/k)+(a-j/k)]/1+a\\
&\quad = -\ln(1+a)(1+a)/(1+a) = -\ln(1+a) = -\ln q
\end{aligned} \tag{9.5.29}$$

This term involves only the equilibrium partition function.

The remaining terms in the entropy expression are

$$(1+a)^{-1}\,[(1+j/k)\ln(1+j/k)+(a-j/k)\ln(a-j/k)] \qquad (9.5.30)$$

This expression can be further divided into separate terms for equilibrium and stationary states; the logarithms are expanded as

$$\ln(1+j/k)= +(j/k)+(j/k)^2-\cdots \qquad (9.5.31)$$

$$\ln(1-j/k)= -(j/k)+(j/k)^2+\cdots \qquad (9.5.32)$$

The expression becomes

$$\begin{aligned}&(1+a)^{-1}\,((1+j/k)[j/k+(j/k)^2]+(a-j/k)\{\ln a+[-j/ka+(j/ka)^2]\})\\ &\quad=(1+a)^{-1}\,[a\ln a-(j/k)\ln a+(j/k)^2\,(1+1/a)+\cdots]\end{aligned} \qquad (9.5.33)$$

The entropy is

$$-S/k_b= -\ln q+q^{-1}[a\ln a-(j/k)\ln a+(j/k)^2\,(1+1/a)+\cdots] \qquad (9.5.34)$$

The first two terms in the series are the equilibrium entropy,

$$\begin{aligned}S^{\text{eq}}/k_b&=[1/(1+a)]\ln[1/(1+a)]+[a/(1+a)]\ln[a/(1+a)]\\ &=[1/(1+a)]\ln 1+[a/(1+a)]\ln a-\ln(1+a)\\ &=q^{-1}a\ln a-\ln q\end{aligned} \qquad (9.5.35)$$

The third term in the summation,

$$-(j/k)\ln(a)/(1+a) \qquad (9.5.36)$$

is related to the stationary-state energy perturbation. Since

$$\begin{aligned}a&=\exp(-zeV/kT)\\ \ln a&= -zeV/kT\end{aligned} \qquad (9.5.37)$$

Equation 9.5.36 becomes

$$-(j/k)(-zeV/kT)/(1+a) \qquad (9.5.38)$$

Comparing this with Equation 9.5.25 gives

$$E^d/kT=(ze/kT)\,V^d \qquad (9.5.39)$$

This term is equal to the change in internal energy when the system is in the stationary state.

The entropy can now be written in terms of other thermodynamic properties:

$$
\begin{aligned}
-S/k &= a \ln(a)/1 + a - \ln(1+a) - (j/k)\ln(a) + (j/k)^2 (1 + 1/a) + \cdots \\
&= -E^{\text{eq}}/kT + A^{\text{eq}}/kT - E^d/kT - A^d/kT
\end{aligned} \tag{9.5.40}
$$

If the entropy is divided into equilibrium and perturbed components,

$$S/k = S^{\text{eq}}/k + S^d/k \tag{9.5.41}$$

the equilibrium terms can be collected to give

$$A^{\text{eq}} = E^{\text{eq}} - TS^{\text{eq}} \tag{9.5.42}$$

The remaining terms give the supplemental definition,

$$A^d = E^d - TS^d \tag{9.5.43}$$

with

$$A^d/k_b T = -(j/k)^2 (1 + 1/a) + \cdots \tag{9.5.44}$$

$$E^d/k_b T = (j/k) \ln a \tag{9.5.45}$$

The energy input due to the flux j increases the energy as the first power of (j/k). However, the free energy increases as the second power of (j/k). The ordering effect of the flux through the system places a proportionately greater fraction of its energy into the form of available free energy.

9.6. Fluctuations about Equilibrium

The kinetic systems which have been described in Chapter 6 and in the previous sections of this chapter ultimately reduce to a single temporal curve for each distinct component in the kinetic scheme. For example, the reaction

$$\mathrm{A} \rightarrow \mathrm{B}$$

with $\mathrm{A}(0) = A_0$, $B(0) = 0$ will decay to zero as

$$A(t) = A_0 \exp(-kt) \tag{9.6.1}$$

This equation is deterministic; at time t, some exact number of unreacted A molecules remain. In reality, reaction is a random process and different concentrations of A may exist at time t if the reaction is repeated several times. The concentration of A at time t will form a distribution, as shown in Figure 9.11. There will be a similar distribution of A for each time t. A single reaction will then show deviations about the deterministic solution, as shown in Figure 9.12.

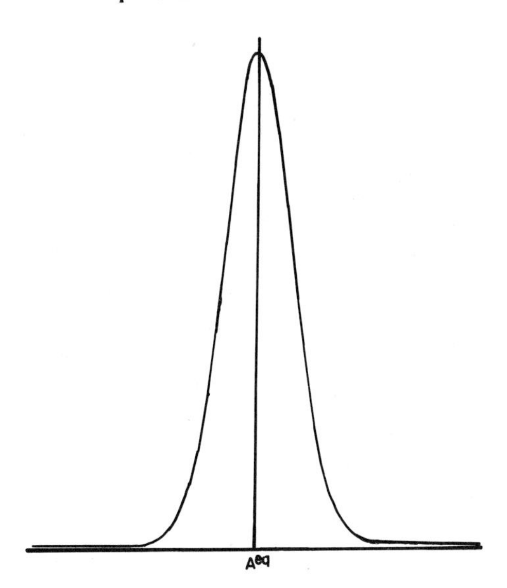

Figure 9.11. The distribution of A molecules about the equilibrium population of A due to fluctuations.

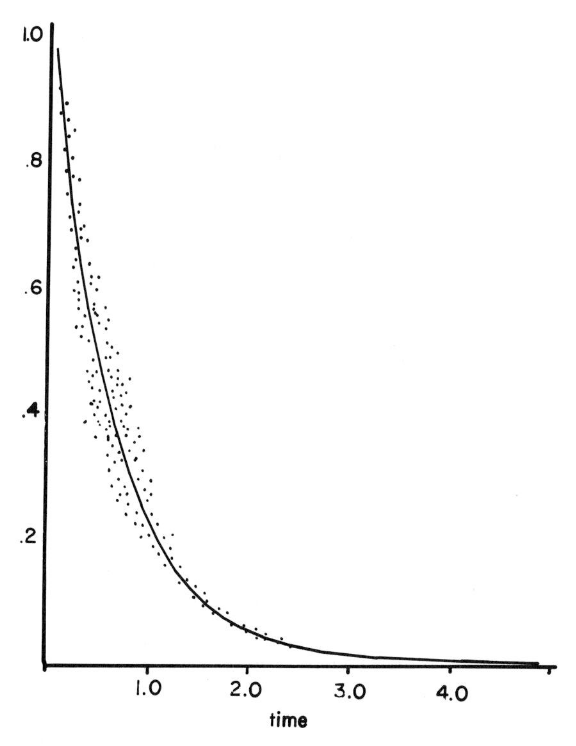

Figure 9.12. Fluctuations in a first-order reaction system. The solid line is the deterministic equation determined from repetitions of the experiment.

These deviations will exist at equilibrium as well if the reaction is reversible. Since the deterministic value of A^{eq} is constant, equilibrium systems provide a good starting point for the study of these fluctuations.

The nature of fluctuations about equilibrium is first demonstrated with the short-circuited capacitor of Figure 9.13. Because the plates are short-circuited, the potential difference between the plates must be zero. However, if the potential is monitored in time, the potential will deviate from zero and small fluctuations of potential are observed. The electrons in the circuit move randomly. If an excess should build up on the left plate, the potential on this plate will become negative relative to the remaining plate. This transient is succeeded by other fluctuations which make the plate either positive or negative. Over an extended time period, these fluctuations will average to zero.

The probability of observing a given fluctuation will be proportional to the total energy associated with this fluctuation. For a potential excursion V from equilibrium ($V=0$), the total energy stored in the capacitor is

$$E = CV^2/2 \tag{9.6.2}$$

and the Boltzmann probability is proportional to

$$\exp(-CV^2/2kT) \tag{9.6.3}$$

The partition function is determined as an integral over all possible voltages, i.e.,

$$q = \int_{-\infty}^{\infty} \exp(-CV^2/2kT)\, dV \tag{9.6.4}$$

Since

$$\int_{-\infty}^{\infty} \exp(-ax^2)\, dx = \sqrt{\pi/a} \tag{9.6.5}$$

the partition function is

$$q = \sqrt{2\pi kT/C} \tag{9.6.6}$$

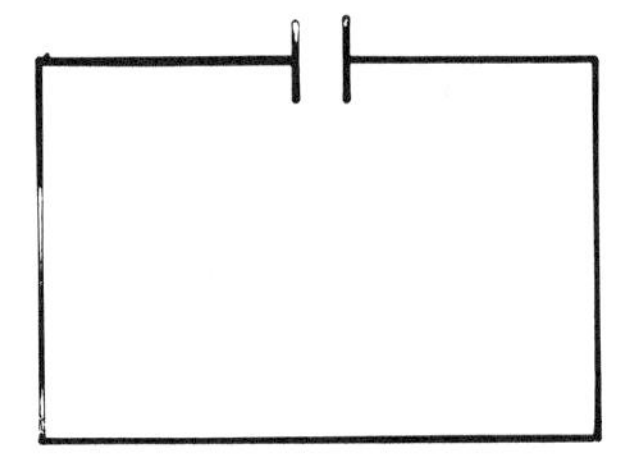

Figure 9.13. A short-circuited capacitor to illustrate voltage fluctuations due to random motions of the electrons.

and the Boltzmann probability for a voltage V is

$$\exp(-CV^2/2kT)/q = \exp(-CV^2/2kT)/\sqrt{2\pi kT/C} \tag{9.6.7}$$

The Boltzmann probability can be used to find a number of average values. For example, the average potential observed for this system is

$$\int_{-\infty}^{\infty} V \exp(-CV^2/2kT)\, dV/q = (2kT/Cq) \exp(-CV^2/2kT)\Big|_{-\infty}^{\infty} = 0 \tag{9.6.8}$$

Since positive and negative voltage fluctuations of the same magnitude are equally possible, the average of the fluctuations is zero.

Since deviations from equilibrium may be positive or negative, it is convenient to use the square of the deviations as the average value of interest. The function which must be averaged is then

$$(V - V^{\text{eq}})^2 \tag{9.6.9}$$

Since $V^{\text{eq}} = 0$ for this case, the average square of deviations from the average is

$$\int_{-\infty}^{\infty} V^2 \exp(-CV^2/2kT)\, dV/q \tag{9.6.10}$$

With the definition,

$$a = C/2kT \tag{9.6.11}$$

the numerator is related to a derivative of the denominator,

$$-dp/da = \int_{-\infty}^{\infty} V^2 \exp(-aV^2)\, dV \tag{9.6.12}$$

and the average square deviation is

$$\langle \Delta V^2 \rangle = -(dq/da)/q = -d(\ln q)/da \tag{9.6.13}$$

Since

$$q = \sqrt{2\pi kT/C} = \sqrt{\pi/a} \tag{9.6.14}$$

$$\ln q = \frac{1}{2} \ln \pi - \frac{1}{2} \ln a$$

$$= \text{constant} - \frac{1}{2} \ln a \tag{9.6.15}$$

Equation 9.6.13 becomes

$$\langle \Delta V^2 \rangle = -d(-\ln a^{1/2})/da = 1/2a = kT/C \tag{9.6.16}$$

The magnitude of fluctuations about equilibrium are larger for smaller capacitors. If the capacitor is smaller, it can be charged more easily and the fluctuations become more important. The factor kT reflects the fact that the energy of the electrons is responsible for producing the fluctuations.

Although the fluctuations about equilibrium for the capacitor were described by a function continuous in V, these voltage fluctuations are really caused by the accumulation of an excess of electrons on a plate. Thus, the equations could also be written for the number of discrete electrons present on the plate. Viewed from this perspective, the problem is equivalent to the problem of a simple isomerization,

$$A \rightleftharpoons B$$

The isomers are analogues of the capacitor plates. A reaction fluctuation which increases A at the expense of B is the analogue of a fluctuation which places excess electrons on the left plate of the capacitor.

To determine fluctuations in electrons in the capacitor, the potential associated with these fluctuations was used as the experimental variable. For the reaction system, the chemical potential for the isomerization reaction can be used as the energy variable (Problem 9.13). However, it is also possible to develop a discrete particle approach to fluctuations, and this is introduced for the isomerization of a set of N molecules.

For an isomerization in which $k_f = k_r$, the equilibrium numbers of A and B molecules will be $N/2$, i.e., the concentrations are equal at equilibrium. Two aspects of this equilibrium can be clarified by considering a set of 100 molecules. The equilibrium deviates from 50 A molecules by units. It may rise to 51 or 55 at the expense of B molecules but not to 52.2. In addition, the 50 molecules which appear as the A isomer do not remain as this isomer. If the 100 molecules are labeled, the set of A isomers might have labels 1–50 at one time, 1–25 and 76–100 at a second time, etc., because of the dynamics of the reacting system.

The number of particles will fluctuate about the equilibrium value of 50 A and 50 B molecules. The equilibrium probabilities are each 0.5. Using the indices 0 and 1 for A and B molecules, the equilibrium probabilities are

$$\begin{aligned} p_0 &= 0.5 \\ p_1 &= 0.5 \end{aligned} \tag{9.6.17}$$

Each fluctuation will create a new pair of probabilities which will reflect the ability to select a given isomer from the mixture during that time. At all times, each molecule must be in one isomeric form:

$$p_0 + p_1 = 1 \tag{9.6.18}$$

Equation 9.6.18 includes the possibility of selecting an A molecule or a B molecule by randomly choosing one molecule. The probability of choosing two A molecules in sequence is

$$p_0^2 \tag{9.6.19}$$

Consider the product

$$(p_0 + p_1)^2 = 1^2 = 1 \tag{9.6.20}$$

This expression is expanded to give

$$(p_0 + p_1)^2 = p_0^2 + 2p_0 p_1 + p_1^2 \tag{9.6.21}$$

Each of the terms in the sum corresponds to some joint arrangement of the two molecules. For example, p_0^2 corresponds to the case where both molecules are the A isomer. The factor 2 multiplying $p_0 p_1$ appears because there are two distinct ways which permit one A and one B: molecule 1 is A and molecule 2 is B or molecule 2 is A and molecule 1 is B.

The technique can be duplicated for arbitrary N. The expansion will always produce products of the probabilities and a binomial coefficient which gives the total number of molecular arrangements. The joint probability for three A isomers and one B isomer is

$$p_0^3 p_1^1 \tag{9.6.22}$$

and there are

$$4!/3!1! = 4 \tag{9.6.23}$$

different arrangements of the four molecules so that three are A isomers (Figure 9.14). For the general case in which there are N molecules and i of them are A molecules and $N - i$ are B molecules, the joint probability is

$$[N!/i!(N-i)!]\, p_0^i p_1^{N-i} \tag{9.6.24}$$

The binomial coefficient becomes increasingly important with increasing N. For this case, the number of distinct arrangements, i.e., the binomial coefficient, becomes extremely large for the 50–50 mixture. For 50 molecules, there are

$$50!/49!1! = 50 \tag{9.6.25}$$

1(A) 2(A) 3(A) 4(A)
1(B) 2(B) 3(B) 4(B)
1(C) 2(C) 3(C) 4(C)
1(D) 2(D) 3(D) 4(D)

Figure 9.14. Different molecular arrangements of four labeled molecules which yield three A isomers and one B isomer.

different arrangements for 49 A molecules and 1 B molecule. Each of the 50 labeled molecules could be the B isomer. By contrast, the 50–50 mixture has

$$50!/25!25! = 3 \times 10^{64}/2.4 \times 10^{50} = 1.25 \times 10^{14} \tag{9.6.26}$$

different arrangements. A sampling of the system would yield the 50–50 mixture most of the time since there are so many different ways to create this mixture. As N increases and most arrangements are 50–50 (equilibrium), the magnitude of fluctuations from this equilibrium will also decrease.

The system requires probabilities, p_i, for single molecules and joint probabilities. The probability p_1 is the probability of selecting a B molecule from the reaction vessel. The probability p_1^4 is the joint probability that all four molecules in the vessel happened to be B molecules when the sampling took place. Some samples might yield four B molecules; others might yield four A molecules. When all the samples are averaged, the equilibrium configuration of two A and two B molecules will result.

The averaging procedure using the joint probabilities can be illustrated for a system with four isomers. The probabilities for each configuration can be generated from the binomial expansion

$$(p_0 + p_1)^4 = p_0^4 + 4p_0^3 p_1^1 + 6p_0^2 p_1^2 + 4p_0^1 p_1^3 + p_1^4 \tag{9.6.27}$$

Each joint probability term in the sum represents the probability of finding some number of B molecules in the system. For example, the probability that all four molecules are B isomers is

$$p_0^4/(p_0 + p_1)^4 = p_0^4 \tag{9.6.28}$$

The average number of B molecules is determined by multiplying the number of B molecules for a given configuration by the probability for that configuration. For the four-molecule case,

$$\langle B \rangle = (0)\, p_0^4 + (1)\, 4p_0^3 p_1^1 + (2)\, 6p_0^2 p_1^2 + (3)\, 4p_0^1 p_1^3 + (4)\, p_0^4 \tag{9.6.29}$$

The equation is verified by substituting $p_0 = p_1 = 0.5$. The average value for the number of B molecules is

$$\langle B \rangle = (0)\left(\frac{1}{16}\right) + (1)\left(\frac{4}{16}\right) + (2)\left(\frac{6}{16}\right) + (3)\left(\frac{4}{16}\right) + (4)\left(\frac{1}{16}\right) = \frac{32}{16} = 3 \tag{9.6.30}$$

Since each molecule can be A or B with equal probability, half the molecules are B molecules.

To determine the average number of B molecules, the number of B molecules for each term is multiplied by the probability. This probability contains the number of B molecules as a power of p_1. A differentiation of each term with respect to p_1 generates the expression for average B. In practice, a dummy variable s is introduced as the differentiation variable. It appears as a factor with each p_1 in the equation. For example, the probability for four B molecules is

$$(p_1 s)^4/(p_0 + p_1 s)^4 \tag{9.6.31}$$

The dummy index can always be equated to 1 to restore the original equations. Differentiation of the numerator in this term with respect to s gives the factor of 4 required in the average value expression. The expression can be multiplied by s to generate the proper power of s.

The expression for the average value for four molecules is

$$s\, d/ds(p_0 + sp_1)^4/(p_0 + sp_1)^4 \Big|_{s=1} = s\, d(\ln q)/ds \Big|_{s=1} \tag{9.6.32}$$

where

$$q = (p_0 + sp_1)^4 \tag{9.6.33}$$

For N molecules, q is

$$q = (p_0 + sp_1)^N \tag{9.6.34}$$

The equation can be checked by determining $\langle B \rangle$ for $N = 4$:

$$\begin{aligned} \langle B \rangle &= s\, d[\ln(p_0 + p_1)^4]/ds \Big|_{s=1} \\ &= 4s(p_0 + sp_1)^{-1} p_1 \Big|_{s=1} \\ &= 4p_1 = 2 \end{aligned} \tag{9.6.35}$$

The variance is the average value of the deviations from equilibrium squared and serves as the measure of fluctuations. For the four-molecule system, the average value of B is 2. The probability for one B and three A would then be multiplied by the factor

$$(1-2)^2 \tag{9.6.36}$$

to determine its contribution to the variance. The total variance is

$$\begin{aligned} \text{var} = {} & (0-2)^2\, p_0^4 + (1-2)^2\, 4p_0^3 p_1 + (2-2)^2\, 6p_0^2 p_1^2 \\ & + (3-2)^2\, 4p_0^1 p_1^3 + (4-2)^2\, p_1^4 \end{aligned} \tag{9.6.37}$$

Substituting for p_0 and p_1 gives

$$\text{var} = (4)\left(\frac{1}{16}\right) + (1)\left(\frac{4}{16}\right) + (0)\left(\frac{6}{16}\right) + (1)\left(\frac{4}{16}\right) + (4)\left(\frac{1}{16}\right) = 1 \tag{9.6.38}$$

The ith term in this expression for N molecules can be written as

$$(i-\langle i\rangle)^2\,[N!/i!(N-i)!]\;p_0^{N-i}p_1^i s^i \tag{9.6.39}$$

where the dummy variable s is included with p_1. This equation can be generalized for N particles.

The factor

$$(i-\langle i\rangle)^2 \tag{9.6.40}$$

is expanded to give the terms

$$i^2-2\langle i\rangle i+\langle i\rangle^2 \tag{9.6.41}$$

The last term is a constant. Since it is common to all terms, it is the average value. The central term is the constant $\langle i\rangle$ multiplied by the single variable i. This variable averages to $\langle i\rangle$. The average of the first term will give the average of the square of i. The equation becomes

$$\begin{aligned}\text{var}&=\langle i^2\rangle-2\langle i\rangle^2+\langle i\rangle^2\\&=\langle i^2\rangle-\langle i\rangle^2\end{aligned} \tag{9.6.42}$$

Since $\langle i\rangle$ is known (Equation 9.6.29), only $\langle i^2\rangle$ is still required for the variance.

The second derivative of q with respect to s is

$$s^2d^2q/ds^2=\sum_i i(i-1)[N!/i!(N-i)!]\;p_0^{N-i}p_1^i \tag{9.6.43}$$

When all terms are summed, this derivative gives

$$\langle i^2\rangle-\langle i\rangle \tag{9.6.44}$$

The variance is determined by adding the expression for $\langle i\rangle$ to this result to cancel $-\langle i\rangle$ and then subtracting the expression for $\langle i\rangle^2$:

$$\text{var}=s^2\,d^2q/ds^2+s\,dq/ds-(s\,dq/ds)^2\Big|_{s=1} \tag{9.6.45}$$

The variance for an N-molecule system is determined by evaluating each term. The expression for $\langle i^2\rangle$ is

$$\begin{aligned}s^2\,d^2q/ds^2&=s^2\,d/ds[(p_0+p_1)^{N-1}\,p_1]\\&=s^2(N)(N-1)(p_0+p_1s)^{N-2}\,p_1^2\\&=(N)(N-1)\,p_1^2\end{aligned} \tag{9.6.46}$$

The remaining two terms are

$$s\,dq/ds = Np_1 \tag{9.6.47}$$

$$(s\,dq/ds)^2 = N^2 p_1^2 \tag{9.6.48}$$

The variance is

$$\begin{aligned} \text{var} &= N(N-1)\,p_1^2 + Np_1 - N^2 p_1^2 \\ &= Np_1 - Np_1^2 = Np_1(1-p_1) \end{aligned} \tag{9.6.49}$$

Since

$$p_0 = 1 - p_1 \tag{9.6.50}$$

the variance is proportional to the product of the two probabilities:

$$\text{var} = Np_1\,p_0 \tag{9.6.51}$$

The fluctuations in the system are greatest when the average probabilities for the A and B isomers are equal. The system moves easily between the two states and large fluctuations are possible. If the system favors B, e.g., 0.1 A, 0.9 B, the fluctuations will be smaller since p_0 is small.

9.7. Fluctuations during Reactions

The isomerization reaction of Section 9.6 was observed when the system had reached equilibrium. The average number of B molecules was the equilibrium value of B while the variance of B gave a measure of the magnitude of fluctuations about this equilibrium. Fluctuations are also observed during the evolution to equilibrium. The deterministic solution for B will give an average of B at each time t. However, if the actual experiments are repeated several times, the observed values of $B(t)$ will cluster around the deterministic value of $B(t)$. Two questions must now be posed: (1) will the magnitudes of the fluctuations differ at different times during the reaction? and (2) how do the equations for the fluctuations during reaction compare to those for equilibrium?

Consider a first-order irreversible isomerization,

$$\text{A} \rightarrow \text{B}$$

for which

$$k_f = k \qquad k_r = 0 \tag{9.7.1}$$

The deterministic solutions for this system are an exponential decay of [A] to zero and a concurrent exponential rise of [B]:

$$[\text{A}] = [\text{A}_0]\exp(-kt) \tag{9.7.2}$$

$$[\text{B}] = [\text{A}_0][1 - \exp(-kt)] \tag{9.7.3}$$

This deterministic solution will be equivalent to the solution which would be found by examining the system at each time for several experiments and finding the average value of [A] or [B] at this time, i.e., the deterministic solution is equivalent to the average values of [A] or [B] at each time t.

It is more convenient to deal with some number A of A molecules rather than a concentration [A]. At time zero, there are A_0 A molecules and no B molecules During the reaction, the probabilities of selecting an A or a B molecule will change. These time-dependent probabilities must then be used to calculate the average value and variance at each time t.

At equilibrium, both the average value and the variance for a set of A_0 molecules could be determined from a generating function,

$$q = (p_0 + p_1 s)^{A_0} \tag{9.7.4}$$

where the equilibrium probabilities p_0 and p_1 were known. For each time t, there are a pair of probabilities $p_0(t)$ and $p_1(t)$, and these can be substituted in the generating function. The generating function is then used to find the average and the variance at each time t.

The generating function at time t is

$$q = [p_0(t) + p_1(t)s]^{A_0} \tag{9.7.5}$$

The probability of A will be 1 at time zero since all molecules are known to be A. This probability will decay exponentially to zero with time. The time-dependent probability for A is

$$p_0 = \exp(-kt) \tag{9.7.6}$$

while the time-dependent probability for B is

$$p_1 = 1 - \exp(-kt) \tag{9.7.7}$$

These two expressions are substituted into the generating function expression to produce a time-dependent generating function:

$$q = \{\exp(-kt) + s[1 - \exp(-kt)]\}^{A_0} \tag{9.7.8}$$

The average value of $B(t)$, $\langle B(t) \rangle$, is determined using

$$\langle B(t) \rangle = s\, d(\ln q)/ds \Big|_{s=1} \tag{9.7.9}$$

This gives the result (Problem 9.15)

$$\langle B \rangle = A_0[1 - \exp(-kt)] \tag{9.7.10}$$

The average number of B molecules as a function of time t is then

$$B_{\text{av}} = A_0 p_1(t) \tag{9.7.11}$$

This result is not particularly exciting. The average probability for finding a B molecule, $p_1(t)$, was given. The analysis simply states that for A_0 molecules, the average number of B molecules is simply the probability for one molecule multiplied by the total molecules present. This is actually a manifestation of the independence of the molecules during the reaction process. The decay for a first-order process is independent of the number of molecules present, i.e., the fact that other molecules in the system are reacting has no effect on the reaction of a single molecule.

The analysis for an average highlights a simple fact. The time-dependent probabilities have been substituted into the generating function. However, the derivatives of this generating function required to obtain the average or variance never involve time. Because of this, the expression for the variance will be identical to the expression for this quantity at equilibrium. The equilibrium probabilities are replaced by the time-dependent probabilities. At equilibrium, the variance was proportional to the total number of particles and the product of p_0 and p_1:

$$\text{var} = A_0 p_0 p_1 \tag{9.7.12}$$

The variance at some time t is determined by introducing the time-dependent probabilities into this equation:

$$\begin{aligned} \text{var} &= A_0 p_0(t)\, p_1(t) \\ &= A_0 \exp(-kt)[1 - \exp(-kt)] \end{aligned} \tag{9.7.13}$$

Since $p_0 = 1$ and $p_1 = 0$ at time zero and $p_0 = 0$ and $p_1 = 1$ for long times, the variance, i.e., fluctuations, at these extremes will be zero. At $t = 0$, all the molecules are known to be A and fluctuations have not had time to occur. At $t = \infty$, the system has evolved irreversibly to B, and there is no pathway to generate A molecules and the fluctuations which accompany the transitions between A and B.

The irreversible system provides answers to some questions raised earlier. The variance is zero at equilibrium (all B), but it is finite at other times during reaction. The magnitude of the fluctuations can vary during the reaction. Plots of the variance and average for a system of 10 particles are shown in Figure 9.15. The variance reaches a maximum when $p_0 = p_1 = 0.5$.

The maximum in the variance at $p_0 = p_1 = 0.5$ reflects the variety of different ways a 50–50 mixture of A and B can be generated, i.e., it reflects the large binomial coefficient for this term. The probabilities will change with time and the variance will always approach its maximum as the probabilities of the two molecules approach each other.

The irreversible reaction illustrates the magnitude of fluctuations as a function of A_0. Although the variance describes the squared deviation of the number of particles from their average, the final expression for this variance involves only the first power of A_0. The square root of the variance provides a

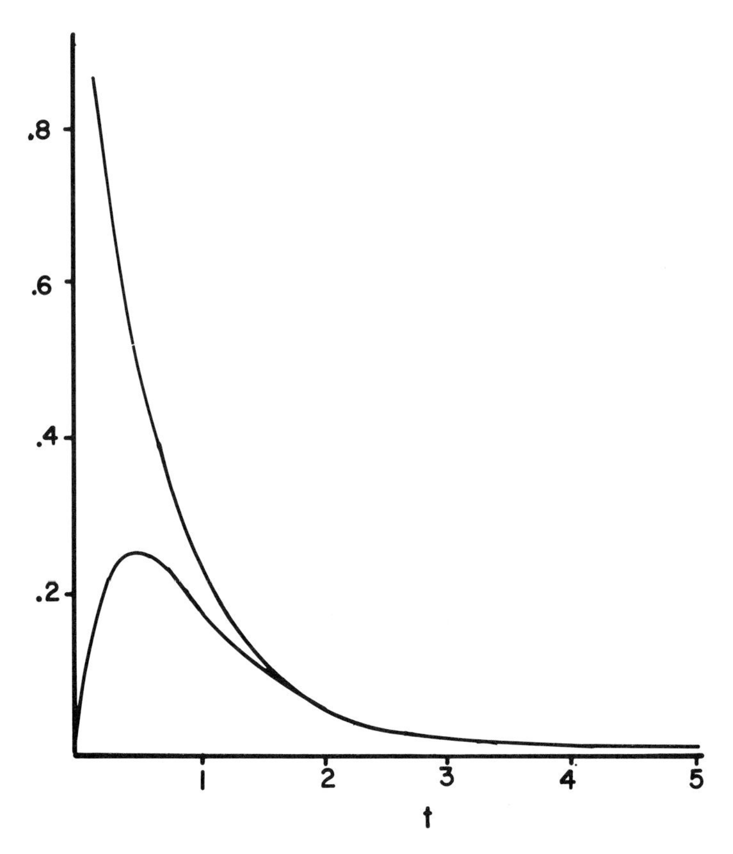

Figure 9.15. (a) The average behavior of an irreversible first-order system as a function of time. (b) Variance of the irreversible first-order reaction as a function of time.

measure of deviations about the average value. These deviations depend on the square root of the number of molecules. As this number increases, the average number of B molecules will increase with A_0 while the variance about this average will increase as the square root of A_0. The fluctuations relative to the average value will decrease as the number of molecules increases. Systems with a minimal number of molecules are ideal for the observations of fluctuations about equilibrium.

The extension of average values and variances to time-dependent kinetic systems is also valid for reversible reactions. If the A–B isomerization is reversible with forward rate constant k and reverse rate constant m,

$$\mathrm{A} \underset{m}{\overset{k}{\rightleftharpoons}} \mathrm{B}$$

and

$$A(0) = A_0, \qquad B(0) = 0 \tag{9.7.14}$$

the time-dependent probabilities are determined from the coupled differential equations:

$$\begin{aligned} dp_0/dt &= -kp_0 + mp_1 \\ dp_1/dt &= kp_0 - mp_1 \end{aligned} \tag{9.7.15}$$

The time-dependent probabilities are

$$p_0 = (m/m+k) + (k/m+k)\exp(-k't) \tag{9.7.16}$$

$$p_1 = (k/m+k) - (k/m+k)\exp(-k't) \tag{9.7.17}$$

where

$$k' = k + m \tag{9.7.18}$$

The generating function is

$$q = [p_0(t) + sp_1(t)]^{A_0} \tag{9.7.19}$$

The average value of B from this equation at any time t is

$$\langle B \rangle = A_0 p_1(t) = A_0[k/(m+k) - [k/(m+k)]\exp(-k't)] \tag{9.7.20}$$

The variance is also determined directly from the final expression,

$$\text{var} = A_0 p_0(t)\, p_1(t) \tag{9.7.21}$$

Substituting for p_0 and p_1 gives

$$\begin{aligned} \text{var} = A_0[m/(m+k) + [k/(m+k)]\exp(-k't)] \times \\ [k/(m+k) - [k/(m+k)]\exp(-k't)] \end{aligned} \tag{9.7.22}$$

As for the irreversible system, the maximal fluctuations will be observed when $p_0 = p_1$. This maximum would be observed at equilibrium only if $k = m$. The locations for the maximal variance for reactions with different k and m are illustrated in Figure 9.16.

The average and variance have been determined using the deterministic time-dependent probabilities $p_0(t)$ and $p_1(t)$ if their kinetic evolution is known. The alternative approach starts from time zero and considers the options open to the molecules in each time interval. Although this technique is preferable for more complicated chemical systems, it is introduced for the irreversible first-order isomerization. The loss of A molecules will be monitored; this will lead to a probability generating function in which s is associated with p_0.

The system has A_0 A molecules at $t = 0$. During some interval of time following $t = 0$, the system has two choices. One molecule may react to form a B

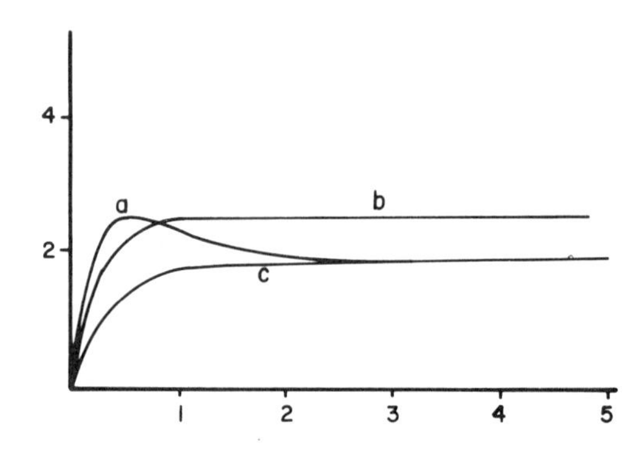

Figure 9.16. Variance as a function of time during a reversible reaction for different combinations of forward and reverse rate constants: (a) $k \gg m$; (b) $k = m$; (c) $k \ll m$.

molecule or no reaction at all occurs. The probability of reaction in the first time interval, Δt, is proportional to

$$kA_0 \, \Delta t \tag{9.7.23}$$

For some later time interval (t to $t + \Delta t$), the probability of reaction depends on the number of A molecules which remain, i, and the interval size, Δt:

$$ki \, \Delta t \tag{9.7.24}$$

The probability of no reaction during this interval is

$$1 - ki \, \Delta t \tag{9.7.25}$$

The probability that one molecule changes in the interval from t to $t + \Delta t$ is now dictated by two possible occurrences. If there are $i + 1$ molecules and one reacts in the interval, it will leave the required i molecules. On the other hand, if there are already i molecules at t and nothing happens in the interval, there will also be i molecules at the finish of the interval. The resultant equation must then include the probabilities that there are $i + 1$ or i molecules at the start of the interval. The equation involves joint probabilities, the probability at time t and the probability for the interval. The probability $p_i(t + \Delta t)$ that there are i molecules at the end of the interval is then the sum of two probability expressions:

$$p_i(t + \Delta t) = [k(i + 1) \, \Delta t] \, p_{i+1}(t) + (1 - ki \, \Delta t) \, p_i(t) \tag{9.7.26}$$

The first term states the joint probability of $i + 1$ molecules at t and i molecules after the interval Δt. The second term states the joint probability of i molecules at t and i molecules after the interval $t + \Delta t$. The factor $i+1$ appears in the first term while i appears in the second term.

The equation is converted to a differential equation by expanding the left-hand side in a Taylor series and allowing Δt to approach a limit of zero,

$$dp_i(t)/dt = k(i+1)\, p_{i+1}(t) - kip_i(t) \tag{9.7.27}$$

with initial condition

$$i = A_0 \quad \text{at} \quad t = 0 \tag{9.7.28}$$

This equation could be solved step by step but the average and variance were determined from a generating function. Since p_i is the probability of i A molecules at time t, the differential equation in p_i can be converted into a differential equation in terms of $q(s, t)$, the probability generating function at time t with dummy index s. The differential equation is formed by multiplying each term by s^i and summing to form

$$q(s, t) = \sum_{i=1}^{p_0} p_i s^i \tag{9.7.29}$$

The equation is

$$d/dt \left\{ \sum_i p_i s^i \right\} = dq/dt = \sum_i k(i+1)\, p_{i+1} s^i - \sum_i kip_i s^i \tag{9.7.30}$$

The first term is obtained by taking the derivative of $q(s, t)$ with s (Problem 9.16):

$$dq(s, t)/ds = \sum (i+1)\, p_i s^i \tag{9.7.31}$$

The second term requires an additional factor of s (Problem 9.16):

$$s\, dq(s, t)/ds = \sum ip_i s^i \tag{9.7.32}$$

These expressions produce a partial differential equation for the probability generating function $q(s, t)$,

$$\partial q/\partial t = k(1-s)\, \partial q/\partial s \tag{9.7.33}$$

with initial conditions

$$q(s, 0) = s^{A_0} \tag{9.7.34}$$

The solution to this differential equation can be verified by substitution in Equation 9.7.33 as

$$q(s, t) = [1 + (s-1)\exp(-kt)]^{A_0} \tag{9.7.35}$$

This probability generating function is equivalent to

$$q(s, t) = (sp_0 + p_1)^{A_0} \tag{9.7.36}$$

and gives the average and variance for A.

For the irreversible bimolecular reaction

$$2A \xrightarrow{k} B$$

a differential equation for the probability p_i for i particles is determined using the same technique. However, in this case, the solutions for both p_i and $q(s, t)$ are more difficult. The probability of reaction of two of $i+2$ molecules in the time interval Δt is

$$(k/2)(i+2)(i+1)\,\Delta t \tag{9.7.37}$$

since the $(i+2)$th molecule can only react with one of the remaining $i+1$ molecules. The probability for no reaction for i molecules is

$$[1 - ki(i-1)\,\Delta t] \tag{9.7.38}$$

The probability of a bimolecular transition in the time $t + \Delta t$ is

$$p_i(t+\Delta t) = (k/2)(i+2)(i+1)\,p_{i+2}\,\Delta t + [1 - ki(i-1)\,\Delta t]\,p_i \tag{9.7.39}$$

As $t \to 0$, the equation becomes

$$dp_i/dt = \frac{1}{2}k(i+2)(i+1)\,p_{i+2} - (k/2)\,i(i-1)\,p_i \tag{9.7.40}$$

The probability generating function for this equation is determined as the solution of the partial differential equation

$$\partial q(s, t)/\partial t = (k/2)(1-s^2)\,\partial^2 q(s, t)/\partial s^2 \tag{9.7.41}$$

The determination of averages and variances then rests on the solution of this differential equation. Methods of solution can be found in texts for differential equations.

9.8. The Kinetics of Single Channels in Membranes

Although fluctuations will occur for all reacting systems, such studies are usually limited by the experimental system. A large number of molecules must be present to permit experimental detection, but the magnitude of the fluctuations generally decreases with increasing numbers of molecules.

There is one particular case in which the fluctuations themselves become the major observable. A planar lipid bilayer has a low ionic conductance. However, certain molecules can diffuse into this bilayer membrane to form a transmembrane ionic channel. The formation of this channel is marked by a quantal increase in the transmembrane current or conductance for a constant transmembrane potential. The current is associated with ion flow through a single membrane channel. These channel currents will disappear if the channel molecule changes conformation or dissociates from the membrane. The appearance and disappearance of gramicidin-A channels is illustrated in the current record of Figure 9.17.

The appearance of a quantal step in membrane current is associated with the formation of a single channel while the cessation of this current should be associated with some kinetic event for this channel molecule. The temporal behavior of such channels should contain detailed information on the kinetic parameters for these channels and these parameters can then be used to describe the macroscopic behavior of ensembles of such channels.

Figure 9.18 illustrates a hypothetical case where the channels form in the membrane but cannot dissociate. Each new channel adds an equal increment of current to the total current, but the time between formation for two successive channels is not a constant. The times form a distribution about some time t_m as shown in Figure 9.19. The average time should be associated with some kinetic parameter and the half-width of the distribution should provide some measure of fluctuations.

The system is an excellent example of a Poisson process. The bathing solution contains a large number of channel molecules relative to the channels on the membrane. Thus, the concentration of channel molecules in solution is essentially constant and channel molecules will reach the membrane and form channels at some constant rate r with units

$$\text{(open channels/time)} \tag{9.8.1}$$

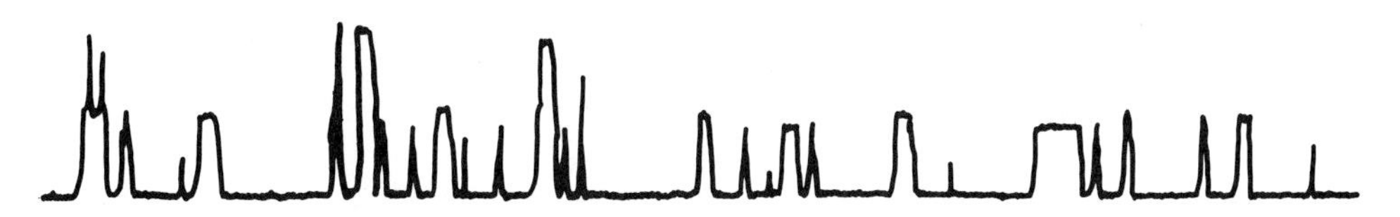

Figure 9.17. The appearance and disappearance of quantal currents associated with the formation of gramicidin-A channels as a function of time.

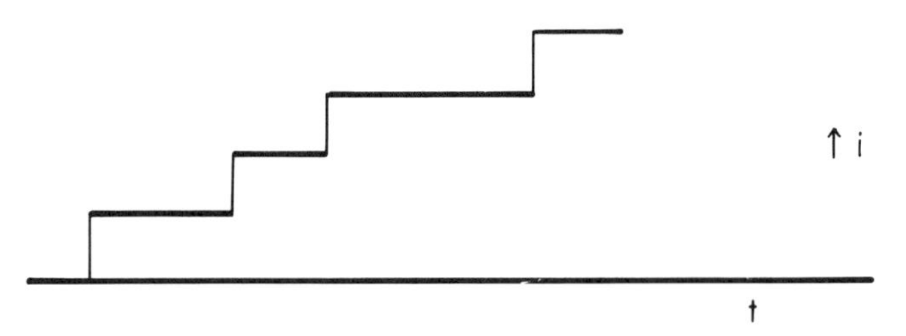

Figure 9.18. Single channel formation as a function of time for a channel which forms irreversibly in the membrane.

The inverse of r has the units of time per open channel. The total number of channels formed in an interval from 0 to t is

$$r(t-0)=rt \tag{9.8.2}$$

The rate r defines the number of channels which form each second. This does not mean channels will appear after equal time intervals. If $r=1$ open channel per second, the first channel might appear in 0.9 s because channel formation is random. Since this event has no effect on other events, the clock is restarted from this point. A second channel may appear 1 s after the first channel. The clock is again reset and the timing is repeated to generate the distribution of Figure 9.19. The average value of t_m for this distribution is 1 s/open channel. The constant rate r of channel formation is the inverse of the average time for channel formation:

$$t_m=1/r \tag{9.8.3}$$

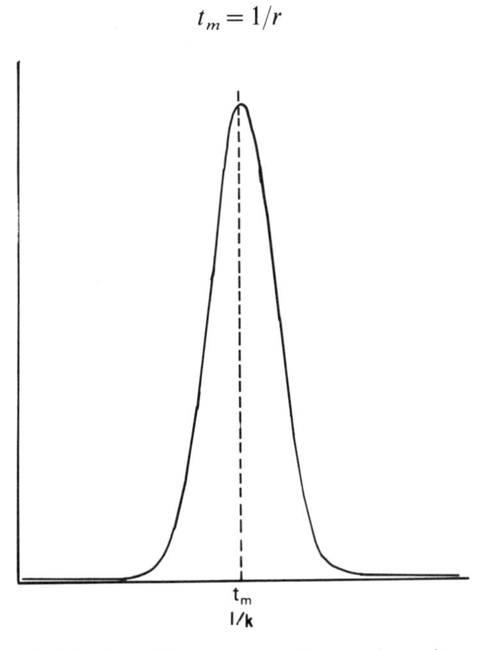

Figure 9.19. The distribution of times between the openings of successive channels.

The kinetics of channel formation is zero order rather than first order. A first-order kinetic rate is proportional to the concentrations of molecules remaining:

$$\text{Rate} = -kc \tag{9.8.4}$$

As the concentration of molecules decreases, so does the rate. For a Poisson process, the rate is constant,

$$\text{Rate} = r \tag{9.8.5}$$

because there is always an overwhelming supply of channel-forming molecules in the bathing solution. In any time interval, the same number of bath molecules are available to form channels. The few that do form channels will not affect the rate r.

The system can be analyzed by considering a series of possibilities. What is the chance that no new channel forms in some interval t? What is the chance that one channel forms in the interval; that two channels form, etc. If the rate is one channel per second, the probability that there is no channel after one second should be quite low. The probability that two or more channels have appeared in one second will also be low but the probability that one channel has appeared after one second should be quite high. The probabilities for no channel or two channels after one second are finite and can be calculated. These probabilities must be time dependent. After two seconds, the probability for the appearance of two channels should reach its maximum. A series of time-dependent probabilities, $p_i(t)$, where i is the number of channels formed, must be determined.

The development of the probabilities $p_i(t)$ follows the pattern established for first- and second-order kinetic processes in Section 9.7. The probability that a channel has not appeared after some time $t + \Delta t$ is determined in two steps. For the interval 0 to t, the channel remained closed. In the interval $t \to t + \Delta t$, the probability that a channel appears is

$$r\,\Delta t \tag{9.8.6}$$

Since there are only two choices (appearance or not), the probability for no appearance is

$$1 - r\,\Delta t \tag{9.8.7}$$

The probability $p(t + \Delta t)$ that no channel has appeared in the full interval $t + \Delta t$ is the joint probability for the intervals t and Δt:

$$p_c(t + \Delta t) = p_c(t)(1 - r\,\Delta t) \tag{9.8.8}$$

The left-hand side of this equation is expanded in a Taylor series to give

$$p_c(t) + (dp_c/dt)\,\Delta t = p_c(t) - rp_c(t)\,\Delta t \tag{9.8.9}$$

Canceling common terms gives

$$dp_c/dt = -rp_c \tag{9.8.10}$$

Since there is definitely no channel at $t = 0$, the initial probability is unity:

$$p_c(0) = 1 \tag{9.8.11}$$

Equation 9.8.10 highlights an important property of such Poisson processes. Because the process is zero order, the deterministic differential equation is

$$dc/dt = r \tag{9.8.12}$$

where c is the number of channels. The solution for this equation is

$$c = rt \tag{9.8.13}$$

By contrast, the differential equation for the probability that the channel has not appeared, p_c, is first order.

The solution of Equation 9.8.10 with the initial condition is

$$p_c(t) = (1)\exp(-rt) \tag{9.8.14}$$

Since $r = 1$, there is a high probability that a channel will appear as t approaches 1 s. The probability p_c that no channel appears will decrease rapidly. Since p_c describes no channel appearance, it is convenient to define it as p_0.

The probability that only one channel will open in the time interval $t + \Delta t$ depends on two different situations:

1. The channel has opened between 0 and t and no other channel appears in Δt.
2. The channel had not appeared at t but does appear in Δt.

The probability of one channel in $t + \Delta t$ is the sum of both probabilities:

$$p_1(t + \Delta t) = p_0(t)(r\,\Delta t) + p_1(t)(1 - r\,\Delta t) \tag{9.8.15}$$

Expanding and canceling common terms leads to

$$dp_1/dt = rp_0 - rp_1 \tag{9.8.16}$$

The equation can be solved using the integrating factor

$$\exp(+rt) \tag{9.8.17}$$

and the time-dependent solution for p_0. Since there is no open channel at the start of the time interval,

$$p_1(0) = 0 \tag{9.8.18}$$

The differential equation becomes

$$\exp(rt)(dp_1/dt + rp_1) = d[p_1 \exp(rt)]/dt$$
$$= r \exp(-rt) \exp(rt) = r \qquad (9.8.19)$$

The solution to this equation with the initial condition is

$$p_1(t) = \exp(-rt)\, rt \qquad (9.8.20)$$

The probability of observing one open channel increases with time (rt) as expected. However, this probability is multiplied by a decaying exponential. As time increases, there is an increased probability that a second channel may also appear. The probability of seeing "just one" reaches a maximum and then decreases.

The solution for p_1 is equivalent to that for the second species in a sequence of consecutive reactions,

$$A \rightarrow B \rightarrow C \rightarrow \cdots$$

where all rate constants are identical. The probability p_0 falls as p_1 rises; p_1, in turn, falls as p_2 rises.

The differential equation for the probability p_i is identical to that for p_1:

$$dp_i/dt = rp_{i-1} - rp_i \qquad (9.8.21)$$

However, in order to solve the equation for p_i using an integrating factor, p_{i-1} must be known. The equations must be solved in sequence: p_1 is substituted into the equation for p_2; the solution for p_2 is then used in the equation for p_3, etc. The probability for observing i channels in some time t will be (Problem 9.18)

$$p_i(t) = (rt)^i \exp(-rt)/i! \qquad (9.8.22)$$

The term $(rt)^i$ produces a delay in the rise of the probability with time. The maximal probability for observing two open channels will occur at a later time. The probabilities p_0, p_1, and p_2 are illustrated in Figure 9.20 for $r=1$. The maxima for p_1 and p_2 are located at 1 and 2, respectively.

The index i defines the number of open channels and is an analogue of the index i which gave the number of B isomers in the reaction systems of Section 9.6. The set of Poisson probabilities can be combined to form a probability generating function by forming the products

$$p_i(t)\, s^i \qquad (9.8.23)$$

and summing them:

$$q(s, t) = \exp(-rt) \sum (rts)^i/i! \qquad (9.8.24)$$

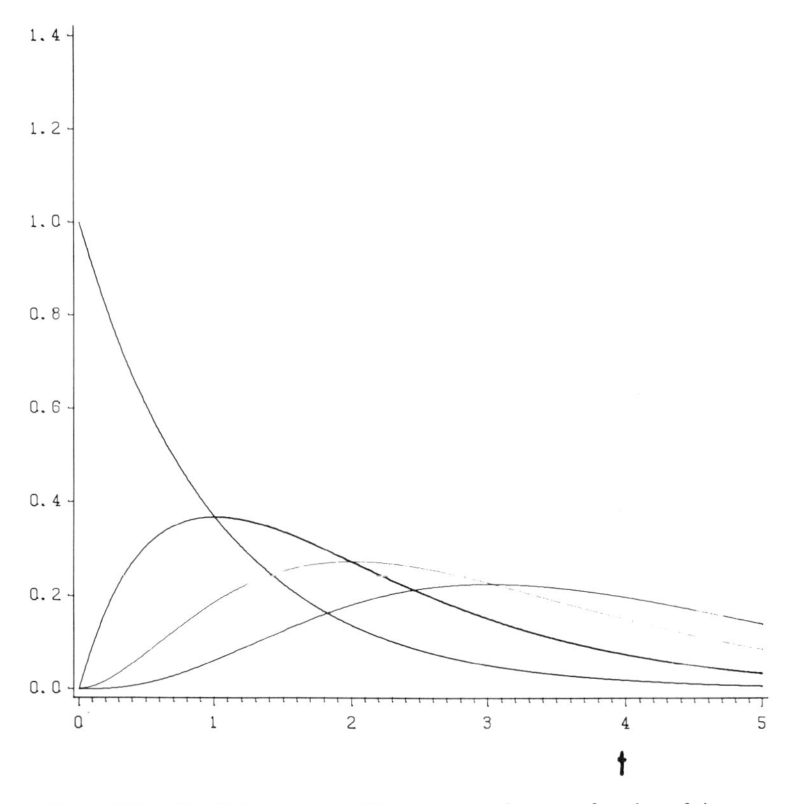

Figure 9.20. The Poisson probabilities p_0, p_1, and p_2 as a function of time.

The summation is the exponential of the argument (rts) and the probability generating function is

$$q(s, t) = \exp(-rt) \exp(rts) = \exp[rt(s-1)] \tag{9.8.25}$$

The average number of channels observed at time t can be obtained directly from the expression for the average value of i,

$$\langle i \rangle = s\, d(\ln q)/ds \Big|_{s=1} \tag{9.8.26}$$

The sum of all probabilities at any time t is 1 (Problem 9.19). $\langle i \rangle$ is

$$s\, d \ln\{\exp[rt(s-1)]\}/ds \Big|_{s=1} = s\, d(rts - rt)/ds \Big|_{s=1} = rt \tag{9.8.27}$$

This is the deterministic equation for the appearance of channels. On average, the number of channels increase with a rate r.

The variance of the system is a measure of fluctuations about the average value, i.e., fluctuations about the linear increase of channels with time. These fluctuations are expected to increase with time since the observed result at longer times will be the cumulative input of many random inputs. The variance is

$$\text{var} = s^2\, d^2\{\exp[rt(s-1)]\}/ds + s\, dq/ds - (s\, dq/ds)^2 \tag{9.8.28}$$

The last two terms defined $\langle i \rangle$ and $\langle i \rangle^2$, respectively. The first term is

$$s^2\, d^2\{\exp[rt(s-1)]\}/ds^2 \Big|_{s=1} = s^2 \exp[rt(s-1)](rt)^2 \Big|_{s=1} = (rt)^2 \tag{9.8.29}$$

Its variance is

$$\text{var} = (rt)^2 + (rt) - (rt)^2 = rt \tag{9.8.30}$$

which also increases linearly with time. At short times, only one or two channels may be open and the variance is relatively small. At longer times, there may be many open channels and the possibility of different numbers of channels increases. The variance increases because there are more possible observable states.

The distribution of times for channel openings (Figure 9.19) correlates with the Poisson distribution. Since the maximum of the experimental distribution is easily determined, it can be compared with the maximum predicted for the Poisson probability, p_1:

$$dp_1(t)/dt = 0 \tag{9.8.31}$$

$$d/dt[(rt)\exp(-rt)] = 0 = r\exp(-rt) - (r^2 t)\exp(-rt) \tag{9.8.32}$$

The time to maximum, t_m, is

$$t_m = 1/r \tag{9.8.33}$$

The most probable time between channels is related to the constant rate of channel formation. It also represents the point on the deterministic equation when

$$rt = 1 \tag{9.8.34}$$

Channels in bilayer membranes appear with the observable rise in current and then disappear (Figure 9.17). The time that the channels remain in their open configuration provides some information on the kinetics of channel closing. If there were an appreciable number of open channels, the kinetics of channel disappearance would depend on the number of open channels. However, the concentration of channel molecules in the bath can be controlled. For low concentrations, the rate of channel formation is small and no more than one channel will be open at any one time. In this limit, the times required for channel disappearance once the channel has appeared will also satisfy a Poisson distribution. In this case, the t_m will equal the inverse of a constant rate, r':

$$t_m = 1/r' \tag{9.8.35}$$

These statistics will remain valid while the system is restricted to no more than a single channel at any time. As the number of channels in the membrane increases, the disappearance kinetics must be modified to include the number of channels present. The analysis is then similar to that developed for first-order isomerizations in Section 9.7.

Problems

9.1. The excited singlet state of a molecule is excited by a light pulse of intensity I which is on for T seconds. The excited singlet is assumed to have two vibrational levels. Each level has a fluorescence rate constant k_f. The upper vibrational level is deactivated to the lower vibrational level by collisions with rate constant z. Find the fluorescence intensity for the system. How does it compare to fluorescence from an excited state with one vibrational level?

9.2. Population inversion is possible for a three-level molecular system. The populations of the ground state (N_0), upper excited state (N_u), and lower excited state (N_l) must be determined. The molecule passes from state u to state l with an irreversible rate constant k_i. The transition from state l back to the ground state is irreversible with rate constant k.

a. Write the matrix form of this equation.
b. Write a formal matrix solution for continuous light intensity I.

9.3. A molecule is excited to its excited singlet state by a short pulse of light whose time course is approximated by a delta function, i.e., the rate of excitation is $kI\delta(t)$. The light excites the upper vibrational level and only down transitions are permitted; the upper excited state is depopulated via rate constants z, k_f, and k_2 while the lower excited state is depopulated via k_f and k_1.

a. Set up the matrix equation for the system.
b. Develop a formal solution of (a).
c. Solve for the fluorescence intensity $I = k_f f_1 + k_f f_2$ as a function of time.

9.4. The Green's function is used to translate the temporal response of an excitation input into a system response. However, the instrument itself will also produce a response to the input function. Develop an integral equation which includes Green's functions for both the molecular and instrumental responses.

9.5. For an excited singlet with two vibrational levels such as that developed in Section 9.2, develop the solutions for a steady excitation input J when the reverse reaction from the lower to the upper level is allowed.

9.6. Use graph theory to determine the stationary-state populations of a three-level laser. An upper level is excited and populates an intermediate level. Lasing takes place between the intermediate and ground states.

9.7. Set up the kinetic equations for the four-level laser model when all the steps are reversible.

9.8. The ground state of a molecule is depopulated via a rate depending on the intensity of light $I(\omega)$ and the population N_0. The excited state, by contrast, depopulates via either stimulated emission with rate

$$kI(\omega)\,N_1$$

or fluorescence (spontaneous emission) with rate

$$k_f N_1$$

a. Show that the rate equation for the excited level is

$$dN_1/dt = kI(\omega)\,N_0 - [kI(\omega) + k_f]\,N_1$$

b. Show, at equilibrium, that

$$N_1/N_0 = kI(\omega)/[k_f + k'I(\omega)]$$

c. Use the result of (b) and the Boltzmann condition,

$$N_1/N_0 = \exp(-\hbar\omega/kT)$$

to show that

$$I(\omega) = k_f \exp(-\hbar\omega/kT)/[k - k' \exp(-\hbar\omega/kT)]$$

d. Show that this equation is consistent with Planck's law,

$$I(\nu) = (8\pi h\nu^3/vc^3)\exp(-h\nu/kT)/[1 - \exp(-h\nu/kT)]$$

$$\hbar\omega = h\nu$$

only if

$$k = k' \quad \text{and} \quad k_f = (8\pi h\nu^3/vc^3)\,k'$$

9.9. Use the chemical potential

$$\tilde{\mu} = RT \ln c + zFV$$

to prove that, at equilibrium, the concentration ratio of the two bathing solutions of the channel permeation model in Section 9.5 is

$$c_2/c_1 = \exp[F(V_1 - V_2)/RT]$$

9.10. For a two site model, show the differences between a nonconservative system where rate constants lead from the two membrane sites to the baths and a conservative system where inputs and outputs from the membrane are controlled entirely by a conservative flux.

9.11. Define a stationary-state entropy in terms of J/λ_1 where λ_1 is the lowest nonzero eigenvalue. Ignore other eigenvalues in the system.

9.12. Prove that only odd eigenvectors contribute to the stationary-state distribution for a conservative membrane system.

9.13. Derive the variance for an equilibrium set of isomers using the chemical potential difference between isomers A and B. The problem is the chemical analogue of the capacitor problem in Section 9.6.

9.14. Derive the expression

$$\langle i \rangle = d \ln q(s, t)/d \ln s$$

9.15. Show that

$$\langle B \rangle = A_0[1 - \exp(-kt)]$$

9.16. Prove the identities

$$dq/ds = \sum (i+1)\, p_{i+1} s^i$$

$$s\, dq/ds = \sum i p_i s^i$$

9.17. In Section 9.7, a differential equation for the p_i was developed for the loss of A. Derive the corresponding differential equation for the probabilities of the formation of B and show that it leads to the probability generating function for B.

9.18. Derive the equation for $p_i(t)$, the Poisson probability for the appearance of i molecules at some time t,

$$p_i(t) = (rt)^i \exp(-rt)/i!$$

9.19. Show that the sum of all the Poisson probabilities at any time t is always 1.

10

Other Techniques: Perturbation Theory and Direct Products

10.1. Development of Perturbation Theory

The matrices used to describe various physical phenomena in previous chapters had eigenvalues and eigenvectors which could be determined exactly. In general, matrix descriptions of physical systems will not be so tractable. However, the complicated matrices can often be analyzed when their elements differ slightly from those of a second matrix whose eigenvalues and eigenvectors are known. This is the purpose of perturbation theory.

The nature of such techniques can be illustrated with the 2×2 matrix

$$\begin{pmatrix} -1 & 1.1 \\ 1 & -1.1 \end{pmatrix} \tag{10.1.1}$$

The eigenvalues for this matrix are determined exactly from the characteristic equation

$$(1+\lambda)(1.1+\lambda) - 1.1 = 0 \tag{10.1.2}$$

as

$$\lambda = 0 \qquad \lambda = -2.1 \tag{10.1.3}$$

The matrix is close to the matrix

$$\begin{pmatrix} -1 & 1 \\ 1 & -1 \end{pmatrix} \tag{10.1.4}$$

with eigenvalues

$$\lambda = 0 \qquad \lambda = -2 \tag{10.1.5}$$

Both have a zero eigenvalue and the second eigenvalue differs only by 0.1.

The eigenvectors of both matrices are also nearly equal. The matrix in Equation 10.1.4 with eigenvalues 0 and -2 has the corresponding eigenvectors

$$|0\rangle = \left(\frac{1}{\sqrt{2}}\right)\begin{pmatrix}1\\1\end{pmatrix} \qquad |-2\rangle = \left(\frac{1}{\sqrt{2}}\right)\begin{pmatrix}1\\-1\end{pmatrix} \tag{10.1.6}$$

while the matrix in Equation 10.1.1 has eigenvectors

$$|0\rangle = (1/\sqrt{2.21})\begin{pmatrix}1.1\\1\end{pmatrix} \qquad |-2.1\rangle = (1/\sqrt{2.21})\begin{pmatrix}1.1\\-1\end{pmatrix} \tag{10.1.7}$$

Only one of the two vector components changes since the set of vector components are linearly dependent and all components are expressed as ratios of one of the components. An $N \times N$ matrix will require $N-1$ independent components.

The eigenvalues and eigenvectors of the matrix

$$\mathbf{K} = \begin{pmatrix}-1 & 1\\1 & -1\end{pmatrix} \tag{10.1.8}$$

are used in perturbation theory as the starting point for estimating the eigenvalues of the matrix

$$\mathbf{K}' = \begin{pmatrix}-1 & 1.1\\1 & -1.1\end{pmatrix} \tag{10.1.9}$$

The $\mathbf{K}'$ matrix can be written as a sum of the $\mathbf{K}$ matrix and a perturbation matrix $\mathbf{R}$ which contains deviations from the elements of $\mathbf{K}$, i.e.,

$$\mathbf{K}' = \mathbf{K} + \mathbf{R}$$

$$\begin{pmatrix}-1 & 1.1\\1 & -1.1\end{pmatrix} = \begin{pmatrix}-1 & 1\\1 & -1\end{pmatrix} + \begin{pmatrix}0 & 0.1\\0 & -0.1\end{pmatrix} \tag{10.1.10}$$

The elements in $\mathbf{R}$ should be small relative to those in $\mathbf{K}$ for effective perturbation analysis. The corrections required by $\mathbf{R}$ are then expressed as some series which converges rapidly. The sequence is developed by introducing the dummy parameter s and a $\mathbf{K}'$ matrix defined as

$$\mathbf{K}' = \mathbf{K} + s\mathbf{R} \tag{10.1.11}$$

The dummy parameter generates the appropriate series in increasing powers of s.

The eigenvalues and eigenvectors of $\mathbf{K}$ serve as the unperturbed, starting parameters. The unperturbed eigenvalue, λ_i, and its eigenvector $|i\rangle$ for the $\mathbf{K}$ matrix must satisfy the equation

$$\mathbf{K}\,|i\rangle = \lambda_i\,|i\rangle \tag{10.1.12}$$

The similarity transform matrices for **K** are formed from the column eigenvectors

$$\mathbf{S} = (|1\rangle \quad |2\rangle \quad \ldots \quad |i\rangle \quad \ldots \quad |N\rangle) \tag{10.1.13}$$

and row eigenvectors

$$\mathbf{S}^{-1} = \begin{pmatrix} \langle 1| \\ \langle 2| \\ \vdots \\ \langle i| \\ \vdots \\ \langle N| \end{pmatrix} \tag{10.1.14}$$

for an $N \times N$ matrix. For the 2×2 **K** matrix, these are

$$\begin{aligned} \mathbf{S} &= (|1\rangle \quad |2\rangle) \\ &= \begin{pmatrix} 1 & 1 \\ 1 & -1 \end{pmatrix} \end{aligned} \tag{10.1.15}$$

and

$$\mathbf{S}^{-1} = \left(\frac{1}{2}\right)\begin{pmatrix} 1 & 1 \\ 1 & -1 \end{pmatrix} \tag{10.1.16}$$

When the **K** matrix operates from the left on the **S** matrix, each column eigenvector in the matrix is regenerated with its eigenvalue. To keep each eigenvalue with its eigenvector, the eigenvalues must be incorporated into a diagonal matrix where λ_1 appears as the a_{11} element, λ_2 appears as the a_{22} element, etc. The diagonal matrix for an $N \times N$ **K** matrix is

$$\mathbf{\Lambda} = [0, \lambda_i, 0] \tag{10.1.17}$$

and the general equation is

$$\mathbf{KS} = \mathbf{S\Lambda} \tag{10.1.18}$$

The $\mathbf{\Lambda}$ matrix must appear to the right of **S** (Problem 10.1).

Equation 10.1.18 is illustrated with the 2×2 **K** matrix:

$$\mathbf{KS} = \begin{pmatrix} -1 & 1 \\ 1 & -1 \end{pmatrix}\begin{pmatrix} 1 & 1 \\ 1 & -1 \end{pmatrix} = \begin{pmatrix} 0 & -2 \\ 0 & 2 \end{pmatrix} \tag{10.1.19}$$

$$\mathbf{K\Lambda} = \begin{pmatrix} 1 & 1 \\ 1 & -1 \end{pmatrix}\begin{pmatrix} 0 & 0 \\ 0 & -2 \end{pmatrix} = \begin{pmatrix} 0 & -2 \\ 0 & 2 \end{pmatrix} \tag{10.1.20}$$

The $\mathbf{K}'$ matrix must also be characterized by still unknown eigenvalues which are the elements of a new diagonal matrix, $\mathbf{\Lambda}'$,

$$\mathbf{\Lambda}' = [0, \lambda_i', 0] \tag{10.1.21}$$

and a new similarity transform,

$$\mathbf{S}' = (|1'\rangle \quad |2'\rangle) \tag{10.1.22}$$

which satisfies the equation

$$\mathbf{K}'\mathbf{S}' = \mathbf{S}'\mathbf{\Lambda}' \tag{10.1.23}$$

The unknown $\mathbf{\Lambda}'$ and $\mathbf{S}'$ matrices must be determined from $\mathbf{K}$, $\mathbf{S}$, and $\mathbf{\Lambda}$.

If two functions $f(x)$ and $g(y)$ have polynomial expansions

$$f(x) = a_0 + a_1 x + a_2 x^2 + \cdots \tag{10.1.24}$$

$$g(y) = b_0 + b_1 y + b_2 y^2 + \cdots \tag{10.1.25}$$

their product is also a polynomial series in the variables x and y. The terms of the product can be ordered by introducing a dummy variable s which is inserted for each x and y. The terms of the product are then grouped as s^0, s^1, ... terms. The product

$$f(x)\, g(y) \tag{10.1.26}$$

is

$$f(x)\, g(y) = a_0 b_0 + s(b_0 a_1 x + a_0 b_1 y) + s^2(b_0 a_2 x^2 + a_0 b_2 y^2 + a_1 b_1 xy) + \cdots \tag{10.1.27}$$

Each set of terms has the same net power as the dummy index s.

The ordering technique is applied to the matrix

$$\mathbf{K}' = \mathbf{K} + s\mathbf{R} \tag{10.1.28}$$

when it operates on $\mathbf{S}'$ to give $\mathbf{S}'\mathbf{\Lambda}'$. Both $\mathbf{S}'$ and $\mathbf{\Lambda}'$ must also be expanded using the common dummy variable:

$$\mathbf{\Lambda}' = \mathbf{\Lambda} + s\mathbf{L} + s^2\mathbf{L}' + \cdots \tag{10.1.29}$$

$$\mathbf{S}' = \mathbf{S}(\mathbf{I} + s\mathbf{X} + s^2\mathbf{X}' + \cdots) \tag{10.1.30}$$

Substituting these expansions into the equation for $\mathbf{K}'\mathbf{S}'$ gives

$$\begin{aligned}(\mathbf{K} + s\mathbf{R})(\mathbf{S})(\mathbf{I} + s\mathbf{X} + s^2\mathbf{X}' + \cdots)\\ = \mathbf{S}(\mathbf{I} + s\mathbf{X} + s^2\mathbf{X}' + \cdots)(\mathbf{\Lambda} + s\mathbf{L} + s^2\mathbf{L}' + \cdots)\end{aligned} \tag{10.1.31}$$

Expansion of the left-hand side gives

$$\mathbf{KSI} + s(\mathbf{RSI} + \mathbf{KSX}) + s^2(\mathbf{KSX}' + \mathbf{RSX}) + s^3(\ldots) + \cdots \tag{10.1.32}$$

while expansion of the right-hand side gives

$$\mathbf{SI\Lambda} + s(\mathbf{SX\Lambda} + \mathbf{SIL}) + s^2(\mathbf{SX}'\mathbf{\Lambda} + \mathbf{SXL} + \mathbf{SL}') + s^3(\ldots) + \cdots \tag{10.1.33}$$

Since each side is ordered in s, perturbations in s on each side of the equation must be the same. The perturbation separates into a series of separate equations, one for each power of s. For example, the terms in s^0 on both sides are equated to give the first equation,

$$\mathbf{KSI} = \mathbf{KS} = \mathbf{S\Lambda} \tag{10.1.34}$$

This is simply the equation for the unperturbed matrix. The matrix $\mathbf{S}$ contains the unperturbed eigenvectors and the $\mathbf{\Lambda}$ matrix contains the unperturbed eigenvalues.

The first perturbation is associated with s^1. Equating terms from each side gives

$$\mathbf{RS} + \mathbf{KSX} = \mathbf{SX\Lambda} + \mathbf{SL} \tag{10.1.35}$$

This equation contains the unknown correction matrix $\mathbf{L}$ for the eigenvalues and the matrix $\mathbf{X}$. $\mathbf{L}$ and $\mathbf{X}$ can be expressed in terms of $\mathbf{R}$ and $\mathbf{S}$. Applying $\mathbf{S}^{-1}$ to each side gives

$$\begin{aligned}\mathbf{S}^{-1}\mathbf{RS} + \mathbf{S}^{-1}\mathbf{KSX} &= \mathbf{S}^{-1}\mathbf{SX\Lambda} + \mathbf{S}^{-1}\mathbf{SL} \\ &= \mathbf{X\Lambda} + \mathbf{L}\end{aligned} \tag{10.1.36}$$

The combination

$$\mathbf{S}^{-1}\mathbf{KS} \tag{10.1.37}$$

is the similarity transform for the unperturbed system and gives the matrix of unperturbed eigenvalues $\mathbf{\Lambda}$:

$$\mathbf{\Lambda} = \mathbf{S}^{-1}\mathbf{KS} \tag{10.1.38}$$

The equation can now be solved for $\mathbf{L}$ as

$$\mathbf{L} = \mathbf{S}^{-1}\mathbf{RS} + \mathbf{\Lambda X} - \mathbf{X\Lambda} \tag{10.1.39}$$

Although $\mathbf{X}$ is still not known, the diagonal $\mathbf{L}$ matrix requires only diagonal elements from the last two terms. These two terms make a net contribution of zero as illustrated with a general 2×2 matrix,

$$\begin{pmatrix} a & b \\ c & d \end{pmatrix} \tag{10.1.40}$$

and a diagonal $\mathbf{\Lambda}$ matrix,

$$\begin{pmatrix} e & 0 \\ 0 & f \end{pmatrix} \tag{10.1.41}$$

The difference of the last two terms in Equation 10.1.39 is

$$\begin{aligned} \mathbf{X\Lambda} - \mathbf{\Lambda X} &= \begin{pmatrix} a & b \\ c & d \end{pmatrix}\begin{pmatrix} e & 0 \\ 0 & f \end{pmatrix} - \begin{pmatrix} e & 0 \\ 0 & f \end{pmatrix}\begin{pmatrix} a & b \\ c & d \end{pmatrix} \\ &= \begin{pmatrix} ae & bf \\ ce & df \end{pmatrix} - \begin{pmatrix} ea & eb \\ fc & df \end{pmatrix} = \begin{pmatrix} 0 & bf - cb \\ ce - fc & 0 \end{pmatrix} \end{aligned} \tag{10.1.42}$$

Even though the off-diagonal elements of this product difference are nonzero, the diagonal elements are identical for each matrix and their difference is zero. The eigenvalue correction matrix $\mathbf{L}$ is

$$\mathbf{L} = \mathbf{S}^{-1}\mathbf{RS} \tag{10.1.43}$$

$\mathbf{L}$ is defined by the perturbation matrix $\mathbf{R}$ and $\mathbf{S}$ and $\mathbf{S}^{-1}$ for the unperturbed $\mathbf{K}$ matrix. Only its diagonal elements are used. Higher-order corrections can be determined from the expansion. However, the first-order expression will provide corrections for both the eigenvalues and eigenvectors. These corrections are considered in Sections 10.2 and 10.3. Corrections generated by the s^2 terms are considered in Sections 10.4 and 10.5.

10.2. First-Order Perturbation Theory—Eigenvalues

The matrix $\mathbf{K}'$ (Equation 10.1.1) is separated into a $\mathbf{K}$ matrix with known eigenvalues and eigenvectors and a perturbation matrix $\mathbf{R}$,

$$\mathbf{K}' = \mathbf{K} + \mathbf{R}$$

$$\begin{pmatrix} -1 & 1.1 \\ 1 & -1.1 \end{pmatrix} = \begin{pmatrix} -1 & 1 \\ 1 & -1 \end{pmatrix} + \begin{pmatrix} 0 & 0.1 \\ 0 & -0.1 \end{pmatrix} \tag{10.2.1}$$

The similarity transform matrices for the $\mathbf{K}$ matrix,

$$\mathbf{S} = \begin{pmatrix} 1 & 1 \\ 1 & -1 \end{pmatrix} \qquad \mathbf{S}^{-1} = \left(\frac{1}{2}\right)\begin{pmatrix} 1 & 1 \\ 1 & -1 \end{pmatrix} \tag{10.2.2}$$

generate the first-order eigenvalue correction matrix $\mathbf{L}$:

$$\begin{aligned} \mathbf{L} &= \mathbf{S}^{-1}\mathbf{RS} \\ &= \left(\frac{1}{2}\right)\begin{pmatrix} 1 & 1 \\ 1 & -1 \end{pmatrix}\begin{pmatrix} 0 & 0.1 \\ 0 & -0.1 \end{pmatrix}\begin{pmatrix} 1 & 1 \\ 1 & -1 \end{pmatrix} \\ &= \left(\frac{1}{2}\right)\begin{pmatrix} 0 & 0 \\ 0.2 & -0.2 \end{pmatrix} = \begin{pmatrix} 0 & 0 \\ 0.1 & -0.1 \end{pmatrix} \end{aligned} \tag{10.2.3}$$

The **L** matrix is not diagonal because **S** and $\mathbf{S}^{-1}$ form a similarity transform for **K**, not **K**′ or **R**. However, if **R** and **K** commute, the eigenvectors for **K** and **K**′ are identical.

The corrected eigenvalues for **K**′ are now

$$\begin{aligned}\mathbf{\Lambda}' &= \mathbf{\Lambda} + \mathbf{L} \\ &= \begin{pmatrix} 0 & 0 \\ 0 & -2 \end{pmatrix} + \begin{pmatrix} 0 & 0 \\ 0 & -0.1 \end{pmatrix} = \begin{pmatrix} 0 & 0 \\ 0 & -2.1 \end{pmatrix}\end{aligned} \tag{10.2.4}$$

where only the diagonal elements of **L** are used.

For this simple case, the first-order eigenvalue perturbation correction yields the exact eigenvalues for the **K**′ matrix. These exact values were determined directly from the characteristic polynomial (Section 10.1). Although this will not occur in general, the first eigenvalue correction is small if the elements in **R** are small. The **K**′ matrix must be separated so the **K** matrix has known eigenvalues and eigenvectors while the correction elements in **R** are minimal.

Equation 10.2.3 gives a matrix **L** containing all eigenvalue corrections. The equation also determines single eigenvalue corrections. These are the diagonal elements of **L**. The similarity transforms for an $N \times N$ **K** matrix are

$$\mathbf{S} = (|1\rangle \quad |2\rangle \quad \dots \quad |N\rangle) \tag{10.2.5}$$

$$\mathbf{S}^{-1} = \begin{pmatrix} |1\rangle \\ |2\rangle \\ |3\rangle \\ \vdots \\ |N\rangle \end{pmatrix} \tag{10.2.6}$$

The diagonal *ii* element of

$$\mathbf{S}^{-1}\mathbf{R}\mathbf{S} \tag{10.2.7}$$

corrects the λ_i eigenvalue,

$$L_{ii} = \langle i|\, \mathbf{R}\, |i\rangle \tag{10.2.8}$$

since

$$\mathbf{L} = \begin{pmatrix} \langle 1|\, \mathbf{R}\, |1\rangle & 0 & \dots & \\ 0 & \langle 2|\, \mathbf{R}\, |2\rangle & \dots & \\ & & \vdots & \\ & & \dots & \langle N|\, \mathbf{R}\, |N\rangle \end{pmatrix} \tag{10.2.9}$$

The correction λ_i' in first-order perturbation theory is

$$\lambda_i' = \lambda_i + \langle i|\, \mathbf{R}\, |i\rangle \tag{10.2.10}$$

and $|i\rangle$ is the eigenvector for λ_i of the **K** matrix.

In function space, where the eigenfunctions are continuous, the first-order perturbation is still

$$L_{ii} = \langle i| \mathbf{R} |i\rangle \tag{10.2.11}$$

In this case, the $|i\rangle$ are the eigenfunctions for an unperturbed operator **A**. **R** is the perturbing operator. Equation 10.2.11 corrects the ith eigenvalue and is determined by the integral over the appropriate space.

In quantum mechanics, the energy for a system must often be estimated even if the eigenvectors are unknown. In this case, the unknown Hamiltonian, $\mathbf{H}'$, is developed as the sum of a Hamiltonian operator, **H**, whose eigenvalues and eigenvectors are known and a perturbation operator, **R**:

$$\mathbf{H}' = \mathbf{H} + \mathbf{R} \tag{10.2.12}$$

If the eigenvalues for **H** are ε_n and the corresponding eigenfunctions are $|n\rangle$, the energy correction due to the perturbation operator **R** is

$$\varepsilon'_n = \langle n| \mathbf{R} |n\rangle \tag{10.2.13}$$

The technique can be illustrated for an electron in a one-dimensional box of length L which is subject to an electric field $\mathscr{E}$ directed along the axis of the box (x axis). The potential energy for the electron due to the electric field is

$$V(x) = \mathscr{E}x \tag{10.2.14}$$

and the total Hamiltonian is

$$\mathbf{H}' = \mathbf{H} + \mathbf{R} = -(\hbar^2/2m)\, d^2/dx^2 + \mathscr{E}x \tag{10.2.15}$$

If the electric field is zero, the Hamiltonian reduces to the Hamiltonian operator for a particle in a box,

$$\mathbf{H}' = \mathbf{H} = (-\hbar^2/2m)\, d^2/dx^2 \tag{10.2.16}$$

which has eigenvalues

$$E_n = (\hbar n\pi)^2/2mL^2 = (hn)^2/8mL^2 \tag{10.2.17}$$

and eigenfunctions

$$|n\rangle = \sin(n\pi x/L) \tag{10.2.18}$$

These eigenfunctions are not eigenfunctions for the **R** operator (Problem 10.2) and perturbation theory is required.

The perturbation energy is

$$E'_n = \langle n| \mathbf{R} |n\rangle = \langle n| \mathscr{E}x |n\rangle = \mathscr{E} \int_0^L \sin(n\pi x/L)x \sin(n\pi x/L)\, dx \tag{10.2.19}$$

The integral is written

$$(2/L)(L/n\pi)^2 \int_0^{n\pi} y \sin^2 y \, dy \tag{10.2.20}$$

$$y = n\pi x/L \tag{10.2.21}$$

with limits

$$\begin{aligned} y &= n\pi(0)/L = 0 \\ y &= n\pi(L)/L = n\pi \end{aligned} \tag{10.2.22}$$

Evaluation of the integral gives

$$\int_0^{n\pi} y \sin^2 y \, dy = (n\pi)^2/4 \tag{10.2.23}$$

and the perturbation is

$$\varepsilon_n' = (2/L)(L/n\pi)^2 \, (n\pi^2/4)\mathscr{E} = L\mathscr{E}/2 \tag{10.2.24}$$

The perturbation energy is proportional to $\mathscr{E}L$ which is the total potential energy difference, $\mathscr{E}(L-0)$, along the length of the box. Division by 2 gives the average potential energy for the electron in the entire box of length L.

First-order perturbation theory will also give corrected rate constants for a rate constant matrix **K**. A three-level kinetic system with reaction from the highest state (Chapter 6) has the matrix

$$\mathbf{K}' = \begin{pmatrix} a & -1 & 0 \\ -a & 1+a & -1 \\ 0 & -a & 1+a \end{pmatrix} \tag{10.2.25}$$

The matrix is tridiagonal, and it can be related to a diagonalizable matrix **K** in which all diagonal elements are equal, i.e.,

$$\mathbf{K}' = \mathbf{K} + \mathbf{R}$$

$$\mathbf{K}' = \begin{pmatrix} a & -1 & 0 \\ -a & 1+a & -1 \\ 0 & -a & 1+a \end{pmatrix} = \begin{pmatrix} 1+a & -1 & 0 \\ -a & 1+a & -1 \\ 0 & -a & 1+a \end{pmatrix} + \begin{pmatrix} -1 & 0 & 0 \\ 0 & 0 & 0 \\ 0 & 0 & 0 \end{pmatrix} \tag{10.2.26}$$

The perturbation matrix has a single nonzero element (11). The eigenvalues for the unperturbed **K** matrix are

$$\begin{aligned} \lambda_1 &= 1 + a + \sqrt{2a} \\ \lambda_2 &= 1 + a \\ \lambda_3 &= 1 + a - \sqrt{2a} \end{aligned} \tag{10.2.27}$$

Since the matrix is not symmetric, the column eigenvectors for the $\mathbf{S}$ matrix will be different from the row eigenvectors for the $\mathbf{S}^{-1}$ matrix. The correct row and column eigenvectors (Problem 10.11)

$$
\begin{aligned}
|1\rangle &= (1/4a)(a \quad \sqrt{2a} \quad 1) \\
|2\rangle &= (1/2a)(a \quad 0 \quad -1) \\
|3\rangle &= (1/4a)(a \quad -\sqrt{2a} \quad 1)
\end{aligned}
\tag{10.2.28}
$$

$$
|1\rangle = \begin{pmatrix} 1 \\ \sqrt{2a} \\ a \end{pmatrix} \qquad |2\rangle = \begin{pmatrix} 1 \\ 0 \\ -a \end{pmatrix} \qquad |3\rangle = \begin{pmatrix} 1 \\ -\sqrt{2a} \\ a \end{pmatrix} \tag{10.2.29}
$$

give similarity matrices

$$
\mathbf{S} = \begin{pmatrix} 1 & 1 & 1 \\ \sqrt{2a} & 0 & -\sqrt{2a} \\ a & -a & a \end{pmatrix} \tag{10.2.30}
$$

$$
\mathbf{S}^{-1} = \begin{pmatrix} a & \sqrt{2a} & 1 \\ 2a & 0 & -2 \\ a & -\sqrt{2a} & 1 \end{pmatrix} (4a)^{-1} \tag{10.2.31}
$$

The eigenvalue corrections are the diagonal elements of

$\mathbf{L} = \mathbf{S}^{-1}\mathbf{R}\mathbf{S}$

$$
= (1/4a) \begin{pmatrix} a & \sqrt{2a} & 1 \\ 2a & 0 & -2 \\ a & -\sqrt{2a} & 1 \end{pmatrix} \begin{pmatrix} -1 & 0 & 0 \\ 0 & 0 & 0 \\ 0 & 0 & 0 \end{pmatrix} \begin{pmatrix} 1 & 1 & 1 \\ \sqrt{2} & 0 & -\sqrt{2a} \\ a & -a & a \end{pmatrix}
$$

$$
= (1/4a) \begin{pmatrix} -a & -a & -a \\ -2a & -2a & -2a \\ -a & -a & -a \end{pmatrix} = \begin{pmatrix} -\frac{1}{4} & \cdot & \cdot \\ \cdot & -\frac{1}{2} & \cdot \\ \cdot & \cdot & -\frac{1}{4} \end{pmatrix} \tag{10.2.32}
$$

The diagonal perturbation matrix for the eigenvalues is independent of a since the perturbation corrects for only the integer 1 in the $\mathbf{R}$ matrix. Although this element appears only in the 11 position, it produces a correction for all three eigenvalues.

The eigenvalue matrix $\mathbf{\Lambda}'$ is

$$\mathbf{\Lambda}' = \mathbf{\Lambda} + \mathbf{L}$$

$$= \begin{pmatrix} 1+a-\sqrt{2a} & 0 & 0 \\ 0 & 1+a & 0 \\ 0 & 0 & 1+a+\sqrt{2a} \end{pmatrix} + \begin{pmatrix} -\frac{1}{4} & 0 & 0 \\ 0 & -\frac{1}{2} & 0 \\ 0 & 0 & -\frac{1}{4} \end{pmatrix}$$

$$= \begin{pmatrix} \frac{3}{4}-\sqrt{2a} & 0 & 0 \\ 0 & \frac{1}{2}+a & 0 \\ 0 & 0 & \frac{3}{4}+a+\sqrt{2a} \end{pmatrix} \tag{10.2.33}$$

The new eigenvalues, i.e., decay constants, are smaller than those of the unperturbed matrix.

The analysis of a multilevel system is simple if the perturbation "drains" molecules from an equilibrium distribution. The N-level truncated harmonic oscillator of Figure 10.1 has equally spaced levels. Using detailed balance and the definition

$$a = \exp(-\varepsilon/kT) \tag{10.2.34}$$

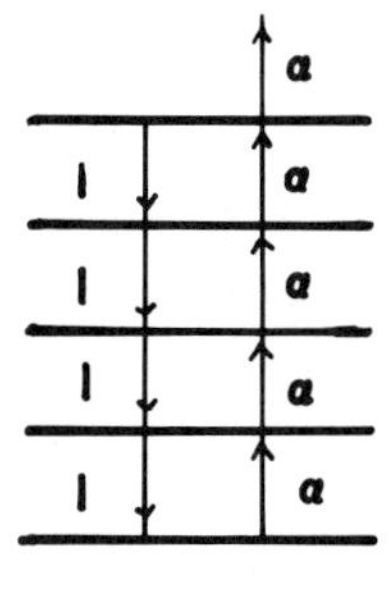

Figure 10.1. Transitions for the truncated harmonic oscillator.

where ε is the energy separation of the levels, the **K** matrix is

$$\begin{pmatrix} -a & 1 & 0 & 0 & \cdots & & \\ a & -(1+a) & 1 & 0 & \cdots & & \\ & & & & \vdots & & \\ & & & \cdots & -(1+a) & 1 \\ & & \cdots & 0 & a & -1-k' \end{pmatrix} \tag{10.2.35}$$

The zero eigenvalue gives the equilibrium eigenvector. A loss of molecules via the Nth level requires a nonzero eigenvalue which is deduced as a correction to the λ_0 eigenvalue.

The components for the $\lambda=0$ eigenvector are the Boltzmann factors for each level:

$$|0\rangle_i = \exp(-i\varepsilon/kT) = a^i \tag{10.2.36}$$

Since the columns of the matrix sum to zero, the row eigenvector for this non-symmetric matrix has components

$$\langle 0|_i = 1 \tag{10.2.37}$$

and the normalization is the partition function for the N-level system,

$$q = \sum a^i \tag{10.2.38}$$

If the equilibrium is disturbed by an irreversible transition from level N, the perturbation matrix **R** will have all zero elements except for the NN element. The perturbation is

$$\langle 0|\, \mathbf{R}\, |0\rangle \tag{10.2.39}$$

Since $k_{nn} = k'$ is the only nonzero element, operation with the **R** matrix gives an N-component vector in which all components equal zero except the Nth, which equals

$$k'a^N \tag{10.2.40}$$

The row vector sums all N components to give

$$k'a^N \tag{10.2.41}$$

i.e., the perturbation produces a steady drain from the equilibrium population of the Nth level. The rate constants within the manifold maintain the equilibrium distribution of the system.

10.3. First-Order Perturbation Theory—Eigenvectors

The first-order perturbation correction for the eigenvalues of a matrix used the matrix products from the first power of the dummy variable s. The matrix equation is

$$\mathbf{RS} + \mathbf{KSX} = \mathbf{SX\Lambda} + \mathbf{SL} \tag{10.3.1}$$

This same equation also determines the first-order correction for the eigenvectors. The analysis is presented for a single eigenvector.

The unknown eigenvector $|i'\rangle$ for the perturbed $\mathbf{K}'$ matrix satisfies the equation

$$\mathbf{K}' |i'\rangle = \lambda' |i'\rangle \tag{10.3.2}$$

The perturbation starts with the known eigenvector $|i\rangle$ for a $\mathbf{K}$ matrix, i.e.,

$$\mathbf{K} |i\rangle = \lambda |i\rangle \tag{10.3.3}$$

and

$$\mathbf{K}' = \mathbf{K} + \mathbf{R} \tag{10.3.4}$$

The dummy variable s is used to expand both λ' and $|i'\rangle$:

$$\lambda' = \lambda_0 + s\lambda_1 + s^2\lambda_2 + \cdots \tag{10.3.5}$$

$$|i'\rangle = |i_0\rangle + s |i_1\rangle + s^2 |i_2\rangle + \cdots \tag{10.3.6}$$

and $\mathbf{K}'$:

$$\mathbf{K}' = \mathbf{K} + s\mathbf{R} \tag{10.3.7}$$

Equation 10.3.2 becomes

$$(\mathbf{K} + s\mathbf{R})(|i_0\rangle + s |i_1\rangle + s^2 |i_2\rangle + \cdots)$$
$$= (\lambda_0 + s\lambda_1 + s^2\lambda_2)(|i_0\rangle + s |i_1\rangle + \cdots) \tag{10.3.8}$$

Expansion produces terms in increasing powers of s. Collecting terms for each power gives the following sequence of equations:

$$\mathbf{K} |i_0\rangle = \lambda_0 |i_0\rangle \tag{10.3.9}$$

$$\mathbf{K} |i_1\rangle + \mathbf{R} |i_0\rangle = \lambda_1 |i_0\rangle + \lambda_0 |i_1\rangle \tag{10.3.10}$$

$$\mathbf{K} |i_2\rangle + \mathbf{R} |i_1\rangle = \lambda_0 |i_2\rangle + \lambda_1 |i_1\rangle + \lambda_2 |i_0\rangle \tag{10.3.11}$$

The equations for a single eigenvector show the developing pattern for higher-order perturbations. The first equation involves only 0 subscripts, the second equation (first-order perturbation) involves only subscripts 0 and 1, and the third (second-order perturbation) involves subscripts 0, 1, and 2. The equation for second-order perturbations will utilize the results from first-order perturbations.

Equation 10.3.10 generates the first-order correction for the eigenvector:

$$\mathbf{K}\ |i_1\rangle + \mathbf{R}\ |i_0\rangle = \lambda_1\ |i_0\rangle + \lambda_0\ |i_1\rangle \tag{10.3.12}$$

The eigenvalue correction is found by applying the row vector $\langle i_0|$ from the left:

$$\langle i_0|\ \mathbf{K}\ |i_1\rangle + \langle i_0|\ \mathbf{R}\ |i_0\rangle = \lambda_1 \langle i_0 | i_0\rangle + \lambda_0 \langle i_0 | i_1\rangle \tag{10.3.13}$$

Although $\mathbf{K}\ |i_1\rangle$ is not known, the **K** matrix can operate to the left on the known bra eigenvector, $\langle i_0|$, to give

$$\langle i_0|\ K = \langle i_0|\ \lambda_i \tag{10.3.14}$$

The equation becomes

$$\lambda_0 \langle i_0 | i_1\rangle + \langle i_0|\ \mathbf{R}\ |i_0\rangle = \lambda_1 + \lambda_0 \langle i_0 | i_1\rangle \tag{10.3.15}$$

Elimination of equal terms on each side gives

$$\lambda_1 = \langle i_0|\ \mathbf{R}\ |i_0\rangle \tag{10.3.16}$$

as determined in Section 10.2.

The eigenvector correction $|i_1\rangle$ is determined from the nondiagonal elements generated with the **R** matrix. The first-order perturbation equation for $|i_0\rangle$ and $|i_1\rangle$ is left multiplied with $\langle j_0|$ with $j \neq i$ so that

$$\langle j_0 | i_0\rangle = 0 \tag{10.3.17}$$

and

$$\langle j_0|\ \mathbf{K} = \langle j_0|\ \lambda_{0,j} \tag{10.3.18}$$

The equation is

$$\langle j_0|\ \mathbf{K}\ |i_1\rangle + \langle j_0|\ \mathbf{R}\ |i_0\rangle = 0 + \lambda_{0,i} \langle j_0 | i_1\rangle \tag{10.3.19}$$

where the subscript 0, i indicates an unperturbed eigenvalue i:

$$\lambda_{0,j} \langle j_0 | i_1\rangle + \langle j_0|\ \mathbf{R}\ |i_0\rangle = 0 + \lambda_{0,i} \langle j_0 | i_1\rangle \tag{10.3.20}$$

The projection

$$\langle j_0 | i_1\rangle \tag{10.3.21}$$

gives the projection of the corrected eigenvector $|i_1\rangle$ on each of the known eigenvectors $\langle j_0|$. The perturbation has rotated the original $|i_0\rangle$ so that it now has finite projections on the remaining vectors $|j_0\rangle$ in the space. The $\langle j_0|$ were orthogonal to the original $|i_0\rangle$. For example, for $\langle j_0|i_1\rangle$:

$$(\lambda_{0,i}-\lambda_{0,j})\langle j_0|i_1\rangle = \langle j_0|\,\mathbf{R}\,|i_0\rangle \tag{10.3.22}$$

$$\langle j_0|i_1\rangle = \langle j_0|\,\mathbf{R}\,|i_0\rangle/(\lambda_{0,i}-\lambda_{0,j}) \tag{10.3.23}$$

The correction projection is defined entirely by the zeroth order eigenvalues and eigenvectors. $|i_0\rangle$ does not appear since the perturbation describes only projections on the remaining basis vectors.

Since each $\langle j_0|$ gives a projection for the corrected eigenvector $|i_1\rangle$, the correction to the original eigenvector is the sum of each scalar projection multiplied by its direction in the space, i.e., $|j_0\rangle$:

$$|i_{\text{corr}}\rangle = \sum_{j_0} |j_0\rangle\langle j_0|\,\mathbf{R}\,|i_0\rangle/(\lambda_{0,i}-\lambda_{0,j}) \tag{10.3.24}$$

The original perturbed eigenvector was written as a summation of zeroth, first, etc., perturbation eigenvectors. With $s=1$, the correction eigenvector of Equation 10.3.24 must be added to $|i_0\rangle$ to give the new eigenvector for a first-order perturbation:

$$|i_1\rangle = |i_0\rangle + \sum_{j_0} |j_0\rangle\langle j_0|\,\mathbf{R}\,|i_0\rangle/(\lambda_{0,i}-\lambda_{0,j}\rangle \tag{10.3.25}$$

The technique is easily illustrated using the simple 2×2 system introduced in Section 10.1. The **R** matrix for this system is

$$\begin{pmatrix} 0 & 0.1 \\ 0 & -0.1 \end{pmatrix} \tag{10.3.26}$$

The unperturbed **K** matrix (Equation 10.1.4) had eigenvalues 0 and -2. The $|0\rangle$ eigenvector,

$$\left(\frac{1}{\sqrt{2}}\right)\begin{pmatrix} 1 \\ 1 \end{pmatrix} \tag{10.3.27}$$

is corrected using a projection on the remaining $|-2\rangle$ eigenvector for this two-dimensional system. This eigenvector,

$$\left(\frac{1}{\sqrt{2}}\right)\begin{pmatrix} 1 \\ -1 \end{pmatrix} \tag{10.3.28}$$

gives the correction for the $|0\rangle$ eigenvector as

$$|0_{\text{corr}}\rangle = |2\rangle\langle 2|\,\mathbf{R}\,|0\rangle/[0-(-2)] \tag{10.3.29}$$

where

$$\langle 2|\,\mathbf{R}\,|0\rangle = \left(\frac{1}{2}\right)(1 \quad -1)\begin{pmatrix} 0 & 0.1 \\ 0 & -0.1 \end{pmatrix}\begin{pmatrix} 1 \\ 1 \end{pmatrix} = 0.1 \tag{10.3.30}$$

so that

$$|0_{\text{corr}}\rangle = (0.1/\sqrt{2})\begin{pmatrix} 1 \\ -1 \end{pmatrix}(+2)^{-1} \tag{10.3.31}$$

and the corrected eigenvector, $|0_1\rangle$, is

$$\begin{aligned} |0_1\rangle &= |0_0\rangle + |0_{\text{corr}}\rangle \\ &= \left(\frac{1}{\sqrt{2}}\right)\begin{pmatrix} 1 \\ 1 \end{pmatrix} + \left(\frac{1}{\sqrt{2}}\right)\begin{pmatrix} 0.05 \\ -0.05 \end{pmatrix} \\ &= \left(\frac{1}{\sqrt{2}}\right)\begin{pmatrix} 1.05 \\ 0.95 \end{pmatrix} \end{aligned} \tag{10.3.32}$$

If the second component is set equal to 1, the vector becomes

$$\begin{pmatrix} 1.1 \\ 1 \end{pmatrix} \tag{10.3.33}$$

The components are identical to those determined via a direct solution of the $\mathbf{K}'$ matrix in Section 10.1.

The first-order perturbation for $|-2\rangle$ is

$$|2_{\text{corr}}\rangle = |0\rangle\langle 0|\,\mathbf{R}\,|2\rangle/(2-0) = 0 \tag{10.3.34}$$

since

$$\langle 0|\,\mathbf{R}\,|2\rangle = 0 \tag{10.3.35}$$

There is no first-order perturbation correction even though the actual eigenvector $|-2'\rangle$ for $\mathbf{K}'$ is different from $|-2\rangle$ for $\mathbf{K}$ (Equation 10.1.6).

Perturbation theory can be used when a chain of carbon atoms includes one N atom, e.g.,

$$\text{C—C—C} \rightarrow \text{N—C—C}$$

The atomic orbital associated with the nitrogen will differ from the carbon atomic orbitals and will require the definitions of new integrals α' and β' associated with the nitrogen atom. The matrix elements are

$$\begin{aligned} \alpha &= \langle C_i|\,\mathbf{H}\,|C_i\rangle \\ \beta &= \langle C_j|\,\mathbf{H}\,|C_i\rangle = \langle C_i|\,\mathbf{H}\,|C_j\rangle \\ \alpha' &= \langle N_i|\,\mathbf{H}\,|N_i\rangle \\ \beta' &= \langle N_i|\,\mathbf{H}\,|C_j\rangle = \langle C_i|\,\mathbf{H}\,|N_j\rangle \end{aligned} \tag{10.3.36}$$

C_i is the carbon atomic orbital while N_i is the nitrogen atomic orbital.

The Hückel matrix is

$$\begin{pmatrix} \alpha' & \beta' & 0 \\ \alpha' & \alpha & \beta \\ 0 & \beta & \alpha \end{pmatrix} \tag{10.3.37}$$

If the nitrogen atomic orbital is not orthogonal to the carbon atomic orbital, the **S** matrix must be included in the calculation. For this example, however,

$$\mathbf{S} = \mathbf{I} \tag{10.3.38}$$

If α' and β' are close to α and β, the three-carbon chain serves as the starting matrix **K**, and α' and β' elements are written in terms of α and β and two perturbations, c and d, respectively:

$$\begin{aligned} \alpha' &= \alpha + c \\ \beta' &= \beta + d \end{aligned} \tag{10.3.39}$$

The matrices are now

$$\begin{pmatrix} \alpha' & \beta' & 0 \\ \beta' & \alpha & \beta \\ 0 & \beta & \alpha \end{pmatrix} = \begin{pmatrix} \alpha & \beta & 0 \\ \beta & \alpha & \beta \\ 0 & \beta & \alpha \end{pmatrix} + \begin{pmatrix} c & d & 0 \\ d & 0 & 0 \\ 0 & 0 & 0 \end{pmatrix} \tag{10.3.40}$$

The eigenvalues and eigenvectors for the three-carbon system are (Chapter 8)

$$\alpha + \beta\sqrt{2} = \left(\frac{1}{2}\right)\begin{pmatrix} 1 \\ \sqrt{2} \\ 1 \end{pmatrix} = |1\rangle \tag{10.3.41}$$

$$\alpha = \left(\frac{1}{\sqrt{2}}\right)\begin{pmatrix} 1 \\ 0 \\ -1 \end{pmatrix} = |2\rangle \tag{10.3.42}$$

$$\alpha - \beta\sqrt{2} = \left(\frac{1}{2}\right)\begin{pmatrix} 1 \\ -\sqrt{2} \\ 1 \end{pmatrix} = |3\rangle \tag{10.3.43}$$

The corrections to the eigenvectors and eigenvalues are

$$\lambda_1' = \lambda_1 + \langle 1 | \mathbf{R} | 1 \rangle$$

$$= \alpha + \beta\sqrt{2} + \left(\frac{1}{4}\right) (1 \quad \sqrt{2} \quad 1) \begin{pmatrix} c & d & 0 \\ d & 0 & 0 \\ 0 & 0 & 0 \end{pmatrix} \begin{pmatrix} 1 \\ \sqrt{2} \\ 1 \end{pmatrix}$$

$$\lambda_1 = \alpha + \beta\sqrt{2} + c/4 + \sqrt{2}\, d/2 \tag{10.3.44}$$

$$\lambda_2' = \lambda_2 + \langle 2 | \mathbf{R} | 2 \rangle$$

$$= \alpha + \left(\frac{1}{2}\right) (1 \quad 0 \quad -1) \begin{pmatrix} c & d & 0 \\ d & 0 & 0 \\ 0 & 0 & 0 \end{pmatrix} \begin{pmatrix} 1 \\ 0 \\ -1 \end{pmatrix}$$

$$\lambda_2' = \alpha + c/2 \tag{10.3.45}$$

$$\lambda_3' = \lambda_3 + \langle 3 | \mathbf{R} | 3 \rangle$$

$$= \alpha - \beta\sqrt{2} + \left(\frac{1}{4}\right) (1 \quad -\sqrt{2} \quad 1) \begin{pmatrix} c & d & 0 \\ d & 0 & 0 \\ 0 & 0 & 0 \end{pmatrix} \begin{pmatrix} 1 \\ -\sqrt{2} \\ 1 \end{pmatrix}$$

$$\lambda_3' = \alpha - \beta\sqrt{2} + c/4 - d/\sqrt{2} \tag{10.3.46}$$

For this 3×3 matrix, each eigenvector correction must have projections for the remaining two eigenvectors, i.e., each correction will have two terms. For example,

$$|1'\rangle = |1\rangle + \frac{|2\rangle\langle 2| \mathbf{R} |1\rangle}{\alpha + \beta\sqrt{2} - \alpha} + \frac{|3\rangle\langle 3| \mathbf{R} |1\rangle}{\alpha + \beta\sqrt{2} - (\alpha + \beta\sqrt{2})} \tag{10.3.47}$$

The relevant off-diagonal elements are

$$\langle 2 | \mathbf{R} | 1 \rangle = (1 \quad 0 \quad -1) \begin{pmatrix} c & d & 0 \\ d & 0 & 0 \\ 0 & 0 & 0 \end{pmatrix} \begin{pmatrix} 1 \\ \sqrt{2} \\ 1 \end{pmatrix} [1/(2\sqrt{2})] = (c + d\sqrt{2})/4 \tag{10.3.48}$$

$$\langle 3 | \mathbf{R} | 1 \rangle = (1 \quad -\sqrt{2} \quad 1) \begin{pmatrix} c & d & 0 \\ d & 0 & 0 \\ 0 & 0 & 0 \end{pmatrix} \begin{pmatrix} 1 \\ \sqrt{2} \\ 1 \end{pmatrix} \left(\frac{1}{4}\right) = c/4 \tag{10.3.49}$$

The eigenvector $|1'\rangle$ is

$$|1'\rangle = \begin{pmatrix} 1 \\ \sqrt{2} \\ 0 \end{pmatrix} \left(\frac{1}{2}\right) + \begin{pmatrix} 1 \\ 0 \\ -1 \end{pmatrix} [(c + d\sqrt{2})/4\sqrt{2}\beta] + \begin{pmatrix} 1 \\ -\sqrt{2} \\ 1 \end{pmatrix} (c/8\sqrt{2}b) \tag{10.3.50}$$

Two correction vectors are required to generate the final eigenvector $|1'\rangle$. The procedure must be repeated for $|2'\rangle$ and $|3'\rangle$.

The change in the energies and wavefunctions for the heteroatomic chain is illustrated with some numerical values for the parameters. The unperturbed eigenvectors were independent of α and β. The parameters are

$$\begin{aligned} \alpha &= 2, \qquad & c &= 0.2 \\ \beta &= -1, \qquad & d &= 0.1 \end{aligned} \tag{10.3.51}$$

The corrected eigenvalues are

$$\begin{aligned} 1' &= 2 - 1.414 + 0.05 + 0.0717 = 0.586 + 0.1217 \\ 2' &= 2 + 0.1 \\ 3' &= 2 + 1.414 + 0.05 - 0.0717 = 3.414 - 0.0217 \end{aligned} \tag{10.3.52}$$

The perturbations are small relative to the contribution from the unperturbed eigenvalues.

The eigenvector $|1'\rangle$ is

$$|1'\rangle = \begin{pmatrix} 0.5 \\ 0.707 \\ 0.5 \end{pmatrix} + \begin{pmatrix} -0.0854 \\ 0 \\ +0.0854 \end{pmatrix} + \begin{pmatrix} -0.0088 \\ +0.0124 \\ -0.0088 \end{pmatrix} = \begin{pmatrix} 0.406 \\ 0.719 \\ 0.577 \end{pmatrix} \tag{10.3.53}$$

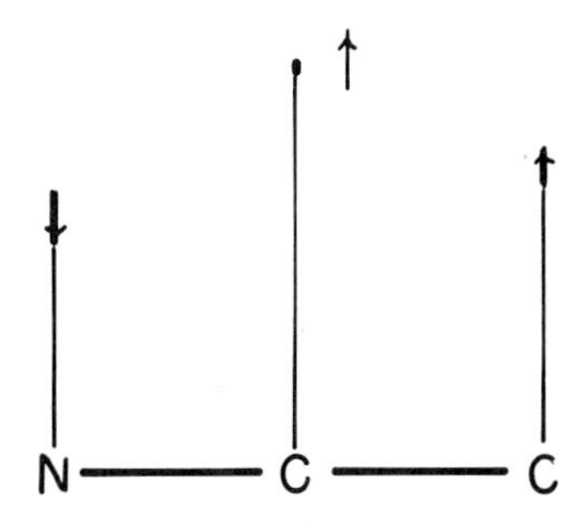

Figure 10.2. Corrected molecular orbitals for the molecule NCC using first-order perturbation theory.

The first-order eigenvectors for the remaining two eigenvalues are

$$|2'\rangle = \begin{pmatrix} 0.671 \\ 0.05 \\ -0.7427 \end{pmatrix} \qquad |3'\rangle = \begin{pmatrix} 0.514 \\ -0.702 \\ 0.493 \end{pmatrix} \tag{10.3.54}$$

(Problem 10.3). The three eigenvectors are compared with the unperturbed eigenvectors in Figure 10.2.

10.4. Second-Order Perturbation Theory—Eigenvalues

Second-order perturbations are determined from the s^2 terms in the perturbation expansion, i.e.,

$$\mathbf{K}\,|i_2\rangle + \mathbf{R}\,|i_1\rangle = \lambda_{i,0}\,|i_2\rangle + \lambda_{i,1}\,|i_1\rangle + \lambda_{i,2}\,|i_0\rangle \tag{10.4.1}$$

Corrections for a single eigenvalue and its eigenvector are developed; the full matrix solution is left as an exercise. The vectors $|i_1\rangle$ and $|i_2\rangle$ are first- and second-order corrections for the original eigenvector $|i_0\rangle = |i\rangle$.

The second-order eigenvalue correction is determined by operating with the bra eigenvector $\langle i_0|$ from the left:

$$\langle i_0|\,\mathbf{K}\,|i_2\rangle + \langle i_0|\,\mathbf{R}\,|i_1\rangle = \lambda_{i,0}\langle i_0 | i_2\rangle + \lambda_{i,1}\langle i_0 | i_1\rangle + \lambda_{i,2}\langle i_0 | i_0\rangle \tag{10.4.2}$$

If $|i_1\rangle$ and $|i_2\rangle$ correct $|i_0\rangle$, they must be orthogonal to it:

$$\begin{aligned} \langle i_0 | i_1\rangle &= 0 \\ \langle i_0 | i_2\rangle &= 0 \end{aligned} \tag{10.4.3}$$

The equation reduces to

$$\langle i_0|\,\mathbf{K}\,|i_2\rangle + \langle i_0|\,\mathbf{R}\,|i_1\rangle = 0 + 0 + \lambda_{i,2} \tag{10.4.4}$$

Since $|i_0\rangle$ is an eigenvector for $\mathbf{K}$, the first term on the left becomes

$$\langle i_0|\,\mathbf{K}\,|i_2\rangle = \lambda_{i,0}\langle i_0 | i_2\rangle = 0 \tag{10.4.5}$$

and the equation reduces to a simple second-order correction for this eigenvalue:

$$\lambda_{i,2} = \langle i_0|\,\mathbf{R}\,|i_1\rangle \tag{10.4.6}$$

Equation 10.4.6 is deceptively simple. It depends on $|i_1\rangle$, the correction eigenvector for $|i\rangle$ in first order.

The eigenvector correction $|i_1\rangle$ using first-order perturbation theory was determined in Section 10.3 as

$$|i_1\rangle = \sum |j_0\rangle \frac{\langle j_0| \mathbf{R} |i_0\rangle}{(\lambda_{i,0} - \lambda_{j,0})} \tag{10.4.7}$$

Equation 10.4.7 is substituted into Equation 10.4.6 to give

$$\lambda_{i,2} = \langle i_0| \mathbf{R} |i_1\rangle = \sum_j \langle i_0| \mathbf{R} |j_0\rangle\langle j_0| \mathbf{R} |i_0\rangle/(\lambda_{i,0} - \lambda_{j,0}) \tag{10.4.8}$$

The equation requires a summation over all $|j\rangle$ eigenvectors which are orthogonal to $|i_0\rangle$. The terms in the summation are products of quadratic forms using the perturbation matrix **R** and the eigenvectors for the unperturbed matrix **K**. The matrix with *ji* element,

$$R_{ji} = \langle j_0| \mathbf{R} |i_0\rangle \tag{10.4.9}$$

i.e.,

$$\mathbf{S}^{-1}\mathbf{R}\mathbf{S} \tag{10.4.10}$$

contains all the information required to determine both first- and second-order perturbations.

The second-order eigenvalue perturbation for a matrix **K**′ is determined by using a tractable **K** matrix,

$$\mathbf{K}' = \mathbf{K} + \mathbf{R}$$

$$\begin{pmatrix} 2.1 & 1.2 & 0 \\ 0.9 & 2.1 & 1.1 \\ 0 & 1.2 & 2 \end{pmatrix} = \begin{pmatrix} 2 & 1 & 0 \\ 1 & 2 & 1 \\ 0 & 1 & 2 \end{pmatrix} + \begin{pmatrix} 0.1 & 0.2 & 0 \\ -0.1 & 0.2 & 0.1 \\ 0 & 0.2 & 0 \end{pmatrix} \tag{10.4.11}$$

The symmetric **K** matrix has eigenvectors and eigenvalues,

$$|1_0\rangle = 2^{-1}\begin{pmatrix} 1 \\ \sqrt{2} \\ 1 \end{pmatrix} \qquad \lambda_{0,1} = 2 + \sqrt{2} \tag{10.4.12}$$

$$|2_0\rangle = (1/\sqrt{2})\begin{pmatrix} 1 \\ 0 \\ -1 \end{pmatrix} \qquad \lambda_{0,2} = 2 \tag{10.4.13}$$

$$|3_0\rangle = (1/2)\begin{pmatrix} 1 \\ -\sqrt{2} \\ 1 \end{pmatrix} \qquad \lambda_{0,3} = 2 - \sqrt{2} \tag{10.4.14}$$

which form the $\mathbf{S}$ and $\mathbf{S}^{-1}$ matrices. The similarity transform

$$\mathbf{S}^{-1}\mathbf{R}\mathbf{S} \tag{10.4.15}$$

gives all the required matrix elements.

The $\mathbf{S}$ and $\mathbf{S}^{-1}$ matrices are identical:

$$\mathbf{S} = \left(\frac{1}{4}\right)\begin{pmatrix} 1 & \sqrt{2} & 1 \\ \sqrt{2} & 0 & -\sqrt{2} \\ 1 & -\sqrt{2} & 1 \end{pmatrix} \tag{10.4.16}$$

and the perturbation matrix (Equation 10.4.15) is

$$\left(\frac{1}{4}\right)\begin{pmatrix} 1 & \sqrt{2} & 1 \\ \sqrt{2} & 0 & -\sqrt{2} \\ 1 & -\sqrt{2} & 1 \end{pmatrix}\begin{pmatrix} 0.1 & 0.2 & 0 \\ -0.1 & 0.2 & 0.1 \\ 0 & 0.2 & 0 \end{pmatrix}\begin{pmatrix} 1 & \sqrt{2} & 1 \\ \sqrt{2} & 0 & -\sqrt{2} \\ 1 & -\sqrt{2} & 1 \end{pmatrix}$$

$$= \left(\frac{1}{4}\right)\begin{pmatrix} 0.5+0.4\sqrt{2} & 0.1\sqrt{2}-0.4 & -0.4\sqrt{2}-0.3 \\ 0.1\sqrt{2} & 0.2 & 0.1\sqrt{2} \\ 0.4\sqrt{2}-0.3 & 0.1\sqrt{2}+0.4 & 0.5-0.4\sqrt{2} \end{pmatrix}$$

$$= \begin{pmatrix} 0.27 & -0.065 & -0.216 \\ 0.035 & 0.05 & 0.038 \\ 0.066 & 0.135 & -0.01 \end{pmatrix} \tag{10.4.17}$$

for

$$\begin{pmatrix} \langle 1|\,\mathbf{R}\,|1\rangle & \langle 1|\,\mathbf{R}\,|2\rangle & \langle 1|\,\mathbf{R}\,|3\rangle \\ \langle 2|\,\mathbf{R}\,|1\rangle & \langle 2|\,\mathbf{R}\,|2\rangle & \langle 2|\,\mathbf{R}\,|3\rangle \\ \langle 3|\,\mathbf{R}\,|1\rangle & \langle 3|\,\mathbf{R}\,|2\rangle & \langle 3|\,\mathbf{R}\,|3\rangle \end{pmatrix} \tag{10.4.18}$$

The first-order eigenvalue corrections give

$$\begin{aligned} \lambda_{1,1} &= 2+\sqrt{2}+0.27 \\ \lambda_{1,2} &= 2+0.05 \\ \lambda_{1,3} &= 2-\sqrt{2}-0.01 \end{aligned} \tag{10.4.19}$$

The relative size of the correction terms is small for each eigenvalue. The eigenvalue subscript $1, i$ represents a first-order correction on the ith eigenvalue.

The eigenvector corrections with first-order perturbation theory are

$$|i_1\rangle = |i_0\rangle + \sum |j_0\rangle \frac{\langle j_0|\,\mathbf{R}\,|i_0\rangle}{(\lambda_{i,0}-\lambda_{j,0})} \tag{10.4.20}$$

Each summation for the 3×3 matrix has two terms. The $|1_1\rangle$ eigenvector is

$$|1_1\rangle = |1_0\rangle + |2_0\rangle \frac{\langle 2|\,\mathbf{R}\,|1\rangle}{2+\sqrt{2}-2} + |3_0\rangle \frac{\langle 3|\,\mathbf{R}\,|1\rangle}{2+\sqrt{2}-2+\sqrt{2}} \tag{10.4.21}$$

The quadratic forms $\langle j|\,\mathbf{R}\,|i\rangle$ are selected from the perturbation matrix (Equation 10.4.18):

$$\begin{aligned} \langle 2|\,\mathbf{R}\,|1\rangle &= 0.035; &\quad \langle 2|\,\mathbf{R}\,|1\rangle/\sqrt{2} &= 0.025 \\ \langle 3|\,\mathbf{R}\,|1\rangle &= 0.066; &\quad \langle 3|\,\mathbf{R}\,|1\rangle/2\sqrt{2} &= 0.023 \end{aligned} \tag{10.4.22}$$

to give the corrected eigenvector

$$\begin{aligned} |1_1\rangle &= |1_0\rangle + 0.025\,|2_0\rangle + 0.023\,|3_0\rangle \\ &= \begin{pmatrix} 0.5 \\ 0.707 \\ 0.5 \end{pmatrix} + \begin{pmatrix} 0.018 \\ 0 \\ -0.018 \end{pmatrix} + \begin{pmatrix} 0.012 \\ -0.018 \\ 0.012 \end{pmatrix} = \begin{pmatrix} 0.53 \\ 0.69 \\ 0.49 \end{pmatrix} \end{aligned} \tag{10.4.23}$$

The perturbation enhances the first component at the expense of the others. However, the components are no longer symmetric, i.e., the 1 and 3 components are now different. The procedure can be repeated for the remaining two eigenvectors (Problem 10.12).

Since $|1_1\rangle$ is known, the second-order eigenvalue correction is determined by substituting $|1_1\rangle$ into

$$\lambda_{2,1} = \langle i_0|\,\mathbf{R}\,|i_1\rangle \tag{10.4.24}$$

It may also be determined directly by substituting the elements of Equation 10.4.18 into

$$\lambda_{2,i} = \sum \frac{\langle i_0|\,\mathbf{R}\,|j_0\rangle\langle j_0|\,\mathbf{R}\,|i_0\rangle}{\lambda_{i,0} - \lambda_{j,0}} \tag{10.4.25}$$

The second-order correction to the first eigenvalue of the 3×3 matrix is

$$\lambda_{2,1} = \frac{\langle 1|\,\mathbf{R}\,|2\rangle\langle 2|\,\mathbf{R}\,|1\rangle}{\lambda_{0,1} - \lambda_{0,2}} + \frac{\langle 1|\,\mathbf{R}\,|3\rangle\langle 3|\,\mathbf{R}\,|1\rangle}{\lambda_{0,1} - \lambda_{0,3}} \tag{10.4.26}$$

The appropriate matrix elements are

$$\begin{aligned} \langle 1|\,\mathbf{R}\,|2\rangle\langle 2|\,\mathbf{R}\,|1\rangle &= (-0.065)(0.035) = -0.0023 \\ \langle 1|\,\mathbf{R}\,|2\rangle\langle 2|\,\mathbf{R}\,|1\rangle/(2+\sqrt{2}-2) &= -0.0016 \\ \langle 1|\,\mathbf{R}\,|3\rangle\langle 3|\,\mathbf{R}\,|1\rangle &= (-0.216)(0.066) = -0.014 \\ \langle 1|\,\mathbf{R}\,|3\rangle\langle 3|\,\mathbf{R}\,|1\rangle/(2+\sqrt{2}-2+\sqrt{2}) &= 0.005 \end{aligned} \tag{10.4.27}$$

The corrected eigenvalue is the sum of the original eigenvalue and the first- and second-order corrections:

$$\lambda_1' = 3.414 + (0.27) + (0.005 - 0.016) = 3.673 \tag{10.4.28}$$

The corrections to the eigenvalue become smaller with increasing perturbation order. If the initial choice of **K** matrix is good, accurate eigenvalues for **K**′ can be determined using only first- and second-order perturbation theory.

10.5. Second-Order Perturbation Theory—Eigenvectors

The sequence of calculations for corrected eigenvalues and eigenvectors proceeded in a regular order. Diagonal elements of a matrix generated from the expression for a first-order perturbation generated the first-order eigenvalue corrections. The off-diagonal elements of this matrix gave the first-order eigenvector corrections. The expression for second-order perturbations of the eigenvalues is written as a sum of products of these matrix elements. The second-order perturbation for the eigenvectors uses products of these same matrix elements.

The second-order perturbation equation is

$$\mathbf{K}\,|i_2\rangle + \mathbf{R}\,|i_1\rangle = \lambda_{0,i}\,|i_2\rangle + \lambda_{1,i}\,|i_1\rangle + \lambda_{2,i}\,|i_0\rangle \tag{10.5.1}$$

The eigenvector perturbation requires application of a bra vector $\langle j_0| = \langle i_0|$ to all terms in the equation:

$$\langle j_0|\,\mathbf{K}\,|i_2\rangle + \langle j_0|\,\mathbf{R}\,|i_1\rangle = \lambda_{0,i}\langle j_0\,|\,i_2\rangle + \lambda_{1,i}\langle j_0\,|\,i_1\rangle + \lambda_{2,i}\langle j_0\,|\,i_0\rangle \tag{10.5.2}$$

Since

$$\langle j_0|\,\mathbf{K}\,|i_2\rangle = \lambda_{0,j}\langle j_0\,|\,i_2\rangle \tag{10.5.3}$$

$$\langle j_0\,|\,i_0\rangle = 0 \tag{10.5.4}$$

the equation becomes

$$(\lambda_{0,i} - \lambda_{0,j})\langle j_0\,|\,i_2\rangle = \langle j_0|\,\mathbf{R}\,|i_1\rangle - \lambda_{1,i}\langle j_0\,|\,i_1\rangle - 0 \tag{10.5.5}$$

The equation is solved for the projection of the second-order eigenvector correction, $|i_2\rangle$, on $\langle j_0|$:

$$\langle j_0\,|\,i_2\rangle = \frac{\langle j_0|\,\mathbf{R}\,|i_1\rangle - \lambda_{1,i}\langle j_0\,|\,i_1\rangle}{\lambda_{0,i} - \lambda_{0,j}} \tag{10.5.6}$$

Both $\langle j_0|\,\mathbf{R}$ and $\langle j_0|$ operate on $|i_1\rangle$, not $|i_0\rangle$. The correction for the eigenvector requires a sum of such projections including their orientation in the space, i.e.,

$$\sum_j |j_0\rangle\langle j_0\,|\,i_2\rangle = \sum_j |j_0\rangle \frac{\langle j_0|\,\mathbf{R}\,|i_1\rangle - \lambda_{1,i}\langle j_0\,|\,i_1\rangle}{\lambda_{0,i} - \lambda_{0,j}} \tag{10.5.7}$$

The first-order correction is also a sum requiring a second index k:

$$|i_1\rangle = \frac{\sum |k\rangle\langle k| \mathbf{R} |i\rangle}{\lambda_{0,i} - \lambda_{0,k}} \tag{10.5.8}$$

The subscript 0 is dropped from matrix elements with zeroth-order eigenvectors.

If Equation 10.5.8 for $|i_1\rangle$ is substituted into Equation 10.5.7, each $\langle j|$ component is

$$\langle j|i_2\rangle = \sum_k \frac{\dfrac{\langle j| \mathbf{R} |k\rangle\langle k| \mathbf{R} |i\rangle}{(\lambda_{0,i} - \lambda_{0,k})} - \dfrac{\lambda_{1,i}\langle j|k\rangle\langle k| \mathbf{R} |i\rangle}{(\lambda_{0,i} - \lambda_{0,k})}}{(\lambda_{0,i} - \lambda_{0,j})} \tag{10.5.9}$$

The second term is "simplified" by substituting the expression for the first-order eigenvalue perturbation,

$$\lambda_{1,i} = \langle i_0| \mathbf{R} |i_0\rangle = \langle i| \mathbf{R} |i\rangle \tag{10.5.10}$$

The summation over k gives a nonzero term only when $j = k$ since

$$\langle j|k\rangle = 1 \quad \text{if} \quad j = k \tag{10.5.11}$$

With Equation 10.5.10, the second term is

$$\langle i| \mathbf{R} |i\rangle\langle j| \mathbf{R} |i\rangle/(\lambda_{0,i} - \lambda_{0,j})^2 \tag{10.5.12}$$

The projection of $|i_2\rangle$ on a single $\langle j|$ vector is

$$\langle j|i_2\rangle = \sum_k \frac{\langle j| \mathbf{R} |k\rangle\langle k| \mathbf{R} |i\rangle}{(\lambda_{0,i} - \lambda_{0,k})(\lambda_{0,i} - \lambda_{0,j})} - \sum_j \frac{\langle i| \mathbf{R} |i\rangle\langle j| \mathbf{R} |i\rangle}{(\lambda_{0,i} - \lambda_{0,j})^2} \tag{10.5.13}$$

Only the first term sums over the index k. Each term depends only on matrix elements from the similarity transform

$$\mathbf{S}^{-1}\mathbf{RS} \tag{10.5.14}$$

Products of these elements divided by the appropriate eigenvalue differences give the projections on each $\langle j|$. The total eigenvector correction is a summation over all $|j\rangle$, i.e.,

$$|i_{2,\text{corr}}\rangle = |j\rangle\langle j|i_2\rangle \tag{10.5.15}$$

Perturbation theory has been applied to matrices of low order; it may also be applied to larger matrices in which the elements of the $\mathbf{S}^{-1}\mathbf{RS}$ matrix are minimized. In the following, the energy eigenvalues and eigenvectors for a particle confined in a circular orbit of constant radius will be perturbed by an applied electric field (Figure 10.3) and determined using first- and second-order perturbation theory.

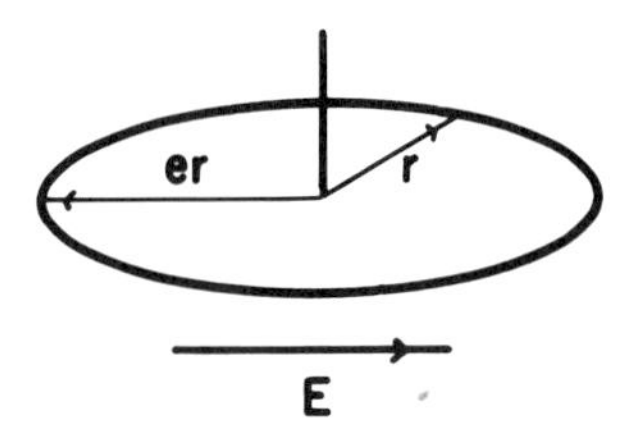

Figure 10.3. An electron confined on a ring of radius r, showing dipole moment and electric field vectors.

In the absence of the applied field, the particle on the ring has eigenfunctions

$$|m\rangle = \exp(im\varphi)/2\pi \tag{10.5.16}$$

for a Hamiltonian operator

$$\mathbf{H} = -(\hbar^2/2\mathscr{I})\, d^2/d\varphi^2 \tag{10.5.17}$$

where $\mathscr{I} = mr^2$ is the moment of inertia for the particle. The energy for the electron with quantum number m $(-l \leqslant m \leqslant l)$ is determined by applying the Hamiltonian to the wavefunction:

$$\begin{aligned} \mathbf{H}\,|m\rangle &= E_m\,|m\rangle \\ &= -(\hbar^2/2\mathscr{I})\, d^2/d\varphi^2[\exp(im\varphi)] \\ &= (\hbar^2 m^2/2\mathscr{I}) \exp(im\varphi) = E_m \exp(im\varphi) \end{aligned} \tag{10.5.18}$$

The wavefunctions are a valid basis set for the particle on a ring. The energies for each $|m|$ are doubly degenerate since positive and negative m are allowed. Since the perturbation expressions require differences of eigenvalues in the denominator, such degeneracies can create a singularity in the expression. However, there are a limited number of nonzero matrix elements in the $\mathbf{S}^{-1}\mathbf{RS}$ matrix. The singularities are associated with terms having zero matrix elements, and these terms are ignored in this analysis.

This system can be perturbed by applying a steady electric field which is directed along the x axis (in the x–y plane) for convenience. The energy perturbation operator is the dot product of the dipole moment

$$\boldsymbol{\mu} = e\mathbf{r} \tag{10.5.19}$$

and the electric field $\mathscr{E}$,

$$-\mu \cdot \mathscr{E} = -\mu\mathscr{E} \cos\varphi \tag{10.5.20}$$

where φ is the angle between the dipole moment along $\mathbf{r}$ and the field.

The new Hamiltonian is

$$\mathbf{H}' = \mathbf{H} + \mathbf{R} = -(\hbar^2/2\mathscr{I})\, d^2/d\varphi^2 - \mu\mathscr{E} \cos\varphi \tag{10.5.21}$$

The perturbation **R** is used to find all matrix elements

$$\langle m|\, \mathbf{R}\, |m\rangle \qquad \langle m'|\, \mathbf{R}\, |m\rangle \tag{10.5.22}$$

for application of perturbation theory. Although the system has an infinite number of wavefunctions $|m\rangle$, evaluation of the quadratic forms for arbitrary $|m'\rangle$ and $|m\rangle$ will clarify the form for the entire matrix.

The first-order perturbation correction for the eigenvalues is

$$E'_m = \langle m|\, \mathbf{R}\, |m\rangle \tag{10.5.23}$$

which is the integral

$$\begin{aligned} \langle m|\, \mathbf{R}\, |m\rangle &= \int_0^{2\pi} \psi_m^*(-\mu\mathscr{E} \cos\varphi)\, \psi_m\, d\varphi \\ &= -(\mu\mathscr{E}/2) \int_0^{2\pi} \exp(-im\varphi) \cos(\varphi) \exp(im\varphi)\, d\varphi \\ &= -(\mu\mathscr{E}/2\pi) \int_0^{2\pi} \cos\varphi\, d\varphi = -(\mu\mathscr{E}/2\pi)(+\sin\varphi)\Big|_0^{2\pi} = 0 \end{aligned} \tag{10.5.24}$$

For any m, the first-order correction to the eigenvalue (the diagonal elements of the **R** matrix) will be zero.

The electron and the dipole it creates will be aligned and opposed to the field during each rotation and the average first-order energy perturbation must be zero.

The remaining perturbations require the off-diagonal elements of the perturbation matrix. A general matrix element

$$\langle j|\, \mathbf{R}\, |m\rangle \tag{10.5.25}$$

requires the integral

$$(\mu\mathscr{E}/2\pi) \int_0^{2\pi} \exp(-ij\varphi) \cos(\varphi) \exp(im\varphi)\, d\varphi \tag{10.5.26}$$

which is evaluated using the identity

$$\cos\varphi = \frac{1}{2}\,[\exp(i\varphi) + \exp(-i\varphi)] \tag{10.5.27}$$

The integral is

$$\mu\mathscr{E}/+\pi \left\{ \int_0^{2\pi} \exp[i(m-j)\varphi] \exp(i\varphi)\, d\varphi + \int_0^{2\pi} \exp[i(m-j)\varphi] \exp(-i\varphi)\, d\varphi \right\} \tag{10.5.28}$$

These integrals are zero unless the exponentials are eliminated from the integral (Problem 10.4). The first integral is nonzero when

$$m - j + 1 = 0 \tag{10.5.29}$$

while the second integral is nonzero when

$$m - j - 1 = 0 \tag{10.5.30}$$

For a given m, nonzero matrix elements exist only when

$$j = m + 1 \quad \text{or} \quad j = m - 1 \tag{10.5.31}$$

respectively. These finite matrix elements are each

$$-\mu\mathscr{E}/2 \tag{10.5.32}$$

A summation over the full set of basis functions $|j\rangle$ for a given value of m has only two nonzero terms:

$$\langle m-1|\,\mathbf{R}\,|m\rangle = \langle m+1|\,\mathbf{R}\,|m\rangle = -(\mu\mathscr{E}/2) \tag{10.5.33}$$

Since there are nonzero off-diagonal elements, all perturbations after the first-order eigenvalue perturbation are finite. The second-order perturbations are built on $|m_1\rangle$, which depends on these off-diagonal elements. The first-order eigenvector correction is

$$|m_{1,\text{corr}}\rangle = \sum \frac{|m'\rangle\langle m'|\,\mathbf{R}\,|m\rangle}{E_{0,m} - E_{0,m'}} \tag{10.5.34}$$

This summation has only two nonzero terms:

$$|m_{1,\text{corr}}\rangle = \frac{|m+1\rangle\langle m+1|\,\mathbf{R}\,|m\rangle}{(\hbar^2/2\mathscr{I})[m^2-(m+1)^2]} + \frac{|m-1\rangle\langle m-1|\,\mathbf{R}\,|m\rangle}{(\hbar^2/2\mathscr{I})[m^2-(m-1)^2]} \tag{10.5.35}$$

Substituting the expressions for the matrix elements gives

$$+|m+1\rangle(\mu\mathscr{E}/2)(2\mathscr{I}/\hbar^2)/(2m+1) \tag{10.5.36}$$

$$-|m-1\rangle(\mu\mathscr{E}/2)(2\mathscr{I}/\hbar^2)/(2m-1) \tag{10.5.37}$$

Defining

$$C = (\mu\mathscr{E}\mathscr{I}/\hbar^2) \tag{10.5.38}$$

produces the final expression for the first-order correction to $|m\rangle$:

$$|m+1\rangle[C/(2m+1)] - |m-1\rangle[C/(2m-1)] \tag{10.5.39}$$

The eigenvector $|m_1\rangle$ is

$$|m_1\rangle = |m\rangle + C[|m+1\rangle/(2m+1) - |m-1\rangle/(2m-1)] \quad (10.5.40)$$

The factor C in Equation 10.5.40 is essentially the ratio of the perturbation energy to the unperturbed energy of the state $|m\rangle$. Since the perturbation appears in the numerator, the perturbation is small for small fields. Equation 10.5.40 can be simplified by combining the two terms using

$$\exp[i(m+1)\varphi] = \exp(im\varphi)\exp(i\varphi) = \exp(i\varphi)\,|m\rangle \quad (10.5.41)$$

and

$$\exp[i(m-1)\varphi] = \exp(im\varphi)\exp(-i\varphi) = \exp(-i\varphi)\,|m\rangle \quad (10.5.42)$$

Factoring $|m\rangle$ from each term gives

$$|m_1\rangle = |m\rangle\{1 + C[\exp(i\varphi)/(2m+1) - \exp(-i\varphi)/(2m-1)]\} \quad (10.5.43)$$

The quantum numbers now appear only in the denominator. It is apparent that the perturbations will become smaller as m increases, i.e., when the number of maxima and minima in the wavefunction increases and the spatial distribution of electron density becomes more homogeneous.

This perturbation is valid for all m except $m=1$ where there is no $m-1$ component. The perturbation is

$$\begin{aligned}|1'\rangle &= |1\rangle + |2\rangle[-(\mu\mathscr{E}/2)]/(\hbar^2/2\mathscr{I})(1^2-2^2)\\ &= |1\rangle + |2\rangle C/3 = \exp(i\varphi)[1+\exp(i\varphi)(C/3)] \end{aligned}\quad (10.5.44)$$

Second-order perturbation summations are also limited by the small number of nonzero matrix elements. The second-order eigenvalue correction, E_2'', for $|m\rangle$ is

$$E_2'' = \frac{\langle m|\,\mathbf{R}\,|m'\rangle\langle m'|\,\mathbf{R}\,|m\rangle}{E_m - E_{m'}} \quad (10.5.45)$$

The sum contains only two nonzero terms:

$$\begin{aligned}E_2'' &= \frac{\langle m|\,\mathbf{R}\,|m+1\rangle\langle m+1|\,\mathbf{R}\,|m\rangle}{E_m - E_{m+1}} + \frac{\langle m|\,\mathbf{R}\,|m-1\rangle\langle m-1|\,\mathbf{R}\,|m\rangle}{E_m - E_{m-1}}\\ &= (\mu\mathscr{E}/2)^2\,(2\mathscr{I}/\hbar^2)[1/(-2m-1) + 1/(2m-1)]\end{aligned} \quad (10.5.46)$$

The terms in the summation can be combined to give

$$-1/(2m+1) + 1/(2m-1) = 2/(4m^2-1) \quad (10.5.47)$$

and the second-order perturbation energy is

$$E_2'' = (\mu\mathscr{E}/2)^2\,(2\mathscr{I}/\hbar^2)[2/(4m^2-1)] = (\mu\mathscr{E}\mathscr{I}/\hbar^2)/(4m^2-1) \tag{10.5.48}$$

The second-order perturbation for the eigenvectors is

$$\langle j|i_2\rangle = \sum_k \frac{\langle j|\,\mathbf{R}\,|k\rangle\langle k|\,\mathbf{R}\,|i\rangle}{(E_i-E_j)(E_i-E_k)} - \frac{\langle i|\,\mathbf{R}\,|i\rangle\langle j|\,\mathbf{R}\,|i\rangle}{(E_i-E_j)^2} \tag{10.5.49}$$

Since

$$\langle i|\,\mathbf{R}\,|i\rangle = 0 \tag{10.5.50}$$

i.e., the diagonal elements of the **R** matrix are zero, the second term must be zero. The remaining summation has only two finite terms:

$$\langle j|m_2\rangle = (E_m-E_j)^{-1} \times \left(\frac{\langle j|\,\mathbf{R}\,|m-1\rangle\langle m-1|\,\mathbf{R}\,|m\rangle}{E_m-E_{m-1}} + \frac{\langle j|\,\mathbf{R}\,|m+1\rangle\langle m+1|\,\mathbf{R}\,|m\rangle}{E_m-E_{m+1}}\right) \tag{10.5.51}$$

The full eigenvector correction requires a summation over all $|j\rangle$. The only finite element in the first sum is observed when $j=m-2$. The only finite element in the second sum occurs when $j=m+2$. The double summation reduces to two terms

$$\frac{|m-2\rangle\langle m-2|\,\mathbf{R}\,|m-1\rangle\langle m-1|\,\mathbf{R}\,|m\rangle}{(E_m-E_{m-2})(E_m-E_{m-1})} + \frac{|m+2\rangle\langle m+2|\,\mathbf{R}\,|m+1\rangle\langle m+1|\,\mathbf{R}\,|m\rangle}{(E_m-E_{m+2})(E_m-E_{m+1})} \tag{10.5.52}$$

The two matrix element products are

$$(\mu\mathscr{E}/2)^2 \tag{10.5.53}$$

while the energy difference products are proportional to

$$(\hbar^2/2\mathscr{I})^2 \tag{10.5.54}$$

Defining a constant,

$$C'' = (\mu\mathscr{E}/2)^2/(\hbar^2/2\mathscr{I})^2 \tag{10.5.55}$$

gives the second-order eigenvector correction,

$$|m-2\rangle\, C''[1/(4m-4)(2m-1)] + |m+2\rangle\, C''[1/(-4m-4)(-2m-1)] \tag{10.5.56}$$

This eigenvector correction $|m_2\rangle$ must be added to $|m\rangle$ and $|m_1\rangle$ to give the eigenvector

$$|m_2\rangle = |m\rangle + |m_{1,\text{corr}}\rangle + |m_{2,\text{corr}}\rangle \tag{10.5.57}$$

The first-order correction, C, depends on the first power of the perturbation energy while the second-order correction depends on its square. For small perturbations, the corrections will converge quickly.

10.6. Direct Sums and Products

Two matrices can be added only if they have the same order since the addition must proceed element by element, i.e.,

$$\mathbf{A} + \mathbf{B} = \mathbf{C} \tag{10.6.1}$$

is identical to

$$[A]_{ij} + [B]_{ij} = [C]_{ij} \tag{10.6.2}$$

where

$$1 \leqslant i, j \leqslant N \tag{10.6.3}$$

for a square matrix of order N. The matrices need not be square matrices to perform an addition. However, both matrices must have the same size, e.g., $N \times M$, so there is a one-to-one match of elements which must be added.

Matrix multiplication is also applicable to $N \times M$ matrices if the number of columns in the left matrix is equal to the number of rows in the matrix on the right. For example, the two matrices

$$\begin{pmatrix} a_{11} a_{12} \\ a_{21} a_{22} \\ a_{31} a_{32} \end{pmatrix} \qquad \begin{pmatrix} b_{11} b_{12} b_{13} \\ b_{21} b_{22} b_{23} \end{pmatrix} \tag{10.6.4}$$

can be multiplied in the order **AB** (only) to give

$$\begin{pmatrix} a_{11} b_{11} + a_{12} b_{21} & a_{11} b_{12} + a_{12} b_{22} & a_{11} b_{13} + a_{12} b_{23} \\ a_{21} b_{11} + a_{22} b_{21} & a_{21} b_{12} + a_{22} b_{22} & a_{21} b_{13} + a_{32} b_{23} \\ a_{31} b_{11} + a_{32} b_{21} & a_{31} b_{12} + a_{32} b_{22} & a_{31} b_{13} + a_{32} b_{23} \end{pmatrix} \tag{10.6.5}$$

The final matrix has the same number of rows as the **A** (left) matrix and the same number of columns as the **B** (right) matrix. These sizes limit the total size of the final product matrix **C**. Two square ($N \times N$) matrices produce a matrix sum and matrix product as $N \times N$ matrices.

The direct sum of two matrices is used when **A** and **B** are two independent matrices which must be combined to give a single matrix with full information on **A** and **B**. The direct sum of **A** and **B** is

$$\mathbf{A} \oplus \mathbf{B} = \begin{pmatrix} \mathbf{A} & 0 \\ 0 & \mathbf{B} \end{pmatrix} \tag{10.6.6}$$

The **A** and **B** matrices become block diagonal elements of the direct sum matrix. The matrices **A** and **B** can have any order, but only square matrices will be used in this chapter. For example, if **A** is a 2×2 matrix and **B** is a 3×3 matrix, the direct sum matrix is a 5×5 matrix,

$$\begin{pmatrix} a_{11} & a_{12} & 0 & 0 & 0 \\ a_{21} & a_{22} & 0 & 0 & 0 \\ 0 & 0 & b_{11} & b_{12} & b_{13} \\ 0 & 0 & b_{21} & b_{22} & b_{23} \\ 0 & 0 & b_{31} & b_{32} & b_{33} \end{pmatrix} \tag{10.6.7}$$

The block matrices of zero which complete the matrix are not square matrices if the two matrices have different order.

For a direct sum, each matrix retains its eigenvalues and eigenvectors. For example, an eigenvector for the 2×2 **A** matrix can be expanded to the five-dimensional system by adding three zero components:

$$|c_i\rangle = \begin{pmatrix} |a_i\rangle_1 \\ |a_i\rangle_2 \\ 0 \\ 0 \\ 0 \end{pmatrix} \tag{10.6.8}$$

Since there is no coupling between the **A** and **B** matrices in the direct sum matrix (**C**), **C** applied to the five-component vector gives

$$\mathbf{C}\,|c_i\rangle = a_i\,|c_i\rangle \tag{10.6.9}$$

where a_i is the eigenvalue for $|a_i\rangle$ or $|c_i\rangle$.

The eigenvectors for **B** can be used as the final three components of the five-component **C** vector; the first two components are equated to zero. In this case, the matrix **C** returns this $|c_j\rangle$ eigenvector with eigenvalue b_j:

$$\mathbf{C}\,|c_j\rangle = b_j\,|c_j\rangle \tag{10.6.10}$$

The 5×5 matrix will have two eigenvalues from **A** and three eigenvalues from **B**.

The direct sum notation can be used for any matrix which can be resolved into block diagonal form. The appearance of the block diagonal regimes indicates that the system itself has been resolved into a set of independent subspaces. Each of these subspaces can be analyzed independently of the remaining subspaces. If these block diagonal subspaces are designated $\mathbf{M}_i$, then the full matrix for the system is

$$\mathbf{M}_1 \oplus \mathbf{M}_2 \oplus \cdots \tag{10.6.11}$$

A primary example of this resolution into subspaces occurs with the symmetry operations of a group (Chapter 11). The matrix for each symmetry operation in the group diagonalizes as a set of irreducible representations. Each irreducible representation is a subset of the full symmetry matrix. The full matrix is constructed as a direct sum of all its irreducible representations.

If a Hamiltonian operator for a two-particle system resolves into Hamiltonians $\mathbf{H}_1$ and $\mathbf{H}_2$ which operate exclusively on $|\psi_1\rangle$ and $|\psi_2\rangle$, respectively, then each particle can be analyzed independently. The two independent operators are combined via the direct sum

$$\mathbf{H}_1 \oplus \mathbf{H}_2 \tag{10.6.12}$$

The direct product of two matrices is formed when one matrix is multiplied times each element of the second matrix. The two 2×2 matrices

$$\mathbf{A} = \begin{pmatrix} a_{11} & a_{12} \\ a_{21} & a_{22} \end{pmatrix} \qquad \mathbf{B} = \begin{pmatrix} b_{11} & b_{12} \\ b_{21} & b_{22} \end{pmatrix} \tag{10.6.13}$$

have a direct product

$$\mathbf{A} \otimes \mathbf{B} = \begin{pmatrix} \mathbf{A}b_{11} & \mathbf{A}b_{12} \\ \mathbf{A}b_{21} & \mathbf{A}b_{22} \end{pmatrix} \tag{10.6.14}$$

Since each element of this matrix actually has four elements, the direct product has converted the two 2×2 matrices into a new 4×4 matrix,

$$\begin{pmatrix} a_{11}b_{11} & a_{12}b_{11} & a_{11}b_{12} & a_{12}b_{12} \\ a_{21}b_{11} & a_{22}b_{11} & a_{21}b_{12} & a_{22}b_{12} \\ a_{11}b_{21} & a_{12}b_{21} & a_{11}b_{22} & a_{12}b_{22} \\ a_{21}b_{21} & a_{22}b_{21} & a_{21}b_{22} & a_{22}b_{22} \end{pmatrix} \tag{10.6.15}$$

This format must be used with care since it obscures the fact that the full $\mathbf{A}$ matrix information is present for each element of the $\mathbf{B}$ matrix.

Vectors for the direct product matrix are constructed in a similar manner. If $|a\rangle$ is an eigenvector for $\mathbf{A}$ and $|b\rangle$ is an eigenvector for $\mathbf{B}$, the direct product vector is

$$|a\rangle \otimes |b\rangle = \begin{pmatrix} |a\rangle\, b_1 \\ |a\rangle\, b_2 \end{pmatrix} = \begin{pmatrix} a_1 b_1 \\ a_2 b_1 \\ a_1 b_2 \\ a_2 b_2 \end{pmatrix} \tag{10.6.16}$$

where a_i and b_j are the ith and jth components of eigenvectors $|a\rangle$ and $|b\rangle$.

The operation of a direct product matrix on the direct product vector constructed from the eigenvectors of **A** and **B** gives a product of the eigenvalues associated with these eigenvectors since

$$\begin{aligned}
&\begin{pmatrix} \mathbf{A}b_{11} & \mathbf{A}b_{21} \\ \mathbf{A}b_{21} & \mathbf{A}b_{22} \end{pmatrix} \begin{pmatrix} |a\rangle\, b_1 \\ |a\rangle\, b_2 \end{pmatrix} \\
&\quad = \lambda_a \begin{pmatrix} b_{11} & b_{12} \\ b_{21} & b_{22} \end{pmatrix} \begin{pmatrix} |a\rangle\, b_1 \\ |a\rangle\, b_2 \end{pmatrix} \\
&\quad = \lambda_a \mathbf{B}\, |a\rangle\, |b\rangle = \lambda_a \lambda_b\, |a\rangle \otimes |b\rangle
\end{aligned} \tag{10.6.17}$$

To verify that the direct product of the two vectors will produce the proper eigenvalue product, i.e., without interference from the products of components, consider the direct product of the matrices

$$\begin{pmatrix} 1 & 1 \\ 1 & 1 \end{pmatrix} \quad \begin{pmatrix} 2 & 1 & 0 \\ 1 & 2 & 1 \\ 0 & 1 & 2 \end{pmatrix} \tag{10.6.18}$$

i.e.,

$$\begin{pmatrix} 2 & 2 & 1 & 1 & 0 & 0 \\ 2 & 2 & 1 & 1 & 0 & 0 \\ 1 & 1 & 2 & 2 & 1 & 1 \\ 1 & 1 & 2 & 2 & 1 & 1 \\ 0 & 0 & 1 & 1 & 2 & 2 \\ 0 & 0 & 1 & 1 & 2 & 2 \end{pmatrix} \tag{10.6.19}$$

acting on the direct product of the eigenvectors,

$$|a\rangle \otimes |b\rangle = \begin{pmatrix} 1 \\ 1 \end{pmatrix} \otimes \begin{pmatrix} 1 \\ 2 \\ 1 \end{pmatrix} = \begin{pmatrix} 1 \\ 1 \\ 2 \\ 2 \\ 1 \\ 1 \end{pmatrix} \tag{10.6.20}$$

The result is

$$\begin{pmatrix} 2 & 2 & 1 & 1 & 0 & 0 \\ 2 & 2 & 1 & 1 & 0 & 0 \\ 1 & 1 & 2 & 2 & 1 & 1 \\ 1 & 1 & 2 & 2 & 1 & 1 \\ 0 & 0 & 1 & 1 & 2 & 2 \\ 0 & 0 & 1 & 1 & 2 & 2 \end{pmatrix} \begin{pmatrix} 1 \\ 1 \\ 2 \\ 2 \\ 1 \\ 1 \end{pmatrix} = (4 + 2\sqrt{2}) \begin{pmatrix} 1 \\ 1 \\ 2 \\ 2 \\ 1 \\ 1 \end{pmatrix} \tag{10.6.21}$$

The eigenvector is regenerated and the eigenvalue is the product of the eigenvalues for **A** ($\lambda=2$) and **B** ($\lambda=2+\sqrt{2}$).

Since the 2×2 matrix has two eigenvalues and the 3×3 matrix has three, there are six eigenvalues for the direct product matrix, representing all possible products between the two sets of eigenvalues:

$$\begin{matrix} 2\times(2+\sqrt{2}), & 2\times 2, & 2\times(2-\sqrt{2}) \\ 0\times(2+\sqrt{2}), & 0\times 2, & 0\times(2-\sqrt{2}) \end{matrix} \qquad (10.6.22)$$

The $\lambda=0$ eigenvalue will be threefold degenerate in the direct product matrix. The appropriate eigenvectors can be constructed as the direct products of the eigenvectors from the **A** and **B** matrices.

If the order of the matrices in the direct product is reversed, a new matrix and a new set of eigenvectors are produced; the eigenvalue products are not changed. The direct product $B\otimes A$ for the given **A** and **B** matrices is

$$\begin{pmatrix} 2 & 1 & 0 & 2 & 1 & 0 \\ 1 & 2 & 1 & 1 & 2 & 1 \\ 0 & 1 & 2 & 0 & 1 & 2 \\ 2 & 1 & 0 & 2 & 1 & 0 \\ 1 & 2 & 1 & 1 & 2 & 1 \\ 0 & 1 & 2 & 0 & 1 & 2 \end{pmatrix} \qquad (10.6.23)$$

with eigenvectors

$$|b\rangle\otimes|a\rangle = \begin{pmatrix} 1 \\ \sqrt{2} \\ 1 \\ 1 \\ \sqrt{2} \\ 1 \end{pmatrix} \begin{pmatrix} 1 \\ 0 \\ -1 \\ 1 \\ 0 \\ -1 \end{pmatrix} \begin{pmatrix} 1 \\ -\sqrt{2} \\ 1 \\ 1 \\ -\sqrt{2} \\ 1 \end{pmatrix} \begin{pmatrix} 1 \\ \sqrt{2} \\ 1 \\ -1 \\ -\sqrt{2} \\ -1 \end{pmatrix} \begin{pmatrix} 1 \\ 0 \\ -1 \\ -1 \\ 0 \\ 1 \end{pmatrix} \begin{pmatrix} 1 \\ -\sqrt{2} \\ 1 \\ -1 \\ \sqrt{2} \\ -1 \end{pmatrix} \qquad (10.6.24)$$

These eigenvectors must be prepared in the order $|b\rangle\otimes|a\rangle$.

The direct products are important when complicated matrices must be analyzed to determine their eigenvalues and eigenvectors. In such cases, it is necessary to determine if the matrix can be resolved as the direct product of two simpler matrices for which the eigenvalues and eigenvectors are known. Patterns within the larger matrix are used to deduce the form of the constituent matrices.

The matrix

$$\begin{pmatrix} 2 & 2 & 1 & 1 \\ 2 & 4 & 1 & 2 \\ 1 & 1 & 2 & 2 \\ 1 & 2 & 2 & 4 \end{pmatrix} \tag{10.6.25}$$

is complicated but the 2×2 matrix

$$\begin{pmatrix} 1 & 1 \\ 1 & 2 \end{pmatrix} \tag{10.6.26}$$

is present in four 2×2 blocks. Each element in the 11 and 22 blocks is multiplied by 2 so the matrix can be generated from the direct product

$$\begin{pmatrix} 1 & 1 \\ 1 & 2 \end{pmatrix} \otimes \begin{pmatrix} 2 & 1 \\ 1 & 2 \end{pmatrix} \tag{10.6.27}$$

Since the eigenvalues for the first matrix are

$$(3+\sqrt{5})/2, \qquad (3-\sqrt{5})/2 \tag{10.6.28}$$

while the eigenvalues for the second matrix are

$$3, \qquad 1 \tag{10.6.29}$$

the four eigenvalues for the direct product matrix of Equation 10.6.25 are

$$(3+\sqrt{5})/2, \qquad (3-\sqrt{5})/2, \qquad 3(3+\sqrt{5})/2, \qquad 3(3-\sqrt{5})/2 \tag{10.6.30}$$

The corresponding eigenvectors are easily determined as the direct products of the eigenvectors for these 2×2 matrices.

Although a single direct product is a convenient construct for producing larger matrices with known eigenvalues and eigenvectors, the matrices which can be produced in this manner are limited. The range of the technique is extended by considering a variation. Two matrices can be summed only if they have the same order. The 2×2 **A** matrix and the 3×3 **B** matrix of Equation 10.6.18 can be converted to the same order with the judicious use of identity operators and the direct product. The 2×2 **A** matrix forms a direct product with a 3×3 **I** matrix:

$$\begin{pmatrix} 1 & 1 \\ 1 & 1 \end{pmatrix} \otimes \begin{pmatrix} 1 & 0 & 0 \\ 0 & 1 & 0 \\ 0 & 0 & 1 \end{pmatrix} = \begin{pmatrix} 1 & 1 & 0 & 0 & 0 & 0 \\ 1 & 1 & 0 & 0 & 0 & 0 \\ 0 & 0 & 1 & 1 & 0 & 0 \\ 0 & 0 & 1 & 1 & 0 & 0 \\ 0 & 0 & 0 & 0 & 1 & 1 \\ 0 & 0 & 0 & 0 & 1 & 1 \end{pmatrix} \tag{10.6.31}$$

The **B** matrix forms a direct product with a 2×2 **I** matrix on the left to give

$$\begin{pmatrix} 1 & 0 \\ 0 & 1 \end{pmatrix} \otimes \begin{pmatrix} 2 & 1 & 0 \\ 1 & 2 & 1 \\ 0 & 1 & 2 \end{pmatrix} = \begin{pmatrix} 2 & 0 & 1 & 0 & 0 & 0 \\ 0 & 2 & 0 & 1 & 0 & 0 \\ 1 & 0 & 2 & 0 & 1 & 0 \\ 0 & 1 & 0 & 2 & 0 & 1 \\ 0 & 0 & 1 & 0 & 2 & 0 \\ 0 & 0 & 0 & 1 & 0 & 2 \end{pmatrix} \tag{10.6.32}$$

These two 6×6 matrices can now be added element by element to give

$$\begin{pmatrix} 3 & 1 & 1 & 0 & 0 & 0 \\ 1 & 3 & 0 & 1 & 0 & 0 \\ 1 & 0 & 3 & 1 & 1 & 0 \\ 0 & 1 & 1 & 3 & 0 & 1 \\ 0 & 0 & 1 & 0 & 3 & 1 \\ 0 & 0 & 0 & 1 & 1 & 3 \end{pmatrix} \tag{10.6.33}$$

The eigenvalues for the 6×6 matrix are the sums of each eigenvalue of **A** with each eigenvalue of **B**, i.e.,

$$\begin{aligned} \lambda_a + \lambda_b = 2+2+\sqrt{2}, \quad & 2+2, \quad 2+2-\sqrt{2}, \\ 0+2+\sqrt{2}, \quad & 0+2, \quad 0+2-\sqrt{2} \end{aligned} \tag{10.6.34}$$

The sum of direct products with the matrices reversed,

$$\mathbf{B} \otimes \mathbf{I}_2 + \mathbf{I}_3 \otimes \mathbf{A} \tag{10.6.35}$$

gives a different matrix,

$$\begin{pmatrix} 3 & 1 & 0 & 1 & 0 & 0 \\ 1 & 3 & 1 & 0 & 1 & 0 \\ 0 & 1 & 3 & 0 & 0 & 1 \\ 1 & 0 & 0 & 3 & 1 & 0 \\ 0 & 1 & 0 & 1 & 3 & 1 \\ 0 & 0 & 1 & 0 & 1 & 3 \end{pmatrix} \tag{10.6.36}$$

whose eigenvalues are also the sums of eigenvalues from **A** and **B**.

In general, any matrix which can be generated as the sum

$$\mathbf{A} \otimes \mathbf{I}_M + \mathbf{I}_N \otimes \mathbf{B} \tag{10.6.37}$$

where matrix **B** has order M and matrix **A** has order N and $\mathbf{I}_N$ is an $N \times N$

identity matrix, will have eigenvalues equal to the sums of eigenvalues from **A** and **B**,

$$\lambda_a + \lambda_b \tag{10.6.38}$$

If the two terms in Equation 10.6.37 are subtracted, the eigenvalues of the final matrix are equal to all differences of the **A** and **B** eigenvalues,

$$\lambda_a - \lambda_b \tag{10.6.39}$$

The eigenvectors for the matrices generated as a sum of direct products (Equations 10.6.33 and 10.6.36) are the direct products of the eigenvectors for the **A** and **B** matrices. For the combination

$$\mathbf{A} \otimes \mathbf{I}_3 + \mathbf{I}_2 \otimes \mathbf{B} \tag{10.6.40}$$

the direct products

$$|a\rangle \otimes |b\rangle \tag{10.6.41}$$

must be used, i.e., the left vector in the direct product must be an eigenvector of the matrix on the left in Equation 10.6.40. The six eigenvectors for the matrix in Equation 10.6.33 are

$$\begin{pmatrix} 1 \\ -1 \\ \sqrt{2} \\ -\sqrt{2} \\ 1 \\ -1 \end{pmatrix} \begin{pmatrix} 1 \\ -1 \\ 0 \\ 0 \\ -1 \\ +1 \end{pmatrix} \begin{pmatrix} 1 \\ -1 \\ -\sqrt{2} \\ +\sqrt{2} \\ 1 \\ -1 \end{pmatrix} \begin{pmatrix} 1 \\ 1 \\ \sqrt{2} \\ \sqrt{2} \\ 1 \\ 1 \end{pmatrix} \begin{pmatrix} 1 \\ 1 \\ 0 \\ 0 \\ -1 \\ -1 \end{pmatrix} \begin{pmatrix} 1 \\ 1 \\ -\sqrt{2} \\ -\sqrt{2} \\ 1 \\ 1 \end{pmatrix} \tag{10.6.42}$$

This can be verified with direct application of the matrix (Problem 10.8).

The direct product provides a convenient way of determining eigenvalues and eigenvectors of some complicated matrix systems. The problems of diagonalization and eigenvector determination are replaced by a search for matrices of lower order which will generate the matrix of interest via a direct product or combination of direct products. The utility of the technique is demonstrated for coupled oscillator systems in the next section.

10.7. A Two-Dimensional Coupled Oscillator System

In Chapter 5, the normal mode vibrations of chains of coupled atoms were determined by generating a force constant matrix for all the atoms of the system.

The analysis could be extended to a chain with N atoms because the tridiagonal force constant matrix for the one-dimensional system had known eigenvalues and eigenvectors for any value of N. With the direct product analyses introduced in Section 10.6, these analyses can be extended to two- and three-dimensional systems of coupled oscillators. The new force equations generate more complicated force constant matrices, but these matrices can be reduced to direct products of matrices whose eigenvalues and eigenvectors are known. The analysis is illustrated for a two-dimensional net of coupled oscillators. Each atom is assigned an x_i and a y_i spatial coordinate, and the system will resolve into two independent sets of equations, one set for all x_i and one set for all y_i.

The coupled system is shown in Figure 10.4 for a set of nine atoms arranged in three rows of three atoms each. The end atom in each row and column is coupled to a fixed surface. The variable x_{ij} gives the displacement of the atom located in the ith row and jth column of the network in the x direction. Similarly, y_{ij} describes the (ij)th atom's motion in the y direction. The force constants for all bonds are equal to k. Since the y and x motions are independent, only the x_{ij} equations of motion are considered. The y_{ij} equations are determined in exactly the same manner.

The force on an atom in row 1 is dictated by the net displacement between it and the two neighboring atoms, i.e.,

$$x_{1,i-1} - x_{1,i}; \qquad x_{1,i+1} - x_{1,i} \tag{10.7.1}$$

If these were the only horizontal displacements possible for atom $1i$, the force equation would be equivalent to that of a single atom in a one-dimensional chain of coupled oscillators,

$$m\, d^2x_{1,i}/dt^2 = -k(-x_{1,i-1} + 2x_{1,i} - x_{1,i+1}) \tag{10.7.2}$$

The two-dimensional problem must also include displacements of the atom produced by displacements of atoms in the rows below and above the atom of interest.

The atom 12 in the net will experience forces proportional to the displacements

$$x_{12} - x_{11}, \qquad x_{12} - x_{13} \tag{10.7.3}$$

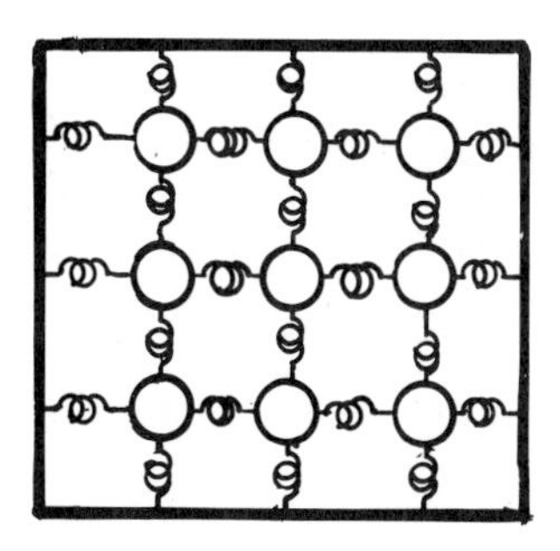

Figure 10.4. A set of nine coupled atoms arranged in a two-dimensional 3×3 network.

in the first row where the atom is located. In addition, the atoms directly above and below atom 12 will produce a net x displacement if they move relative to atom 12. This constitutes the coupling between adjacent rows of the network. Since atom 12 is attached directly to the surface above it, this connection is defined with the dummy displacement variable x_{02}. The atom also couples with the displacement of the actual atom below it in row 2, i.e., x_{22}.

The displacement forces produced by the atoms in row 1 on atom 12 are

$$\begin{aligned} F &= k[(-x_{11}+x_{12})+(x_{12}-x_{13})] \\ &= k(-x_{11}+2x_{12}-x_{13}) \end{aligned} \tag{10.7.4}$$

while the displacement forces for the surface above and the atom below atom 12 are

$$\begin{aligned} F_a &= -k[(x_{12}-x_{02})+(x_{12}-x_{22})] \\ &= -k(-x_{02}+2x_{12}-x_{22}) \end{aligned} \tag{10.7.5}$$

These two forces are summed and substituted into the force equation for atom 12:

$$m\,d^2x_{12}/dt^2 = -k[(-x_{11}+2x_{12}-x_{13})+(-x_{02}+2x_{12}-x_{22})] \tag{10.7.6}$$

The force equation for atom 12 illustrates the pattern required for each atom in the system. The first group of terms includes displacements for atoms on either side of the atom of interest in the same row. The second group of terms includes the displacements of the atoms directly above and below the atom of interest. The index 0 is used when the atom is attached to the upper or left-hand surfaces. For the 3×3 coupled system, the index 4 signifies coupling with the lower or right-hand fixed surfaces. The pattern is apparent with this notation so the system is easily expanded to nets with an arbitrary number of rows and columns. The nine force equations for the nine atoms in the 3×3 net are

$$\begin{aligned} d^2x_{11}/dt^2 &= -k[(-x_{10}+2x_{11}-x_{12})+(-x_{01}+2x_{11}-x_{21})] \\ d^2x_{12}/dt^2 &= -k[(-x_{11}+2x_{12}-x_{13})+(-x_{02}+2x_{12}-x_{22})] \\ d^2x_{13}/dt^2 &= -k[(-x_{12}+2x_{13}-x_{14})+(-x_{03}+2x_{13}-x_{23})] \\ d^2x_{21}/dt^2 &= -k[(-x_{20}+2x_{21}-x_{22})+(-x_{11}+2x_{21}-x_{31}) \\ d^2x_{22}/dt^2 &= -k[(-x_{21}+2x_{22}-x_{23})+(-x_{12}+2x_{22}-x_{32}) \\ d^2x_{23}/dt^2 &= -k[(-x_{22}+2x_{23}-x_{24})+(-x_{13}+2x_{23}-x_{33}) \\ d^2x_{31}/dt^2 &= -k[(-x_{30}+2x_{31}-x_{32})+(-x_{21}+2x_{31}-x_{41})] \\ d^2x_{32}/dt^2 &= -k[(-x_{31}+2x_{32}-x_{33})+(-x_{22}+2x_{32}-x_{42})] \\ d^2x_{33}/dt^2 &= -k[(-x_{32}+2x_{33}-x_{34})+(-x_{23}+2x_{33}-x_{43})] \end{aligned} \tag{10.7.7}$$

The inclusion of the variables for coupling to the fixed surfaces, i.e., indices 0 and 4, generates the most general form of these equations. If the atoms are indeed coupled to the fixed surfaces, the displacements x_{01}, x_{34}, etc. are equated to zero but the displacements such as x_{11} are multiplied by the factor of 2 which appears in the equations to include coupling to both the adjacent atoms and the fixed surfaces. The equations reduce as

$$\begin{aligned}
d^2x_{11}/dt^2 &= -k(4x_{11}-x_{12}-x_{21})\\
d^2x_{12}/dt^2 &= -k(-x_{11}+4x_{12}-x_{13}-x_{22})\\
d^2x_{13}/dt^2 &= -k(-x_{12}+4x_{13}-x_{23})\\
d^2x_{21}/dt^2 &= -k(-x_{11}+4x_{21}-x_{22}-x_{31})\\
d^2x_{22}/dt^2 &= -k(-x_{12}-x_{21}+4x_{22}-x_{23}-x_{32})\\
d^2x_{23}/dt^2 &= -k(-x_{22}+4x_{23}-x_{13}-x_{33})\\
d^2x_{31}/dt^2 &= -k(-x_{21}+4x_{31}-x_{32})\\
d^2x_{32}/dt^2 &= -k(-x_{22}-x_{31}+4x_{32}-x_{33})\\
d^2x_{33}/dt^2 &= -k(-x_{23}-x_{32}+4x_{33})
\end{aligned} \tag{10.7.8}$$

These equations in matrix format are

$$d^2\,|x\rangle/dt^2 = (k)\begin{pmatrix} 4 & -1 & 0 & -1 & 0 & 0 & 0 & 0 & 0\\ -1 & 4 & -1 & 0 & -1 & 0 & 0 & 0 & 0\\ 0 & -1 & 4 & 0 & 0 & -1 & 0 & 0 & 0\\ -1 & 0 & 0 & 4 & -1 & 0 & -1 & 0 & 0\\ 0 & -1 & 0 & -1 & 4 & -1 & 0 & -1 & 0\\ 0 & 0 & -1 & 0 & -1 & 4 & 0 & 0 & -1\\ 0 & 0 & 0 & -1 & 0 & 0 & 4 & -1 & 0\\ 0 & 0 & 0 & 0 & -1 & 0 & -1 & 4 & -1\\ 0 & 0 & 0 & 0 & 0 & -1 & 0 & -1 & 4 \end{pmatrix}\begin{pmatrix} x_{11}\\ x_{12}\\ x_{13}\\ x_{21}\\ x_{22}\\ x_{23}\\ x_{31}\\ x_{32}\\ x_{33} \end{pmatrix} \tag{10.7.9}$$

Although the force constant matrix is complicated, it can be expressed as a direct product of two smaller matrices and this facilitates the determination of its eigenvalues and eigenvectors. The forms for the vertical and horizontal coupling equations for the x direction were the same and the matrix is the direct product of two identical 3×3 matrices,

$$\mathbf{B} = \mathbf{A} = \begin{pmatrix} 2 & -1 & 0\\ -1 & 2 & -1\\ 0 & -1 & 2 \end{pmatrix} \tag{10.7.10}$$

The direct product is the summation

$$\mathbf{A} \otimes \mathbf{I}_3 + \mathbf{I}_3 \otimes \mathbf{B} \tag{10.7.11}$$

which gives

$$\begin{bmatrix} 2 & -1 & 0 & 0 & 0 & 0 & 0 & 0 & 0 \\ -1 & 2 & -1 & 0 & 0 & 0 & 0 & 0 & 0 \\ 0 & -1 & 2 & 0 & 0 & 0 & 0 & 0 & 0 \\ 0 & 0 & 0 & 2 & -1 & 0 & 0 & 0 & 0 \\ 0 & 0 & 0 & -1 & 2 & -1 & 0 & 0 & 0 \\ 0 & 0 & 0 & 0 & -1 & 2 & 0 & 0 & 0 \\ 0 & 0 & 0 & 0 & 0 & 0 & 2 & -1 & 0 \\ 0 & 0 & 0 & 0 & 0 & 0 & -1 & 2 & -1 \\ 0 & 0 & 0 & 0 & 0 & 0 & 0 & -1 & 2 \end{bmatrix}$$

$$\oplus \begin{bmatrix} 2 & 0 & 0 & -1 & 0 & 0 & 0 & 0 & 0 \\ 0 & 2 & 0 & 0 & -1 & 0 & 0 & 0 & 0 \\ 0 & 0 & 2 & 0 & 0 & -1 & 0 & 0 & 0 \\ -1 & 0 & 0 & 2 & 0 & 0 & -1 & 0 & 0 \\ 0 & -1 & 0 & 0 & 2 & 0 & 0 & -1 & 0 \\ 0 & 0 & -1 & 0 & 0 & 2 & 0 & 0 & -1 \\ 0 & 0 & 0 & -1 & 0 & 0 & 2 & 0 & 0 \\ 0 & 0 & 0 & 0 & -1 & 0 & 0 & 2 & 0 \\ 0 & 0 & 0 & 0 & 0 & -1 & 0 & 0 & 2 \end{bmatrix}$$

$$= \begin{bmatrix} 4 & -1 & 0 & -1 & 0 & 0 & 0 & 0 & 0 \\ -1 & 4 & -1 & 0 & -1 & 0 & 0 & 0 & 0 \\ 0 & -1 & 4 & 0 & 0 & -1 & 0 & 0 & 0 \\ -1 & 0 & 0 & 4 & -1 & 0 & -1 & 0 & 0 \\ 0 & -1 & 0 & -1 & 4 & -1 & 0 & -1 & 0 \\ 0 & 0 & -1 & 0 & -1 & 4 & 0 & 0 & -1 \\ 0 & 0 & 0 & -1 & 0 & 0 & 4 & -1 & 0 \\ 0 & 0 & 0 & 0 & -1 & 0 & -1 & 4 & -1 \\ 0 & 0 & 0 & 0 & 0 & -1 & 0 & -1 & 4 \end{bmatrix} \tag{10.7.12}$$

The eigenvalues and eigenvectors for the direct product matrix are constructed directly from the eigenvalues and eigenvectors for the 3×3 **A** matrix. Since the eigenvalues for **A** and **B** are

$$2-\sqrt{2}, \qquad 2, \qquad 2+\sqrt{2} \tag{10.7.13}$$

the nine eigenvalues for the full system are the nine sums possible if each eigenvalue from **A** is added to each of the three (identical) eigenvalues from **B**. The nine eigenvalues are

$$\begin{array}{ccccc} 4-2\sqrt{2}, & 4-\sqrt{2}, & 4, & 4-\sqrt{2}, & 4, \\ 4+\sqrt{2}, & 4, & 4+\sqrt{2}, & 4+2\sqrt{2} & \end{array} \tag{10.7.14}$$

Although the original three eigenvalues for **A** are not degenerate, degenerate eigenvalues are present in the coupled system. The normal mode frequencies for x motions are

$$\omega^2 = k/m \tag{10.7.15}$$

A second set of normal mode frequencies for y is developed in the same way.

The normal mode eigenvectors for the system are the direct products of the eigenvectors from the matrices **A** and **B**. The eigenvectors for these identical 3×3 matrices are

$$|2+\sqrt{2}\rangle = \begin{pmatrix} 1 \\ +\sqrt{2} \\ 1 \end{pmatrix} \qquad |2\rangle = \begin{pmatrix} 1 \\ 0 \\ -1 \end{pmatrix} \qquad |2-\sqrt{2}\rangle = \begin{pmatrix} 1 \\ -\sqrt{2} \\ 1 \end{pmatrix} \tag{10.7.16}$$

and the nine nine-component eigenvectors for the full system are

$$\begin{bmatrix} 1 \\ \sqrt{2} \\ 1 \\ \sqrt{2} \\ 2 \\ \sqrt{2} \\ 1 \\ \sqrt{2} \\ 1 \end{bmatrix} \quad \begin{bmatrix} 1 \\ \sqrt{2} \\ 1 \\ 0 \\ 0 \\ 0 \\ -1 \\ -\sqrt{2} \\ -1 \end{bmatrix} \quad \begin{bmatrix} 1 \\ \sqrt{2} \\ 1 \\ -\sqrt{2} \\ -2 \\ -\sqrt{2} \\ 1 \\ \sqrt{2} \\ 1 \end{bmatrix} \quad \begin{bmatrix} 1 \\ 0 \\ -1 \\ \sqrt{2} \\ 0 \\ -\sqrt{2} \\ 1 \\ 0 \\ -1 \end{bmatrix} \quad \begin{bmatrix} 1 \\ 0 \\ -1 \\ 0 \\ 0 \\ 0 \\ -1 \\ 0 \\ 1 \end{bmatrix} \quad \begin{bmatrix} 1 \\ 0 \\ -1 \\ -\sqrt{2} \\ 0 \\ \sqrt{2} \\ 1 \\ 0 \\ -1 \end{bmatrix}$$

$$\begin{bmatrix} 1 \\ -\sqrt{2} \\ 1 \\ -\sqrt{2} \\ -2 \\ -\sqrt{2} \\ 1 \\ -\sqrt{2} \\ 1 \end{bmatrix} \quad \begin{bmatrix} 1 \\ -\sqrt{2} \\ 1 \\ 0 \\ 0 \\ 0 \\ -1 \\ +\sqrt{2} \\ 1 \end{bmatrix} \quad \begin{bmatrix} 1 \\ -\sqrt{2} \\ 1 \\ -\sqrt{2} \\ 2 \\ -\sqrt{2} \\ 1 \\ -\sqrt{2} \\ 1 \end{bmatrix} \tag{10.7.17}$$

These vectors describe the nine displacements in the order x_{11}, x_{12}, x_{13}, x_{21}, x_{22}, x_{23}, x_{31}, x_{32}, x_{33}. The normal modes described by the first, second, and fifth eigenvectors listed in Equation 10.7.17 are illustrated in Figure 10.5. The two vectors which constitute the direct product carry the amplitudes both horizontally and vertically along the grid to give the final amplitudes for each x displacement.

The full set of equations for the nine atoms in the network must be altered if the atoms are released from the fixed surfaces. This problem is more complicated because the environment around the terminal atoms is no longer symmetric (for the fixed surface system, each atom was always surrounded by four bonds). The four atoms on the corners of the network now have only two bonds. Atoms in the center of outside rows have three bonds while those in the interior retain the symmetric four-bond configuration. The factor of 4 which appeared in every diagonal element of the matrix when the atoms were tied to the surface is reduced to 2 for the corner atoms and 3 for the inner atoms of the outside row. The x_i variables are used to determine the final equation. For example, if x_{01} and x_{10} appear in the equation for x_{11}, the factor for x_{11} must be reduced by 2. The force constant matrix for the free net generated in this manner is

$$\begin{bmatrix} 2 & -1 & 0 & -1 & 0 & 0 & 0 & 0 & 0 \\ -1 & 3 & -1 & 0 & -1 & 0 & 0 & 0 & 0 \\ 0 & -1 & 2 & 0 & 0 & -1 & 0 & 0 & 0 \\ -1 & 0 & 0 & 3 & -1 & 0 & -1 & 0 & 0 \\ 0 & -1 & 0 & -1 & 4 & -1 & 0 & -1 & 0 \\ 0 & 0 & -1 & 0 & -1 & 3 & 0 & 0 & -1 \\ 0 & 0 & 0 & -1 & 0 & 0 & 2 & -1 & 0 \\ 0 & 0 & 0 & 0 & -1 & 0 & -1 & 3 & -1 \\ 0 & 0 & 0 & 0 & 0 & -1 & 0 & -1 & 2 \end{bmatrix} \tag{10.7.18}$$

Although the variation in diagonal elements suggests a very complicated analysis, the one-dimensional version of a three-atom chain provides a starting point. Its force constant matrix is

$$\begin{pmatrix} 1 & -1 & 0 \\ -1 & 2 & -1 \\ 0 & -1 & 1 \end{pmatrix} \tag{10.7.19}$$

Since the coupling is identical for horizontal and vertical coupling, the matrix should be a sum of direct products of the matrix of Equation 10.7.19, i.e.,

$$\mathbf{A} \otimes \mathbf{I}_3 + \mathbf{I}_3 \otimes \mathbf{A} \tag{10.7.20}$$

(Problem 10.10). The nine eigenvalues are now the sums of the eigenvalues for $\mathbf{A}$,

$$0,\ 1,\ 3 \tag{10.7.21}$$

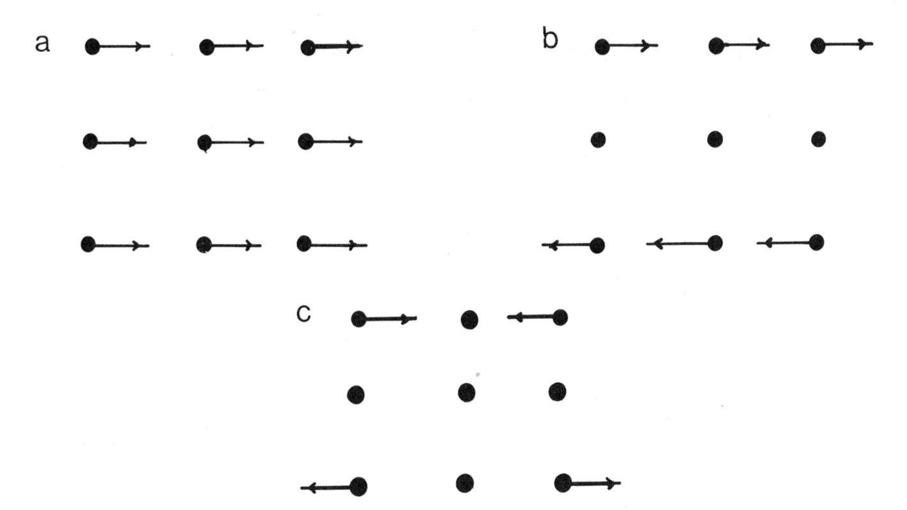

Figure 10.5. Three normal modes for the 3×3 coupled atom network: (a) $n=1$ normal mode; (b) $n=2$ normal mode; (c) $n=5$ normal mode.

i.e.,

$$0, 1, 3, 1, 2, 4, 3, 4, 6 \tag{10.7.22}$$

and the corresponding eigenvectors are formed as the direct products of the three eigenvectors

$$\begin{pmatrix} 1 \\ 1 \\ 1 \end{pmatrix} \quad \begin{pmatrix} 1 \\ 0 \\ -1 \end{pmatrix} \quad \begin{pmatrix} 1 \\ 2 \\ 1 \end{pmatrix} \tag{10.7.23}$$

For example, the direct product of the first eigenvector with itself will give a nine-component vector with identical components. This corresponds to a translation along x of the entire network.

Although a 3×3 network of coupled atoms was selected for analysis, the direct product technique can be used for an arbitrary network of N rows and M columns. The general formula

$$\mathbf{A} \otimes \mathbf{I}_M + \mathbf{I}_N \otimes \mathbf{B} \tag{10.7.24}$$

will then permit generation of the full matrix for the system and give its eigenvalues and eigenvectors directly from the **A** and **B** matrices.

Problems

10.1. Show that $\mathbf{KS} = \mathbf{S\Lambda}$.

10.2. Develop the first- and second-order perturbation eigenvalues and eigenvectors for a particle in a one-dimensional box of length L oriented along the x axis when an electric field is directed along this axis.

10.3. Determine $|2'\rangle$ and $|3'\rangle$ for the molecule NCC using the parameters given in Section 10.3 (Equations 10.3.54).

10.4. (a) Sin $m\varphi$ and cos $m\varphi$ are eigenvectors for the particle confined to a ring of radius r. Use these eigenvectors to develop the matrix $\mathbf{S}^{-1}\mathbf{RS}$ for the system in an electric field. (b) Develop the first- and second-order perturbation corrections for the eigenvalues and eigenvectors of this system.

10.5. The kinetic sequence

$$A \underset{k_r}{\overset{k_f}{\rightleftarrows}} B$$

with forward and reverse rate constants equal to 2 and 1 s^{-1}, respectively, is perturbed by the slow irreversible loss of B to a molecule C with rate constant 0.1 s^{-1}.

a. Determine the change in the equilibrium and kinetic concentrations of A and B when the perturbing reaction is present.
b. Determine the change in equilibrium and kinetic concentrations if the perturbing rate constant has a value of 0.5 s^{-1}.

10.6. Find the eigenvalues and eigenvectors for the matrices

$$\begin{pmatrix} -2.1 & 1 \\ -1 & -2.1 \end{pmatrix} \quad \begin{pmatrix} -1 & 2 \\ 1 & -2.3 \end{pmatrix}$$

using first- and second-order perturbation theory.

10.7. The partition function for an Ising model of a long-chain polymer is characterized by the transfer matrix

$$\begin{pmatrix} qq_b & q \\ 1 & 1 \end{pmatrix}$$

Develop a tractable starting matrix and use first-order perturbation theory to develop a partition function for this system.

10.8. Show that the eigenvectors of Equation 10.6.42 are eigenvectors for the matrix of Equation 10.6.33.

10.9. Show that the eigenvectors of Equation 10.7.17 are eigenvectors of the matrix of Equation 10.7.12.

10.10. Determine the direct products of the eigenvectors of Equation 10.7.19 and show that they are eigenvectors for the matrix of Equation 10.7.18.

10.11. Determine the row and column eigenvectors for the **K** matrix of Equation 10.2.26.

10.12. Determine the numerical values of eigenvector components for the matrix of Equation 10.4.11 using first- and second-order perturbation theory.

11

Introduction to Group Theory

11.1. Vectors and Symmetry Operations

When the axis of a homonuclear diatomic molecule is confined to a single direction in space, the molecule has a total of two degrees of freedom. These degrees of freedom are expressed using the coordinates x_1 and x_2 for each atom or combinations of these displacements, i.e., the normal modes. When the atomic components x_1 and x_2 are assigned positions 1 and 2, respectively, in a two-component vector, the normal modes for the molecule are

$$(1, 1) \tag{11.1.1}$$

$$(1, -1) \tag{11.1.2}$$

The first vector describes a translation of the entire molecule while the second describes a vibration of the molecule. For the translation, both atoms move in phase. For the vibration, they move 180° out of phase.

The normal modes are a set of relative coordinates. If the actual displacement of one atom is known at some time, the normal mode components give the corresponding displacements for all the other atoms at this time. For the two-atom molecule, these normal modes can also be generated using symmetry operations.

If the displacement for atom 1 is known, then it must certainly appear as one of the components of the normal mode. The symmetry operation which "creates" this particular component is the identity E. E is used in place of I by convention. The identity symmetry operation reproduces the known component x_1:

$$Ex_1 = x_1 \tag{11.1.3}$$

The normal mode contains two components and the second is obtained by rotating the atom with its vector through 180°. The rotation axis is perpendicular to the internuclear axis and midway between the two atoms so this rotation effectively moves the atom with its vector to the position of the second

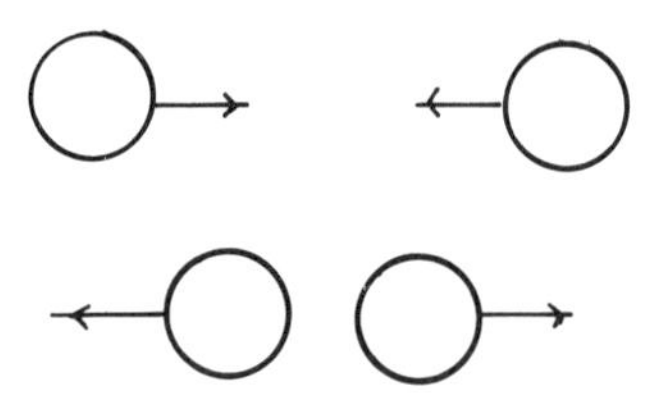

Figure 11.1. The vibrational normal mode for a diatomic molecule. The two position vectors always have the same magnitude and face in opposite directions.

atom (Figure 11.1). The resulting normal mode is the symmetric stretch vibration since the vectors generated are exactly 180° out of phase.

The rotation operation is called a C_2 operation. C is used for rotation and the subscript 2 indicates that two rotations of 180° are required to restore the molecule to its original position. In general, a rotation of $180°/n$ will be defined as C_n, indicating that n such rotations are required to restore the original vector. Written formally, the C_2 operation moves the x_1 component to the x_2 position:

$$C_2 x_1 = x_2 \tag{11.1.4}$$

The normal mode for this symmetric stretch can be written using these symmetry operators:

$$(Ex_1, C_2 x_1) \tag{11.1.5}$$

The symmetry elements themselves are now associated with the specific components of the normal mode vector.

The translation normal mode is also generated using the E and C_2 symmetry operations. However, both vectors must now face in the same direction. The x_1 component must be rotated to the second atom and then multiplied by -1 to change its direction by 180° (Figure 11.2). The elements appear with the same components as for the vibration but the C_2 operation is multiplied by -1:

$$(Ex_1, -C_2 x_1) \tag{11.1.6}$$

These ordered arrangements can be compared with the ordered sets of components in a regular vector. The column vector contains only the magnitude of the vector in each component direction, not the direction itself. This suggests

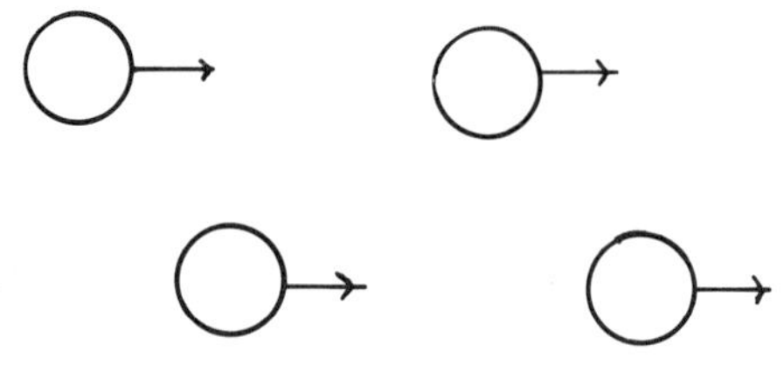

Figure 11.2. The translation mode for the diatomic molecule in one dimension. The two position vectors have equal magnitudes and face in the same direction.

Figure 11.3. The normal mode vibrations for 1,3-butadiene. $\mathbf{x}_1$ and $\mathbf{x}_2$ have different magnitudes.

that the set of two symmetry-generated vectors for the diatomic molecule can be expressed by their components. For the symmetric stretch, the ordered vector components are

$$(1, 1) \tag{11.1.7}$$

The translation vector is

$$(1, -1) \tag{11.1.8}$$

These vectors are identical to the normal mode vectors for this system but they now describe the components in a symmetry space. The first component is the scalar projection of a vector along an "identity" axis. The second scalar component is the projection of this same vector on a C_2 symmetry axis. The symmetry elements now define the components in this space. In this example with only two atoms, the normal mode eigenvectors and the "symmetry space" vector components are identical but the two spaces are different.

The difference between this "C_2" symmetry space and the space for the normal modes can be illustrated for 1,3-butadiene. The molecule will have the four normal modes shown in Figure 11.3. These modes require four components, i.e., a four-dimensional space.

The symmetry space still has only two components and can only develop relationships between the left and right sides of the vibrating molecule. A vector (1, 1) in the symmetry space will create the remaining two normal mode vector components if two are known. The two possible operations are shown in Figure 11.4. In Figure 11.4a, the identity operation reproduces the two in-phase vectors on the left. The C_2 operation rotates the atoms with the two vectors through 180° to produce vectors on atoms 3 and 4 which are 180° out of phase with respect to the first two. This is the symmetric stretch vector for the molecule if the vectors on atoms 1 and 2 have the proper magnitudes for a symmetric

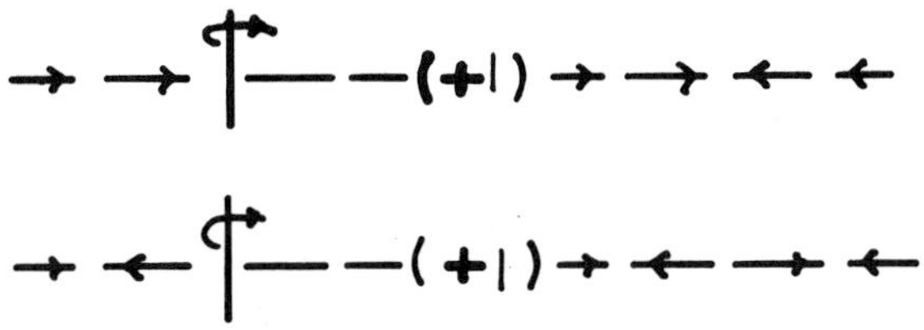

Figure 11.4. The generation of vibrational normal modes for 1,3-butadiene using the operations E and $(+1)\,C_2$.

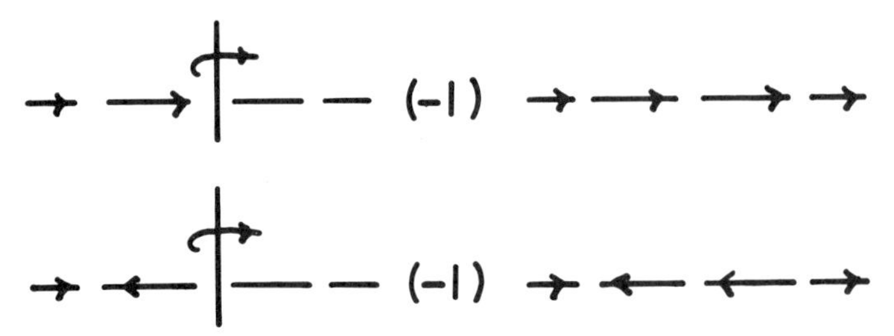

Figure 11.5. The generation of (a) a molecular translation vector and (b) a normal vibrational mode using the operations E and $(-1)\,C_2$.

stretch normal mode. The same symmetry operations will generate a second normal mode vector (Figure 11.4b) if the vectors on atoms 1 and 2 are opposed.

The remaining two normal modes are generated if the symmetry space vector $(1, -1)$ is used. In Figure 11.5a, the in-phase vectors on atoms 1 and 2 are maintained by the identity operation. The C_2 operation rotates the vectors while the -1 component reverses their direction to create four in-phase vectors, i.e., a translation. This translation is generated only if the initial two vectors have equal magnitudes; the symmetry space cannot give all the detailed information on the normal mode. It only selects symmetry properties involving the left and right sides of the molecule. The final normal mode vector generated with the $(1, -1)$ vector is shown in Figure 5b.

Symmetry operations are not restricted to normal mode vectors. A molecular orbital for ethene is generated starting from a p_z orbital on carbon 1. The E operation reproduces this orbital while the C_2 operation rotates atom and orbital to the location of carbon 2. The $+1$ component for C_2 completes the bonding molecular orbital for the molecule (Figure 11.6a). The vector $(1, -1)$ rotates and then negates the orbital from carbon 1 to produce the antibonding orbital of Figure 11.6b.

The operations E and C_2 constitute a C_2 symmetry group. There was insufficient information in this two-dimensional system to reproduce the four normal modes for 1,3-butadiene since the vectors for carbons 1 and 2 were required. If a system has additional symmetry elements, the symmetry space will have

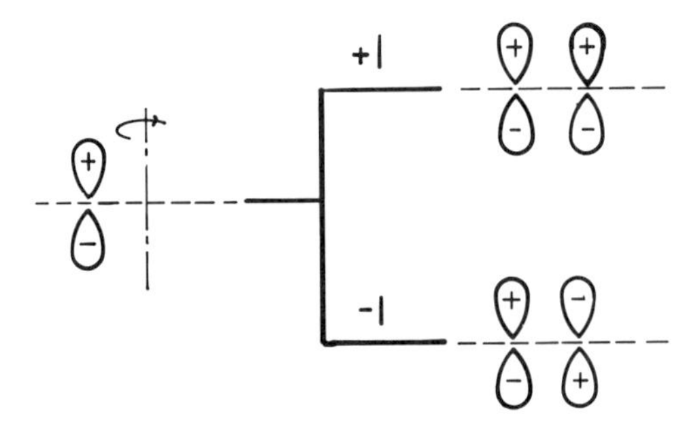

Figure 11.6. Generation of the molecular orbitals of ethene from p_z orbitals using the E and C_2 symmetry operations.

more components and more detailed vectors are possible. For example, the 1,3-cyclobutadiene molecule is symmetric with respect to rotations of 90°, 180°, and 270° (−90°). These rotations bring carbon 1 to the locations for carbons 2, 3, and 4, respectively. With the identity E, they constitute a four-dimensional symmetry space. The molecular orbitals from p_z orbitals were determined in Section 8.5 as

$$\begin{aligned}&(1, 1, 1, 1)\\&(1, -1, 1, -1)\\&(1, 1, -1, -1)\\&(1, -1, -1, 1)\end{aligned} \tag{11.1.9}$$

These four-component vectors are sufficient to reproduce the molecular orbitals involving the four p_z orbitals. They also appear in the symmetry space. The rotation of 90° is called C_4. Since the rotation of 270° can be generated via three applications of the C_4 operation, it is written C_4^3. The rotation of 180° is again C_2. If the four symmetry elements are ordered as E, C_4, C_2, and C_4^3, then the vectors of Equation 11.1.9 constitute a set of orthogonal vectors in this space as well. The individual orbitals are generated using the symmetry operations and their scalar components. The totally symmetric orbital is generated starting with the p_z orbital on carbon 1, $|1\rangle = p_{z1}$, as

$$\begin{aligned}|1\rangle &= (+1)E\,|1\rangle + (+1)\,C_4\,|1\rangle + (+1)\,C_2\,|1\rangle + (+1)\,C_4^3\,|1\rangle\\&= |1\rangle + |2\rangle + |3\rangle + |4\rangle\end{aligned} \tag{11.1.10}$$

The remaining three molecular orbitals are generated using the scalar components for the remaining three vectors of Equation 11.1.9.

Although the vector components for the vectors in the symmetry space have been identical to those generated via Hückel or normal mode analysis, not every orthogonal vector in the symmetry space will generate a corresponding motion or configuration for the molecule. The C_{3v} symmetry group includes rotations of 120° and 240°, which are defined as C_3 and C_3^2, respectively. In addition, this group contains three separate reflection operators, σ_1, σ_2, and σ_3. All these symmetry operations can be illustrated for a three-atom triangular molecule confined to a plane. The C_3 and C_3^2 symmetry operations are shown in Figure 11.7. Atom 1 is rotated to the positions for atoms 2 and 3 via the symmetry operations C_3 and C_3^2, respectively. The reflection σ_1 takes place through a

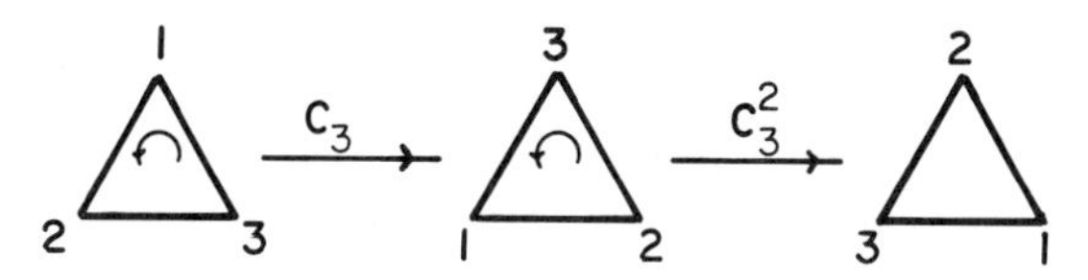

Figure 11.7. The C_3 and C_3^2 rotation operations.

reflection plane perpendicular to the molecular plane which passes through atom 1 and bisects the bond between atoms 2 and 3 (Figure 11.8). Reflection σ_2 has a perpendicular reflection plane which passes through atom 2 while reflection σ_3 has its reflection plane through atom 3 (Figure 11.8). Each of these six symmetry operations moves atoms to the locations for other atoms in the molecule or maintains the atom in its original position. No new atomic positions can be created by the symmetry operations.

Since there are six symmetry elements, the symmetry space vectors must have six components. The triangular molecule confined to the plane should have six independent coordinates (two per atom) and a set of six orthogonal vectors for the normal modes of the molecule might be expected. However, this is not the case. The acceptable six-component vectors in the symmetry space are

$$(1, 1, 1, 1, 1, 1) \tag{11.1.11}$$

$$(1, 1, 1, -1, -1, -1) \tag{11.1.12}$$

$$(2, -1, -1, 0, 0, 0) \tag{11.1.13}$$

for the ordering E, C_3, C_3^2, σ_1, σ_2, and σ_3. Note that the C_3 and C_3^2 elements have the same scalar component value in each of the vectors. The scalar components for the three reflections are also identical within each vector. This pattern will be clarified later.

There are only three six-component vectors in the symmetry space. However, these three orthogonal vectors will not generate the three molecular orbitals expected for this system. The fully symmetric molecular orbital is generated starting with the p_z orbital on carbon 1, $|1\rangle$. The six symmetry operations in conjunction with the six scalar components of the first vector give

$$\begin{aligned}|\psi\rangle &= (+1)E\,|1\rangle + (+1)\,C_3\,|1\rangle + (+1)\,C_3^2\,|1\rangle \\ &\quad + (+1)\,\sigma_1\,|1\rangle + (+1)\,\sigma_2\,|1\rangle + (+1)\,\sigma_3\,|1\rangle \\ &= |1\rangle + |2\rangle + |3\rangle + |1\rangle + |3\rangle + |2\rangle\end{aligned} \tag{11.1.14}$$

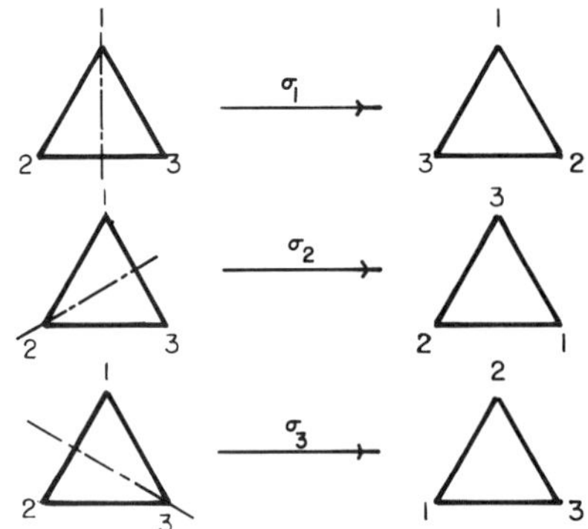

Figure 11.8. The σ_1, σ_2, and σ_3 reflection operations.

The reflections regenerate the original orbital if the orbital lies in the reflection plane. If not, it becomes a new orbital. For example, $|1\rangle$ becomes $|3\rangle$ during a reflection in σ_2 (Figure 11.8).

The six operations can only reproduce the three orbitals available, and the final unnormalized molecular orbital is just

$$|\psi\rangle = (+1)\,|1\rangle + (+1)\,|2\rangle + (+1)\,|3\rangle \tag{11.1.15}$$

The $|1\rangle$ orbital in conjunction with the $(2, -1, -1, 0, 0, 0)$ symmetry vector generates one of the degenerate molecular orbitals for cyclopropene. However, the vector with components $(1, 1, 1, -1, -1, -1)$ will not generate a nonzero molecular orbital. The symmetry operations give

$$\begin{aligned} |\psi\rangle &= (+1)E\,|1\rangle + (+1)\,C_3\,|1\rangle + (+1)\,C_3^2\,|1\rangle \\ &\quad + (+1)\,\sigma_1\,|1\rangle + (+1)\,\sigma_2\,|1\rangle + (+1)\,\sigma_3\,|1\rangle \\ &= |1\rangle + |2\rangle + |3\rangle - |1\rangle - |3\rangle - |2\rangle = 0 \end{aligned} \tag{11.1.16}$$

Although this vector is acceptable in symmetry space, it is unable to create a valid molecular orbital.

The group theory in this chapter will focus on two main questions. (1) How are the eigenvectors for the symmetry elements in a symmetry space determined? (2) How are the specific eigenvectors which generate valid molecular orbitals or normal modes determined?

11.2. Matrix Representations of Symmetry Operations

When a single atomic orbital and the symmetry of a molecule are known, the full molecular orbital for the molecule is generated by using the valid symmetry operations to move the known atomic orbital to other "consistent" locations. The coefficients for the new component orbitals are components of a vector which can be deduced from matrices which describe the symmetry operations.

Simple molecular orbitals for ethene can be developed using one p_z orbital from each carbon atom. The identity operator reproduces these atomic orbitals:

$$\begin{aligned} \mathbf{E}\varphi_1 &= \varphi_1 \\ \mathbf{E}\varphi_2 &= \varphi_2 \end{aligned} \tag{11.2.1}$$

If the two orbitals are orthogonal, each atomic orbital constitutes one direction in a two-dimensional space. The orbitals are expressed as two-dimensional vectors

$$\begin{pmatrix} 1 \\ 0 \end{pmatrix} \quad \begin{pmatrix} 0 \\ 1 \end{pmatrix} \tag{11.2.2}$$

and the identity matrix operator in this space

$$\mathbf{E}=\begin{pmatrix}1 & 0\\ 0 & 1\end{pmatrix} \tag{11.2.3}$$

reproduces each atomic orbital.

The matrix for the C_2 symmetry operation must be able to rotate either atomic orbital or an arbitrary combination of these orbitals in the two-dimensional space. If it operates on φ_1, the rotation of 180° must generate φ_2; if it operates on φ_2, it generates φ_1:

$$\begin{aligned}\mathbf{C}_2\varphi_1 &= \varphi_2\\ \mathbf{C}_2\varphi_2 &= \varphi_1\end{aligned} \tag{11.2.4}$$

The $\mathbf{C}_2$ matrix which describes both these transforms is

$$\mathbf{C}_2=\begin{pmatrix}0 & 1\\ 1 & 0\end{pmatrix} \tag{11.2.5}$$

For example, a $\mathbf{C}_2$ rotation of φ_2 is

$$\begin{pmatrix}0 & 1\\ 1 & 0\end{pmatrix}\begin{pmatrix}0\\ 1\end{pmatrix}=\begin{pmatrix}1\\ 0\end{pmatrix} \tag{11.2.6}$$

and gives the two-dimensional vector for φ_1.

The process can be repeated for the **E** and $\mathbf{C}_2$ operations on the four atomic orbitals of 1,3-butadiene. If the atomic orbitals φ_1, φ_2, φ_3, and φ_4 are assigned vector components 1 through 4, respectively, the **E** and $\mathbf{C}_2$ matrices are (Problem 11.2)

$$\mathbf{E}=\begin{pmatrix}1 & 0 & 0 & 0\\ 0 & 1 & 0 & 0\\ 0 & 0 & 1 & 0\\ 0 & 0 & 0 & 1\end{pmatrix} \qquad \mathbf{C}_2=\begin{pmatrix}0 & 0 & 0 & 1\\ 0 & 0 & 1 & 0\\ 0 & 1 & 0 & 0\\ 1 & 0 & 0 & 0\end{pmatrix} \tag{11.2.7}$$

There are only two symmetry matrices even though the system now has four atomic orbital components. Both the 2×2 and 4×4 matrices are unitary with eigenvalues of $+1$ or -1.

The **E** and $\mathbf{C}_2$ matrices commute in each case. For the 2×2 system,

$$\begin{gathered}\mathbf{E}\mathbf{C}_2=\mathbf{C}_2\mathbf{E}\\ \begin{pmatrix}1 & 0\\ 0 & 1\end{pmatrix}\begin{pmatrix}0 & 1\\ 1 & 0\end{pmatrix}=\begin{pmatrix}0 & 1\\ 1 & 0\end{pmatrix}\begin{pmatrix}1 & 0\\ 0 & 1\end{pmatrix}\\ \begin{pmatrix}0 & 1\\ 1 & 0\end{pmatrix}=\begin{pmatrix}0 & 1\\ 1 & 0\end{pmatrix}\end{gathered} \tag{11.2.8}$$

All rotations about a common axis commute. For example,

$$\mathbf{C}_4\mathbf{C}_2 = \mathbf{C}_2\mathbf{C}_4 \tag{11.2.9}$$

i.e., an initial rotation of 180° followed by a rotation of 90° gives a net rotation of 270°. An initial rotation of 90° followed by a rotation of 180° will also give a 270° rotation.

Commuting matrices have a common set of eigenvectors and similarity transforms. The eigenvectors for one of these matrices can be used to diagonalize all the remaining symmetry matrices. This is illustrated with the $\mathbf{E}$ and $\mathbf{C}_2$ matrices for the ethene system. The identity matrix will regenerate any eigenvector so the $\mathbf{C}_2$ matrix must be used to develop the common set in this case. The characteristic determinant and polynomial for the 2×2 matrix are

$$\begin{vmatrix} 0-\lambda & 1 \\ 1 & 0-\lambda \end{vmatrix} = 0 \tag{11.2.10}$$

$$\lambda^2 - 1 = 0 \tag{11.2.11}$$

and the eigenvalues are

$$\lambda = +1 \qquad \lambda = -1 \tag{11.2.12}$$

The associated eigenvectors are the now familiar

$$|1\rangle = \left(\frac{1}{\sqrt{2}}\right)\begin{pmatrix} 1 \\ 1 \end{pmatrix} \tag{11.2.13}$$

$$|-1\rangle = \left(\frac{1}{\sqrt{2}}\right)\begin{pmatrix} 1 \\ -1 \end{pmatrix} \tag{11.2.14}$$

and the similarity transforms are

$$\mathbf{S} = \begin{pmatrix} 1 & 1 \\ 1 & -1 \end{pmatrix} \tag{11.2.15}$$

$$\mathbf{S}^{-1} = \left(\frac{1}{2}\right)\begin{pmatrix} 1 & 1 \\ 1 & -1 \end{pmatrix} \tag{11.2.16}$$

The identity matrix $\mathbf{E}$ is transformed into itself:

$$\mathbf{S}^{-1}\mathbf{E}\mathbf{S} = \mathbf{E} \tag{11.2.17}$$

The similarity transform of $\mathbf{C}_2$ is

$$\mathbf{S}^{-1}\mathbf{C}_2\mathbf{S} = \left(\frac{1}{2}\right)\begin{pmatrix} 1 & 1 \\ 1 & -1 \end{pmatrix}\begin{pmatrix} 0 & 1 \\ 1 & 0 \end{pmatrix}\begin{pmatrix} 1 & 1 \\ 1 & -1 \end{pmatrix} = \begin{pmatrix} 1 & 0 \\ 0 & -1 \end{pmatrix} \tag{11.2.18}$$

The two diagonal matrices partition into scalar regions. The corresponding scalar regions from each symmetry matrix give the scalar components for each eigenvector in the symmetry space:

$$\left(\begin{array}{c|c} 1 & 0 \\ \hline 0 & 1 \end{array}\right) \qquad \left(\begin{array}{c|c} 1 & 0 \\ \hline 0 & -1 \end{array}\right) \tag{11.2.19}$$

i.e., they give the scalar components needed to construct the molecular orbitals for ethene.

When the 4×4 $\mathbf{E}$ and $\mathbf{C}_2$ matrices are diagonalized with the common similarity transform (Problem 11.3)

$$\mathbf{S} = \mathbf{S}^{-1} = \frac{1}{2}\begin{pmatrix} 1 & 1 & 1 & 1 \\ 1 & 1 & -1 & -1 \\ 1 & -1 & -1 & 1 \\ 1 & -1 & 1 & -1 \end{pmatrix} \tag{11.2.20}$$

the diagonal matrices are

$$\mathbf{E} = \begin{pmatrix} 1 & 0 & 0 & 0 \\ 0 & 1 & 0 & 0 \\ 0 & 0 & 1 & 0 \\ 0 & 0 & 0 & 1 \end{pmatrix} \qquad \mathbf{C}_2 = \begin{pmatrix} 1 & 0 & 0 & 0 \\ 0 & -1 & 0 & 0 \\ 0 & 0 & 1 & 0 \\ 0 & 0 & 0 & -1 \end{pmatrix} \tag{11.2.21}$$

For this four-dimensional system, the pairs $(\mathbf{E}, \mathbf{C}_2)$ of diagonal elements are either $(1, 1)$ or $(1, -1)$. The two orthogonal eigenvectors which generate the molecular orbitals appear as diagonal elements in the diagonalized $\mathbf{E}$ and $\mathbf{C}_2$ matrices. Each eigenvector generated by selecting components from 1×1 diagonal blocks, i.e., diagonal elements, from each of the symmetry matrices is called an irreducible representation. The 1×1 matrices will have symmetry properties like those of the full matrices.

The 1,3-cyclobutadiene system requires 4×4 matrices for the four symmetry operations E, C_4, C_2, and C_4^3. These matrices are (Problem 11.4)

$$\mathbf{E} = \begin{pmatrix} 1 & 0 & 0 & 0 \\ 0 & 1 & 0 & 0 \\ 0 & 0 & 1 & 0 \\ 0 & 0 & 0 & 1 \end{pmatrix} \tag{11.2.22}$$

$$\mathbf{C}_4 = \begin{pmatrix} 0 & 1 & 0 & 0 \\ 0 & 0 & 1 & 0 \\ 0 & 0 & 0 & 1 \\ 1 & 0 & 0 & 0 \end{pmatrix} \tag{11.2.23}$$

$$\mathbf{C}_2 = \begin{pmatrix} 0 & 0 & 1 & 0 \\ 0 & 0 & 0 & 1 \\ 1 & 0 & 0 & 0 \\ 0 & 1 & 0 & 0 \end{pmatrix} \tag{11.2.24}$$

$$\mathbf{C}_4^3 = \begin{pmatrix} 0 & 0 & 0 & 1 \\ 1 & 0 & 0 & 0 \\ 0 & 1 & 0 & 0 \\ 0 & 0 & 1 & 0 \end{pmatrix} \tag{11.2.25}$$

The unit diagonal elements move one unit to the right for each 90° of rotation.

These four matrices all commute and determination of the similarity transform for one of them suffices for all four. The matrix for C_4 gives a characteristic determinant

$$\begin{vmatrix} 0-\lambda & 1 & 0 & 0 \\ 0 & 0-\lambda & 1 & 0 \\ 0 & 0 & 0-\lambda & 1 \\ 1 & 0 & 0 & 0-\lambda \end{vmatrix} = 0 \tag{11.2.26}$$

Expansion in the first row of the determinant gives

$$\lambda^4 - 1(1) = 0 \tag{11.2.27}$$

which reduces to

$$\begin{gathered} (\lambda^2+1)(\lambda^2-1) = 0 \\ (\lambda+i)(\lambda-i)(\lambda+1)(\lambda-1) = 0 \end{gathered} \tag{11.2.28}$$

and the four eigenvalues for this matrix are 1, -1, i, and $-i$.

The four eigenvectors $|1\rangle$, $|-1\rangle$, $|i\rangle$, and $|-i\rangle$ are applied in this order to produce the similarity transform matrices:

$$\mathbf{S} = \begin{pmatrix} 1 & 1 & 1 & 1 \\ 1 & -1 & i & -i \\ 1 & 1 & -1 & -1 \\ 1 & -1 & -i & i \end{pmatrix} \tag{11.2.29}$$

$$\mathbf{S}^{-1} = \left(\frac{1}{4}\right) \begin{pmatrix} 1 & 1 & 1 & 1 \\ 1 & -1 & 1 & -1 \\ 1 & -i & -1 & i \\ 1 & i & -1 & -i \end{pmatrix} \tag{11.2.30}$$

The $\mathbf{S}$ and $\mathbf{S}^{-1}$ matrices have different elements because the C_4 matrix is not symmetric.

The four symmetry matrices can now be transformed with these matrices. The identity transform is

$$\mathbf{S}^{-1}\mathbf{E}\mathbf{S} = \mathbf{E} \tag{11.2.31}$$

The transform for $\mathbf{C}_4$ is

$$\begin{aligned}\mathbf{S}^{-1}\mathbf{C}_4\mathbf{S} &= \left(\frac{1}{4}\right)\begin{pmatrix}1 & 1 & 1 & 1\\ 1 & -1 & 1 & -1\\ 1 & -i & -1 & i\\ 1 & i & -1 & -i\end{pmatrix}\begin{pmatrix}0 & 1 & 0 & 0\\ 0 & 0 & 1 & 0\\ 0 & 0 & 0 & 1\\ 1 & 0 & 0 & 0\end{pmatrix}\begin{pmatrix}1 & 1 & 1 & 1\\ 1 & -1 & i & -i\\ 1 & 1 & -1 & -1\\ 1 & -1 & -i & +i\end{pmatrix}\\ &= \begin{pmatrix}1 & 0 & 0 & 0\\ 0 & -1 & 0 & 0\\ 0 & 0 & i & 0\\ 0 & 0 & 0 & -i\end{pmatrix}\end{aligned} \tag{11.2.32}$$

The four eigenvalues appear as diagonal elements.

The $\mathbf{C}_2$ transform gives

$$\mathbf{S}^{-1}\mathbf{C}_2\mathbf{S} = \begin{pmatrix}1 & 0 & 0 & 0\\ 0 & 1 & 0 & 0\\ 0 & 0 & -1 & 0\\ 0 & 0 & 0 & -1\end{pmatrix} \tag{11.2.33}$$

while

$$\mathbf{S}^{-1}\mathbf{C}_4^3\mathbf{S} = \begin{pmatrix}1 & 0 & 0 & 0\\ 0 & -1 & 0 & 0\\ 0 & 0 & -i & 0\\ 0 & 0 & 0 & i\end{pmatrix} \tag{11.2.34}$$

Since all four matrices are diagonal, corresponding *ii* elements from each symmetry matrix are collected to form the four possible eigenvectors for the symmetry space:

$$|11\rangle = \begin{pmatrix}1\\ 1\\ 1\\ 1\end{pmatrix} \quad |22\rangle = \begin{pmatrix}1\\ -1\\ 1\\ -1\end{pmatrix} \quad |33\rangle = \begin{pmatrix}1\\ i\\ -1\\ -i\end{pmatrix} \quad |44\rangle = \begin{pmatrix}1\\ -i\\ -1\\ i\end{pmatrix} \tag{11.2.35}$$

The first two vectors generated in this way are identical to the components which form the molecular orbitals for 1,3-cyclobutadiene. The remaining two vectors correspond to the doubly degenerate energy state and can be combined to produce two mutually orthogonal vectors with real components:

$$\begin{pmatrix} 1 \\ 0 \\ -1 \\ 0 \end{pmatrix} \quad \begin{pmatrix} 0 \\ 1 \\ 0 \\ -1 \end{pmatrix} \tag{11.2.36}$$

The 2×2 matrices for **E** and $\mathbf{C}_2$ and the 4×4 matrices for **E**, $\mathbf{C}_4$, $\mathbf{C}_2$, and $\mathbf{C}_4^3$ can all be diagonalized so that all the matrices can be separated into 1×1 submatrices along the diagonal. These 1×1 matrices are called the irreducible representations of the symmetry matrices. Since they are diagonal elements, they couple only with diagonal elements of the same row when the symmetry matrices are multiplied, i.e., the rows are uncoupled. The single diagonal element in each row is the scalar coefficient for the orbital generated with that symmetry operation.

The irreducible representations generated by diagonalizing the symmetry matrices provide a clear example of the great orthogonality theorem for such matrix systems. The vector composed of ordered ii elements from the ith row is orthogonal to the vector composed of ordered jj elements from the jth row. The scalar product of the vector of ii elements equals h, the order of the group. Formally,

$$\sum_i \gamma_i(\mathbf{R})\,\gamma_j(\mathbf{R}) = h\delta_{ij} \qquad \delta_{ij} = \begin{cases} 1 & i = j \\ 0 & i \neq j \end{cases} \tag{11.2.37}$$

where γ_i is the scalar value ($+1$ or -1) for the ith row of the symmetry matrix **R**. For example, the elements from the first row generate the vector components

$$(1 \quad 1 \quad 1 \quad 1) \tag{11.2.38}$$

while the diagonal elements of the second row are

$$(1 \quad -1 \quad 1 \quad -1) \tag{11.2.39}$$

The vectors are clearly orthogonal. The scalar product of the second row vector (Equation 11.2.39) is

$$\langle 22|22 \rangle = (1 \times 1) + (-1 \times -1) + (1 \times 1) + (-1 \times -1) = 4 = h \tag{11.2.40}$$

There are four symmetry elements in the group and the scalar product equals this number. The procedure can be repeated with any of the vectors formed from the diagonal elements of different rows. The complex components must be multiplied by their complex conjugate in the scalar products.

In Section 11.1, a set of scalar components were used in conjunction with the different symmetry operations to generate molecular orbitals from a single atomic orbital. One set of such scalar components formed an eigenvector which was perpendicular to other eigenvectors formed in the same manner. For the examples considered, a set of N symmetry operations gave an N-component vector. Although N such eigenvectors were required to span an N-dimensional space and this set could be generated using a Gram–Schmidt orthogonalization, not all such eigenvectors gave an acceptable molecular orbital. In this section, the components for these eigenvectors appeared as the diagonal elements for specific rows of diagonalized rotation matrices. Only "valid" eigenvectors appeared as diagonal elements. The particular matrix set selected valid irreproducible representations from all possible eigenvectors in the symmetry space. The simultaneous diagonalization of the symmetry matrices produced the orthogonal symmetry space eigenvectors required for that specific system. If a second system had the same symmetry elements but different rotation matrices, e.g., a larger number of atomic orbitals, these new matrices would have to be diagonalized with different similarity transforms. However, the diagonal elements still yield 1×1 representations, i.e., eigenvectors, which are consistent with that particular symmetry.

The molecular orbitals can be determined once the vectors for the irreducible representations are known. These representations were determined by diagonalization in this section. However, additional group theoretical techniques will facilitate diagonalization by determining the number of each irreducible representation vector present in the symmetry matrices for a system of arbitrary size. A set of $N \times N$ symmetry matrices are then resolved into $n_1\gamma_1$ irreducible representations, $n_2\gamma_2$ irreducible representations, etc. The components of these vectors are then used to generate the full $N \times N$ diagonal matrix set without similarity transforms.

11.3. Group Operations and Tables

In Section 11.2, all symmetry matrices for a group were diagonalized with a common set of eigenvectors. The elements of the diagonalized matrices served as the components of orthogonal vectors. In addition, the 1×1 matrices which constitute the eigenvector will represent all the properties of the actual symmetry operation. For example, two symmetry operations in sequence will generally equal a single symmetry operation. The product of the corresponding 1×1 matrices for each symmetry element will give a result equal to the 1×1 matrix for the symmetry element which results from the two operations. If the irreducible representation is $+1$ for symmetry operation A in the group and -1 for symmetry operation B and a third symmetry operation C for the group is $C = \text{BA}$, then its irreducible representation must be $(+1)(-1) = -1$. The irreducible "representations" are a simple way of expressing the symmetry properties of the group elements.

A group of symmetry elements must regenerate some group element for any combination of group operations. The group with elements $\mathbf{E}$ and $\mathbf{C}_2$ only illustrates this clearly. All possible pairs of symmetry operations are equivalent to either **E** or $\mathbf{C}_2$, the elements of the group:

$$\begin{aligned} \mathbf{E}\mathbf{C}_2 &= \mathbf{C}_2 \\ \mathbf{C}_2\mathbf{E} &= \mathbf{C}_2 \\ \mathbf{E}\mathbf{E} &= \mathbf{E} \\ \mathbf{C}_2\mathbf{C}_2 &= \mathbf{E} \end{aligned} \tag{11.3.1}$$

When the **E** and $\mathbf{C}_2$ matrices for ethene were diagonalized, the two irreducible representations from the diagonal elements had components

$$\begin{gathered} (1, 1) \\ (1, -1) \end{gathered} \tag{11.3.2}$$

Since these scalars represent the corresponding symmetry operation, products of these scalars give results which are consistent with the products of symmetry operators. For example, the products for the first irreducible representation with both **E** and $\mathbf{C}_2$ assigned a scalar $+1$ and performed in the same order as in Equation 11.3.1 are

$$\begin{aligned} (+1)(+1) &= +1 \\ (+1)(+1) &= +1 \\ (+1)(+1) &= +1 \\ (+1)(+1) &= +1 \end{aligned} \tag{11.3.3}$$

The irreducible representations obey the same product rules as the symmetry operations for the group.

The second vector in Equation 11.3.2 also satisfies the group multiplication conditions. The equations in order are

$$\begin{aligned} (+1)(-1) &= -1 \\ (-1)(+1) &= -1 \\ (+1)(+1) &= +1 \\ (-1)(-1) &= +1 \end{aligned} \tag{11.3.4}$$

The -1 "represents" C_2 and appears each time the product of symmetry elements gives the single element C_2.

The symetry element products are often written as a group table where the elements in the array are the product elements generated by the symmetry elements for the corresponding row and column:

	E	C_2
E	E	C_2
C_2	C_2	E

(11.3.5)

Either of the vectors satisfy the same table. For example, the vector (1, −1) has a product table,

	1	−1
1	1	−1
−1	−1	+1

(11.3.6)

A +1 entry appears for each E in the table and −1 appears for each C_2. The scalars have the same group properties as the group elements.

The correlation between groups of scalars and group element product tables persists as the number of symmetry elements increases. A planar water molecule has C_{2v} symmetry with four symmetry operations. A C_2 rotation about a vertical axis which bisects the HOH angle moves H_1 to the coordinates of H_2 while rotating H_2 to the coordinates of H_1. A reflection through the yz plane (σ) also shifts H_1 to the coordinates of H_2 and vice versa. The triatomic molecule lies in the xz plane. Since the three atoms lie in this plane, a reflection in this plane (σ') leaves all atoms in the original positions. These operations and the identity operation are the four elements for the C_{2v} group.

The group table is generated by considering the effects of all pairs of symmetry operations. The three coordinate vectors on each of the atoms (Figure 11.9) must be monitored to establish that the correct result has been obtained. The net operation σC_2 is a good example. C_2 is performed first for operations to the right. The direction of each z vector is unchanged but the

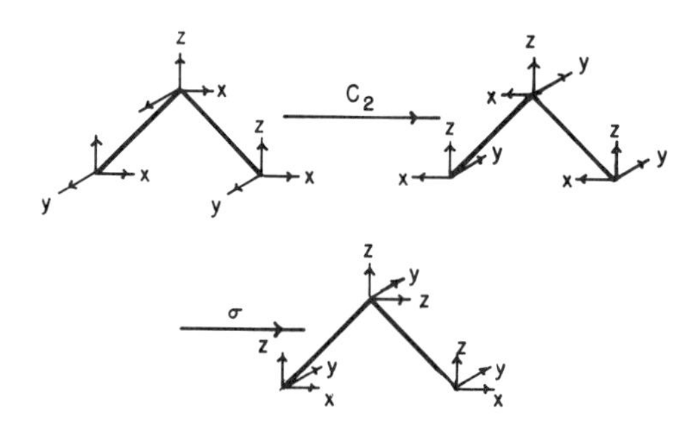

Figure 11.9. Changes in the coordinate components x, y, and z under a C_2 operation followed by a σ operation.

directions of the x and y vectors are reversed. The operation then reflects this altered vector arrangement through the yz plane. The x vectors revert to their original positions but the y vectors remain reversed. The product of symmetry elements is equivalent to a single σ' operation through the plane of the molecule (xz). Only the y vectors are reversed in this operation. The group product operation is

$$\sigma C_2 = \sigma' \tag{11.3.7}$$

The full group table can be generated in this manner as

(11.3.8)

	E	C_2	σ	σ'
E	E	C_2	σ	σ'
C_2	C_2	E	σ'	σ
σ	σ	σ'	E	C_2
σ'	σ'	σ	C_2	E

The first operation is tabulated as a row element on the left while the second operation appears as a column element in this format.

Although this group table is more complicated, it is clear that replacement of each symmetry element by $+1$ satisfies the multiplication in the group table. Each product must be $+1$. This will happen for all groups and there will always be an irreducible representation which contains only $+1$'s. Since other irreducible representations are orthogonal to this representation via the great orthogonality theorem, the group property can be verified for any vector orthogonal to this vector. The vector with components

$$(E, C_2, \sigma, \sigma') = (1, -1, 1, -1) \tag{11.3.9}$$

forms the group table

(11.3.10)

	1	-1	1	-1
1	1	-1	1	-1
-1	-1	1	-1	1
1	1	-1	1	-1
-1	-1	1	-1	1

which could also be formed by substituting the scalars for the symmetry elements in the table in Equation 11.3.8. The two remaining orthogonal vectors, i.e., irreducible representations, are

$$\begin{gathered}(1, 1, -1, -1)\\(1, -1, -1, 1)\end{gathered} \tag{11.3.11}$$

(Problem 11.5). When four symmetry matrices for a system of any size are diagonalized, the corresponding diagonal elements will be identical to the components of one of these four vectors since the vectors represent the four distinct ways in which the system can change while satisfying the group product properties.

The C_{2v} group has some properties of interest. Each of the elements in the group is its own inverse, e.g.,

$$\mathbf{C}_2\mathbf{C}_2 = \mathbf{E} \tag{11.3.12}$$

The diagonal elements in the table are all **E**. Each symmetry element appears only once in each row of the table and the matrix is symmetric in these elements. The order of operations is irrelevant for this group, e.g.,

$$\mathbf{C}_2\boldsymbol{\sigma} = \boldsymbol{\sigma}\mathbf{C}_2 = \boldsymbol{\sigma}' \tag{11.3.13}$$

and the group elements all commute with each other, i.e.,

$$[\mathbf{R}_i, \mathbf{R}_j] = 0 \tag{11.3.14}$$

where $\mathbf{R}_i$ is any symmetry operation in the group. The matrices which describe these symmetry operations must also commute and are diagonalized with a common similarity transform. The group table produces a set of four orthogonal irreducible representations. Each distinct symmetry element in this case constitutes a class of the group; a symmetry element must commute with every other element to constitute a single class. More than one element may occupy a class if symmetry operations do not commute.

The ammonia molecule of Figure 11.10 is an example of C_{3v} symmetry. Rotations of 120° and 240° (C_3 and C_3^2) about the z axis will displace H_1 to H_2 or H_3, respectively. H_1 and the N atom lie in the yz plane (σ_1) so a reflection in this plane will transfer H_2 to the coordinates of H_3 and H_3 to the coordinates of H_2. There are two additional reflections through N and atoms H_2 and H_3, respectively. A reflection plane through N and H_2 (σ_2) will switch H_1 and H_3 while a reflection plane (σ_3) through N and H_3 will switch H_1 and H_2. The group table now requires these six distinct symmetry elements. In this case, the symmetry element products are determined by monitoring the positions

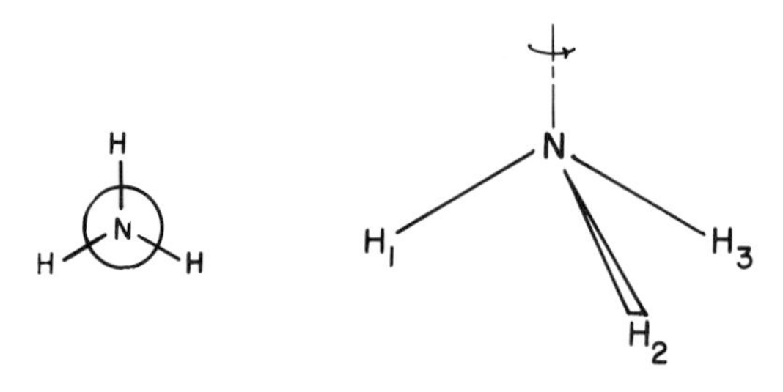

Figure 11.10. The ammonia molecule.

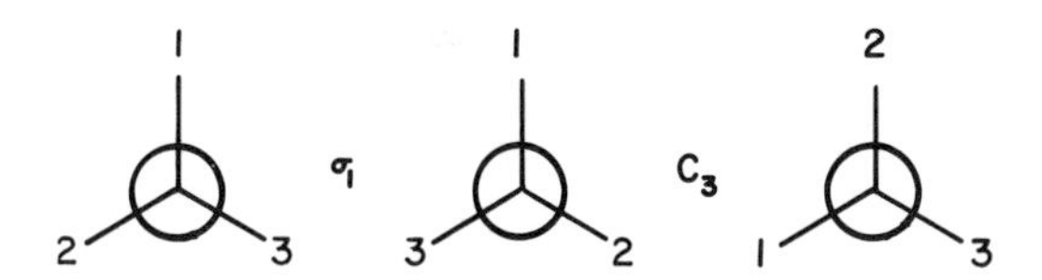

Figure 11.11. The $C_3\sigma_1$ combined symmetry operation for the NH_3 molecule, yielding a net σ_3 symmetry operation for the product table.

of the four atoms for each symmetry operation and the products of these symmetry operations.

The procedure parallels that for the C_{2v} elements. The $C_3\sigma_1$ operation is described in Figure 11.11 and gives

$$C_3\sigma_1 = \sigma_3 \tag{11.3.15}$$

The result of the two operations is always identical to a single operation. In this case, σ_1 switches H_2 and H_3 while leaving N and H_1 unchanged. A rotation of 120° then moves the labeled atoms in their new positions to the position of the atom on their right, e.g., H_1 to the present H_3 position. The final result leaves H_3 and N in their original positions but H_1 and H_2 are switched. This result can be obtained directly with the σ_3 reflection operation.

The product of two σ operations gives

$$\sigma_1\sigma_2 = C_3 \tag{11.3.16}$$

as illustrated in Figure 11.12. During each operation, the labeled reflection planes retain their original orientation. Some product operations are shown in Figure 11.13.

The full group table for the C_{3v} group is

	E	C_3	C_3^2	σ_1	σ_2	σ_3
E	E	C_3	C_3^2	σ_1	σ_2	σ_3
C_3	C_3	C_3^2	E	σ_2	σ_3	σ_1
C_3^2	C_3^2	E	C_3	σ_3	σ_1	σ_2
σ_1	σ_1	σ_3	σ_2	E	C_3^2	C_3
σ_2	σ_2	σ_1	σ_3	C_3	E	C_3^2
σ_3	σ_3	σ_2	σ_1	C_3^2	C_3	E

(11.3.17)

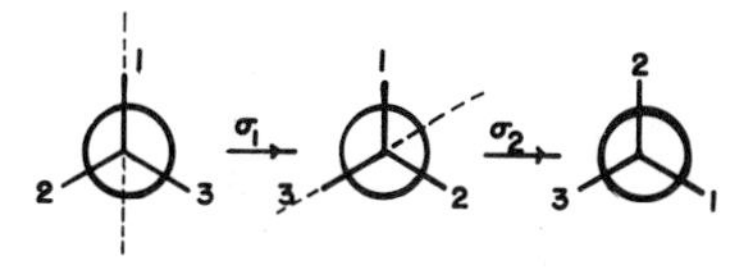

Figure 11.12. The $\sigma_1\sigma_2$ combined symmetry operation for NH_3 which is equivalent to a single C_3 symmetry operation.

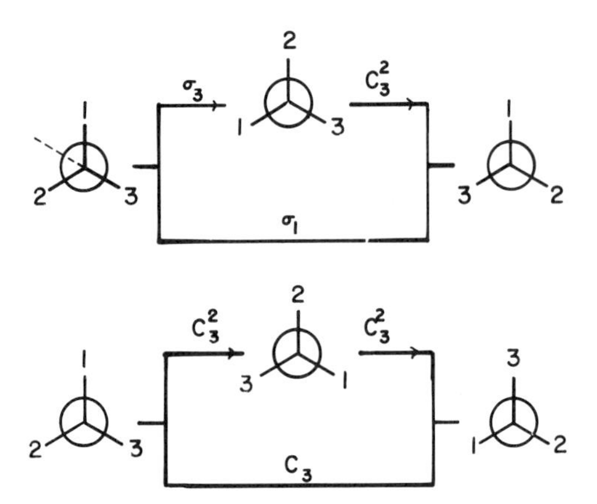

Figure 11.13. Some product transformations for the NH_3 molecule.

The elements of the table are no longer symmetrically arranged and E does not appear on each diagonal. However, each element appears only once in a given row of the table.

The group table gives the commutation properties of the elements. The element E must commute with all the elements and it will always exist as a distinct class. The element C_3 must be tested by checking its commutation with every other element in the group. The five commutators are evaluated with the group table as

$$\begin{aligned}
C_3E - EC_3 &= C_3 - C_3 = 0 \\
C_3C_3^2 - C_3^2C_3 &= E - E = 0 \\
C_3\sigma_1 - \sigma_1C_3 &= \sigma_3 - \sigma_2 \neq 0 \\
C_3\sigma_2 - \sigma_2C_3 &= \sigma_1 - \sigma_3 \neq 0 \\
C_3\sigma_3 - \sigma_3C_3 &= \sigma_2 - \sigma_1 \neq 0
\end{aligned} \tag{11.3.18}$$

The C_3 and C_3^2 elements commute with each other but do not commute with the σ elements.

The same procedure can be applied to a reflection operation, e.g., σ_1. The commutation relations are

$$\begin{aligned}
\sigma_1E - E\sigma_1 &= 0 \\
\sigma_1C_3 - C_3\sigma_1 &= \sigma_2 - \sigma_3 \neq 0 \\
\sigma_1C_3^2 - C_3^2\sigma_1 &= \sigma_3 - \sigma_2 \neq 0 \\
\sigma_1\sigma_2 - \sigma_2\sigma_1 &= C_3 - C_3^2 \neq 0 \\
\sigma_1\sigma_3 - \sigma_3\sigma_1 &= C_3^2 - C_3 \neq 0
\end{aligned} \tag{11.3.19}$$

Although the operators do not commute, a pattern is apparent. The nonzero commutators for C_3 always gave a difference of reflection operators. The nonzero commutators for σ_1 always gave a difference of rotation operators. If C_3 and C_3^2 are combined to form a second class while the three σ operations are combined as the third and final class, then these classes do "commute" with each other. In Equation 11.3.19, the commutation operations which give a difference of two different reflection operators are zero in terms of the classes since both operators are in the same class. In practice, the division of elements into classes is done using a variation of the commutation relation. The difference

$$\sigma_1 C_3 - C_3^2 \sigma_1 \tag{11.3.20}$$

is rewritten like a similarity transform:

$$\begin{aligned} \sigma_1 C_3 \sigma_1^{-1} &= \sigma_1 C_3 \sigma_1 \\ &= C_3^2 \end{aligned} \tag{11.3.21}$$

Every element and its inverse are defined, for example, so that

$$\sigma_1 \sigma_1^{-1} = E \tag{11.3.22}$$

must be applied to a single group element. The operations will generate every element which is in the class with C_3.

The irreducible representations which are associated with the C_{3v} group are developed for a specific example in Section 11.4. However, the group table provides some information on these irreducible representations already. An eigenvector with all components equal to $+1$ must always satisfy the group table. The vector with components

$$(1, 1, 1, -1, -1, -1) \tag{11.3.23}$$

is also consistent with the group table

	1	1	1	−1	−1	−1
1	1	1	1	−1	−1	−1
1	1	1	1	−1	−1	−1
1	1	1	1	−1	−1	−1
−1	−1	−1	−1	1	1	1
−1	−1	−1	−1	1	1	1
−1	−1	−1	−1	1	1	1

(11.3.24)

since σ elements appear only in the 3×3 blocks at the upper right and lower left.

Each of these eigenvectors has identical scalar components for elements of the same class. This is a general result which permits a reduction of the eigenvector into classes:

$E(1)$	$C_3, C_3^2(1)$	$\sigma_1, \sigma_2, \sigma_3(-1)$
1	1	-1

(11.3.25)

Because elements in the same class have the same scalar components, the total number of orthogonal eigenvectors for this space is restricted. Only three eigenvectors are consistent with this requirement. The third such eigenvector must be

$$(2, -1, -1, 0, 0, 0) \tag{11.3.26}$$

This result is particularly interesting since the irreducible representation for an **E** matrix, i.e., the first component, is 2, not 1, as expected for the diagonal matrix. This is related to the fact that not all symmetry elements commuted for the C_{3v} group and all matrices might not diagonalize with a common similarity transform. In fact, for such systems, some 2×2 blocks of elements may remain on the diagonals of the symmetry matrices. This is illustrated in the next section.

11.4. Properties of Irreducible Representations

The group table for the C_{3v} group was used to find two irreducible representations which satisfied the group properties. These were

$$(1, 1, 1, 1, 1, 1) \tag{11.4.1}$$

$$(1, 1, 1, -1, -1, -1) \tag{11.4.2}$$

These representations gave the components of two orthogonal vectors. These vectors can also be obtained via a simultaneous diagonalization of a set of symmetry matrices for a C_{3v} group. These matrices will not diagonalize completely and the matrices will have a block diagonal form. Some 2×2 blocks of nonzero elements will remain along the diagonal and they are irreducible. The C_{3v} group contains both 1×1 and 2×2 irreducible representations.

The p_z orbitals of the cyclopropene molecule provide a simple example of a C_{3v} system. Atomic orbital φ_1 can be rotated to the coordinates of φ_2 with a rotation of 120°. The reflection operators are identical to those introduced for NH_3 in the last section, i.e., σ_1 is a reflection in a plane through carbon 1 and the midpoint of the carbon 2–carbon 3 bond (Figure 11.14).

The basis vectors for this system are the three atomic orbitals, which are ordered as φ_1, φ_2, and φ_3. The identity does not change the orbitals and is

$$\mathbf{E} = \begin{pmatrix} 1 & 0 & 0 \\ 0 & 1 & 0 \\ 0 & 0 & 1 \end{pmatrix} \tag{11.4.3}$$

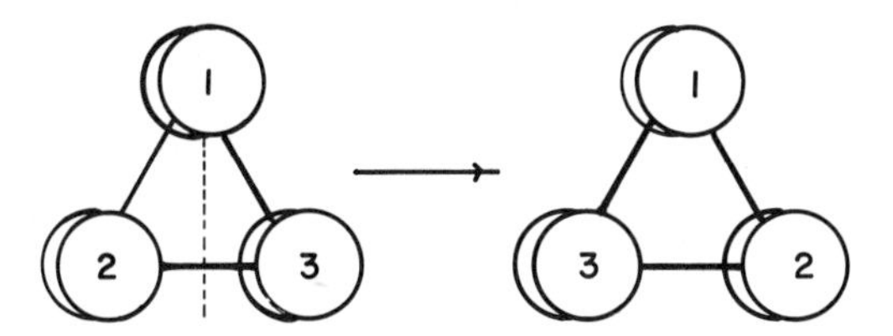

Figure 11.14. Generation of the full set of cyclopropene orbitals with the symmetry elements of C_{3v}.

$\mathbf{C}_3$ is a counterclockwise rotation of each of the orbitals. With the numbering scheme of Figure 11.14, φ_1 moves to the coordinates of φ_2, φ_2 moves to the coordinates of φ_3, and φ_3 moves to the coordinates of φ_1. The $\mathbf{C}_3$ matrix equation is

$$\begin{pmatrix} \varphi_1 \\ \varphi_2 \\ \varphi_3 \end{pmatrix} = \begin{pmatrix} 0 & 1 & 0 \\ 0 & 0 & 1 \\ 1 & 0 & 0 \end{pmatrix} \begin{pmatrix} \varphi_1 \\ \varphi_2 \\ \varphi_3 \end{pmatrix} \tag{11.4.4}$$

The identity elements move one position to form $\mathbf{C}_3$. The element in each row is moved one position to the right.

The $\mathbf{C}_3^2$ matrix requires motion of φ_1 through the coordinates of φ_2 to the coordinates of φ_3. The matrix equation is

$$\begin{pmatrix} \varphi_3 \\ \varphi_1 \\ \varphi_2 \end{pmatrix} = \begin{pmatrix} 0 & 0 & 1 \\ 1 & 0 & 0 \\ 0 & 1 & 0 \end{pmatrix} \begin{pmatrix} \varphi_1 \\ \varphi_2 \\ \varphi_3 \end{pmatrix} \tag{11.4.5}$$

The reflections through a plane will change the positions of two orbitals which do not lie in the reflection plane. The reflection $\boldsymbol{\sigma}_1$ will reverse the positions of $\boldsymbol{\sigma}_2$ and $\boldsymbol{\sigma}_3$. The matrix for $\boldsymbol{\sigma}_1$ is

$$\begin{pmatrix} \varphi_1 \\ \varphi_3 \\ \varphi_2 \end{pmatrix} = \begin{pmatrix} 1 & 0 & 0 \\ 0 & 0 & 1 \\ 0 & 1 & 0 \end{pmatrix} \begin{pmatrix} \varphi_1 \\ \varphi_2 \\ \varphi_3 \end{pmatrix} \tag{11.4.6}$$

The matrix equations for $\boldsymbol{\sigma}_2$ and $\boldsymbol{\sigma}_3$ are

$$\boldsymbol{\sigma}_2 = \begin{pmatrix} 0 & 0 & 1 \\ 0 & 1 & 0 \\ 1 & 0 & 0 \end{pmatrix} \tag{11.4.7}$$

$$\boldsymbol{\sigma}_3 = \begin{pmatrix} 0 & 1 & 0 \\ 1 & 0 & 0 \\ 0 & 0 & 1 \end{pmatrix} \tag{11.4.8}$$

The commutation properties of these matrices, like those of the corresponding symmetry operators, limit complete diagonalization. The eigenvectors for $\mathbf{C}_3$ are determined and used to develop a similarity transform which is then applied to all the matrices in the group to determine the effects of noncommutation.

The characteristic equation for $\mathbf{C}_3$ is

$$\lambda^3 - 1 = 0 \tag{11.4.9}$$

$\lambda = 1$ is one solution. The remaining two eigenvalues are exponential functions,

$$\begin{aligned} \lambda_2 &= \exp(i2\pi/3) = \varepsilon \\ \lambda_3 &= \exp(i4\pi/3) = \exp(-i2\pi/3) = \varepsilon^* \end{aligned} \tag{11.4.10}$$

These results can be verified by direct substitution in Equation 11.4.9. The last two eigenvalues are complex conjugates.

The column eigenvectors for $|1\rangle$, $|\varepsilon\rangle$, and $|\varepsilon^*\rangle$ are combined in this order to give the similarity transform

$$\mathbf{S} = \begin{pmatrix} 1 & 1 & 1 \\ 1 & \varepsilon & \varepsilon^* \\ 1 & \varepsilon^* & \varepsilon \end{pmatrix} \tag{11.4.11}$$

Since the $\mathbf{C}_3$ matrix is not symmetric, the set of row vectors $\langle 1|$, $\langle \varepsilon|$, and $\langle \varepsilon^*|$ must be developed separately:

$$\mathbf{S}^{-1} = \begin{pmatrix} 1 & 1 & 1 \\ 1 & \varepsilon^* & \varepsilon \\ 1 & \varepsilon & \varepsilon^* \end{pmatrix} \tag{11.4.12}$$

The $\mathbf{C}_3$ matrix transforms as

$$\left(\frac{1}{2}\right) \begin{pmatrix} 1 & 1 & 1 \\ 1 & \varepsilon^* & \varepsilon \\ 1 & \varepsilon & \varepsilon^* \end{pmatrix} \begin{pmatrix} 0 & 1 & 0 \\ 0 & 0 & 1 \\ 1 & 0 & 0 \end{pmatrix} \begin{pmatrix} 1 & 1 & 1 \\ 1 & \varepsilon & \varepsilon^* \\ 1 & \varepsilon^* & \varepsilon \end{pmatrix} = \begin{pmatrix} 1 & 0 & 0 \\ 0 & \varepsilon & 0 \\ 0 & 0 & \varepsilon^* \end{pmatrix} \tag{11.4.13}$$

The off-diagonal elements are all zero because

$$1 + \varepsilon + \varepsilon^* = 1 + 2\cos(2\pi/3) = 1 - 1 = 0 \tag{11.4.14}$$

The similarity transform of $\mathbf{C}_3^2$ is

$$\left(\frac{1}{3}\right) \begin{pmatrix} 1 & 1 & 1 \\ 1 & \varepsilon^* & \varepsilon \\ 1 & \varepsilon & \varepsilon^* \end{pmatrix} \begin{pmatrix} 0 & 0 & 1 \\ 1 & 0 & 0 \\ 0 & 1 & 0 \end{pmatrix} \begin{pmatrix} 1 & 1 & 1 \\ 1 & \varepsilon & \varepsilon^* \\ 1 & \varepsilon^* & \varepsilon \end{pmatrix} = \begin{pmatrix} 1 & 0 & 0 \\ 0 & \varepsilon^* & 0 \\ 0 & 0 & \varepsilon \end{pmatrix} \tag{11.4.15}$$

The eigenvalues for this matrix are equal to those for $\mathbf{C}_3$. However, they now appear in conjunction with different eigenvectors in each diagonalized matrix.

The $\mathbf{C}_3$ and $\mathbf{C}_3^2$ matrices commute but they do not commute with the $\boldsymbol{\sigma}$ matrices (Problem 11.6). Although the $\boldsymbol{\sigma}$ matrices are not diagonalized with the similarity matrices from $\mathbf{C}_3$, they do reduce to a block diagonal form. The similarity transform for $\boldsymbol{\sigma}_1$ is

$$\left(\frac{1}{3}\right)\begin{pmatrix}1 & 1 & 1\\ 1 & \varepsilon^* & \varepsilon\\ 1 & \varepsilon & \varepsilon^*\end{pmatrix}\begin{pmatrix}1 & 0 & 0\\ 0 & 0 & 1\\ 0 & 1 & 0\end{pmatrix}\begin{pmatrix}1 & 1 & 1\\ 1 & \varepsilon & \varepsilon^*\\ 1 & \varepsilon^* & \varepsilon\end{pmatrix}=\begin{pmatrix}1 & 0 & 0\\ 0 & 0 & 1\\ 0 & 1 & 0\end{pmatrix} \tag{11.4.16}$$

The similarity transform does not change $\boldsymbol{\sigma}_1$. The elements of the lower two rows remain off the diagonal.

The similarity transform for $\boldsymbol{\sigma}_2$ is

$$\left(\frac{1}{3}\right)\begin{pmatrix}1 & 1 & 1\\ 1 & \varepsilon^* & \varepsilon\\ 1 & \varepsilon & \varepsilon^*\end{pmatrix}\begin{pmatrix}0 & 0 & 1\\ 0 & 1 & 0\\ 1 & 0 & 0\end{pmatrix}\begin{pmatrix}1 & 1 & 1\\ 1 & \varepsilon & \varepsilon^*\\ 1 & \varepsilon^* & \varepsilon\end{pmatrix}=\begin{pmatrix}1 & 0 & 0\\ 0 & 0 & \varepsilon\\ 0 & \varepsilon^* & 0\end{pmatrix} \tag{11.4.17}$$

and the similarity transform for $\boldsymbol{\sigma}_3$ is

$$\left(\frac{1}{3}\right)\begin{pmatrix}1 & 1 & 1\\ 1 & \varepsilon^* & \varepsilon\\ 1 & \varepsilon & \varepsilon^*\end{pmatrix}\begin{pmatrix}0 & 1 & 0\\ 1 & 0 & 0\\ 0 & 0 & 1\end{pmatrix}\begin{pmatrix}1 & 1 & 1\\ 1 & \varepsilon & \varepsilon^*\\ 1 & \varepsilon^* & \varepsilon\end{pmatrix}=\begin{pmatrix}1 & 0 & 0\\ 0 & 0 & \varepsilon^*\\ 0 & \varepsilon & 0\end{pmatrix} \tag{11.4.18}$$

For each of the reflection matrices, the 11 element is diagonal and constitutes a 1×1 matrix in each reflection. Since the $\mathbf{C}$ and $\mathbf{E}$ matrices are completely diagonal, the upper row of each of the six symmetry matrices has a 1×1 diagonal matrix with a value of $+1$. This is one of the vectors which satisfied the group multiplication property.

The last two rows of the matrices are in block diagonal form. The 2×2 blocks are uncoupled from the 1×1 blocks of the first row. Since each matrix should contribute one component of a vector, these vector components must be the 2×2 matrices. The "vector" components for this irreducible representation are the matrices tabulated below.

E	C_3	C_3^2	σ_1	σ_2	σ_3
$\begin{pmatrix}1 & 0\\ 0 & 1\end{pmatrix}$	$\begin{pmatrix}\varepsilon & 0\\ 0 & \varepsilon^*\end{pmatrix}$	$\begin{pmatrix}\varepsilon^* & 0\\ 0 & \varepsilon\end{pmatrix}$	$\begin{pmatrix}0 & 1\\ 1 & 0\end{pmatrix}$	$\begin{pmatrix}0 & \varepsilon\\ \varepsilon^* & 0\end{pmatrix}$	$\begin{pmatrix}0 & \varepsilon^*\\ \varepsilon & 0\end{pmatrix}$

(11.4.19)

This irreducible representation in 2×2 matrices is different from the 1×1 matrices determined previously but it has the same properties. For example, it is orthogonal to the first vector, $\langle A|$,

$$(1, 1, 1, 1, 1, 1) \tag{11.4.20}$$

which is verified by multiplying component by component and adding:

$$1\begin{pmatrix}1 & 0\\ 0 & 1\end{pmatrix}+1\begin{pmatrix}\varepsilon & 0\\ 0 & \varepsilon^*\end{pmatrix}+1\begin{pmatrix}\varepsilon^* & 0\\ 0 & \varepsilon\end{pmatrix}$$

$$+1\begin{pmatrix}0 & 1\\ 1 & 0\end{pmatrix}+1\begin{pmatrix}0 & \varepsilon\\ \varepsilon^* & 0\end{pmatrix}+1\begin{pmatrix}0 & \varepsilon^*\\ \varepsilon & 0\end{pmatrix}$$

$$=\begin{pmatrix}1+\varepsilon+\varepsilon^*+0+0+0 & 0+0+0+1+\varepsilon+\varepsilon^*\\ 0+0+0+1+\varepsilon^*+\varepsilon & 1+\varepsilon^*+\varepsilon+0+0+0\end{pmatrix}=0 \quad (11.4.21)$$

The orthogonality result is even more general than that of Equation 11.4.21. The great orthogonality theorem states that corresponding elements from each 2×2 matrix create a vector which is also orthogonal to the one-dimensional vector (Equation 11.4.20). For example, the 21 elements from each 2×2 matrix give the vector

$$(0, 0, 0, 1, \varepsilon^*, \varepsilon) \quad (11.4.22)$$

which is orthogonal to the vector of Equation 11.4.20 since

$$1+\varepsilon+\varepsilon^*=0 \quad (11.4.23)$$

Each set of elements from the 2×2 matrices will be orthogonal to this vector. They are also orthogonal to the vector

$$(1, 1, 1, -1, -1, -1) \quad (11.4.24)$$

which satisfied the group product in Section 11.3.

Although any element of the 2×2 matrices can be used with the great orthogonality theorem, it is more convenient to use a sum of elements. Since the trace is an invariant for these matrices, the traces of each matrix are selected and used as the vector components. For this case, the vector is

$$(2, \varepsilon+\varepsilon^*, \varepsilon^*+\varepsilon, 0, 0, 0)$$
$$=(2, -1, -1, 0, 0, 0) \quad (11.4.25)$$

The components of the vectors for the C_3 group can be arranged in classes:

	E	C_3, C_3^2	$\sigma_1, \sigma_2, \sigma_3$
$\lvert 1\rangle$	1	1	1
$\lvert 2\rangle$	1	1	-1
$\lvert 3\rangle$	2	-1	0

(11.4.26)

There are three classes and three vectors of three components each rather than the six orthogonal vectors of six components which might be expected. However, the same component will appear for each element of the class and this separation by classes makes the vector operations more tractable. The traces of the 2×2 irreducible matrices generate the components for the third vector. The table of irreducible representations is called a character table.

The orthogonality condition for the vectors must include the number of elements in each class. This class orthogonality requires the condition

$$\langle i| \mathbf{B} |j\rangle \tag{11.4.27}$$

where the diagonal matrix **B** has the number of elements in each of its classes as diagonal elements. For C_{3v}, the matrix is

$$\mathbf{B} = \begin{pmatrix} 1 & 0 & 0 \\ 0 & 2 & 0 \\ 0 & 0 & 3 \end{pmatrix} \tag{11.4.28}$$

The orthogonality condition for $|1\rangle$ and $|3\rangle$ is

$$\begin{aligned} \langle 3| \mathbf{B} |1\rangle &= (2 \quad -1 \quad 0) \begin{pmatrix} 1 & 0 & 0 \\ 0 & 2 & 0 \\ 0 & 0 & 3 \end{pmatrix} \begin{pmatrix} 1 \\ 1 \\ 1 \end{pmatrix} \\ &= 2 - 2 + 0 = 0 \end{aligned} \tag{11.4.29}$$

In addition,

$$\begin{aligned} \langle 3| \mathbf{B} |3\rangle &= (2 \quad -1 \quad 0) \begin{pmatrix} 1 & 0 & 0 \\ 0 & 2 & 0 \\ 0 & 0 & 3 \end{pmatrix} \begin{pmatrix} 2 \\ -1 \\ 0 \end{pmatrix} \\ &= 4 + 2 + 0 = 6 \end{aligned} \tag{11.4.30}$$

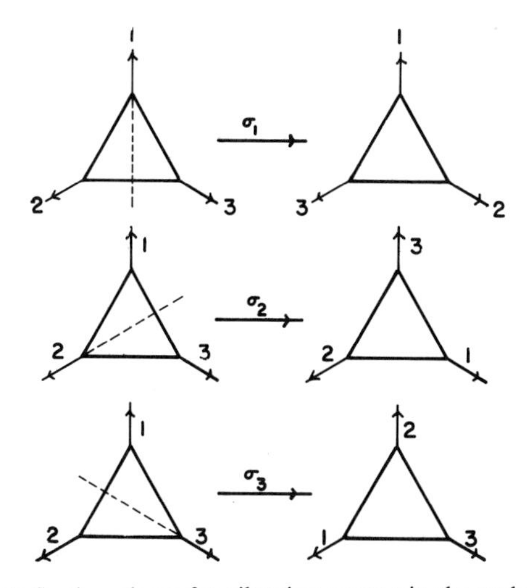

Figure 11.15. Reflection planes for vibration vectors in the cyclopropene system.

The scalar product gives the total elements in the group and this number will normalize the vectors.

The simultaneous diagonalization of the symmetry matrices was performed for matrices generated by rotation of atomic orbitals. However, the same technique can be used to monitor the change of three position vectors centered on some atom. The rotation matrices

$$\mathbf{C}_3 = \begin{pmatrix} -\frac{1}{2} & \frac{\sqrt{3}}{2} & 0 \\ -\frac{\sqrt{3}}{2} & -\frac{1}{2} & 0 \\ 0 & 0 & 1 \end{pmatrix} \tag{11.4.31}$$

$$\mathbf{C}_3^2 = \begin{pmatrix} -\frac{1}{2} & -\frac{\sqrt{3}}{2} & 0 \\ \frac{\sqrt{3}}{2} & -\frac{1}{2} & 0 \\ 0 & 0 & 1 \end{pmatrix} \tag{11.4.32}$$

describe the rotation of unit vectors x, y, and z through 120° and 240°.

The reflection matrices for these vectors are placed at 0°, 120°, and 240° perpendicular to the xy plane (Figure 11.15). The z component remains constant but x and y unit vectors are mixed during the reflections. Thc planc along x leaves x unchanged but converts y to $-y$ (Figure 11.16a). The matrix is

$$\boldsymbol{\sigma}_1 = \begin{pmatrix} 1 & 0 & 0 \\ 0 & -1 & 0 \\ 0 & 0 & 1 \end{pmatrix} \tag{11.4.33}$$

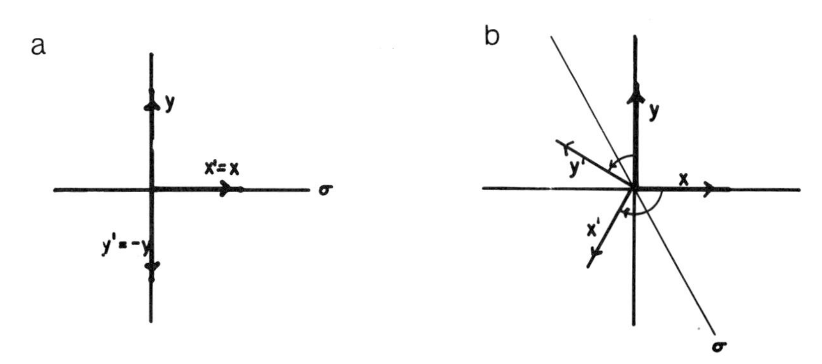

Figure 11.16. (a) The reflection plane and new x' and y' coordinates for the σ_1 reflection for cyclopropene. (b) The reflection plane and new x' and y' coordinates for the σ_2 reflection for cyclopropene.

The reflection σ_2 for the two vectors is illustrated in Figure 11.16b. The final vector has components

$$\begin{pmatrix} -x/2+\dfrac{\sqrt{3}}{2}y \\ -\dfrac{\sqrt{3}}{2}x+y/2 \\ 1 \end{pmatrix} \tag{11.4.34}$$

and the matrix is

$$\begin{pmatrix} -\dfrac{1}{2} & -\dfrac{\sqrt{3}}{2} & 0 \\ -\dfrac{\sqrt{3}}{2} & \dfrac{1}{2} & 0 \\ 0 & 0 & 1 \end{pmatrix} \tag{11.4.35}$$

The matrix for the reflection through the plane at 240° is

$$\begin{pmatrix} \dfrac{1}{2} & \dfrac{\sqrt{3}}{2} & 0 \\ \dfrac{\sqrt{3}}{2} & -\dfrac{1}{2} & 0 \\ 0 & 0 & 1 \end{pmatrix} \tag{11.4.36}$$

All these matrices can be diagonalized with the same eigenvectors and the transformed matrices will all contain 1×1 diagonal blocks (irreducible representations) with the components of $|1\rangle$ or $|2\rangle$ or 2×2 blocks with traces equal to the components of $|3\rangle$. Since the trace of each of these matrices is invariant, a direct examination of the trace for each of the six matrices can provide some indication of which of the three irreducible representations are appropriate for this example. The vector formed with the traces of the six elements is

$$(3, 0, 0, 1, 1, 1) \tag{11.4.37}$$

The traces for matrix elements in each class are identical and this vector can be written as a three-component class vector,

$$(3, 0, 1) \tag{11.4.38}$$

There is only one way to recreate these traces. The similarity transforms will produce the $|1\rangle$ and $|3\rangle$ vectors. Their sum is

$$(1, 1, 1) + (2, -1, 0) = (3, 0, 1) \tag{11.4.39}$$

The matrices do not have to be diagonalized. Knowledge of the traces of each of the matrices dictates the appropriate vectors and shows that the 11 elements of each vector will be 1. The second diagonal elements will be 2×2 blocks which will have traces of 2, -1, and 0.

A symmetry matrix can be any size. For example, a symmetry operation on a triatomic molecule in three dimensions will require nine spatial vectors and a 9×9 symmetry matrix. Diagonalization of each of these matrices is difficult. However, the traces of these matrices are easily determined and can be used to determine the irreducible representations which will appear in these diagonal matrices. The techniques for the determination of the numbers of each irreducible representation for matrices of arbitrary order are introduced in the next section.

11.5. Applications of Group Theory

A group of symmetry elements for a system can be a set of formal operators or a set of matrices which perform the symmetry transformations on atomic orbitals or vibration vectors. Although these matrices have an order equal to the total number of orbitals or vectors in the system of interest, these matrices will always reduce to 1×1, 2×2, or 3×3 block diagonal matrices. These block diagonal matrices are the irreducible representations for that specific group of elements, and the block diagonal form of the matrices can be constructed once the numbers of each irreducible representation are known.

Since the scalar components of the irreducible representations can be used to construct a molecular orbital or normal mode (Section 11.1), the number and type of irreducible representations in a set of symmetry matrices must be determined. The trace of the matrix is invariant and constitutes one component of a reducible representation. If the matrices were diagonalized, the traces would be formed as a sum of the irreducible components which lay along the diagonal. This suggests the technique required to establish the diagonal form of each symmetry matrix without a similarity transform. The reducible representation formed from the traces of the $N \times N$ symmetry matrices will be a linear combination of traces of the irreducible representations which would appear in the diagonal matrices. Projection operator techniques are used to establish the number of each irreducible representation in the diagonalized $N \times N$ matrix.

The determination of the traces for the symmetry matrices is simplified by noting an important property of such matrices. When vectors or orbitals move during a symmetry operation, they change their location in space and acquire new coordinates. This requires matrix elements in off-diagonal positions. In other words, any orbital or vector which moves during the symmetry operation will have no diagonal elements and will contribute zero to the trace.

This observation is illustrated with the atomic orbitals for cyclopropene. The $\mathbf{C}_3$ matrix for the set of three atomic orbitals is

$$\begin{pmatrix} 0 & 1 & 0 \\ 0 & 0 & 1 \\ 1 & 0 & 0 \end{pmatrix} \tag{11.5.1}$$

Each of the three orbitals moves during the symmetry operation. Since no orbital is restored to its original position, there are no elements along the diagonal and the trace for this $\mathbf{C}_3$ operation must be zero. The $\mathbf{C}_3^2$ operation, which is in the same class, also moves each orbital from its original position and its trace will also be zero. By contrast, the reflection matrix $\boldsymbol{\sigma}_1$ is

$$\begin{pmatrix} 1 & 0 & 0 \\ 0 & 0 & 1 \\ 0 & 1 & 0 \end{pmatrix} \tag{11.5.2}$$

One orbital does not move in the operation and only it contributes to the trace for this reducible representation. The remaining two reflection coefficients in this class will also have net traces of 1.

The same effect is observed with sets of position vectors. The H_2 molecule is a homonuclear system which must be described with six spatial components, x_1, y_1, z_1, x_2, y_2, and z_2, as shown in Figure 11.17. The identity matrix for this system is 6×6 and its trace is then 6. The C_2 operation on this set of vectors will move atom 1 to the coordinates of atom 2 and atom 2 to the coordinates of atom 1 with appropriate changes in the vector directions due to the rotation. The symmetry operation is

$$\begin{pmatrix} -x_2 \\ -y_2 \\ z_2 \\ -x_1 \\ -y_1 \\ z_1 \end{pmatrix} = \begin{pmatrix} 0 & 0 & 0 & -1 & 0 & 0 \\ 0 & 0 & 0 & 0 & -1 & 0 \\ 0 & 0 & 0 & 0 & 0 & 1 \\ -1 & 0 & 0 & 0 & 0 & 0 \\ 0 & -1 & 0 & 0 & 0 & 0 \\ 0 & 0 & 1 & 0 & 0 & 0 \end{pmatrix} \begin{pmatrix} x_1 \\ y_1 \\ z_1 \\ x_2 \\ y_2 \\ z_2 \end{pmatrix} \tag{11.5.3}$$

With this ordering of vectors, the C_2 operation simply moves the entire 3×3

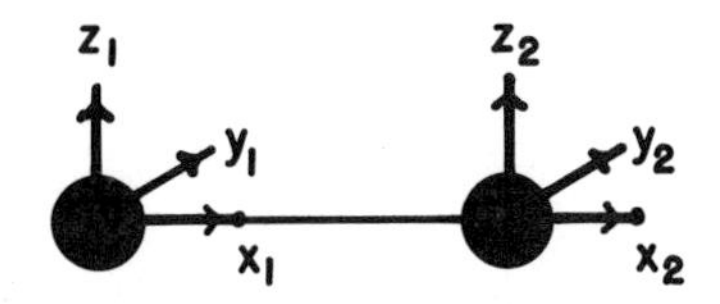

Figure 11.17. The six spatial coordinates required for H_2 in three-dimensional space.

block for the transform of a single atom off the diagonal of the matrix. Since each atom moves in the operation, the blocks which remain on the diagonal have only zero elements. If the atom did not move, the C_2 operation would contribute -1 to the total trace since this is the trace of each of the 3×3 matrices for the C_2 rotation of one atom.

The linear CO_2 molecule also has C_2 symmetry. In this case, the oxygens are assigned the spatial vectors x_1, y_1, and z_1 and x_2, y_2, and z_2 while the carbon atom is assigned x_3, y_3, and z_3 for a total of nine independent components. The trace of the identity matrix is now 9. The $\mathbf{C}_2$ symmetry matrix again contains 3×3 blocks but only the block for C remains on the diagonal since it is not moved by the C_2 operation:

$$\begin{pmatrix} 0 & 0 & 0 & -1 & 0 & 0 & 0 & 0 & 0 \\ 0 & 0 & 0 & 0 & -1 & 0 & 0 & 0 & 0 \\ 0 & 0 & 0 & 0 & 0 & 1 & 0 & 0 & 0 \\ -1 & 0 & 0 & 0 & 0 & 0 & 0 & 0 & 0 \\ 0 & -1 & 0 & 0 & 0 & 0 & 0 & 0 & 0 \\ 0 & 0 & 1 & 0 & 0 & 0 & 0 & 0 & 0 \\ 0 & 0 & 0 & 0 & 0 & 0 & -1 & 0 & 0 \\ 0 & 0 & 0 & 0 & 0 & 0 & 0 & -1 & 0 \\ 0 & 0 & 0 & 0 & 0 & 0 & 0 & 0 & 1 \end{pmatrix} \tag{11.5.4}$$

The total trace for C_2 is -1.

The traces of the $\mathbf{E}$ and $\mathbf{C}_2$ matrices for H_2 form a vector $|\Gamma\rangle$ which is a reducible representation,

$$|\Gamma\rangle = (6, 0) \tag{11.5.5}$$

The reducible representation is now expressed as a linear combination of irreducible representations. Since the two irreducible representations are eigenvectors which span the two-dimensional symmetry space, the reducible representation can be projected onto each of these eigenvectors. The magnitude of the projection will give the number of times each irreducible representation appears in the reducible representation. The two eigenvectors for the C_2 group are generally assigned the letters A and B:

$$\begin{aligned} \langle A| &= (1, 1) \\ \langle B| &= (1, -1) \end{aligned} \tag{11.5.6}$$

The symbols A and B are used for one-dimensional irreducible representations. A is used when the irreducible representation has a component of $+1$ for C_2. B is used when this component is -1. For groups with reflection elements, the subscripts 1 and 2 define the irreducible representations with reflection components of +1 and −1, respectively.

Since $|\Gamma\rangle$ must project onto a unit vector to establish the actual number of irreducible representations of each type, the projection is

$$|A\rangle\langle A|\Gamma\rangle = \begin{pmatrix}1\\1\end{pmatrix}\begin{pmatrix}1\\\frac{1}{2}\end{pmatrix}(1 \quad 1)\begin{pmatrix}6\\0\end{pmatrix} = 3\begin{pmatrix}1\\1\end{pmatrix} \tag{11.5.7}$$

There are three irreducible $\langle A|$ representations and

$$|B\rangle\langle B|\Gamma\rangle = \begin{pmatrix}1\\-1\end{pmatrix}\begin{pmatrix}1\\\frac{1}{2}\end{pmatrix}(1 \quad -1)\begin{pmatrix}6\\0\end{pmatrix} = 3\begin{pmatrix}1\\-1\end{pmatrix} \tag{11.5.8}$$

three $|B\rangle$ irreducible representations. This was required by the trace of 0 for C_2. The diagonal C_2 matrix would have three diagonal elements of +1 and three of −1 to give the net trace of zero. The projections on the irreducible vectors give this information directly.

The normalization factor for these projections is equal to the number of elements in the group. When it is included in the projection expression for each irreducible representation, it always gives the integral number of this representation which appears on the diagonal. This number represents the magnitude of the projection on that particular irreducible representation.

The projections for the CO_2 molecule with reducible representation

$$|\Gamma\rangle = \begin{pmatrix}9\\-1\end{pmatrix} \tag{11.5.9}$$

are

$$\begin{aligned} |A\rangle\langle A|\Gamma\rangle &= |A\rangle 4 \\ |B\rangle\langle B|\Gamma\rangle &= |B\rangle 5 \end{aligned} \tag{11.5.10}$$

The $\mathbf{C}_2$ trace of −1 requires one extra $|B\rangle$ representation.

The projection procedure must be modified when the classes of group elements have more than one element. The traces for the symmetry matrices for cyclopropene give the reducible representation

$$(3, 0, +1) \tag{11.5.11}$$

which must be projected onto the three irreducible representations for the C_{3v} group:

	E	C_3, C_3^2	$\sigma_1, \sigma_2, \sigma_3$
$\langle A_1\|$	1	1	1
$\langle A_2\|$	1	1	−1
$\langle E\|$	2	−1	0

(11.5.12)

E is the symbol for the identity but it is also used for the 2×2 irreducible representation. The subscripts 1 and 2 are introduced for σ components of $+1$ and -1, respectively. Each vector uses A because the sign for each C_3 component is positive.

The projection operator for $|A\rangle$ using all six group elements is

$$|A_1\rangle\langle A_1|\Gamma\rangle = |A_1\rangle\left(\frac{1}{6}\right)(1 \quad 1 \quad 1 \quad 1 \quad 1 \quad 1)\begin{pmatrix}3\\0\\0\\1\\1\\1\end{pmatrix} = |A_1\rangle(1) \tag{11.5.13}$$

A single $|A_1\rangle$ irreducible representation will appear in the diagonalized symmetry matrices.

Since the components for each class are identical, the projection notation can be compressed into a three-component vector for the three classes. The **B** matrix for C_{3v} is incorporated into the scalar product. The projection for $|A_2\rangle$ is

$$|A_2\rangle\langle A_2|\,\mathbf{B}\,|\Gamma\rangle = |A_2\rangle\left(\frac{1}{6}\right)(1 \quad 1 \quad -1)\begin{pmatrix}1 & 0 & 0\\0 & 2 & 0\\0 & 0 & 3\end{pmatrix}\begin{pmatrix}3\\0\\1\end{pmatrix} = |A_2\rangle 0 \tag{11.5.14}$$

Although $|A_2\rangle$ is an acceptable irreducible representation for the C_{3v} group, it is absent for this particular set of symmetry matrices. By contrast, the projection for $|E\rangle$ is

$$|E\rangle\langle E|\,\mathbf{B}\,|\Gamma\rangle = |E\rangle 1 \tag{11.5.15}$$

The 3×3 diagonalized matrices have a single 1×1 block (A_1) and a 2×2 block (E). This result was determined earlier by simultaneous diagonalization of the symmetry matrices.

When the scalar projections, i.e., the irreducible representations, for a given system are known, the actual molecular orbital or normal mode can be constructed by performing the symmetry operations and multiplying the results of each operation by the corresponding component of the irreducible representation. The $|E\rangle$ representation will generate doubly degenerate wavefunctions or normal modes.

The generation technique is illustrated for the cyclopropene molecule since its molecular orbitals are known. The system will have a single $|A_1\rangle$ representation. The molecular orbital for this representation is generated by selecting a single atomic orbital, e.g., φ_1. The remaining orbitals are generated using the symmetry operations. If the ith scalar component is γ_i and $\mathbf{R}_i$ is the ith symmetry element for the $|A_1\rangle$ representation, the generation formula is

$$\sum \gamma_i \mathbf{R}_i \tag{11.5.16}$$

where the sum is over all elements rather than classes.

The cyclopropene molecular orbital for $|A_1\rangle$ is generated from φ_1 as

$$\begin{aligned} |\psi_A\rangle &= (+1)\,E\varphi_1 + (+1)\,C_3\varphi_1 + (+1)\,C_3^2\varphi_1 + (+1)\,\sigma_1\varphi_1 \\ &\quad + (+1)\,\sigma_2\varphi_1 + (+1)\,\sigma_3\varphi_1 \\ &= +\varphi_1 + \varphi_2 + \varphi_3 + \varphi_1 + \varphi_3 + \varphi_2 \\ &= 2(\varphi_1 + \varphi_2 + \varphi_3) \end{aligned} \tag{11.5.17}$$

This is the symmetric molecular orbital expected for this irreducible representation.

The $|E\rangle$ irreducible representation generates

$$\begin{aligned} |\psi_E\rangle &= (+2)\,E\varphi_1 + (-1)\,C_3\varphi_1 + (-1)\,C_3^2\varphi_1 \\ &= 2\varphi_1 - \varphi_2 - \varphi_3 \end{aligned} \tag{11.5.18}$$

which is one of the molecular orbitals for the doubly degenerate energy eigenvalue. Generation of the second eigenvector is deferred to the next section.

The C_{2v} group for H_2 with its six spatial coordinates gave a reducible representation

$$|\Gamma\rangle = (6, 0) \tag{11.5.19}$$

which resolved into

$$|A\rangle 3 + |B\rangle 3 \tag{11.5.20}$$

Each of the six irreducible representations describes some spatial displacement of the six coordinates of the molecule. If the molecule moves in the z direction, both atoms have equal displacements in this direction. The single vector $|z_1\rangle$ is selected as the starting vector. Equation 11.5.16 generates the full normal mode with its $|A\rangle$ symmetry properties:

$$(+1)E\,|z_1\rangle + (+1)\,C_2\,|z_1\rangle = +|z_1\rangle + |z_2\rangle \tag{11.5.21}$$

This is the z translation of the entire molecule.

If the $|B\rangle$ irreducible representation is applied to the $|z_1\rangle$ vector, the result is

$$(+1)\,E\,|z_1\rangle + (-1)\,C_2\,|z_1\rangle = |z_1\rangle - |z_2\rangle \tag{11.5.22}$$

The two z components must be displaced in the opposite directions and this corresponds to a rotation about any axis in the xy plane.

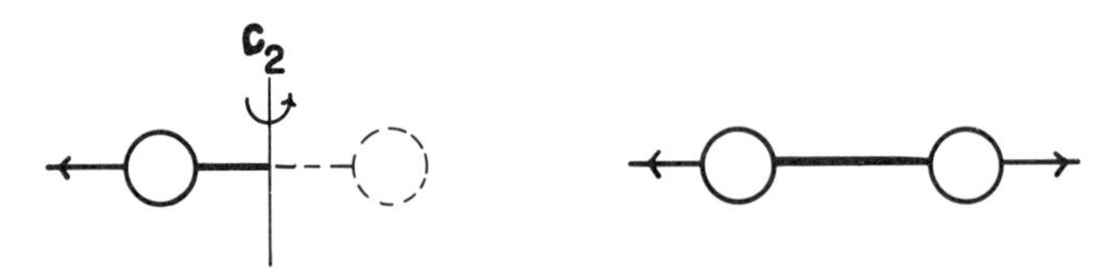

Figure 11.18. The C_2 operation used to generate the spatial vectors on atom 2 using the spatial vectors of atom 1.

With $|x_1\rangle$ as the initial vector, the $|A\rangle$ representation gives

$$(+1)E\,|x_1\rangle + C_2\,|x_1\rangle = |x_1\rangle + |x_2\rangle \tag{11.5.23}$$

The positive sign, in conjunction with the group operation C_2, reverses the direction of the $|x_1\rangle$ vector when it moves it to atom 2 (Figure 11.18). The $|A\rangle$ operation has generated the vibration of the two atoms along the x axis. The $|B\rangle$ representation will give the translation on this axis:

$$(+1)E\,|x_1\rangle + (-1)\,C_2\,|x_1\rangle = |x_1\rangle - |x_2\rangle \tag{11.5.24}$$

The normal modes generated with the vector $|y_1\rangle$ will be translation in the y direction and rotation about the z axis.

The molecular orbitals and normal modes presented here are simple because the total symmetry of the system is relatively low. As the number of symmetry elements in the group increases, more elaborate molecular orbitals and normal modes can be generated from minimal initial information.

11.6. Generation of Molecular Orbitals

When the symmetry matrices are resolved into their irreducible representations, these representations dictate the allowed molecular orbitals or normal modes for that specific system. The generation of these orbitals or modes requires an interesting interpretation of projection operators. A projection operator is normally composed of column and row eigenvectors. The column defines the direction in space while the row establishes the magnitude of the projection. In group theory, a symmetry operation replaces the column eigenvector and operates on an atomic orbital to create a new atomic orbital, i.e., the new "direction." The projection in this direction is dictated by the components of the irreducible representation. They represent the contribution of this new atomic orbital to the total molecular orbital.

Two molecular orbitals for cyclopropene were generated in Section 11.5 using the A_1 and E irreducible representations. The coefficients for these orbitals are

$$(1, 1, 1) \tag{11.6.1}$$

$$(2, -1, 0) \tag{11.6.2}$$

The E irreducible representation requires 2×2 matrices and two molecular orbitals which must have the same energy. The pair of molecular orbitals is generated by starting with any atomic orbital. If φ_2 is selected as the starting atomic orbital, the molecular orbital is

$$|\psi\rangle = -\varphi_1 + 2\varphi_2 - \varphi_3 \tag{11.6.3}$$

The coefficients are the same but the orbital is 120° out of phase with respect to the orbital created starting with $|\varphi_1\rangle$, $(2\varphi_1 - \varphi_2 - \varphi_3)$. Because the rotation is 120°, the two orbitals are not orthogonal:

$$(2 \quad -1 \quad -1)\begin{pmatrix} -1 \\ 2 \\ -1 \end{pmatrix} = -3 \tag{11.6.4}$$

The orthogonal pair is easily found using a Gram–Schmidt orthogonalization:

$$\begin{pmatrix} -1 \\ 2 \\ -1 \end{pmatrix} - \left(\frac{1}{6}\right)(-3)\begin{pmatrix} 2 \\ -1 \\ -1 \end{pmatrix} = \begin{pmatrix} 0 \\ \frac{3}{2} \\ -\frac{3}{2} \end{pmatrix} \tag{11.6.5}$$

The orthogonal eigenvector is

$$|\psi\rangle = 0 + \varphi_2 - \varphi_3 \tag{11.6.6}$$

Although the C_{3v} group generates the molecular orbitals for cyclopropene, the C_3 group with elements E, C_3, and C_3^2 contains sufficient information to generate the molecular orbitals from a single atomic orbital. The irreducible representations for C_3 are three-dimensional with one symmetry element per class:

	E	C_3	C_3^2
A	1	1	1
1	1	ε	ε^*
2	1	ε^*	ε

(11.6.7)

Although the last two vectors are one-dimensional, the appearance of and ε^* in these vectors suggests it might ultimately be convenient to combine them to produce molecular orbitals with real coefficients.

Since the 3×3 symmetry matrices for these orbitals give traces of 3, 0, and 0, the reducible representation is

$$|\Gamma\rangle = (3, 0, 0) \tag{11.6.8}$$

and the projections on each irreducible representation vector in the table are

$$|A\rangle \left(\frac{1}{3}\right) \langle A|\Gamma\rangle = |A\rangle(1) \tag{11.6.9}$$

$$|1\rangle\langle 1|\Gamma\rangle = |1\rangle \left(\frac{1}{3}\right)(3+0+0) = |1\rangle(1) \tag{11.6.10}$$

$$|2\rangle\langle 2|\Gamma\rangle = |2\rangle \left(\frac{1}{3}\right)(3+0+0) = |2\rangle(1) \tag{11.6.11}$$

The three molecular orbitals are

$$\begin{gathered} \varphi_1 + \varphi_2 + \varphi_3 \\ \varphi_1 + \varepsilon\varphi_2 + \varepsilon^*\varphi_3 \\ \varphi_1 + \varepsilon^*\varphi_2 + \varepsilon\varphi_3 \end{gathered} \tag{11.6.12}$$

If ε and ε^* are expanded in their real and imaginary parts, the coefficients in the last two orbitals will be imaginary:

$$\begin{gathered} \varphi_1 + \left(-\frac{1}{2} + i\frac{\sqrt{3}}{2}\right)\varphi_2 + \left(-\frac{1}{2} - i\frac{\sqrt{3}}{2}\right)\varphi_3 \\ \varphi_1 + \left(-\frac{1}{2} - i\frac{\sqrt{3}}{2}\right)\varphi_2 + \left(-\frac{1}{2} + i\frac{\sqrt{3}}{2}\right)\varphi_3 \end{gathered} \tag{11.6.13}$$

They can be converted to orbitals with real coefficients by addition and subtraction, i.e., combining them into a two-dimensional subspace. The molecular orbital sum is

$$|\psi\rangle = 2\varphi_1 - \varphi_2 - \varphi_3 \tag{11.6.14}$$

while the difference is

$$|\psi\rangle = i3\varphi_2 - i\sqrt{3}\varphi_3 \propto \varphi_2 - \varphi_3 \tag{11.6.15}$$

The two degenerate molecular orbitals appear directly in this way.

The benzene molecule group D_{6h} contains rotations about axes in the plane of the molecule as well as the C_6, C_3, and C_2 rotations about the z axis perpendicular to the molecular plane. All these symmetry elements would be useful for a normal mode analysis since there are a total of 3×12 spatial coordinates for the molecule and each element would establish connections between these coordinates. The molecular orbital for benzene is created from six atomic orbitals, and the C_6 group with the six symmetry operations E, C_6, C_3, C_2, C_3^2, and C_6^5 is sufficient. Since each of these rotation matrices commute, six 1×1 irreducible representations are possible for the group. However, since these representations contain imaginary components, it is convenient to eliminate the imaginaries by combining pairs of 1×1 irreducible representations to create 2×2 irreducible representations. The character table is

	E	C_6	C_3	C_2	C_3^2	C_6^5
A	1	1	1	1	1	1
B	1	-1	1	-1	1	-1
E_1	1	ε	$-\varepsilon^*$	-1	$-\varepsilon$	ε^*
	1	ε^*	$-\varepsilon$	-1	$-\varepsilon^*$	ε
E_2	1	$-\varepsilon^*$	$-\varepsilon$	1	$-\varepsilon^*$	$-\varepsilon$
	1	$-\varepsilon$	$-\varepsilon^*$	1	$-\varepsilon$	$-\varepsilon^*$

(11.6.16)

The parameters ε and ε^* are

$$\varepsilon = \exp(i2\pi/6) = +\frac{1}{2} + i\frac{\sqrt{3}}{2} \tag{11.6.17}$$

$$\varepsilon^* = \exp(-i2\pi/6) = \frac{1}{2} - i\frac{\sqrt{3}}{2} \tag{11.6.18}$$

Since each orbital moves for every rotation in the group, the vector $|\Gamma\rangle$ is

$$|\Gamma\rangle = (6, 0, 0, 0, 0, 0) \tag{11.6.19}$$

The nondegenerate molecular orbitals are

$$|\psi_A\rangle = \varphi_1 + \varphi_2 + \varphi_3 + \varphi_4 + \varphi_5 + \varphi_6 \tag{11.6.20}$$

$$|\psi_B\rangle = \varphi_1 - \varphi_2 + \varphi_3 - \varphi_4 + \varphi_5 - \varphi_6 \tag{11.6.21}$$

The $|\psi_{E_1}\rangle$ orbitals are formed as the sum and difference of the two one-dimensional vectors for E_1:

$$\begin{aligned} |\psi_{E_1}\rangle &= 2\varphi_1 + \varphi_2 - \varphi_3 - 2\varphi_4 - \varphi_5 + \varphi_6 \\ |\psi'_{E_1}\rangle &= \varphi_2 + \varphi_3 - \varphi_5 - \varphi_6 \end{aligned} \tag{11.6.22}$$

The $|\psi_{E_2}\rangle$ orbitals are the sum and difference of the final two one-dimensional vectors:

$$\begin{aligned} |\psi_{E_2}\rangle &= 2\varphi_1 - \varphi_2 - \varphi_3 + 2\varphi_4 - \varphi_5 - \varphi_6 \\ |\psi'_{E_2}\rangle &= \varphi_2 - \varphi_3 + \varphi_5 - \varphi_6 \end{aligned} \tag{11.6.23}$$

These orthogonal orbitals have real coefficients and are illustrated in Figure 11.19.

The cyclic cyclopropene and benzene systems are convenient because all the atomic orbitals in the system can be generated from a single atomic orbital. The situation is different for the 1,3-butadiene molecule. To generate the orbitals for the molecule using C_2 symmetry, two atomic orbitals are combined to form the starting orbital. The C_2 operation then gives the remaining two orbitals.

The atomic orbitals φ_1 and φ_2 are selected as starting orbitals and used to generate two new wavefunctions. Since the C_2 operation moves each orbital, the $|\Gamma\rangle$ vector is

$$|\Gamma\rangle = (2, 0) \tag{11.6.24}$$

which resolves into 1 A and 1 B representation. The atomic orbitals φ_1 and φ_2 are selected to generate two combination orbitals in the B representation:

$$\begin{aligned} |B_1\rangle &= \frac{1}{\sqrt{2}}(\varphi_1 - \varphi_4) = \frac{1}{\sqrt{2}}(|1\rangle - |4\rangle) \\ |B_2\rangle &= \frac{1}{\sqrt{2}}(\varphi_2 - \varphi_3) = \frac{1}{\sqrt{2}}(|2\rangle - |3\rangle) \end{aligned} \tag{11.6.25}$$

These "symmetry-adapted" vectors are used to form the characteristic determinant:

$$\begin{vmatrix} \langle B_1|\,\mathbf{H}\,|B_1\rangle - E & \langle B_1|\,\mathbf{H}\,|B_2\rangle \\ \langle B_2|\,\mathbf{H}\,|B_1\rangle & \langle B_2|\,\mathbf{H}\,|B_2\rangle - E \end{vmatrix} \tag{11.6.26}$$

The off-diagonal elements in this case are

$$\begin{aligned} \beta &= \frac{1}{2}(\langle 1| - \langle 4|)\,\mathbf{H}(|2\rangle - |3\rangle) \\ &= \frac{1}{2}(\langle 1|\,\mathbf{H}\,|2\rangle - \langle 1|\,\mathbf{H}\,|3\rangle - \langle 4|\,\mathbf{H}\,|2\rangle + \langle 4|\,\mathbf{H}\,|3\rangle) \end{aligned} \tag{11.6.27}$$

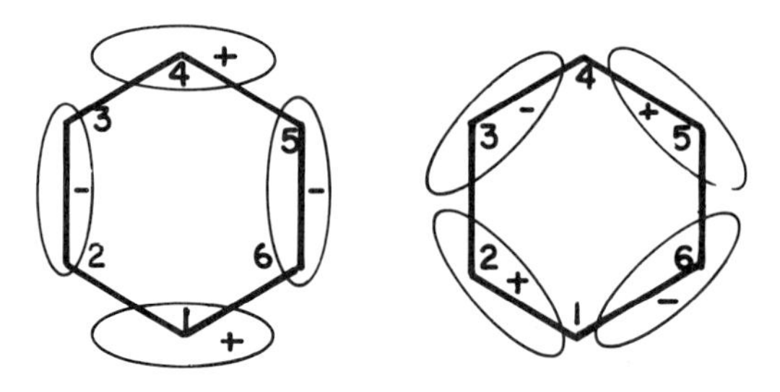

Figure 11.19. Molecular orbitals for the benzene molecule.

Since β is finite only for nearest neighbors, the entire expression reduces to β. The a_{11} element of the matrix is

$$\frac{1}{2}(\langle 1| - \langle 4|)\,\mathbf{H}(|1\rangle - |4\rangle)$$

$$= \frac{1}{2}(\langle 1|\,\mathbf{H}\,|1\rangle - \langle 1|\,\mathbf{H}\,|4\rangle - \langle 4|\,\mathbf{H}\,|1\rangle + \langle 4|\,\mathbf{H}\,|4\rangle)$$

$$= \frac{1}{2}(\alpha - 0 - 0 + \alpha) = \alpha \tag{11.6.28}$$

as normally observed for the Hückel matrices. However, the a_{22} element is

$$\frac{1}{2}(\langle 2| - \langle 3|)\,\mathbf{H}(|2\rangle - |3\rangle)$$

$$= \frac{1}{2}(\langle 2|\,\mathbf{H}\,|2\rangle - \langle 2|\,\mathbf{H}\,|3\rangle - \langle 3|\,\mathbf{H}\,|2\rangle - \langle 3|\,\mathbf{H}\,|3\rangle)$$

$$= \alpha - \beta \tag{11.6.29}$$

since φ_2 and φ_3 are coupled. The Hückel matrix is

$$\begin{vmatrix} \alpha - E & \beta \\ \beta & \alpha - \beta - E \end{vmatrix} = (\alpha - \beta - E)(\alpha - E) - \beta^2 = 0 \tag{11.6.30}$$

with eigenvalues

$$\begin{aligned} E &= \alpha - \beta(1 + \sqrt{5})/2 = \alpha - 1.62\beta \\ E &= \alpha - \beta(1 - \sqrt{5})/2 = \alpha - 0.62\beta \end{aligned} \tag{11.6.31}$$

The eigenvectors for this system show how the two orbital combinations couple. The first eigenvector is

$$\begin{gathered} \begin{pmatrix} 1.62 & 1 \\ 1 & +2.62 \end{pmatrix}\begin{pmatrix} \varphi_1 \\ \varphi_2 \end{pmatrix} = 0 \\ \begin{pmatrix} \varphi_1 \\ \varphi_2 \end{pmatrix} = \begin{pmatrix} 1 \\ -1.62 \end{pmatrix} = \begin{pmatrix} 0.588 \\ -0.95 \end{pmatrix} \end{gathered} \tag{11.6.32}$$

Because these two coupled orbitals belong to the B representation, the full orbital is generated using this irreducible representation:

$$\begin{aligned} |\psi\rangle &= E(0.588\varphi_1 - 0.95\varphi_2) - C_2(0.588\varphi_1 - 0.95\varphi_2) \\ &= 0.588\varphi_1 - 0.95\varphi_2 - 0.588\varphi_4 + 0.95\varphi_3 \end{aligned} \tag{11.6.33}$$

The $|B\rangle$ representation was responsible for the correct signs but the Hückel analysis was required to establish the relative coefficients of the φ_1 and φ_2 atomic orbitals.

The second eigenvalue for B gives the eigenvector

$$\begin{pmatrix} \varphi_1 \\ \varphi_2 \end{pmatrix} = \begin{pmatrix} 1.6 \\ 1 \end{pmatrix} = \begin{pmatrix} 0.95 \\ 0.588 \end{pmatrix} \tag{11.6.34}$$

and the components of the full eigenvector are

$$\begin{pmatrix} 0.95 \\ 0.588 \\ -0.588 \\ -0.95 \end{pmatrix} \tag{11.6.35}$$

For each of the B representation eigenvectors, the coefficients of orbitals 3 and 4 are equal and opposite to those orbitals 2 and 1, respectively.

The remaining two orbitals are generated from the A representation. They are

$$\begin{gathered} \frac{1}{\sqrt{2}}(\varphi_1 + \varphi_4) \\ \frac{1}{\sqrt{2}}(\varphi_2 + \varphi_3) \end{gathered} \tag{11.6.36}$$

The characteristic determinant for these orbitals is

$$\begin{vmatrix} \alpha - E & \beta \\ \beta & \alpha + \beta - E \end{vmatrix} = 0 \tag{11.6.37}$$

and the eigenvalues are

$$\begin{aligned} E &= \alpha + 1.62\beta \\ E &= \alpha - 0.62\beta \end{aligned} \tag{11.6.38}$$

The two-component eigenvectors for the matrix are

$$\begin{pmatrix} 0.588 \\ 0.95 \end{pmatrix} \qquad \begin{pmatrix} 0.95 \\ -0.588 \end{pmatrix} \tag{11.6.39}$$

The character table for the A representation now generates these molecular orbitals (Figure 11.20):

$$\begin{pmatrix} 0.588 \\ 0.95 \\ 0.95 \\ 0.588 \end{pmatrix} \qquad \begin{pmatrix} 0.95 \\ -0.588 \\ -0.588 \\ 0.95 \end{pmatrix} \tag{11.6.40}$$

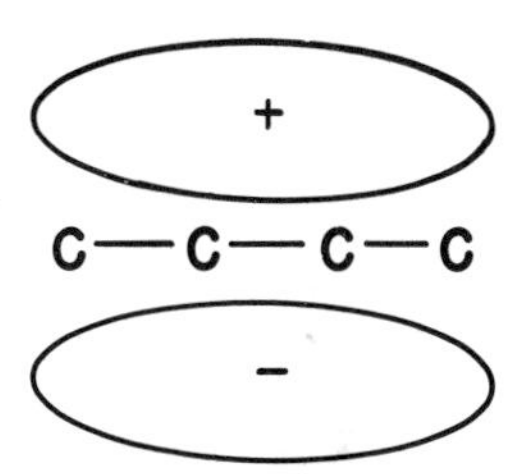

Figure 11.20. The fully symmetric molecular orbital for 1,3-butadiene.

The group theory results can be verified because the 4×4 **H** matrix for 1,3-butadiene can be solved directly using the techniques introduced in Chapter 8. The eigenvectors for the matrix

$$\begin{pmatrix} \alpha & \beta & 0 & 0 \\ \beta & \alpha & \beta & 0 \\ 0 & \beta & \alpha & \beta \\ 0 & 0 & \beta & \alpha \end{pmatrix} \tag{11.6.41}$$

are determined using the general formula for the jth component of the ith eigenvector,

$$|i\rangle_j = \sin\left[i\pi j/(N+1)\right] \tag{11.6.42}$$

and these eigenvectors are identical to those generated using group theory (Problem 11.7).

11.7. Normal Vibrational Modes

In Section 11.6, p_z orbitals of a carbon skeleton were combined into a molecular orbital using the characters of the proper irreducible representations and the symmetry operations which were used to generate each of the atomic orbitals for the molecular orbital. The procedure for the generation of normal modes of a molecule is identical. For a given vibrational mode described by a specific irreducible representation, a single vector can be used to generate the vectors on the remaining atoms of the molecule. The simultaneous excursions of all these vectors then constitutes the normal mode.

The procedure is illustrated by generating the normal modes for the H_2O molecule. The molecule has the symmetry of elements of the C_{2v} group, i.e., E, C_2, σ, and σ'. Since each atom has three vibration vectors in the three-dimensional space, the dimension of the system is 9 and the trace of **E** is also 9. The C_2 rotation moves both H_1 and H_2 and their vectors will not contribute to the trace. The O atom lies on the rotation axis and its symmetry matrix is

$$\begin{pmatrix} \cos 180^\circ & -\sin 180^\circ & 0 \\ \sin 180^\circ & \cos 180^\circ & 0 \\ 0 & 0 & 1 \end{pmatrix} = \begin{pmatrix} -1 & 0 & 0 \\ 0 & -1 & 0 \\ 0 & 0 & 1 \end{pmatrix} \tag{11.7.1}$$

with a trace of -1. For any rotation of θ, each atom which does not change its spatial position will contribute

$$1+2\cos\theta \tag{11.7.2}$$

to the total trace.

The σ reflection perpendicular to the molecular plane again moves the two H atoms. The O atom remains in place and the x vector on O is reversed during the reflection. The matrix for the O atom reflection is

$$\begin{pmatrix} -1 & 0 & 0 \\ 0 & 1 & 0 \\ 0 & 0 & 1 \end{pmatrix} \tag{11.7.3}$$

with trace $+1$. The $\sigma'(\Gamma_{xz})$ reflection requires a symmetry matrix

$$\begin{pmatrix} 1 & 0 & 0 \\ 0 & -1 & 0 \\ 0 & 0 & 1 \end{pmatrix} \tag{11.7.4}$$

for each atom. The three atoms in this plane then contribute a total of $+3$ to the trace. The reducible representation $|\Gamma\rangle$ is

$$|\Gamma\rangle = (9, -1, +1, +3) \tag{11.7.5}$$

The C_{2v} group has the character table

	E	C_2	σ	σ'	
A_1	1	1	1	1	T_z
A_2	1	1	-1	-1	R_z
B_1	1	-1	1	-1	T_x, R_y
B_2	1	-1	-1	1	T_y, R_x

(11.7.6)

and $|\Gamma\rangle$ can be resolved into irreducible representations as

$$3\,|A_1\rangle + 1\,|A_2\rangle + 2\,|B_1\rangle + 3\,|B_2\rangle \tag{11.7.7}$$

Six of the nine representations in Equation 11.7.7 describe translations and rotations of the entire molecule. Although these constants of the motion can be generated using the group properties, the vibrations are generally more interesting. For this reason, the irreducible representations for translations and rotations are categorized by including a column after the character table. This is included in Equation 11.7.6 using the notation T_i for a translation in the i direction and R_i for a rotation about i. The z, x, and y translations transform as the A_1, B_1, and B_2 representations, respectively, while the R_z, R_y, and R_x rotations

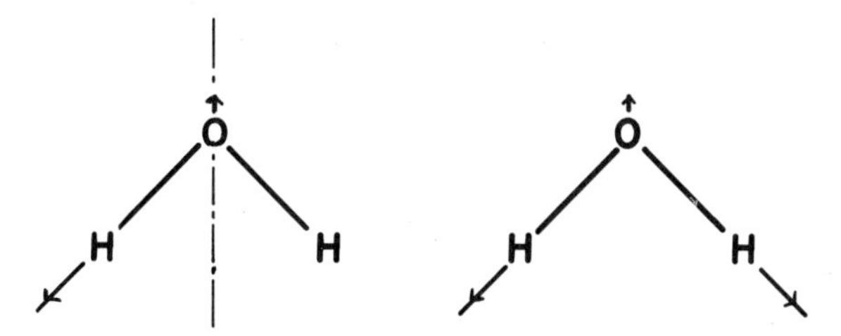

Figure 11.21. A starting vector for H_2O which generates the proper symmetric stretch (A_1) vibration.

transform as A_2, B_1, and B_2, respectively. These representations can be eliminated from the nine representations of Equation 11.7.7. The remaining three vibrational normal modes then have the representations

$$2\,|A_1\rangle + 1\,|B_2\rangle \tag{11.7.8}$$

The normal modes are generated from a single vector using the generating equation

$$\sum \gamma_i R_i\,|1\rangle \tag{11.7.9}$$

The water molecule has nine spatial components. However, instead of performing symmetry operations on each of these components, it is often convenient to infer vectors which will be consistent with the symmetry of the system and use them to generate the full normal mode. An A_1 mode is generated by selecting a single vector for H_1 along the $O-H_1$ axis (Figure 11.21). This vector is selected because rotation and reflection in either plane will not change its orientation as required by the $+1$ components of A_1. The symmetry operations generate the vectors for each H in the normal mode. The x and y vectors on O will change direction during the reflections and rotations and are not consistent with A_1. The only acceptable vector for O lies along the z axis and is one-eight the magnitude of the H vectors to maintain the center of mass constant.

The second A_1 vibration uses a starting vector on H_1 which is perpendicular to the H vectors of Figure 11.21. This is illustrated in Figure 11.22. The C_2 and σ operations generate the vector on H_2. In this case, the O vibration again lies

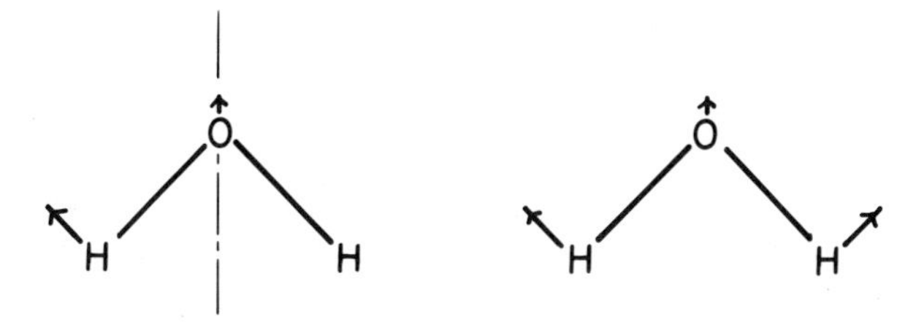

Figure 11.22. A starting vector perpendicular to the starting vector of Figure 11.21 which gives the second A_1 normal mode for H_2O.

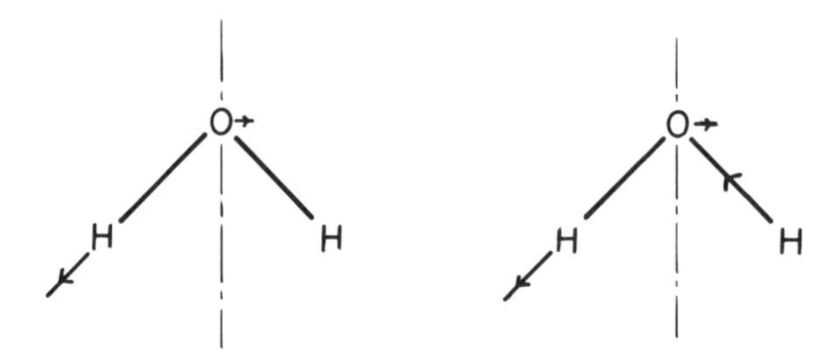

Figure 11.23. The B_2 normal mode for the H_2O molecule.

along the z axis with negative direction to maintain the center of mass of the molecule. This normal mode is a bending vibration.

The final B_2 mode requires a change in the direction of the vector on H_2 for both rotation and reflection σ. The generation process is illustrated in Figure 11.23. The vector on H_1 is directed along the OH axis as it was for the symmetric stretch of A_1 (Figure 11.21). When this vector is rotated or reflected, however, its direction must be reversed because of the -1 character for each of these symmetry operations. Since both H atoms move to the left, the O atom must move to the right. A vector x will change its direction for either C_2 or σ and it is also consistent with B_2. The three vectors of Figure 11.23 are sufficient for a full description of the nine-dimensional system. The symmetry conditions have permitted this simplification.

In Chapter 5, the normal modes for a homonuclear triangular model were developed directly from the force equations for this system. The same results are developed more rapidly using group theory. The molecule and its coordinates are shown in Figure 11.24 and are identical to those introduced in Chapter 5. The coordinates along the medians of the molecule are labeled q_1, q_2, and q_3 for atoms 1, 2, and 3, respectively. The remaining three vectors for the two-dimensional analysis of this system are perpendicular to each of the first three vectors and are labeled q_4, q_5, and q_6 for atoms 1, 2, and 3, respectively.

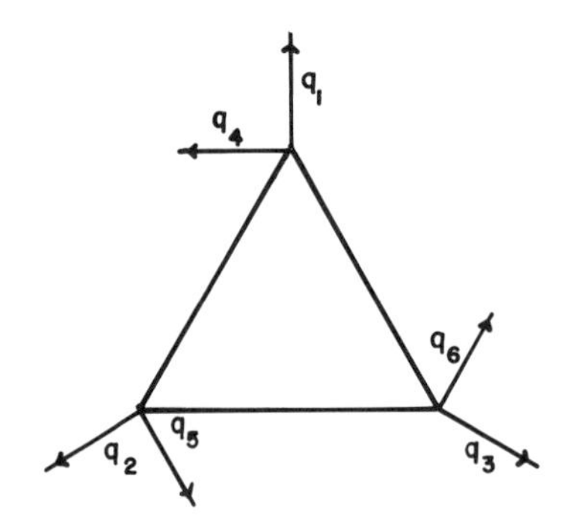

Figure 11.24. A special coordinate system for the triangular molecule. Three of the vectors lie on the medians of the three atoms and the remaining three are perpendicular to the first three and directed in a counterclockwise direction.

The molecule has C_{3v} symmetry. In addition, the particular basis vectors chosen are already "symmetry-adapted." For example, C_3 rotations of q_1, q_2, and q_3 remain as a subset during the symmetry operations; q_4, q_5, and q_6 constitute the second subset.

The first three vectors produce a trace of 3 for the **E** matrix. Since each vector moves during the C_3 and C_3^2 rotations, the trace for this class is 0. The reflections move two of the vectors to new atoms while the remaining vector is unchanged. The trace for each reflection is +1 and $|\Gamma\rangle$ is

$$|\Gamma\rangle = (3, 0, +1) \tag{11.7.10}$$

The projections for each irreducible representation are

$$1\,|A_1\rangle + 1\,|E\rangle \tag{11.7.11}$$

The A_1 irreducible representation for the molecule is generated by starting with q_1. The six symmetry operations reproduce vectors q_1, q_2, and q_3 with coefficients of +1, i.e.,

$$\begin{aligned} |v_1\rangle &= [(+1)E + (+1)\,C_3 + (+1)\,C_3^2 + (+1)\,\sigma_1 + (+1)\,\sigma_2 + (+1)\,\sigma_3]\,q_1 \\ &= 2(q_1 + q_2 + q_3) \end{aligned} \tag{11.7.12}$$

This is the symmetric stretch normal mode for this triangular molecule. Each atom is displaced equally and in phase along the three medians for the triangular molecule as illustrated in Figure 11.25.

One vector for the E irreducible representation is generated starting with q_1. The normal mode is

$$\begin{aligned} |v_{E1}\rangle &= [(2)E + (-1)\,C_3 + (-1)\,C_3^2 + 0 + 0 + 0]\,q_1 \\ &= 2q_1 - q_2 - q_3 \end{aligned} \tag{11.7.13}$$

The second vector in this E representation has already been determined for the molecular orbitals of cyclopropene as

$$(0, 1, -1) \tag{11.7.14}$$

and the second normal mode in this representation is

$$[(0)\,q_1 + (+1)\,q_2 + (-1)\,q_3] = q_2 - q_3 \tag{11.7.15}$$

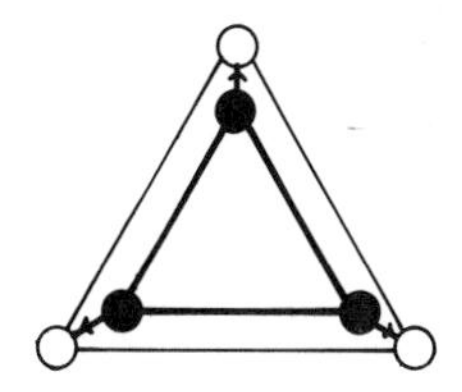

Figure 11.25. The symmetric stretch normal mode for the triangular molecule.

The second set of three vectors also have a trace of 3 for E and a trace of 0 for C_3 and C_3^2 since each vector moves. The trace for reflections is best illustrated with the reflection plane through atom 2. The sign of vector q_5 is reversed on reflection while the remaining two vectors move to the new coordinates. The trace is -1 in this case and

$$|\Gamma\rangle = (3, 0, -1) \tag{11.7.16}$$

The vector resolves into the irreducible representations

$$(1)\,|A_2\rangle + (1)\,|E\rangle \tag{11.7.17}$$

For this coordinate subset, the A_2 representation generates a rotation in the plane,

$$\begin{aligned}
&[(+1)E + (+1)\,C_3 + (+1)\,C_3^2 + (-1)\,\sigma_1 + (-1)\,\sigma_2 + (-1)\,\sigma_3]\,q_4 \\
&\quad = q_4 + q_5 + q_6 + q_4 + q_5 + q_6 \\
&\quad = q_4 + q_5 + q_6
\end{aligned} \tag{11.7.18}$$

since each atom is moving in the same direction and each vector is tangent to a circle with its center at the center of mass of the molecule. The elements reproduce these same vectors. For example, σ_1 reverses the direction of q_4 but the negative coefficient for the operation restores the original orientation. The reflection σ_2 generates the negative of q_5 which is also multiplied by the scalar -1 to give the vector $+q_5$. Each reflection produces one of the three vectors with proper sign.

The E representation for vectors q_4, q_5, and q_6 gives the two degenerate modes

$$\begin{gathered}
2q_4 - q_5 - q_6 \\
q_5 - q_6
\end{gathered} \tag{11.7.19}$$

In this coordinate system, the four different E normal modes describe z and x translations and two additional vibrational modes. Since the vectors do not all point in the z and x directions, the vectors within the E representations must be combined to demonstrate these motions. The translation will require vectors with all six components. The normal mode vector

$$2q_1 - q_2 - q_3 \propto q_1 - \left(\frac{1}{2}\right) q_2 - \left(\frac{1}{2}\right) q_3 \tag{11.7.20}$$

has projections on the z axis. The three vectors have positive z components of different magnitude. A second vector with positive z components for q_5 and q_6 must be added to this vector to produce the translation. The vector

$$q_5 - q_6 \tag{11.7.21}$$

is multiplied by $-\frac{\sqrt{3}}{2}$ to give a net component of 1 in the z direction (Figure 11.26). The six-component z translation is

$$q_1-\left(\frac{1}{2}\right)q_2-\left(\frac{1}{2}\right)q_3+0+\frac{\sqrt{3}}{2}q_5-\frac{\sqrt{3}}{2}q_6 \tag{11.7.22}$$

or

$$\begin{bmatrix} 1 \\ -\frac{1}{2} \\ -\frac{1}{2} \\ 0 \\ 0 \\ 0 \end{bmatrix} + \begin{bmatrix} 0 \\ 0 \\ 0 \\ 0 \\ -\frac{\sqrt{3}}{2} \\ \frac{\sqrt{3}}{2} \end{bmatrix} = \begin{bmatrix} 1 \\ -\frac{1}{2} \\ -\frac{1}{2} \\ 0 \\ -\frac{\sqrt{3}}{2} \\ \frac{\sqrt{3}}{2} \end{bmatrix} \tag{11.7.23}$$

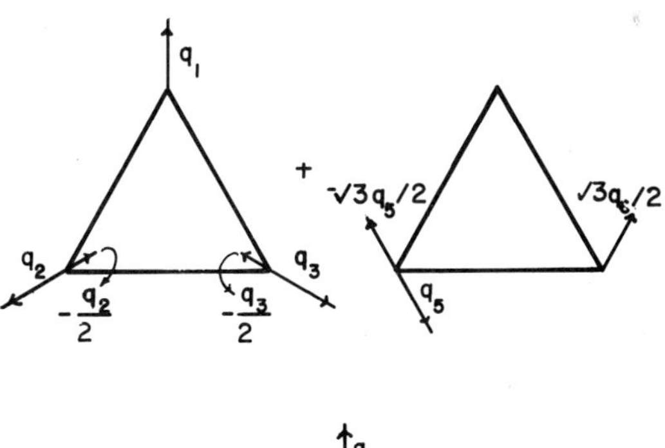

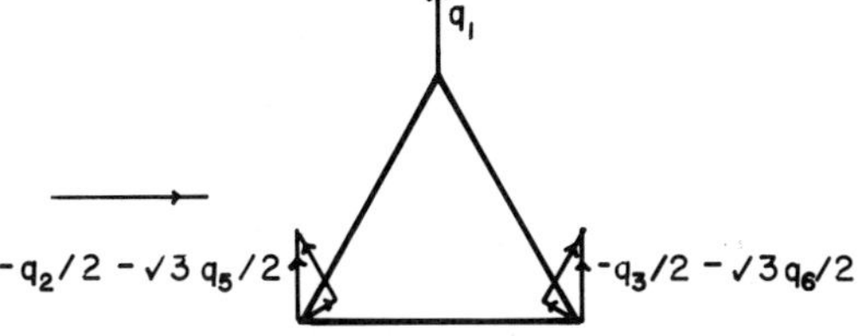

Figure 11.26. The linear combination of vector components for the triangular molecule which generate a net translation in the z direction.

When two orthogonal vectors are summed to obtain a new vector, a second vector results from the difference of the two vectors. In this case, the difference vector is

$$\begin{bmatrix} 1 \\ -\frac{1}{2} \\ -\frac{1}{2} \\ 0 \\ 0 \\ 0 \end{bmatrix} - \begin{bmatrix} 0 \\ 0 \\ 0 \\ 0 \\ -\frac{\sqrt{3}}{2} \\ \frac{\sqrt{3}}{2} \end{bmatrix} = \begin{bmatrix} 1 \\ -\frac{1}{2} \\ -\frac{1}{2} \\ 0 \\ \frac{\sqrt{3}}{2} \\ -\frac{\sqrt{3}}{2} \end{bmatrix} \tag{11.7.24}$$

This vibrational normal mode is developed in Figure 11.27. The components of

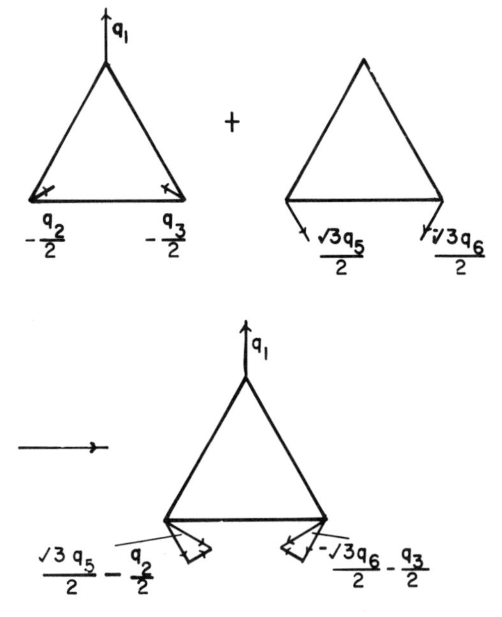

Figure 11.27. A linear combination of basis vectors for the triangular molecule which leads to a normal mode vibration.

the second vector are reversed and now give resultant vectors directed toward the reflection plane through atom 1 at an angle of 30° to the base.

The final two E representation vectors are

$$\begin{bmatrix} 0 \\ 0 \\ 0 \\ 1 \\ -\frac{1}{2} \\ -\frac{1}{2} \end{bmatrix} \qquad \begin{bmatrix} 0 \\ \frac{\sqrt{3}}{2} \\ -\frac{\sqrt{3}}{2} \\ 0 \\ 0 \\ 0 \end{bmatrix} \tag{11.7.25}$$

The factor $\frac{\sqrt{3}}{2}$ has been introduced for the second vector by analogy with the combination of the other two E vectors. The sum of these two vectors is

$$\begin{bmatrix} 0 \\ \frac{\sqrt{3}}{2} \\ -\frac{\sqrt{3}}{2} \\ 1 \\ -\frac{1}{2} \\ -\frac{1}{2} \end{bmatrix} \tag{11.7.26}$$

This is the x translation; the development of the directional vectors on each atom is illustrated in Figure 11.28.

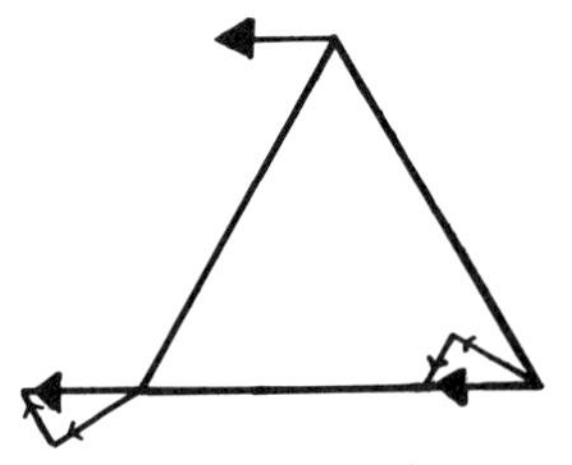

Figure 11.28. The linear combination of basis vectors for the triangular molecule which yields an x translation of the molecule.

The final vibration normal mode is formed as the difference of these two vectors to give

$$\begin{pmatrix} 0 \\ -\frac{\sqrt{3}}{2} \\ \frac{\sqrt{3}}{2} \\ 1 \\ -\frac{1}{2} \\ -\frac{1}{2} \end{pmatrix} \tag{11.7.27}$$

This vibration is illustrated in Figure 11.29.

The analysis for the triangular model emphasizes the difference between one- and two-dimensional irreducible representations. The one-dimensional representations generated for this particular set of basis vectors gave the sym-

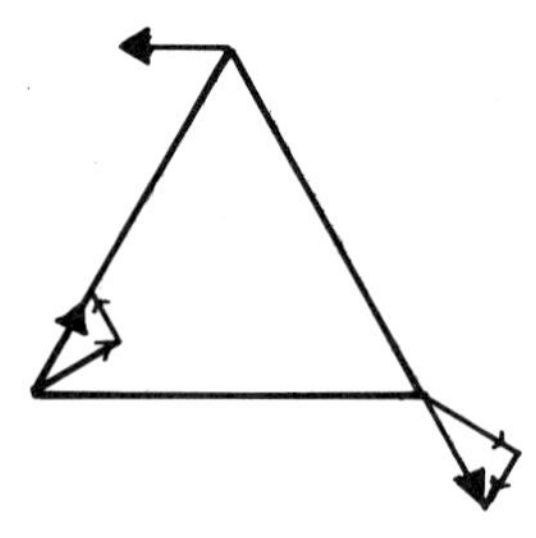

Figure 11.29. A linear combination of basis vectors for the triangular molecule which produces a normal mode vibration.

metric stretch and rotation modes directly. The vectors in the E representations had to be combined to produce an orthogonal set corresponding to molecular parameters of interest, e.g., x translation. An alternate six-vector basis set can be used, e.g., standard Cartesian x and z vectors on each atom. In this case, the x and z translations will give the two irreducible one-dimensional representations and the normal vibrational modes and the rotation modes must be constructed as a set of orthogonal vectors in the E representation (Problem 11.8).

11.8. Ligand Field Theory

Molecular orbitals and normal modes are generated from the symmetry operations of the group and the irreducible representations consistent with the system. The symmetry operator moved a vector on an orbital in space. The orbital systems were selected so that the orbitals superposed exactly. In general, however, the orbital is a wave. The symmetry operation must transfer the amplitude at every spatial location $\psi(x, y, z)$ to some new set of spatial locations. In other words, the symmetry operation must relocate entire functions. The symmetry operation on the function is

$$\mathbf{P_R} \tag{11.8.1}$$

where $\mathbf{P_R}$ is some functional operator which moves each coordinate for the function to its new location in space via the symmetry operation $\mathbf{R}$. This approach was straightforward when the entire wavefunction was described by a vector directed to its maximal amplitude. However, in other cases, a specific symmetry operation may produce some linear combination of orbitals. In such cases, it is more convenient to transform the function by rotating the three-dimensional coordinate system in the opposite direction, i.e.,

$$\mathbf{P_R} f(|x\rangle) = f(\mathbf{R}^{-1}|x\rangle) \tag{11.8.2}$$

where $|x\rangle$ is the set of three variables x, y, and z in the space. The transformed function is determined by replacing the x, y, and z with the x', y', and z' from

$$\mathbf{R}^{-1}|x\rangle = |x'\rangle \tag{11.8.3}$$

The properties of orbitals can be influenced by their surroundings and their intrinsic symmetry. For example, an atom with p_x, p_y, and p_z orbitals rests in an environment with three other B atoms symmetrically arranged in the xy plane (Figure 11.30). The environment can perturb the three p orbitals in different ways. The p_z orbital has a node in the xy plane so the B atom ligands will perturb it minimally. The orbital p_x faces one ligand and should experience the largest perturbation. However, the choice of the x and y axes and the locations of the ligands are completely arbitrary. The p_y orbital might have faced a ligand. In fact, it is impossible to assign p_x or p_y a larger share and they will experience

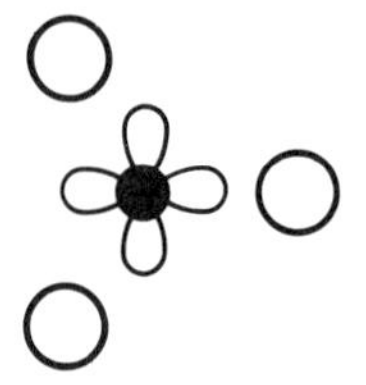

Figure 11.30. The C_{3v} orientation of 3 B atoms about a central A atom containing three degenerate p orbitals.

the same perturbation from the ligand field while p_z always experiences a different perturbation. The ligand field theory which follows will demonstrate that the degeneracy of the three p orbitals will be broken in the ligand field: p_z will have one energy while p_x and p_y will have a second energy.

The p_x and p_y orbitals will experience the maximal effects of the ligand field every 120°. However, the p orbitals are oriented 90° to each other. When atom A is rotated 120°, both p_x and p_y will experience a perturbation by the ligand field. The C_3 rotation will give some weighted sum of p_x and p_y.

The new orbital combinations created by C_3 are found by rotating the coordinates by $-120°$. The transform matrix is

$$\mathbf{C}_3 = \begin{pmatrix} \cos(-120°) & -\sin(-120°) & 0 \\ \sin(-120°) & \cos(-120°) & 0 \\ 0 & 0 & 1 \end{pmatrix}$$

$$= \begin{pmatrix} -\frac{1}{2} & +\frac{\sqrt{3}}{2} & 0 \\ -\frac{\sqrt{3}}{2} & -\frac{1}{2} & 0 \\ 0 & 0 & 1 \end{pmatrix} \tag{11.8.4}$$

The original x, y, and z become

$$\begin{aligned} x' &= -x/2 + \sqrt{3}y/2 \\ y' &= -\sqrt{3}x/2 - y/2 \\ z' &= z \end{aligned} \tag{11.8.5}$$

The rotated function is determined by substituting these new coordinates for the original x, y, and z in the function.

A p_x orbital with normalization constant A becomes

$$
\begin{aligned}
p_x &= Ax/r \\
\mathbf{P}_C p_x &= Ax'/r = (A/r)(-x/2 + \sqrt{3}y/2) \\
&= -\frac{1}{2}(Ax/r) + \frac{\sqrt{3}}{2}(Ay/r) \\
&= -p_x/2 + \sqrt{3}p_y/2
\end{aligned}
\tag{11.8.6}
$$

This weighted sum of orbitals determines the portion of each orbital which experiences the ligand field at $\theta = 120°$.

The transformed p_y and p_z orbitals for a C_3 rotation are determined using the same $|x'\rangle$ as

$$
\begin{aligned}
\mathbf{P}_C p_y &= (A/r)(-\sqrt{3}x/2 - y/2) \\
&= -\sqrt{3}p_x/2 - p_y/2
\end{aligned}
\tag{11.8.7}
$$

$$\mathbf{P}_C p_z = Az'/r = Az/r \tag{11.8.8}$$

The three equations can be combined into a matrix operation for the C_3 transform:

$$
\begin{pmatrix} p'_x \\ p'_y \\ p'_z \end{pmatrix} = \begin{pmatrix} -\frac{1}{2} & +\frac{\sqrt{3}}{2} & 0 \\ -\frac{\sqrt{3}}{2} & -\frac{1}{2} & 0 \\ 0 & 0 & 1 \end{pmatrix} \begin{pmatrix} p_x \\ p_y \\ p_z \end{pmatrix}
\tag{11.8.9}
$$

Since this matrix is the orbital transform matrix, its trace will contribute to the reducible representation as the C_3 component. This trace is 0.

The C_3^2 rotation belongs to the same class and should also have a trace of 0. The transformed coordinates are (Problem 11.9)

$$
\begin{pmatrix} p'_x \\ p'_y \\ p'_z \end{pmatrix} = \begin{pmatrix} -\frac{1}{2} & -\frac{\sqrt{3}}{2} & 0 \\ \frac{\sqrt{3}}{2} & -\frac{1}{2} & 0 \\ 0 & 0 & 1 \end{pmatrix} \begin{pmatrix} p_x \\ p_y \\ p_z \end{pmatrix}
\tag{11.8.10}
$$

For these axis-oriented orbitals, the rotation matrix is identical to that for a rotation of x, y, and z vectors through 120°.

The transform matrix for reflections (σ) requires σ^{-1} but this is identical to σ. σ_1 through atom B_1 lies in the xz plane and leaves the x and z components unchanged. The y axis reverses direction and the matrix is

$$\begin{pmatrix} 1 & 0 & 0 \\ 0 & -1 & 0 \\ 0 & 0 & 1 \end{pmatrix} \tag{11.8.11}$$

The observed trace of $+1$ will be the same for other reflections of this class. Since the transform matrices for the remaining reflections are more complicated, the reflection σ_2 through B_2 at an angle of 120° with respect to the x axis is illustrated. The reflections of the x and y coordinates are shown in Figure 11.31. The matrix is

$$\begin{pmatrix} -\frac{1}{2} & -\frac{\sqrt{3}}{2} & 0 \\ -\frac{\sqrt{3}}{2} & +\frac{1}{2} & 0 \\ 0 & 0 & 1 \end{pmatrix} \tag{11.8.12}$$

and the trace is $+1$ as expected.

Since the traces for each symmetry transform are known, the reducible representation vector is

$$|\Gamma\rangle = (3, 0, +1) \tag{11.8.13}$$

which resolves into the irreducible representations as

$$1\,|A_1\rangle + 1\,|E\rangle \tag{11.8.14}$$

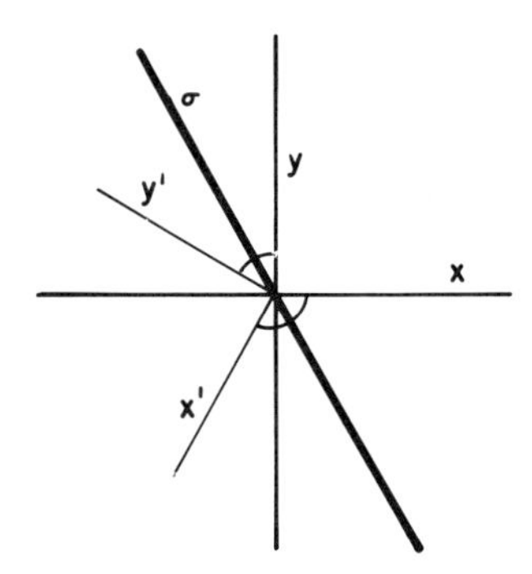

Figure 11.31. Reflections of the x and y coordinates to x' and y' through the reflection plane σ_2, showing the components for x' and y'.

The nature of these irreducible representations is easily seen with the p orbitals. The p_z orbital is perpendicular to the plane which contains the three ligands. None of the symmetry operations will change its orientation relative to these ligands and it is consistent with the A_1 irreducible representation. The p_x and p_y orbitals are mixed by the symmetry operations of the C_{3v} group and they must be included as a two-dimensional irreducible representation under the C_{3v} operations. The ligand field will break the threefold degeneracy of the p orbitals to produce two distinct representations. The energies for the p_z orbital and the p_x and p_y orbitals differ in the C_{3v} ligand field.

The coordinate transform technique is applicable to any sets of degenerate orbitals. For example, the Eu^{3+} ion has an excited D state which is fivefold degenerate. In the presence of a ligand field, this degeneracy can be broken. Different orbitals or groups of orbitals within the manifold will transform with a different irreducible representation in the ligand field, and each distinct representation will produce a distinct energy level in the spectrum.

Since the spatial transforms for C_{3v} are already known, the problem is illustrated with the C_{3v} ligand symmetry of Figure 11.32. Water molecules are arranged in groups of three in three planes symmetric about the equatorial plane. This arrangement contains both the rotation and reflection elements required for C_{3v}.

The D_{z^2} orbital is oriented on the z axis and should be unaffected by the rotation and reflection symmetry operations. It will be consistent with an A_1 irreducible representation.

To determine the traces for the reducible representation, the 5×5 matrices which describe the transforms of the five D orbitals must be known. These transforms are developed by determining which orbitals are produced when the coordinates $|x'\rangle$ are substituted for $|x\rangle$ in the expressions for each of the D orbitals. The inverse spatial transforms for the C_{3v} group elements were determined for the p-orbital transformations. The $|x'\rangle$ spatial transforms for C_3 are

$$\begin{aligned} x' &= -x/2 - \sqrt{3}y/2 \\ y' &= \sqrt{3}x/2 - y/2 \\ z' &= z \end{aligned} \tag{11.8.15}$$

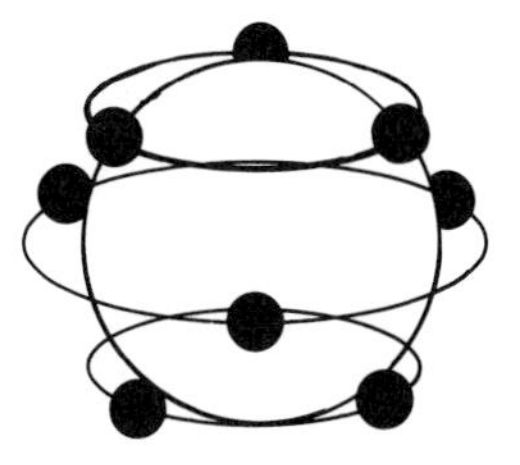

Figure 11.32. A C_{3v} arrangement of water molecules about a Eu^{3+} ion.

The transformed orbitals for the C_3 transform are determined by substituting these values into the expressions for the five D orbitals:

$$\begin{aligned}
\varphi_{yz} &= \sqrt{15/4\pi}(A/r^2)\ yz = Cyz \\
\varphi_{xz} &= \sqrt{15/4\pi}(A/r^2)\ xz = Cxz \\
\varphi_{xz} &= \sqrt{15/4\pi}(A/r^2)\ xy = Cxy \\
\varphi_{z^2} &= \sqrt{45/16\pi}(A/r^2)(3z^2 - r^2) \\
&= \frac{\sqrt{3}}{2} C(3z^2 - r^2) \qquad (11.8.16) \\
\varphi_{x^2-y^2} &= \sqrt{15/16\pi}(A/r^2)(x^2 - y^2) \\
&= \frac{1}{2} C(x^2 - y^2) \\
C &= \sqrt{15/4\pi}(A/r^2)
\end{aligned}$$

The C_3 operation gives five transform operations for the five D orbitals:

$$\begin{aligned}
C_3\varphi_{yz} &= Cx'z' = Cz(\sqrt{3}x/2 - y/2) \\
&= C\left(\frac{\sqrt{3}}{2} xz - \frac{1}{2} yz\right) = C\left(\frac{\sqrt{3}}{2} \varphi_{xz} - \frac{1}{2} \varphi_{yz}\right) \\
C_3\varphi_{xz} &= Cx'z' = Cz(-x/2 - \sqrt{3}y/2) \\
&= C\left(-\frac{1}{2} xz - \frac{\sqrt{3}}{2} yz\right) = C\left(-\frac{1}{2} \varphi_{xz} - \frac{\sqrt{3}}{2} \varphi_{yz}\right) \\
C_3\varphi_{xy} &= Cx'y' = C(-x/2 - \sqrt{3}y/2)(\sqrt{3}x/2 - y/2) \\
&= C\left(-\frac{\sqrt{3}}{4} x^2 + \frac{\sqrt{3}}{4} y^2 + \frac{1}{4} xy - \frac{3}{4} xy\right) \\
&= C\left[-\frac{1}{2} xy - \frac{1}{2}\left(\frac{\sqrt{3}}{2}\right)(x^2 - y^2)\right] \\
&= -\frac{1}{2} \varphi_{xy} - \frac{\sqrt{3}}{2} \varphi_{x^2-y^2} \qquad (11.8.17) \\
C_3\varphi_{z^2} &= \frac{\sqrt{3}}{2} C(3z^2 - r^2) \\
&= \frac{\sqrt{3}}{2} C(3z^2 - r^2) = \varphi_{z^2}
\end{aligned}$$

$$C_3\varphi_{x^2-y^2} = \frac{1}{2}C(x'^2 - y'^2)$$

$$= \frac{1}{2}C[(-x/2 - \sqrt{3}y/2)^2 - (\sqrt{3}x/2 - y/2)^2]$$

$$= \frac{1}{2}C[x^2/4 + 3y^2/4 + 3xy/2 - (3x^2/4 + y^2/4 - \sqrt{3}xy/2)]$$

$$= +\frac{\sqrt{3}}{2}\varphi_{xy} - \frac{1}{2}\varphi_{x^2-y^2}$$

The 5×5 matrix for the rotation of the five D orbitals under C_3 is

$$\begin{pmatrix} -\frac{1}{2} & \frac{\sqrt{3}}{2} & 0 & 0 & 0 \\ -\frac{\sqrt{3}}{2} & -\frac{1}{2} & 0 & 0 & 0 \\ 0 & 0 & -\frac{1}{2} & -\frac{\sqrt{3}}{2} & 0 \\ 0 & 0 & \frac{\sqrt{3}}{2} & -\frac{1}{2} & 0 \\ 0 & 0 & 0 & 0 & 1 \end{pmatrix} \begin{pmatrix} xz \\ yz \\ xy \\ x^2 - y^2 \\ z^2 \end{pmatrix} \tag{11.8.18}$$

The trace for the C_3 operation on the five D orbitals is -1. The same procedure can be repeated to determine the trace for C_3^2. However, since this operation is the same class as C_3, its trace must also be -1. The arrangement in classes can reduce the total calculations significantly.

The simplest reflection is σ_{xz}. Since the x and z vectors lie in this plane, the inverse reflection involves only a change in y. The transform equations are

$$\begin{aligned} x' &= x \\ y' &= -y \\ z' &= z \end{aligned} \tag{11.8.19}$$

The transformed orbitals are

$$\begin{aligned} \varphi'_{xz} &= \varphi_{xz} = \varphi_{xz} \\ \varphi'_{yz} &= \varphi_{yz} = -\varphi_{yz} \\ \varphi'_{xy} &= \varphi_{xy} = -\varphi_{xy} \\ \varphi'_{x^2-y^2} &= \varphi_{x^2-y^2} = \varphi_{x^2-y^2} \\ \varphi'_{z^2} &= \varphi_{z^2} = \varphi_{z^2} \end{aligned} \tag{11.8.20}$$

The two orbitals containing y^1 change sign during the transform; the remainder are unchanged. The σ transform matrix is

$$\begin{pmatrix} 1 & 0 & 0 & 0 & 0 \\ 0 & -1 & 0 & 0 & 0 \\ 0 & 0 & -1 & 0 & 0 \\ 0 & 0 & 0 & 1 & 0 \\ 0 & 0 & 0 & 0 & 1 \end{pmatrix} \begin{pmatrix} xz \\ yz \\ xy \\ x^2 - y^2 \\ z^2 \end{pmatrix} \tag{11.8.21}$$

which has a trace of $+1$. The remaining two reflections in the class will have the same trace.

The $|\Gamma\rangle$ vector for the five D orbitals is

$$|\Gamma\rangle = (5, -1, 1) \tag{11.8.22}$$

which gives the irreducible representations

$$1\,|A_1\rangle + 0\,|A_2\rangle + 2\,|E\rangle \tag{11.8.23}$$

With the actual information on the required irreducible representations available, the form of the diagonalized matrices of the symmetry matrices is clear. One orbital will transform separately with A_1 symmetry. The remaining four orbitals will transform in groups of two with E symmetry. The orbitals for each representation are determined directly from the C_3 matrix which had the block diagonal form. The φ_{z^2} orbital is unaffected by the C_{3v} symmetry of the ligand field and transforms as the A_1 irreducible representation. The φ_{xz} and φ_{yz} orbitals are perpendicular to the xy plane and the ligand field mixes them, i.e., they transform as an E representation. The φ_{xy} and $\varphi_{x^2-y^2}$ orbitals also transform together; both lie essentially in the xy plane of the system.

The approach can be used for any set of degenerate orbitals in a ligand field of any symmetry. For example, the five D orbitals may separate in a different manner if the waters surrounding the ion are arranged with a different symmetry. If the ion binds to some molecular binding site, determination of the splitting using spectroscopic techniques gives a measure of the local symmetry of the ion at this binding site.

11.9. Direct Products of Group Elements

The direct product of two matrices produces a new matrix of higher order. The eigenvalues and eigenvectors of this new matrix are related to those of the generating matrices. For the matrices or operators for symmetry operations in group theory, direct products can be used to generate new symmetry operations. Such operations differ from group table operations, which involve elements within a group. The direct product combines elements from one symmetry group

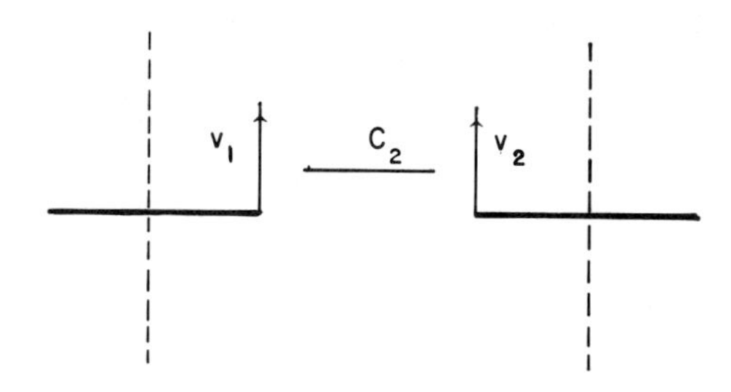

Figure 11.33. The generation of vectors $\mathbf{v}_1$ and $\mathbf{v}_2$ using E and C_2 group operations.

with those of a second symmetry group to produce a new group with symmetry elements from both groups plus additional symmetry elements which do not appear in the original groups.

The direct product groups are illustrated by starting with two simple groups. The C_2 group with elements E and C_2 is applied to a vector parallel to the rotation axis (Figure 11.33). The C_2 rotation generates a second vector in the same direction as the first. The elements E and i of a C_i group also operate to produce vectors (Figure 11.34). The inversion element i changes any vector $|r\rangle$ to $-|r\rangle$. Since the vector of Figure 11.34 has coordinates $(x, 0, z)$, the inverted vector has coordinates $(-x, 0, -z)$. The inversion could be generated by a C_2 rotation followed by a reflection in the xy plane. However, there is no σ in the two starting groups.

C_2 and C_i are true groups since any pair of operations give an operation within the group. This situation changes when the operations from each group act on the same vector. For example, the two operations, iC_2, act on the vector $\mathbf{v}_1$ as shown in Figure 11.35. The first operation moves the vector to the left side with no direction change. The i operation then returns it to the right with a net direction change, i.e., it now faces downward. A new group operation must be defined to convert $\mathbf{v}_1$ directly to $\mathbf{v}_4$; this is a reflection, σ, in the xy plane. The group constructed from C_2 and C_i must then contain the four group elements E, C_2, i, and σ.

The four-element group will generate a total of four distinct vectors from the vector $\mathbf{v}_1$. The matrices which describe these symmetry operations can be constructed as direct products of the matrices which describe the symmetry

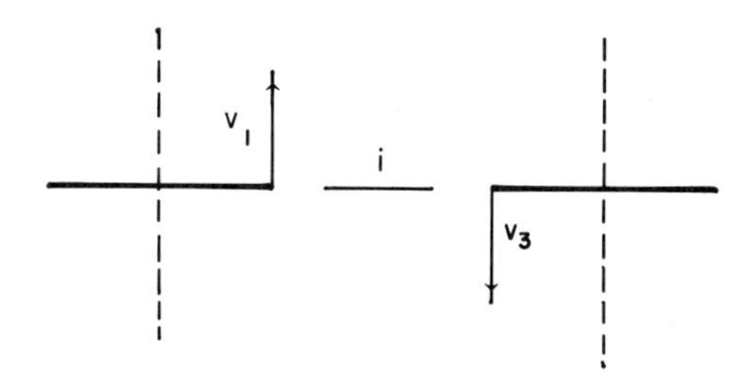

Figure 11.34. The generation of vectors $\mathbf{v}_1$ and $\mathbf{v}_3$ using E and i group operations.

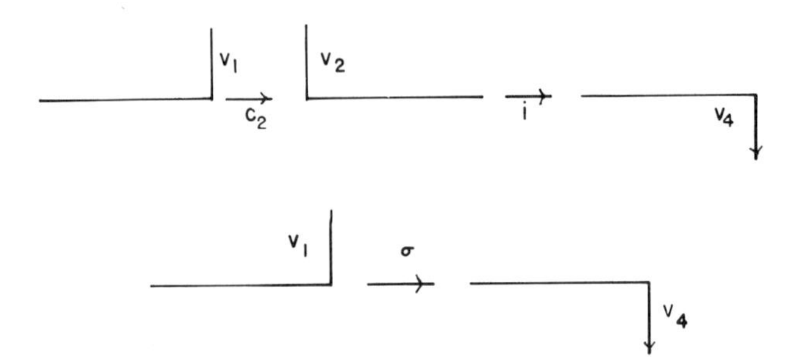

Figure 11.35. The generation of a new vector $\mathbf{v}_4$ using the sequence of symmetry operations iC_2 on the vector $\mathbf{v}_1$. The two operations are equivalent to a reflection in the xy plane.

operations in the C_2 and C_i groups. Since all vectors remain parallel to the z axis, 2×2 matrices describe each of the four symmetry operations. The 2×2 matrices for the C_2 group are

$$\mathbf{E}=\begin{pmatrix}1 & 0\\ 0 & 1\end{pmatrix} \qquad \mathbf{C}_2=\begin{pmatrix}0 & 1\\ 1 & 0\end{pmatrix} \tag{11.9.1}$$

The corresponding 2×2 matrices for the C_i group are

$$\mathbf{E}=\begin{pmatrix}1 & 0\\ 0 & 1\end{pmatrix} \qquad \mathbf{i}=\begin{pmatrix}0 & -1\\ -1 & 0\end{pmatrix} \tag{11.9.2}$$

The four symmetry matrices for the new group are formed as the four possible direct products:

$$\mathbf{E}\otimes\mathbf{E}=\begin{pmatrix}1 & 0\\ 0 & 1\end{pmatrix}\otimes\begin{pmatrix}1 & 0\\ 0 & 1\end{pmatrix} =\begin{pmatrix}1 & 0 & 0 & 0\\ 0 & 1 & 0 & 0\\ 0 & 0 & 1 & 0\\ 0 & 0 & 0 & 1\end{pmatrix} \tag{11.9.3}$$

$$\mathbf{C}_2\otimes\mathbf{E}=\begin{pmatrix}0 & 1\\ 1 & 0\end{pmatrix}\otimes\begin{pmatrix}1 & 0\\ 0 & 1\end{pmatrix} =\begin{pmatrix}0 & 1 & 0 & 0\\ 1 & 0 & 0 & 0\\ 0 & 0 & 0 & 1\\ 0 & 0 & 1 & 0\end{pmatrix} \tag{11.9.4}$$

$$\mathbf{E}\otimes\mathbf{i}=\begin{pmatrix}1&0\\0&1\end{pmatrix}\otimes\begin{pmatrix}0&-1\\-1&0\end{pmatrix}$$

$$=\begin{pmatrix}0&0&-1&0\\0&0&0&-1\\-1&0&0&0\\0&-1&0&0\end{pmatrix} \qquad (11.9.5)$$

$$\mathbf{C}_2\otimes\mathbf{i}=\begin{pmatrix}0&1\\1&0\end{pmatrix}\otimes\begin{pmatrix}0&-1\\-1&0\end{pmatrix}$$

$$=\begin{pmatrix}0&0&0&-1\\0&0&-1&0\\0&-1&0&0\\-1&0&0&0\end{pmatrix} \qquad (11.9.6)$$

The results can be verified by operating on the $\mathbf{v}_1$ vector. For example, the final matrix (Equation 11.9.6) gives

$$\begin{pmatrix}0&0&0&-1\\0&0&-1&0\\0&-1&0&0\\-1&0&0&0\end{pmatrix}\begin{pmatrix}1\\0\\0\\0\end{pmatrix}=\begin{pmatrix}0\\0\\0\\-1\end{pmatrix} \qquad (11.9.7)$$

The order of components is $\mathbf{v}_1$, $\mathbf{v}_2$, $\mathbf{v}_3$, and $\mathbf{v}_4$ so the operation reproduces the results illustrated in Figure 11.35. This is the σ operation. Equations 11.9.3, 11.9.4, and 11.9.5 are symmetry matrices for E, C_2, and i, respectively. They create all the vectors which could be created by the original groups plus $\mathbf{v}_4$.

All these matrices commute and they can be diagonalized with a single set of eigenvectors. The similarity transform matrices which do this are

$$\mathbf{S}=\begin{pmatrix}1&1&1&1\\1&1&-1&-1\\1&-1&-1&1\\1&-1&1&-1\end{pmatrix} \qquad \mathbf{S}^{-1}=\begin{pmatrix}1&1&1&1\\1&1&-1&-1\\1&-1&-1&1\\1&-1&1&-1\end{pmatrix} \qquad (11.9.8)$$

The diagonalized symmetry matrices for this four-element group are

$$\mathbf{E}=\begin{pmatrix}1&0&0&0\\0&1&0&0\\0&0&1&0\\0&0&0&1\end{pmatrix} \qquad (11.9.9)$$

$$\mathbf{C}_2 = \begin{pmatrix} 1 & 0 & 0 & 0 \\ 0 & 1 & 0 & 0 \\ 0 & 0 & -1 & 0 \\ 0 & 0 & 0 & -1 \end{pmatrix} \tag{11.9.10}$$

$$\mathbf{i} = \begin{pmatrix} -1 & 0 & 0 & 0 \\ 0 & 1 & 0 & 0 \\ 0 & 0 & -1 & 0 \\ 0 & 0 & 0 & 1 \end{pmatrix} \tag{11.9.11}$$

$$\boldsymbol{\sigma} = \begin{pmatrix} -1 & 0 & 0 & 0 \\ 0 & 1 & 0 & 0 \\ 0 & 0 & 1 & 0 \\ 0 & 0 & 0 & -1 \end{pmatrix} \tag{11.9.12}$$

The four matrices have resolved into the four possible 1×1 irreducible representations for this group. The characters (traces) for these representations can be read from the diagonal elements of the four matrices as

C_{3h}	E	C	i	σ
A_g	1	1	1	1
A_u	1	1	−1	−1
B_g	1	−1	1	−1
B_u	1	−1	−1	1

(11.9.13)

The g (gerade, even) and u (ungerade) subscripts specify the character for i. The character vectors produced in this manner have the same components as the eigenvectors which were used in the similarity transform to diagonalize the matrices.

The characters for the irreducible representations could be obtained directly as direct products of diagonalized matrices for the C_2 and C_i groups. The diagonalized matrices for these groups are

$$\mathbf{E} = \begin{pmatrix} 1 & 0 \\ 0 & 1 \end{pmatrix} \quad \mathbf{C}_2 = \begin{pmatrix} 1 & 0 \\ 0 & -1 \end{pmatrix} \quad \mathbf{i} = \begin{pmatrix} -1 & 0 \\ 0 & 1 \end{pmatrix} \tag{11.9.14}$$

and the relevant direct products are

$$\mathbf{E} \otimes \mathbf{E} = \mathbf{E} = \begin{pmatrix} 1 & 0 \\ 0 & 1 \end{pmatrix} \otimes \begin{pmatrix} 1 & 0 \\ 0 & 1 \end{pmatrix}$$

$$= \begin{pmatrix} 1 & 0 & 0 & 0 \\ 0 & 1 & 0 & 0 \\ 0 & 0 & 1 & 0 \\ 0 & 0 & 0 & 1 \end{pmatrix} \tag{11.9.15}$$

$$\mathbf{C}_2 = \mathbf{C}_2 \otimes \mathbf{E} = \begin{pmatrix} 1 & 0 \\ 0 & -1 \end{pmatrix} \otimes \begin{pmatrix} 1 & 0 \\ 0 & 1 \end{pmatrix}$$

$$= \begin{pmatrix} 1 & 0 & 0 & 0 \\ 0 & -1 & 0 & 0 \\ 0 & 0 & 1 & 0 \\ 0 & 0 & 0 & -1 \end{pmatrix} \tag{11.9.16}$$

$$\mathbf{i} = \mathbf{E} \otimes \mathbf{i} = \begin{pmatrix} 1 & 0 \\ 0 & 1 \end{pmatrix} \otimes \begin{pmatrix} -1 & 0 \\ 0 & 1 \end{pmatrix}$$

$$= \begin{pmatrix} -1 & 0 & 0 & 0 \\ 0 & -1 & 0 & 0 \\ 0 & 0 & 1 & 0 \\ 0 & 0 & 0 & 1 \end{pmatrix} \tag{11.9.17}$$

$$\boldsymbol{\sigma} = \mathbf{C}_2 \otimes \mathbf{i} = \begin{pmatrix} 1 & 0 \\ 0 & -1 \end{pmatrix} \otimes \begin{pmatrix} -1 & 0 \\ 0 & 1 \end{pmatrix}$$

$$= \begin{pmatrix} -1 & 0 & 0 & 0 \\ 0 & 1 & 0 & 0 \\ 0 & 0 & 1 & 0 \\ 0 & 0 & 0 & -1 \end{pmatrix} \tag{11.9.18}$$

The direct product of each of these diagonal 2×2 matrices is also diagonal. The irreducible representations for the four-element group appear as the block diagonal elements of these matrices.

The diagonal elements of the 2×2 matrices gave the irreducible representations for the groups. Their direct product gave the irreducible representations for the product group. Thus, the character table for the larger group is generated as a direct product of the character tables of the constituent groups.

The technique can be illustrated for the direct product of the D_3 group and the C_i group. The D_3 group differs from the C_3 group since it contains three C_2 operations about axes which are perpendicular to the z axis for the C_3 operations. The symmetry is illustrated in Figure 11.36. Rotation about any of the three axes in the plane will reproduce the vectors. The character table for D_3 has characters identical to those for C_{3v}, i.e.,

	E	$2C_3$	$3C_2$
A_1	1	1	1
A_2	1	1	-1
E	2	-1	0

(11.9.19)

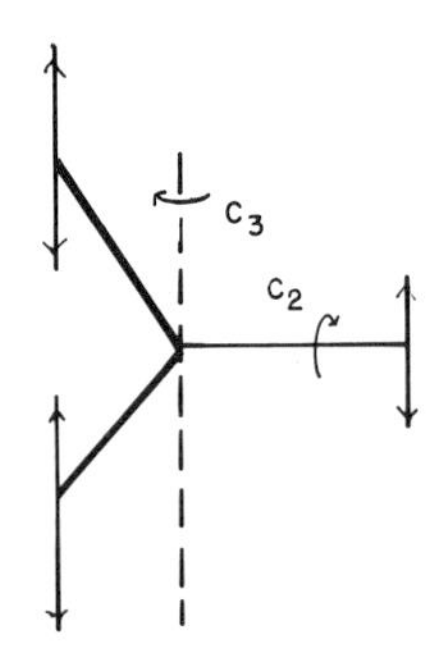

Figure 11.36. A set of vectors which satisfy the symmetry operations of D_3. The operations include C_3 and C_3^2 operations about z and three C_2 rotations about the three axes in the xy plane.

The subscripts 1 and 2 define the character of C_2. The elements are listed by class, i.e., two C_3 operations and three C_2 operations. The character table for C_i is

	E	i
A_g	1	1
A_u	1	−1

(11.9.20)

The direct product of the two tables is

$$\mathbf{D}_3 \otimes \mathbf{C}_i = \begin{pmatrix} 1 & 1 & 1 \\ 1 & 1 & -1 \\ 2 & -1 & 0 \end{pmatrix} \otimes \begin{pmatrix} 1 & 1 \\ 1 & -1 \end{pmatrix}$$

$$= \begin{pmatrix} 1 & 1 & 1 & 1 & 1 & 1 \\ 1 & 1 & -1 & 1 & 1 & -1 \\ 2 & -1 & 0 & 2 & -1 & 0 \\ 1 & 1 & 1 & -1 & -1 & -1 \\ 1 & 1 & -1 & -1 & -1 & 1 \\ 2 & -1 & 0 & -2 & 1 & 0 \end{pmatrix} \tag{11.9.21}$$

Although this is the correct character table for the D_{3d} group, the symmetry elements for each column must be determined. Some belong to the original groups. Since D_3 occurs first in the direct product, its three classes are coupled with E from C_i and they are repeated in the new table (E, $2C_3$, and $3C_2$). The fourth column couples E from D_3 with i from C_i and gives the i symmetry

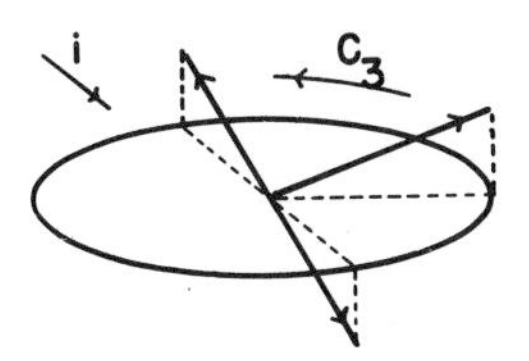

Figure 11.37. Combination of the C_3 and i operations to produce a new S_6^5 operation.

element. The fifth and sixth columns will contain new symmetry elements generated from the coupling of the C_3 and C_2 operations, respectively, with the i operation. Since the classes C_3 and C_2 respectively contain two and three elements while the class for i has one element, columns 5 and 6 are classes with two and three elements, respectively.

The first new symmetry operation results from a direct product of C_3 or C_3^2 and i. The three vectors of Figure 11.37 are 120° apart. A C_3 rotation, followed by an inversion, places the vector at $-60°$. This is the S_6^5 operation in which the vector is rotated $-60°$ and then reflected through a plane perpendicular to the rotation axis (Figure 11.37). The direct product of the C_3^2 and i operations gives the S_6 operation (Figure 11.38). Two S_6 operations in sequence are equivalent to a C_3 operation.

The final set of three symmetry operations are the direct product of the C_2 operations and the inversion operation. These operations are equivalent to a reflection through a plane perpendicular to the xy plane and oriented halfway between the C_2 rotation axes (Figure 11.39). The reflection plane is not present for D_3 and the presence of the inversion element in the group is required for its appearance.

The direct product functions in two ways with groups. The direct product of two symmetry operations from the constituent subgroups generates new symmetry elements for the group. The direct product of the two character tables for the constituent subgroups generates the full character table for the direct product group. When the full character table is known, it can be used to establish the irreducible representations present for a molecule which contains all symmetries tabulated for the group.

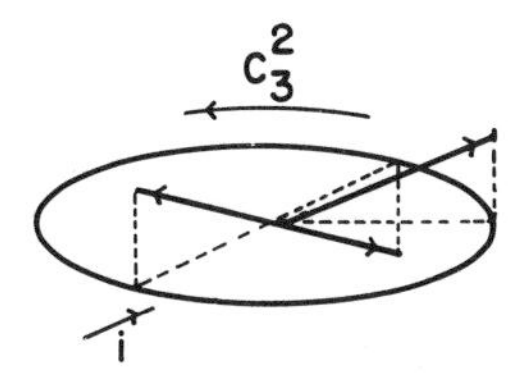

Figure 11.38. Combination of the C_3^2 and i operations to produce the S_6 operation.

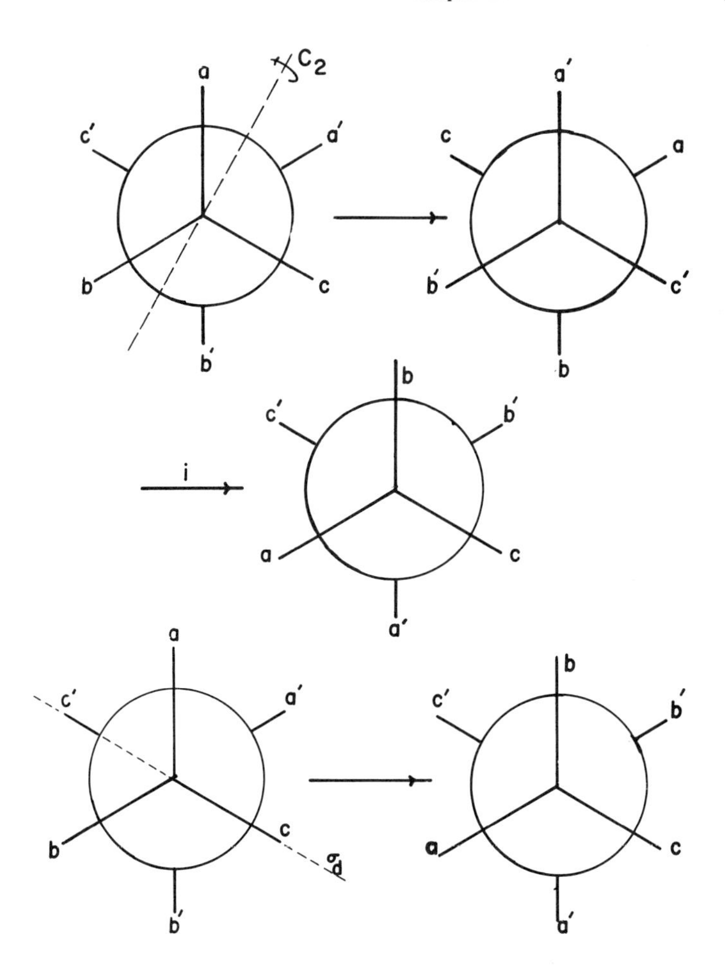

Figure 11.39. The σ_d reflection operation generated as a combination of C_2 and i operations.

11.10. Direct Products and Integrals

If a molecule has an intrinsic symmetry, normal modes and molecular orbitals can be generated using the appropriate irreducible representations for that group. Since the irreducible representations are mutually orthogonal, molecular orbitals generated from the symmetry operations will also be orthogonal. Integrals of products of such orthogonal orbitals must be zero. For this reason, it becomes possible to evaluate integrals involving products of functions by examining the irreducible representations of these functions.

The role of group theory for the evaluation of integrals is illustrated with the simple, one-dimensional function $f(x) = x$. The function is defined on the interval $0 \leqslant x \leqslant 1$ and must be defined on the remaining interval $-1 \leqslant x \leqslant 0$

using symmetry operations. The full function can be defined with a C_2 group; the actual function will depend on the irreducible representation chosen. The two possibilities are shown in Figure 11.40. The A representation generates an even function while the B representation generates an odd function. Each of these functions can be integrated over the full interval. The even A function gives a nonzero integral while the odd B function gives a total integral of zero since the areas for the two halves of the function are equal and opposite. The nonzero result is expected for the even function. However, there is now an additional correlation between a nonzero integral and the A (symmetric) representation of the function.

The correlation persists for more complicated functions. For example, the function

$$f(x, y) = xy \tag{11.10.1}$$

can be defined for values of x and y such that

$$0 \leqslant x, y \leqslant 1 \tag{11.10.2}$$

in the first (positive) xy quadrant. The function lies above the xy plane. The symmetry operations C_4, C_2, and C_4^3 can then extend this function to the remaining four x–y quadrants. The choice of irreducible representation selected for the extension is now crucial. If the A (totally symmetric) representation is chosen, the function will be identical in each quadrant and the total integral is just four times the integral determined for the first quadrant. However, if the B irreducible representation is selected, the function will give a positive integral in the first and third quadrants but a negative integral in the second and fourth quadrants. The total integral will be zero. A total integral of zero will also be observed if the E representation is used to generate the function in all four quadrants.

These simple examples illustrate a general property of functions generated using the symmetry operations and irreducible representations. The integrals will be zero unless the integrand is generated using the totally symmetric irreducible representation, i.e., A_1. If the function or product of functions under the integral has a finite projection on the A_1 irreducible representation, then the integral will be nonzero.

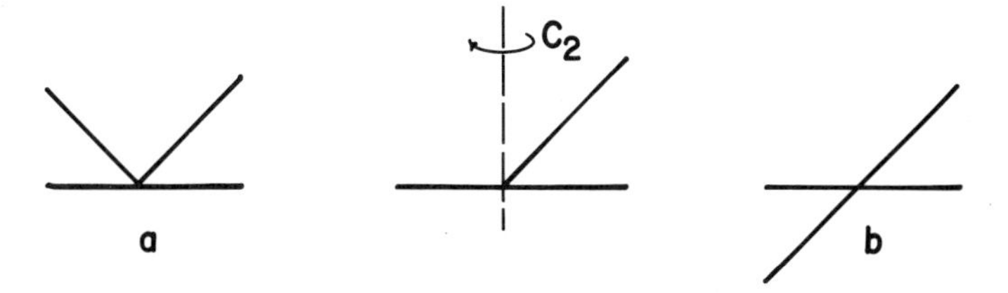

Figure 11.40. The generation of even and odd functions by application of the (a) A and (b) B irreducible representation characters to the function $f(x) = x$, $0 \leqslant x \leqslant 1$.

The utility of the correlation between irreducible representations and the values of integrals becomes significant when the integral contains products of functions which are generated using the irreducible representations for a group. This situation arises frequently in studies of transitions between states in quantum mechanics where such transitions are dictated by matrix elements

$$\langle f|\, \mathbf{O}\, |i\rangle \tag{11.10.3}$$

The operator **O** perturbs the initial state $|i\rangle$. If the perturbation generates some $\langle f|$ wavefunction character, there is a finite projection on $\langle f|$ and the integral of Equation 11.10.3 will be nonzero. The result of the integral can be inferred from a knowledge of the irreducible representations of $\langle f|$ and $|i\rangle$ and the operator **O**.

Two functions which are generated from irreducible representations in a group will have a product which can be correlated with the direct product generated from these irreducible representations. Consider the two functions $f(x)=x$ and $g(x)=x^2$. The function $f(x)$ is extended over the interval $-1 \leqslant x \leqslant 1$ using the A irreducible representation of the C_2 group. The function $g(x)$ is extended over this interval using the B irreducible representation. The two functions (Figure 11.41) are introduced as projection operators,

$$[(+1)E+(+1)\,C_2]\,f(x)[(+1)E+(-1)\,C_2]\,g(x)\,dx \tag{11.10.4}$$

The individual projections must generate the independent functions so the two group projections must be combined as a direct product. Because the functions are generated as irreducible representations, each "matrix" for the group multiplication is a 1×1 block diagonal matrix. The direct product for such matrices is the product of the block diagonal elements. The direct product is then the product of the two characters for E (1×1) and the product of the characters for C_2 with each function (1×-1). The direct product has a representation

$$(1,\, -1) \tag{11.10.5}$$

Because this product corresponds to the B irreducible representation, the function product under the integral corresponds to a net B irreducible representation and the integral must be zero.

Figure 11.41. (a) The generation of the function $f(x)=x$ using the A irreducible representation. (b) The generation of the function $g(x)=x^2$ using the B irreducible representation.

The basic process which occurred here is quite simple. The irreducible representation for $f(x)$ made it even on the interval. The irreducible representation for $g(x)$ made it odd on this interval. The direct product of the two irreducible representations indicated that the product of the two functions was odd and their integral was zero.

The technique can be illustrated with two molecular orbitals generated from the atomic orbital φ_1 using the A_1 and A_2 irreducible representations of the group C_{3v}. The vector $|\Gamma\rangle$ for the product of the two functions is the direct product of their irreducible representations, i.e.,

$$\mathbf{A}_1 \otimes \mathbf{A}_2 = (1\times 1 \quad 1\times 1 \quad 1\times -1) = (1 \quad 1 \quad -1) \tag{11.10.6}$$

The direct product has the same characters as the A_2 representation. The symmetry of the product of the two wavefunctions is then A_2 and the integral of the product must be zero since only the A_1 representation will give a nonzero integral.

If the two functions are generated as two E irreducible representations, their direct product is

$$\mathbf{E} \otimes \mathbf{E} = (2\times 2 \quad -1\times -1 \quad 0\times 0) = (4 \quad 1 \quad 0) \tag{11.10.7}$$

This $|\Gamma\rangle$ vector does not correspond to a simple irreducible representation. However, it can be resolved into such representations. The projections of this vector on the three irreducible representations give

$$\begin{aligned} n_{A_1} &= \frac{1}{6}(1\times 4 + 2\times 1\times 1 + 3\times 1\times 0) = 1\ |A_1\rangle \\ n_{A_2} &= \frac{1}{6}(1\times 4 + 2\times 1\times 1 + 3\times -1\times 0) = 1\ |A_2\rangle \\ n_E &= \frac{1}{6}(2\times 4 + 2\times -1\times 1 + 3\times 0\times 0) = 1\ |E\rangle \end{aligned} \tag{11.10.8}$$

The vector resolves into each of the irreducible representations. However, A_1 is the prime concern. Because there is a finite projection on this representation, the total integral for the product of the two E representation functions is nonzero.

It is convenient to express the irreducible representations produced via the direct products as a table. For the C_{3v} group, this table is

C_{3v}	A_1	A_2	E
A_1	A_1	A_2	E
A_2	A_2	A_1	E
E	E	E	$A_1 \oplus A_2 \oplus E$

(11.10.9)

The direct sum notation is used to indicate that the direct product of two E representation resolves into three distinct irreducible representations.

The direct product technique can be extended to an arbitrary number of functions in the integrand. The direct product in this case is the product of the corresponding characters for each symmetry element. Such direct products can now be used to determine transition moments. Dipole moment-allowed transitions require an operator proportional to the dipole moment of the molecule since the electron distribution of the molecule must be changed by the incoming light if a light-induced transition is to take place. The dipole moment operator can be resolved into three components in a Cartesian coordinate system:

$$\mu_x = ex$$
$$\mu_y = ey \qquad (11.10.10)$$
$$\mu_z = ez$$

Each of these dipole moment components carries an intrinsic symmetry which must be included in the direct product for the factors in the integrand.

The transition moments for an electron in a one-dimensional box can be determined using the simple symmetry of this system. The wavefunctions for the electron are

$$|\psi\rangle = \sqrt{2/L}\sin(n\pi x/L) \qquad (11.10.11)$$

where L is the length of the one-dimensional box. If a rotation axis is introduced at $x = L/2$, wavefunctions with odd n will be even with respect to this C_2 rotation. The wavefunctions with even n are odd with respect to this rotation. Thus, the odd-n wavefunctions belong to the A representation in the C_2 group while the even-n wavefunctions belong to the B representation.

Although the symmetry for the dipole moment operator can be determined directly for this simple case, the character table for the C_2 group tabulates this information. The character table has the form

	E	C_2		
A	1	1	T_z, R_z	xx, yy, zz, xy
B	1	-1	T_x, T_y, R_x, R_y	yz, xz

(11.10.12)

For dipole moment operators, the first supplemental column is required. It states that a translation along z transforms as an A representation of the group. The direction of a z vector will not be affected by the E or C_2 operations. By contrast, x and y vectors will change sign under the C_2 operation.

For the particle in a one-dimensional box, the electron wave is distorted only by a dipole operator in the x direction. Consider a transition from a wavefunction with odd n to one with even n with a change in the dipole moment along x induced by the electric field of an incoming light wave. The direct product of the irreducible representations of the three functions under the integral is

$$\langle n \text{ even}|\, ex\, |n \text{ odd}\rangle = \mathbf{B} \otimes \mathbf{B} \otimes \mathbf{A} = (1 \times 1 \times 1 \quad -1 \times -1 \times 1) = (1 \quad 1) \tag{11.10.13}$$

The product of the three irreducible representations is the A irreducible representation. This totally symmetric representation indicates that the product of the three functions is even and the integral is nonzero. The transition from n odd to n even is allowed by symmetry considerations.

A transition between two odd or two even wavefunctions is not permitted. The direct product of the irreducible representations for the ex dipole moment component is

$$\langle n \text{ even}|\, ex\, |n \text{ even}\rangle = \mathbf{B} \otimes \mathbf{B} \otimes \mathbf{B} = (1 \times 1 \times 1 \quad -1 \times -1 \times -1) = (1 \quad -1) = \mathbf{B} \tag{11.10.14}$$

$$\langle n \text{ odd}|\, ex\, |n \text{ odd}\rangle = \mathbf{A} \otimes \mathbf{B} \otimes \mathbf{A} = (1 \times 1 \times 1 \quad 1 \times -1 \times 1) = (1 \quad -1) = \mathbf{B} \tag{11.10.15}$$

Both sets of products are odd on the interval and the integrals are zero.

The second supplementary column for the table gives the irreducible representations associated with products of functions. The polarizability α_{xx} indicates a volume change of the molecule associated with the x spatial direction. This column permits evaluation of integrals for Raman-allowed transitions since such transitions require a net change in volume of the molecule in the electric field of an incoming light wave. For a Raman-allowed transition, integrals such as

$$\langle \psi|\, \alpha_{xx}\, |\psi\rangle \tag{11.10.16}$$

must be nonzero. In this case, for example, transitions between two even functions for a particle in a one-dimensional box are allowed:

$$\langle n \text{ odd}|\, \alpha_{xx}\, |n \text{ odd}\rangle = \mathbf{A} \otimes \mathbf{A} \otimes \mathbf{A} = (1 \times 1 \times 1 \quad 1 \times 1 \times 1) = (1 \quad 1) = \mathbf{A} \tag{11.10.17}$$

Different levels can be probed using dipole-allowed transitions and Raman-allowed transitions.

Problems

11.1. Use the similarity transforms of Equation 11.9.8 to generate diagonal or block diagonal matrices for $\mathbf{E}$, $\mathbf{C}_2$, $\mathbf{i}$, and $\boldsymbol{\sigma}$ (Equations 11.9.9, 11.9.10, 11.9.11, and 11.9.12).

11.2. Determine the $\mathbf{E}$ and $\mathbf{C}_2$ matrices for the 1,3-butadiene molecule (Equation 11.2.7).

11.3. Determine the similarity transform matrices and diagonalize the $\mathbf{E}$ and $\mathbf{C}_2$ matrices of Problem 11.2.

11.4. Develop the matrices for $\mathbf{E}$, $\mathbf{C}_2$, $\mathbf{C}_4$, and $\mathbf{C}_4^3$ for 1,3-cyclobutadiene (Equations 11.2.22–11.2.25).

11.5. Show that Equation 11.3.11 is consistent with the group multiplication table for C_{2v}.

11.6. Show that the $\mathbf{C}_3$ and $\mathbf{C}_3^2$ matrices do not commute with the $\boldsymbol{\sigma}_1$, $\boldsymbol{\sigma}_2$, and $\boldsymbol{\sigma}_3$ matrices.

11.7. Show that the eigenvectors determined from Equation 11.6.42 are identical to those derived using group theory.

11.8. Use regular Cartesian coordinates on each atom of a triangular molecule to determine the normal modes of the molecule.

11.9. Show the orbitals which result from a rotation of p_x, p_y, and p_z orbitals by 120° about the z axis.

Index

(8)

P,236